Agricultural insect
and their control

last issue
2004

This book is dedicated to Jeremy and
to the memory of my mother

# Agricultural insect pests of the tropics and their control

## SECOND EDITION

**DENNIS S. HILL** M.Sc., Ph.D., F.L.S., M.I. Biol.

*Formerly Senior Lecturer in Entomology and Ecology
Department of Zoology, University of Hong Kong
Senior Lecturer in Entomology, Department of Crop Science
and Production, Makerere University, Uganda*

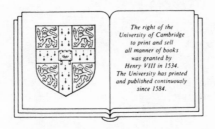

*The right of the
University of Cambridge
to print and sell
all manner of books
was granted by
Henry VIII in 1534.
The University has printed
and published continuously
since 1584.*

**CAMBRIDGE UNIVERSITY PRESS**

*Cambridge*

*New York    New Rochelle    Melbourne    Sydney*

Published by the Press Syndicate of the University of Cambridge
The Pitt Building, Trumpington Street, Cambridge CB2 1RP
32 East 57th Street, New York, NY 10022, USA
10 Stamford Road, Oakleigh, Melbourne 3166, Australia

First published 1975
Reprinted 1978
Second edition 1983
First paperback edition 1987

Printed in Great Britain at the Alden Press
Oxford London and Northampton

Library of Congress catalogue card number: 81–24216


*British Library cataloguing in publication data*

Hill, Dennis S.
Agricultural insect pests of the tropics and
their control. – 2nd ed.
1. Tropical crops – Diseases and pests
2. Insect control
I. Title
632'.7'0913 SB935.T7

ISBN 0 521 24638 5 hard covers
ISBN 0 521 28867 3 paperback

(First edition
ISBN 0 521 20261 2 hard covers
ISBN 0 521 29441 X paperback)

# Contents

# *Foreword*

*by Professor J. L. Nickel*

Agricultural systems in the tropics will undergo major changes in the next several decades. In order to meet increasing needs for food and fibres farming will need to be greatly intensified. Such intensification will inevitably lead to increasing pest problems. Without sufficient biological information to design pest management systems, or adequate numbers of qualified personnel to operate them, simplistic solutions, such as blanket calendar spray schedules, are likely to be utilized. Such practices are even more likely to result in ecological imbalances in the tropics than they have in the temperate zones, and are often too expensive for the limited resources of small farmers in developing countries. It is essential, therefore, that the training of practically oriented entomologists be accelerated in the tropical countries. This book will provide a useful tool for this purpose.

One of the major problems facing agricultural education in the tropics has been the paucity of textbooks with a tropical bias from which students could learn the agricultural sciences using local examples. Too often they have had to learn about soils, insects, diseases or livestock from textbooks written for the temperate areas, giving examples from the United States or Europe. Fortunately, the number of books on tropical agriculture is increasing rapidly. This volume is a welcome member of that group.

Dr Hill, soon after his arrival in Uganda to teach Entomology in the Faculty of Agriculture of Makerere University, saw the need for a textbook based on East African insect pests. As I was Dean of that Faculty at that time I encouraged him to do so. While the information and plates were being assembled, they already enhanced greatly the teaching of Entomology in the Faculty. Before the book was complete Dr Hill left Uganda, and sub-

sequently expanded the scope of the book well beyond the modest initial beginnings. By adding sections on control principles, materials, and methods, and by expanding the species coverage to include agricultural pests of the tropics around the world, he has produced a book which can serve as a useful text throughout the warmer regions.

I believe teachers and students of Agricultural Entomology will find this a highly useful text and reference.

John L. Nickel
Director General
Centro Internacional de Agricultura
    Tropical (CIAT)
Palmira, Colombia                 November 1974

# Preface to second edition

In this second edition the number of insect pests treated in the pest descriptions chapter has been very considerably increased by the addition of about 70 species of agricultural and horticultural pests from tropical and sub-tropical Asia, and some from the tropical parts of the New World. The object of this increase is to remove the original bias towards African pests that prevailed in the first edition. Similarly, the examples of biological control have been increased by addition of Asian and Australian examples of note. The total number of insect and mite pests dealt with in some detail now totals more than 310 species.

The chapter dealing with crops and their pest spectra has also been very extensively enlarged and many more data have been made available. The total number of crops included now has reached 104 including stored products pests, and the pest spectra species lists for each crop are more comprehensive.

In the production of the first edition an unforeseen delay in publication had unfortunate repercussions in several respects. Firstly, at the time the manuscript was completed the insect growth regulators (IGR) and juvenile hormones (JH) used as biological insecticides were both still largely at the experimental stage of development and had not been proved effective in field trials. Also insect pheromones were just beginning to be synthesized experimentally and their value in pest management programmes were still to be properly assessed. Lastly, at that time DDT, dieldrin, and the other organochlorine compounds were still being used widely throughout the world, although insect resistance was developing quite rapidly in a number of different pests, but the environmental 'hue and cry' as exemplified by the legislation in the USA which established the Environmental Protection Agency had not then been raised.

Unfortunately a delay of a couple of years in publication saw very rapid developments in all these aspects of pest control, which meant that in these aspects the book was already out-of-date at the time of publication.

In the sections dealing with pesticide recommendations, where a pest was formerly controlled successfully by the use of DDT or dieldrin, or another organochlorine compound, these chemicals are still referred to, even though in most countries their use now is at least restricted, if not completely prohibited. The object of the pesticide recommendations is to indicate which chemicals either have been or are still effective against a particular pest. Also it must be borne in mind that for some pests no suitable alternative chemical to dieldrin, etc., has yet been found, and some pests are not resistant to it; in some countries there is no legal ban yet on the use of the organochlorine compounds. It must be stressed that the pesticide recommendations are for general information only, and that any pest control programme must be planned in accordance with local pesticide regulations.

Drawings of pest species added to the second edition were made by Karen Phillipps, of Hong Kong, and photographs were taken by the author. Photographs of immature stages of *Chilades* and *Lampides* were taken by Mrs. G. Johnston, and for *Cephanodes* by Mrs F. Bascombe. Dr D. K. Butani drew the illustrations for *Homona, Stephanitis,* and *Diaphorina citri.* Dr J. M. Cherrett provided the photographs of *Atta* spp. and *Acromyrmex* spp.

Assistance and the provision of data for inclusion in the second edition have been provided by the following people, to whom my gratitude is now expressed: Dr B. Banerjee, Dr M. J. Bascombe,

Dr D. K. Butani, Dr J. M. Cherrett, Dr R. H. Gonzalez,
Dr S. M. Hammad, Dr H. Y. Lee, Dr Li, Li-ying,
Dr H. E. Ostmark, Dr D. L. Struble, Dr K. Yasumatsu,
Mr R. Winney, Mr R. Wong, Mrs F. Bascombe and
Mrs G. Johnston. Thanks are also due to Professor
B. Lofts for his interest and support, in whose
department the additional work was done.

Additional information and assistance was also
given by the staff of the Commonwealth Institute of
Entomology, London, and the Entomology Department of the British Museum (Natural History).

A few of the more temperate crop pests have
been removed from the second edition (such as
Codling Moth, Colorado Beetle, etc.) as it is now
intended that a companion volume, titled *'Agricultural Insect Pests of Temperate Regions and their
Control'*, be completed in the near future; publication is planned for 1983.

Dennis S. Hill
August, 1981

'Haydn House',
20, Saxby Avenue,
Skegness,
Lincs.,
ENGLAND.

# Preface to first edition

The need for such a book as this became apparent during the course of first lectures and practicals on 'Crop pests of East Africa', and 'Principles and practice of crop protection', which were given to the senior students in the Faculty of Agriculture at Makerere University, Uganda. That semester, in January 1968, at a meeting of the East African Specialist Entomology and Insecticide Committee held in Arusha, Tanzania, it also became obvious that a good handbook on crop pests for agricultural field entomologists was long overdue. Financial support for this project was provided out of the Agriculture Faculty Rockefeller Grant for which grateful acknowledgement is made.

At the time this project was started there was already published an excellent series of books dealing with various aspects of tropical crop pests and pesticides, but the number of books required for reference purposes for the two courses referred to came to a total of about twelve. What was required was a suitable amalgam of the published data available in the form of a single text, which would be the basis for courses on 'Tropical crop pests' and 'Crop protection'. Although the immediate requirement was for a textbook for the three E. African universities, since the crops grown there and the pests which attack them are generally of very wide distribution throughout the tropics and the warmer parts of the world, it was decided that the book would be more useful if it gave an outline of the insect pest situation throughout the tropical parts of the world. For reasons of space this book only includes insect and mite pests of tropical crops. The mollusc, myriapod, bird and mammal pests tend to be more specific to an area or country, rather than to a crop or range of crops, as is the case with insects and mites, and it would be very difficult to generalize about these other pests on a worldwide basis. Because of the complexity and scope of nematology in relation to tropical crops this book does not include any plant parasitic nematodes.

So far as possible, references have been made to the original sources of data, but wherever a monograph had already been written about a crop or a particular pest, or group of pests, this was used as the primary source of information. The main sources of information used in the compilation of the text are referred to in the general bibliography as well as in the text.

Originally this book was to be produced jointly by myself and T. J. Crowe, then Senior Entomologist, Kenya; but on his transfer as FAO Entomologist to Addis Ababa, Ethiopia, he was obliged to relinquish his part in the production. His initial support and encouragement were however particularly important.

About half of the drawings used in this book were done by Mrs Hilary Broad; the others were by M. Sanders, and several other artists in Nairobi. The assistance of the Ministry of Agriculture, Kenya, in the provision of some drawings, especially some of the coffee and cotton pests, is gratefully acknowledged. Specimens of insect pests for drawing purposes were loaned from the British Museum (Natural History) through the Keeper of Entomology, Dr P. Freeman, and the Trustees are accordingly thanked.

The staff at the Commonwealth Institute of Entomology, London, kindly checked the list of names for the major pests, and provided important synonymy.

I am most grateful for the support and assistance received from a large number of colleagues and fellow entomologists, including Professors J. L. Nickel and K. Oland, Drs R. G. Fennah, R. H. Le Pelley,

D. J. Williams, V. F. Eastop, D. V. Alford, J. B. Muthamia, M. E. A. Materu, C. A. Edwards, D. J. Greathead and J. C. Davies, Mrs S. D. Feakin and Mrs A. Thompson, Mr R. Gair, L. Mound, R. J. A. W. Lever, K. M. Harris, D. Macfarlane, G. Rose, H. Stroyan, D. N. McNutt, and the other staff of the Commonwealth Institute of Entomology and Entomology Department of the British Museum (Natural History), and C. Myram of Bayer Agrochemicals Ltd.

*Cambridge, 1972*                    Dennis S. Hill

# 1 *Introduction*

This book is intended for use as a student text for courses in 'Applied Entomology', 'Crop pests' and 'Crop protection', at both undergraduate and postgraduate level. It presupposes a basic knowledge of entomology to the level of that in Imms, A.D. (1967): *Outlines of Entomology*, (1960): *A General Textbook of Entomology*, or alternatively Borror, D.J. & D.M. Delong (1971): *An Introduction to the Study of Insects*. In other words, the reader should be acquainted with the major groups of insects and their characteristics, which may mean Order, Suborder or Superfamily in some cases, but in the more economically important Orders this would mean familiarity with Superfamilies or Families, for example in the Hemiptera, Lepidoptera, Coleoptera, and Diptera.

Although the title is *Agricultural Insect Pests of the Tropics*, there are crops and pests included that are found in many warmer parts of the world outside the tropics, namely in S. Africa, Australia, N. New Zealand, China, Korea, Japan, S. USA, and the Mediterranean region. Certain tropical crops, such as rice and citrus fruit, can be cultivated in countries outside the tropics. A recent trend in many tropical countries is to make an effort to diversify local agricultural crops, and in particular various temperate crops are now being successfully cultivated at higher altitudes in the tropics. Examples of such crops are potato, wheat, apple, peach, various brassicas and other vegetables. Hence the scope of this book is really insect and mite pests of agricultural crops in warmer climates.

In some parts of chapter 2, on the principles of pest control, it has not been possible to provide tropical examples, for it is only in the long-established agricultural systems of Europe and N. America where examples are available.

The examples of biological control were initially restricted to Africa, partly for reasons of space and convenience, and also because there were sufficient examples to illustrate adequately the main problems connected with biological control introductions. But in the second edition more examples are included from Asia and Australasia because of their importance and effectiveness.

The section on pesticides was compiled from data published by Martin (1970), from Worthing (1979) and from various original data sheets provided by the firms concerned, and that part dealing with application equipment is largely from G. Rose (1963). It is not feasible to generalize extensively about persistence, efficiency, pre-harvest intervals, toxicity, and tolerance levels, for not only do these characteristics vary considerably according to local climatic conditions, but each country has its own requirements with regard to residues and toxicity. Some countries are more concerned with operator safety, whereas others regard consumer hazards more important. Thus the same chemical may have a pre-harvest interval of seven days in one country and as many as 28 in another; or alternatively an approved pesticide in one country may be banned in another.

In chapter 7, on pest descriptions, biology and control measures, the original scheme was to illustrate all the important stages of the major insect pests and to show the damage done to the host. But it was not possible to provide all stages and damage for more than 310 major pests, and so in some cases only the adult insect is drawn. Unfortunately, some of the earlier drawings were designed more to give an impression of the pest and the crop plant rather than accurate detail of the insect. In the more recent drawings by Mrs Broad and Karen Phillipps we have endeavoured to reproduce morphological details which are taxonomically specific.

The species here designated as major pests have, in a few instances, been chosen for academic reasons or to demonstrate a point of particular biological interest, rather than always being primarily economic pests. I have attempted to include a well-balanced range of pests, most of which are important on major crops, and widely distributed throughout the warmer parts of the world. The denotation of the term 'major pest' to a species is necessarily somewhat arbitrary when dealing with 104 crops grown throughout the warmer parts of the world. However, this term has usually only been applied to species which are economically important over a wide part of the range in which the crop is cultivated. A small number of pests restricted to E. Africa have been included because of the original nature of the project.

The common names used are either those used on the Commonwealth Institute of Entomology (CIE) distribution maps, or else those most frequently employed in international entomological journals. The distributions given are summarized from the maps produced by the CIE, and in the cases where a map has not been produced for a particular species the appropriate distribution data have been made available from the CIE card index system. Reference to the CIE map, where one is available, is made at the end of each summary of distribution.

In the section on control, emphasis has been placed on methods of cultural control whenever these are available, and so far as pesticides are concerned no details as to rates, etc., are included. Pesticide recommendations vary extensively from country to country, and also from season to season, so only the barest details of pesticide recommendations are included. For full details of these for local crops in each country, the appropriate Ministry or Department of Agriculture or Regional Entomologist should be consulted. It would be quite impossible to provide adequate pesticide detail suitable for practical use in all the different parts of the tropical world.

According to figures provided by Dr. R.G. Fennah for Wilson (1971) it can be said that there are some 30 000 insect pest species, but Fletcher (1974) referred to there being only about 1000. Later in the book he mentioned that the total number of insect and mite pest species recorded from several major crops ranges from 1400 on cocoa and cotton to 838 on coffee. It seems reasonable to assume that on a world-wide basis there are something in the region of 1000 species of 'serious' crop pest species, including pests of forests and ornamentals, and maybe up to 30 000 minor pest species. In chapter 8, under the headings of the 104 crops considered, are listed the more important major pests, many of which were included in chapter 7, and in addition a selection of the minor pests recorded from each crop. In some of the more restricted crops the number of recorded pest species is very small, whereas the widespread crops may have more than 1000 recorded minor pests. In these cases the list of minor pests has been restricted to the more important, more interesting, or more widespread of the minor pest species.

## Problems relating to agricultural entomology in the tropics

Successful pest control anywhere is not easy. Most of the time we have to use insecticides and there are problems with chemical formulation, method of application, precise timing, hitting the target area, and retention on the target, to name but a few. But in tropical situations it is even more difficult. In the past the main source of pest control recommendations in the tropics were the usual insecticide measures employed in the western temperate countries against similar or closely related pests. These recommendations were often singularly unsuccessful because the basic agroecological systems in the two regions were dissimilar, sufficiently so that the proven western recommendations were ineffective in the hot and usually wet tropical situation.

In recent years, however, the situation has improved considerably as most major pesticide firms have testing stations in the tropics, and for some time many pesticides have been developed specifically for the extensively grown tropical crops, such as cotton, rice, maize, coffee, etc. But until very recently most of the basic research in crop protection was still carried out in the temperate western countries. The situation in the USA was less drastic as the southern states are semitropical in climate, so that tropical and temperate crops have been grown in the same area for many years. Within the last ten years or so there has been the establishment of several important international centres for agricultural research in tropical countries, concerned with indigenous local crops, and these centres are providing the basic research into all aspects of crop production and protection. These centres include IRRI at Manila (Philippines) for rice; IITA at Ibadan (Nigeria); CIAT at Cali (Columbia); ICRISAT at Hyderabad (India) for dryland tropical crops.

The main problems arise because of the climatic differences, in that the tropics are hot and usually wet, and these factors affect both insecticide performance and insect pest biology.

## Insecticide performance in the tropics: weathering

(a) Torrential tropical rain may wash off insecticide residues from crop plant foliage at a greater rate than the more gentle temperate rainfall.

(b) Chemical degradation may be rapid under tropical conditions because of the high ambient temperatures and the high level of insolation on exposed crops.

## Insect pest physiology/ecology in the tropics
### (a) *Continuous breeding*

In the absence of a cold winter or hot, dry season most insect pests will breed continuously, given host plant availability, often with a complete lack of synchronization of life-cycles, so that at any one time all instars and life-history stages may be present and there may be many (overlapping) generations per year. The Diamond-back Moth, for example, in the UK has 2–3 generations per year, in eastern Ontario (Canada) with its hot summer there may be 6 generations, but in S.E. Asia there may be as many as 15 generations per year. Many Lepidoptera are notoriously difficult to kill with insecticides in their later instars, and species such as stem borers or bollworms have to be attacked during the first instar stage prior to their penetrating the host plant tissues. With unsynchronized life-cycles such pests are very difficult to control with insecticides without recourse to continually repeated applications. In a temperate situation, with diapause over the cold winter period, more or less simultaneous emergence in the spring, and most pests being uni- or bi-voltine, pest management is a somewhat simpler affair.

### (b) *Development of resistance to insecticides*

Because of the greater biological productivity in the tropics, development of resistance to insecticides by pests may be more rapid. It is generally recognized that 10–15 generations are required in most insect species for resistance to manifest itself. Thus, with continuous breeding in the tropics as opposed to perhaps bi-voltinism in a temperate country, insecticide resistance can become apparent far more rapidly. Diamond-back Moth is the major pest of cultivated brassicas in S.E. Asia, in part because it has developed resistance to almost all of the insecticides used against caterpillars.

3

### Agricultural systems in the tropics

#### (a) *Complexity of agroecosystems*

With the absence of a cold winter the continuous breeding of many insect pests results, in general, in an enhanced importance of natural predators and parasites as population controlling factors, particularly in the case of the longer-term plantation crops, such as oil palm, coconut, coffee and cocoa. In countries where plantation crops are grown alongside tracts of native vegetation the complexity is increased due to the presence of a reservoir of crop pest species on alternative hosts in the forest, and also a reservoir of predators and parasites. This situation has been found in several parts of S.E. Asia where rice is cultivated in paddies surrounded by native grassland. In tropical forest ecosystems, or rather agroecosystems, the insect pests and their natural enemies usually coexist in a delicate balance which can easily be upset by indiscriminate or careless insecticide use. Good examples of this type of situation were seen in Malaysia, on both oil palm and cocoa, as reported a few years ago by Wood (1971) and Conway. Thus, in many tropical pest situations the use of cultural methods of control and biological control may be far more appropriate than insecticides.

#### (b) *Diversity of agricultural systems*

Most tropical countries have large areas of long-term plantation crops, and large tracts of some short-term or annual crops such as maize, rice and sorghum, where pest management programmes can be carefully executed. But adjacent to the plantations may be many small farms or smallholdings where subsistence farming is practised, and where plantation-type crops such as coffee, cocoa, cotton and rubber are grown in small plots as cash crops. On the shambas there is often little or no pest control because of the cost of insecticides and equipment. In some countries (for example Uganda) smallholder production of crops such as coffee and cotton may represent a sizeable proportion of the national production, often without any rational pest control programme being employed.

Thus, in some tropical countries it is necessary for more integrated research into rural systems of cultivation, social structures, life-styles, and marketing practices to be undertaken in order for pest management programmes to be formulated sensibly. Simplistic remedies based upon western technology may not meet the requirements for pest control in these situations.

#### (c) *Ignorance of farmers*

Surveys conducted in the Philippines by IRRI have revealed that many farmers do not know what is meant by the terms 'pest' and 'pesticide', and those that spray their crops do so solely because they are told to, and in order to do the same as neighbouring farmers. Consequently, they often use far too little water for proper coverage, and frequently use too little insecticide, because of its high cost, to be effective. This situation could well be prevalent amongst peasant farmers in the tropics, especially where the literacy level is low.

#### (d) *Water shortage*

In parts of India, Africa, Indonesia and the Philippines, water for spraying purposes is not readily available and conventional high-volume drenches, as required for some soil-inhabiting pests (e.g. cutworms and chafer grubs), are quite impractical. This imposes a serious restriction upon the range of insecticide (pesticide) practices available against any particular pest.

### 'Vicious circle' of subsistence farming in the tropics (Winney, *in litt.*)

Most subsistence farming is on a very small scale, and often in isolated situations. A shortage of money makes the farmers use local varieties of crops, with little fertilizer, and the yield is low. However, pest damage levels are also usually low. In order to increase the yield it would be necessary to use introduced crop varieties, at a greater cost, which require increased use of fertilizer, and are more susceptible to pest attack so that pesticides would be required to protect the crop. Thus an increased crop yield could be achieved on the same land area, but at the cost of extra outlay and extra effort. Most farmers are not able to afford the additional cash outlay in order to achieve this extra yield. Some could afford the outlay provided they sell the surplus crop afterwards, but this is often not practical because of the isolated situation and the small quantities of produce involved.

Thus the value of the crop in relation to the cost of insecticide in subsistence farming constitutes the 'vicious circle', and is the reason why much subsistence farming is practised with little or no insecticide usage. The farmers simply cannot afford the extra money to spend on a low-yielding crop, and if they buy high-yielding varieties with their additional fertilizer and pesticide requirements they often cannot find a market for the surplus yield.

Most countries with an agricultural economy are now sponsoring local co-operatives which will buy seed, agrochemicals and machinery on behalf of local farmers and then arrange for collective disposal and sale of surplus crops to the nearest markets, thus breaking the 'vicious circle'.

# 2  *Principles of pest control*

## Definition of the term 'pest'

Before contemplating taking any control measures against an insect species in a crop, the species must be correctly identified; then, presuming its biology is known, it should be clearly established that the species in this particular context is a pest, and that it could be profitable to attempt population control.

**Pest.** The definition of a pest can be very subjective, varying according to many criteria, but in the widest sense any animal (or plant) causing harm or damage to man, his animals, his crops or possessions, even if just causing annoyance, qualifies for the term pest. From an agricultural point of view, an animal or plant out of context is regarded as a pest (individually) even though it may not belong to a pest species. Thus an elephant in a shamba is a pest, but next-door in a game park it is not, and is in fact a valuable national asset there. Similarly, volunteer maize plants growing in a field with cotton have to be regarded as 'weed' pests.

Many insects belong to generally accepted *pest species*, as listed in chapter 8, but individual populations are not necessarily always pests; that is, of course, not necessarily *economic pests*.

As pointed out by Norton & Conway (in Cherrett & Sagar, 1977), we are often somewhat over-preoccupied at the present time with the state of the 'pest' population, whereas probably the most important aspect of a pest species is the damage (or illness) caused by the pest and the value placed upon these consequences by human society.

**Economic pest.** On an agricultural basis, we are concerned when the crop damage caused by insects leads to a loss in yield or quality, resulting in a loss of profits by the farmer. When the yield loss reaches certain proportions the pest can be defined as an economic pest. Clearly the value of the crop is of paramount importance in this case, and it is difficult to generalize, but as a general guide for most crops it is agreed that most species reach *pest status* when there is a 5–10% loss in yield. Obviously a loss of 10% of the plant stand in a cereal or sugarcane field (note that this is not quite the same as a 10% loss in yield!) is not particularly serious, whereas the loss of a single mature tree of *Citrus*, macadamia or mango is important.

**Economic damage.** This is the amount of damage done to a crop that will financially justify the cost of taking artificial control measures, and will clearly vary from crop to crop according to its basic value, the actual market value at the time and other factors. In practice, many peasant farmers in the tropics engaged in subsistence farming feel that they cannot justify use of pesticides at all.

**Economic injury level (EIL).** This is the lowest population density that will cause economic damage, and will vary between crops, seasons and areas. But it is of basic agricultural importance that it is known for all the major crops in an area.

**Economic threshold.** It is desirable that control measures be taken to prevent a pest population from actually causing economic injury. So the economic threshold (Stern, Smith, van den Bosch & Hagen, 1959) is the population density of an increasing pest population, at which control measures should be started to prevent the population from reaching the economic injury level.

**Pest complex.** The normal situation in a field or plantation crop is that it will be attacked by a number of insects, mites, birds and mammals, nematodes and pathogens which together form a complicated interacting pest complex. The control of a pest complex is complicated and requires careful assessment, especially as to which are the *key pests*, and careful integration of the several different methods of control which may be required. This, of course, makes the process of evaluation difficult, and generally, in the past, much money was wasted on uneconomic pest control, either through carelessness or lack of knowledge.

**Pest spectrum.** This is the total range of different types and species of pests recorded attacking any particular crop, and especially of concern in one particular area. The total number of insects (and mites) recorded from the major crop species are considerable; these records, incidentally, are of insects feeding or egg-laying on the plant, and do not include casual observations when the insect might just be resting, or when, for example, caterpillars have crawled up on to the plant just to pupate (e.g. Sweet Potato Butterfly). Simmonds & Greathead (in Cherrett & Sagar, 1977) listed the numbers of pest species, on a world basis, recorded from sugarcane as 1300, cotton 1360, coffee 838, and cocoa 1400. Fortunately, for the practising entomologist, and the farmers, these numbers reflect the situation globally and many of these pests are restricted geographically to one part of the world. For example, wherever cocoa is grown it will be attacked by a capsid complex (Miridae) but the actual species differ from region to region. Only a few major pests are cosmopolitan (e.g. *Myzus persicae, Agrotis ipsilon*) or pantropical (e.g. *Maruca testulalis*) in distribution (in point of fact, their widespread distribution is one reason for their being regarded as major pests).

**Pest load.** This is the actual number of different species (and numbers of individuals) of pests found on either a crop or an individual plant at any one time, and, as already mentioned, this would usually be a pest complex but could also be a monospecific population.

**Pest species accumulation.** Long-term stable habitats generally exhibit an extensive species diversity, both in host plants and in phytophagous arthropods. This is shown typically in old forests, and also some plantation crops such as cocoa, rubber, sugarcane and tea (Banerjee, 1981). The pest species accumulation in the monocultures is in part a reflection of the area under cultivation; other important factors are the type of plant, the geographical location of the habitat (area) and its natural species richness, and the age of the actual plants and of the community (crop). It appears that in some cases the age of the plant community is not particularly important, but Banerjee observed that in tea plantations (from the Old World tropics, excluding China) age appeared to be a major factor and pest species saturation was apparently reached (in plantations in N.E. India) at the plantation age of about 35 years. After this age there was no further increase in pest species number.

**Pest species recruitment.** As mentioned elsewhere, each major crop has a more or less clearly defined area of origin in one particular part of the world, and during historical times most of the crops have gradually been transported by travellers and commerce to other parts of the world where suitable climatic conditions prevail.

With some crops, such as *Citrus,* it appears that most of the species in the pest spectrum in most parts of the world are allochthonous in that they have originated in the S. China/Indo-China region where *Citrus* is endemic, and the pests have been gradually spread to the new areas of cultivation. Thus at the present time most of the important *Citrus* pests are pantropical in distribution.

Other crops such as rubber, sugarcane and tea (Banerjee, 1981) have apparently been subjected to autochthonous pest recruitment in their new areas of cultivation. Tea, according to Banerjee, has only about 3% of its total number of pest species common to the different areas of cultivation, the other pests being recruited locally and thus being different in each area.

As would be expected, it appears that with many crops pest recruitment has been in part allochthonous and partly autochthonous. The origin of the crop pest spectrum may have an important effect so far as pest control strategy is concerned, particularly with regard to natural and biological control.

**Serious pest.** This is a species that is both a major pest and an economic pest of particular importance, being very damaging and causing considerable harm to the crop plants and a large loss in yield. It almost invariably occurs in large numbers.

**Major pest.** In this book these are the species of insects and mites that are either serious pests of a crop (or crops) in a restricted locality, or are economic pests over a large part of the distributional range of the crop plant(s). Thus the species here regarded as major pests usually require controlling over a large part of their distributional (geographical) range, most of the time. As mentioned in chapter 1 however, some species of insects have been included as 'major pests' in this book because of their widespread and frequent occurrence, biological interest, wide range of host plants, or other aspects of academic interest. In any one crop, in one location, at one time, there is usually only a rather small number (say 4-8) of major pests in the complex that actually require controlling. For example, although the pest spectrum for cotton

worldwide is 1360 species, on any one cotton crop there will probably only be about five species requiring population control. Usually for most crops in most localities the major pest species remain fairly constant from year to year, but several entomologists have commented recently that in some areas they have observed that the major pest species complex has been gradually changing over a long period of time. Soehardjan (1980) reported that in Indonesia the 8–10 major pests of rice have largely changed over the period of time 1929–79, although there are some differences within different parts of the island of Java. As mentioned in chapter 7, the Brown Planthopper (BPH) of Rice has risen in 10 years from obscurity to becoming the most serious pest of rice in most parts of S.E. Asia. And it is reported from IRRI that there are two new major rice pests in tropical Asia (Pathak, 1980); these are the Sugarcane Leafhopper and Rusty Plum Aphid. So over a period of time (some 10–50 years) it is expected that the complement of major pests for a crop will change. It must be remembered that evolution continues all the time, even though it is not often obvious, and that in an artificial environment, such as agriculture, it can be expected that evolution will be accelerated.

**Minor pests.** These are the species that are recorded feeding or ovipositing on the crop plant(s) but usually do not inflict damage of economic importance; often their effect on the plant is indiscernible. They may be confined to particular crop plants or may prefer other plants as hosts. Many (but not all) pests listed as minor pests are potentially major pests (viz. BPH of Rice). Many species that are major pests of one crop will occur in a minor capacity on other crops. And sometimes a major pest of a particular crop in one part of the world (e.g. Africa) will be a minor pest on the same crop in a different part (e.g. Asia or the New World).

**Potential pest.** This term is used occasionally in the literature and refers to a minor pest species that could become a major pest following some change in the agroecosystem. Only a relatively small proportion of the species listed as minor pests are really potential pests in this sense, because of their basic biology.

**Secondary or sporadic pest.** Defined by Coaker (in Cherrett & Sagar, 1977) as a species whose numbers are usually controlled by biotic and abiotic factors which occasionally break down, allowing the pest to exceed its economic injury threshold.

**Key pests.** In any one local pest complex it is usually possible to single out one or two major pests that are the most important; these are defined as key pests, and are usually perennial and dominate control practices. A single crop may have one or more key pests, which may or may not vary between areas and between seasons. It is of course necessary to establish economic thresholds for these key pests in order to be certain when to apply control measures, for it has been often observed that the mere presence of a few individuals of a key pest species in a crop may cause undue alarm and lead to unnecessary pesticide treatment. Key pests owe their status to several factors, including their usually high reproductive potential, and the type of damage they inflict on the host plant.

## Pest populations

The important point to remember about any pest is that it is only an economic pest at or above a certain population density, and that usually the control measures employed against it are designed only to lower the population below the density at which the insect is considered to be an economic pest; only very rarely is complete eradication of the pest aimed at. The schematic representation of the growth of a popu-

lation in fig. 1 (adapted from Allee *et al*, 1955) has had four separate population levels indicated; these are represented by the numbers 1 to 4. These population levels indicate purely hypothetical densities at which any particular insect species may be designated an economic pest. Population level 1 might well represent the economic pest level for such an insect as *Antestiopsis* spp., for in this case control measures are recommended when the population density reaches more than two bugs per coffee bush, whereas observations on isolated unsprayed arabica coffee bushes

*Fig. 1.* The growth of populations (after Allee *et al*., 1955).
STAGE    I   Period of positive, sigmoid growth; population increasing
     *A*   Establishment of population
     *B*   Period of rapid growth (exponential growth)
    II   Equilibrium position (asymptote); numerical stability
    III   Oscillations and fluctuations
     *A*   Oscillations – symmetrical departures from equilibrium
     *B*   Fluctuations – asymmetrical departures
    IV   Period of population decline (negative growth)
    V   Extinction

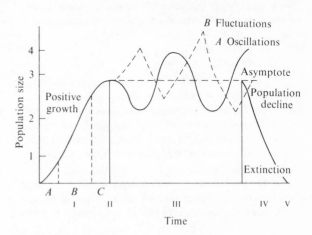

indicate that the normal asymptote may be in the region of 20—40 bugs per bush. At the other extreme population level 4 could well apply to insects such as the Desert Locust which are only economic pests in E. Africa at irregular intervals at times of population irruption. Most of the more common pests would come into the categories which reach pest density at population levels 2 and 3. The growth of a population can be expressed very simply in the equation:

$$P_2 \rightleftharpoons P_1 + N - M \pm D,$$

where $P_2$ = final population, $P_1$ = initial population, $N$ = natality, $M$ = mortality, $D$ = dispersal.

To simplify this equation, natality can be regarded as synonymous with birthrate, mortality with deathrate, and dispersal is either regarded as movement out of the population (emigration), or movement into the population from outside (immigration). The object of pest control is to lower $P_2$, which quite clearly can be done by either lowering the birthrate of the pest, increasing the deathrate, or inducing the pest to emigrate away from the area concerned.

Four hypothetical pest populations are illustrated graphically in Stern *et al.* (1959), in relation to their equilibrium position, economic threshold, and economic injury levels; these graphs are illustrated in fig. 2.

### Life-table

The examination of a pest population and its separation into the different age-group components, i.e. eggs, larvae, pupae and adults, enables a life-table for that pest population to be compiled. The construction of life-tables for a pest species is an important component in the understanding of its population dynamics. However, this is a complicated procedure requiring mathematical ability and considerable ecological knowledge. The growth of an insect population, especially the recruitment and the survival of the different stages, varies considerably according to the type of insect concerned. One result of this variation is that there are half-a-dozen different methods for the construction of a *budget* (for further details see chapters 10 and 11 in Southwood (1978)). As pointed out by Harcourt (1969) it is necessary to be careful in the choice of the appropriate method for compiling a life-table budget when planning the sampling methods to be used.

### Resurgence

The term resurgence is used to express a sudden increase in population numbers. It occurs when the target species, which was initially suppressed by insecticidal treatment, undergoes rapid recovery after the decline of the treatment effect.

It may also occur as a result of the development of a new biotype of the pest, or if the insecticide treatment kills a disproportionate number of the natural enemies of the pest species.

### Population dynamics theory

(After Southwood, 1977.) Applied biologists have for a long time been concerned with two basic aspects of animal numbers: firstly that population numbers may change greatly, as pointed out by Andrewartha & Birch (1954); and secondly, that most animal populations are relatively stable in comparison to their prodigious powers of increase. It now seems that certain species of animals belong to the one category and others to the second. Southwood pointed out that the change of the population fluctuation to a state of stability is associated with an increasing duration of stability in the habitat, and may be conveniently equated with the *r—K continuum. r-strategists* are opportunists, living in temporary (ephemeral) habitats and adapted to obtain maximum food intake in a short time; they are generally small, mobile and migratory, and have a short generation time.

*Fig. 2.* Schematic graphs of the fluctuations of four theoretical arthropod populations in relation to their general equilibrium position, economic threshold and economic injury levels (from Stern *et al.,* 1959).

1. Non-economic species whose general equilibrium position and highest fluctuations are below the economic threshold, e.g. *Aphis medicaginis* on alfalfa in California, USA.
2. Occasional pest whose general equilibrium position is below the economic threshold, but whose highest population fluctuations exceed the economic threshold, e.g. *Cydia molesta* on peaches in California, USA.
3. Perennial pest whose general equilibrium position is below the economic threshold, but whose population fluctuations frequently exceed the economic threshold, e.g. *Lygus* spp. on seed alfalfa in western USA.
4. Serious pest whose general equilibrium position is above the economic threshold, usually requiring insecticide application to prevent economic damage, e.g. *Musca domestica* in milking sheds of dairy farms.

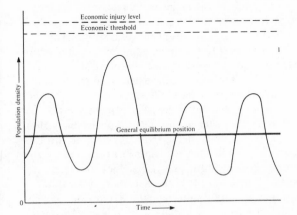

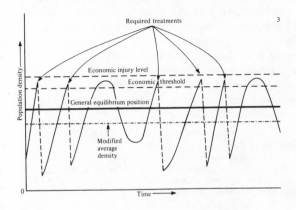

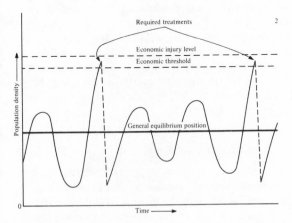

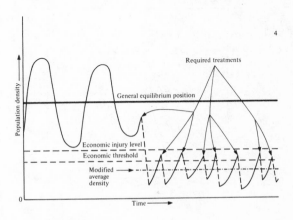

11

*K-strategists* live in stable habitats, often in crowded conditions, with their population size near the carrying capacity of their habitat; they are usually larger in size, less migratory and have a long generation time.

Fig. 3 is a synoptic population model. Three regions can be recognized. The r-strategists, whose habitats are ephemeral and whose numbers are characteristically 'boom and bust'; this strategy is dominated by large-scale migration, massive population losses, and new populations continually developing from a handful of colonizers. Secondly, the K-strategists represent the other extreme, maintaining a steady population at or near the carrying capacity of the habitat, basically in equilibrium with their resources; recruitment, mortality. and migration are low, so there is less opportunity to adapt to changed environments. These animals are special-ized to their particular environment, and if their numbers are reduced to a low level they are liable to become extinct.

Finally, the middle region recognized is the 'natural enemy ravine'. Both kinds of strategists have a stable equilibrium point, the upper one at the population density of the carrying capacity of the habitat. Where the 'natural enemy ravine' dips below the zero population growth contour there is a second equilibrium point, and where it rises through the contour on the other side of the ravine is the release, or escape point from natural enemies. Above this point, in the absence of density independent catastrophes, the population rises to the upper equilibrium point where intraspecific competition mechanisms (disease, etc.) operate. These two levels have been referred to as the endemic (lower) level and the epidemic (upper) level.

*r-pests* include species such as locusts, army-worms, leafhoppers, aphids, planthoppers, many flies, and, in plants, the ruderals (weeds) belong to this category. *K-pests* include elephants in Africa, tapeworms, Codling Moth, ants, tsetse flies, and many beetles. Obviously the r- and K-pests represent the extremes of a continuum, and there is correspondingly a large group of *intermediate pests*. It is with this large group that natural enemies have most population impact.

Applied biologists generally appreciate that habitat characters are important indicators for IPM strategies, and Conway (in May, 1976) has shown that as the r–K continuum is related to habitat characteristics it is relevant to decisions on the choice of a particular control strategy.

*Fig. 3.* The synoptic population model (after Southwood & Comins, 1976).
K = carrying capacity (as in K-selected).
Note: equilibrium points only occur where an 'east facing slope' cuts a zero population growth contour, as only here does negative feedback operate.

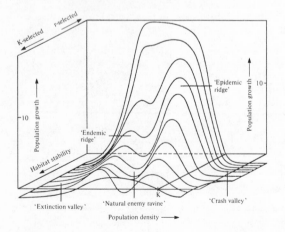

## Ecology and pest control

The earliest recorded attempts at pest control were mainly concerned with the biology of the pests and their ecology, and attempts were made to make the environment less favourable for the pests by cultural and physical means.

The complex and interacting system comprising all the living organisms of an area, and their physical environment (soil, water, climate, shelter etc.) is termed the *ecosystem*, and the study of ecosystems is called *ecology*. Definitions of ecology vary according to the speciality of the definer; botanists often have a different viewpoint from zoologists, and agriculturalists may have a third view. In its simplest form ecology can be defined as 'the total relationships of the plants and animals of an area (habitat) to each other, and to their environment'.

Environment has been defined by Andrewartha & Birch (1961) as being composed of four main factors: weather, food, other animals and plants, and shelter (a place in which to live).

For convenience it is customary to lump together environmental factors into two broad categories, biotic (i.e. organic) and physical (i.e. abiotic or inorganic). Weather and shelter (usually) are clearly physical factors, although shelter for a parasite could be regarded as biotic. Other animals and plants clearly constitute a biotic factor. Food is a biotic factor for animals which are holozoic (heterotrophic) in their feeding habits, but could possibly be more suitably described as physical for plants, which are holophytic (autotrophic) in their nutrition.

The *environmental factors* can be further defined as follows.

## Weather

(a)  Temperature – ranges defined as tropical, temperate, arctic or boreal.
(b)  Humidity – ranges from moist, moderate, to dry conditions.
(c)  Water – includes groundwater, rainfall, etc.
(d)  Light – intensity important for many organisms.
(e)  Wind – important for dispersal, and drying effects.

## Food

(a)  For animals
    (i) Organic remains – detritivores
    (ii) Plant material – herbivores (phytophagous)
    (iii) Other animals – carnivores and parasites
(b)  For plants
    (i) Organic remains – saprophytes (mostly fungi and bacteria
    (ii) Other plants – parasites and pathogens
    (iii) Animals – insectivores (carnivores)
    (iv) Sunlight, water, $CO_2$, minerals, chlorophyll – autotrophs

**Other animals and plants** (i.e. the community)
(a) Competition – intraspecific (within the species)
                  – interspecific (between different species)
(b) Predation
(c) Parasitism
(d) Pathogens causing diseases

**Shelter** (a place in which to live; habitat)
(a) For animals (insects) and pathogens – frequently a plant, and often a specific location on the plant, e.g. in the cases of a leaf-miner, stem-borer, bollworm and leaf-roller. Some insects are soil-dwellers (e.g. termites, crickets, beetle larvae), and adult, winged insects may not be very habitat-specific.
(b) For plants – usually a physical location (habitat), including the soil (e.g. hilltop, valley, field) together with the other plants that constitute the community. Two basic ecological terms should perhaps be included here for reference.

**Habitat.** The place where the plants and animals live; usually with a distinctive boundary, e.g. a field, pond, stream, sand-dune or rocky crevice. Often initially broadly subdivided into terrestrial, marine, and freshwater habitats.

**Community.** The collection of different species and types of plants and animals, in their respective niches, within the common habitat, e.g. a lake community, mangrove community or ravine community. The basic plan for all communities is the same, i.e.they are composed of saprophytes, autotrophic plants, detritivores, herbivores, carnivores, parasites, etc.

With the general disillusionment that followed the widespread continual use of synthetic chemical pesticides, especially the early organochlorine compounds, the situation has changed so that now attention is focussed on biological and ecological understanding, linked with careful application of selected pesticides. This approach was initially called *integrated control* but was more recently redefined as *pest management* (PM), and is now finally referred to as *integrated pest management* (IPM).

As indicated above, the number of different factors operating in an insect pest/host plant relationship is large and hence the different possibilities available for ecological manipulation are considerable. But, of course, a vital prerequisite is a detailed knowledge of the insect's life-history and biology, and especially its relationship with the host-plant.

As already mentioned, an insect species is only a pest (that is an economic pest) at or above a certain population density, and in any pest ecosystem any one (or more) aspect of the environment may be of over-riding importance. In the study of pest populations the key to control will inevitably lie in the understanding of the complex of environmental factors and their relative importance. However, our knowledge at present of most pest situations falls short of this ideal, and much basic ecological study is still required. Too frequently pest control still consists of hastily and ill-considered applications of chemical pesticides, which sometimes wreak ecological havoc, especially in the tropics, often without controlling the pest at which they were aimed. Progress is gradually being made though, as evidenced by the ever-growing number of IPM programmes for different crops in different parts of the world.

## Agroecosystems

An agroecosystem is basically the ecosystem of an area as modified by the practice of agriculture, horticulture or animal rearing. Agriculture consists of methods of soil management and plant cultivation so as to maintain a continuous maximum yield of crop produce, in the shortest time possible. This is achieved by manipulation of the environment so as to make growing conditions for the crop plants as near ideal as possible, and also to minimize damage to the crop by pest and disease attacks. Obvious manipulations are listed below, under the appropriate environmental headings.

### Weather
(a) Temperature control by shading (lowering) or use of greenhouse (raising).
(b) Humidity control by spraying or altering plant density.
(c) Irrigation (below or above-ground) and drainage.
(d) Light increased by use of ultra-violet lamps, or reduced by shade trees, shelter, etc.
(e) Wind protection by growing shelter-belts, windbreaks, tall trees and hedgerows.

### Food
(a) Animal feedstuffs, grazing leys, dietary supplements.
(b) Plants are 'fed' by addition of fertilizers, minerals, and trace elements; sometimes increased radiation by use of extra illumination.

**Competition** — intraspecific, reduced by careful crop spacing.
— interspecific, reduced by weeding and use of herbicides.
Predation, parasitism and disease reduced by crop protection procedures.

### Shelter
(a) Animal houses

(b) Windbreaks, shelter-belts, greenhouses, polythene shelters, protected seedbeds, etc. Also soil improvement by drainage, irrigation, liming, fertilizers, deep ploughing, hardpan breaking, manuring etc.

Thus it is clear that every aspect of the environment can be (and usually is) manipulated in the course of modern sophisticated agriculture. Generally though, only practices that show a definite economic profit are indulged.

The major ecological modifications that are made during the process of agriculture (*sensu lato*) that affect pest populations are as follows.

**Monoculture** — the extensive growth of a single plant species, with a simplification of the flora, partly by weed destruction.

**Increased edibility of crop plants** — the crop plants are more succulent, larger and generally more attractive to pests than the wild progenitors.

**Multiplication of suitable habitats** — the habitat and the microclimate becomes uniform over a large area.

**Loss of competing species** — may lead to the formation of new pests.

**Change of host/parasite relationships** — will lead to the development of secondary pests.

**Spread of pests by man** — as crops are grown in more parts of the world, the pests are also eventually spread around by accident.

These and other topics will be looked at in more detail in the next section of this chapter.

It should be stressed at this point that the vast majority of crop pests are in fact human-created through the ancient practice of agriculture. Completely 'natural' serious crop pests are very few and are limited to locusts, possibly a few tropical armyworms, and to some extent a few defoliating caterpillars (e.g.

Spruce Budworm) that occur in the extensive natural semi-monocultures of the northern taiga in N. Europe, N. Asia, and N. America.

# Pollination

The value of insect pollination of crops to man is really inestimable, although the annual yield of insect-pollinated crops in the USA has been estimated to be in the region of US $5000 million (at present values).

The crops pollinated by insects (entomophilous) include top fruit (e.g. citrus, apple, pear, peach, plum, cherry, almond, mango); bush and cane fruits (e.g. currants, raspberry, blackberry, gooseberry); ground fruit (e.g. strawberry); some Leguminosae (e.g. pulses, clovers); all Cruciferae (brassicas and other vegetables); other vegetables such as Cucurbitaceae and onions; cotton, cocoa, tea, some coffees; most flowers and some trees (e.g. lime).

The crops that are anemophilous (wind-pollinated) are mostly the Gramineae (cereals and sugarcane), and many trees, particularly the Gymnospermae (pines, spruces and other conifers). A few crop plants are partly entomophilous and partly anemophilous, such as beet, spinach, carrot, parsnip, white mustard, charlock and chrysanthemum. With a surprisingly large number of crops there is apparently uncertainty as to the precise manner of natural fertilization. Some crops are grown in cultivated clones with fruit-set by parthenocarpy (e.g. banana, and Smyrna fig), and others are propagated vegetatively like sugarcane and potato. Some crops are self-pollinating, such as *Coffea arabica,* pea, groundnut, some beans and many Solanaceae, but apparently sometimes cross-pollination by insects is effected.

One result of widespread use of chemical insecticides has been the great reduction of insect pollinator populations (mostly bees) in many parts of the world, and at the present time many crops are grown under conditions of inadequate pollination. However, for many tropical crops which are at least partially self-pollinating the precise value of increased insect pollination is not known (Free & Williams, 1977). As mentioned later (page 63), some crops typically over-produce flowers and fruit so increased pollination of these crops may be of little value; but it does seem quite likely that many crops are now suffering from under-pollination. An example of the difference in yield that can be achieved by increased pollination in some crops was shown on red clover in Ohio (USA) where the average yield reported was about $0.1 \, m^3/ha$ of seed; after an increase in the local bee population the field yield rose to about $0.4 \, m^3/ha$, and when a plot was enclosed and subjected to maximum pollination by bees the yield was raised to $1.1 \, m^3/ha$.

Bee destruction in Japan has been serious since World War II due to the very intensive nature of agriculture there, and in some orchards, by 1980, growers had to resort to hand-pollination — both time-consuming and labour-expensive. It has been estimated that about 25% of labour-time in fruit orchards in Japan is spent on pollination of the crop. An interesting development in Japan has been to use mason bees (in particular *Osmia cornifrons*) in fruit orchards for pollination. Artificial nest sites (usually consisting of small bundles of open canes or narrow tubes, about 5–6 mm diameter) are manufactured and situated in suitable sheltered locations in the orchards, often under the eves of the houses and buildings. These sites are readily colonized by the mason bees. Research has shown the *Osmia* bees to be good pollinators; they forage in Japan from 08.00 h to 18.00 h daily, visiting some 15 flowers per minute. A local population of 500–600 bees per hectare will give adequate pollination without recourse to outside pollinators, and there will be a 50% fruit-set within 65 m of the nest site (Maeta & Kitamura, 1980).

The main groups of insects responsible for flower pollination are
(a) Hymenoptera
    Apidae — Honey Bee (*Apis mellifera*), cosmopolitan
    — Bumble Bees (*Bombus* spp.), Holarctic only
    Megachilidae — Leaf-cutting Bees (*Megachile* spp.)
    — Mason Bees (*Osmia* spp.)
(b) Diptera
    Syrphidae — Hover Flies
    Muscidae — House Fly, Bluebottles, etc.

Some Lepidoptera (butterflies and moths) may be of importance, but mostly for ornamentals with flowers having a long tubular corolla. A few beetles and some thrips pollinate some crops, and a few somewhat bizarre tropical plants (e.g. orchids) are pollinated by hummingbirds or bats.

The most important wild pollinators are probably bumble bees and flies, but *Bombus* is only Holarctic in distribution. In the tropics, honey bees, Megachilidae and flies are probably the most important pollinators, but flies (Muscidae) can only pollinate the open-type flowers such as in the Cruciferae. In the cooler temperate regions *Apis mellifera* is domesticated and kept in hives which are easily handled and may be transported to orchards specifically for crop pollination. In the tropics however *Apis mellifera*, although domesticated and kept in 'hives', also occurs widely as wild colonies nesting in hollow trees, etc.

After the early catastrophes, when insecticide spraying in orchards and crops in flower resulted in large-scale bee destruction, most chemical companies are now including toxicity testing against bees as part of their regular pesticide screening programmes, and in the UK there are elaborate arrangements made to ensure that apiarists (bee-keepers) are warned before any major local insecticide applications are made. Also care is taken to avoid spraying particular crops in flower, so that bumble bee populations are safeguarded. In the tropics, however, there is seldom any warning given prior to spraying, but in these regions the majority of bees are wild anyway.

---

## Insect pheromones in relation to pest control

---

Pheromones, originally referred to as ectohormones, are complex chemical compounds, basically long-chain hydrocarbons such as alcohols, esters, ketones, aldehydes and sometimes ethers. Many have now been successfully identified, and some synthesized; a number are now available commercially from some chemical companies, for pest monitoring or control purposes.

They are secretions from several different types of glands, on different parts of the insect body, which open directly to the exterior. The secretory product is usually airborne for its distribution. Their basic function is for communication of a specific type between individuals of the same species, and the chemical elicits a specific reaction in the receiving individual. Within the large, complex, social colonies of ants, bees, wasps and termites, apparently quite sophisticated systems of communication have evolved, mostly based upon the use of pheromones.

The term 'pheromone' is usually regarded in a behavioural context in that it is a chemical or chemical complex that elicits a specific behavioural response in the receiving insect. Apparently some glands secrete a single chemical, whereas others secrete several which appear to act in concert. Generally there has been no overall agreement for a scheme of classification, but most workers favour the basis to be the type of behaviour released in the receiving insect. One approach

is to consider them to be of two basic types, those which give a releaser effect (this entailing a more or less immediate and reversible effect on the behaviour of the recipient), the others having a primer effect (this starting a chain of physiological events in the receiving insect). The latter group are usually gustatory in operation and typically control the social behaviour in Hymenoptera and Isoptera. Behaviour-releasing pheromones are typically odorous and their action is direct upon the central nervous system of the recipient, usually through the chemoreceptors in the antennae.

Recent work has demonstrated that most insect pheromones are in fact a complex of chemicals, and the individual chemicals are referred to as *components*. For example, the Smaller Tea Tortrix Moth sex pheromone has four components, two are major components and two are minor. Current opinion is that further research will reveal that almost all pheromones are actually chemical complexes of several compounds, and that the original idea of there being only one chemical present is completely incorrect, and originated because at the time the methods of chemical analysis were insufficiently sensitive to detect the minor components.

It is now apparent that much of the early work on insect pheromones is largely worthless (or, at best, of limited value) in that the researchers did not appreciate that almost invariably each pheromone was in fact a complex of major and minor components acting in concert on the receiving insect. Experimentation using only some of the chemical components of a particular pheromone inevitably led to anomalous results.

The types of behaviour (pheromones) used by Shorey (1976) were aggregation (including aerial and ground trail-following); dispersion; sexual; oviposition; alarm and specialized colonial behaviour. Some pheromones appear to have more than one function so these categories are of somewhat limited application.

## Aggregation

The reasons for aggregation are numerous and varied, and include collection around or to a food source, a shelter site, a site for oviposition, or colonization, recruitment of a sexual partner, and aggregation for swarming or dispersal purposes.

One of the most obvious cases can be seen by watching ants on their foraging trails, where the scouts are clearly laying scent trails. Trail-following by ants has been well studied. The scouts, after having located a food source, deposit droplets of pheromone on the ground and this stimulates trail-following behaviour amongst other workers. While the food source persists the ant workers continually reinforce the scent trail, but after depletion trail reinforcement ceases and the trail eventually disappears. Some trails are ephemeral but others persist for weeks or months. The Imported Fire Ant (*Solenopsis saevissima*) in the USA uses short-lived trails near the nest, but some species of Leaf Cutting Ant (*Atta* spp.) lay persistent trails to leaf sources up to 100 m distant, which may last for months. Similar trails are laid by termites when foraging.

Bees and wasps (*Vespa vulgaris*) leave a scent trail from their feet which is important in delimiting the entrance to their nests.

Bark Beetles (Scolytidae) use aggregation pheromones to designate host trees suitable for colonization, as these beetles only flourish when present in quite dense populations. The pheromones are released from the hind-gut of the beetle mixed with the various terpenoid compounds of the host tree which initially attracted the first invaders to the tree. These aggregation pheromones can be released by either sex and serve to attract individuals of both sexes. Colonization is usually succeeded by mating which involves the use of sexual pheromones. This type of aggregation can also be seen in the Japanese Beetle (*Popillia japonica*) and the Cotton Boll Weevil (*Anthonomus grandis*).

Certain mosquitoes release pheromones into the water at oviposition which attracts other females. And Sheep Blowfly (*Lucilia cuprina*) females apparently use an aggregation pheromone to form dense populations at sheep carcases for oviposition.

Aggregation at a suitable resting site has been demonstrated for the Bed Bug (*Cimex lectularis*) and some other cryptozoic species.

The mechanism for aggregation at a chemical source is usually chemotaxis where the insect can detect the gradient of odour molecules, and it often involves orientation by anemotaxis, that is positive orientation to air currents, particularly in the case of flying insects.

## Dispersal

Dispersal is clearly the opposite of aggregation, but is not encountered very often. However, some Bark Beetle males produce pheromones after mating which repel other males, and some female beetles release a repelling pheromone when they are unwilling to mate. *Tribolium confusum* females release a pheromone in the foodstuff they infest which repels other females and ensures a uniform population distribution throughout the available space. It is thought that the female Apple Fruit Fly leaves a pheromone trace on the apple surface after oviposition for she can be seen to drag her ovipositor over the surface and generally other females do not lay their eggs in the same fruit.

It seems that some dispersal secretions are the same as the defence secretions; for instance, nymphs of *Dysdercus* produce stinking coxal gland secretions when disturbed (thought to be a defence against predators) which causes the gregarious bug nymphs to scatter. Certain species of ants have alarm pheromones which in some circumstances (in the nest) induce aggregation but under other conditions (away from the nest) result in dispersal.

## Sexual behaviour

Sex pheromones may be produced in either sex and stimulate a series of behavioural sequences that usually results in mating. There appears to be typically a hierarchy of behavioural responses with increasing stimulation by sex pheromones. Once the two sexes are in proximity there is usually a close-range series of behavioural reactions, referred to as courtship behaviour.

The most usual situation is that a receptive virgin female insect will announce her availability through release of aerial sex pheromones, known as 'calling', and these cause a flight response and approach by receiving males. The night-flying moths (especially Saturniidae, Geometridae, and Noctuidae) are best known for their nocturnal emission of sex pheromones, which reputedly can attract males from as far as 5 km downwind. Males may produce a pheromone (sometimes called an 'aphrodisiac') when in the immediate vicinity of the female, which operates by inhibiting the female's tendency to fly away.

Sex pheromones are commonly referred to as 'sex attractants' or 'sex lures', which is misleading in that it implies that the odorous chemicals simply cause attraction, which is a great oversimplification. As will be discussed later, the male response to a female pheromone is complicated and sequential involving half-a-dozen or more separate stages.

Most of the sex pheromones that have been isolated, identified and synthesized are from the Lepidoptera, and include Red Bollworm, Spiny Bollworm, Pink Bollworm, *Heliothis* spp., *Spodoptera littoralis*, *Chilo* spp., *Prays citri*, *Prays oleae*, Gypsy Moth, *Bombyx mori*, Cabbage Looper, Codling Moth, and Honey Bee queen. Pheromones of some of the important fruit flies (Tephritidae) such as *Dacus* and *Ceratitis* spp. have also been synthesized, as have some for Scarab Beetles and Scolytidae. Many of the sex pheromones are either difficult to synthesize or else expensive to produce, and this has led to the development of chemi-

cal pheromone mimics for large-scale management programmes. These chemicals are discussed in the section on 'attractants' which follows later.

## Oviposition

As already mentioned, Sheep Blowfly females release an aggregation pheromone when they oviposit on a suitable sheep carcase in Australia, and the result is the formation of a dense population. Bark Beetles (Scolytidae) aggregate on suitable trees as a result of use of aggregation pheromones, but the ultimate purpose of the aggregation is for oviposition and breeding. Thus, functionally many of the aggregation pheromones are also used to stimulate oviposition upon the correct host plant. This point is mentioned later under the heading of 'plant odours', as there appears to be probable interaction between plant volatiles and pheromones connected with the oviposition of many phytophagous insects.

## Alarm behaviour

This is characteristic of social Hymenoptera and Isoptera, and may be seen most dramatically when field workers disturb a nest of Paper Wasp (*Polistes* spp.) or arboreal ants in plantation trees. The wasps (and bees) produce alarm pheromones both when they sting and when gripping with their mandibles. The pheromone is released from glands in the stinging apparatus and from the mandibular glands, but apparently the worker can open the sting chamber and emit the pheromone without the necessity of stinging.

The strictly social (colonial) aspects of pheromone use are not considered here.

## Attractants

Initial experimental studies on female sex pheromones were conducted using live virgin females that had been laboratory-reared, or else a chemical extract was made from the abdomen tips of many young females and this was used instead of the live insects. It was soon apparent that these sex pheromones have considerable potential application in pest management programmes and so many organic chemists in government and industrial establishments started work in this field.

One approach was to carefully analyze the tiny quantities of natural pheromone produced by virgin females, and, when the chemical components were isolated and identified, to attempt to synthesize the same chemical compounds in the laboratory.

The second approach was to try and synthesize closely related chemical compounds which might possess the behavioural qualities of the natural pheromone, but were easier and cheaper to manufacture. In this way pheromone homologues and analogues have been produced commercially. A *pheromone homologue* is a very closely related compound, which differs only from the natural pheromone by chain lengthening or shortening following, for example, addition or removal of a methylene group. A *pheromone analogue* is a less closely related compound that has major basic differences in structure, such as the change of a functional group or its position, for example an alcohol, ester or ketone.

The third approach was to use a vast range of organic chemicals which it was thought might possibly function in a manner similar to sex pheromones and to use them in laboratory and field trials in a purely empirical manner to see if they did possess such qualities. The organic chemicals that have been found to be successful in attracting certain male insects are collectively referred to as *sex attractants*. Obviously for monitoring purposes it does not matter how the chemical attracts the insects so long as it does attract them sufficiently well, and a great deal of research effort is being expended in this field at present. Sometimes these chemicals are termed *pheromone mimics*, for the obvious reason that they produce a similar reaction in the receiving male insect. 'Hexalure' is a chemical attractant produced commercially in the USA for use with Pink Bollworm on cotton, in a disruptive technique to prevent mating.

Recent work has demonstrated that many pheromone complexes are subjected to *synergism* in one way or another. In some cases it appears that some of the minor components have a synergistic effect on the major components or else on the pheromone complex as a whole. Sometimes the synergist may be a chemical released by the host plant; this may be of more importance in aggregation or oviposition behaviour, for example in scolytid infestations of forest trees various terpenoids are released by the injured tree which interact with the aggregation pheromones released by the beetles. Empirical chemical testing has discovered a number of synergists for use with sex pheromones that appreciably enhance their performance.

The sex pheromones of insects must, by their very nature, be quite specific to each species but some of the attractants have a very useful and much broader response, for example 'Cu-lure', developed initially for Melon Fly (*Dacus cucurbitae*), and methyl eugenol both attract all species of *Dacus* and some other fruit flies in addition, which makes these chemicals very useful for survey studies.

### Plant odours

Most plants release volatile odorous chemicals into the atmosphere (although the majority are undetectable by human sense) and phytophagous insects react to these chemical stimuli when locating host plants. Monophagous and oligophagous phytophagous insects usually react to specific volatile odorous chemicals in, and emitted from, the host plant. It is thought that polyphagous insects either have no olfactory chemical response, or else react to general plant chemicals. The olfactory chemoreceptors are mostly situated in the antennae, but in some Diptera they are located on the tarsi (feet).

It has very recently been demonstrated that very

strong plant odours can inhibit sex pheromone reception; the interaction of plant odours and sex pheromones is thought to be complementary under natural conditions, so that mating is more likely to be successful on appropriate host plants, whereas the chances of mating taking place on inappropriate host plants are reduced. As this is a very recent discovery, as yet little work has been carried out, but future studies might well give rise to a greater understanding of the general phenomenon of host-specificity in phytophagous insects.

The chemical attractants in plants, when identified, are usually a mixture of many different compounds, for example, cruciferin in the brassicas is a complex mixture of glucosides, amines, and other chemicals.

The major chemical repellents in plants seem to be terpenes, tannins and various alkaloids. Tannins are mostly found in horsetails, ferns, gymnosperms, and some angiosperms, and they are quite antibiotic to many pathogens. Alkaloids are mostly found in angiosperms and are thought to be of more recent origin. It is thought that the tannins were developed initially in the process of evolution as a deterrent to grazing reptiles, and the alkaloids similarly evolved as a protective mechanism in angiosperms to repel grazing mammals. It seems unlikely that the Insecta were at all involved in the evolution of feeding repellents in plants, although these may now be of considerable importance with respect to phytophagous insect feeding behaviour. The insect biotypes that come to feed on 'repellent' varieties of crop plants (and other plants) usually develop biochemical detoxification mechanisms, so that the poisonous compounds are broken down into non-toxic degradation products.

### Sex attraction in Lepidoptera

Most of the work on sex pheromones and attraction has been done on the Lepidoptera, and the greatest potential for pheromones in pest management is in this group.

The 'calling' female moth emits her sex pheromones from the genital opening on the abdomen tip and the chemical complex is carried downwind as a plume. The odour plume is basically cone-shaped, but is flattened ventrally if the moth is close to the ground; the plume widens and the chemicals disperse as they are carried from the source. The shape of the plume is clearly controlled by wind speed and direction, the ground contours, and the presence of tall vegetation such as trees. Most virgin female moths 'call' at dusk or at night when there may be strong temperature gradients, or even inversions, over the ground, and these will obviously have effects on the spread of the pheromone.

The male moth responds to the pheromone by anemotaxis, in that it flies upwind in a zig-zag pattern. At first (assuming the male to be some distance from the female) the flight pattern of the male moth diverges from the plume of pheromone quite often, but as it approaches the female the pheromone concentration increases and the flight of the male moth becomes more direct.

The patterned response by a resting male moth to female sex pheromone can generally be described in half-a-dozen sequential stages, as follows:
(a) reception — antennal elevation or twitching,
(b) activation — wing fanning or fluttering,
(c) active flight,
(d) orientation to the source of pheromone — i.e. anemotaxis,
(e) alighting — landing in the immediate vicinity of the female moth,
(f) courtship — including gland extrusion and release of male pheromone,
(g) mating.

Thus it is clear that the behavioural response by a male moth to a 'calling' female is a complex sequential series of events and is not just a simple attraction.

The initial responses are made to a low concentration of pheromone, but the later events require an ever-increasing concentration of pheromone. It is now thought that the different components of the sex pheromone are responsible for different parts of the behaviour sequence. Thus if in an experiment one minor component is missing this will result in the behaviour sequence being broken, and the experimental results confusing.

In *Adoxophyes orana* it has been demonstrated that the female sex pheromone has two components and that there are three different types of chemoreceptors on the male antennae (Den Otter, 1980). One type of sensillum reacts to the first component in electroantennagram studies, the second type reacts to the second component in the female pheromone, and it is suspected that the third type of sensillum may react to the male pheromone at close range, but this has not yet been demonstrated.

---

## Development of pest status

---

This takes place through a number of different agencies that fall into two main categories, as follows.
(a) Ecological changes — as already mentioned, these are largely results from the artificial innovations of widespread agriculture.
(b) Economic changes — which are related to the human social context.

An important publication relating to this problem is the 18th Symposium of the British Ecological Society (April 1976) on *'Origins of Pest, Parasite, Disease and Weed Problems'*, edited by J.M. Cherrett and G.R. Sagar (1977).

### Ecological changes

There are really only three aspects of importance in considering crop pests, and these are

(a) state of the pest population (i.e. numbers of pests present);
(b) nature of the damage done to the crop;
(c) value of the damage as assessed by human society. Thus each aspect of the way in which pests arise through ecological changes will relate to population numbers, for by definition a species is only a pest at or above a particular population density.

**Increase in numbers.** This is the most common way in which an insect species attains pest status and it usually follows because of one of the factors listed below. But often insect populations can be seen to change, sometimes drastically from a few individuals per plant to literally thousands on the same plant the following season, for no obvious reason at all; probably the most spectacular population fluctuations of this sort are seen with some Diaspididae, Aleyrodidae and Aphidoidea. Presumably there is a subtle population control being effected but often we are not able to understand it.

Most insect species will have a natural phenological cycle whereby the relatively small number of overwintering (or equivalent) or immigrant adults lay large numbers of eggs and the population develops in a regular, stepped cycle, increasing through each generation of the growing season; finally, there will be a population 'crash' or large-scale emigration at the time of harvest, or the onset of winter (or dry season). The population growth cycle is 'stepped' because of natural predation on the different instars, the earliest instars usually being the most heavily preyed upon.

**Population resurgence.** The effect of most chemical pesticides is short-lived nowadays, and once the suppressive effect declines the pest population will naturally resurge, possibly up to the economic injury level.

There may also be resurgence of a secondary pest due to insecticidal destruction of natural enemies, or from some other ecological upset.

**Migration.** Immigration is the movement of a pest population into an area from elsewhere and in certain parts of the world is a very important source of major pests. In most of the tropics there is little insect migration (apart from locusts) because of the general stability of climate. Migration is typically an animal phenomenon of the colder parts of the world, where animals move away from northern areas after the short warm summer, and before the onset of the long cold winter. There are some tropical migrations, including the spectacular ungulate migrations of eastern Africa, where animals move to new food sources away from arid areas suffering from their annual dry season; they are basically following seasonal rains and the new grass growth that is promoted across part of the continent.

Animal migrations and dispersal movements are natural phenomena characteristic of all phyla in the animal kingdom and many species now regarded as pests will have generally dispersed during the millenia from their endemic areas (areas of origin). This innate tendency will still be present in all animal populations, but in some cases it operates slowly and may not be at all obvious. Flying insects can effect their own dispersal, influenced by winds and air currents of course. Some of the more novel means of dispersal include riding on floating flotsam (for rats, insects etc.), concealment in the feathers of migrating birds, and aerial transport by typhoon and hurricane. Migration and natural succession was clearly demonstrated on the island of Krakatoa after its total devastation by a volcanic eruption.

Locusts and armyworms (*Spodoptera* spp.) are migratory tropical species of considerable economic importance; they breed in areas of hazardous climate which compels them to disperse on their sometimes lengthy journeys.

The Brown Planthopper of Rice (BPH) has recently become established as a major pest of rice in Japan through its migratory behaviour. The climate of S. Japan (Kyushu) is subtropical or tropical in the summer, but the winter is cold with snow and ice and the BPH cannot survive the winter. The White-backed Planthopper (*Sogatella furcifera*) is likewise an annual migrant into Japan. These planthoppers (Delphacidae) live along the coastal regions of E. Asia (and elsewhere in India and S.E. Asia) and each spring they migrate northwards, usually entering Kyushu in Japan in the period mid-June to mid-July when they are caught in wind traps along the coast. They then breed on the rice crops and usually have four generations each summer in Japan before dying out in the cooler weather of the autumn after harvest.

The climate of Kyushu is such that many tropical crops can be grown in the summer, either annuals such as rice, or hardy perennials like *Citrus* and peach. So, climatically and floristically Kyushu lends itself to exploitation by migratory insect pests. Several other Delphacidae and some Cicadellidae are also probably migratory here, as also is the Rice Leaf-roller which has recently become a serious pest in Japan.

**Character of food supply.** Plants grown for agricultural purposes have usually been selected for their nutritive value and are typically large and succulent with especially large fruits, leaves and lush foliage. Thus cabbage and lettuce are far more attractive as food for caterpillars than are wild crucifers (such as *Capsella bursa* and *Cardamine flexuosa*), and *Solanum nigrum* hardly compares with egg-plant, tobacco or tomato as a food source; similarly maize and sorghum are far more attractive than wild grasses. A large field containing such a crop represents literally an inexhaustible food supply (for at least the duration of the crop) for a potential pest species, so it is little wonder that pests evolved concurrently with the practice of cultivation.

This topic is scarcely distinguishable from the following one in practical terms, especially concerning the use of new varieties and clonal propagation. The growing of new varieties of crop plant can lead to the development of a new pest species. In S.E. Asia new high-yielding varieties of rice were introduced some years ago from IRRI, and in order to achieve the high yields possible the farmers had to use extra nitrogenous fertilizers; the use of these fertilizers produced a lush vegetative growth which seems to be more favourable for BPH reproduction. Thus as already mentioned, *Nilaparvata lugens* rose in status from a minor pest in the 1960s to its first record as a major pest in 1970, to a generally serious pest in 1975 and a widespread serious pest by 1980. In many parts of S.E. Asia it is now regarded as the 'number one' rice pest. This new-found fecundity has also led to a very rapid biotype formation and resistance to many pesticides and resistance-breaking of many new rice varieties.

**Monoculture.** The growing of a single crop species over a large area provides an unlimited source of food for pest species, especially when the crop plant is particularly succulent. The present worldwide tendency is towards mechanised agriculture which requires larger fields and fewer hedgerows which traditionally delimited each field, and so the crop monoculture becomes even more extensive. This practice will encourage some insect species to become more abundant, and hence important as pests, as was observed about a century ago with Colorado Beetle on potatoes in the USA.

As well as monoculture becoming more extensive some new crop varieties have become more specialised in their growing conditions, and there is a tendency to reduce the extent of crop rotation.

As mentioned before, the taiga of northern Asia and America with its coniferous forest dominated by only a few tree species, and the northern temperate oak/beech forests, are the natural equivalents of an agricultural monoculture. Thus it is not surprising to learn that *Quercus* is attacked by 1000 different species of insects!

Farmers also often specialize in particular crops for reasons of convenience, both agriculturally and economically, so that, for example, in parts of eastern England and eastern Canada where the fenland peats (and mucklands) occur, it is common practice to grow carrots, rotated with celery and parsnip, because they all require similar soil conditions, but from a pest point of view this is scarcely a rotation as all belong to the Umbelliferae and have a similar pest spectrum. The more extensive a monoculture becomes then the greater is the pest problem, generally.

*Fig. 4.* Schematic graph of the change in general equilibrium position of the Colorado Potato Beetle (*Leptinotarsa decemlineata*) following the development of widespread potato culture in the USA (from Stern *et al.*, 1959).

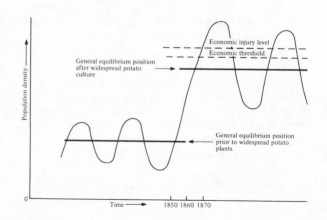

Plantation and orchard crops are of necessity long-term monocultures which suffer heavy pest depredation, but there is the advantage in these cases that biological control may establish a long-term pest reduction.

A recent trend in agriculture is towards extreme standardization of product, for a number of different reasons connected with both crop production and product utilization. A good example is rubber in Malaysia. After World War II it was thought that synthetic rubbers (produced from oil) had destroyed the natural rubber (NR) market. However, it was found that natural rubber possesses some qualities that have not been matched by synthetic rubber. Then in 1965 new cultivation methods and processing produced the new Standard Malaysia Rubber (SMR) constant in quality to tight specifications. There is now a tremendous world market for natural rubber, and this industry is booming throughout S.E. Asia, mostly through new improved methods of production.

These new production methods include *cloning*, the vegetative mass-production of ideal genetic stock through mist-propagation and other techniques. Clonal propagation of oil palm in Malaysia and S.E. Asia is proceeding rapidly at present as this industry expands throughout the region. To have large areas of genetically identical trees is very useful from both crop production and product processing points of view, but these trees are vulnerable to disease and pest epidemics by virtue of their uniformity, and the most serious threat would be from a new major pest or disease which might arise. Clonal cultivation clearly represents the most extreme form of monoculture.

Oddly enough, in the very large-scale monocultures, such as the wheat fields on the Canadian prairies, there is the rather anomalous situation whereby the wheat monoculture is regarded as a highly stable habitat with a diminished pest problem. On reflection this situation is not so strange because the extent of the wheat monoculture is very great. It seems that each extreme of habitat diversity (from tropical rain forest to prairie wheat) is stable with a reduced pest problem, whereas it is the intermediate stages in the process of simplification that suffer the greatest pest problems.

Mixed stands (*mixed cropping*) are being recognized as a cultural method of alleviating pest situations (Way; in Cherrett & Sagar, 1977), and it is being developed successfully for use in barrios in the Philippines by IRRI. For large-scale agriculture, with the present reliance on mechanization, mixed cropping is generally not feasible, but for small peasant farmers (barrios, shambas, etc.) in the tropics it can be a valuable means of combating pest infestations. In the tropics by far the greatest area of crops is actually grown on smallholdings and not the huge agricultural estates characteristic of the northern temperate regions.

**Continuous cropping.** The plantation and orchard crops are all very long-term and because of this they suffer from particular pest problems, but in compensation their pests are to some extent controlled by natural enemies. Field crops are typically characterized by their short duration (an ephemeral habitat) as at most they are annuals, and sometimes by careful timing the crop may be grown before the pest population catches up. However, there is pressure of late to produce a continuous supply of vegetables, and new varieties tolerant of different climates have been bred. In the vicinity of large urban conurbations many vegetables are now being grown almost all year round. For example in S. China and S.E. Asia, by varying the varieties grown, it is possible to produce onions, tomatoes, and several species of *Brassica* continuously for the local markets. This encourages the build-up of the pest populations; for instance, in S.E. Asia, Diamondback Moth and Turnip Mosaic Virus have now both become serious pests of *Brassica* crops.

**Minimum cultivation techniques.** A recent agricultural technique in ground preparation is known as minimum cultivation, and it consists essentially of a chemical destruction of old crop remains and weeds, followed by a subsequent planting of the new crop into the undisturbed soil. Ploughing and harrowing normally reduce the population of soil pests by exposing them to sunlight and desiccation, and to predators and parasites. Many Coleoptera, Lepidoptera and Diptera feed on the aerial parts of plants as larvae but pupate in the soil. These and other soil-inhabiting pests would normally be depleted in numbers by normal cultivation methods, and so in areas where minimum cultivation techniques are employed there is often a build-up in numbers of soil pests.

**Multiplication of suitable habitats.** Farming leads to a simplification of the flora by a selection of plants suitable for husbandry. Thus insects associated with these plants have a more attractive and concentrated food supply, as well as a greater total host number. The most outstanding example of this is shown in the storage of grain and foodstuffs; many storage pests exist in small populations in the field but increase enormously in numbers in the favourable micro-climate and abundant food of the grain store. Examples are to be seen in *Sitophilus* on maize cobs, *Sitotroga* on sorghum, and the Bean and Cowpea Bruchids.

**Loss of competing species.** Under conditions of monoculture an area harbours fewer insect species than under 'natural conditions', and many insects now become pests which were not pests under the natural conditions. Sometimes specific pest control measures may remove one pest, but another insect released from competitive pressure may increase in numbers and become a new pest. Once the new pest is established mutations make the relationship between it and the crop even closer, for as numbers increase

more mutants appear and can be selected to consolidate their niche as pests.

**Change of host/parasite relationships.** Most insects are kept in check by their predators and parasites, although when a pest species increases in number there is typically a time-lag between its increase and the parasite and predator numbers. Parasites are generally quite specific but predators less so, and the time-lag of the parasite population build-up is generally less than that for a predator. The greater the time-lag between the pest population increase and that of the parasite or predator, then the more likely is the species to be a serious pest. Generally agricultural operations involving large-scale insecticide applications may affect parasites and predators more than the pests. One of the classical cases is that of the Red Spider Mite (*Metatetranychus ulmi*) which became a pest on fruit trees after widespread use of DDT in orchards. A good African example is the Giant Looper Caterpillar (*Ascotis selenaria*) on coffee; this is normally a minor pest at most, but has become serious in places where parathion has been used regularly over a long time.

**Spread of insects and crops by man.** Almost every major crop was once endemic only to one particular part of the world, but during the centuries of world exploration and trade most crops were distributed throughout the regions which had a suitable climate for their cultivation. This state of affairs is generally highly desirable, especially since some crops flourish in their new habitats (e.g. rubber in Malaysia and coffee in S. America) and most countries now aim at agricultural diversification. Unfortunately some of the endemic crop pests were accidentally (or otherwise) distributed along with the crops (or afterwards) and became serious pests in the new locations, for they were usually without the restraint of their endemic natural enemies. Several aspects of this problem are reviewed in the book by Elton (1958). The main means of pest dispersal by human agency are as follows.

*Dispersal by unknown means.* Some pests (and crops) appear to have been dispersed since antiquity, though it is not clear how this dispersal was effected. It must be remembered that in some parts of the world there have been human population movements ever since the early days of agriculture. Also, of course, all animal species have an innate tendency to disperse, or migrate, of their own accord. However, there are numerous records of crop pests having been spread by man, either accidentally or intentionally.

*Transport on the host (plant).* The symbionts that normally live with man and his animals have been transported to all parts of the world, and likewise the transport of live plants provides an easy mechanism for the dispersal of many pests, especially scale insects, mealybugs, and aphids on the plants, and beetles, nematodes, moth pupae, etc. in the soil around the roots. It is only recently that phytosanitation measures have been adopted. Classical cases include the spread of *Icerya purchasi* around the Mediterranean on nursery stock, and also several scale insects on *Citrus* and other fruit stocks from S. China to California, Hawaii and Australia. Hessian Fly (*Mayetiola destructor*) is thought to have been transported to N. America from Europe in wheat straw. Colorado Beetle was brought from the USA to Europe (France) on stored potatoes.

*Dispersal through trade.* Rats, many weeds, and some insects have been dispersed over the international trade routes, usually accidentally, in either produce (as above), containers or ballast. A recent case of note is the spread of Rice Water Weevil from the USA into Japan, where unfortunately it is now established. It is thought that weevil pupae were present in hay exported to Japan for use by dairy farmers, and they were then carried into rice fields in the manure from dairy farms. About half the immigration records for Colorado Beetle in the UK are from produce (seldom potatoes) in ships arriving at coastal ports in southern and eastern England.

*Stowaways in air transport.* With the recent development of international airways and rapid transport from one part of the world to another, there is a constant danger of insect pests being transported live from one region to another. This threat applies also to medical pests (especially vectors) which is why aeroplane cabins are sprayed with aerosol insecticides prior to passenger departures. Several of the records of Colorado Beetle introductions into the UK are clearly cases where the beetles were stowaways in the undercarriage of the planes from Spain and France, and the beetles were dropped along the airport approach line as the undercarriage was lowered.

*Deliberate introduction by man.* The most notorious examples are without doubt the introduction of prickly pear cactus (*Opuntia* spp.) and rabbit into Australia. Many other examples are listed by Simmonds & Greathead (in Cherrett & Sagar, 1977) and include the Gypsy Moth (*Lymantria dispar*) imported into the USA in about 1870 by an amateur entomologist to test its value for silk production. Some moths escaped and became established as one of the most serious forest pests in the north-eastern USA.

### Economic changes

The definition of a pest relates clearly to the value of the damage done by the insect (or animal) as assessed by human society, and so any changes in the value of the crop will affect the importance of the pest. Damage that is not important when prices are low can be very serious when prices are high. Sometimes the converse situation is true, if an important food crop is in short supply then some pest damage may be tolerated.

**Change in demand.** If some crops are replaced by others, the pests of the former crops become less important. Greater demand for a crop increases its value and the incentive to grow it. The demand may be for increased quantity and quality, both of these factors affecting the importance of the pests. If the crop is in short supply the consumers are less selective than if it is abundant. Wireworms do not greatly affect the yield of potatoes, but their tunnels spoil the appearance and keeping qualities. If the supply is short, consumers overlook a little damage, but with the recent demand for packaged vegetables such potatoes are generally unsaleable, thus making wireworms a much more serious pest.

**Change in production costs.** A pest may become economically important when agricultural practices change. If a new high-yielding variety is developed, minor pests which attack it may become of economic importance.

The case of the Brown Planthopper becoming a serious pest on new rice varieties is complicated in that, in order to achieve the high yield possible from these new varieties, it is necessary to use more fertilizer on the crop. The extra fertilizer produces a lush foliage which induces the BPH to a higher level of fecundity, making it a serious pest on a now more expensive crop.

Crops for export are more valuable than those for local consumption. In Europe it is now possible to buy fresh peppers, aubergines, avocados, and many other tropical fruits and vegetables that are being air-freighted from Africa, the Middle East and other tropical regions. These crops are of course now much more expensive to produce because of air freight but are sold at correspondingly higher prices.

---

## Pest damage

---

This is considered in more detail later (chapter 4) but is mentioned here briefly because of its implications in connection with choosing a pest control strategy. The single most important aspect is the relation between the damage done by the pest and the part of the plant harvested. *Direct damage* is when the part of the plant to be harvested is the part attacked, such as the leaves of tobacco, fruits of tomato and *Citrus*, tubers of potato and sweet potato; in these cases clearly the damage is more important. *Indirect damage* is when the part of the plant damaged is not the part to be harvested. Examples include the roots of tobacco and sugarcane, leaves of tomato, sweet potato, and *Citrus*. In these cases it is usually possible to ignore quite surprisingly high levels of pest infestation, as these infestations may only have a marginal effect on the crop yield.

### Direct effects of insect feeding

Biting insects may damage plants as follows.
(a) Reduce the amount of leaf assimilative tissue and hinder plant growth; examples are leaf-eaters, such as adults and nymphs of *Zonocerus* and *Epilachna* and larvae of *Plutella* and *Athalia*.
(b) Tunnel in the stem and interrupt sap flow, often destroying the apical part of the plant; these are stem borers and shoot flies, such as *Dirphya* in coffee, *Earias* in cotton, *Atherigona* in maize and sorghum.
(c) Ring-bark stems, for example *Anthores* on coffee.
(d) Destroy buds or growing points and cause subsequent distortion or proliferation, as with *Earias* in cotton shoots.
(e) Cause premature fruit-fall, as with Mango Fruit Fly.
(f) Attack flowers and reduce seed production, as with Maize Tassel Beetle on maize.
(g) Injure or destroy seeds completely, or reduce germination due to loss of food reserves; examples are Sorghum Midge, Maize Weevil, Coffee Berry Borer and Pea Pod Borer.
(h) Attack roots and cause loss of water and nutrient absorbing tissue; as in Black Maize Beetle, and various Chafer larvae (Scarabaeidae).

(i) Remove stored food from tubers and corms, and affect next season's growth; examples are *Cylas* weevils (both adults and larvae) in sweet potato tubers, Potato Tuber Moth larvae, and Yam Beetle.

Insects with piercing and sucking mouthparts may damage plants as follows.

(a) Cause loss of plant vigour due to removal of excessive quantities of sap, in extreme cases wilting results, as in the stunting of cotton by *Bemisia* (Whitefly), and Aphids on various plants.

(b) Damage floral organs and reduce seed production, for example Coffee Lygus Bug on coffee.

(c) Cause premature fruit-fall; example Coconut Bug (*Pseudotheraptus*).

(d) Cause premature leaf-fall, as do many diaspidid scales.

(e) Inject toxins into the plant body, causing distortion, proliferation (galls), or necrosis; examples are seen in Lygus Bug damage on cotton leaves and shoots, and the stem necrosis on cashew by *Helopeltis anacardii*, and cotton boll abortion by *Calidea* bugs.

(f) Provide entry points for pathogenic fungi and bacteria, as does *Dysdercus* on cotton bolls (for fungus *Haemospora*), and *Calidea*, also on cotton bolls.

**Indirect effects of insects on crops**

(a) Insects may make the crop more difficult to cultivate or harvest; they may distort the plant, as do *Earias* larvae on cotton when they cause the plant to develop a spreading habit which makes weeding and spraying more difficult. They may delay crop maturity, as do the bollworms on cotton, and grain in cereals may become distorted or dwarfed.

(b) Insect infestation results in contamination and loss of quality in the crop; the quality loss may be due to reduction in nutritional value or marketability (lowering of grade). Loss of yield in a crop is obvious but a nutritional quality loss is easily overlooked; this is the type of damage done to stored grain by *Ephestia cautella* and *Tribolium*. A more common loss of quality is the effect of insects on the appearance of the crop, for example skeletonized or discoloured cabbages have a lower market value than intact ones. Attacked fruit is particularly susceptible to this loss in quality, as seen by skin blemishes and hard scales on citrus fruit. Contamination by insect faeces, exuviae, and corpses all reduce the marketability of a crop, as do black and sooty moulds growing on the honeydew excreted by various homopterous bugs.

(c) *Transmission of disease organisms*

1. Mechanical transmission, also termed passive transmission, takes place through feeding lesions in the cuticle. Sometimes the pathogen (usually fungi or bacteria) is carried on the proboscis of the bug or sometimes it is on the body of the tunnelling insects. Examples are seen in the case of the platygasterid wasp which transmits Coffee Leaf Rust, and *Dysdercus* and *Nezara* on cotton but in this case the spores are carried in the saliva of the bugs.

2. Biological transmission. Most viruses depend upon the activity of an insect vector for transmission. The vector is usually also an intermediate host, as is the case with most aphid and whitefly hosts. Diseases transmitted in this manner include Leaf-curl of cotton, Tobacco Mosaic, and Cassava Mosaic.

---

# Economics of pest attack and control

---

Economics in this sense is the relation of crop losses to production costs. Sometimes due to pest outbreaks a particular crop may be scarce and sell for a higher price, so that the individual producer may not suffer financially. From the economic point of view, however, one must consider the overall (national) picture, not just individual producers.

**Decision to take action**

This decision is based upon an accurate estimation of the cost of control measures in relation to possible profits likely to accrue as a result of control. The *economic threshold* in pest control is the point at which a particular pest can be controlled at a cost of less than the expected market value of the expected yield increase. Often the cost of control may be known very accurately but the possible profit is a subjective assessment based upon past experience and guesswork. Because many pest control costs are relatively low, against a serious pest attack it is usual for no calculations to be made; none are needed! However, sometimes costs are critical, for example Coffee Berry Disease in E. Africa and the spray Difolitan which is prohibitive in cost to the small farmers and generally can only be afforded by the large estates. Impressions of the degree of infestation of a crop are often wrong or else very misleading; temperate examples include the infestation of Mangold Fly on sugar beet, where the damage is often striking and unsightly, but experiments have shown that 70% defoliation reduces the yield by only 5%, and a 50% defoliation has no measurable yield loss. Similarly Black Bean Aphid infestation on bean stems is misleading in that an early infestation which looks only light can reduce yield by as much as 30%; in this case damage is more serious than would appear at first sight.

In some areas certain crops are always *at high risk* from particular pests, as shown by years of empirical observations, and in these cases preventative pesticide applications are usually made, often in the form of granule application at the time of sowing or else seed dressings.

There is a definite danger though that in some areas the mere sight of a particular key pest in a crop will precipitate panic action by a farmer who will immediately spray with insecticides when in fact their application might not be necessary.

### Cost/potential benefit ratio in decision

In practice, cost/potential benefit ratios are only known for valuable crops like citrus and apples, and usually then apply only to the benefit likely to result from control of the heaviest infestations that may occur. Generally little account is taken of differences in infestation and prices; thus spray programmes planned using this concept are only applicable to comparatively steady rates of infestation and cash return. When the ratio is high, uncertainty about the likelihood of attack can be ignored. Seed dressings are a cheap form of protection and can give a cost/potential benefit ratio as high as 1:10, when used against severe attacks of Wheat Bulb Fly or wireworms. Doubt as to whether they are needed on each occasion wheat is planted after fallow (danger from Wheat Bulb Fly) or ploughed up pasture (wireworm) is usually discounted because the ratio of 1:10 need only be reached once in ten times for the operation to break even with regard to costs. Surveys for use of insecticides against wireworms on potato lands in the British Isles by the National Agricultural Advisory Service (now ADAS) have shown that in only 25% of fields is there likelihood of wireworm damage. Therefore the cost/potential benefit ratio must be at least 1:4 for treatment to be profitable on average, but attacks are seldom severe enough to give a ratio of 1:10, and so an average of 1:4 or more for treated fields is unlikely. Thus treatment of all potato fields is not only unnecessary three times out of four, but the total effect is to increase the cost of production of potatoes and probably to decrease growers' profits. Surveys have also shown that potato damage due to wireworms has decreased in recent years, but the use of insecticide by farmers continues to rise; an obvious waste of time and money! Experience has shown that often growers continue a treatment long after the need for it has finished.

### Relation between yield increase and pesticide dosage

If a graph is drawn showing the relationship between rate of application of pesticide and percentage increase in crop yield, it is apparent that the curve is sigmoid but slightly asymmetrical. At the lower application rates the yield increases are low; at higher application rates the increase becomes high and then gradually decreases; finally, the yield increase decreases rapidly. It is possible to calculate a dosage response curve for a particular pest at different population densities, which shows what control measures are necessary, and also shows the rate of application at which profits no longer increase. However, if the dosage reponse curve is to be used to decide upon the correct pesticide use, the relationship between yield and infestation must be known, and the populations likely to cause economic damage must be predicted. Unfortunately, this relationship and prediction cannot yet be made very accurately for any insect pest, and can only be made at all for a few.

### Economics of national pest control programmes

Most of the cost of pest control programmes is borne by the consumer in the price which he pays for food. However, the cost of quarantine, eradication of imported pests, biological control, and much of the cost of research into pest biology is met through government expenditure. The cost of the UK quarantine laws against Colorado Beetle is probably in the order of £50 000 per year, or 10p per acre to be added to the cost of the potato crop; this is a very good investment when compared with the losses which a heavy attack would cause. Attempts to eradicate a newly established pest by chemical control can be very costly. After the Mediterranean Fruit Fly was accidently introduced into Florida a total of 2 million hectares of citrus land and adjoining land was sprayed at a total cost of some £3–4 million, yet the entire operation cost no more than about 5% of the annual value of the citrus crop and saved the industry. The cost of biological control programmes is very difficult to estimate but DeBach (1964) said that, for the Department of Biological Control of the University of California for the period 1923–59, a total expenditure of £1½ million has resulted in a saving on five projects of £40 million, with a recurrent benefit of £10½ million. Thus it can be seen that the cost of biological control does compare very favourably with chemical control.

### Pest assessment

In sampling and assessment programmes it is found that the relationship between pest numbers and damage is often logarithmic rather than linear, and it is usually more effective to take many samples and separate them quickly into different, easily distinguished, categories of infestation, than to carefully count numbers of pests on a few samples, if the results are to be statistically analysable.

It is highly advisable to seek advice from a biological statistician, when planning a sampling and assessment programme, in order to ensure that the results obtained will be suitable for statistical treatment. The publication by FAO/CAB (1971) includes details of damage assessment for about 48 major pests of 14 tropical crops (see chapter 4 for more detail).

### Efficiency of pest control measures

Methods of pest control tend to operate with a characteristic *efficiency* in that they reduce pest populations by a fixed proportion, more or less regardless of the number of pests involved. The efficiency of a treatment can be expressed as the percentage reduction of the pest population, or of the damage. It therefore follows that the numbers of a pest surviving a treatment depend upon the numbers attacking the crop, as well as the efficiency of the control measure. The efficiency of an insecticide treatment invariably declines with time since application, and is closely related to residue persistence, the growth of the crop, type of soil, sunshine, and moisture level.

The *effectiveness* of a control measure, as distinct from its efficiency, is reflected by the number of pests surviving after treatment, or, from a practical point of view, the amount of damage occurring after treatment. Thus a control measure will be most effective when dealing with relatively light infestations. Consequently, any actions which reduce a pest population will automatically improve the effectiveness of a control treatment, even though the efficiency of that treatment remains unchanged. Conversely, any practice encouraging a high population density in an area will immediately reduce the effectiveness of the control treatment, often giving the false impression that the treatment no longer works. The kinds of action which encourage a pest population build-up are destruction of predators and parasites, insufficient crop rotation, presence of weeds and alternative hosts in the vicinity of the crop, a low standard of crop hygiene, etc.

# Forecasting pest attack

The ultimate goal of pest control programmes is the accurate forecasting of pest attacks before they actually take place, so that control measures can be planned with maximum efficiency. Successful forecasting techniques should be as simple as possible, and will be based upon detailed knowledge of the biology and ecology of the pests concerned. The types of detailed studies required to give the basic information are as follows.

*Quantitative seasonal studies* which must be made over several years to determine seasonal range, variability in numbers, and geographical distribution. Such studies must use sampling methods appropriate to the pest and its abundance, and the seasonal counts should be related to climate and topographical data.

*Life history studies*, to find the length of the life cycle, number of eggs laid, amount of food eaten, maturation period for the females, are aspects that can be studied both in the field and in the laboratory. Behaviour of different larval instars and possible number of generations under different conditions can most suitably be studied in the laboratory. The expected range in relative humidity, temperature, etc., can be considered in relation to limits of survival of the insect under study.

*Field studies* to find the effects of weather on the pest. Either directly or indirectly climatic factors control pest numbers, affecting not only the pests themselves but also their predators and parasites. The ways in which pests spread from crop to crop are largely influenced by weather. Until population dynamics of insects are more fully understood, accurate forecasting is very difficult, and generally the many forecasting schemes in different countries have met with widely differing successes. The National Agricultural Advisory Service (now ADAS) in Great Britain is usually quite good at making pest outbreak forecasts. The essential point of forecasting is to predict the timing of critical pest populations, or populations reaching the economic threshold. In this section the term forecasting is used in the wide sense to include simple spray warnings based upon first occurrence records, as well as more complicated forecasting by prediction. Unfortunately, most of the examples given here are European or Japanese, for in tropical countries at the present time there are not many pest forecasting schemes in operation, with the exception of Armyworm and Locust forecasting in Africa.

## Emergence or occurrence warnings

In temperate regions these are basically emergence warnings as the first of the overwintering eggs hatch, or the first adults emerge from the overwintering pupae. Because of the climatic regulation most emergences take place over a relatively short period of time and are not too difficult to monitor. In the tropical parts of the world, where weather conditions permit continuous breeding of the pests most of the time, the warning is basically for the first occurrence of the pest in the crop, or sometimes the recording of immigrants from an adjoining area. In southern Japan on the island of Kyushu there are several serious rice pests which are unable to over-winter and which arrive from E. China each summer as aerial immigrants. In the areas at risk from these pests large wind-traps are mounted along the coast to sample the anemoplankton (small insects carried on the wind or air currents) to detect the arrival of these visitors. In these different cases, this type of warning is applicable only where the pest is known to be a serious one, and when economic damage is to be expected, based upon years of previous empirical experience. Such a crop is referred to as being at *high risk*.

In its simplest form, this type of warning consists of a visual record of adult insects in the vicinity of the crop, otherwise the use of different types of traps is required. To make this method more reliable it is preferable to use a trap that actively attracts the insects and so will be effective at lower pest densities; if electricity is available in the field then an ultra-violet (u.v.) mercury-vapour light-trap (sometimes called 'black-light') can be used for some species such as moths and leafhoppers. Black-light traps are used to monitor the arrival of migratory rice leafhoppers in Japan, and are used to such an extent throughout China for pest monitoring that electricity cables run throughout all the agricultural areas of eastern China. In some cases the u.v. traps are of a sufficient density to be also used for the '*trapping-out*' of several pest species. Otherwise pheromone traps are very good, but pheromones are only available commercially for a small number of pest species. Codling Moth warnings are based upon both pheromone and u.v. light trap catches; fruit flies of the genus *Dacus* are easily trapped using 'Cu-lure' or methyl eugenol baited traps. Warnings of danger from locusts and armyworms usually rely on sight records of hopper or adult swarms. Suction traps are suitable for small flying insects such as aphids and midges, but the more active fliers can elude the trap easily. Sticky traps are cheap to make and easy to handle, and by adjusting the level of stickiness it is possible to catch only the correct sized insects. But the efficiency of such traps is rather dubious and may require some trial runs to establish whether it is a suitable method for any particular pest. For some small insects water traps can be quite effective. Some traps such as sticky and water traps may be more effective if coloured yellow, as it seems that a number of pests find the colour yellow attractive.

Emergence traps fixed on the soil between the crop plants are very suitable for species that pupate in the soil or litter, such as some Diptera, Lepidoptera and Thysanoptera, but care has to be taken for such traps may heat the soil and cause precocious emergence.

The use of traps as an ecological tool for the assessment of population size is highly problematical, and requires great care and forethought, but for forecasting use most of the time all that is required is a record of the first emergence or occurrence in the crop. In some temperate situations empirical studies over many years can link certain levels of trap catches (e.g. Codling Moth) with particular levels of infestation and crop damage. For details concerning methods of trapping see Southwood (1978), and in McNutt (1976) various types of traps are illustrated.

Spray warnings based on such trapping methods have to make due allowance for the female maturation period, rate of oviposition and time required for egg development.

**Forecasting by sampling**
Strictly speaking there is no real difference be-

tween this category of forecasting and the previous one, for in both cases the pest population is being sampled, but for practical purposes there is some merit in treating them as distinct. The study of the development of a pest population is often referred to as *pest monitoring*.

By sampling immature stages of insect pests it is possible to arrive at approximate estimations of numbers expected in later stages. The further back in the life history one samples, the less reliable will be the method, because of the difficulty in estimating natural mortality rates. Thus, it is rather unreliable to sample fly pupae in the soil in the autumn (for example in the UK); it is more useful to sample eggs in the soil (around the plants) in the following spring. In the UK the taking of soil cores for insect eggs of Carrot Fly and Cabbage Root Fly is quite successful for estimating later population of root maggots, and hence determining whether or not to apply insecticides.

With pests that have alternative hosts, they may be sampled while on the other host, so that an estimate of their probable pest density on the crop can be made; this method has most application in temperate regions where the pest overwinters as eggs on another plant (e.g. Peach—potato Aphid).

With many Lepidoptera it may be feasible to determine the best spraying date by the finding of eggs on the crop. This is done with various pulse moths (Pea Moth, etc.), some stem borers and many of the bollworms of cotton. In Malawi and the Sudan especially, the major cotton bollworms are *Diparopsis* and *Heliothis* and crops at risk are examined either in the field or else samples taken back to the laboratory to search for eggs and/or young larvae. Generally good control has resulted if pesticide sprays are applied when field surveys reveal either 5 larvae per 100 plants, or 10 eggs per 100 terminal shoots.

### Forecasting by prediction

Temperature is the single most important factor controlling insect development and hence population numbers. A simple method using mean temperatures for two months has been developed to predict the date of emergence of the adult of the Rice Stem Borer (*Chilo simplex*) in Japan. It has also been used in the USA for the prediction of outbreaks of the European Corn Borer (*Ostrinia nubilalis*). A base temperature of $10°C$ is used, and the amount of development the pest attains daily is indicated by the number of degrees above $10°C$ each daily mean temperature reaches. The accumulation of these departures from the base temperature in one season is expressed as *degree-days*. Observations over several years have given the number of degree-days required and have equated these with each stage of development.

Rice Leaf-roller has recently become a serious pest of rice in Japan, and here it is recorded that there are five larval instars on young plants and six instars on the ripening crop, usually with three generations each season in Japan. The second generation requires 210 degree-days for development, whereas the later, third generation requires 300 degree-days.

Rainfall has also been used to forecast the likelihood of pest attack. In Tanzania outbreaks of the Red Locust (*Nomadacris septemfasciata*) have been forecast from an index of the previous year's rainfall, and in the Sudan the amount of pre-sowing rains frequently enables Jassid damage to the cotton crops to be predicted.

A simple diagram of the climate or weather characteristic of an area can suggest the likelihood of outbreaks within it. If the monthly means of temperature are plotted against relative humidity, a polygonal diagram results — this is called a *climatograph*. If temperature and rainfall are used then the diagram is a *hythergraph*. The temperature and humidity requirements of the different instars of the pest must be known, and can be plotted on to the diagram. Then the relationship between the basic pest requirements and the prevailing conditions in the area in question will give an indication as to the likelihood of the pest becoming established in that area. Such studies have formed the basis for many of the international phytosanitary regulations when it has been shown that there are areas where a particular pest could become established, because the climatic conditions are suitable and if introductions were permitted.

The final type of prediction is based upon observation of climatic areas, for on this basis some areas where critical infestations are likely to occur can be predicted for some pests. The principal factors controlling a pest population build-up may be climatic, biotic and topographical, although a combination of temperature and relative humidity (or rainfall) is probably the most important. Geographical distribution of many pests is controlled by some limiting climatic factor. The distribution of a pest can be divided into three zones.

(a) Zone of natural abundance — here the insect is always present in detectable numbers, and it is regularly a pest.

(b) Zone of occasional abundance — here the population is kept low by climatic conditions and only sometimes can it rise to pest proportions.

(c) Zone of possible abundance — sometimes the climate permits an outbreak to occur, but rarely. The insects often move into this zone from zones (a) and (b), and it may here be a pest for a while until the climate controls it.

Knowing the temperature and relative humidity requirements for the different instars of an insect species and knowing the climatic conditions of an area, enables the likelihood of an outbreak of that pest in the area to be estimated.

An alternative aspect of climatic effects on a pest can be seen in the rate of development, this being a clear reflection of the climate as measured by temperature and relative humidity. Fig. 5 shows the effects of temperature and relative humidity on the rate of development of the Cotton Boll Weevil, the three inner areas roughly representing the conditions in the three natural zones of abundance for this pest.

*Fig. 5.* Time of development of the Cotton Boll Weevil (in days) in relation to two climatic factors (after Uvarov, 1931).

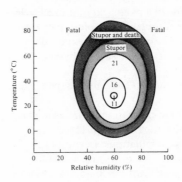

35

# 3  *Methods of pest control*

There is a large number of different methods of pest (including disease) control available to the crop protectionist, but careful deliberation is required in making a choice of methods. In general the orientation of the control project is towards the crop plant population rather than individual plants, so that low levels of pest infestation are acceptable provided that damage levels are low. Obviously with some expensive horticultural crops the welfare of each individual plant is of concern.

The choice of method(s) to be used depends on several factors.

(a) Degree of risk – some crops in some fields are at *high risk*, because in that area serious pests are invariably present in large populations. In such situations *preventative* measures (sometimes called insurance measures) may be justified. Similarly a high risk area may be predicted by sampling and forecasting techniques.

(b) Nature of pest and disease complex –usually several (or many) different pests and pathogens will be interacting on the crop in the form of a pest complex. Key pests will dominate the control strategy. With many crops between four and eight major pests will require control at any one time. Ideally the method(s) used will control several pests (and sometimes pathogens also) simultaneously.

(c) Nature of the crop and agricultural system – e.g. height of crop and spacing.

(d) Economic factors – e.g. cost of chemicals and specialized equipment.

(e) Ecological factors – e.g. extent and type of natural control, and availability of water.

Obviously it is vitally important that the pests be correctly identified and that their general biology be known.

The system of classifying control measures which follows is based upon the type of mode of action, and is widely used by plant pathologists.

(a) Exclusion – including quarantine, use of disease-free seed and planting material; designed to keep (new) pests and diseases out of an area or crop.

(b) Avoidance – uses cultural control methods and sites free of infection, and resistant crops.

(c) Protection – use of chemicals mostly, as protectants, therapeutants and disinfectants; physical protection may be included.

(d) Eradication – for an outbreak of a pest or disease in a new area; uses soil sterilization, fumigation, heat treatment, etc.

Generally it is more useful, from a pest viewpoint, to regard control measures according to their basic nature as follows:

(a) legislative methods
(b) physical methods
(c) cultural control
(d) crop plant resistance to pest attack
(e) biological control
(f) chemical control
(g) integrated control
(h) pest management (integrated pest management)
(i) eradication.

## Legislative methods

These are obviously methods of control where government legislation (laws) has been passed so that certain control measures are mandatory, with failure to comply being a legal offence. These are extreme measures and only apply to certain very serious pest situations of national importance.

### Phytosanitation (Quarantine)

When the major crops were distributed around the world from their indigenous areas, initially they may have been free of their native pests and diseases (particularly crops grown from seed). But over the years many reintroductions have been made and gradually the native pests and pathogens have spread also, until now some pests and diseases are completely sympatric with their host crops. However, there is still a large number of pests and diseases that have not yet spread to all parts of the crops' areas of cultivation. For example, the early success of coffee in Brazil (S. America) was due to the absence of Coffee Rust, Antestia Bugs, and Coffee Berry Borer, which remained behind in E. Africa; but this advantage has now been lost as in recent years both Rust and Berry Borer have become established in S. America. Similarly the success of rubber as a crop in S.E. Asia is due in part to the absence of its native South American Leaf Blight.

FAO (Food and Agricultural Organization of the United Nations) have organized a system of international plant protection with respect to the import and export of plant material. The world is divided into a number of different geographical zones for the basis of phytosanitation; each zone has its own regional organization for co-ordination. There is now an International Phytosanitary Certificate, an essential document required for importation of plant material into almost every country of the world.

Disease-free and pest-free plants can usually be imported, provided they are accompanied by appropriate documentation from the country of export, but certain plants (and fruits) are completely prohibited because of the extreme likelihood of their carrying specific noxious pests or pathogens. Other categories of plants are allowed to be imported (and exported) after routine treatment to eradicate possible pests;

such treatment usually consists of fumigation. Sometimes the plants have to be kept in quarantine isolation for a period of time to check that no symptoms develop (as with domestic pets and rabies quarantine). Specific import regulations will vary from country to country according to the nature of the main agricultural crops; thus California and Florida are very concerned about *Citrus* fruit import, and Hawaii has rigorous regulations concerning orchids. Importation of pome fruit into many African countries from Asia and America is very rigorously controlled because of the danger of importation of San José Scale (*Quadraspidiotus perniciosus*); this is potentially the most destructive orchard pest and caused tremendous damage to commercial orchards in the southern USA at the turn of the century, attacking all types of pome fruit, plums and cherries, in addition to shade trees and ornamentals. The terminal stages of infestation usually resulted in the death of the trees. California Red Scale (*Aonidiella aurantii*) is a serious pest of *Citrus* that is not yet quite pantropical, so that citrus fruit importation is subjected to careful scrutiny in several tropical countries.

For further information on international phytosanitation see Caresche *et al.* (1969) and Hill & Waller (1982).

### Prevention of spread

In all the cotton growing areas of Africa and the Middle East the Red Bollworms (*Diparopsis castanea* – south of E. Africa; *D. watersi* – north of E. Africa) are serious pests which cause considerable damage to developing bolls. They are, however, monophagous pests whose diet is restricted solely to cotton plants, and they have restricted powers of dispersal. The only area adjacent to E. Africa where Red Bollworm occurs is Malawi, but by the very strict enforcement of the law maintaining a cotton-free zone in S. Tanzania the pest is prevented from extending its range into E. Africa.

*Trogoderma granarium* is probably the most serious international pest of stored grains, by nature of its biology. Regular introductions accidentally occur in the harbour areas of Dar es Salaam and Mombasa from cargo ships, and very strict control measures are taken to ensure that there is no spread of these infestations beyond the dockside areas. Usually, of course, eradication measures are then taken against the beetle in the dockside godowns and infected ships are usually fumigated.

A serious temperate and subtropical pest is the Colorado Beetle, which as yet has failed to become established in the UK, largely due to the effect of the Colorado Beetle Order (1933). The legislation ensures that all farmers and growers must immediately report any infestation that appears to be by this pest, under threat of legal prosecution. Most years there is an accidental introduction of Colorado Beetles somewhere in southern or eastern England, but so far the precautions taken by MAFF (Ministry of Agriculture, Fisheries and Food) staff have proved effective.

Fruit flies (*Dacus* and *Ceratitis* spp.; Tephritidae) are very important pests of most tropical fruits and a serious threat to *Citrus* (and other fruit) cultivation in both Florida and California (USA). There are regular accidental introductions of Medfly (*Ceratitis capitata*) into both areas and on each occasion legislation has ensured an immediate eradication programme involving compulsory intensive widespread insecticide spraying both from the ground and from the air. To date, these campaigns have proved successful and the Medfly (and some other Tephritidae) have been denied establishment. The cost of eradication of a pest such as the Medfly (caught before it has spread very far afield) is quite infinitesimal in comparison with the value of the $14 000 million fruit industry in California.

# Physical methods

These refer to methods of mechanical removal or destruction of the pest, and are usually unimportant in most countries owing to the high cost of labour. They are still of use in some countries and for particularly valuable crops.

## Mechanical

Hand-picking of pests was probably one of the earliest methods of pest control and is still a profitable method for the removal of *Papilio* caterpillars from young *Citrus* trees. The killing of *Dirphya* larvae and *Apate* adults boring in branches of coffee bushes is recommended by the pushing of a springy wire (e.g. bicycle spoke) up the bored hole and spiking the insect. The use of mechanical drags, which crush insects on the ground, has been made against armyworms (*Spodoptera* larvae) but this practice is now generally outmoded. Banding on fruit trees is particularly effective against caterpillars and ants, which gain access to the tree by crawling up the trunk. Spray-banding of coffee bush trunks is practised against White Borer (*Anthores leuconotus*) and the adults of Tip Borer (*Eucosma nereidopa*) which typically rest on the shaded lower parts of the trunk. Much of the Tsetse (*Glossina* spp.) field sampling and catching in E. Africa is still done by hand, using cattle as bait, and by the use of special bicycles with decoy plates on the front. An earlier method of locust control was to herd the hoppers into a large pit which was afterwards filled in with soil. In northeastern Thailand in 1980 it was reported that local villagers were employed to hand-collect Bombay Locust (*Patanga succincta*) from the maize crops, and apparently they collected 80 tons of locusts in only a few days!

In parts of S.E. Asia, especially in gardens and on smallholdings, it is common practice to place bags around large fruits (e.g. grapefruit, pomelo, pomegranate and jackfruit) to deter fruit flies (Tephritidae) from oviposition. In rural areas and in the forests the bag is usually woven from grass or raffia, but in more surburban situations paper bags and polythene may be employed (fig. 6). Occasionally a single fruit may be left unprotected so that the local fruit flies will concentrate their egg-laying upon this one fruit which can later be destroyed (fig. 7).

A greenhouse is essentially a structure to provide physical shelter for a crop, but its main purpose is of course to provide a suitable microclimate, often for a tropical crop grown in a temperate country or a local temperate crop being grown early or out of season. In the tropics greenhouses (and polythene tunnels) are of less use, largely because of the dangers of overheating, but structures of a wooden or metal frame covered with fine nylon or plastic netting are extremely useful. They provide some physical shelter from strong winds, frost, heavy rain and extreme insolation, but allow free air circulation to prevent overheating. The fine insect mesh keeps most pests at bay

*Fig. 6.* Guava fruits protected by polythene bags against fruit flies and other pests in Thailand.

*Fig. 7.* Pomelo attacked by fruit flies.

and a crop may be grown (with care) under insect-free conditions. Such structures are invaluable for rearing very costly crops, or vegetable crops for seed (provided pollinators are released inside), and for seed-beds. If the netting material is too dense, however, plant growth may be somewhat retarded due to reduced sunlight levels.

### Use of physical factors

The use of lethal temperatures, both high and low, for insect pest destruction is of importance in some countries but not particularly so in Africa. However, the use of cool storage in insulated stores for maize grain is practised locally in E. Africa. The purpose of this method is not the actual destruction of the pests, which may occur when lethal temperatures are employed, but the drastic retardation of development following the reduction of the metabolic rate.

Kiln treatment of timber for control of timber pests is very widely practised in many countries. Plant bulbs are often infested with mites, fly larvae (Syrphidae) or nematodes, and hot-water treatment (dipping) can be a very successful method of control if carefully carried out. The drying of grain, which is widely practised for a reduction in moisture content, usually results in lower infestation rates by most pests. In the Sudan the heating of cotton seed to kill the larvae of Pink Bollworm (*Pectinophora gossypiella*) is an effective control. In different parts of Kenya a hermetic storage of grain is being developed as a standard long-term storage method. The stores are Cyprus bins (of 1 000 000-bag capacity), some 70 in number originally, and the principle involved is that only a small quantity of air is enclosed within the sealed bin, the oxygen in which is quickly used up by the respiration of the pests and the subsequent carbon dioxide accumulation quickly results in the death of all contained pests, both arthropod and microbial.

On-farm storage of grain is being carried out in some areas using butyl silos; the addition of small quantities of diatomite fillers increases the effectiveness of this control, as the abrasive effect removes the outer waxy covering of the epicuticle of the insects, resulting in greater water loss (and possible dehydration) and greater ease of insecticide penetration through the cuticle.

### Use of electromagnetic energy

The radio-frequency (long wavelength radiations) part of the spectrum has been extensively studied in the development of radio communications, radar, etc. and it has been known for a long time that absorption of radio-frequency energy by biological material results in the heating of the tissues. Control of insect pests by such heating is only practicable in enclosed spaces of small or moderate size (food stores, warehouses, timber stores). The nature of absorption of radio-frequency energy by materials in a high-frequency electrical field is such that for certain combinations of hosts and insect pests their dielectric properties are favourable for differential absorption of energy, hence the insects can be killed without damaging the host material. Timber beetles in wood blocks have been killed in this manner, but whether this treatment offers any real advantage over normal kiln treatment is doubtful.

Use of infrared radiation for heating purposes is very much in its infancy.

Many insects show distinct preferences for visible radiation of certain wavelengths (i.e. certain colours), as well as the long recognized attraction of ultra-violet radiation for various nocturnal insects, especially Lepidoptera. Ultra-violet light traps have on occasions significantly lowered pest populations in various crops, but have also failed when used against other pests.

Aphids and some other plant bugs are attracted to yellow colours; this is possibly because most aphids feed either on young or senescent leaves, presumably because these are the plant parts where active transport of food material occurs. The young leaves are photosynthesizing rapidly and the sugars formed are transported away in the phloem system to be stored as starch grains in older leaves, tubers, etc. As leaves become senescent, the stored starch is reconverted into soluble sugars for transportation prior to leaf dehision. Senescent leaves are usually yellowish in colour and young foliage is often a pale yellowish-green. So the attraction of aphids to yellow colours seems fairly obvious, but why so many flies (especially Anthomyiidae) and some moths are similarly attracted to yellow is not obvious.

Conversely, many flying insects are repelled by blue colours and by reflective material. This has been exploited by using strips of aluminium foil, or metallicized plastic, between the rows and around the periphery of the crop, the result being that fewer flying aphids and other insects settle in the crop than would otherwise. Thus, reflective strips also result in crops having far less aphid-borne virus diseases. Another effective method is to surround the seed-beds by flooded furrows, as practised in S. China with vegetable crops.

The ionizing radiations (X-rays, $\gamma$-rays) are sterilizing at lower dosages but lethal at higher. The use of these radiations in controlling stored product pests, particularly in grain, is being quite extensively studied in various countries.

## Cultural control

These are regular farm operations, that do not require the use of specialized equipment or extra skills, designed to destroy pests or to prevent them causing economic damage. Often these are by far the best methods of control since they combine effectiveness with minimal extra labour and cost.

### Optimal growing conditions

A healthy plant growing vigorously has considerable natural tolerance to pests and diseases (as with a healthy animal), both physically and physiologically. Good plant vigour is a result of sound genetic stock and optimal growing conditions. Obviously the farmer attempts to provide such growing conditions so that the crop yield will be maximal. Many diseases are more severe, and the damage by pests more serious, if the plant is suffering from water-stress (drought), unfavourable temperature, imbalance of nutrients or nutrient deficiency, etc. This predisposition to pest attack and disease can be very serious when crops are grown on marginal land. This is one main reason, together with reduced yield, why the cultivation of marginal land is generally not very successful.

### Avoidance

Empirical observations will reveal that certain areas (and fields) are constantly 'at risk' from particular pests and conversely others are pest-free. Clearly, if a crop can be grown in areas of the latter category it can be expected to remain free from that particular pest. This practice is particularly effective against certain soil-borne diseases and nematodes, but less so against most insects because of their greater mobility. This is one of the advantages of shifting cultivation.

### Time of sowing

By sowing early (or sometimes late) it may be possible to avoid the egg-laying period of a pest, or else the vulnerable stage in plant growth may have passed by the time the insect numbers have reached pest proportions. Early sowing is regularly practised against Cotton Lygus (*Taylorilygus vosseleri*) and Sorghum Midge (*Contarinia sorghicola*) in Africa. In N. Thailand it has been shown that early transplanting of paddy rice reduced the level of Rice Gall Midge attack appreciably. Another important aspect of the time of sowing is that of simultaneous sowings of the same crop over a wider area, to avoid successive plantings which often permit the build-up of very large pest populations.

### Deep sowing (planting)

Some seeds are less liable to damage and pest attack if rooted deep, but of course if planted too deep germination will be impaired. Many root crops are also less liable to attack by pests if they are deeper in the soil. This is true for sweet potato in that the deeper tubers always have fewer weevils (*Cylas* spp.) boring inside, and the deeper potatoes have fewer infestations of Tuber Moth larvae. Even if no special attempts at deeper planting of root crops are made then care should be taken to ensure that no tubers are allowed to grow too close to the soil surface, earthing-up should be done when required.

### Time of harvesting

Prompt harvesting of maize and beans may prevent these crops from becoming infested by Maize Weevil (*Sitophilus zeamais*) and Bean Bruchid (*Acanthoscelides obtectus*) respectively. Both of these pests infest the field crops from neighbouring stores but are generally not able to fly more than about half a mile; so an added precaution is to always grow these crops at least half a mile away from the nearest grain store.

New varieties of crops which mature early may enable a crop to be harvested early, before pest damage is serious.

### Close season

In E. Africa legislation has been passed to ensure that there is a close season for cotton growing in order to prevent population build-up of Pink Bollworm (*Pectinophora gossypiella*), which is oligophagous on Malvaceae. This legislation stresses that all cotton plants should be uprooted and destroyed (or burned) by a certain date and quite clearly no seed would be planted until the following rains arrive. However, it is clear that many farmers do not bother to destroy the old plants by the appointed date and so in some areas there is considerable survival of diapausing Pink Bollworm larvae. Similarly there is a natural close season for beans against the Bean Fly (*Ophiomyia phaseoli*) and it is strongly recommended that this crop should not be grown during the dry season, for at this time pest damage is maximal.

### Secondary hosts

Most pests are not monophagous and so will live on other plants in addition to the crop. Sometimes, in point of fact, the crop itself is not the preferred host! For example, Cotton Lygus is often more abundant on sorghum, and sometimes on wild Malvaceae, than on the cotton plants. In many cases the pests build up their numbers on wild hosts and then invade the crop when the plants are at the appropriate stage of development. Many Cicadellidae and Delphacidae that are crop pests (on rice and other cereals) feed and breed on wild grasses in the vicinity of the paddy fields so that when the young rice is planted out there is a large bug population waiting to infest the young tender shoots of the rice. The destruction of alternative hosts (or the control of insects on them) may be an important part of an IPM programme. Often the alternative hosts are native (or introduced) trees and shrubs, although they may be herbaceous plants. Solanaceous weeds, for example, are important alternative hosts for pests of tomato, tobacco, egg-plant and potato. Morning glory in its various species is one of the most widespread tropical plants (climbers) and all the important sweet potato pests feed on this plant. The pest status of Cotton Stainers (*Dysdercus* spp.) in a particular area is entirely related to the presence and abundance of wild alternative hosts in the bush (Pearson, 1958).

In temperate situations a number of pests and diseases have an alternation of generations on quite different hosts, e.g. *Myzus persicae* on peach and potato. The removal of the alternative host can effectively reduce such pests to insignificance, but in the case cited, of course, the alternative host is itself a crop plant.

With monophagous pests it is important to remove any host plants between crops (if feasible for the species concerned). A monophagous insect would normally be confined to the species of a single genus for host, though it would be unusual for it to be restricted to a single species of one genus. Thus, for sweet potato pests it is not feasible to attempt to remove all wild species of *Ipomoea* as they are too abundant. But, for the Brown Planthopper of Rice (*Nilaparvata lugens*) which is restricted to rice (*Oryza* spp.) as a host plant, the alternative host plants are either wild rice or volunteer rice; by destruction of wild rice weeds, volunteer plants and crop residues, this pest can be controlled.

### Deep ploughing

Many Lepidoptera (particularly Noctuidae, Sphingidae and Geometridae), Coleoptera and Diptera pupate in the soil, and a large number of their larvae live there. The bulk of the soil insect population lie in the top 20 cm of the soil (most are in the top 10 cm). Deep ploughing will bring these insects to the surface, to be exposed to hot sunlight (insolation), desiccation and predators. In many tropical areas a farmer ploughing a field will be followed by a flock of cattle egrets, little egrets, crows or starlings, in a seaside locality there might be a flock of gulls; all these birds will feed on the exposed worms, slugs and insects.

### Fallow

Allowing a field to lie fallow almost invariably reduces pest and pathogen populations, but care must be taken to ensure that there are no volunteer crop plants or important secondary host weed species. Fallowing may be done as *bare fallowing* when the soil surface is left bare, or *flood fallowing* when the field is flooded with water for a while. Sometimes, as an alternative, a cover crop of legumes is grown as *green manure* which is then ploughed under.

### Crop rotation

In olden times a period of fallow was an essential part of all crop rotations, but nowadays economic pressures mean that fields can seldom be left fallow. Instead, basic crop rotation is usually practised for the obvious reasons that continuous cultivation of one crop depletes the minerals and trace elements in the soil quite rapidly, and also induces disease and pest build-up. However, some agricultural crops require rather specialized growing conditions, and these, combined with the practice of large-scale cultivation, result in some areas growing crops such as sugarcane, pineapple, maize or cotton, almost continuously. Obviously the plantation crops (coffee, rubber, coconut, oil palm, etc.) are also very long-term monocultures.

The alternation of completely different crops in a field has very obvious advantages from the pest and disease control aspects. But in a rotation it is necessary to remove a particular crop quite a distance away; having the same crop in an adjacent field is not really 'rotation' so far as active insects are concerned, although it might be adequate for soil nematodes and some soil-borne diseases. So, for effective pest control crop rotation has to separate crops both spatially and in time.

Against monophagous and oligophagous pests crop rotation can be effective, especially with beetle larvae that may take a year or more to develop, but it is not effective against migratory pests or those with effective powers of dispersal. The alternation of cereals with non-cereals may be an important method of curtailing *Nematocerus* weevils in Africa. A common type of rotation is the alternation of a legume crop with cereals; this is effective against some pests, but others (e.g. *Colaspis, Diabrotica*) can utilize both host types.

### Weeds

Most cultivation practices require destruction of weeds because of their competition with the crop plants, and their interference with different aspects of cultivation. But weeds can also be important from the viewpoint of pests and diseases; sometimes the weeds may be alternative hosts, as already mentioned. Weeds often belong to the same family as the crop plant because of the selective nature of many post-emergence herbicides. Some pests seem to prefer weeds as oviposition sites; for example, some cutworms (*Agrotis* spp.) and some beetles (Scarabaeidae) lay most of their eggs on, or in the immediate vicinity of, weeds in the crop. Weed removal at the appropriate time may result in many potential pests being destroyed.

### Trap crops

The use of trap plants to reduce pest infestation of various crops is based upon the knowledge that many pests actually prefer feeding upon plants other than those on which they are the most serious pests. This preference may be exploited in two different ways: either the pests are just lured from the crop on to the trap plants where they stay and feed, or else, because of the greater concentration of pests on the trap plants, only the trap plants need be sprayed with pesticides. In the latter case, since the trap plants are either grown as a peripheral band or else interplanted at about every fifth to tenth row, the saving represented is considerable. At Namulonge Cotton Growing Corporation Research Station, Uganda, it has been shown very clearly that *Cissus* (Ampelidaceae) is a very attractive trap plant for Lygus Bugs on cotton, as little as one row in ten of *Cissus* producing significantly lower Lygus infestations on the cotton plants. Work is also in progress at Makerere University to determine the effectiveness of maize and other plants as traps for late bollworms of cotton (False Codling Moth) in Uganda.

### Intercropping

This practice obviously has various drawbacks for large-scale agriculture, but can be of particular application for the small farmers, who often use little insecticide. Intercropping can certainly reduce a pest population on a crop, and without doubt reduces the visual and olfactory stimuli that attract insects to a particular crop species. As a method of control it is most effective against exogenous pests, such as locusts, which enter the crop for only part of their life-cycle. In N.E. Thailand in 1980 groundnuts were intercropped with maize, and nymphs of Bombay Locust (*Patanga succincta*) were induced to leave the maize (particularly on hot days) for the lower foliage of the groundnuts where they were eaten by ducks. Work is in progress on different schemes of intercropping at IRRI in the Philippines, for use particularly in that country with its large proportion of peasant farmers.

### Crop sanitation

This is a rather general term, and hence is used to include the following different aspects of crop cultivation.

(a) Destruction of diseased or badly damaged plants —

the roguing of such plants is an important agricultural practice, but of course requires hand labour.

(b) Removal and destruction of rubbish – old crop remnants, fallen leaves, branches, dead trunks, also weeds, etc. Some pests use rubbish heaps for breeding purposes, for example *Oryctes* larvae are to be found in rotting palm trunks and in rotting vegetation and soil, especially in rubbish heaps. There are, in fact, laws to promote crop hygiene on the east coast of Kenya and Tanzania and, when enforced, this method of control is effective. On oil palm plantations in Malaysia crop hygiene is recognized as a major method of suppression of *Oryctes*.

(c) Removal and destruction of fallen fruits – for the control of many fruit flies and boring caterpillars this is important, as the insects will continue to develop in the fallen fruit and will pupate either there or in the soil. This is, in fact, still the most successful method of suppression of Coffee Berry Borer (*Hypothenemus hampei*) on *robusta* coffee.

(d) Destruction of crop residues – this is often vitally important in order to kill the resting stages (pupae, etc.) of many pests after harvest. Many stalk borers (Pyralidae, Noctuidae) pupate in the lower parts of the cereal stems and as such will be left in the stubble even if the main parts of the stalks are removed. (Of course, more will be left in a tall stubble than in a short one.) With some crops of maize, sorghum and millets most of the actual stem is left. The stems should be burnt immediately after harvest. In some parts of S.E. Asia it is customary to leave rice stubble tall; so then many pupae (Lepidoptera and Diptera) may survive in the stems. For diseases, crop residue destruction may be even more important. Ploughing in of crop stubble may kill a small proportion of pupae but most will survive; burning is most effective.

The recommended method of destruction of all weeds, crop residues, rubbish, rogued plants, fallen fruits, etc., is by burning; other methods may not kill the pests.

# Crop plant resistance to pest attack

In most growing crops it may be observed that some individual plants either harbour far fewer pests than the others or else show relatively little sign of pest damage. These individuals usually represent a different genetic variety from the remainder of the crop, and this variety is said to show *resistance* to the insect pest. Also when different varieties of the same crop are grown side by side, differences in infestation level may be very marked. Resistance to pest attack is characterized by the resistant plants having a lower pest population density, or fewer damage symptoms, than the other plants which are termed *susceptible*. Conversely, there will be some plants that appear to be preferred by the pests and these especially susceptible plants will bear very large pest populations. Frequently these plants will actually be destroyed by the pests and so will not breed and pass on their disadvantageous genetic material.

Varietal resistance to insect pests was broadly classified by Painter (1951) into three categories: non-preference, antibiosis, and tolerance; but Russell (1978) suggests the use of a fourth category: pest avoidance. Some workers restrict the use of the term varietal resistance to antibiosis, but this view is rather narrow and not practical. In fact it is often very difficult to distinguish between some cases of non-preference and antibiosis.

### Types of resistance
1. Pest avoidance
2. Non-preference (= non-acceptance)
3. Antibiosis
4. Tolerance

The basis of these types of resistance is slight variations in genetic material; as defined by Russell (1978), 'Resistance is any inherited characteristic of a host plant which lessens the effect of parasitism.' The term *parasitism* is used in a broad sense to include the attack of insect pests, mites, vertebrates, nematodes, and pathogens (fungi, bacteria, viruses) on the host plant. The feeding of phytophagous insects on plants is not generally regarded as parasitism by most zoologists, but rather as ecological grazing.

Genetically there are three main types of resistance. *Monogenic* resistance is controlled by a single gene, usually a *major gene* which has a relatively large effect. This type of resistance (often biochemical, involving phytoalexins) is fairly easily incorporated into a breeding programme, and it usually gives a high level of resistance; unfortunately this resistance is just as easily 'broken' by new pest 'biotypes' (new races or strains).

*Oligogenic* resistance is the term used when the character is controlled by several genes acting in concert.

*Polygenic* resistance is the result of many genes, and is clearly more difficult to incorporate into a plant breeding programme. It may be either morphological or biochemical, and it is generally less susceptible to biotype resistance ('breaking'). Many of the genes will be *minor genes* which individually only have a small effect genetically.

In epidemiological terms, resistance is classified as either *horizontal resistance* (alternatively, durable resistance), with a long-lasting effect and effective against all genetic variants of a particular pest, or *vertical resistance* (alternatively, transient resistance), effective for a short period and against certain variants only.

There are a few other terms which are in use in plant breeding for pest resistance. *Field resistance* is the term used commonly to describe resistance which gives effective control of a pest under natural conditions in the field, but is difficult to characterize in laboratory tests; usually it is a complex kind of resistance giving only partial control. *Passive resistance* is

when the resistance mechanism is already present before the pest attack, for example an especially thick cuticle, or hairy (pubescent) foliage. *Active resistance* is a resistance reaction of the host plant in response to attack by a parasite, more usually applicable to attack by pathogens rather than pests (insects etc.); for example, the formation of phytoalexins or other antibiotics (antifungal compounds) by some host plants in response to attack by some pathogenic fungi. This reaction is not unlike the human production of antibodies in response to foreign matter in the blood or tissues. *Qualitative resistance* applies when the frequency distribution of resistant and susceptible plants in the crop population is discontinuous, and the plants are individually easily categorized as either resistant or susceptible. *Quantitative resistance* is the term used when a crop shows a continuous gradation between resistant plants and susceptible plants within the population, with no clear-cut distinction between the two types.

Varietal resistance has been shown in a broad range of crop plants ranging from cereals and herbaceous plants to trees, against an equally broad range of insect, mite and vertebrate pests. The insect groups involved include Orthoptera (locusts), Hemiptera (bugs), Lepidoptera (caterpillars), Diptera (fly maggots), Thysanoptera (thrips), and Coleoptera (beetles), in both temperate and tropical parts of the world.

In many respects crop plant breeding for pest resistance is the most ecologically desirable method to be employed in pest management programmes. The time factor is quite considerable in that a breeding programme is slow to conduct, and the cost may be exorbitant, but the end result could be very long-term control without the use of pesticides and their associated problems. At present there are a number of plant breeding programmes in operation in different parts of the world for different crops and different pests, some at major international research institutes. Some of the better known stations include IRRI, Philippines (rice), CIAT, Colombia (cassava, etc.), IITA, Nigeria (many tropical crops), Cotton Growing Corporation, Africa, India, etc. (cotton), PBI, England (potatoes, wheat), NVRS, England (vegetables), and many other stations specializing in fruit (UK, USA), sorghum (East Africa), maize (India), pulses (E. Africa and UK), and so on. The scope of some of these plant breeding programmes is often very large, for example at IRRI all 10 000 varieties of rice, representing the world germ plasm collection, have been screened for resistance to rice stalk borers. More than 20 varieties show high levels of resistance to *Chilo suppressalis* caterpillars by non-preference and/ or antibiosis mechanisms, and some of these varieties are also resistant to leafhoppers and planthoppers. Some of these varieties of rice show tolerance towards these pests, as well as definite antibiosis and non-preference.

It is, however, now clear that, in response to the development of new resistant plants, it must be expected that new insect biotypes will arise to which the existing plants are not resistant. Fortunately, it appears that most insect resistance in plants is complex and polygenic in nature which may help to discourage development of insect biotypes.

### Pest avoidance

This is when the plant escapes infestation by the pest by not being at a susceptible stage when the pest population is at its peak. Some varieties of apple escape infestation by several different pest species in the spring by having buds which do not open until after the main emergence period of the pests, thus reducing the final amount of damage inflicted.

### Non-preference

The term 'non-acceptance' has been proposed as a more suitable alternative, but has never gained general acceptance. Insects are noticeably reluctant to colonize some individual plants, or some particular strain of host-plant, and these plants seem to be less attractive to the pest by virtue of their texture, colour, odour or taste. Non-preference usually is revealed when the bug or caterpillar either refuses to feed on the plant or takes only very small amounts of food, or when an ovipositing female insect refrains from laying eggs on the plant. In the Philippines *Chilo suppressalis* females laid about 10–15 fewer egg masses on resistant rice varieties than on susceptible ones. Also at IRRI, the Brown Planthopper of Rice (*Nilaparvata lugens*) punctures the tissues of a certain rice variety but apparently feeds only little, probably because of the reduced amount of a particular amino acid (asparagine) in the sap of that variety. In Uganda, experiments at Namulonge (Stride, 1969) showed that Cotton Lygus Bugs (*Taylorilygus vosseleri*) found some red-coloured varieties of cotton less acceptable for feeding purposes than the usual green plants, and this difference in preference was attributed to the presence of certain aromatic compounds in the sap of the red-coloured plants.

Non-preference based upon the absence or presence of certain chemicals in the plant tissues or sap may equally well be categorized as a form of antibiosis.

There are two temperate examples of this type of resistance. Raspberry Aphid (*Amphorophora idaei*) placed on to leaves of resistant plants exhibit a reaction so strong that they will quickly walk off the plants completely. On sugar beet aphids do not actually walk off the resistant plants, but they do feed for noticeably shorter periods of time and are quite restless whilst on resistant plants. The precise details of the resistance in these cases is not yet known, but the insect reactions are easily observed.

### Antibiosis

In this case the plant resists insect attack, and has an adverse effect on the bionomics of the pest by causing the death of the insects or decreasing their rate of development or reproduction. The resistant plants are generally characterized by anatomical features such as thick cuticle, hairy stems and leaves, a thickened stem (cereals), a narrower diameter of the hollow pith in cereal stems, compactness of the panicle in sorghum, tightness of the husk in maize, and tightness of leaf sheaths in rice. Biochemical aspects usually involve the presence of various toxic or distasteful chemicals in the sap of the tissues of the plant which effectively repel feeding insects, sometimes to the extent that the odour is sufficient to completely deter them from feeding. Alternatively, there may be a chemical which normally functions as a feeding stimulant missing from the body of the resistant plant, or else at a sufficiently low concentration that it fails to stimulate the insect into feeding behaviour.

Cotton Jassids (*Empoasca* spp.) have ceased to be important pests of cotton in Africa and India since the post-war development of pubescent strains which the bugs find quite unacceptable as host plants. In a similar manner, hairy-leaved varieties of wheat in North America are attacked significantly less often by the Cereal Leaf Beetle (*Oulema melanopus*); the females lay fewer eggs on the leaves and, of the larvae that hatch, fewer survive. It is also recorded that pubescent foliage apparently deters oviposition by many species of Lepidoptera, but this situation is complicated in that some bollworms will apparently lay more eggs on the foliage of some pubescent varieties of cotton.

The tightness of the husk in some maize varieties will deter feeding on the cobs by larvae of *Heliothis zea* (Corn Earworm) in the USA, and should also apply to field infestations of the drying grain by Maize Weevil (*Sitophilus zeamais*). At IRRI it has been shown that the tightness of the leaf cleavage in rice varieties is closely correlated with resistance to stem borers. If the leaf sheath is tight and closed and covers the entire stem internode, the young caterpillars usually fail to establish themselves between the leaf sheath and the stem where they would normally spend some six days feeding before boring into the stem. Varieties of sorghum in Africa, with an open panicle, suffer far less damage by False Codling Moth (*Cryptophlebia leucotreta*) and other caterpillars. Wheat varieties with solid stems (i.e. very reduced pith) are noticeably resistant to Wheat Stem Sawfly (*Cephus cinctus*) in that growth and development of the larvae is retarded. Some species of pyralid and noctuid stem-borers are closely restricted to cereal hosts with stems of a particular thickness, and varieties of cereals with thicker or thinner stalks may harbour fewer caterpillars.

An anatomical factor of considerable importance is the development of silica deposits in leaves and stems of various graminaceous crops. The Gramineae as a group are basically semixerophytic and so many species have silica deposits in the leaves and stems; not only the truly xerophytic species such as marram grass (*Ammophila* spp.), but also species which have become secondarily adapted as hydrophytes such as rice, still possess some deposits. A number of cereals are indigenous to semi-arid areas, and one example is sorghum. In these plants silica deposits are apparently formed in some resistant varieties at about the fourth leaf stage. Up to this stage most varieties of sorghum are attacked by Sorghum Shoot Fly (*Atherigona soccata*), but in the resistant varieties infestations are not usually recorded once these silica deposits are formed, although infestations will still occur in susceptible varieties up to the sixth leaf stage. This is a simple type of resistance which is easily incorporated into all sorghum breeding programmes.

At IRRI rice varieties have been developed with a high silica content in the leaves and stems, and these are resistant to larvae of *Chilo suppressalis* in that the caterpillars' mandibles become worn down by the abrasive nature of the silica deposits.

The Hessian Fly (*Mayetiola destructor*) was in-

48

troduced into the USA in the mid-18th century and rapidly became a major pest of wheat; the larvae attack the wheat stem, and either destroy the shoot or weaken the stem so that lodging subsequently occurs. Heavy attacks cause a serious drop in wheat yield. Since 1914 a search for resistance was carried out in Kansas. Several resistant varieties were developed, and since 1965 the widespread growing of resistant varieties has resulted in the virtual extinction of Hessian Fly in Kansas. Some 25 different resistant varieties of wheat are being grown in the USA now, and the annual value of the wheat crop has increased by many millions of dollars. The mechanism of resistance has not been established but it is known that these resistant wheats have unusually large deposits of silica in the leaf sheaths.

The biochemical factors involved in plant resistance arise from differences in the chemical constituents of the plants. The differences may be restricted to different parts of the plant body and/or particular stages in the growth of the plants. It is thought that in some resistant plants the pest concerned suffers nutritional deficiencies resulting from the absence of certain essential amino-acids. Some maize varieties show direct physiological inhibition of larvae of the European Corn Borer (*Ostrinia nubilalis*) for they possess biochemical growth inhibitors at various stages. Some pests are influenced in their host selection by aromatic compounds present in the plant tissues, as well as various sugars, amino-acids, and vitamins in the sap. Resistance to the Brown Planthopper of Rice by a rice variety at IRRI was attributed to a low concentration of asparagine, thought to be a feeding stimulant for this pest. Cabbage Aphid (*Brevicoryne brassicae*) is stimulated by sinigrin (a mustard oil glucoside) in the leaves, and it has been shown that a high level of resistance to Cabbage Aphid is associated with low foliar concentrations of sinigrin. Some cotton varieties have a poisonous polyphenolic pigment, gossypol, in subepidermal glands, and they show a strong resistance to several different insect pests (Pink Bollworm, Spiny Bollworm, etc.), although a direct causal relationship has not been established.

Members of the family Solanaceae often have peculiar glandular hairs on the leaves and stems, and recently new potato varieties have been obtained by crosses with various wild stocks from S. America which have an abundance of these glandular hairs. The sticky exudates easily trap small insects like aphids which soon die, and even such large insects as larvae of Colorado Beetle have been trapped. This easily inherited character is hoped to be of value in combatting potato aphid attack, for these sap-sucking bugs are vectors of several major virus diseases.

### Tolerance

Tolerance is the term used when host plants suffer little actual damage in spite of supporting a sizeable insect pest population. This is characteristic of healthy vigorous plants, growing under optimum conditions, that heal quickly and show compensatory growth. In fact many plants bear more foliage than they actually need, and can usually suffer a fair amount of defoliation with no discernible loss in crop yield.

Tolerance is frequently a result of the greater vigour of a plant, and this may result from the more suitable growing conditions rather than from the particular genetic constitution of the plant. For example, sorghum growing vigorously will withstand considerable stalk borer damage with no loss of yield. Some varieties of crop plant (e.g. rice) may show both tolerance to a pest as well as antibiosis; this is true for several stalk borers.

Sometimes pest attack on a tolerant variety can actually increase the crop yield; this occurs quite frequently with the tillering of cereals following shoot fly, stem borer, or cutworm destruction of the initial shoot in the young seedling.

From a pest management point of view the use

49

of a tolerant variety could in theory be a disadvantage in that it could support a larger population of the pest and so encourage a local population build-up rather than a decline.

Many cases of clearcut resistance to insect pests have been recorded, but they have not been investigated sufficiently for the mechanism of resistance to be evident. This is particularly the case in respect to the aphids *Myzus persicae* and *Aphis fabae* on sugar beet, and to Carrot Fly (*Psila rosea*) on carrots (Hill, 1974*b*).

### Breaking of host plant resistance

In some agricultural situations there develop physiological races of the insect pest, known as *biotypes*, some of which are not susceptible to the host plant resistance. These are *resistance-breaking biotypes*. In nematology this type of variant is known as a *pathotype*, in virology it is a *strain* and, applied to fungi, it is a *race*.

The development of resistance-breaking biotypes has been known for a long time in the Hessian Fly, and several biotypes can attack wheat varieties that are quite resistant to other biotypes. The Brown Planthopper of Rice (BPH) in S.E. Asia has recently become notorious in that for several reasons it has changed status from a minor to a serious major pest on rice. However, rice varieties resistant to this bug were developed at IRRI, and have been widely grown throughout the area; they have given such good control that sometimes insecticides have not been required. In some localities, however, resistance-breaking biotypes of BPH have developed to such an extent as to threaten local rice production. The present situation is that as fast as the research workers at IRRI produce resistant varieties of rice to BPH, the insects correspondingly produce new resistance-breaking biotypes. Detailed biosystematic and ecological studies of the biotypes of *Nilaparvata lugens* have very recently

been initiated at IRRI and through the auspices of the ODM in the UK.

The breaking of host plant resistance is generally less common amongst insect pests than pathogens; this is thought to be because insects produce far fewer propagules than the fungi, bacteria and viruses, and thus far less genetic variation can be expressed.

The main use of resistant varieties of crop plants in agriculture has been against plant diseases (Russell, 1978), and in general a high level of success has been achieved. Plant-parasitic nematodes (eelworms) in some ways behave rather like soil pathogens, and the development of resistant varieties of potato and wheat have been very successful in combating Potato Cyst Eelworm and Cereal Root Eelworm. Against insect pests plant breeding for resistance has not had the same success, but in some instances good control has been achieved with enough success to encourage further work in this area. On the whole it can be said that many resistant varieties of crop plants have given quite good control of insect pests, albeit only partial, against a very wide range of insect species. Many varieties of crop plants showing good resistance to important pest species have not been fully exploited because their yield is less, or of inferior quality, than the usual susceptible varieties.

## Biological control

In the broad sense (*sensu lato*) this can include all types of control involving the use of living organisms, so that, in addition to the use of predators, parasites, and disease-causing pathogens (biological control – *sensu stricta*), one can include sterilization, genetic manipulation, use of pheromones, and use of resistant varieties of crop plant.

As already indicated in this book, the use of resistant crop varieties is being dealt with separately as it is an aspect of control of such importance, and plant breeding is a very specialized subject in its own rights. In chapter 5 biological control (*sensu stricta*) is considered in more detail, and here it is only intended to be introduced in its broadest aspects.

The main attraction of biological control is that it obviates the necessity (or at least reduces it) of using chemical poisons, and in its most successful cases gives long-term (permanent) control from one introduction. This method of control is most effective against pests of exotic crops which often do not have their full complement of natural enemies in the introduced locality. Then the most effective natural enemies usually come from their native locality, for the local predators/parasites/pathogens are usually in a state of delicate ecological balance in their own environment and cannot be expected to exercise much population control over the introduced pests. On rare occasions a local predator or parasite will successfully control an introduced pest, but this is rare!

### Natural control

This is the existing population control already being exerted by the naturally occurring predators and parasites (and diseases) in the local agroecosystem, and it is vitally important in agriculture not to upset this relationship. Because it is not readily apparent, the extent of natural control in most cases is not appreciated. It is only after careless use of very toxic, broad-spectrum, persistent insecticides which typically kill more predators and parasites than the less sensitive crop pests, and which is then followed by a new, more severe pest outbreak, that the extent of the previously existing natural control may be appreciated. In summary, the importance of natural control of pests in most agroecosystems cannot be overemphasized.

### Predators

The animals that prey and feed on insects are very varied, as are their effects on pest populations. The main groups of entomophagous predators are as follows:

Mammalia – (Man), Insectivora, Rodentia
Aves – Passeriformes (many families), many other groups
Reptilia – small snakes, lizards, geckos, chamaeleons
Amphibia – most Anura (frogs and toads)
Pisces – *Gambusia* etc. (control mosquito larvae)
Arachnida – Spiders, harvestmen, chelifers, scorpions, etc.
Acarina – mostly family Phytoseiidae
Insecta – Odonata (adults, and nymphs in water), Mantidae, Neuroptera, Heteroptera (Miridae, Anthocoridae, Reduviidae, Pentatomidae), Diptera (Cecidomyiidae, Syrphidae, Asilidae, Therevidae, Conopidae, etc.), Hymenoptera (Vespidae, Scoliidae, Formicidae), Coleoptera (Cicindelidae, Carabidae, Staphylinidae, Histeridae, Lampyridae, Hydrophilidae, Cleridae, Meloidae, and Coccinellidae).

A few predators are quite host-specific; for example, the larvae of Meloidae feeding on the egg-pods of Acrididae in soil, and Scoliidae feeding on scarab larvae in soil and on rubbish dumps. But most predators are not particularly confined to any specific host. Some of the predators live in rather specialized habitats; for example, all the fish are aquatic, as are some insect larvae (e.g. Odonata), and so only prey on aquatic insects (such as mosquito larvae); some live in soil or leaf-litter so their prey is restricted to certain types of insects (more details are given in chapter 5).

### Parasites

These are almost entirely other insects and belong to two large groups (Diptera and Hymenoptera) and one small group (Strepsiptera), together with a few species of entomophilic nematodes. The Diptera include the large family Tachinidae which parasitize Lepidoptera (larvae), Coleoptera, Hemiptera and Orthoptera. Other parasitic families include Phoridae, Pipunculidae, Bombylinidae, and there are some parasites in the Sarcophagidae and Muscidae.

The Hymenoptera include the very important Chalcidoidea and Ichneumonoidea, and some Bethylidae, Scelionidae and Proctotrupidae. Some groups of parasitic wasps are miniscule (being the tiniest insects known, about 0.2 mm in body length) and are all egg-parasites. Almost all groups of insects (as well as spiders and ticks) are parasitized by the Hymenoptera Parasitica, at all stages of development from egg to adult. Many species of parasitic wasps lend themselves to exploitation in biological control projects.

## Pathogens

Control by pathogens is sometimes referred to as *microbial control*. There are three main groups concerned; bacteria, fungi and viruses, and some other groups of entomophagous micro-organisms which are rather obscure and little studied. There are several types of *Bacillus*, which are specific to caterpillars or beetle larvae, responsible for natural epizootics, and several species are now commercially formulated and very important in pest control projects.

Fungi are responsible for producing antibiotics and apparently about 300 antibiotics do show some promise as pesticides; these act directly as killing agents or inhibitors of growth or reproduction.

Viruses are quite commonly found attacking insects in wild populations of caterpillars and beetle larvae, as well as some temperate sawfly larvae. They have long been used as biological insecticides, by finding dead larvae in the field and making an aqueous suspension of their macerated bodies. But now a few commercial preparations are available.

Most of these new biological insecticides using insect pathogens are however only easily available in the USA as yet.

### Sterilization

This usually refers to the sterilization of males by X-rays or $\gamma$-rays and is called the *sterile-male technique* – control of a pest by this technique is termed *autocide*. Sterilization can be effected by exposure to various chemicals and this practice is called *chemosterilization*. The rationale behind this method is that male-sterilization is effective in species where females only mate once and are unable to distinguish or discriminate against sterilized males. The classical case was in about 1940 on the island of Curaçao against Screw-worm (*Callitroga*) on goats – the male flies were sterilized by exposure to $\gamma$-rays, and dropped from planes at a rate of 400/square mile/week. The whole pest population was eradicated in twelve months. The life-cycle took only about four weeks to complete, and the females only mated once in their lifetime. Generally, autocide is most effective when applied to restricted populations (islands, etc.), but can be effective on parts of continents. The Screw-worm eradication campaign was extended to the southern part of the USA where the pest is very harmful to cattle. In Texas 99.9% control was achieved in only three years. Male-sterilization trials were effective against Mediterranean Fruit Fly (*Ceratitis capitata*) on part of the island of Hawaii in 1959 and 1960, but immigration from untreated parts of the island prevented control from being long-lived.

This method of control could quite possibly be effective against *Orcytes* spp. attacking the coconuts along the coastal strips of Kenya and Tanzania, and work is in progress on the feasibility of autocide as a method of controlling Tsetse.

Chemosterilization has now advanced from a theoretical technique to a practical one, a variety of chemicals have been demonstrated to interrupt the reproductive cycles of a large number of insect species.

### Genetic manipulation

In reality this method is an extension of the previous one, in that the electromagnetic radiations (X-rays, γ-rays) induce *dominant lethal mutations* in the germ cells of the insects. These mutations in insect sperm have been used successfully in several eradication programmes. Lethal mutations are not lethal to the treated cell — they are lethal to its descendant, in that the zygote fails to develop to maturity. These mutations arise as a result of chromosome breakages in the treated cells.

### Potential uses for pheromones in pest control

Pheromones are reviewed in more detail in chapter 2 (page 17), but here they are considered in their actual roles in pest control.

The two obvious ways in which pheromones may be used in a pest control programme are firstly, in pest population surveys or for population monitoring (for emergence warnings, and spray warnings), and secondly for direct behavioural modification control. It is clear now from the work that has been done in recent years that pheromone traps are extremely useful in monitoring projects and this use is likely to be increased in the future. But to date there has not yet been a good example of pheromone use actually achieving a significant level of population control in a pest management programme; though, as already mentioned (page 20), there has been a behaviour disruption trial in S. America with Pink Bollworm on cotton with very encouraging results.

**Insect population monitoring.** The presence or absence of a particular insect species in an area can be established through the use of attractant pheromones, so that control measures may then be exercised, if necessary, with precise timing. Previously field population monitoring relied largely on either light-trapping, which requires a source of electricity, or the finding of eggs on the crop plants. The finding of the first eggs on a particular crop is a very tedious and time-consuming process requiring a great deal of labour and is not particularly efficient. The examination of a few pheromone traps for the presence of male insects is relatively very easy, and much more efficient. Alternatively, the pheromone traps can be used to monitor the effectiveness of a pest control programme, even though not directly employed in the programme themselves.

Emergence of male Codling Moth in apple orchards in the spring in Europe and N. America is now regularly monitored by the use of small paper (waterproof) pheromone traps with a sticky interior. Pink Bollworm, and other bollworms, on cotton crops in many parts of the tropics are likewise monitored with the use of these sticky pheromone traps, with considerable success. Various species of fruit flies (*Dacus* spp.) are monitored, sometimes using sticky pheromone traps and sometimes in traps with insecticides inside, in citrus and peach orchards to determine whether or not insecticide spraying is required, and if so just when. As more and more sex pheromones are being synthesized, and more chemical attractants are being discovered, it seems likely that the use of these chemicals in monitoring programmes will increase and will play a constant role in many pest management programmes.

**Insect behavioural control.** Bark Beetles have been induced to fly to inappropriate host trees by aggregation pheromones; the host tree was either resistant and killed the boring beetles or they were unable to breed successfully. Carefully designed traps, incorporating the required visual stimuli, have been very successfully used in destroying Bark Beetle populations in forests in N. America.

Orientation to a trap baited with sex pheromones (*trapping-out*) has proved a feasible method of reducing populations of various moths and some fruit flies (Tephritidae). Destruction of males responding to a female sex pheromone-baited trap will, however, only effectively control a pest population if enough males respond and are destroyed to result in most females not being inseminated. Using theoretical population models it is suggested that such a trapping technique could be effective at low population densities, but would be unlikely to be effective at high densities. Such traps can kill the male insects either by adhesion (sticky traps) or by insecticides.

Male nocturnal moths have been observed to be additionally attracted to light in the presence of female pheromones. An experiment in America in 1966 using Cabbage Looper moths showed that when a cage of virgin females was placed on to an ultra-violet (blacklight) light trap the catch of males overnight was increased twenty-fold! With the recent synthesizing of Cabbage Looper sex pheromone this method of population reduction could be feasible agriculturally.

The *communication disruption technique* uses pheromones or sex attractants to prevent orientation of males to virgin females. This method could give direct population control of a pest species in a crop, but as yet no really convincing results have been obtained. The basic idea is to saturate the air with pheromone or sex attractant, or at least to make the concentration high enough so that the pheromone released by the wild females is imperceptible to the males; thus the males would not find the females in the crop and the females would not get inseminated. A recent experiment in the New World gave encouraging results. On cotton crops grown in southern USA, C. and S. America, Pink Bollworm is a major pest which builds up through six generations during the warmer part of the year, the last two or three generations generally being above the economic threshold for this crop and causing economic damage. A sex attractant 'Gossyplure', was used in the communi-

cation disruption technique to saturate the air over the cotton fields with pheromone mimic so that the males would be disoriented by the odour and would fail to find and mate with many of the virgin females. This technique was timed to be used against the fourth and fifth Pink Bollworm generations. Following the use of 'Glossyplure', the level of cotton boll attack was significantly reduced, demonstrating that this technique can be agriculturally successful (Brooks, 1980). However, most experiments to date have not been particularly successful.

It has been shown by experiments that most male moths show adaptation in their response to sex pheromones as it is weakened following previous recent exposure. In addition, it was shown that some non-pheromone chemicals may react on the antennal sense cells or central nervous system to cause adaptation to the natural pheromone, but conclusive studies have not yet been carried out, although there would seem to be promise for eventual agricultural use.

Recent work by Cherrett and others has involved the addition of trail pheromones to baits for Leafcutting Ants in C. and S. America; this renders the bait pellets more attractive to the foraging ants which pick them up and take them back to the nest, making the poison bait more effective.

**Pheromone release.** The precise method of release of the pheromone, or mimic, is of importance since for trapping purposes a steady, controlled release over a known period of time is most desirable. Some preparations are sold as impregnated rubber or polythene caps, or as a 'wick', each usually containing 1 mg of the chemical, which are generally effective for several (often four) weeks. However, experimentation has shown that there does appear to be a considerable difference in success of pheromone release, sometimes when a different matrix is used. This is a factor that requires consideration when experimentation or monitoring programmes are planned, rather like the importance of the base (matrix) of a bait in a baiting project.

A recent development described by Brooks (1980) is a hollow fibre formulation for pheromone release. Each tiny fibre is 1½ cm in length, sealed at one end, and with the lumen filled with pheromone which evaporates from the open end. In the experiment against Pink Bollworm two pheromone components were used in a 1 : 1 ratio and the release was steady over a period of 2–3 weeks.

Another method of delaying evaporation of the chemicals is by microencapsulation, but after some encouraging results ICI found that pheromones in gelatine capsules used in experiments in the Mediterranean region rapidly degraded in daylight because of the ultra-violet radiation.

---

## Chemical methods

---

### Pesticides

Insecticides and their methods of application will be dealt with more fully in chapter 6. Only rarely does chemical application kill all the pests, and the few which survive usually soon give serious problems by the development of resistance. Chemical control is essentially repetitive in nature and has to be applied anew with each pest outbreak. However, this method is very quick in action and, for the majority of pest outbreaks, chemical control remains the method by which the surest and most predictable results are obtained. The different modes of action of insecticides are briefly listed below.

(a) *Repellants and antifeedants* — designed to keep the insects away; usually employed against mosquitoes and other medical pests.

(b) *Fumigants* — gases and smokes.

(c) *Stomach poisons* — often mixed with baits to encourage ingestion, or as foliar application.

(d) *Ephemeral contact poisons* — absorbed through the cuticle; usually foliar application.

(e) *Residual poisons* — persistent; remain active for a long period of time; foliar or soil application.

(f) *Systematic poisons* — watered into the soil, sprayed on to the plant, or applied to the trunk; absorbed and translocated by the plant and effective against sapsuckers especially. May be applied as sprays or granules, to either soil or foliage.

Pesticide application still remains the major weapon in the pest war, for obvious reasons, but we now find that often, sometimes usually, there are three definite post-application effects.

(a) Resurgence of treated pest — this always occurs as the target species initially suppressed by the insecticidal treatment shows a rapid population recovery after the decline of the treatment effect.

(b) Resurgence of the target species due to either development of a resistant biotype (as with the Brown Planthopper of Rice) and/or destruction of natural enemies.

(c) Outbreak of a secondary pest or pests, due to the alteration of the agroecosystem, usually by the destruction of natural enemies.

### Insect resistance to pesticides

This topic is now of sufficient importance that a small section of text be devoted to it, for it is a major factor to be considered in many tropical pest management programmes.

In the 1940s the many new synthetic organic pesticides, mostly the organochlorine group, became widely available throughout the world for crop protection. Pesticides such as DDT, BHC and dieldrin were in many respects thought to be ideal for crop protection: they were highly toxic to insects (most groups), broad-spectrum and persistent. Control levels (kills) of major insect pests were consistently at the level of 98–99% or even higher. Many serious pests were controlled effectively for the first time, and a heavy dependence upon these pesticides resulted. Invariably, though, despite a high kill there was always a small proportion of individuals that possessed a natural resistance to such poisons, and of course during successive generations this genetically based, inherited, natural resistance spread throughout the insect (and mite) populations. After a number of years (generations) this resistance became manifestly obvious and a cause for concern. Eventually the entire local pest population became resistant to the chemical involved. These groups of individuals, with slight genetic differences from the main stock of the insect species, are known as *biotypes*. In disease-causing pathogens these races are known as *pathotypes*, and in a few cases a widespread pathogen is known to have as many as 20–30 pathotypes throughout its geographical range. Insects generally have far fewer biotypes, but the Brown Planthopper of Rice probably has a dozen biotypes in the area from India, through S.E. Asia, up to Japan. As already mentioned on page 46, there are resistance-breaking biotypes associated with varieties of resistant crop plants.

Sometimes resistance to one chemical compound leads to resistance to other closely related compounds in the same group. For instance, after initial development of resistance to DDT and the other organochlorines, there followed resistance to the organophosphorous compounds, and now there is also resistance to some carbamates.

To date, more than 300 major pest species (insects and mites) have developed resistance to one or more major pesticides. An extreme case is that of Diamond-back Moth in S.E. Asia which is now showing marked resistance to most of the pesticides used against caterpillars, with the exception of BTB; whether it is now developing resistance to pyrethroids is not quite clear at present. One of the most spectacular cases of biotype development is that shown by the Brown Planthopper of Rice (*Nilaparvata lugens*) in India and S.E. Asia. Prior to the widespread use of diazinon on paddy rice it was only a very minor pest. Diazinon was the most widely used insecticide for general rice pest control in S.E. Asia for many years, and in 1970 this planthopper was first recorded as a major rice pest in a few localities; by 1975 it was recorded as a serious pest, and by 1980 was both serious and widespread. Recent work at IRRI has shown that the situation now is such that if diazinon is used on rice it invariably causes a resurgence of the planthopper.

It is generally thought that some 10–15 generations are required for the development of manifest resistance, and, as already mentioned, a widespread crop pest may show the development of several distinct biotypes over different parts of its distributional range. In temperate regions where many pests are uni- or bi-voltine, resistance develops slowly (6–10 years), but in the tropics, with the higher temperatures and no cold winter, insect breeding may be more or less continuous, and resistance may be strongly apparent after only 2–4 years in extreme cases. The Brown Planthopper of Rice can apparently develop a new biotype in 18 months on rice, that is about 18 generations.

To date, field resistance when encountered has been dealt with by just increasing the amount of pesticide applied, or using available chemical substitutes. The former remedy is generally useless (except for a very short time) and only adds to the general ecological disturbance as well as causing residue problems; it may also accelerate resistance development. Resistance problems are expanding and intensifying to such an extent that they are outstripping the development of new chemical pesticides.

Most cases of resistance have been shown by the Arthropoda, but in recent years resistance has been also shown by rats, some fungi, and some weeds.

## Antifeedants

Certain chemicals possess the properties of inhibiting the feeding of insect pests – these can be classed as antifeedants. The first chemicals noticed with these properties were initially referred to as repellants, but they are not repellants in that they do not merely drive the insect away to another plant but actually inhibit feeding on that plant. In laboratory tests insects have remained on treated plants indefinitely and eventually starved to death without eating the leaves. In field tests the insects were free to wander elsewhere seeking food – they either found weed plants to feed on or died of predation and starvation.

The most successful source of natural antifeedant is the Neem tree (Indian lilac; *Azadirachta indica*) and the closely related Persian lilac (*Melia azedarach*). Extracts from the Neem have long been known to have germicidal properties, but for the last ten years research in India and Germany has concentrated on its pesticide potential. Extracts from leaves and fruits (Neem oil) act as an antifeedant for many different insects (locusts, caterpillars and Hemiptera) and they also upset insect development and inhibit gravid females from oviposition. The active ingredient is a chemical called azadirachtin, and it is thought to be related to the ecdysonoids. The first Neem Conference was held in Germany in 1980 and was reported in *International Pest Control* (1981) No. 3, pp. 68–70 (25 references are listed).

The earliest antifeedant used in agriculture was ZIP (a complex compound zinc salt) used to keep rodents and deer from feeding on the bark and twigs of trees in the winter. The first recognized antifeedant for use in insect pest control was introduced by Cyanamid in 1959. Since then a number of compounds have been shown to possess antifeedant properties, but a commercially successful antifeedant has not yet been produced, although work is still progressing along this line, and it appears that Neem oil may be the first such product.

# Integrated control

The original concept of integrated control was developed to stress the need for understanding the complicated and antagonistic relationship between biological control and the use of chemical pesticides. Early pest control measures, up until about the turn of the century, were mainly concerned with the biology and ecology of the pests, and in particular those aspects relating to their population numbers. Attempts were made to make the crop environment less favourable to the pests by a combination of cultural and biological methods. The chemical poisons available at that time were rather simple, such as kerosene, sulphur and some inorganic salts (lead, arsenic and mercury), and were not very effective as insecticides. The earliest attempts at biological control were in the early 1900s when various insect predators and parasites were imported into California and Hawaii to try to control some of the pests of the newly established *Citrus* industry. It was soon appreciated that the use of chemical poisons was basically inimical to biological control in that the poisons were not selective, and in fact the natural enemies were more susceptible to poisons than the biologically very robust pests. In 1940 DDT was discovered to have insecticidal properties and was soon available commercially. It was capable of killing a broad range of insects and mites in small doses, and had a long-lasting residual activity. Literally almost overnight insect pest control was revolutionized, and a series of new synthetic organic insecticides was rapidly discovered and made available commercially. The organochlorine compounds (DDT, BHC, dieldrin, etc.) gave consistently high kills (98–99%) of a wide range of pests, as seed dressings, sprays and powders; persistence was lengthy and application was not difficult.

But there was no universal panacea after all; it

soon transpired that pest problems often continued, and indeed sometimes worsened, and many undesirable side-effects were noted. These were the accidental destruction of natural enemies of the pests, and the development of resistance to the insecticides by the pests, and soon various other ecological disruptions became evident. Now, after considerable ecological damage and widespread resistance has developed, the culmination has been the extensive banning of the organochlorines in most parts of the world, particularly the total ban imposed by the Environmental Protection Agency (EPA) in the USA. The cycle of events is now completed as we are turning back to the original approach whereby biological and ecological understanding assumes predominance. The judicious use of the synthetic organic pesticides is a most important weapon in pest control, but must be used in conjunction with other appropriate methods.

The term *integrated control* was originally coined to describe the combining of biological control with compatible chemical application. (In this sense biological control included natural control and biological control *sensu stricta*.) The basic idea was to use chemical pesticides judiciously so as to avoid disrupting the existing natural control by killing the predators and parasites in the crop community. This can be done in several ways; by using specific, carefully screened pesticides only, by careful timing of the treatment, using minimal dosages, by reducing spray drift, and so on. This attitude developed because these two basic approaches to pest control are our primary resources, and as used in many instances in the past they have been in direct conflict with each other (Smith & van den Bosch, in Kilgore & Doutt, 1967).

Eventually accumulated experience and logic made it clear that it is necessary to integrate not only chemical and biological control, but all available techniques and procedures, into a single pattern aimed at profitable crop production together with minimal environmental disturbance. This realization led to the concept of *pest management* (PM) with its broad ecological approach to pest situations.

During the last decade the literature on pest control has been very confusing for in some instances integrated control was regarded in its original context and in many others it was virtually synonymous with pest management. After all, there is no basic qualitative difference between the two approaches, the difference is essentially quantitative. This confusion in terminology was widespread, but finally there appears to have been some international agreement to remove this confusion and the new term for PM is now accepted as *integrated pest management* (IPM) as formulated by Glass (1975) on behalf of the Entomological Society of America. The term integrated control is regarded as an historical term, now superseded.

The control of Coffee Leaf Miners (*Leucoptera* spp.) at Ruiru in Kenya has been, for many years, a carefully balanced, integrated programme using selected insecticides together with natural control by parasites. A similar situation prevails with regard to Coffee Tip Borer (*Eucosma nereidopa*) where the shaded tree trunks are sprayed to kill the resting adults. This avoids the application of chemicals to the leaves and shoots, so that the quite effective natural control by parasites may continue without interference. Conway (1972) found that, in Sabah, excessive use of pesticides was severely aggravating the local pest situation on cocoa and oil palm, and the most effective immediate measure was to curtail spray applications to allow survival of the predators and parasites which normally keep most of the pest populations in check.

## Pest Management (PM)
### *now Integrated Pest Management (IPM)*

In 1967 the FAO panel of experts on integrated pest control defined *integrated control* as 'a pest management system that, in the context of the associated environment and the population dynamics of the pest species, utilizes all suitable techniques and methods in as compatible a manner as possible and maintains the pest population at levels below those causing economic injury'.

This definition incorporates the concept of pest management as defined by the Entomological Society of America, now expressed as IPM (Glass, 1975).

The concept of PM is now well established. One of the earliest definitions was by Rabb & Guthrie (1970); they commented that originally integrated control generally referred to the modification of insecticidal control in order to protect and enhance the activities of beneficial insects (predators and parasites). Subsequently, however, integrated control interpretations have become more comprehensive until, now, some definitions of integrated control embody most of the essentials of pest management. Rabb preferred the term pest management because it connotes a broader ecological basis and a wider variety of opinions in devising solutions to pest problems.

Pest management can be defined as the reduction of pest problems by actions selected after the life systems of the pests are understood and the ecological as well as economic consequences of these actions have been predicted, as accurately as possible, to be in the best interests of mankind. In developing a pest management programme, priority is given to understanding the role of intrinsic and extrinsic factors in causing seasonal and annual changes in pest populations. Such an understanding implies a conceptual model of the pests' life system functioning as a part of the ecosystem involved. Ideally such a model would be mathematical, but a word or pictorial model may be useful in predicting effects of environmental manipulations.

Five of the most characteristic features of the population management approach to pest problems are as follows.

(a) The *orientation* is to the entire pest population, or a relatively large portion of it, rather than to localized infestations. The population to be managed is not contiguous to an individual farm. county, state or country, but is more often international – hence a high degree of co-operation, both nationally and internationally, is a prerequisite for success.

(b) The *immediate objective* is to lower the population density of the pest so that the frequency of fluctuations, both spatially and temporally, above the economic threshold is reduced or eliminated.

(c) The *method*, or combination of methods, is chosen to supplement the effects of natural control agents where possible and is designed to give the maximum long-term reliability of protection, the minimum expenditure of effort and money, and the least objectionable effects on the ecosystem.

(d) The *significance* is that alleviation of the problem is general and long-term rather than localized and temporary, and that harmful side effects are minimized or eliminated.

(e) The *philosophy* is to manage the pest population rather than to attempt to eradicate it. The real significance of the concept is seen in relation to serious pest problems which defy solution through the more traditional approaches.

As previously stated, this broad ecological approach to pest control problems is still rather new in concept and, as yet, is more developed theoretically than in practice, particularly with regard to the use of computer models.

Throughout the world there is now an IPM programme devised for many different crops in different locations, for example tobacco in N. Carolina (Rabb, Todd & Ellis, 1976), sugarcane in Louisiana (Hensley, 1980), sugarcane in Taiwan (Chu, 1980) and others in Apple & Smith (1976). Clearly some of these programmes are in the category of first attempts, and as yet rather crude, whereas others are quite elaborate, very sophisticated and obviously very effective. Presumably, the basic problems with a particular crop remain much the same wherever the crop is grown, but the pest and disease spectra will be different in each locality where the crop is important (i.e. site-specific). The ultimate aim for crop protectionists is clearly an IPM programme for all the more important crops in all the major agricultural regions of the world.

## Eradication

In most cases, pest control is undertaken to reduce the population density of an insect to a point at which the damage done is not of economic significance; very rarely is complete eradication the goal, and even more seldom is it achieved. The more usual cases of complete eradication are directed against pests of medical importance, and very successful campaigns have been carried out in many areas against diseases such as malaria, yellow-fever and dengue. Against agricultural pests about the only time eradication is aimed at is when a new pest, which is potentially very serious, has been introduced into a country and has not spread too far. The campaign is often very costly and difficult but can be won if the pest is still restricted to a relatively small area.

Some eradication programmes have been successful against Screw-worm (*Callitroga*) and fruit flies (*Drosophila* spp.), when sterile-male techniques were employed. This technique is particularly suitable for eradication programmes.

California and Florida in the USA, have multi-

million dollar *Citrus* industries, as well as for peach, fig, guava and many other tropical and subtropical fruits. Some very damaging fruit flies are still not established here, as are some other pests such as San José Scale and California Red Scale (Florida). Some years ago Medfly (*Ceratitis capitata*) was accidentally introduced into Florida; a total of 0.7 million hectares of *Citrus* orchards and adjoining land was sprayed with insecticides, at a total cost of US $8 million; but the project was successful and the Medfly population exterminated. The total value of the Florida fruit industry was at that time about $180 million. Very recently (1981) Medfly was accidentally established in California and presented the Governor with a difficult decision. With a $14 000 million fruit industry in California at stake it was imperative that an aerial spraying eradication programme be started quickly, but in recent years the environmentalists in California have become a very powerful political lobby and they objected strongly to the idea of aerial spraying of suburban homes and gardens. The Medfly was, however, too serious a threat so the eradication project went ahead.

Colorado Beetle is accidentally introduced into the UK in most years, but this very damaging pest of potatoes is usually destroyed quite rapidly. Since the 1933 Colorado Beetle Order all farmers and growers are aware of this pest and the danger it presents, and the odd infestation is usually reported promptly. On receipt of a report, the Ministry of Agriculture regional staff take immediate action to collect and kill adults and larvae on the potato foliage, and also to fumigate the soil in the infested area in order to kill any pupae that might be present.

The decision to eradicate a pest is a grave biological responsibility and should not be made unless careful study, involving diverse perspectives, has produced convincing evidence that the benefits to be accrued more than balance the ecological impoverishment represented by removing the pest species.

# 4  *Pest damage to crop plants*

The agricultural entomologist has a distinct advantage over his colleagues in nematology and plant pathology in that the damaging organisms with which he is concerned are relatively large and usually to be found in the vicinity of the damage on the crop plant. This helps to make the identification of insect pests a relatively simple matter, at least to the level of family and genus.

For most control purposes pest identification to family, or preferably genus, is often adequate, for most members of most insect (and mite) families produce the same type of damage on the crop plant and are likewise controlled by the same practices. After all, this is to be expected, since the object of insect systematics is to place closely related species together in the same taxa. However, there are occasions when the specific identity of the pest is important, especially in the case of aphids and other Homoptera that are virus vectors. For example, *Toxoptera citricidus* is the vector of Citrus Die-back disease (Tristeza) in Africa and S. America, whereas the partially sympatric and morphologically very similar *T. aurantii* is apparently not a vector. At such times a high level of taxonomic expertise is demanded, but these occasions are relatively few. Generally however, it is necessary to identify the pest correctly (to the appropriate systematic level) in order that the economic significance of the infestation be accurately assessed, and the most appropriate control measures be applied.

On occasions only the damaged plant may be found, or else there may be several similar pests on the crop and it may not be evident which insects are responsible for which damage. Some pests are nocturnal or crepuscular in habits and during daylight hours remain hidden in the plant foliage or the soil. The damage inflicted on the various parts of the plant body is, however, sometimes characteristic of a specific pest, or a group of closely related pests. Then the experienced entomologist will usually be able to make a fairly accurate determination (identification) of the identity of the damaging animal.

In practice the situation is often more complicated. Several quite unrelated insect groups produce almost identical attack symptoms on the plants; for example, leaf miners may be the larvae of Diptera (Anthomyiidae, Ephydridae, Agromyzidae), or Lepidoptera (Gracillariidae, Lyonetidae) or Coleoptera (Chrysomelidae, Hispinae). 'Dead-hearts' in graminaceous seedlings may be produced by larvae of Lepidoptera (Pyralidae, Noctuidae), or Diptera (Muscidae, Anthomyiidae), or Coleoptera (Scarabaeidae). The more generalized defoliation resulting from the browsing and grazing of herbivores can be very difficult to identify without extensive practice, and even then it is often not possible. Leaf-eating is the normal method of feeding of some mammals (ungulates, pigs, rabbits, rodents), birds (sparrows, ducks, etc.), molluscs (slugs, snails), most grasshoppers, locusts, caterpillars, and some beetles (both larvae and adults). If one is presented with such damaged plant material in the laboratory or regional 'Plant Clinic', then identification of the causal organism may be quite impossible. But if the plants are examined *in situ* in the field various clues such as footprints, scats, hair, feathers, slime trails, etc. may permit the entomological detective to identify the culprit. Detailed ecological knowledge is of great assistance in such cases, especially knowledge of local animal migrations and behaviour.

The situation is actually even more complicated in reality because damage similar to that produced by some pests may also result from adverse weather conditions, such as drought, flooding or waterlogging of the soil, lightning-strike, hail, frost, sun-scorch and strong wind. Excessive fertilizer application, herbicide damage and use of insecticide on susceptible crops also produce distortions and damage to the plant body. Disease organisms (bacteria, fungi, viruses) and nematodes (eelworms) produce some symptoms reminiscent of insect and mite attack, and mineral deficiencies (manganese, magnesium, etc.) result in symptoms similar to various virus infections.

The growing crop will be subjected to the vagaries of the weather, prevailing soil conditions, and the ravages of maybe a vast pest and disease complex, which altogether will influence the basic genetic control of the development of the plant body. All of these constituent ecological factors will be interacting, and the various agricultural advisers (agronomist, horticulturalist, pathologist, entomologist) will be confronted by the end-result of this complicated interaction. Obviously, in some situations the adviser cannot hope to rectify the situation, but only aim at minimizing damage in an attempt to secure a crop yield of economic proportions.

It should finally be noted that various pests and disease organisms interact by predisposition, so that an infected plant may be more susceptible to pest attack, and a slight infestation may produce unusually severe symptoms. This also applies to climatic and edaphic factors; a water-stressed plant is invariably more susceptible to pest attack.

Pest damage is probably most conveniently considered according to the part of the plant body attacked, and in the following account it is studied under these headings: damaged leaves, flowers and buds, fruits and seeds, stems, roots and tubers, sown seeds and seedlings.

## Pest damage assessment and crop yields

The ultimate aim of agriculture is to produce a sustained economic yield of crop produce, so it becomes of prime importance to understand the effect of the insect pest population on the subsequent yield or harvest. Obviously, if the pests are causing no crop loss their presence on the plants and the damage they cause may be ignored, and in the context of ecological stability they should be left alone! However, most pest populations produce some damage of significance, but the damage assessment in relation to possible or expected yield loss is difficult. The total number of interacting factors responsible for determining crop yield is quite overwhelming, and any decision as to the probable effect of any single factor, such as the population of one insect pest species, is problematical. However, the gradual accumulation of empirical data over many years has resulted in our being able to make various generalizations about some pest populations and their probable effect on crop yield. These results are used to define economic injury levels (and economic thresholds) for some pests on some crops in different parts of the world. But in general, many more data are required for many more pests on the more important tropical crops.

Some types of damage are obviously more important than others, depending upon the part of the plant body damaged and the part harvested; if the two are the same then clearly the damage is more serious. A single fruit fly larva can effectively destroy a single mango, peach or orange fruit, and a relatively small number could ruin an entire crop. On the other hand, a mango tree can accommodate a large number of foliage-eating caterpillars and root-eating termites or beetle larvae, with no discernible loss of yield. Root crops generally can withstand quite heavy leaf damage without appreciable yield loss; cereals can likewise tolerate leaf-eating, stem sap-sucking and also root-eating; palms and pulses can tolerate quite heavy defoliation. Vegetables such as brassicas and lettuce may have their outside leaves removed at harvest, prior to sale, so damage to the older leaves is unimportant. In summary, damage which can be ignored on one crop may be of considerable economic importance on another, so damage assessment is different for each crop grown.

Most plants produce far more leaves and often more flowers than actually required, so that partial defoliation should result in no loss of yield, and there will often be a natural bud-drop (e.g. *Hibiscus* and cotton); cotton generally sheds about 70% of the young squares naturally. Similarly, some plants produce excessive numbers of flowers followed, in most recent years, by a subsequent natural premature fruit-fall, as seen in coconut and apple. Coconut Bug is thus a debatable pest in that 'early-nutfall' symptoms of its attack may really be insignificant as it is anticipated that 70% of the young nuts will fall naturally due to natural overproduction. It must be remembered that in 'the natural scheme of things' (i.e. the ecological situation) one major function of green plants is to have their foliage eaten by herbivores, as the bases for the myriad terrestrial food chains and webs, so natural foliage and floral overproduction would be an obvious evolutionary adaptation.

Studies on cotton in Queensland by the Integrated Pest Management Unit of the University found that looper caterpillars (*Anomis flava*) were often serious defoliators. However, a large population of larvae was required before any significant reduction in leaf area index (LAI) was recorded. In experiments on artificial defoliation it was found that in general the critical LAI was 3.0, and the critical growth time was between the first flower to the first split boll, but even then more than 25% defoliation was required to produce a significant yield loss. Usually the loopers

attack the lower leaves, which are the most shaded and thus the least important photosynthetically. It is even thought that such bottom defoliation may be advantageous in allowing more air and light to the lower bolls and reducing boll rots which are often a problem if the LAI is high.

Casual inspection of pest infestations can be very misleading in relation to actual damage done. Pests such as aphids are relatively serious on young, actively growing plants; a light infestation of aphids on some young plants may be very damaging and cause severe stunting and poor yield, whereas a heavy infestation on a woody plant (shrub or tree) or a mature herbaceous plant may be very unsightly (especially if associated with sooty mould) but have a negligible effect. In fact, in the case of large (heavy) infestations of scales or whiteflies on the leaves of trees (such as *Citrus*) there may be an overall beneficial effect in permitting a build-up of parasite populations.

The two basic ways in which pest infestation or damage is assessed are, first, the *incidence* of the pest or damage symptoms, which is the proportion of plants in a sample which show symptoms or are host to pests, and second, the *severity* of the infestation which is a measure of the size of pest population on the plants or the extent of the damage done, which is often measured as so many insects per bush, or so many egg masses per bush, per fruit, etc. The most important crops have well-studied and easily recognized growth stages, only some of which are critical in relation to pest and disease attack vulnerability (FAO/CAB, 1971). It is not possible to generalize further than this for damage assessment will remain different for each different crop. Insect infestation severity is generally broken down into about six different levels for easy recognition in the field by non-experts, and it is usually recommended that a large number of small samples be taken (for rapid assessment) rather than a small number of large samples; this will accommodate an uneven pest distribution throughout the crop which is the more usual situation.

Accumulated observations reveal that for some crops the effect of a particular pest infestation level will vary according to the usual yield of the crop. Thus losses for a high-yielding crop will be relatively less than the loss caused by the same size of pest population on a low-yielding crop. So for some crops it is necessary to distinguish between high-, medium- and low-yielding crops in damage assessments, e.g. sorghum in W. Africa.

In some cases the economic injury levels and the damage assessments can be conveniently expressed in simple terms as so many insects per plant, per shoot, or per trap, whereas in others the methods involve detailed analysis and computer programmes. For further details see the publication by FAO/CAB (1971) which presents loss assessment methods for a total of 80 pests on 27 crops. About 44 of the pests can be regarded as tropical, on 14 tropical crops as follows (number of pests in brackets): banana (1), beans (1), citrus (1), cotton (4), cocoa (1), coffee (1), cowpea (3), maize (12), opium poppy (1), rice (12), sorghum (3), soybean (1), sugarcane (2) and sunflower (1).

Examples of economic injury levels (economic thresholds) for some pests on some tropical crops are given below.
(a) Coffee (*arabica*) in E. Africa — two Antestia Bugs per bush.
(b) Maize in Thailand: Asian Corn Borer — 15 egg-masses per 100 plants.
(c) Cocoa in Ghana — 45 capsid bugs per acre.
(d) Sorghum in W. Africa — c. 3% spikelets infested with larvae of Sorghum Midge.
(e) Banana in Honduras — 15–20 Banana Weevils per disc-on-stump trap.
(f) Cotton in Malawi and Sudan: bollworms — 5 larvae per 100 plants, or 10 eggs per 100 terminal shoots.

Some examples of damage assessment levels are as follows.

(a) Cotton in Queensland (Australia) — the critical LAI is 3.0; if defoliation by Cotton Looper results in a lower LAI, control spraying is recommended; defoliation is most serious in the period between first flower and first split boll.
(b) Cowpea in Nigeria: *Maruca testulalis* — for high-yielding crops (average maximum yield 1120 kg dry seed/ha), an increase from 0–40% flowers attacked by caterpillars gives a 30% reduction in seed yield; for low-yielding crops (average maximum yield 560 kg dry seed/ha), an increase from 0–40% flowers attacked by caterpillars gives a 70% reduction in seed yield.
(c) Sorghum in W. Africa: *Contarinia sorghicola* — loss is proportional to the percentage of spikelets attacked; for a 1% increase in spikelets attacked, the reduction of yield is:
high-yielding crop (1390 kg grain/ha) — 1.34%
medium-yielding crop (870 kg grain/ha) — 1.63%
low-yielding crop (270 kg grain/ha) — 3.48%.

---

# Types of pest damage to crop plants

---

### Damaged leaves
(a) Margin with regular notches (fig. 8–1): many adult broad-nosed weevils (Coleoptera: Curculionidae) feed on the leaf margins of many different plants throughout the world, producing characteristic notches.
(b) Margin with large, clean, semicircular (or sub-circular) pieces of lamina missing; adults of leaf-cutting ants (Formicidae: Attinae) in C. and S. America, and leaf-cutting bees of the genus *Megachile*.
(c) Margin irregularly eaten (figs. 8–14 and 9–7): the commonest form of leaf damage by defoliating pests, caused by grasshoppers, locusts, many caterpillars, leaf-beetles, sawfly larvae, some slugs and snails, some birds; in a severe attack the entire leaf lamina may be

eaten away, sometimes leaving only the main veins intact.

(d) Lamina skeletonized (fig. 8−12): caterpillars of several families (Epiplemidae, Bombycidae, etc.) especially when young, eat part of the way through the leaf lamina but leave the veins and one epidermis intact; some beetles (adults and larvae of *Epilachna* (fig. 9−8), adult Citrus Flea Beetle) and some slug sawflies (Tenthredinidae).

(e) Lamina holed (fig. 8−7): many tiny holes are made by adult flea beetles (Chrysomelidae, Halticinae); fewer larger holes are made by adults of tortoise beetles and some other Chrysomelidae, and some caterpillars.

(f) Lamina windowed (fig. 8−13): a window is a small hole in the lamina with one epidermis left intact, but after a while the thin epidermis dries and ruptures leaving a small hole; such damage is characteristic of larvae of the Diamond-back Moth.

(g) Leaves and shoot webbed: some caterpillars (Tortricidae, Pyralidae, Lasiocampidae, etc.) web leaves and shoots with silk and make a 'tent' or web in which they live and eat the leaves, often the web traps a large number of faecal pellets. An aerial, webbed-leaf nest is made by red tree ants of the genus *Oecophylla*.

(h) Lamina cut and rolled (figs. 8−5 and 9−9): caterpillars of Hesperiidae (Skippers), some Tortricidae and Pyralidae cut the lamina and make a leaf-roll, binding the edges with strands of silk; commonest examples are Cotton Leaf-roller (*Sylepta derogata*) on cotton and *Hibiscus*, Banana Skippers (*Erionota* spp.) and the several Rice Skippers; the leaf lamina is eaten within the protection of the roll; leaf-rolling can be induced (but no silk used) by some Aphididae, Aleyrodidae, many thrips, some sawfly larvae, and nesting spiders (fig. 9−5).

(i) Leaf curled under (fig. 8−4) or generally distorted, sometimes completely distorted into a bunchy lump of tissue (fig. 8−10): the former damage is done by

Aphididae, Psyllidae, Aleyrodidae, Cicadellidae and some other Homoptera, the latter damage is by thrips of the family Phlaeothripidae.

(j) Leaf folded (fig. 9−1,2): some adult Pyralidae fold over the edges of graminaceous leaves to provide a protected egg-site, and some pyralid caterpillars also fold cereal and grass leaves longitudinally; *Gynaikothrips ficorum* fold the young leaves of the Chinese banyan (*Ficus microcarpa*) by their feeding activities.

(k) Lamina pitted (fig. 8−6): Psyllidae in the group Triozinae cause ventral leaf-pits at the sites where the nymphs sit and feed, young leaves sometimes may be considerably deformed.

(l) Lamina scarified (fig. 8−3): adults and nymphs of some thrips (Thysanoptera) and some Spider Mites (Tetranychidae) make tiny epidermal feeding lesions which give the lamina a silvery, bronzed or scarified appearance; soft leaves will be caused to wilt, as shown by Onion Thrips on onions (fig. 9−17).

(m) Lamina with erinia (fig. 8−11): on the lower surface (usually) of some leaves are found wart-like outgrowths (erinia) inhabited by microscopic gall mites (Eriophyidae).

(n) Leaf galls (fig. 8−15): round or elongate galls, small or large, few or numerous, found on either upper or lower leaf surface, sometimes arranged randomly, sometimes distributed peripherally or alongside veins; made by feeding larvae of gall midges (Diptera: Cecidomyiidae), gall mites (Acarina: Eriophyidae), gall wasps (Hymenoptera: Chalcidoidea, Cynipoidea and Symphyta), and some Psyllidae.

(o) Lamina tattered with irregular holes and tears (fig. 8−16): margin usually intact; many Miridae and other Heteroptera feed on young leaves, and their toxic saliva results in small necrotic spots. As the leaf grows and expands the dead areas enlarge and tear, resulting in the characteristic tattering of the expanded leaf.

(p) Lamina mined (figs. 8−17,18; 9−15): the mines may be either tunnel mines or blotches, sometimes starting as a tunnel and ending as a blotch mine; tunnel mines with a central line of faecal pellets usually belong to caterpillars (Lepidoptera: Gracillariidae, Lyonetidae, etc.); tunnels without evident faecal pellets are made by maggots (Diptera: Agromyzidae, Anthomyiidae, Ephydridae) and beetle larvae (Hispinae, Halticinae).

(q) With bubble-froth (fig. 8–2): either on leaves or in leaf axils; spittle mass built by nymphs of Cercopidae for protection; Spittle Bugs.

(r) With black sooty mould (fig. 9–11): infestations by aphids, mealybugs, and soft scales are often associated with sooty moulds that develop on the sugar excreted as 'honey-dew'.

(s) Leaf petiole galled: several species of temperate woolly aphids (Pemphigidae) make large galls in the leaf petioles of various woody shrubs and trees; also Bean Fly larvae mine and swell petioles of beans.

(t) Cereal leaves with longitudinal feeding scars (fig. 8–9): made by adult leaf-beetles, especially Hispinae, and both adults and larvae of some Criocerinae; hispine larvae mine inside the leaves.

(u) Cereal leaves with serial holes (fig. 8–19): these were the feeding sites of young cereal stem borers made when the young leaf was folded (Lepidoptera: Pyralidae, Noctuidae) prior to their penetrating the stem; as the leaves grow and expand, the tiny feeding sites become expanded into a series of matched holes.

(v) Dangling cases constructed of leaf fragments (fig. 9–3,4); small cases on rice leaves are made by larvae of the Rice Caseworm (*Nymphula depunctalis* – Pyralidae); small or large cases are made by bagworms (Lepidoptera: Psychidae) on many types of tree and bush (sometimes herbaceous plants), but abundant on palms.

(w) Cereal leaves with longitudinal streaking (adjacent pale and dark areas) (fig. 9–16), and dicotyledonous leaves with a conspicuous regular blotching; these are symptoms of virus infection; all plant viruses are trans-

*Fig. 8.* Damaged leaves (no.1)

1. Leaf margin notched (adult broad-nosed weevils: Coleoptera; Curculionidae).
2. Leaf axils with frothy spittle mass (spittlebugs: larvae of Cercopidae; Hemiptera).
3. Leaf lamina scarified (adults and nymphs of thrips: Thysanoptera. Red Spider Mites: Acarina; Tetranychidae).
4. Leaf edges curled under (aphids, jassids, psyllids and thrips: Hemiptera; Aphididae, Cicadellidae, Psyllidae, and Thysanoptera).
5. Dicotyledonous leaf rolled longitudinally (leaf-rollers – larvae of some Pyralidae and Tortricidae; Lepidoptera; Cotton Leafroller (*Sylepta derogata*) on cotton and *Hibiscus*).
6. *Citrus* leaf lamina with ventral pits (some pysllids (*Trioza* spp.): Hemiptera; Psyllidae).
7. Leaf lamina with many, small (shot) holes (adult flea beetles: Coleoptera; Chrysomelidae; Halticinae).
8. Leaf lamina with larger, regular-shaped holes (adults and some larvae of leaf beetles, especially tortoise beetles: Coleoptera; Chrysomelidae; Cassidinae).
9. Cereal leaf with elongate, deep scarification (adult and larvae of some leaf beetles: Coleoptera; Chrysomelidae) or elongate mines (larvae of hispid beetles: Coleoptera; Chrysomelidae, Hispinae).

10. Leaves completely folded, rolled or distorted into a bunchy lump (adult and nymphal thrips: Thysanoptera; Phlaeothripidae).
11. Wart-like outgrowths on underside of leaves (gall mite erinia: Acarina; Eriophyidae).
12. Leaf lamina extensively skeletonized (leaf skeletonizers: larvae of various Lepidoptera; Epiplemidae, Bombycidae, etc.)
13. *Brassica* leaf lamina with many, small, round holes and windows (Diamond-back Moth larvae (*Plutella xylostella*): Lepidoptera; Yponomeutidae).
14. Vegetable leaf with large pieces eaten away (leafworms, etc. : Lepidoptera; Noctuidae).
15. Leaf lamina with many round or elongate galls (gall midges and gall wasps: Diptera; Cecidomyiidae: Hymenoptera; Symphyta and Chalcidoidea, or by some Psyllidae, or gall mites (Eriophyidae).
16. Leaf lamina tattered with many irregular tears and holes (capsid bugs: Hemiptera; Miridae, etc.)
17. Leaf-mine starting as a tunnel and ending as a blotch-mine (larvae of Lyonetidae and Gracillariidae; Lepidoptera. Anthomyiidae;Diptera).
18. Leaf-mine with broad tunnel and central row of continuous faecal pellets, and edge of lamina turned over dorsally for pupation site (larvae of Gracillariidae; Lepidoptera).

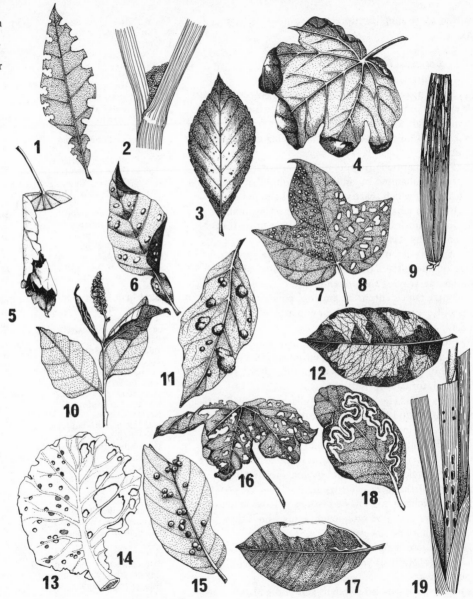

19. Graminaceous leaf with regular series of small holes in expanding lamina (feeding holes of young stalk borers prior to penetration of the plant stalk: Lepidoptera; Noctuidae and Pyralidae).

mitted by feeding Hemiptera, or sometimes by nematodes.

## Damaged flowers and buds
*(Illustrated in fig. 9 for convenience)*

(a) Petals gnawed (fig. 9—13,14): adult blister beetles (Meloidae) chew petals of many plants, often common on Malvaceae; adult flower beetles (Scarabaeidae) make small holes in petals, *Popillia* being especially injurious.

(b) Petals scarified: flowers of Leguminosae, Compositae, etc. inhabited by adults and nymphs of thrips (Thripidae) which scarify the bases of the petals.

(c) Flowers inhabited by tiny black beetles, making feeding scars at the base of the petals: legume flowers inhabited by *Apion* weevils; worldwide.

(d) Anthers eaten: Pollen Beetles (*Coryna* spp. etc.) feed on the anthers of many flowers, especially Malvaceae, destroying the pollen sacs.

(e) Maize tassels eaten: by grasshoppers, or Maize Tassel Beetle in East Africa, or Japanese Beetle in the USA.

(f) Flowers inhabited by tiny maggots: gall midge larvae (Diptera: Cecidomyiidae), either white, yellow, orange or red in colour, usually causing shoot and flower deformation (nettlehead).

(g) Buds bored (fig. 9—6): caterpillars of some Tortricidae bore into large buds of shrubs and trees (called 'budworms' in the USA).

(h) Buds gnawed, with large holes (fig.9—12): eaten by large caterpillars (e.g. Cotton Semi-looper, *Anomis flava*: Noctuidae); sometimes by long-horned grasshoppers (Orthoptera: Tettigoniidae).

(i) Buds webbed and gnawed: some Tortricidae have caterpillars that feed on opening buds which they cover with a fine silk webbing.

(j) Buds pierced and dying: feeding adult and nymphal jassids (Miridae) and other Heteroptera, with their toxic saliva.

(k) Buds enlarged and swollen: gall mites of the fam-

*Fig. 9.* Damaged leaves (no.2; including flowers)

1. Graminaceous leaf cut laterally and rolled longitudinally (larvae of some skippers: Lepidoptera; Hesperiidae).

2. Graminaceous leaf with edges folded inwards (larvae and pupae of skippers, and egg site of some Pyralidae: Lepidoptera).

3. Small leaf-cases on rice leaves (Rice Caseworm (*Nymphula depunctalis*): Lepidoptera; Pyralidae).

4. Palm leaves eaten and small bagworm cases (bagworms: Lepidoptera; Psychidae).

5. Leaf edge rolled dorsally (nymphs of Aleyrodidae, thrips, and spiders nests).

6. Large buds bored (budworms: Lepidoptera; Tortricidae).

7. Leaf lamina extensively eaten away (large caterpillars: Lepidoptera; Noctuidae, Sphingidae, etc. Grasshoppers and locusts: Orthoptera; Acrididae. Adult scarab beetles: Coleoptera; Scarabaeidae).

8. Leaf lamina with ladder-like windowing leaving veins intact (adults and larvae of epilachna beetles (*Epilachna* spp.): Coleoptera; Coccinellidae).

9. Banana leaf with lamina cut and rolled (banana skippers (*Erionota* spp.): Lepidoptera; Hesperiidae).

10. Shoot telescoped, small distorted leaves, reduced flowers (mealybugs: Hemiptera; Pseudococcidae).

11. Leaf lamina covered dorsally with black growth (sooty mould, grows on honey-dew excreted by various Homoptera; Aphididae, Coccidae, Pseudococcidae, etc.).

12. Flower bud eaten into, leaving large jagged hole (various caterpillars: Lepidoptera; Noctuidae, Pyralidae, etc. Grasshoppers: Orthoptera; Tettigoniidae).

13. Flower petals with small holes (adult flower beetles: Coleoptera; Scarabaeidae; Rutelinae).

14. Flower perianth largely destroyed (adult blister beetles: Coleoptera; Meloidae).

15. Tunnel leaf mine with no central line of faecal pellets (dipterous leaf miners: Diptera; Agromyzidae, Anthomyiidae, Ephydridae, etc.).

16. Graminaceous leaves with longitudinal streaking (streak viruses, transmitted by Cicadellidae, Aphididae, etc.).

17. Plant leaves silvered and wilting (thrips, especially Onion Thrips (*Thrips tabaci*): Thysanoptera.

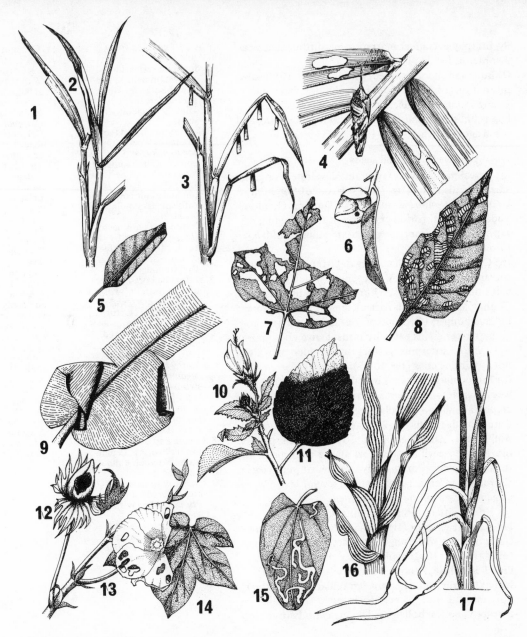

ily Eriophyidae infest buds of some shrubs causing a disorder called 'big-bud'.

(l) Buds and shoots stunted/wilting (fig.9–10): heavy infestations of aphids, scales, mealybugs, can stunt young shoots during active growth; Hibiscus Mealybug is unusual in causing severe shoot telescoping and often shoot death.

### Damaged fruits and seeds

(a) Cereal panicle with few grains developed, or grains small and distorted (fig. 10–1): larvae of Sorghum Gall Midge (*Contarinia sorghicola*; Diptera: Cecidomyiidae); small distorted grains result from feeding of sap-sucking Heteroptera with toxic saliva (Miridae, Coreidae, etc.).

(b) Cereal panicle with missing or bitten grains: young cereal heads attacked by caterpillars which eat the soft milky grains (e.g. sorghum attacked by False Codling Moth larvae in Africa); similar damage done by grasshoppers (Orthoptera; Acrididae and Tettigoniidae); riper grains eaten by birds; maize cobs grazed by *Heliothis* caterpillars (Noctuidae) which may also bore up inside cobs (fig. 10–5).

(c) Fruits, nuts and seeds bored; bored internally by weevil larvae, and hole made by emerging adult, examples are Mango Weevils (*Cryptorrhynchus* spp., Curculionidae) (fig. 10–2), Hazel-nut Weevil, Cotton Boll Weevil, Maize Weevil (fig. 10–12,18); in small ovaries one weevil larva will eat all the endosperm and thus each ovary contains only one weevil pupa (e.g. Clover Seed Weevils: *Apion* spp.); several Maize Weevils can develop in one maize kernel.

(d) Fruits with large internal tunnels, sometimes large holes to the exterior (fig. 10–5,7,8,14,16,17); sometimes infected with fungal and bacterial rots, sometimes frass expelled; made by caterpillars of the families Tortricidae, Pyralidae and Noctuidae; eggs are laid externally, and the tiny caterpillars (first instar) bore into the fruit; sometimes infested fruits fall prematurely; cotton bolls bored by many bollworms; top-

*Fig. 10.* Damaged fruits and seeds

1. Sorghum grains destroyed in panicle (larvae of Sorghum Midge (*Contarinia sorghicola*): Diptera; Cecidomyiidae).
2. Mango fruit bored (larvae of Mango Weevil (*Cryptorrhynchus* spp.): Coleoptera; Curculionidae).
3. Cocoa pod with necrotic spots (feeding of nymphs and adult cocoa capsids: Hemiptera; Miridae).
4. Cotton boll with necrotic spots (caused by the toxic saliva of feeding stink bugs: Hemiptera; Pentatomidae).
5. Maize cob bored (larvae of American Bollworm (*Heliothis armigera*): Lepidoptera; Noctuidae).
6. Coffee berries webbed and gnawed (caterpillars of Coffee Berry Moth (*Prophantis smaragdina*): Lepidoptera; Pyralidae).
7. Apple bored (Codling Moth larvae (*Cydia pomonella*): Lepidoptera; Tortricidae).
8. Cotton boll bored (Cotton Bollworms: Lepidoptera; Noctuidae and Tortricidae).
9. Orange with necrotic area (caused by Fruit Fly maggots: Diptera; Tephritidae).
10. Coffee berries bored (adult Coffee Berry Borer (*Hypothenemus hampei*): Coleoptera. Scolytidae).
11. Coconut scarred (by toxic saliva from feeding Coconut Bug (*Pseudotheraptus wayi*): Hemiptera; Coreidae).
12. Maize cob with bored seeds (Maize Weevil (*Sitophilus zeamais*): Coleoptera; Curculionidae): holes made by emerging adults.
13. Orange fruit distorted (Citrus Bud Mite (*Aceria sheldoni*): Acarina, Eriophyidae).
14. Sesame pod bored (caterpillars of Sesame Webworm (*Antigastra catalaunalis*): Lepidoptera; Pyralidae).
15. Bean seeds windowed and holed (bruchids: Coleoptera; Bruchidae).
16. Pea (pulse) pod bored and seeds eaten (Pea Pod Borer (*Etiella zinckenella*): Lepidoptera; Pyralidae).
17. Bean (pulse) pod bored with large holes by the seeds (the large larvae of American Bollworm (*Heliothis armigera*): Lepidoptera;Noctuidae, and other Noctuidae).
18. Dried maize seeds bored (Maize Weevil (*Sitophilus zeamais*): Coleoptera; Curculionidae).

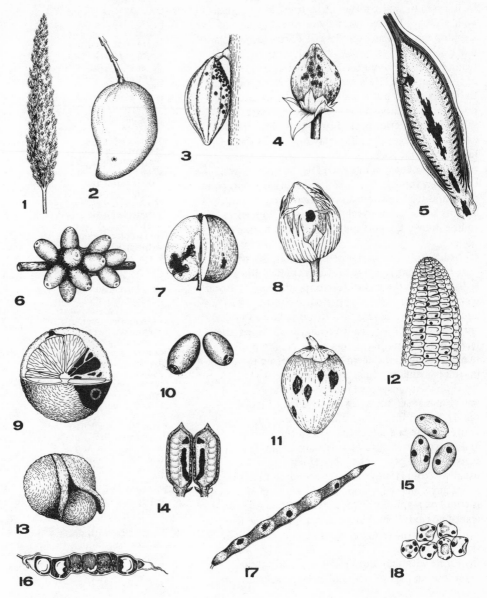

fruit bored by many different tortricid larvae; pulse pods attacked by two basically different groups of Lepidoptera: small caterpillars of Tortricidae, Lycaenidae and Pyralidae, which live inside the pods and feed on the developing seeds and the larger caterpillars of the Noctuidae (*Heliothis* etc.) which are too large to live inside the pulse pods, so bore holes into the pods in the vicinity of the seeds which they eat by pushing the anterior part of their body into the pod cavity. One large caterpillar may bore a number of holes in several pods.

(e) Fruits webbed and gnawed (fig. 10–6): some caterpillars (mostly Pyralidae) will spin silk over young fruits and feed on the fruits (Coffee Berry Moth: *Prophantis smaragdina*; Pyralidae); Sesame Webworm spins silk over the pod and shoot before boring the pod.

(f) Fruits tunnelled and bored: adults of Scolytidae bore host plants to make breeding tunnels and galleries in which the larvae develop; Coffee Berry Borer (*Hypothenemus hampei*) bore coffee berries in Africa and South America (fig. 10–10); Bruchidae (adults) enter ripened pulse pods which have split open; the larvae develop inside the ripe seeds, finally pupating beneath a thin translucent 'window' before the emerging adult makes the final exit hole (fig. 10–15).

(g) Fruits with maggots inside, a necrotic area, and sometimes small holes (fig. 10–9): fruit flies of the family Tephritidae attack almost all larger types of fruit; eggs are laid under the skin of the young fruit and the fly larvae develop internally as the fruit ripens; the fruit often falls prematurely, pupation usually takes place in the soil; secondary infection of the tunnels and oviposition site by bacteria and fungi is common; typically there is neither frass hole, entrance nor exit hole evident in the skin of the fruit while still on the tree, but sometimes there may be sap exudation (see fig. 7).

(h) Fruits with numerous small necrotic patches: heteropteran bugs of the families Miridae, Coreidae

*Fig. 11.* Damaged stems
1. Apical part of stem bored and dying (spiny bollworms in cotton (*Earias* spp.). larvae of some longhorn beetles: Coleoptera; Cerambycidae).
2. Sweet potato vine bored (caterpillars of clearwing moths; Sesiidae, and plume moths; Pteryphoridae).
3. Tree trunk or branch with eaten bark and deep holes into the wood, with a frass and silk tube (Wood-borer Moth (*Indarbela* spp.) : Lepidoptera; Metarbelidae).
4. Tree trunk or branch with single round emergence hole from 4–25 mm diameter (longhorn beetles: Coleoptera; Cerambycidae). With a pupal exuvium protruding from the hole (clearwing moths: Lepidoptera; Sesiidae. Goat or leopard moths: Lepidoptera; Cossidae).
5. Sisal stem bored internally (larvae of Sisal Weevil (*Scyphophorus interstitialis*): Coleoptera; Curculionidae).
6. Graminaceous seedling with 'dead-heart' (larvae of root fly: Diptera; Anthomyiidae. Muscidae, etc. Or stem borer caterpillar: Lepidoptera; Pyralidae, Noctuidae).
7. Woody stem with oval emergence hole (jewel beetle: Coleoptera; Buprestidae).
8. Woody stem with small circular holes in the bark (bark beetles: Coleoptera; Scolytidae).
9. Twig with round or irregular galls, old galls with small emergence holes (gall midges: Diptera; Cecidomyiidae. Gall wasps: Hymenoptera; Cynipidae, Eurytomidae, etc.).
10. Twig with bagworm case (bagworm: Lepidoptera; Psychidae).
11. Graminaceous stem bored internally, with emergence holes to exterior (stalk borers: Lepidoptera; Pyralidae, Noctuidae).
12. Tree or woody stem with sapwood eaten away in patches and some deep tunnels (longhorn beetle larvae: Coleoptera; Cerambycidae).
13. Tree branches or trunk bored centrally with a line of frass holes to the exterior (longhorn beetle larvae: Coleoptera; Cerambycidae).
14. Tree branch or stem bored with a cylindrical tunnel (Black Borer adults (*Apate* spp.): Coleoptera; Bostrychidae. Larvae of leopard or goat moths: Lepidoptera; Cossidae, or clearwing moth; Sesiidae)
15. Banana 'stem' extensively bored and tunnelled (larvae and adults of Banana Stem Weevil (*Odoiporus longicollis*): Coleoptera; Curculionidae).
16. Twig or woody stem with elongate swelling, sometimes with a single emergence hole (gall weevils (several species) : Coleoptera; Curculionidae).

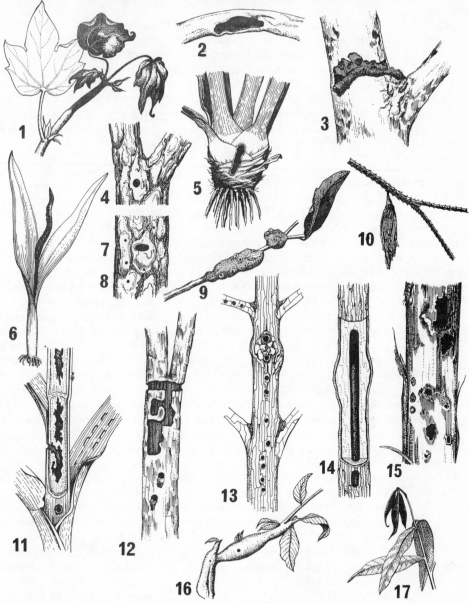

17. Woody stem with apical shoots killed and wilting (coreid bugs with their toxic saliva: Hemiptera; Coreidae).

73

and Pentatomidae have toxic saliva, and when feeding on fruits they cause necrosis at the feeding sites, which usually become infected with fungi and bacteria, resulting in rotting, death, and premature fruit-fall, e.g. Cocoa Capsids (fig. 10–3), Cotton Stink Bugs (fig. 10–4), and Coconut Bug (fig. 10–11).

(i) Fruit deformed (fig. 10–13): various phytophagous mites (Eriophyidae) feeding on flowers and young fruits cause distortion and deformation of the fruits.

### Damaged stems

These fall into four main categories for practical purposes: shoots of graminaceous seedlings; stems of cereals and grasses; shoots and twigs of shrubs; trunks and branches of trees (and large woody shrubs).

(a) Cereal shoots with 'dead-hearts' (fig. 11–6): caused by boring larvae of shoot-flies (Agromyzidae, Muscidae, Anthomyiidae), and caterpillars of Pyralidae.

(b) Cereal stems galled and distorted: larvae of Diptera (Opomyzidae, Chloropidae); cereal stem gall midges (Cecidomyiidae) are mostly temperate in distribution.

(c) Cereal and grass stems bored (fig. 11–11): caterpillars of the family Pyralidae generally bore rice and grass stems, while the larger caterpillars of the Noctuidae bore stalks of maize, sorghum and other larger species of Gramineae; tunnels in sugarcane are usually very short because the stem is solid and pithless. Stem sawflies (*Cephus*) bore apically in temperate cereals, especially wheat.

(d) Herbaceous stems and woody shoots bored, swollen often with a single emergence hole: larvae of gall weevils (Curculionidae) (fig. 11–16), also Sisal Weevil (fig. 11–5), and the temperate Cabbage Stem Weevil and Turnip Weevil.

(e) Banana pseudostem bored (fig. 11–15): Banana Stem Weevil (*Odoiporus longicollis*) larvae make extensive tunnel galleries in which they pupate and adults may also be found.

(f) Sweet potato stem (vine) bored (fig. 11–2): caterpillars of some clearwing moths (Sesiidae) and plume moths (Pteryphoridae).

(g) Twigs galled (fig. 11–9): made by feeding larvae of gall midges (Diptera: Cecidomyiidae), some gall wasps (Cynipidae, Eurytomidae, Torymidae), old galls have multiple exit holes; a few twig galls are made by weevil larvae and some woolly aphids (Pemphigidae).

(h) Twig-covered bags dangling from twigs and thin branches (fig. 11–10): bagworms (Lepidoptera: Psychidae) usually feed on the leaves but pupate with the bag firmly attached to twigs by a silken thong.

(i) Shoot and distal part of stem wilted and dying: the shoot may be bored by a caterpillar (Spiny Bollworm on cotton), or a longhorn beetle larva (Coleoptera: Cerambycidae) (fig. 11–1); some Heteroptera (Coreidae, etc.) feed on young shoots of woody shrubs and their toxic saliva enters the vascular system and kills the stem distally (fig. 11–17); shoots on trees are killed terminally in some regions by ovipositing cicadas, and also by some long-horned grasshoppers.

(j) Tree trunk and branches bored, sometimes bark eaten externally: timber borers belong to two orders of insects, the Lepidoptera and Coleoptera; Metarbelidae feed externally at night on the bark under a silken tube coated with frass but retire during the day to a deep tunnel in the heartwood (fig. 11–3); larvae of Cossidae and some Sesiidae bore in the heartwood of trunks and usually down the centre of smaller branches (fig. 11–4,14); the circular adult emergence hole typically contains the pupal exuvium projecting. Tree-boring beetles belong mainly to the families Cerambycidae, Buprestidae, Bostrychidae, and Scolytidae, with a few weevils (Curculionidae) and a few small families such as the Lymexylidae. Cerambycidae are long-horn beetles and may be very large (*Batocera* spp.) with larvae that bore in trees for up to 3–4 years before pupating; they mostly eat sapwood just under the bark and may leave frass holes to the exterior; the emergence hole is circular (fig.11–4, 12,13); the adults sometimes chew patches of bark. Buprestidae are called jewel beetles, and the larvae, known as flat-headed borers, and the emerging adults leave an oval exit hole in the tree bark (fig. 11–7). Bostrychidae are black, cylindrical beetles, completely circular in cross-section and the tunnels are bored by the adult beetles (fig.11–14) called black-borers. Scolytidae (shot-hole and twig borers) belong to the group of ambrosia or fungus beetles; the adults bore into trees (fig. 11–8) and make extensive breeding galleries under the bark; they carry fungus spores on their bodies with which to inoculate the new fungus galleries where the larvae feed; some species bore down the centre of twigs on bushes such as tea.

### Damaged roots and tubers

(a) Fine roots eaten: by subterranean caterpillars, fly larvae, beetle larvae and termites.

(b) Tubers tunnelled (fig. 12–1): usually infected with secondary rots; Potato Tuber Moth larvae bore down the stem into the tubers.

(c) Tubers with narrow tunnels (fig. 12–2): bored by wireworms (larvae of Elateridae) and some small species of slugs.

(d) Tubers with wide, sometimes shallow tunnels (fig. 12–3,11): bored by cutworms (Noctuidae), chafer grubs (Scarabaeidae), and some larger species of subterranean slugs.

(e) Yam tubers with wide tunnels (fig. 12–11): bored by adults and larvae of yam beetles (Scarabaeidae).

(f) Sweet potato tubers bored (fig. 12–4): infected with secondary rots; larvae and adults of Sweet Potato Weevils (*Cylas* spp., Apionidae).

(g) Root stock encrusted: root mealybugs form a hard layer of encrustation over plant roots; ants are often found in attendance.

(h) Roots with waxy agglomeration: root aphids (usually Pemphigidae) found as a waxy or mealy agglomeration on roots of some herbaceous plants; mostly temperate.

(i) Roots with small globular cysts (fig. 12–7): although nematodes are not included in this book, mention of them cannot be avoided here; small globular cysts are produced by female *Heterodera* spp. (cyst eelworms) on a wide range of host plants; probably more temperate than tropical.

(j) Roots with large swellings (fig. 12–5): root-knot nematodes (eelworms), *Meloidogyne* spp. cause extensive root swellings on many crop plants, especially in the tropics.

(k) Roots stunted but bushily prolific (fig. 12–6): Citrus Root Eelworm (*Tylenchulus* spp.) a free-living nematode.

(l) Roots galled: larvae of several weevils and other beetles live inside root galls, sometimes inside the 'nodules' on roots of Leguminosae.

(m) Taproot eaten or hollowed (fig. 12–8): cutworms (Noctuidae) and chafer grubs (Scarabaeidae) will eat both fine roots and the main taproot of herbaceous plants, especially vegetables.

(n) Taproot tunnelled and eaten (fig. 12–9): symptoms are wilting and plant collapse (fig. 13–5); larvae of Anthomyiidae (root maggots), and other Diptera such as Carrot Fly maggots (*Psila rosae*); with root crops, plant wilting does not normally occur.

(o) Bulb bored and tunnelled: Onion Fly (*Delia antiqua*) larvae (temperate and sub-tropical), and the temperate Narcissus Flies (Syrphidae).

(p) Rhizome bored (fig. 12–10): Banana Weevil (*Cosmopolites sordidus*) larvae bore the rhizome extensively, and both pupae and adults may also be found in the tunnels; various fly maggots belonging to the Chloropidae and other families bore the rhizomes of ginger.

### Damage to sown seeds and seedlings

(a) Cotyledons of large seeds bored and eaten (fig. 13–6): larvae of Bean Seed Fly (Corn Seed Maggot in the USA), *Delia platura*, bore into cotyledons, epicotyl and hypocotyl and prevent germination; similar

*Fig. 12.* Damaged roots and tubers

1. Potato tuber with extensive, wide tunnelling (Potato Tuber Moth larvae (*Phthorimaea operculella*): Lepidoptera; Gelechiidae).
2. Potato tuber with small entrance tunnels (wireworms: Coleoptera; Elateridae).
3. Potato tuber with wide, sometimes shallow, hole (cutworms: Lepidoptera; Noctuidae. Chafer grubs: Coleoptera; Scarabaeidae. Slugs: Mollusca; Limacidae, etc.).
4. Sweet potato tuber tunnelled (larvae and adults of Sweet Potato Weevils (*Cylas* spp.): Coleoptera; Apionidae).
5. Roots with extensive swellings (larvae of some weevils: Coleoptera; Curculionidae. Root Knot Eelworms (*Meloidogyne* spp.): Nematoda).
6. Roots stunted and bushy (Citrus Root Nematode (*Tylenchulus* spp.).
7. Roots with small round cysts (Root Cyst Eelworms (*Heterodera* spp.): Nematoda).
8. Tap root eaten or hollowed (cutworms: Lepidoptera; Noctuidae. Chafer grubs: Coleoptera; Scarabaeidae).
9. Carrot root tunnelled (Carrot Fly larvae (*Psila rosae*): Diptera; Psilidae). Other roots tunnelled (root fly larvae: Diptera; Anthomyiidae).
10. Banana rhizome bored (larvae and adults of Banana Weevils (*Cosmopolites* spp.): Coleoptera; Curculionidae).
11. Yam tubers bored (larvae of yam beetles: Coleoptera; Scarabaeidae).

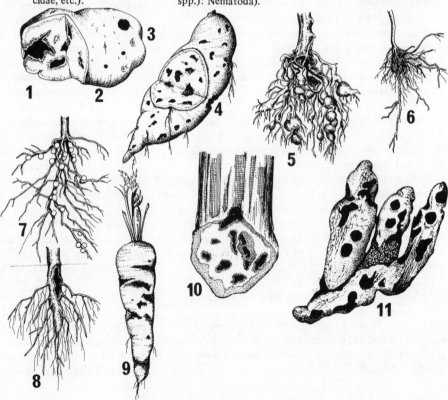

**Fig. 13.** Damage to sown seeds and seedlings

1. Cereal or grass with 'deadheart' (larvae of shoot flies, grass flies, etc.: Diptera; Muscidae, Anthomyiidae, Opomyzidae, Chloropidae, etc.). Sometimes young caterpillars of lepidopterous stalk borers.
2. Sugarcane or cereal seedling with stem eaten below ground level (adult Black Maize Beetles (*Heteronychus* spp.): Coleoptera; Scarabaeidae).
3. Legume seedling with swollen hollowed stem (larvae of Bean Fly (*Ophiomyia phaseoli*): Diptera; Agromyzidae).
4. Earth tube up side of woody stem, and young plant wilting; woody stem eaten away under earth tube (termites: Isoptera or ants).
5. Whole seedling or young plant wilting and dying; root/stem eaten (larvae of root fly: Diptera; Anthomyiidae. Cutworms: Lepidoptera; Noctuidae. Chafer grubs: Coleoptera; Scarabaeidae).
6. Seed cotyledons bored and tunnelled (larvae of Bean Seed Fly (*Delia platura*): Diptera; Anthomyiidae).
7. Seedling stem severed and wilting plant lying on ground (cutworms: Lepidoptera; Noctuidae. Adult crickets: Orthoptera; Gryllidae. Adult Surface Weevil (*Tanymecus indicus*): Coleoptera; Curculionidae).

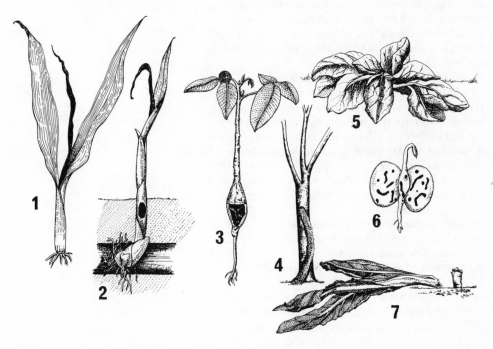

damage done by small species of slugs.

(b) Seeds dug up and eaten: various species of birds (sparrows etc.) and mice.

(c) Cereal seedling stem with 'dead-heart' (fig. 13–1); maggots of various Diptera (Agromyzidae, Anthomyiidae, Muscidae, Opomyzidae, Chloropidae, and Oscinellidae) bore into the young stem, usually killing the growing point and making the apical leaf turn brown and die; some caterpillars (Crambidae, Pyralidae, and Noctuidae) also bore seedling stems of graminaceous plants, but typically they attack older plants.

(d) Stem bored, usually swollen (fig. 13–3): several species of Agromyzidae (Diptera) have larvae that bore in the stem of various seedlings, best known is probably the Bean Fly (*Ophiomyia phaseoli*) widespread and abundant on legumes, it will also bore into leaf petioles on beans.

(e) Stem severed with shoot lying alongside (fig. 13–7): typical cutworm (Noctuidae) and cricket (Orthoptera: Gryllidae) damage; often several consecutive seedlings are cut through; with crickets the cut plant is left for a day to wither and the next night is pulled down into the nest; such damage may be done by surface slugs.

(f) Stem severed and plant removed: several species of termites; can also be done by some leaf-cutting ants and harvester ants.

(g) Stem gnawed at about ground level (fig. 13–2) or below ground: a few beetles belonging to small families, and a few adult scarabs.

(h) Cotyledons or first leaves pitted and eaten (fig. 8–7): adult flea beetles (Halticinae) make a shot-hole effect on seedlings of Cruciferae, cotton, and other crops, frequently stunting and killing the seedling.

(i) Seedling or young plant wilting and dying (fig. 13–2,5) as a result of the underground stem being eaten: typical damage by temperate Cabbage Root Fly (*Delia brassicae*), cutworms, and African Black Maize Beetles (*Heteronychus* spp.) on maize and sugarcane seedlings.

(j) Woody seedling or young plant wilting, with earth-tube up part of the stem; under the tube the bark is eaten away by termites (fig. 13–4), or ants.

# 5 Biological control of crop pests

The initial enthusiasm with which the first modern insecticides, the chlorinated hydrocarbons, were greeted in the late 1940s and early 1950s has long since begun to wane. This is, in part, the result of the development of resistance by so many pests to many of the pesticides, and also because of the growing awareness of the dangers involved in a gradual poisoning of the whole environment. One result is that many governments are now introducing legislation designed to curb indiscriminate use of the more toxic and persistent insecticides (mostly chlorinated hydrocarbons) and indeed to prohibit the use of the most dangerous pesticides. Another result of this awareness is the acknowledgement of the desirability of trying to control pests by use of cultural or biological means. The single most promising aspect of cultural control is the development and use of resistant varieties of crops. In some texts this is regarded as a biological method of control, but this view is not generally held, since whilst it involves the reactions of living organisms, it is normally regarded as a distinct branch of agricultural science.

The main advantages of biological control lie in:

(a) the absence of toxic effects;
(b) no development of resistance by the pests;
(c) no residues of poison in the soils and rivers, etc;
(d) no build-up of toxins in food chains;
(e) no killing of pollinators, or development of secondary pests through the destruction of their natural enemies;
(f) the permanence of successful biological control programmes where repeated application of chemicals would be required;
(g) the fact that biological control is self-adjusting and does not require the careful timing and organization which should be given to pesticide applications, and which often make it impracticable on small peasant holdings in underdeveloped areas.

Quite often natural control is effective for many pest populations most of the time, but when a population outbreak does occur and methods of control are required, care should be taken so that the *disturbance of natural control factors should be kept to a minimum*. Some of the ways in which this can be attained are:

(a) prediction of pest outbreaks so that insecticidal control can be well planned in advance and used only when necessary;
(b) use of selective insecticides — the more specific the insecticide the less damage it is likely to do to natural control;
(c) by avoiding 'blanket' treatments with insecticides as an insurance measure;
(d) the use of cultural or biological control measures wherever possible;
(e) the encouragement of the increase of natural enemies of the pest;
(f) the growing of pest-resistant crops where possible.

In this chapter biological control (BC) is viewed in the strict sense as the control (lowering) of pest populations by predators, parasites and disease-causing pathogens.

## Natural control

First and foremost it must be stressed that it is not possible to overemphasize the importance of the natural control of insect populations. Under natural conditions most insect populations are regulated by a complex of predators, parasites and pathogens causing diseases, all sharing the same habitat and belonging to the same ecological community. This population regulation is sometimes referred to as the *balance of nature* or *natural control*. Under agricultural systems of extensive monoculture, conditions are created that may induce a pest population explosion, but often it is found that the natural enemies increase correspondingly, and so natural mortality rates for the pest remain high and the final pest population does not become very large after all. This is particularly true for perennial orchard and plantation crops, which are present for a long enough time to allow some stabilization of insect and mite populations. The time factor is critical in that parasite populations always lag behind the host populations in their rate of growth (increase), and a fairly lengthy time is required to reach community stability (i.e. some measure of pest population control). With annual crops, such as vegetables, cereals and cotton, there is generally insufficient time available for specific predator or parasite populations to build up to a level where they can exert any controlling effect, although non-specific (polyphagous) predators and parasites may be important in such situations. The spectacular control of greenhouse pests (Red Spider Mites and Whiteflies) by introduced predators and parasites on cucumbers and tomatoes in the UK, northern Europe and the USA is attributable to the greenhouses being very small (in effect tropical) and completely enclosed agroecosystems, and an unnaturally high predator or parasite population is introduced at the start of the control programme. However, although the natural enemies of pests on annual and short-lived crops may seldom completely control the pest populations, their controlling effect is always important. Many crop pests have spectacular fecundity; for example, some female noctuid moths (especially cutworms) lay as many as 2000 eggs per season, and the only reason that the world is not inundated with caterpillars is the high rate of natural mortality due largely to predators, parasites and diseases. In temperate regions the fecundity of viviparous and parthenogenetic aphids is almost unbelievable!

Since natural control is present in all pest ecosystems at all times, it is difficult to appreciate the actual extent of its population controlling effect as it is usually not really noticed until it breaks down. Indiscriminate use of highly toxic, non-specific, persistent pesticides has often resulted in the death of more natural enemies than crop pests. In point of fact, this is now the normal situation with most orchard infestations of scale insects (Coccoidea), in that the mature scales are notoriously difficult to kill with insecticides but the complex of tiny parasitic chalcid and braconid wasps are far more susceptible to the poisons. In these situations pesticide application (with chemicals such as DDT and dieldrin), is almost invariably followed by a population outbreak of the pest. Examples of this phenomenon are now legion!

A nice example of recognized natural control was recorded in Thailand where *Patanga succincta* is a serious maize pest; careful field studies revealed that the egg pods in the soil were heavily parasitized by a scelionid wasp and preyed upon by *Mylabris* beetle larvae; insecticide spraying was carefully restricted to avoid disruption of this natural control.

## Biological control

This is the deliberate introduction of predators, parasites, and/or pathogens into the pest/crop agroecosystem, and is designed to reduce the pest population to a level at which damage is not serious.

Two schematic representations of a pest population being controlled (economically) by natural enemies are shown (from Smith & van den Bosch in Kilgore & Doutt, 1967) in figs. 14 and 15.

The advantages of biological control are several, and have already been listed (see page 79). The predators and parasites may be local species whose natural population is being augmented by the introduction, or they may be exotic species from another country. The most successful examples of biological control are elegant in all respects, but unfortunately most pest situations are not amenable to this method of control. However, suitable situations are more likely to be found in the tropics (and sub-tropics) where the climatic conditions favour continuous breeding of insect populations. To date, there are in the region of 200–300 cases of clear success of biological control (*sensu stricta*) that are well documented. The original work in this field was carried out in the USA in the early part of this century, mostly in California and Hawaii, to find means of controlling pests of introduced exotic fruits (*Citrus*, etc.). At the present time most work is being done by the various stations of the Commonwealth Institute of Biological Control (CIBC), which are situated in Africa, India, Pakistan, Switzerland, the W. Indies, and S. America. This is one of the institutes belonging to the Commonwealth Agricultural Bureaux, Slough, UK, and is a sister institute to the Commonwealth Institute of Entomology, London.

Most crops are indigenous to one particular region of the world, and have been widely transported by man to other regions with a suitable climate in relatively recent times. These exotic crops in their new locations may be attacked by indigenous insect pests quick to seize upon a new food source. Alternatively, their own pests may have either accompanied the crops or else followed soon after, and in the new situation their usual complement of natural enemies is absent, thus permitting a rapid population build-up. The local predators, parasites and pathogens will inevitably be in a state of delicate balance in their own environment and usually cannot be expected to exercise much control over the introduced pests.

The most successful cases of biological control are usually where the predator or parasite has been brought from the area of origin of the crop. For ex-

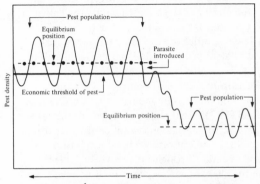

*Fig. 14.* Complete biological control (based on the economic criterion) of a pest by an introduced natural enemy. Note that it is not the economic threshold of the pest that is affected by the parasite, but rather its equilibrium position (e.g. long-term mean density). (After Smith & van den Bosch 1967.)

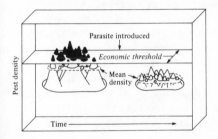

*Fig. 15.* Three-dimensional representation of biological control of a pest species by an introduced natural enemy. The two populations depict the distributional status of the pest species and its density at particular points in time before and after parasite introduction. (After Smith & van den Bosch, in Kilgore & Doutt, 1967.)

81

ample, the most successful citrus scale parasites have all come from S. China where *Citrus* is indigenous. Usually specific parasites are more successful in stable habitats such as long-established orchards. If the habitat is unstable (as with most agricultural crops) then more general parasites are better suited for control as they can better survive the periods when the specific pest host is absent (between crops). An interesting case is recorded from Japan; here the rice stem borers are controlled partially by the use of *Trichogramma* spp. to parasitize the eggs. After harvest, when there are no more lepidopterous egg masses available, the wasps parasitize the eggs of flies of the genus *Sepedon* (Diptera: Sciomyzidae) in the weedy, fallow lands around the paddy fields, and so the parasite population is maintained (in a state of diapause) over winter. The fly larvae prey on aquatic snails of the genus *Lymnaea* and as such are normally part of the fauna of oriental rice paddy fields.

Most of the natural enemies now bred commercially are either non-specific predators or non-specific parasites, which may obviously be employed against a number of different crop pests. For example, Rincon–Vitova Insectaries Inc., California, USA, offer for sale lacewings (as larvae or pupae), ladybird beetle (*Cryptolaemus*) adults, *Nasonia* (Hym., Pteromalidae) fly parasites (as parasitized puparia), and several species of *Trichogramma* for control of lepidopterous eggs. The more specific predators or parasites used in the control of greenhouse Red Spider Mite and Whitefly in temperate parts, which are available commercially, are in a different category, as the two pests concerned usually occur as single species populations, in a very specialized environment. More specific parasites and predators may be obtained from some of the CIBC stations by special arrangement. Some research stations maintain their own stocks of particular parasites for local use, for example as practised in S. China (Guangdong) and GCRI in the UK.

The cost of a successful introduction may be relatively low in comparison with most chemical applications. For example, DeBach (1964) estimated that the Department of Biological Control, University of California, over the period 1923–59 spent a total of US $3.6 million with a resulting annual saving in California of about US $100 million. A recent case by CIBC, Pakistan, was the establishment of *Apanteles ruficrus* (Hym., Braconidae) against the maize pest *Mythimna separata* (also attacking vegetables and other crops) in New Zealand in 1973/75. The cost of supplying the parasites from Pakistan was US $1700 and the costs involved in New Zealand amounted to about US $20 000; the annual saving is estimated at about US $5 million, with the control still being successful. One oddity noticed was that only the Pakistan strain of parasite was particularly effective.

A few examples of some successful tropical biological control projects are listed below in Table 1.

### Fortuitous biological control

It should perhaps be mentioned, as pointed out by DeBach (1971), that there has been considerable, unrecognized, fortuitous biological control resulting from ecesis (accidental dispersal and establishment) of natural enemies throughout the world. This is particularly the case with some scales (especially Diaspididae) and their parasites, because the scales are tiny, sessile, inconspicuous (especially in low numbers) and easily transported on either fruits or planting material (shoots and rootstocks). DeBach uses the specific parasite, *Aphytis lepidosaphes*, of the Purple Scale (*Lepidosaphes beckii*) as a good example. Purple Scale is indigenous to the area of S. China and Indo-China, and it is highly specific to *Citrus.* During the past 100 years or so Purple Scale has gradually (accidentally) spread and invaded most of the major *Citrus* growing areas of the world, from the Mediterranean to S. Africa, Australia, N., C. and S. America, and the W. Indies. DeBach records that *Aphytis lepidosaphes* (first discovered in China in 1949) is

Table 1: *Some successful tropical biological control projects*

| Pest | Crop | Predator or parasite | Country | Date | Estimated annual value |
|---|---|---|---|---|---|
| Banana Weevil | Banana | *Plaesius javanus* (Col., Histeridae) | Fiji | 19 ? | not known |
| *Mythimna separata* (Noctuidae) | Maize | *Apanteles ruficrus* (Hym., Braconidae) | New Zealand | 1973/5 | US $5 mill. |
| *Promecotheca* spp. | Coconut | ? | Sri Lanka | 1972 | US $3 mill. |
| | | *Pediobius parvulus* | Fiji | 1933 | not known |
| *Phthorimaea operculella* | Potato | *Apanteles subandrinus* | Zambia | 1968 | US $70 000 |
| *Diatraea* spp. | Sugarcane | *Glyptomorpha deesae* (Hym., Braconidae) | Barbados | 1965 | US $1 mill. |
| *Chrysomphalus ficus* | Citrus | *Aphytis holoxanthus* (Hym., Aphelinidae) | Israel | 1965/7 | US $1 mill. |
| *Parlatoria oleae* | Olive, etc. | *Aphytis maculicornis* (Hym., Aphelinidae) | USA (California) | 1949 | not known |

slowly spreading around the world although the only deliberate introduction was into California in 1949; apparently, at that time it was (as discovered later) already established in Hawaii, presumably as a result of an earlier accidental introduction. The most recent colonizations by this specific parasite include Spain (1969) and Argentina (1970). In almost every country checked by DeBach, *Aphytis lepidosaphes* is responsible for substantial to complete biological control of Purple Scale on *Citrus*.

### Predators

The importance of insect pest predators is now being reassessed in many countries because of their obvious value in IPM programmes.

The value of wild birds as insect predators is clearly demonstrated in urban situations, where tits (Paridae) and sparrows (Ploceidae) can be seen searching fruit trees, roses, etc., for aphids, caterpillars, and other insects, and thrushes (Turdidae) eat large numbers of slugs and snails. Domesticated birds are being used to control insects in parts of the tropics very successfully. For many years peasant farmers in Africa have been using chickens on their shambas in cotton plots to eat the cotton stainers and other bugs that drop to the ground (as an escape mechanism) when disturbed. About 40 chickens per acre is the accepted number of birds required for adequate control. Recently in S. China (and other parts of India and S.E. Asia) effective use of ducks has been made against rice pests. In Guangdong on the early rice crop, after transplanting and establishment of the plants, 220 000 ducklings (about half-grown) were herded slowly through the paddy fields. Each duckling was estimated to eat about 200 insects per hour, and most of these were inevitably pests. The overall effect of this predation was that the amount of chemical insecticides required on the early rice crop was reduced from 77 000 kg in 1973 to 6700 kg in 1975. Following this early success, ducks are now being used as rice pest predators in many parts of the Far East. In Thailand, in 1980, adult ducks were used successfully to eat adults and large nymphs of the Bombay Locust (*Patanga succincta*) that had been enticed out of the upland maize crop into intercropped soybean foliage and also groundnuts. In this case ducklings would have been useless as the locusts were too large for them to kill. At the same time local villagers were paid to hand-collect adult locusts, and in a few days 80 tons were collected; the locusts were later either roasted and eaten by the villagers or made into a paste for culinary purposes. (There are times when man can be an insect predator of some consequence!)

Within the Vertebrata one amphibian has been used a few times for biological control of insects; this is the giant toad (*Bufo marinus*), which was introduced into Hawaii in 1932–3 in an attempt to control various beetles (Scarabaeidae) attacking sugarcane, and was released in Australia (Queensland) in 1935–6. It is now flourishing in both locations where it generally seems to cause more ecological disturbance than benefit; this is in part because it eats small vertebrates as well as insects, and the toads are poisonous if eaten by other vertebrates. An introduction of small numbers into Tanzania in 1948–9 was unsuccessful in that the toad did not become established; it was thought that predation by *Varanus* lizards destroyed the small population.

It should perhaps be mentioned that the main reason for the tremendous upsurge of rats (*Rattus* spp.) as urban and agricultural crop pests in India and throughout S.E. Asia is largely due to the destruction of their main predators which are snakes, especially the venomous cobra (*Naja naja*; Elapidae), and the non-venomous, but fanged and aggressive, rat snakes (*Ptyas* spp., *Elaphe* spp., etc., Colubridae). For in-

explicable reasons most peasants and farmers in Africa and tropical Asia have a morbid dread of all snakes, most lizards, and salamanders, being convinced that they are all highly venomous and lethal. The consequence is, regrettably, the wholesale and widespread destruction of local snake populations, with a subsequent population explosion of suburban rats. The situation is somewhat aggravated in S.E. Asia as it is the endemic source of the frugivorous arboreal *Rattus rattus* subspecies group. It is estimated by FAO that in India there are five rats per person, and that in recent years 10% of the grain harvests have been lost due to rat attack. In S.E. Asia the main pests of oil palm (and in some areas coconut) are rats; in the Philippines the crops damaged also include sugarcane and maize.

It has long been obvious that spiders were an important part of the natural control community throughout the world, but only recently have they been studied in this role. Work at IRRI in the Philippines and in S. China has shown that many rice pests (especially Cicadellidae and Delphacidae) are heavily predated by spiders. In some experiments mortality rates of the Brown Planthopper of Rice reached 70% due almost entirely to spider predation. In parts of China (Hunan, Zhejiang), since 1975, spiders belonging to more than five different families have been reared and released on rice and cotton crops, against a broad range of insect pests, with considerable success (Li, Li-ying, *in litt.*, see table 2). Both the web-spinning and the webless wolf spiders have been used. At IRRI it was discovered that $\gamma$-BHC was particularly toxic to spiders, both directly by contact and indirectly by their eating poisoned prey; obviously such insecticides should be avoided in any IPM programme where spiders are important natural predators.

Also in the class Arachnida are the mites (order Acarina); in the same way that insects are heavily preyed upon by other insects, the major predators of phytophagous mites are carnivorous mites, most of

Table 2: *Biological control of some insect pests in southern China* (Li, Li-ying, 1980; *in litt.*)

| Pest | Crop | Predator, parasite or pathogen | Date | Locality | Success |
|------|------|-------------------------------|------|----------|---------|
| *Icerya purchasi* | *Citrus* | *Rodolia rufopilosa* (Col., Coccinellidae) | 1954− | Sichuan | Yes |
| *Aphis* spp. | Cotton, Wheat, etc. | *Coccinella septempunctata* (Col., Coccinellidae) | 1971− | Henan | Yes |
| *Heliothis armigera* | Cotton | *Trichogramma* spp. (3) | 1970− | Hubei, Shangsi | Yes |
| *Cnaphalocrocis medinalis* | Rice | *Trichogramma* spp. (3) | 1970− | S. China | Yes |
| *Tessaratoma papillosa* | Litchi | *Anastatus* sp. (Hym., Eupelmidae) | 1964− | Guangdong | Yes |
| *Pectinophora gossypiella* | Cotton | *Pteromalus puparum* (Hym., Pteromalidae) | 1955− | Hubei, Hunan, Shangsi, etc. | Yes |
| *Rhynchocoris humeralis* | *Citrus* | *Oecophylla smaragdina* (Hym., Formicidae) | since ancient times | Guangdong | Yes |
| Various borers | Sugarcane | *Tetramorium guineense* (Hym., Formicidae) | 1961− | Guangdong, Fujian | Yes |
| Various pests | Cotton | *Chrysopa* spp. (Neuroptera, Chrysopidae) | 1975 | Henan | Limited |
| *Panonychus citri* | *Citrus* | *Amblyseius newsami* *Typhlodromus pyri* (Acarina, Phytoseiidae) | 1975− 1975− | Guangdong Sichuan | Yes Yes |
| Various pests | Rice, Cotton | Various Argiopidae, Lycosidae, Tetragnathidae, Micryphantidae, Clubionidae, etc. (Arachinda, Araneida) | 1975− | Hunan, Zhejiang | Yes |
| Corn borers *Grapholitha glycinivorella* | Maize, Soybean, Sweet potato | *Beauveria bassiana* (fungi) | 1965− | most provinces of China | Yes Yes |

which belong to the family Phytoseiidae. The three main genera used in biological control programmes are *Phytoseius, Amblyseius,* and *Typhlodromus.* The original widespread use of these predatory mites against spider mites (Tetranychidae) was in greenhouses in the UK and the USA (on cucurbits, tomatoes, etc.) with spectacular success. Later it was realized that on field and orchard crops in the tropics they could also be successful. Various companies now provide commercial sale of phytoseiid mites for biological control purposes.

Predaceous insects are important population-controlling factors for many insect pests, but their precise roles in natural control have seldom been evaluated and documented. It was however reported from the Solomon Islands that, in 1975 and 1976, the Brown Planthopper of Rice was not a pest on rice (economic pest, that is) because of good control by a predaceous mirid bug (*Cyrtorhinus*). Many members of the Miridae, Anthocoridae, Reduviidae, Pentatomidae, and other Heteroptera, are important natural predators. *Orius minutus* has been used in China since 1976 for the control of various cotton pests in Hubei and Jiansu with some success. *Platymeris laevicollis* (Reduviidae) is an important predator of Rhinoceros Beetle adults (*Oryctes* spp.) and has been introduced from Zanzibar into India, Fiji and New Caledonia (Lever, 1969); each adult bug can destroy one adult beetle per day and may live for four months.

In the Diptera there are some Cecidomyiidae that prey on Coccoidea, and there are a number of rather obscure families, in addition to the obvious Asilidae (robber flies), where the adults prey on other insects. But without doubt the most important group is the Syrphidae (hover-flies) in which many species have predaceous larvae. They feed on aphids, coccids, other small plant bugs and small caterpillars. Being legless, headless and blind it is rather surprising that syrphid larvae are effective predators, but the eggs are laid amidst dense populations of suitable prey so the emerging larvae manage to feed effectively with the aid of their mouth-hooks and rather simple sensilla (chemosensory and tactile). Although very important in natural control, Syrphidae are not much used in biological control programmes as their limited powers of searching make them not really suitable. One oddity is that in the UK they have been recorded as a nuisance in commercial pea (*Pisum sativa*) crops; the pea-viner also threshes and when the fresh peas (seeds) are removed from the pods and collected in the hopper they are accompanied by the small globular syrphid pupae which were dislodged from the foliage. Customers buying a packet of frozen peas are invariably annoyed to find syrphid pupae enclosed.

The Hymenoptera have a few predaceous groups of importance. The Vespidae, the social wasps, are all carnivorous and feed upon other insects. In most parts of the world *Vespa* and *Polistes* are regarded as pests because they damage ripe fruits, and also they nest in crops and attack the field-workers if disturbed. But in China (and some other countries) they are being used as general predators in cotton crops. Similarly ants (Formicidae) include carnivorous species that will prey on crop pests, and *Oecophylla smaragdina* (Red Tree Ant) is successfully controlling *Rhynchocoris humeralis* on *Citrus* in Guangdong, China; this method has in fact been used in China since ancient times. The African Red Tree Ant (*O. longinoda*), when nesting in a coconut palm, keeps the palm free of Coconut Bug. But these ants sometimes guard bug colonies, and as they are quite aggressive they may attack field-workers. So like the wasps (*Vespa* and *Polistes*) it is a moot point whether they are more important as pests or as beneficial predators. Scoliidae prey on beetle larvae, mostly chafer grubs, in the soil, but attempts to use them for the control of *Oryctes* spp. in S.E. Asia have not been very successful.

The Neuroptera as an order are almost entirely predaceous, both as adults and larvae, and clearly they are an important group in the natural control of many insect species. Green Lacewings (*Chrysopa* spp.) are quite easy to rear, and are now available commercially for biological control use. As general (i.e. non-specific) predators they are useful in IPM programmes but will only give limited control of any particular pest.

Coleoptera include many important groups of predators. In some families, both adults and larvae are fiercely predaceous (e.g. Carabidae, Staphylinidae, Cicindelidae, Histeridae and Coccinellidae), but in others only the larvae are (Hydrophilidae, Meloidae, Lampyridae). The Carabidae include the Indian *Pheropsophus hilaris* which has been successfully used in Mauritius to kill larvae of the Rhinoceros Beetle. Most of the predaceous beetles are litter and soil-dwellers and so are mostly used against soil-inhabiting pests, for obvious reasons. The major exception to this is,

however, the Coccinellidae (ladybird beetles) which are arboreal foliage dwellers. Apart from *Epilachna*, the entire family are predators and very effective control agents, both as adults and larvae. They are used mainly against aphids and coccoids, and also kill small caterpillars. Ladybirds are easy to rear and can be bought commercially in large numbers from many establishments. Successful examples of their use in biological control are too numerous to list, but a few are included in the tables. Fig. 16 shows the effect of coccinellid predators on *Icerya purchasi* in California in 1868, one of the earliest documented cases of biological control.

The Meloidae are important in the natural control of many locusts and grasshoppers; the triungulin larvae seek out and prey upon the egg-pods of Acrididae in the soil, and in parts of the tropics their value is considerable. As already mentioned *Mylabris* larvae exert considerable control pressure on the Bombay Locust in N. Thailand. The histerid beetle *Plaesius javanicus* has been used very successfully in Fiji for controlling Banana Weevil, but in other countries its performance has been varied. Attempts to use it in Uganda in 1934—5 failed, but this was most probably due to mismanagement in the handling of the beetle consignments as many of the insects were dead on arrival.

### Parasites

The three outstanding groups of insect parasites are the Tachinidae (Diptera), and the Chalcidoidea and Ichneumonoidea (both Hymenoptera). They are vitally important both in the natural control and biological control of insect crop pests. Within the Diptera are families such as Pipunculidae, Phoridae, Bombyliidae and Sarcophagidae which are of minor importance as natural parasites of insect pests. There are also some anomalous groups like the Sciomyzidae which live in oriental paddy fields and whose eggs provide alternative hosts for *Trichogramma* wasps

*Fig. 16.* Schematic graph of the fluctuations in population density of the Cottony Cushion Scale (*Icerya purchasi*) on citrus from the time of its introduction into California in 1868. Following the successful introduction of two of its natural enemies in 1888, this scale was reduced to noneconomic status except for a local resurgence produced by DDT treatments (from Stern *et al.*, 1959).

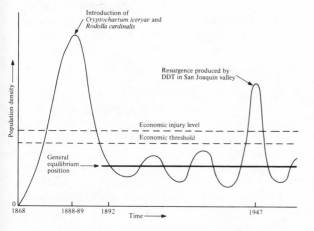

during the period when stalk borers (*Chilo, Tryporyza*, etc.) are absent.

The Tachinidae parasitize a very large range of hosts worldwide, including the larvae of lepidopterans, the larvae and adults of coleopterans, and nymphs and adults of orthopterans and hemipterans. Important African tachinid parasites include *Exorista sorbilans* on *Ascotis* (Coffee Giant Looper), and *Bogosia rubens* on *Antestiopsis* spp.

Ichneumonidae and Braconidae parasitize a wide range of insects, ranging from the wood-boring larvae of Siricidae, to cereal stem borers, aphids and scales. Some species have long ovipositors and are able to parasitize larvae *in situ* in wood or cereal stems. Probably the most important genus is *Apanteles* (Braconidae) with its many species, which parasitize the larvae of Noctuidae and Pyralidae. These species are generally not difficult to rear, and some are bred regularly at various CIBC stations.

The Chalcidoidea includes some 19 families, most of which (excluding the Agaonidae — fig wasps) are parasites of caterpillars, beetle larvae, fly larvae and pupae, and large numbers feed on aphids, scales, mealybugs, psyllids and other Hemiptera. A few species parasitize Orthoptera, spiders, ticks, and other insects. Several groups, especially the Trichogrammatidae, are solely egg-parasites and use the eggs of Hemiptera, Lepidoptera, and some Diptera, Orthoptera and others. Species of *Trichogramma* are being widely used in many IPM programmes in different parts of the tropics, and several species are available commercially in the USA. Some species of *Trichogramma* are especially useful in their being polyphagous and parasitizing a range of similar-sized eggs. It should be mentioned that Scelionidae and Proctotrupidae are also important as egg parasites, largely of Lepidoptera; they are Parasitica but not Chalcidoidea. The total complex of parasitic Hymenoptera (including hyperparasites) that can be reared from scale insect colonies is at times quite bewildering. The most important chalcid families for biological control purposes and their insect hosts are as follows: Chalcididae (pupae of Lepidoptera and Diptera mostly); Pteromalidae (many different insects, some species are polyphagous); Encyrtidae (Coccoidea, also aphids, psyllids, some Diptera, Lepidoptera, Coleoptera; some species are hyperparasites); Eulophidae (eggs, larvae and pupae of many insects), Aphelinidae (Coccoidea, Aleyrodidae, Aphidoidea, some Psyllidae, Cicadoidea and eggs of Orthoptera and Lepidoptera); Trichogrammatidae and Mymaridae (eggs of many different groups).

The Strepsiptera are a small order of the Insecta in which the larvae and adult females are obligatory internal parasites of some bees and Hemiptera, and a few species are regarded as important in the natural control of some heteropteran bugs.

A few species of Nematoda are entomophagous. Species of nematode are being used very successfully in controlling populations of mosquito larvae in ponds and swamps. The nematode—bacterium complex of *Neoaplectana carpocapsae* and *Achromobacter nematophilus* (known as strain DD-136 and available commercially as 'Biotrol NCS' from Nutrilite Products Inc., USA) appears to be effective against some rice stem borers (larvae and sometimes pupae), for example *Chilo, Tryporyza* and *Sesamia*, and seems to be a very promising biotic insecticide for use against caterpillars and some beetle larvae (such as Colorado Beetle).

### Pathogens

**Fungi.** A number of different fungi are of some importance in the natural control of insect pests, but only a few have been used with any success in biological control projects. Some of the better known examples are listed below.
(a) *Entomophthora musci* (more temperate): attacks muscoid flies.
(b) *Botrytis tenella* : attacks white grubs (Coleoptera, Scarabaeidae).

(c)[+] *Metarrhizium anisopliae* (Green Muscarine Fungus): *Oryctes* larvae and other white grubs.
(d) *Beauveria globulifera* (Chinch Bug Fungus): Heteroptera.
(e)[+] *Beauveria bassiana* ('Biotrol FBB'): larvae of Lepidoptera.
(f) *Empusa grylli* (African Locust Fungus): attacks Orthoptera.
[+] Available from International Minerals & Chemicals Corp., (IMC) USA.

A few fungi are now available as commercial preparations (as mentioned above) but are generally only available in the USA.

**Bacteria.** *Bacillus thuringiensis* is now so well-known and generally available that it has been included as a biological insecticide in chapter 6 on the chemical control of pests. *B. popilliae* is now available from the Fairfax Biological Laboratories in the USA under the tradename 'Doom', and a few other species of entomophagous bacteria (all *Bacillus* spp.) are also available in the USA.

One drawback with the use of *B. thuringiensis* is that very often it works extremely well in laboratory experiments but in field trials its performance may be unpredictable. Commercial preparations are generally short-lived and so repeated applications may be required. However, with pesticide resistance becoming an ever-increasing problem, especially in the tropics, BTB is an excellent alternative to be used at regular intervals against resistant caterpillars, especially for Diamond-back Moth caterpillars on brassicas in S.E. Asia.

**Viruses.** The two main types of entomophagous viruses are those described as granulosis viruses (including Codling Moth GV, Summer Fruit Tortrix GV and Cabbage White Butterfly GV) and the polyhedrosis viruses. *Heliothis* PHV has already been mentioned as a biological insecticide (page 150) as it is now widely available commercially. Other PHVs include ones specific to *Neodiprion* larvae (sawfly), *Trichoplusia* and *Spodoptera/Prodenia* larvae. These latter viruses are still undergoing trials, and at present it is not known if they will become generally available.

A natural PHV occurs on Cotton Semi-looper (*Anomis flava*) in Queensland, Australia, and is an important population controlling factor; at times the proportion of the looper population which was infected was as high as 50%. It was suggested by Blood & Bishop (1975), as might be expected with an epizootic, that a certain minimum population of larvae (population abundance threshold) is required for a disease outbreak proper, and they calculated from two seasons' data, using regression analysis, that for *A. flava* the population abundance threshold might be about 130 000 (± 10 000) larvae per hectare before a significant disease outbreak would occur.

A review of recorded cases of insects with virus diseases was published by Hughes (1957).

### Biological control failures

In the past many of the biological control failures appear to have resulted from careless handling of the predators and parasites in shipment, together with accidental delays in transit and distribution, and were not due to intrinsic factors in the host/parasite biology. However, simultaneously, many introductions were made without any real scientific basis for the choice of predator or parasite, except perhaps convenience! Thus, there was often little success achieved. For example, the history of biological control attempts in Australia (F. Wilson, 1960) against insect pests reveals a large number of unsuccessful cases, and a similar situation prevailed in Africa (Greathead, 1971). A selection of these control attempts in Africa are presented in the next section in some detail as they nicely illustrate the various aspects of the complexity of these situations, and in particular show some of the more obvious mistakes that have been made in the past.

A few of the more successful and interesting cases of biological control currently being practised in S. China are listed in table 2.

## Biological control of insect pests in Africa

The history of biological control in Africa is rather disjointed. The earliest efforts took place as long ago as 1924 and there were scattered attempts in the decades following this, but major progress in this field did not start until 1958 when the Commonwealth Institute of Biological Control opened its E. African station at Kawanda Research Station, Kampala.
A series of control programmes were in operation, and in some cases it was too early for a comprehensive assessment of these operations to be made. Some of the major attempts at biological control in Africa are summarized in table 3 (p. 91), but are dealt with here in more detail.

1. Antestia Bug of coffee (*Antestiopsis* spp. – Pentatomidae)
From the earliest days of coffee growing in E. Africa *Antestiopsis* spp. have been recognized as serious pests of *Coffee arabica*. Great confusion arose over the status and names of the forms found in E. Africa as a result of the practice of using only colour patterns in attempting to distinguish species. It is now recognized that *A. orbitalis bechuana* (Kirkaldy) is the principal species in eastern Kenya and Tanzania, with *A. facetoides* Greathead in some areas at lower altitudes, while *A. orbitalis ghesquierei* Carayon and *A. intricata* (Ghesquiere and Carayon) replace them in western Kenya and throughout Uganda (Greathead, 1966). It was discovered in 1930 by Wallace that the

very considerable damage done to coffee beans is due to the introduction of fungi (*Nematospora* spp.) during the feeding of the bugs.

Several local parasites (mostly Scelionidae) were discovered, and an introduction was made on a plantation at Limuru, Kenya, before it was realized that they were widespread and already exerting their maximum controlling effect.

*Corioxenus antestiae* Blair (Strepsiptera) was discovered on Mt. Kilimanjaro in 1935 parasitizing adult *A. orbitalis bechuana*, the entry of the triungulins being made at the previous moult. Despite intensive searching in Kenya and Tanzania it was only found on Mt. Kilimanjaro and Mt. Meru. In 1936 three consigments of parasitized individuals were sent from Tanzania to the National Agricultural Laboratories, Nairobi, but all died.

Later, *C. antestiae* was discovered in western Uganda by Taylor. The first release made by Taylor, in 1940 in Toro District, was quite successful in that two years later the parasite had become well established in the bug population and the latter had declined seriously. By 1946 *Corioxenus* was found to have spread half a mile from the release point, and by 1970 it was to be found two miles away.

In 1942 Taylor made a second release, some 150 parasitized *A. orbitalis ghesquierei,* collected from Butukuru on Mt. Algon. The only species of bug on Mt. Elgon at the time was *A. intricata*. When the two release sites were revisited a year later both the *A. orbitalis ghesquierei* and the parasite (in both hosts) were established at Bumasobo, but neither was found at Sipi. In 1946 *Corioxenus* was still confined to the original plot at Bumasobo. By 1948 it had spread a mile or two and by 1950 a little further, *A. orbitalis ghesquierei* at each stage having spread a little further than the parasite. In subsequent years this slow spread has been continued.

*Bogosia rubens* (Villeneuve) (Tachinidae), a parasite of adult *Antestiopsis* which occurs in parts of

Table 3: *Some biological control attempts in Africa* (from Greathead, 1971)

| Pest | Crop | Predator or parasite | Date | Locality | Success |
|------|------|----------------------|------|----------|---------|
| 1. *Antestiopsis* spp. | Coffee (arabica) | *Bogosia rubens* (Tachinidae) | 1940 | Kenya and Uganda | Limited |
| | | *Corioxenus antestiae* (Strepsiptera) | 1965/6 | Kenya and Uganda | Limited |
| 2. *Planococcus kenyae* | Coffee | *Cryptolaemus* sp. (Coccinellidae) | 1924 | Kenya | No |
| | | *Anagyrus* sp. nr *kivuensis* (Encyrtidae) | 1939–40 | Kenya | Yes |
| 3. Mealybugs | Cocoa | 2 species of Chalcidoidea/ 6 of Coccinellidae | 1948–55 | Ghana | No |
| 4. *Leucoptera* spp. | Coffee (arabica) | *Mirax insularis* (Braconidae) | 1962/3 | Kenya (Ruiru) | No |
| 5. *Pseudotheraptus wayi* | Coconut | *Oencyrtus* sp. (Encyrtidae) | 1959 | Kenya and Zanzibar | ? |
| 6. *Cochliotis melolonthoides* | Sugarcane | *Bufo marinus* (Amphibia) | 1948/9 | Tanzania | No |
| | | *Cordyceps barnsii* (Fungi) | 1968 | Tanzania | Too soon |
| 7. *Scyphophorus interstitalis* | Sisal | *Hololepta* sp. and *Plaesius* sp. (Histeridae) | 1948 | Tanzania | No |
| | | *Dactyloernum* sp. (Hydrophilidae) | 1948 | Tanzania | No |
| 8. *Orthezia insignis* | Jacaranda | *Hyperaspis jocosa* (Coccinellidae) | 1945 | Kenya (Uganda and Tanzania) | Yes |
| 9. *Aonidiella aurantii* | *Citrus* | *Aphytis africanus* (Aphelinidae) | 1964 | S. Africa | Yes |
| 10. *Cosmopolites sordidus* | Bananas | *Plaesius javanus* (Histeridae) | 1934 | Uganda | No |
| 11. *Eriosoma lanigerum* | Apples | *Aphelinus mali* (Aphelinidae) | 1927/8 | Kenya (highlands) | Yes |
| 12. *Aleurocanthus woglumi* | *Citrus* | *Eretmocerus serius* (Aphelinidae) | 1958 | Kenya (coast) | Yes |
| | | *Prospaltella opulenta* (Aphelinidae) | 1966 | Kenya (Kisumu area) | Yes |
| 13. *Tetranychus cinnabarinus* | Cotton | *Phytoseiulus riegeli* (Phytoseiidae) | 1966/7 | Uganda (Namulonge) | ?Yes |
| | | | | Kenya (coast) | ?Yes |

Uganda, western Kenya and the Eukobe District of Tanzania, has also been the subject of attempts at biological control. At the same time that *Corioxenus* was released at Butukuru 30 of the *Antestiopsis* used in the introduction bore eggs of *Bogosia*, but it failed to become established, though it is present in the area now. Another six releases were made in different parts of Uganda, but only the release at Namutamba in Mengo District was successful. *Bogosia* is now very locally distributed in Uganda, apparently far less common than it was in the 1940s, probably due to changing cultural practices and to increased insecticide spraying.

In 1962 the E. African station of CIBC undertook a survey throughout E. Africa which showed that all the egg parasites were widespread but that the two adult parasites, *Corioxenus* and *Bogosia* were restricted in their distribution. Since *Corioxenus* showed little promise on Mt. Elgon, it was decided that an experimental release should be made at the Coffee Research Station, Lyamungu, Tanzania. In 1965 an attempt was made to establish laboratory cultures but without success, and a second collection from Fort Portal was released in the field at Lyamungu in 1966. Only 70 flies were released, but in spite of this small number a survey made the following year showed that the fly was established.

At the same time CIBC (E. Africa) undertook studies in other countries where *Antestiopsis* spp. are found on coffee, in an attempt to discover new parasites. However, surveys in Sudan, S. Africa, India and Madagascar failed to yield any new species of parasites that might have been used against *Antestiopsis* in E. Africa.

2.  **Kenya Mealybug** (*Planococcus kenyae* (Le Pelley) – Pseudococcidae)

*P. kenyae* became such a serious coffee pest in Kenya that it was the subject of a major control effort by the Department of Agriculture, and later the Coffee Research Station, Ruiru, for many years. The early part of the history of its control was bedevilled by confusion as to its identity. Since its description by Le Pelley in 1935 as *Pseudococcus*, it had to be transferred to the new genus *Planococcus* Ferris.

The first outbreak of *P. kenyae* (at the time identified as *Pseudococcus lilacinus* Cockerell) was in 1923 at Thika, from where it quickly spread to Ruiri, Fort Hall and Ngong, and subsequently throughout the Eastern Highlands, not only on coffee but also on many peasant crops. Associated with the mealybug was the ant *Pheidole punctulata* Mayr, and it was believed that the ant was responsible for the outbreaks, but it was later shown conclusively that the ant was present prior to the spread of mealybug outbreaks.

The first control attempts recognized the importance of the role played by the ants in protecting coccoids from their natural enemies, and consisted of recommendations for sticky-banding coffee trees and introducing ladybirds (Coccinellidae) on to them. The most important predators of *P. kenyae* then to be found in Kenya were five species of Coccinellidae (*Chilocorus* sp., *Hyperaspis* – two spp., and *Scymnus* – two spp.). But even with banding these predators were not sufficient to control the mealybug. In 1924 *Cryptolaemus* sp. was introduced from S. Africa, as also were several parasites from other regions. With the failure of *Cryptolaemus* to exercise control, attempts were made to culture and release large numbers of the two native species of *Scymnus* in 1932, but with little success.

In the 1930s natural enemies of *Pseudococcus gahani* were imported from California, and of *P. krauhiniae* from Honolulu and Japan. None of these would breed on the Kenya Mealybug. These failures, and the failures of enemies from Uganda and Tanzania imported in 1935, led to a reappraisal of the situation. Since the origin of *P. lilacinus* was in S.E. Asia, recommendations for a search in that area were made. In the meantime, Le Pelley had been able to show that the Kenya Mealybug was not *P. lilacinus*, although very similar in morphology to it, and he described the new species in 1935. This discovery led to the first investigations in Uganda, but their failure and the doubt still remaining as to the conspecificity of the Kenya Mealybug and the mealybug in Uganda, gave weight to the recommendations which resulted in the despatch of Le Pelley to the Philippines, Indonesia, S. India and Sri Lanka. None of the insects collected by Le Pelley would feed satisfactorily on *P. kenyae*, and later Le Pelley (in 1943) showed that *P. lilacinus* also had biological differences from *P. kenyae*. These discoveries finally narrowed down the search to Uganda, N.W. Tanzania and Zaire, and with this came confirmation that this area was indeed the natural home of the Kenya Mealybug and that it was completely under natural control there.

Two species of *Anagyrus* (Encyrtidae) were discovered in the field parasitizing the mealybugs and they proved easy to rear in the laboratory. The first species to be released was *A.* sp. near *kivuensis*, which was distinct biologically, but morphologically indistinguishable, from *A. kivuensis*. It was first released in 1939 at Ruiru when 244 females and 280 males were liberated. The parasite was detected in the field during the first generation and after only six months it had controlled the mealybug at the release point and was present up to 70 yards away. Two months later it was 300 yards upwind of the release point and the original infestation was almost completely wiped out. By the same time in the following year a total of 104 liberations had been made and widespread control achieved. *Anagyrus* sp. near *kivuensis* soon became the dominant parasite, while the other species, at first indeterminate but later described as *A. beneficians* Compere, also spread throughout, though in coffee estates it was apparently not able to compete with the other parasite species. In peasant areas and under other less artificial conditions it became the chief parasite, but may not be so now. It was also less effective at the very low population densities achieved by the parasitism of *A.* sp. near *kivuensis*. Several other species of Encyrtidae were released in large numbers and became established, but apparently contributed little to the control of the mealybug.

By 1949, control of the mealybug was regarded as good, and any further outbreaks were controlled by the parasites. The ant *Pheidole punctulata* continued to be troublesome where banding was not effectively carried out, but the introduction of dieldrin spray banding in 1951 greatly increased the ease with which ants could be excluded from the coffee trees, and resulted in enhanced control of the mealybug, making further work on natural enemies unnecessary.

The spread of the use of persistent chlorinated hydrocarbon insecticides against other coffee pests in the early 1950s resulted in new outbreaks of *P. kenyae*, which have subsided since the introduction of the widespread use of non-persistent organophosphorous insecticides in their place. These remarks refer mainly to estate coffee; on peasant holdings breakdown of biological control has been rare, and here the growing of yams, which had been virtually eliminated, began again.

This biological control programme is one of the most successful to be carried out in E. Africa. It is not possible to estimate the value of control of *P. kenyae* to the peasant farmer, but estimates were made for the coffee industry; in 1959 Melville estimated that £10 million had been saved, at a control cost of not more than £30 000.

## 3.  Mealybugs on cocoa in Ghana

*Planococcoides njalensis* and other indigenous mealybugs are vectors of Swollen Shoot Virus of cocoa in Ghana. Between 1950 and 1953 a survey of local natural enemies of mealybugs was carried out, and 22 species of Encyrtidae, and two of Aphelinidae, were found as parasites. So far as predators were concerned there were three species of Cecidomyiidae, two of Lycaenidae, one of Tinaeidae, one of Chrysopidae, and several species of Acarina attacking the mealybugs. The survey showed that the natural level of parasitisim of *P. njalensis* and the other species was between 4 and 25%.

In order to supplement this rather low level of parasitism and predation the Cocoa Research Institute imported *Anagyrus* sp. near *kivuensis* from Kenya in 1948, and from 1949 to 1955 a total of 25 species of various Chalcidoidea (mostly Encyrtidae) and six of Coccinellidae were imported from California, the W. Indies, China, Hawaii, and S. Africa. However these enemies did not become established and importation of the mealybug's natural enemies did not continue.

4. **Coffee Leaf Miners** (*Leucoptera* spp. – Lyonetidae)

The two species of *Leucoptera* mining leaves of *Coffea arabica* in E. Africa are *L. meyricki* Ghesquière, formerly confused with the W. Indian *L. coffeella* Guier., and *L. caffina* Washb. which has been misidentified as *L. daricella* Meyr., an Australian species.

Prior to 1950 *Leucoptera* spp. were not serious pests, but since then there have been outbreaks in both Kenya and Tanzania, especially during the dry seasons. The use of persistent insecticides (chlorinated hydrocarbons) aggravated the problem with the accidental destruction of the natural enemies, but since the change to non-persistent organophosphorous insecticides the situation has improved and it has been suggested that these pesticides somehow enhance the effects of the parasites.

In 1948 Notely suggested the use of indigenous parasites for control of *Leucoptera*, suggesting that, as a *Eulophus* sp. was common and effective in the Usambara Mountains and not present elsewhere, it might be worth introducing elsewhere. In recent surveys, however, this parasite has not been found. Notely published, in 1948 and 1956, studies of the parasite complex on Mt. Kiliminajaro, and in 1959 a parasite survey was started in Kenya. This was extended by CIBC to other areas, especially Uganda and N.W. Tanzania where the leaf miners are not pests. This work showed that a complex of eulophids, *Agoniaspis* sp. (Encyrtidae), and braconids, while not all present everywhere, left little scope for inserting other parasites into the complex to any effect. The egg stage seems to be free of parasites and the pupae only lightly parasitized, so a survey was made by CIBC in the W. Indies to see if there were any suitable parasites which could be introduced to fill these gaps.

Comparison of the parasites of *L. coffeella* with those of the E. African species showed that, though the complexes are generically very similar, the genus *Mirax* (Braconidae) was absent from E. Africa. Accordingly in 1962 and 1963 seven shipments (totalling 2040 insects) of *Mirax insularis* Mues. were made from Dominica to the Coffee Research Station, Ruiru. Unfortunately it was found that although *M. insularis* would parasitize *L. coffeella* in cages it would not accept the E. African species of *Leucoptera* as hosts in the laboratory.

5. **Coconut Bug** (*Pseudotheraptus wayi* Brown – Coreidae)

The disease of coconuts known as gumming, gummosis or nut-fall has been noticed since 1924, and in 1950 it was found to be due to the feeding of a coreid bug, later described as *P. wayi*.

It was discovered that ants, especially *Oecophylla longinoda* Latr. can control the bug. Since the ant distribution was uneven, their encouragement was advocated, even to the extent of growing special nesting bushes. However, this ant bites the plantation workers, so this method of control did not become popular. Also, it was often found that competition with other ants reduced the effectiveness of the *Oecophylla*.

Two species of *Ooencyrtus* (Encyrtidae) were found as egg parasites in Zanzibar, and a further species was found on the Kenya coast. Cultures of one of the Zanzibar species and the Kenya species were exchanged in 1959, but it is not known if that sent to Zanzibar became established or if any benefit ensued. In Kenya recoveries were made in 1959 and 1960, but no further searches have been made. Since many workers consider that nut-fall is inevitable as it is the natural method of the palm for regulating its crop size, and since the price of copra has fallen recently, interest in this problem has waned.

6. **Sugarcane Whitegrub** (*Cochliotis melolonthoides* (Gerst.) — Scarabaeidae)

Whitegrub pests of sugarcane have become important on an estate at Arusha Chini, near Moshi, Tanzania. The first outbreak was in 1941 and in subsequent years the pests were studied. The major pest was *Cochliotis melolonthoides*, although there were five other scarabaeoid beetles involved. Four species of *Campsomeris* (Scoliidae) were already present in the area, as well as an elaterid and a predatory beetle larva *Didimus* sp. (Passalidae). As no dipterous parasites of the adults were found at Arusha it was recommended that suitable flies be found and introduced. The giant toad (*Bufo marinus*) was imported in 1948 (71 individuals) and 1949 (80) and released, but failed to become established probably owing to predation, especially by monitor lizards (*Varanus* sp.). In 1962 the possibility of further attempts at biological control was discussed with the estate manager, but owing to the heavy application of aldrin to the soil and herbicides to destroy dicotyledonous weeds, it was decided that this was not worthwhile.

Subsequently the fungus *Cordyceps barnsaii* Thw. was found attacking larvae on the estate, and it has been investigated at TPRI, Arusha, as a promising biological insecticide.

7. **Sisal Weevil** (*Scyphophorus interstitialis* Gylh. — Curculionidae)

Infestation of *Agave* spp. by *Scyphophorus* (both originally from C. America) has followed the development of sisal plantations in the tropics. Sisal was first planted in E. Africa at about the end of the last century and the weevil arrived in about 1930. In 1945 enquiries were made to CIBC as to the possibility of biological control of this pest in E. Africa. A brief survey was made in Mexico which resulted in the discovery of two possible predators, *Morion* sp. (Carabidae) and *Hololepta* sp. (Histeridae), but only five specimens were actually sent to Tanzania. A later survey in S.W. USA failed to reveal any natural enemies.

In 1948 three beetle predators of *Cosmopolites sordidus* were introduced into Tanzania from Trinidad (two Histeridae and one Hydrophilidae). These were sent from CIBC (W. Indies) in the hope that as the two weevils have similar life histories the predators might be the same. None of the beetles became established and there was no evidence that any benefit resulted.

After a survey of the problem in E. Africa in 1952, Dr Simmonds (Director, CIBC) recommended that any further work should take the form of a thorough survey in the countries where the weevil is indigenous and not, as had been done, in merely sending predators of unknown value.

8. **Jacaranda Bug** (*Orthezia insignis* Browne — Orthezidae)

*Othezia insignis,* also known as the Lantana Bug, clearly is of some value when feeding on lantana, but this is heavily outweighed by its nuisance value in severely damaging flowering trees, especially jacaranda, and sometimes coffee and garden flowers. Control was attempted in 1945 with the introduction of *Hyperaspis jocose* Muls. (Coccinellidae) from Hawaii into Kenya by the Forest Department. It became established, but inital results were disappointing; then the culture was taken over by the Agriculture Department and in 1953, as a result of laboratory breeding, 70 000 were released in Nairobi, Kisumu, Embu and Fort Hall, and stocks sent to Uganda and Tanzania. In all, more than 100 000 were released in Kenya by the end of 1954, and a considerable measure of control has resulted. In Tanzania, releases at Moshi are said to have saved the jacaranda there. In Uganda, *Hyperaspis* was also established and *Orthezia* is no longer a pest of any importance.

Shipments of other predators were sent from CIBC (W. Indies – Trinidad) to Kenya but some died during rearing and others failed to become established.

### 9.    Red Scale (*Aonidiella aurantii* (Mask.) – Diaspididae)

Since about 1900 Red Scale has been regarded as the major pest of *Citrus* in S. Africa and, in the belief that there were no adequate natural enemies, emphasis was placed upon a search for suitable pesticides. When DDT and parathion became available and were widely used, the Red Scale became even more numerous, and pesticide research became more intensive. However, a local entomologist (E. C. G. Bedford) believed that local natural enemies were able to exert satisfactory control, provided that ants were excluded from the trees, and his ideas were developed on several estates. He showed that, once ants were removed from the orchards, within a few years the natural enemies were able to control Red Scale to such an extent that commercially acceptable citrus fruit could be grown without the use of chemical sprays. This was chiefly due to the parasite *Aphytis africanus* (Aphelinidae) and also to some extent to *Habrolepis rouxi* (Encyrtidae).

Over the years, various attempts at classical biological control were made by introducing cocci-

nellids from Australia and California, but without success. Parasites from Australia and California were also introduced but none was recorded as having survived. In 1964 three exotic species of *Aphytis* were introduced and have apparently survived, but it appears that they are less successful in controlling the Red Scale than is the indigenous *A. africanus*. The remarkable feature is that here is apparently a case where an indigenous parasite has attacked an imported scale and proved to be superior to its own native parasites in controlling it.

*Compieriella bifasciata* (Encyrtidae) was introduced in 1966 from W. Australia into the Transvaal where it has become established.

### 10.    Banana Weevil (*Cosmopolites sordidus* Germ. – Curculionidae)

*Cosmopolites sordidus* seems to have spread with the cultivation of bananas. It was first recorded in Uganda in 1918, though it had clearly been present for some time. The focus of infestation appears to have been the Kampala/Entebbe area. It may well have been introduced with bananas imported for the Botanic Gardens at Entebbe before 1908. Subsequently there was confirmation of Banana Weevil presence in most other areas of E. Africa. In 1934 the histerid predator *Plaesius javanus* Erichs. was imported into Uganda from Java, where it had been discovered. Three shipments of beetles were received, the first in 1934 consisting of 600 individuals of which 225 were alive on arrival. These were held in quarantine and the survivors released on Kibibi Island in Lake Victoria in January 1935. The second shipment received in 1935 consisted of 820 dead beetles, and in the third of 845 only 27 survived for release on Kibibi. The release site was revisited for the first time in 1944 but no *Plaesius* were found, and it is presumed that the predator did not become established.

### 11.    Woolly Apple Aphid (*Eriosoma lanigerum* (Hausm.) – Pemphigidae)

There is no record of the arrival of *Eriosoma lanigerum* in Kenya, but by the middle of the 1920s it had become sufficiently serious for the introduction of *Aphelinus mali* (Aphelinidae) in 1927/8 from England. The introduction was presumably made in the Eastern Highlands as it was introduced to the Western Highlands at Molo in 1952, using material from Kimonkop, where it was very common. Apple-growing is said to be possible in Kenya only owing to the presence of this parasite.

## 12. Citrus Blackfly (*Aleurocanthus woglumi* Ashby – Aleyrodidae)

*Aleurocanthus woglumi* was first collected on citrus in Kenya in 1913. In the more humid coastal areas it has for many years been a serious pest. A survey in the coastal areas of Kenya in 1958 showed that though coccinellid predators exerted some degree of control, there were no parasites.

*Eretmocerus serius* Silv. (Aphelinidae) a native of Malaya which had been successfully used for biological control, was already present in the Seychelles. The first small shipment in 1958 failed to become established, but the second shipment later in the year enabled 267 adult parasites to be released on a single tree at Matuga (10 miles S. of Mombasa). The parasite became established and was detected in the second generation. Six months after the successful liberation, several hundred *E. serius* were collected from the original release site and were released at Ukundu and Gazi (12 miles N. of Mombasa). Two years later (June 1962) only an estimated 1% of citrus leaves at Matuga bore traces of *A. woglumi* eggs. In March 1963 only a few *A. woglumi* could be found at Ukunda, establishing clearly that commercial control of *A. woglumi* had been achieved on the Kenya coast. The parasite was released in Nairobi and Kitui in July 1965.

As a result of the success of *Prospaltella opulenta* Silv. (Aphelinidae) in the drier parts of the W. Indies it was decided in 1966 to introduce it for trial in the Kisumu area. Two consignments were sent by airmail from Barbados in 1966; they were received in good condition and about 616 parasites were released immediately on infected citrus at the Cotton Research Station, Kibos, where the parasite is now established.

## 13. Red Cotton Mite (*Tetranychus cinnabarinus* (Boisd.) – Tetranychidae)

Control of Spider Mites in greenhouses in N. America and Europe, by the S. American predaceous mite *Phytoseiulus riegeli* (Phytoseiidae), suggested that under tropical conditions it might be effective in the field. *Tetranychus cinnabarinus* causes considerable damage to cotton and is not affected by the insecticides often applied to cotton for the control of insect pests. Accordingly a culture of *P. riegeli* was obtained by the Cotton Research Station, Namulonge (Uganda), from the Glasshouse Crops Research Institute in England in 1966 for trial against *T. cinnabarinus*. Control was successful in the laboratory but unfortunately the predators escaped into the host culture and destroyed it. A second attempt was made in 1967 which was more successful.

In August 1967, a culture of *P. riegeli* was sent to the Kenya coast and released at Msabaha. In October examination of 100 cotton leaves from the release point revealed only a few mites, including six *P. riegeli*.

## Conclusions

A study of the history of biological control attempts in Africa in many ways nicely reflects various developments in biological control techniques, and also demonstrates various problems which are frequently encountered. Firstly, the identity of the pest must be clearly established, and the taxonomists need to be made aware of any biological differences, as well as morphological or anatomical differences between various pest species (cf. *Planococcus kenyae*). Secondly, most serious pests are not indigenous to the area in which they are pests, and so it is necessary to go back to their country of origin for suitable predators and/or parasites for possible introduction (cf. *Planococcus kenyae*). Thirdly, it is generally a waste of time to introduce parasites or predators which usually attack another genus of pest, even though the two pests may have very similiar biologies (cf. *Scyphophorus*). It must be borne in mind, however, that generally predators are less specific in their choice of prey than are parasites. Parasites of closely related species of pests from different parts of the world may be worthwhile introducing sometimes, but more often parasites are species-specific and will not attack the new host (cf. *Leucoptera* spp. and *Mirax*). Fourthly, when the pest is indigenous to the area in which it is a nuisance, it is generally not worthwhile to attempt to breed-up large numbers of one of the local parasites for local release since the usual pest/parasite complex is more often a stable and long-established one which will not permit changes of this nature. Also, if the parasite complex is extensive there is little chance of being able to successfully insert another species into the complex, even if it comes from another area altogether (cf. *Leucoptera* spp.), although sometimes this is successful. Generally it is best to find out if there are any predator/parasite niches which are not filled and then attempt to introduce a species that will effectively fill that niche.

Fifthly, when making the original survey of the pest-parasite complex, it is vital to do the job thoroughly to make certain that all the local parasites are found. Should an introduction be made of a parasite species which is already present in the area, but not detected owing to careless surveying, then this is likely to have little effect and prove to be a waste of effort (cf. *Hypothenemus hampei* in Kenya).

One of the problems associated with early attempts at the introduction of parasites from other countries was the time taken for consignments of insects to be shipped from country to country, and many consignments were found to be dead on arrival. Nowadays, with rapid and regular airline services, such consignments of insects can be sent either by airmail or air freight and arrive at their destination within only a few days.

The facilities required for mass propagation of parasites or predators are not unduly complicated but are generally beyond the scope of most agricultural and entomological research stations. However, in recent years many research stations have become equipped with such facilities. In addition there is the chain of stations and substations of CIBC scattered throughout the world. The result of these developments is that now it is generally possible to obtain suitable predator or parasite species in very large numbers, bred either in the country of origin or in the country of introduction, so that tens or hundreds of thousands of individuals may be released instead of only a few hundred, which was all that was often possible in the past. Clearly the larger the number of individuals that can be released, the greater the chances of the introduction being successful.

As pointed out by F. Wilson (1960) it has been customary to characterize attempts at biological control as either 'successful' or 'unsuccessful'. In reality this oversimplification is undesirable for it is fairly obvious that many different levels of biological control are achieved. For the sake of definition it can be said that a biological control project is 'completely successful' if other forms of control can then be dispensed with. In the oversimplified view the project is regarded as 'unsuccessful' if other methods of control still have to be utilized. This attitude is totally unrealistic; chemical applications normally give many different levels of control, as is only to be expected. Wilson summarizes attempts at biological control of insect pests in Australia by making five categories based on the levels of success achieved:

(a) pests substantially reduced in status;

(b) pests reduced in status;

(c) pests of doubtfully diminished status;

(d) pests of unchanged status;

(e) pests against which the introduced enemies failed to become established.

# 6 Chemical control of insect pests

## Pesticides

### Efficient use of pesticides

To ensure accurate application the label should be read carefully *before* use, and the recommended doses, volumes and times of applications followed. With many of the newer, short-lived chemicals, dosage, placement and timing are critical. Attention should also be paid to maintenance of equipment, for defects such as faulty or worn nozzles can appreciably alter rates of application.

Compatibility of mixed products used in combined spray programmes is important. If no information is given on the labels consult the manufacturer or dealer. Wetting agents are sometimes used as additives to sprays for crops that are difficult to wet (e.g. brassicas, peas), but not all wetting agents are compatible with all proprietary formulations.

Taint and damage to plants are hazards to be avoided. Where chemicals are known invariably to cause damage to certain crops or varieties of plants, or if taint or off-flavours are produced in edible crops, this is mentioned under the specific chemical.

**Pest resistance to pesticides.** Pest resistance to chemicals is one of the major problems in agricultural entomology in all parts of the world now, and is ever increasing. Such cases arise in localized areas, or even on certain holdings, and may remain purely local for some time. Resistance does not usually develop until a particular chemical has been employed in that area for some considerable time (3–10 years). Sometimes resistance to one compound may be closely followed by resistance to other related compounds, e.g. one organochlorine compound and the others in the same group.

Information as to which major local pests may be exhibiting resistance to widely used insecticides can usually be obtained from the local Ministry or Department of Agriculture staff (see page 55).

**Natural control agents.** It is very important that pesticide applications should not adversely affect existing natural control of the pest population. To this end, it is desirable that knowledge of the local predators and parasites of key pests be accumulated so that, whenever possible, existing levels of natural control may be maintained in addition to the artificial control measures that are being applied. Fairly obvious cases where extreme care should be taken are infestations of scale insects on trees, leaf miners, and sometimes the late stages of aphid infestations.

**Storage of chemicals.** Storage of chemicals is important; they should be kept in a dry place protected from extremes of temperature. Generally frost is more harmful to stored liquid formulations than is high temperature. Most formulations may be expected to be used for at least two years without loss of efficiency, but unsatisfactory storage conditions can impair stability and effectiveness.

### Safe use of pesticides

All pesticides should be treated with care whether they are known to be particularly poisonous or not, and the following routine observed.

1. Read the label, especially the safety precautions, carefully before use.
2. Use all products as recommended on the labels and do not use persistent chemicals where there are effective, less persistent alternatives.

3. Avoid drift on to other crops, livestock and neighbouring property and take care to prevent contamination of any water source, whether for drinking or irrigation purposes.

4. Safely dispose of all used containers. Liquid contents must first be washed out thoroughly and the washings added to the spray tank. Packages containing powders or granules must be completely empty before disposal. Burn bags, packets and polythene containers. Puncture non-returnable metal containers (except aerosol dispensers) and bury them in a safe place. Bury glass containers or dispose of them with other refuse. On no account use empty pesticide containers for any other purpose.

5. Return unused materials to store under lock and key.

6. Clean any protective clothing used, and wash exposed parts of the body thoroughly when the job is completed.

Particular attention should be paid to the paragraph headed 'Caution' in the pesticide section of this chapter. It indicates:

1. Whether the chemical is particularly poisonous; if so, certain protective clothing should be worn.

2. Whether there are any special user risks involved, even if not very toxic, such as irritation to the skin. Some protective clothing should be used with these chemicals, and their labels should be consulted for guidance.

3. What precautions should be taken to ensure that unacceptable residues do not remain on edible crops at harvest. Where appropriate, the time at which a chemical may be applied is shown and also the minimum interval that must be observed before last treatment and harvest.

4. Whether there are risks to bees, livestock, fish or other wild life. In the case of bees, the degree of risk is given as *dangerous*, *harmful* and *safe* (see following section).

Fish are susceptible to many chemicals, and great care must be taken to prevent contamination of ponds, waterways and ditches with chemicals or used containers.

## Toxicity to bees

*Dangerous:* highly toxic to bees working the crop or weeds, at the time of treatment, and toxic for 24 hours or more

| | |
|---|---|
| Azinphos-methyl | Fenitrothion |
| BCH (HCH) | Formothion |
| Carbaryl | Heptenophos |
| Chlorpyrifos | Lead arsenate |
| Demeton-*S*-methyl sulphone | Methomyl |
| Diazinon | Mevinphos |
| Dichlorvos | Parathion |
| Dimethoate | Permethrin |
| Endrin | Phosphamidon |
| Ethoate-methyl | Pirimiphos-methyl |
| | Triazophos |

*Harmful:* toxic to bees working a crop at time of treatment, but not hazardous if applied when bees are not foraging.

| | |
|---|---|
| DDT | Menazon |
| DDT with malathion | Nicotine |
| Demephion | Omethoate |
| Demeton-*S*-methyl | Oxydemeton-methyl |
| Endosulfan | Phosalone |
| Malathion | Tetrasul |
| Mercury | TDE |
| Methidathion | Thiometon |
| | Vamidothion |

*Safe:*

| | |
|---|---|
| Binapacryl | Disulphoton granules |
| Derris | Phorate granules |
| Dicofol | Schradan |
| Dinocap | Tetradifon |

### The Agriculture (Poisonous Substances) Regulations, UK

These regulations are laid down in the Act of 1952 and, although not applicable outside the UK, they give a good indication of the toxicity of some of the various chemicals and the protective clothing ideally required. It must be noted though that this clothing is designed for use in a temperate climate and not the tropics.

#### Part I substances

| | |
|---|---|
| Dimefox | *Full protective clothing*, e.g. rubber gloves and boots, respirator and either an overall and rubber apron, or a mackintosh when preparing the diluted chemical. The respirator may be dispensed with when the diluted chemical is applied to the soil. |
| Chloropicrin | |

#### Part II substances

| | |
|---|---|
| Aldicarb | *Full protective clothing*, which includes rubber gloves and boots and, either a face shield or dust mask, and an overall and rubber apron or a mackintosh; or a hood, or a rubber coat and sou'wester, depending on the operation being performed. A respirator is required when applying aerosols or atomizing fluids in glasshouses. |
| Carbofuran | |
| Disulfoton | |
| DNOC | |
| Endosulfan | |
| Endrin | |
| Fonofos | |
| Methomyl | |
| Mevinphos | |
| Oxamyl | |
| Parathion | |
| Phorate | |
| Schradan | |
| Thionazin | When the substances are used in the form of granules certain relaxations in the protective clothing requirements are allowed. |

#### Part III substances

| | |
|---|---|
| Amitraz | Rubber gloves and face shield |

Azinphos-methyl
Chlorfenvinphos
Demephion
Demeton-*S*-methyl
Dichlorvos
Methidathion
Nicotine
Omethoate
Oxydemeton-methyl
Phosphamidon
Thiometon
Triazophos
Vamidothion

when preparing the diluted chemical. Full protective clothing, i.e. overall, hood, rubber gloves and respirator, is needed however when applying aerosols or atomizing fluids in glasshouses.

## Methods of pesticide application

In the application of pesticides to crop plants, the object is to place the chemical in or on the correct part of the plant in order that it might come into suitable contact with the insect pest. For a leaf-eating insect a pesticide should either be on the leaf surface or in the leaf tissues; for a sap-sucker the poison should be in the phloem system. For a leaf-miner the poison must penetrate into the leaf tissues to be effective. A soil-inhabiting, root-eating pest can be attacked either through the tissues of the roots or else by a contact insecticide, introduced into the soil around the roots. Many of the caterpillars (and fly larvae) which in their later instars bore into fruit, or plant stems, hatch from eggs laid on the surface of the plant or on the soil and the first instar larvae spend some time on the plant surface before burrowing into the stem or fruit. Thus, these larvae, which in later instars are almost invulnerable once they are inside plant tissues, can be attacked by carefully timed and placed pesticide application. Seedling pests can be attacked by the use of a seed dressing on the sown seed. With a contact insecticide it is imperative that the chemical comes into contact with the pest during the period of its potency. This point is most important with the increasing use of the more short-lived organophosphorous pesticides.

A recent and highly informative publication on pesticide application methods is by Matthews (1979).

Most pesticides are either crystalline solids or oily liquids in the pure state or the technical product, and they are usually effective in quite small quantities (i.e. a litre or two per hectare). In order to apply a pesticide to a crop it is usually necessary for the chemical to be contained in a carrier fluid or powder. The earliest carrier (or diluent) used was water and the chemical was either dissolved in it or, if it was water-insoluble, as a suspension or emulsion. In the case of emulsions and suspensions it is important that the mixture be stable enough to ensure the application, under practical conditions, of a solution of uniform and known concentration. The physical stability of a spray solution may be conferred by the addition of supplementary materials. Thus the sedimentation of a suspension can be delayed by the addition of protective colloids or dispersing agents. The coalescence of the scattered droplets in an emulsion may be retarded by the addition of an emulsifier.

### Formulations

As previously mentioned most pesticides are crystalline solids or oily liquids and as such are usually not suitable for spraying direct on to a crop. In a few cases the technical product is soluble in water — then the pesticides can be prepared as a very *concentrated solution* (c.s.) which only requires dilution by the farmer to the appropriate strength for spraying. Usually these concentrated solutions have to have wetting agents or detergents added (see later in the chapter). This type of spray solution is typically very homogeneous and spreads a very even level of pesticide over the foliage.

Many solid substances that will not dissolve in water can be ground and formulated as *wettable powders* (w.p.). Wettable powders are powders which can easily be wetted and do not resist the penetration of water, and are miscible in water in which they pass into suspension — others are more accurately termed water-dispersible powders in that when mixed with water they remain as individual particles in suspension for a considerable period of time. Various additives (dispersants) can be included in the formulation of wettable powders to delay the process of sedimentation.

Oils and other water-immiscible liquids, if agitated with water, break up into tiny droplets which on standing rapidly coalesce to form a separate layer. This coalescence may be retarded or prevented by the addition of auxiliary materials known as surfactants or emulsifiers. With mixtures of oil and water, two types of *emulsion* are possible. The oil may be dispersed as fine droplets suspended in water, which is then the continuous phase, giving an oil-in-water (o/w) emulsion, or the water may be the disperse phase giving a water-in-oil (w/o) emulsion. The type of emulsion generally required for crop spraying pratice is the o/w emulsion which, as water is the continuous phase, is readily dilutable with water. Thus pesticides insoluble in water may be dissolved in various organic solvents forming an *emulsifiable concentrate* (e.c.) which can be diluted in water to an appropriate spray strength. The 'breaking' of an emulsion is the usual way in which the toxic dispersed phase comes into play, breaking occurring after the evaporation of most of the water. Various chemical substances can cause the 'inversion' of an emulsion which then becomes useless for spray purposes. Sometimes emulsions are caused to 'cream' (named from the analogous creaming of milk), resulting from differences in specific gravity between the dispersed and continuous phases.

Some pesticides are more suitably formulated as *miscible liquids* (m.l.). In this case the technical product is usually a liquid and is mixed (dissolved) with an organic solvent which is then, on dilution, dissolved in the water carrier.

Other pesticides may, for use against specific pests, be formulated as *seed dressings* (s.d.), both wet and dry, or *granules*, but these types will be dealt with later in the chapter.

At times it will be necessary to know precisely the quantity or proportion of the pure chemical in any formulation; sometimes this may be shown on the pesticide container, but more often is not. However, it will be available on the company data sheets, and in such sources as *The Pesticide Manual*. The usual method of expression of the proportion is as grams of *active ingredient* per kilogram of formulation (g a.i./kg) for powders, and grams per litre (g/l) for liquids.

### Spraying

When water is the carrier, the usual method of application of the spray is by passage under pressure through special nozzles which distribute the chemical in a fine spray over the crop. In general, the type of spray application can for convenience be expressed as high-volume, low-volume, or ultra-low-volume according to the amount of carrier liquid.

**High-volume spraying.** Definitions vary considerably, but according to Maas (1971) the term 'high-volume' applies to spray rates of more than 400 l/ha. With high volume spraying the carrier is invariably water, and the usual quantity involved is in the region of 600–1200 l/ha. In the case of wanting a run-off of spray from the upper parts of the plants on to the lower parts, or on to the soil, then the water volume may be doubled up to the extent of 2400 l/ha. High-volume spraying has several major disadvantages, the first being the problem of transporting the large quantities of water required, especially in areas where piped water is not available. In many parts of the drier tropics obtaining sufficient water for this purpose can be difficult. The cost of the high-volume spraying

equipment is considerable, and the bulk of the equipment is such that its operation often requires a large tractor.

Because of these problems alternative methods of application for the treatment of large areas have been sought. The 'scent spray' principle involves the jet of liquid being dispersed into fine droplets by the force of a copious air-flow. This method has been successful for the dispersal of DDT/petroleum mixtures from aeroplanes where the air speed alone is sufficient to break up the jet of solution into droplets which are then dispersed by the slip stream. A helicopter can be even more effective for this method using the down-draught from the rotor blades. For terrestrial use, the air-stream can be provided by a fan mounted horizontally, or by a turbine fan. These techniques which rely upon the energy for spray dispersion being provided through the air-stream, and not, as in conventional spraying, through the liquid, are referred to as 'atomization'. Typically the droplets produced are much smaller than those produced by the usual high-volume sprayers. Because of the smaller droplet size smaller volumes of spray per hectare are needed and the use of organic solvents such as kerosene, petroleum oil, or fuel oil, in place of water, becomes economically feasible. If water is used as carrier, the concentration of active ingredient can be increased, giving rise to what is termed 'low-volume' spraying.

**Low-volume spraying.** This term usually applies to volumes in the region of 5-400 l/ha. However, the typical rates for ground application are commonly 100-200 l/ha, whereas for aerial application they are 15–75 l/ha. As referred to in the previous section the equipment used for ground application is typically an air-blast machine, and the carrier liquids are frequently organic solvents, although water is used sometimes.

**Ultra-low-volume (u.l.v.) spraying.** This technique

consists essentially of the production of very small droplets (c. 70 $\mu$m) by a rotary atomizer, which are carried in a light oil and blown by a fan in drift spraying. Most u.l.v. work is carried out aerially from light aircraft. When applying u.l.v. sprays the pilot flies higher than in conventional spray application (5–10 m instead of 1–2 m), often flying crosswind so as to use the movement of the ambient air to distribute the spray over the crop. The higher altitude increases the spray swath to about three times that of conventional aerial spray swaths (about 30 m as compared to 10 m). However, a recent trend is for aerial u.l.v. spray applications now being used to apply the spray from heights of 2–4 m using spray swaths of 15–25 m, depending on type of aircraft and equipment used; these conditions seem to be about optimum for aircraft u.l.v. applications. Because of the small amount of liquid to be sprayed the total time required for the operation is usually much less than half that for conventional low-volume aircraft spraying.

U.l.v. ground-spraying was introduced quite recently, and now very diversified equipment is being used. The advantages of changing from conventional methods to u.l.v. for ground spraying are less apparent than with aerial spraying, but there are some advantages, namely in *ultra-low dosage* spraying and in cases where transport of water is a problem.

The use of u.l.v. sprays requires fine droplets to ensure adequate cover but not so fine that losses due to drift and evaporation become too great. Recent development of rotary atomizers has ensured the production of a narrow spectrum of droplet size. Several manufacturers are modifying their ground-spraying equipment for u.l.v. spraying.

### Electrostatic spraying

A recent innovation in crop spraying is the electrostatic method developed by Dr R. Coffee of the University of Hong Kong, now with ICI Ltd., England (Coffee, 1973). The basic principles are that the spray or dust particles are electrostatically charged and emitted from the nozzle region in a directed stream along the electrical field flux lines so that they surround the earthed target (crop plant), and the result is usually a very even cover of particles on both sides of the leaves. Since the particles are similarly charged they settle evenly on the leaves due to mutual repulsion and as there is a positive reaction on the tiny particles to the target leaves spray drift is minimized.

The method was initially developed using insecticide dusts but has since been adapted for use with low-volume sprays, and it would seem that this method of insecticide application has great potential commercially, particularly for small farmers. (For further details, see page 114.)

### Dusting

Sometimes it is more convenient to use a dust instead of a spray, the need for water is obviated; the dust may be bought ready for use and is more easy to handle than spray concentrate; the dusting appliances are generally lighter and easier to manipulate than sprayers. For dusting, the active ingredient is diluted with a suitable finely divided 'carrier' powder, such as talc. The dust is usually applied by introduction into the air-stream of a fan or turbine blower. However, in practice, dusts are often not easy to apply; frequently the powder 'cakes', usually through absorption of atmospheric moisture, or 'balls' in the hopper (through static electrification). Also it is difficult to ensure that the dust is homogeneously mixed.

It is generally found that dusting is only practicable in the calmest weather, and that best results are obtained when the dust is applied to wet or dew-covered plants. When dusts are applied to dry foliage usually not more than 10–15% of the applied material sticks to the foliage. Thus there are not many occasions when dusting is a more suitable application method than spraying.

## Fumigation

The toxicity of a gas to a pest is proportional to its concentration and to the time of exposure against that pest. Research into gas properties has shown that usually fumigation is only successful in completely closed spaces or with special precautions to lengthen the time of exposure. For stored products fumigation, the material can either be treated in special chambers or under large gas-proof sheets. Some field crops are treated by drag sheet techniques in which the fumigant is enclosed below a light impervious sheet dragged at a rate dependent on its length behind the vaporizing appliance.

Soil can be fumigated by the injection of volatile liquids directly into the soil at frequent regular intervals. The 'DD' soil injector is used to control nematodes and other soil pests, but it is a tedious process only suitable for relatively small areas.

**Smoke generators** contain a blend of the pesticides and a combustible mixture which burns in a self-sustained reaction at a low temperature, so that the minimum amount of pesticide is destroyed during volatilization, and the finely dispersed pesticide is carried in the cloud of smoke.

**Aerosols** contain the toxicant dissolved in an inert liquid which is gaseous at ordinary temperatures but liquefiable under pressure. When the pressure is released, the solution is discharged through a fine nozzle, the solvent evaporates and the toxicant is dispersed in a very finely-divided state. Methyl chloride, at 5.5 bar, and dichlorodifluoromethane ('Freon') at 6.2 bar at ordinary temperatures are two widely used aerosol solvents.

## Seed dressings

The earliest form of seed dressing was to steep the seed in liquids such as urine or wine. The object of a seed dressing is to protect the seed in the soil and also to protect the seedling for a period after germination. The dressing will form a protective zone around the seed, and the extent of the zone will depend upon whether the pesticide has any fumigant or systemic action. In the past, seed dressings have been used mainly against smuts and other diseases, but now there are many insecticides which can be successfully formulated as seed dressings against insect pests such as wireworms, chafer larvae and shoot flies, and preparations of systemic compounds will protect against aphids and other sap-sucking insects on the young plants. Seed dressings can be liquids which are adsorbed onto the seed coat, or powders which are either sufficiently adhesive to stick directly on to the seed coat or else have to be stuck to the seed with the aid of a 'sticker' (paraffin, or methyl cellulose).

With the advent of precision seed drilling came the development of *pelleted seed*. The object of seed pelleting is to make an irregular-shaped, rough seed into a smooth spherical shape so that it will pass easily through the drill. Also a very small seed (e.g. *Brassica* seed) is given more bulk for easier handling. A few major seed companies are now supplying an ever-increasing range of pelleted seeds. The pellet is composed of inert material (such as powdered clay, pumice or Fuller's earth), and clearly during the pelleting process it is easy to incorporate pesticides into the pellet. The process is quite expensive though, and at times of soil water shortage pelleted seed suffers from impaired germination.

## Granules

A new method of pesticide formulation is that of granules; these are small solid particles and are becoming widely used for the treatment of seedling crops with systemic organophosphorous and carbamate insecticides. The main advantage of granular formulations is that the insecticide can be placed in such a manner that gives maximum protection to the plant, with minimal danger of large-scale soil pollution and negligible danger to the operator. This is of particular importance with highly toxic chemicals. Another major advantage of granules is that the active ingredient is less affected by the soil than would otherwise be the case. Many pesticides are strongly adsorbed in soil and rapidly become ineffective once they reach it. The rate at which the pesticide escapes from granular formulations is mainly controlled by the rate of leaching by rain water. However, the organophosphorus compounds used in granules are generally of low aqueous solubility. Other major factors controlling rate of pesticide release from granules are temperature, dosage, and size of granule. The six major organophosphorous insecticides at present formulated as granules are aldicarb, dimethoate, phorate, disulfoton, chlorfenvinphos and diazinon, used against fly maggots, beetle larvae, aphids and nematodes. Granules are sometimes applied broadcast, but more typically as row treatments at sowing by bow-wave technique or where labour permits by spot applications round the base of individual plants using hand applicators — 'Rogor' and 'Birlane' applicators are generally available for this purpose. The body of the granule is made of various inert substances; for soil application both phorate and disulfoton granules are made of Fuller's Earth. Also used are coal, rice husks, corn cob grits, gypsum, and other minerals. On occasions it is advantageous to make a foliar application of granules and there is a pumice formulation for this purpose; pumice is more expensive as a base but is lighter, stickier, and more effective for foliar lodging.

### Encapsulation

The most recent development in pesticide formulation is the technique of micro-encapsulation. Work at Rothamsted Research Station, England, has shown that it is possible to encapsulate an insecticide in a non-volatile envelope of cross-linked gelatine in such a way that it is non-toxic by contact, but is toxic to insects ingesting it. By the addition of suitable stickers the formulation can be given considerable

resistance to weathering. This type of formulation would appear to be of great promise for the control of leaf-eating insects in the tropics. The formulation has the advantage of being far safer to handle than other pesticide formulations where very toxic chemicals are being used, and it presents far fewer hazards to beneficial insects such as predators, parasites and pollinators. The capsules are so tiny that the formulation has the appearance of being a slightly coarse powder. However there have been reports that honey bees collected the capsules mistakenly for pollen grains, which are about the same size, and that this caused a number of larval deaths in the hives.

Where a contact kill is required, it is possible to prepare leaking capsules which will release the poison over a period of time. At Rothamsted it was found that under warm conditions a standard wettable powder of DDT lost over 90% from the target area in 35 days, while a leaking capsule formulation lost about 20% over the same period.

### Baits

The use of poison baits in pest control is generally confined to the insect groups Orthoptera (crickets, cockroaches, sometimes locusts), Isoptera (termites) and Hymenoptera (ants), in addition to terrestrial molluscs and vertebrates (especially birds and rodents). These pests are typically gregarious with underground nests. Social insects in nests are particularly difficult to kill by spraying insecticides, and sometimes the nests are hard to locate. Most social insects take food back to the nest where it is shared by trophallaxis amongst other adults and the larvae. In the case of fungus-growing termites (family Termitidae) and ants (tribe Attini), the collected food material is incorporated into the fungus gardens on which the fungus grows and poisons introduced in this manner eventually get distributed throughout the colony.

The earliest poison baits relied upon inorganic stomach poisons such as salts of arsenic, lead, mercury and sodium fluoride, but these have now largely been replaced by certain organochlorine compounds.

Poison baits generally comprise four separate components (Cherrett & Lewis, 1974).

1. *Carrier* or *matrix*; inert material which provides the structure of the bait. This may be an edible or attractive material such as meal (e.g. soyabean, groundnut) or citrus pulp, or a more inert or less attractive substance such as ground rice husks, corn cob grits, clay, vermiculite and gelatin microcapsules.

2. *Attractants*; this may be an integral part of the carrier itself (e.g. soyabean meal, citrus pulp), or may be substances such as sugar, molasses, or vegetable oils added to the inert carrier base. A recent development in leaf-cutting ant baits is the incorporation of a trail pheromone to attract the worker ants to the baits. Obviously any chemical that will induce the insects to pick up, or eat, the bait will make this method of control more successful (see Cherrett & Lewis, 1974).

3. *Toxicant* (*poison*); previously usually a stomach poison, but more recently contact insecticides such as aldrin, dieldrin, and heptachlor are being used.

4. *Additives*; for specific formulation purposes, such as preservatives, materials to bind the bait together in pellets or blocks, and waterproofing agents. The physical properties of baits can be important, especially for use in the tropics with high temperatures, high humidity and torrential monsoon rains. For a broadcast bait to remain effective in the field for several weeks under wet tropical conditions it must not disintegrate in the rain. A standard ant bait is 'Mirex 450', but the compressed pellets apparently break down in heavy rain and become ineffective. Experiments by Cherrett and others have shown that leaf-cutting ant baits composed of citrus pulp with soyabean oil and aldrin can be rendered fairly waterproof by the addition of a hydrophobic surface deposit of siloxane, which prolongs the effective life of the bait under wet conditions.

Baits are often placed by hand, but clearly this is time-consuming and costly at present times, in most countries, and recent trials against leaf-cutting ants in C. America have used aerial application from low altitude with considerable success.

### Systemic pesticides

Certain pesticides are capable of entering the plant body and being translocated to other parts of the plant through either the phloem or xylem systems. These insecticides may be applied as sprays on to the soil, sprays on to the foliage, granules on to the soil, or a foliar application of granules, or in the case of woody plants direct injection into the phloem system can be made using special injectors. Sap-feeding insects, such as aphids, are more readily killed by systemic insecticides than by those with a contact action. Parasites and predators are not affected unless they come into contact with the insecticide, that is if the plants are sprayed. Some of these insecticides are highly poisonous to man, but others are now available (such as malathion, trichlorphon and menazon) which are still effective against the insect pests but with reduced toxicity to man.

A systemic insecticide must persist in the plant body in an active form until the contaminated sap is sucked or eaten by the insect pests. There are, however, no problems of surface adhesion and weathering, although the problem of penetrating the plant has to be overcome. Consequently, these insecticides are sufficiently lipid-soluble to enable them to penetrate the plant cuticle and also soluble enough in water to be easily translocated within the plant body.

They must also resist hydrolysis and enzymatic degradation for a sufficiently lengthy period of time to be effective as a pesticide. Obviously, not many insecticides possess this subtle balance between lipophilic and hydrophilic properties and, of course, non-phytotoxic at insecticidal concentrations.

The term systemic is not always used in the same sense: sometimes the substances taken up by the plant roots are not referred to under this heading.

Some pesticides which can penetrate the plant cuticle and pass through the cells are termed *translaminar* pesticides. These can penetrate leaf cuticle and will pass through the leaf to the other surface: hence pests on the underneath of the leaf can be killed by spraying the pesticide on the upper surface, and it can also kill leaf-miners effectively. These pesticides are not usually transmitted through either phloem or xylem systems but just diffuse through the cells.

## Pesticide persistence

In temperate countries a considerable amount is known about the amount of pesticide reaching the target area, and how long it persists there. The relationship between the amount of pesticide present and the control of the pest obtained is thus fairly well established.

Some work has been done to study the behaviour of pesticides under tropical conditions, but because of the varying climatic conditions from one country to another, and because of the range of crops and the ways in which they are grown, data on this subject are scanty and control measures largely speculative and empirical.

There are also practical application problems in tropical countries due to agronomic methods and environmental conditions. Some of these arise from the practice of growing crops such as coffee and cotton by peasant farmers in small patches, making it impossible or uneconomic to use any but the smallest hand applicators. Even with the crops, such as coffee, which are grown in larger plantations, often the variation in plant growth and pruning methods is such that it is most difficult to spray all parts of the plants effectively.

Apart from difficulties of ensuring that sufficient quantities of the insecticide reach the target area, which may often be much more difficult in tropical areas than in temperate ones, the extent of the differences in the persistence of the deposits may not be sufficiently realized.

**Temperature.** At Rothamsted, experiments have shown that temperatures would be reached on isolated surfaces in the tropics such that no non-systemic insecticide applied in commonly used formulations would persist for more than a few days. Under tropical conditions the majority of even the most persistent insecticides may be lost from the target surface in a few days. For example, dieldrin deposits lost 25% at $20°C$ and 95% at $40°C$, in 24 hours, at the same low wind-speeds (2 mph). However, when the deposit level is very low (about $0.005 \, \mu g/cm^2$ on a glass surface; $0.01-0.02 \, \mu g/cm^2$ on a cotton leaf) it no longer obeys these rules, but remains very firmly bound to the surface for a very long period. These data are likely to apply to insecticides applied as sprays of wettable powders, emulsions, or dusts.

Work done at Beltsville, Md, USA, in 1975 on solar degradation of pesticides in hot weather showed that, after application of dieldrin and heptachlor as sprays on short grassland (5 kg/ha) during hot sunny weather, volatilization was so great that the half-life of dieldrin was 2.7 days and for the heptachlor 1.7 days, and after 30 days there was 10% of the dieldrin left and 4% of the heptachlor.

Most agricultural crops in the tropics are quite exposed to the sun and so the above result can be expected, although for some plantation crops the trees may form a canopy with shade underneath. Dieldrin applied to tree trunks and the underneath of branches for Tsetse Fly control will persist, since it is shaded from direct sunlight, and effectively kill Tsetse resting there for periods of up to four to five months.

Unfortunately, however, not a great deal of research into the persistence of most insecticides has actually been carried out in tropical countries and often spray programmes tend to be too speculative.

**Rain.** Most parts of the wet tropics receive their annual rainfall torrentially and it is expected that insecticide residues on crop plant foliage will be washed off at a far greater rate than similar residues exposed to the more gentle temperate rainfall.

**New recommendations for the tropics.** A recent suggestion from the staff of the Asian Vegetable Research and Development Centre (AVRDC) (reported in *Trop. Pest Man.*, 1981; p. 512) urges that recommendations for pesticide use in the tropics be revised. At present most international pesticide recommendations are made more specifically for temperate situations and then adopted without adaptation to the tropics. After four years of trials in Taiwan, AVRDC staff reconfirmed previous observations that pesticide breakdown under tropical conditions is usually very rapid. The half-life of some persistent chemicals in the surface soil of temperate countries is generally measured in years, whereas in hot humid tropics half-lives may only be a few months. Trials with DDT and dieldrin at 5 kg/ha, and carbofuran, fonofos and phorate at 10 kg/ha (double usual dosage), applied twice yearly, were carried out over a four-year period. Results showed that during the cool winter period there was slight chemical accumulation in the soil but during the hot wet summer the entire spring application was broken down and also part of the winter accumulation. On a year-to-year basis only dieldrin showed any overall accumulation at this dosage, and that was slight; for the other chemicals there was no accumulation at all. Tests apparently showed that the chemicals were actually degraded and were not just leached to a lower level.

### Spray additives

**Spreaders** (sometimes called *wetters* or *surfactants*) are substances added to the spray to reduce the surface tension of the droplets so as to facilitate contact between spray and sprayed surface. Plain water falling on a waxy leaf such as that of a *Brassica* will normally collect in large drops and will then run off, leaving the leaf surface dry. The incorporation of a spreader is now standard production practice in the manufacture of most modern pesticides. For crops with particularly waxy leaves (such as brassicas) or against pests with particularly waxy cuticles (like mealybugs, and Woolly Apple Aphid) it is necessary to add extra spreader to the spray solution. Sometimes when extensive run-off is required to enable the pesticide to penetrate to the lower part of a dense crop, this can be achieved by addition of extra spreader to the spray. From a physical point of view wetting and spreading are not quite the same, but for practical purposes they can be regarded as synonymous. Spreaders and surfactants exist in three forms: non-ionic, anionic and cationic, classified according to their ionizing properties. Surfactants are defined as surface-active components. The non-ionic detergents depend upon a balance between hydrophilic and lipophilic properties throughout the molecule for their wetting properties. The advantages of these substances include the fact that they are incapable of reacting with cations or anions present in other spray components and in hard water, and that they are not hydrolysed in either acidic or alkaline solutions. However, phytotoxicity of supplements has to be taken into account. Anionic spreaders possess a negative charge on the amphipathic ion, and typical examples include soap, sulphated alcohols and sulphonated hydrocarbons. Cationic spreaders carry a positive charge on the amphipathic ion and a negative charge on the gegenion, and examples include the quaternary ammonium and pyridinium salts. The advantage of cationic spreaders is that they cannot react with ions of heavy metals. The incompatibility of anionic and cationic additives must be borne in mind if it is necessary to add several supplements to a spray mixture.

**Dispersants.** Sprays must be of uniform concentration and with suspensions there is always the danger of sedimentation. By the addition of a dispersant (or protective colloid) sedimentation can be effectively delayed. The more effective colloids used as dispersants are the methyl celluloses and the sodium carboxymethyl celluloses.

**Emulsifiers.** These are added to emulsions to modify the properties of the interface between the disperse and continuous phases. Many of the spreaders or surfactants also function as emulsifiers (e.g. soap).

**Penetrants.** Oils may be added to a spray to enable it to penetrate the waxy cuticle of an insect more effectively. Some of the more effective sprays against locusts were solutions of dieldrin in light petroleum oils.

**Humectants.** These are substances added to a spray to delay evaporation of the water carrier, and the more commonly used compounds are glycerol and various glycols. They are more frequently used with herbicides than insecticides.

**Stickers.** Stickers such as methyl cellulose, gelatine, various oils and gums, are used to improve the tenacity of a spray residue on the leaves of the crop. Maximum spray retention is particularly important in the tropics where rainfall is often monsoonal and torrential. Generally a fine particle deposit is more tenacious than one of coarse particles. Spreaders usually enhance the tenacity of a deposit and its retention on the plant, although they may retain their wetting properties and thus cause the deposit to be washed off by rain or dew. Some spreaders break down on drying and form insoluble derivatives and these can greatly enhance retention of the deposit.

**Lacquers.** In order to achieve a slow release of a pesticide in certain locations it is possible to formulate some insecticides into lacquer, varnish or paint. The painted area then releases the insecticide slowly over a lengthy period of time. The insecticides used in this manner were mainly organochlorines and particularly DDT and dieldrin. Incorporation of insecticides into paint is of some value but the lacquers are of limited use owing to the problem of the lacquer remaining after the pesticide has dispersed. This practice is of more use against household and stored products pests than crop pests.

**Waterproofing.** Baits used in the tropics where they are exposed to torrential rainfall are liable to disintegrate in the rain, and hence lose their effectiveness. As mentioned on page 107, Cherrett and others have shown that leaf-cutting ant baits can be rendered sufficiently waterproof by the addition of a hydrophobic surface deposit of siloxane, without reducing the attractiveness of the bait to the foraging ants.

**Synergists.** These are substances which cause a particular pesticide to have an enhanced killing power. They are sometimes called *activators*. The way in which synergists act is not always fully understood, but some operate on a biochemical level inhibiting enzyme systems which would otherwise destroy the toxicant. Usually the synergist itself is not insecticidal. Piperonyl butoxide is a synergist for the pyrethrins and certain carbamates. Some pairs of organophosphorous insecticides have a mutually synergistic action (sometimes called '*potentiation*' — especially in American literature). Other synergists include piprotal, propyl isome, sesamin and sesamex. Many of these are particularly effective on the pyrethrins.

---

# Equipment for application

---

### High- and low-volume spraying
As previously mentioned, the usual amount of carrier liquid (water) for high-volume spraying is

quantities in excess of 400 l/ha, typically 600–1100 l/ha, and occasionally as much as 2200 l/ha.

Whilst low volume spraying uses water volumes in the region of 5–400 l/ha, typically the rates are 15–75 l/ha for aerial application, and 100–200 l/ha for ground application. All spraying systems consist basically of a tank for holding the spray liquid, a device for applying pressure to the liquid, and a nozzle or outlet through which the liquid is forced. Thus the only basic difference between equipment for low volume spraying and that for high-volume is the capacity of the tank. The different types of high- and low-volume sprayers are very numerous but certain types can be categorized as follows. Further details of sprayer types can be obtained from Rose (1963) and a very comprehensive account is given by Matthews (1979).

### Hand sprayers

#### Compression systems

*Atomizers.* These consist of a simple compression cylinder with an inlet at one end for the air and an outlet at the other for the compressed air. A plunger is moved up and down the cylinder to produce the compressed air. There are seldom valves present. The outlet tube is fixed at right-angles to a fine tube leading from the liquid container, and on the compression stroke the air is forced across the open end of the feed tube and creates a vacuum which draws up the spray liquid from the tank. As the liquid is drawn up it is broken up into tiny droplets by the air stream. Hand atomizers are useful for treating individual plants, but they are tiring to operate for long periods. On the more refined atomizers the spray is delivered continuously by means of a pressure build-up system.

*Pneumatic hand sprayers.* These are machines with a tank capacity varying from a half to three litres and the tank acts as a pressure chamber. An air pump is attached to the chamber and it projects inside. The outlet pipe runs from the bottom of the tank and ends in a nozzle externally. Air is pumped into the tank which compresses the liquid and forces it out of the nozzle when the release valve is opened. On release, the spray is forced out by the air pressure in a continuous fine spray. The better machines can deliver a continuous spray for up to about five minutes when fully charged with compressed air. These sprayers are most useful in glasshouses or for treatment of individual bushes under calm conditions. As with the previous type these sprayers use very fine nozzles and as such are more suitable for use with solutions or emulsions than suspensions, which tend to block the aperture.

*Knapsack pneumatic sprayers.* These are basically the same as pneumatic hand sprayers except that they are designed for spraying large quantities of liquid (tank capacity up to 23 litres). The tank is usually carried on the operator's back, suspended on a harness with shoulder straps. The outlet pipe is extended by means of flexible tubing and terminates in some form of hand lance. The lance usually carries from one to four nozzles, and is easily carried in one hand. A hand valve on the lance base controls the flow of liquid. The air pump is operated with the sprayer on the ground, and a high pressure is normally built up, which will last for about ten minutes of operation. These sprayers are manufactured in a wide variety of models, of varying degrees of efficiency. In general they are very useful, especially for the small farmer or for pesticide trial work. They can be very effective for estate work when teams of operators are employed, and individual attention for the plant is required. Since no system of agitation is incorporated, knapsack sprayers are more suitable for use with solutions than with suspended materials. Very long lances can be obtained for orchard and plantation use.

#### Pump systems

*Syringes.* Syringes consist of a cylinder into which the spray liquid is drawn on the return stroke of the plunger, and expelled on the compression stroke. The spray is sucked in through the spray nozzle aperture, or else through a separate ballvalve-controlled inlet near the nozzle. The spray produced is drenching, and the syringe is difficult and tedious to use; but they are useful for spraying small numbers of plants. Most syringes are simple in construction, and will last for years with minimum maintenance.

*Force-pump sprayers.* These are sprayers with a hand-operated pump, with a lance and nozzle outlet and a feed-pipe to draw the spray liquid from a separate container. Although small in size, these sprayers (fitted with a 45 cm double-action pump) can throw a jet of spray up to a height of 12 m. These sprayers are good for spot treatments in orchards, and, provided that the solution is kept stirred, they will spray suspensions as well as solutions and emulsions. This type of sprayer is obviously tiring to use and it is quite difficult to control the rate of application, but due to the double-action hand pump the spray is continuous.

*Stirrup-pump sprayers.* These consist of a double-action pump suspended in a bucket. For support there is a foot stirrup reaching to the ground on the outside. A flexible outlet pipe carries the spray liquid from the pump to the spray lance which may vary in length and arrangement of nozzles. A stirrup-pump sprayer requires two operators, one to hold the lance and direct the spray, the other to stir the solution (if it is a suspension) and to work the pump. They are very useful, all-purpose machines, of robust construction which will withstand hard wear. Bush crops, buildings, and small trees can be easily sprayed with a stirrup-pump sprayer, and it is ideal for team operation. Providing the liquid is kept stirred, then quite coarse suspensions can be sprayed. A large version, mounted on wheels, with a large capacity double-action pump is available for treating larger areas.

*Knapsack sprayers.* These are the all-purpose, very successful, sprayers used throughout the world for spraying pesticides over smaller areas. They consist basically of a spray container which sits comfortably on the back of the operator held by shoulder straps. The pump is of double-action, built either inside or outside the spray container, and is operated by working a lever which projects alongside the operator's body. In some models the pump lever also operates an agitation paddle in the spray tank. The spray liquid is applied through a lance held in the operator's free hand; the lance is connected to the spray tank by a long flexible hose. The tank capacity is usually about 23 litres. Provided that a sufficiently coarse spray nozzle is used, this sprayer can be used with any type of spray. Many of the most recent knapsack sprayers are almost entirely made of plastic, which obviates the problem of metal corrosion by the more corrosive pesticides. With a little practice the rate of spray application can be controlled quite accurately. Knapsack sprayers can be tiring to operate over a long period of time, but they are very versatile, quite robust in construction, very portable, and most useful on small farms or in teams on larger estates, or for pesticide trials.

## Power-operated sprayers

### Compression systems

*Hand guns.* Two types of compression hand-spray guns are made. One type has the spray liquid fed into a pipe through which air from a portable compressor is fed. The other type is similar to the small compressed-air spray that is worked by hand, but the air pump is replaced by an inlet from a portable compressor. Droplet size, and rate of application, can usually be carefully controlled, but the capacity of the spray tank is small and hand guns are only of value where small areas have to be covered with small amounts of spray. Suspensions may block the outlet nozzle, especially if it is of very small aperture. The advantage of these sprayers is that any type of compressor can be used, and they may also be used as paint sprayers.

*Portable sprayers.* Many types of small portable sprayers are manufactured; some can be carried easily by one man and others are larger and mounted on a wheeled chassis. Air is compressed by a small compressor and is forced into the spray container. The container usually holds about 45 litres, is of strong welded construction and is operated at a pressure of about 7 bar. The outlet hose from the spray container may end either in single or multiple lances, or may end in a boom. Provided the spray tank is lined with an anticorrosive material, these sprayers can be used for spraying corrosive liquids, since there is no pump to be corroded. As the compressor can be used for other purposes these machines can be useful to the smaller grower, who of necessity requires versatility in his equipment.

*Large mounted sprayers.* These are basically similar to the smaller portable sprayers, but they have spray tanks of a much larger capacity, hence they require larger compressors. The whole machine is usually pulled by a tractor, and the outlet terminates in a spray boom of varying design, or else in a series of hand lances for the spraying of individual trees. The spray booms may cover 6–9 m or more in a single swath. The better booms can usually be adjusted for height to suit the crop being sprayed. Rates of application can be adjusted by altering the size of the jet aperture in the nozzles. The booms may be positioned vertically for fruit crop spraying. Typically these machines are usually used for low-volume pesticide application, and they may be mounted on aircraft or helicopters for low-volume aerial application.

### Pump systems

*Portable sprayers.* The range of small portable sprayers available is now very extensive. They are all based upon the fundamental units of a power source, a pump and a spray tank. The smallest units consist of a small, simple, double-action reciprocal pump harnessed to a small air-cooled engine upon a framework so that it can be conveniently carried. The largest units may include a tank complete with positive agitator, mounted upon a four-wheeled chassis. The pump outlet pipe may supply a number of hand lances, depending upon the capacity of the pump. The mounted machines may have a small boom mounted so that they can spray ground crops. These sprayers are only really suitable for small areas of orchard or plantation crops, for generally two lances is the effective limit.

*Large sprayers.* These have a range of tank capacity varying from about 180–1800 l and may operate at very high pressures (up to 55 bar) with a high rate of delivery. They are invariably mounted on tractors and often have booms which will deliver from 50–3500 l/ha). Some of the larger booms are built vertically so that the nozzles point upwards, to ensure that the under-leaf is sprayed as well as the top surfaces; these are used particularly for crops such as coffee. These sprayers may be mounted on aircraft for aerial use, usually for low-volume application.

*Low-volume mist blowers.* Low-volume mist blowers and fogging machines are of relatively recent development. Since the water used in high-volume spraying is purely an inert carrier, use is made in the low-volume equipment of air-streams to carry the very finely dispersed spray droplets, and so far less water need be employed. The air-blast systems consist mainly of a series of nozzles which produce a coarse spray through wide apertures which is then broken into fine droplets by a fast-moving air-stream. This basic system has many modifications, mainly in the method in which the liquid is introduced into the air-stream to be broken up. The air-stream is usually produced by a centrifugal fan of large capacity. Mist blowers vary enormously in tank size, from portable machines that can be carried by one man to large, self-mounted, tractor-drawn machines. They can be

113

used to spray in two different methods, either as blast spraying or drift spraying. Blast spraying is usually carried out in orchards, or plantations of such crops as apples, mangoes, coffee or rubber. The air all around each tree is replaced by a mixture of air and spray droplets, and generally a good, even cover is achieved using as little as 100 l/ha. Run-off of insecticide is eliminated, and the small droplets dry out quickly so the chemical becomes firmly attached to the foliage. Drift spraying relies upon the movement of ambient air to carry the insecticide mist through the crop, and hence its use is dependent upon favourable weather conditions. The optimum droplet size is thought to be about $80-120\,\mu m$; very fine droplets either evaporate or are lost in the wind and never settle, and larger droplets settle too rapidly. Mist blowers are expensive to buy, but very efficient, up to 30 ha of orchard can be sprayed by two operators in a day. Drift on to other crops is the main danger when using this type of sprayer.

### Ultra-low-volume spraying

U.l.v. application technique was developed in E. Africa in the control of Desert Locust, shortly after World War II, using solutions of DNOC and dieldrin in diesel oil. Development of u.l.v. for crop spraying started much later. The technique consists essentially of the production of very small droplets (c. $70\,\mu m$) carried in a light oil and blown by a fan in drift spraying.

In the early 1960s much of the development work on u.l.v. was done by the Plant Pest Control Division of USDA in co-operation with American Cyanamid. When applying u.l.v. sprays the pilot generally flew the plane much higher (3–6 m) than in conventional spray application, where the altitude is usually about 1–2 m. The higher altitude increases the spray swath to about three times that of the conventional aerial spray. Generally, with the Cynanamid

u.l.v. method the spray swath was about 30 m compared to 12 m.

The more usual u.l.v. technique now used is to apply the spray from a height of 2–4 m, depending upon the type of aircraft and equipment used. These conditions seem to be about optimum for aircraft u.l.v. applications. Because of the small amount of liquid to be sprayed the total time required for the operation is usually much less than half that for conventional aircraft spraying.

U.l.v. ground spraying was introduced quite recently, and now the very diversified equipment being used makes it virtually impossible for one general technique to be developed. The advantages of changing from conventional methods to u.l.v. for ground spraying are less apparent than with aerial spraying, but there are some advantages, namely in ultra-low-dosage spraying and in cases where transport of water is a problem.

The use of u.l.v. sprays requires fine droplets to ensure adequate cover, but not so fine that losses due to drift and evaporation become too great. The development of rotary atomizers enabled the production of a narrow spectrum of droplet size. The more popular aerial rotary atomizers are the Micronair, the Minispin (developed by the US Department of Agriculture), and Turbair.

**Electrostatic sprayers.** The ICI 'Electrodyne' sprayer is a small, hand-held apparatus with a disposable combined bottle and nozzle ('Bozzle') containing the already formulated insecticide. The efficiency of the application appears to be very high, with minimal drift, but at present only the hand-held sprayer is commercially available; ICI state that they expect it to be widely available in 1982. They hope that tractor-mounted 'Electrodyne' sprayers might be available by 1983/4. For further details see Coffee (1973) or your local ICI representative.

**Droplet size.** The main objective in crop spraying is to

spread the active chemical evenly over all the plant surfaces so that a lethal dose is available for the pest to pick up. When large spray volumes are used the coverage is often far from complete due to the coalescence of drops and run-off. Thus the pesticide should be distributed over the plant surface in spray droplets that are as small as possible to produce a complete coverage. The droplet density required in a given spray operation depends upon various factors: type of pesticide (fungicide or insecticide); mobility of pest to be controlled; mode of action of insecticide (systemic, contact, or stomach poison). Thus when a contact insecticide is sprayed against a sluggish pest, a much higher droplet density is required than for the spraying of a stomach poison for the control of a highly mobile pest. In the case of fungicides an even better coverage is required for effective control.

Generally, the smallest droplet size would be the most effective, but very small droplets do not fall freely often being carried by wind and air currents. It appears that droplets of less that $30 \mu m$ are practically airborne – they will be carried by the surrounding air so that they will not touch large target surfaces such as cotton leaves, but they may settle on small diameter targets such as the needles of coniferous plants, hairs of caterpillars, etc. Thus, for adult mosquito control, the optimum droplet size is $5-25 \mu m$; for tsetse in vegetation, $10-30 \mu m$. For agricultural crop spraying the situation is complicated; for many cotton pests it was found that droplet size of $20-50 \mu m$ was best, but typically many crop areas are relatively small and situated near other crops, which means that the spray must be deposited in a much more limited target area. Consequently the optimum droplet size in many crop-spraying operations will be decided upon mainly by the necessity to avoid spray drift, and thus larger droplet sizes will be used.

In aerial crop spraying for insect control, the optimum droplet size seems to be in the range of $80-120 \mu m$, the lower values being more suitable for the treatment of larger areas only. In ground crop spraying the u.l.v. droplet size will be in the range $60-90 \mu m$, although with systemic insecticides the optimum droplet size may be larger.

*Controlled droplet application* (CDA). This term was coined in 1975 in discussions on herbicide application, in order to emphasize the importance of having a known and uniform size of droplet in sprays of liquid pesticides, but the basic concept had of course been known for a long time. However, awareness of the precise importance of droplet size in pesticide sprays, and methods of achieving both uniform droplet size and pattern is relatively recent (Matthews, 1977; 1979).

**Solvents.** Anyone with an elementary knowledge of applied mathematics will be aware that the volume of a sphere increases according to the cube of its linear measurement (radius or diameter) whereas the surface area increases according to the square. Thus smaller droplets have relatively larger surface areas, which implies that the rate of evaporation of spray droplets is higher with smaller droplets. Consequently, solvents used in u.l.v. formulations must have low evaporation rates, and water should never be used. The more volatile liquids would also lead to evaporation of the solvent in the atomizer and might cause crystallization of the pesticide there. Liquid pesticides are sometimes sprayed undiluted in u.l.v. application, but more often some solvent is required.

The solvent must be non-phytotoxic, of low volatility, of high dissolving power for the pesticide, of low viscosity, and compatible with the pesticide. Not many of the solvents generally available will fit all these categories and so clearly good u.l.v. solvents are difficult to find. Xylene is too volatile, other aromatic hydrocarbons of lower volatility are often highly phytotoxic. Alcohols and ketones show higher phytotoxic effects at lower volatility, and other com-

115

mon solvents show the same variations in properties. Sometimes a mixture of various solvents can be used with reasonable success. In the laboratories of Phillips-Duphar have been developed several compounds called *adjuvants* which dramatically decrease the phytotoxicity of many low-volatile solvents, thus the range of solvents that can be used for u.l.v. formulations has now increased. Conventional emulsifiable concentrates usually have a low flash point, but this seldom represents a fire hazard since the concentrate is mixed with a large volume of water before spraying. However, with u.l.v. techniques the solvent must be of low volatility for with rotary atomizers with electrical systems and many rotating parts the possibility of electrical discharge is always present.

With the use of these adjuvants a wider range of low-volatile solvents can be used for u.l.v. formulations and many of these formulations are now regarded as being special ultra-low-volume (s.u.l.v.) formulations. Their characteristics are high concentration, low volatility, low phytotoxicity, with a much lower viscosity than most u.l.v. preparations, and with a flash point above 75°C.

**Spray residues.** In conventional spraying the spray liquid contains a large amount of water, containing various wetting agents, dispersants and emulsifiers. The pesticide is generally present as a finely dispersed phase, solid in the case of wettable powders, or liquid when using emulsions. In u.l.v. sprays the pesticide is generally present as a true solution in an oil carrier, or sometimes the technical material is sprayed as such. Such differences in the spray make-up will have a profound effect on the behaviour of the spray droplets on the biological target.

After deposition of a spray droplet on a leaf, the droplet will assume a certain shape by spreading over the leaf surface. With conventional aqueous spray liquids this spreading depends greatly on the properties of the leaf — on hydrophilic leaves the

droplet will spread to a thin film, but on hydrophobic leaves the droplets usually retain their spherical shape. The spreading of oils on most smooth leaf surfaces is much better than that of aqueous solutions. The s.u.l.v. formulations spread to a thin film on most leaves, even hydrophobic leaves, with very little run-off. However, there are times when run-off is wanted, and in these cases u.l.v. application is not the most suitable method of application.

The formation of a residue from an emulsion droplet is very complicated; evaporation of the water, breaking of the emulsion, and crystallization of the pesticide from the oil phase can occur simultaneously. Often from u.l.v. sprays the crystalline residue on the leaves is particularly coherent and very resistant to dislocation, and also the rain-fastness of u.l.v. residues seems to be better than that of conventional formulations.

The real establishment of u.l.v. techniques for crop spraying was in 1963 with the good results obtained with malathion for the control of cotton insects. From that time a great many papers have been published on this topic, and a large number of pests have been successfully controlled on a wide range of crops. Further information on u.l.v. techniques can be found in Maas (1971).

### Dusters

All dusters consist basically of a hopper (a container for the dust), a system of agitation to disturb the dust and a feed mechanism to pass the dust into a current of air which is carried through an outlet as a turbulent cloud.

Hand dusters are usually primitive in structure and tedious to use, but can be effective for the small farmer. They are often a crude distributing arrangement built as part of the packaging — either two concentric cylinders free to move within each other, or a cardboard piston or diaphragm for pumping the air. As the air passes through the pack it picks up a small quantity of dust and ejects it through a nozzle.

Hand pump dusters are cheap and easy to operate but do not have very much control over the amount of dust delivered. More advanced pumps have a double-action plunger which maintains a constant and even stream of air.

Bellows-type dusters generate the air-stream by the contraction and expansion of a pair of bellows. The commonest types are worn on the back in knapsack fashion. The dust hopper (containing from 3.5—7 kg of dust) is carried on the back in a metal cradle; the bellows are situated either on the back or top of the hopper and drive a stream of air through a tube into a small mixing chamber by the hopper outlet. The dust is fed into the mixing chamber by a simple agitator, and then the air and dust travel along a flexible pipe running along the side of the operator, and can be controlled easily with one hand. The outlet may vary in shape and design for different purposes. The bellows and agitator are operated by a simple up-and-down movement of a lever worked by the free hand.

Rotary hand dusters produce an air-stream by a fan driven off a hand crank through a reduction gear. They are often mounted either on the chest of the operator or on his back, using a metal frame and system of straps.

Power dusters are manufactured in a variety of forms. Traction dusters derive their motive power from the turning of the land wheels — either as a wheelbarrow type or a two-wheeled trailer type. Wheelbarrow-type dusters are suitable for use on small areas only, whereas the trailer type is usually pulled by a tractor and has a much greater capacity. Power dusters are equipped with independent engines to provide the power for their operation. The smallest types can be strapped to the chest of the operator, but the larger types become very expensive. The larger power boom dusters are effective for treatment of large areas. Some of these dusters work on the drift principle and use the movement of ambient air to distribute the powder over the crop. Aeroplane dusting is carried out in areas where the terrain and crops are suitable, and again the air-stream is used to spread the powder over the crop.

### Filters and nozzles

Two components of sprayers of particular importance are filters and nozzles. All spraying machines are equipped with a series of filters to ensure that no coarse particles are permitted to pass into the feed-pipe and block the nozzles. Filters are vitally important, particularly since the mixing of the spray and the filling of the tank take place in the field — without the main tank filter the spray would frequently become contaminated with insect bodies, leaves, pieces of grass, and other detritus, which would clog the nozzles and ruin the spray programme. At various points in the pipe system additional filters may be placed and the nozzles themselves may also be fitted with gauze filters in various positions according to their design.

Nozzles are the detachable apertures which break up the pesticide liquid into spray. Many different types of nozzles are made, and each gives its own particular spray pattern. The nozzle disc controls the final shape of the spray pattern — in *cone nozzles* (which have a spray pattern in the form of a hollow or solid cone) the disc is perforated by a series of small holes. In *fan nozzles* the perforation is an elongate, horizontal slit. Discs are often made either of tungsten steel or more recently in ceramics of various sorts — these materials being less subject to abrasion and wear. Cheap nozzles made of inferior materials soon abrade and then as the apertures enlarge the rate of spray application increases. Fan nozzles are generally less subject to wear than cone nozzles, but cone nozzles give a better breakdown of the water into droplets. The size of the apertures on the disc control the rate of spray delivery, but below a certain diameter the size does not greatly affect the size of droplet produced. Some nozzles can be adjusted to obtain droplet sizes from a fine mist to a heavy drenching spray. Pesticides in suspension have to be used with particular care for nozzles with small apertures can be easily blocked by them. For many of the better makes of sprayers a range of nozzles is available so that the most suitable nozzle for a particular purpose can be used. This is of great importance in spraying field trials because at the same rate of operation (i.e. pumping and walking with a knapsack sprayer) use of the different nozzles will give spray delivery rates in either the high-volume or low-volume ranges, according to the nozzle used. Using a very fine nozzle (low-volume) to put on a high-volume spray is exhausting to the operator, and conversely with a coarse nozzle it is almost impossible to apply a low-volume spray successfully.

## Pesticides in current use

The data presented here are mostly derived from Martin (1972a), the 6th ed. by Worthing (1979), MAFF (1981) and from the pesticide data sheets provided by the various chemical companies (Bayer, Shell, ICI, Ciba-Geigy, etc.).

Insecticides in current use mostly fall into several different categories, according to their chemical structure and mode of action, but some are unfortunately intermediate in chemical structure, or else somewhat anomalous, and do not lend themselves to easy categorization. For teaching purposes it is desirable to be able to distinguish between these basic groups. However, in this book, for practical purposes the following basic types of insecticides are recognized: chlorinated hydrocarbons, sub-stituted phenols, organophosphorous compounds, carbamates, miscellaneous compounds, natural organic compounds, organic oils, biological compounds, and insect growth regulators. Clearly some of these groups could equally well be subdivided and their arrangement here is somewhat arbitrary.

In the following section only a few of the more widely used trade names for the pesticides are given, but a more extensive list of trade names is included in Appendix A.

### Chlorinated hydrocarbons

These compounds are often referred to as the 'organochlorines' and collectively they are a broad spectrum and very persistent group, which usually kill both by contact and as stomach poisons. Generally they are more effective against insects with biting mouthparts than the sap-suckers. Because of their persistence they get taken up in food chains very easily and accumulate in the body fat of the vertebrate predators at the apex of the food chains. Under normal conditions this build-up in the body fat may not affect the animal, but in times of starvation, when body fat reserves are being utilized, the amount of pesticide released into the blood may be critical or even fatal. Most countries in Europe and N. America are in the process of restricting the use of chlorinated hydrocarbons where *suitable alternatives* are available, because of the long-term contamination dangers to the environment. It can also be observed that in most of these countries the major pests have already become, or are now becoming, resistant to the organochlorine compounds, and so their period of utility is ending. In some cases, though, suitable alternatives are not readily forthcoming.

The organochlorine compounds may be subdivided into several groups — the three most commonly used groups are represented by DDT, BHC and aldrin (cyclodienes). Despite the structural differences between the subgroups they do possess several characteristics in common: they are chemically stable, have a low solubility in water, moderate solubility in organic solvents and lipids, and a low vapour pressure. The properties of stability and solubility make most of the group very persistent, and they also produce similar physiological responses in the insects.

The solubility of DDT in lipids enables the poison to penetrate the insect integument quite readily (as opposed to slow penetration through animal skin). This difference in penetrative ability would account for its selective toxicity to insects. Cuticle thickness does not seem to greatly influence the susceptibility of insects to DDT, although it may penetrate more readily through the flexible intersegmental membranes. Dissolution in the epicuticular wax is apparently the essential prerequisite to toxic action. Penetration rate of the poison increases with temperature, often nearly doubling for a $20^{\circ}$C rise in temperature. The precise mode of action of these poisons on the insect is not at all fully understood as yet.

## 1. Aldrin (*Aldrin, Aldrite, Aldrex, Drinox, Toxadrin, etc.*)

*Properties.* A broad-spectrum, persistent, non-systemic, non-phytotoxic insecticide with high contact and stomach activity, effective against soil insects at rates of 0.5–5 kg/ha.

Aldrin is stable to heat, alkali and mild acids, but the unchlorinated ring is attacked by oxidizing agents and strong acids, and it is oxidized to dieldrin. It is compatible with most pesticides and fertilizers, but is corrosive because of the slow formation of HCl on storage.

The technical product is a brown colourless solid practically insoluble in water but is quite soluble in mineral oils, and readily soluble in acetone, benzene, and xylene.

The acute oral $LD_{50}$ for rats is 67 mg/kg; it is absorbed through the skin.

*Use.* Effective against all soil insects, e.g. termites, beetle adults and larvae, fly larvae, cutworms, crickets, etc.

*Caution*
(*a*) Harmful to fish.
(*b*) Treated seeds should not be used for human or animal consumption.
(*c*) Avoid excessive skin contact.
(*d*) Risks to wildlife are considerable on a long-term basis.

*Formulations.* 30% e.c.; 2.5–5% dust; m.l.; 5 and 20% granules; 20–50% w.p.; or liquid seed dressing; and as an insecticidal lacquer.

## 2. BHC, γ (*Lindane, HCH, BHC, etc.*)

*Properties.* Benzene hexachloride exists as five isomers in the technical form but the active ingredient is the γ-isomer. Lindane is required to contain not less than 99% γ-BHC. It exhibits a strong stomach poison action, persistent contact toxicity, and fumigant action, against a wide range of insects. It is non-phytotoxic at insecticidal concentrations. The technical BHC causes 'tainting' of many crops but there is less risk of this with 'Lindane'.

Lindane is stable to air, light, heat and carbon dioxide; unattacked by strong acids but can be dehydrochlorinated by alkalis.

It occurs as colourless crystals and is practically insoluble in water; slightly soluble in petroleum oils; soluble in acetone, aromatic and chlorinated solvents.

The acute oral $LD_{50}$ for rats is 88 mg/kg.

*Use.* Effective against many soil insects, e.g. beetle adults and larvae, fly larvae, ants, Collembola, and also against many other biting and sucking insects, e.g. aphids, psyllids, whiteflies, capsids, midges, sawflies and thrips.

*Caution*
(*a*) Dangerous to bees.
(*b*) Harmful to fish and livestock.

(*c*) Pre-harvest interval about two weeks on most crops.

(*d*) Treated seeds should not be used for human or animal consumption.

*Formulations.* Many different seed dressings, some with added organomercury compounds or with Captan or Thiram; 50% w.p. with bran as bait; dust; liquid e.c., or suspension.

### 3. Chlorobenzilate (*Akar, Folbex*)

*Properties.* A non-systemic acaricide with little insecticidal action, particularly effective against phtyophagous mites. Some phytotoxicity has been noted on pear, plum, and some apple varieties.

The technical product is a brown liquid, insoluble in water, but soluble in most organic solvents including petroleum oils. It is hydrolysed by alkali and strong acids.

The acute oral $LD_{50}$ for rats is from 700 to 3100 mg/kg.

*Use.* Effective only against phytophagous mites, but will kill all active and inactive stages, including eggs.

*Caution.* Pre-harvest interval for edible crops is not known.

*Formulations.* 25% and 50% e.c.; 25% w.p.; and 'Folbex' fumigation strips for smoke generation.

### 4. DDT (*DDT, many trade names*)

*Properties.* A broad-spectrum stomach and contact poison, of high persistence, non-systemic, and non-phytotoxic except to cucurbits.

The $pp'$-isomer forms colourless crystals, practically insoluble in water, moderately soluble in petroleum oils and readily soluble in most aromatic and chlorinated solvents. The technical product is a waxy solid. DDT is dehydrochlorinated at temperatures above $50°C$, a reaction catalysed by u.v. light. In solution it is readily dehydrochlorinated by alkalis or organic bases; otherwise it is stable, being un-

attacked by acid and alkaline permanganate or by aqueous acids and alkalis.

The acute $LD_{50}$ for male rats is 113 mg/kg. DDT is stored in the body fat of birds and mammals and excreted in the milk of mammals.

*Use.* Effective against most insects, but with little action on phytophagous mites.

*Caution*

(*a*) Harmful to bees, fish and livestock.

(*b*) Pre-harvest interval for edible crops — 2 weeks.

(*c*) Do not use on cucurbits or certain barley varieties, as damage may occur.

*Formulations.* 25% e.c.; 25% m.l.; 50% w.p.; 5% dusts; smokes. May be combined with BHC or malathion in sprays or smokes.

### 5. Dicofol (*Kelthane, Acarin, Mitigan*)

*Properties.* It is a non-systemic acaricide, with little insecticidal activity, recommended for the control of mites on a wide range of crops. Although residues in soil decrease rapidly, traces may remain for a year or more. Safe to bees.

The pure compound is a white solid, practically insoluble in water, but soluble in most aliphatic and aromatic solvents. It is hydrolysed by alkali, but is compatible with all but highly alkaline pesticides. Wettable powder formulations are sensitive to solvents and surfactants, which may affect acaricidal activity and phytotoxicity.

The acute oral $LD_{50}$ for male rats is $809 \pm 33$ mg/kg.

*Use.* Effective only against Acarina; kills eggs and all active stages of the mites.

*Caution.* Pre-harvest interval for edible crops — 2 to 7 days.

*Formulation.* 18.5% and 42% e.c.; 30% dust.

### 6. Dieldrin (*Dieldrex, Alvit, Dilstan etc.*)

*Properties.* A broad-spectrum, persistent, non-systemic, non-phytotoxic insecticide made by oxidation of aldrin. It is of high contact and stomach activity.

It is stable to light, alkali and mild acids, and is compatible with most other pesticides. Dieldrin occurs as white odourless crystals; the technical product is in light brown flakes; practically insoluble in water, slightly soluble in petroleum oils, moderately soluble in acetone, soluble in aromatic solvents.

The acute oral $LD_{50}$ for male rats is 46 mg/kg; it can be absorbed through the skin.

*Use.* Effective against most insects.

*Caution*
(*a*) Harmful to fish.
(*b*) Dangerous to bees.
(*c*) Seed dressings containing organomercury compounds can cause rashes or blisters on the skin. Those containing thiram can be irritating to skin, eyes, nose and mouth.
(*d*) Treated seed should not be used for human and animal consumption.
(*e*) Dressed seeds are dangerous to birds.

*Formulations.* 15% e.c; m.l.; 50% w.p.; dry seed dressings. Some seed dressings made with thiram or organomercuric compounds.

### 7. Endosulfan (*Thiodan, Cyclodan* etc.)

*Properties.* It is a non-systemic contact and stomach insecticide and acaricide.

The technical product is a brownish crystalline solid, practically insoluble in water, but moderately soluble in most organic solvents. It is a mixture of two isomers. It is stable to sunlight but subject to a slow hydrolysis to the alcohol and sulphur dioxide. It is compatible with non-alkaline pesticides.

The acute oral $LD_{50}$ for rats is 110 (55–220) mg/kg.

*Use.* Effective against most crop mites, and some Hemiptera (aphids, capsids) and beetles.

*Caution*
(*a*) This is a poisonous substance (Part II) – full protective clothing should be worn.
(*b*) Extremely dangerous to fish.
(*c*) Dangerous to livestock.
(*d*) Harmful to bees.
(*e*) Pre-harvest interval for fruit – 6 weeks.
(*f*) Pre-access interval to treated areas: of unprotected persons – 1 day; animals and poultry – 3 weeks.

*Formulations.* 17.5% and 35% e.c.; 17.5%, 35% and 50% w.p.; 1%, 3%, 4% and 5% dusts, and 5% granules.

### 8. Endrin (*Endrex, Hexadrin, Mendrin*)

*Properties.* A very toxic, broad-spectrum, persistent insecticide and acaricide isomeric with dieldrin. It is non-systemic, and non-phytotoxic at insecticidal concentrations, but is suspected of damage to maize.

It is a white crystalline solid practically insoluble in water, sparingly soluble in alcohols and petroleum oils, moderately soluble in benzene and acetone. The technical product is a light brown powder of not less than 85% purity. It is stable to alkali and acids but strong acids or heating above 200°C cause a rearrangement to a less insecticidal derivative. It is compatible with other pesticides.

The acute oral $LD_{50}$ for male rats is 17.5 mg/kg.

*Use.* Effective against many insects and mites – used mainly on field crops.

*Caution*
(*a*) A very poisonous pesticide (Part II) – full protective clothing should be worn.
(*b*) Dangerous to bees, fish, livestock, wild birds and animals.
(*c*) Fruit should not be sprayed after flowering.
(*d*) Minimum interval to be observed between last application and access to treated areas: of unprotected persons – 1 day; of animals and poultry – 3 weeks!

*Formulations.* 20% liquid formulation, or dust.

## 9. Heptachlor (*Drinox, Heptamul, Velsicol*)

*Properties.* A broad-spectrum, non-systemic, stomach and contact insecticide with some fumigant action.

It is a white crystalline solid, practically insoluble in water, slightly soluble in alcohol, more so in kerosene; it is stable to light, moisture, air and moderate heat. It is compatible with most pesticides and fertilisers.

The acute oral $LD_{50}$ for male rats is 100 mg/kg.

*Use.* Effective against many different insect species; used mostly against soil-inhabiting fly and beetle larvae, and in ant baits.

*Caution*

*Formulations.* As seed dressings; e.c.; w.p.; dusts and granules of various a.i. contents.

## 10. Mirex (*Mirex, Dechlorane*)

*Properties.* It is a stomach insecticide, with little contact effect, used mainly against ants.

It is a white solid of negligible volatility; insoluble in water but moderately soluble in benzene, carbon tetrachloride and xylene; unaffected by concentrated mineral acids.

Acute $LD_{50}$ for male rats is 306 mg/kg.

*Use.* Mostly used in baits against Fire Ants, Harvester Ants and Leaf-cutting Ants.

## 11. Tetradifon (*Tedion, Duphar*)

*Properties.* A systemic acaricide toxic to the eggs and all stages of phytophagous mites except adults. At acaricidal concentrations it is non-phytotoxic.

It forms colourless crystals, almost insoluble in water, slightly soluble in alcohols and acetone, more soluble in aromatic hydrocarbons and chloroform. It is resistant to hydrolysis by acid or alkali, is compatible with other pesticides, and is non-corrosive.

The acute oral $LD_{50}$ for rats is more than 5000 mg/kg.

*Use.* Effective against eggs, larvae, nymphs of phytophagous mites, but not adults. Recommended for application to top fruit, citrus, tea, cotton, grapes, vegetables, ornamentals and nursery stock.

*Caution*

(*a*) Do not use smokes to young cucumbers or plants that are wet or damage may occur.

(*b*) At correct dosage is harmless to beneficial insects.

(*c*) Safe to bees.

*Formulations.* 20% w.p.; 18% e.c.; may be combined with malathion in smoke foundations.

## 12. Tetrasul (*Animert V-101*)

*Properties.* A non-systemic acaricide, highly toxic to eggs and all stages of phytophagous mites except adults. At the correct dosage it is non-phytotoxic. As it is highly selective it does not pose a hazard to beneficial insects or to wild life.

It is a brown crystalline solid, only slightly soluble in water, moderately soluble in acetone and ether, but soluble in benzene and chloroform. Stable under normal conditions, but should be protected against prolonged exposure to sunlight; it is oxidized to its sulphone, tetradifon. It is non-corrosive, and compatible with most other pesticides.

The acute oral $LD_{50}$ for female rats is 6810 mg/kg.

*Use.* Effective against eggs, larvae and nymphs of most phytophagous mites, but not adults. Recommended for use on fruit and cucurbits at the time when the winter eggs are hatching.

*Caution*

*Formulations.* 18% e.c. and 18% w.p.

## 13. TDE (*DDD, Rhothane*)

*Properties.* It is a non-systemic contact and stomach insecticide which, not of the general high potency of DDT, is of equal and greater potency against certain insects, e.g. leaf-rollers, mosquito larvae, hornworms.

It is non-phytotoxic except possibly to cucurbits.

The pure compound forms colourless crystals practically insoluble in water but soluble in most aliphatic and aromatic compounds. Its chemical properties resemble those of DDT but it is more slowly hydrolysed by alkali.

The acute oral $LD_{50}$ for rats is 3400 mg/kg.

*Use.* Particularly effective against leaf-rollers, mosquito larvae, hornworms — effective against many caterpillars, weevils, capsids, thrips and earwigs.

*Caution*
(*a*) Harmful to bees, fish and livestock.
(*b*) Pre-harvest interval for edible crops — 2 weeks.
(*c*) Pre-access interval for livestock to treated areas — 2 weeks.

*Formulations.* 50% w.p.; e.c. 25%; 5% and 10% dusts.

## 14. Amitraz (*Mitac, Taktic, Triatox, Azaform, Baam*)

*Properties.* An acaricide effective against a wide range of phytophagous mites; all stages are susceptible. It has some insecticidal properties, and is effective against some Hemiptera and eggs of Lepidoptera. Relatively non-toxic to predaceous insects and bees.

It is insoluble in water, but soluble in methyl benzene and propanone. At acid pHs it is unstable, and will slowly deteriorate under moist conditions; generally compatible with most commonly used pesticides.

The acute oral $LD_{50}$ for rats is 800 mg/kg.

*Use.* Effective against phytophagous mites, especially red spider mites, at concentrations from 20–50 g a.i./ 100 l, depending upon the species. Also effective against eggs of *Heliothis* spp., and used against ticks and mites on cattle and sheep.

*Caution*
(*a*) This is a poisonous substance (Part II) — protective clothing should be worn.
(*b*) Pre-harvest interval for fruit is 2 weeks.
(*c*) Harmful to fish.

*Formulations.* e.c. 200 g a.i./l; d.p. 250 and 500 g a.i./kg.

### Substituted phenols

The nitrophenols are not likely to compete with the newer insecticides; they are used mainly as herbicides and for the control of powdery mildews. The dinitrophenols have mammalian toxicity so high as to restrict their usefulness in crop protection.

## 15. Binapacryl (*Morocide, Acaricide, Dapacryl*)

*Properties.* It is a non-systemic acaricide (also effective against powdery mildews) and mainly used against red spider mites. Non-phytotoxic to a wide range of apples, pears, cotton and citrus; some risks of damage to young tomatoes, grapes and roses.

It is a white crystalline powder practically insoluble in water but soluble in most organic solvents. It is unstable in concentrated acids and dilute alkalis, suffers slight hydrolysis on long contact with water and is slowly decomposed by u.v. light. It is non-corrosive, and compatible with w.p. formulations of insecticides and non-alkaline fungicides. With organophosphorous compounds it may be phytotoxic.

Acute oral $LD_{50}$ for rats is 120–165 mg/kg.

*Use.* Effective especially against red spider mites.

*Caution*
(*a*) Harmful to fish and livestock, but safe to bees.
(*b*) Pre-harvest interval for edible crops — 1 week.
(*c*) Pre-access interval for livestock to treated areas — 4 weeks.

*Formulations.* 25% and 50% w.p.; 40% e.c.; 4% dust.

## 16. DNOC in petroleum oil (*Sinox, Dinotrol*)

*Properties.* It is a non-systemic stomach poison and contact insecticide; ovicidal to the eggs of certain insects. It is strongly phytotoxic and its use as an insecticide is limited to dormant sprays or on waste ground, e.g. against locusts. It is also used (not usually in oil) as a contact herbicide for the control of broad-leaved weeds in cereals, and in e.c. formulations for the pre-harvest desiccation of potatoes and leguminous seed crops. It forms yellowish, odourless crystals, only sparingly soluble in water, but soluble in most organic solvents and in acetic acid. The alkali salts are water soluble. It is explosive, and is usually moistened with up to 10% water to reduce the hazard, though it is corrosive to mild steel in the presence of water.

*Use.* Effective against overwintering stages of aphids, capsids, psyllids, scale insects, red spider mites, and various Lepidoptera on top, bush and cane fruit. Products containing DDT also control various weevils and tortrix moths in their overwintering stages. Also used as a herbicide.

*Caution*

(*a*) If the concentrated substance contains more than 5% of DNOC it is a Part II substance, and protective clothing should be worn.

(*b*) Dangerous to fish.

(*c*) Very phytotoxic; use only as dormant spray or on waste ground.

*Formulations.* The insecticide formulation is an e.c. in petroleum oil; it may be formulated with DDT.


## 17. Pentachlorophenol (*Dowicide, Santophen,* etc.)

*Properties.* An insecticide used for termite control, a fungicide used for protection of timber from fungal rots and wood-boring insects. It is strongly phytotoxic, and is used as a pre-harvest defoliant and as a general herbicide.

It forms colourless crystals; volatile in steam; almost insoluble in water; soluble in most organic solvents. It is non-corrosive in the absence of moisture, solutions in oil cause deterioration of natural rubber but synthetic rubbers may be used in equipment and protective clothing.

The acute oral $LD_{50}$ for rats is 210 mg/kg; it irritates mucous membranes and causes sneezing, the solid and aqueous solutions stronger than 1% cause skin irritation.

*Use.* Effective against termites and other wood boring insects; use restricted by strong phytotoxicity.

*Caution*

(*a*) Very phytotoxic.

(*b*) Irritation to skin and mucous membranes.

*Formulations.* Used as such or formulated in oil. *Santobrite* and *Dowicide G* are the technical sodium salt.


### Organophosphorous compounds

These were discovered and developed during the Second World War by a German research team responsible for developing nerve gases; they are amongst the most toxic substances known to man.

These compounds have phosphorus chemically bonded to the carbon atoms of organic radicals, and are effective as both contact and systemic insecticides and acaricides. Nearly 50 compounds are in current use against insects and mites. Many of these compounds are very toxic to mammals and birds and have to be handled with care. Doses may be accumulative. The systemic compounds are very effective against sap-sucking insects. All the organophosphorous compounds are relatively transient and are soon broken down to become non-toxic. In comparison with the persistent chlorinated hydrocarbons great care in timing and application is required in the use of these compounds for effective results. The mode of action in both insects and mammals appears to be inhibition of acetyl cholinesterase (or some very similar enzyme).


## 18. Azinphos-methyl (*Gusathion, Benthion*)

*Properties.* A non-systemic, broad-spectrum, insecticide and acaricide of relatively long persistence, with contact and stomach action. It forms white crystals, almost insoluble in water but soluble in most organic solvents. It is unstable at temperatures above $200°C$ and is rapidly hydrolysed by cold alkali and acid.

The acute oral $LD_{50}$ for rats is 16.4 mg/kg.

*Use.* Effective against Lepidoptera, mites, aphids, whiteflies, leafhoppers, scales, psyllids, sawflies, thrips, grasshoppers, some fly larvae and some beetles.

*Caution*

(*a*) A poisonous substance (Part III) — protective clothing should be worn.

(*b*) Dangerous to bees.

(*c*) Harmful to fish and livestock.

(*d*) Pre-harvest interval for edible crops — 1 to 4 weeks according to crop, and country.

(*e*) Pre-access interval for livestock to treated area — 2 weeks.

*Formulations.* 20% e.c.; 25% and 50% w.p.; 2.5% and 5% dusts; u.l.v. formulations. A formulation with demeton-*S*-methyl sulphone known as *Gusathion MS* has a more systemic action.

### 19. Azinphos-methyl with demeton-*S*-methyl sulphone (*Gusathion MS*)

*Properties.* A mixture with *Metasystox* combining the properties of both pesticides — usually the proportion is 75% to 25%. In practice the mixture acts like azinphos-methyl with a systemic action.

*Use.* Also effective against thrips, aphids, and fly larvae (midges).

### 20. Bromophos (*Brofene, Bromovur, Nexicon, Pluridox*)

*Properties.* A broad-spectrum, contact and stomach insecticide: persists on sprayed foliage for 7—10 days. Non-phytotoxic at insecticidal concentrations, but suspected of damage to plants under glass. No systemic action.

It occurs as yellow crystals, relatively insoluble in water, but soluble in most organic solvents, particularly in tetrachloromethane, diethyl ether, and

methyl benzene. It is stable in media up to pH 9, non-corrosive, and compatible with all pesticides except sulphur and the organometal fungicides.

The acute oral $LD_{50}$ for rats is 3750—7700 mg/kg.

*Use.* Effective against a wide range of insects on crops, at concentrations of 25—75 mg/100 l.

*Caution*

(*a*) Harmful to fish and bees.

(*b*) Possibly phytotoxic under glass.

*Formulations.* e.c. 250, 400 g a.i./l; w.p. 250 g a.i./kg; dusts 20—50 g a.i./kg; atomising concentrate 400 g a.i./l; coarse powder 30 g a.i./kg; granules 50—100 g a.i./kg.

### 21. Bromopropylate (*Acarol, Neoreon*)

*Properties.* A contact acaricide with residual action, effective against mites resistant to organophosphorous compounds.

It is a crystalline solid, only slightly soluble in water, but is readily soluble in most organic solvents.

The acute oral $LD_{50}$ for rats was more than 5000 mg/kg.

*Use.* Effective against all phytophagous mites on crops; recommended for use on fruits, vegetables, cotton, flowers at 37.5—60 g a.i./100 l; on field crops at 0.5—1.0 kg/ha.

*Caution*

(*a*) Dangerous to fish.

(*b*) Harmful to birds.

*Formulations.* e.c. at 250 and 500 g a.i./l.

### 22. Carbophenothion (*Trithion, Garrathion, Dagadip*)

*Properties.* A non-systemic acaricide and insecticide, with a long residual action. It is phytotoxic at high concentrations to some plants. 50% degradation in soil occurs in 100 days or longer, depending upon the soil type.

It is a pale amber liquid; insoluble in water, but

miscible with most organic solvents. Relatively stable to hydrolysis; is oxidized on the leaf surface to the phosphorothiolate; compatible with most pesticides, and is non-corrosive to mild steel.

The acute oral $LD_{50}$ for male rats is 32.2 mg/kg.

*Use.* Used mainly on deciduous fruit, in combination with petroleum oil, as a dormant spray for the control of overwintering mites, aphids and scale insects; on citrus as an acaricide.

*Caution.* Harmful to livestock, birds and wild animals.

*Formulations.* E.c. 0.25, 0.5, 0.75, 1.0 kg/l; 25% w.p.; dusts of 1, 2, and 3%.

## 23. Chlorfenvinphos (*Birlane, Sapecron, Supona*)

*Properties.* A relatively short-lived insecticide, effective against soil insects, non-phytotoxic at recommended dosages. The pure compound is an amber-coloured liquid, sparingly soluble in water but miscible with acetone, ethanol, kerosene and xylene. It is stable when stored in glass or polythene vessels, but is slowly hydrolysed by water. It may corrode iron and brass on prolonged contact and the e.c. formulations are corrosive to tin plate.

Acute oral $LD_{50}$ for rats is 10–39 mg/kg.

*Use.* Particularly effective against rootflies, rootworms and cutworms as soil applications. As a foliage insecticide it is recommended for the control of Colorado Beetle on potato, leafhoppers on rice, and for stem borers on maize, sugarcane and rice. The half-life in soil is normally only a few weeks.

*Caution*
(*a*) This is a poisonous substance (Part III) – protective clothing should be worn.
(*b*) Dangerous to fish.
(*c*) Use seed dressings carefully to avoid risks to birds.
(*d*) Treated seed should not be used for human or animal consumption.
(*e*) Pre-harvest interval for edible crops – 3 weeks.

126

*Formulations.* 24% e.c.; 25% w.p.; 5% dust; 10% granules; seed dressings (liquid) 40% (+ 2% mercury compounds).

## 24. Chlorpyrifos (*Dursban, Lorsban*)

*Properties.* A broad-spectrum insecticide with contact, stomach and vapour action. No systemic action. At insecticidal concentrations it is non-phytotoxic. It is sufficiently volatile to make insecticidal deposits on nearby untreated surfaces. In soil it persists for 2–4 months.

It forms as white crystals; insoluble in water, but soluble in methanol and most other organic solvents. Stable under normal storage conditions. It is compatible with non-alkaline pesticides; corrosive to copper and brass.

The acute oral $LD_{50}$ for male rats is 163 mg/kg.

*Use.* Effective against many soil and foliar insects, and mite pests. Also used domestically against flies, mosquitoes and household pests, as well as ecto-parasites of sheep and cattle. Specifically used for control of root maggots, aphids, capsid bugs, caterpillars and red spider mites.

*Caution*
(*a*) Dangerous to bees, fish and shrimps.
(*b*) Pre-harvest interval for all edible crops is 2–6 weeks.

*Formulations.* w.p. 25%; e.c.; 0.2 and 0.4 kg/l; granules 1–10%.

## 25. Demephion (*Cymetox, Pyracide*)

*Properties.* A systemic insecticide and acaricide effective against sap-feeding insects, and non-phytotoxic to most crops.

It is a straw-coloured liquid (a mixture of two isomers), miscible with most aromatic solvents, chlorobenzene and ketones; immiscible with most aliphatic solvents. It is generally non-corrosive and compatible with most, except strongly alkaline, pesticides.

The acute oral $LD_{50}$ for rats is about 0.015 ml/kg; the acute dermal $LD_{50}$ is about 0.06 ml/kg.

*Use.* Mainly used against aphids — on all crops.

*Caution*

(*a*) This is a poisonous substance (Part III) — protective clothing should be worn.

(*b*) Harmful to bees, livestock, fish, game, wild birds and animals.

(*c*) Pre-harvest interval for edible crops — 3 weeks.

(*d*) Pre-access interval for livestock to treated areas — 2 weeks.

*Formulations.* 30% e.c.

## 26. Demeton (*Systox, Solvirex*)

*Properties.* A systemic insecticide and acaricide with some fumigant action, effective especially against sap-sucking insects and mites. No marked phytotoxicity has been recorded.

The technical product is a light yellow oil, hydrolysed by strong alkali, but is compatible with most non-alkaline pesticides. It is almost insoluble in water but is soluble in most organic solvents.

The acute oral $LD_{50}$ for male rats is 30 mg/kg.

*Use.* Effective against sap-sucking insects and mites.

*Caution*

(*a*) Harmful to bees.

(*b*) Harmful to fish and livestock.

(*c*) Pre-harvest interval is not known.

*Formulations.* E.c. of different oil contents.

## 27. Demeton-S-methyl (*Metasystox 55, Demetox*)

*Properties.* A systemic and contact insecticide and acaricide, metabolized in the plant to the sulphoxide and sulphone; rapid in action; moderate persistence.

It is a colourless oil, only slightly soluble in water, soluble in most organic compounds. It is hydrolysed by alkali. The acute oral $LD_{50}$ for rats is 65 mg/kg of the technical material, 40 mg/kg of the pure.

*Use.* Effective against most sap-sucking pests (aphids, leafhoppers, etc.), sawflies and red spider mites.

*Caution*

(*a*) This is a poisonous substance (Part III) — protective clothing should be worn.

(*b*) Certain ornamentals, especially some chrysanthemums, may be damaged by sprays.

(*c*) Harmful to bees, fish, livestock, game, wild birds and animals.

(*d*) Pre-harvest interval for edible crops — 2 to 3 weeks.

(*e*) Pre-access interval for livestock to treated areas — 2 weeks.

*Formulations.* 25 and 50% e.c. with emulsifier chosen to reduce dermal hazards.

## 28. Diazinon (*Basudin, Diazitol, DBD, Neacide*)

*Properties.* A non-systemic insecticide with some acaricidal action, used mainly against flies, both in agriculture and veterinary practice. At higher dosages it may be phytotoxic.

It is a colourless oil almost insoluble in water, but is miscible with ethanol, acetone, xylene, and is soluble in petroleum oils. It decomposes above $120\degree C$ and is susceptible to oxidation; stable in alkaline media, but is slowly hydrolysed by water and dilute acids. It is compatible with most pesticides but should not be compounded with copper fungicides.

The acute oral $LD_{50}$ for male rats is 108 mg/kg.

*Use.* Especially effective against flies and their larvae (e.g. Anthomyiidae on vegetables and Carrot Fly), also used against mites, thrips, springtails, glasshouse pests, and some bugs (aphids, capsids, etc.).

*Caution*

(*a*) Dangerous to bees.

(*b*) Harmful to fish, livestock, game, wild birds and animals.

(*c*) Overdosage may lead to phytotoxicity on some crops.

127

(d) Pre-harvest interval for edible crops —usually 2 weeks.

(e) Pre-access interval for livestock to treated areas — 2 weeks.

*Formulations.* Aerosol solutions; 25% e.c.; 40 and 25% w.p.; 4% dust; 5% granules.

## 29. Dichlorvos (*Vapona, Nogos, Oko, Mafu, Dedevap, Nuvan*)

*Properties.* A short-lived, wide-spectrum, contact and stomach insecticide with fumigant and penetrant action, non-phytotoxic. It is used as a household and public health fumigant, especially against mosquitoes and other Diptera, in addition to crop protection uses.

It is little soluble in water but is miscible with most organic solvents and aerosol propellants. It is a colourless to amber liquid, stable to heat, but is hydrolysed; corrosive to iron and mild steel but non-corrosive to stainless steel and aluminium.

The acute oral $LD_{50}$ for male rats is 80 mg/kg.

*Use.* Especially effective against flies; often used for glasshouse fumigation — kills most glasshouse pests. Also used on outdoor fruit and vegetables where rapid kill is required close to harvest. Will kill sap-sucking and leaf-mining insects also.

*Caution*

(a) A poisonous substance (Part III) requiring protective clothing to be worn.

(b) Dangerous to bees.

(c) Pre-harvest interval — 1 day.

(d) Pre-access interval to treated areas — 12 hours.

*Formulations.* 50 and 100% e.c.; 'Vapona Pest Strip'; 0.4 to 1.0% aerosols; 0.5% granules.

## 30. Dimefox (*Terra-sytam, Hanane*)

*Properties.* A systemic insecticide and acaricide of very high toxicity, used mainly for soil treatment of

128

hops against aphids and red spider mites; non-phytotoxic at insecticidal concentrations.

It is a colourless liquid, miscible with water and most organic solvents. It is resistant to hydrolysis by alkali but is hydrolysed by acids, slowly oxidized by vigorous oxidizing agents, rapidly by chlorine. Hence, for decontamination, treat with acids followed by bleaching powder. It is compatible with other pesticides, but the technical product slowly attacks metals.

The acute oral $LD_{50}$ for rats is 1—2 mg/kg; the acute dermal $LD_{50}$ for rats is 5 mg/kg; the hazards of vapour toxicity are high.

*Use.* Effective against sap-sucking insects (aphids) and mites, but toxicity hazards are high.

*Caution*

(a) This is a very poisonous substance (Part I) — full protective clothing must be worn.

(b) Dangerous to fish, livestock, game, wild birds and animals.

(c) Pre-harvest interval for picking hops — 4 weeks.

(d) Pre-access interval to treated areas, of unprotected persons — 1 day; of livestock and poultry — 4 weeks.

*Formulations.* *Terra-systam* is a 50% w.v. solution.

## 31. Dimethoate (*Rogor, Roxion, Cygon, Dantox*)

*Properties.* A systemic and contact insecticide and acaricide, used mainly against fruit flies (Olive Fly and Cherry Fly) and aphids.

The pure compound is a white solid, only slightly soluble in water, soluble in most organic solvents except saturated hydrocarbons such as hexane. It is stable in aqueous solution and to sunlight, but is readily hydrolysed by aqueous alkali. It is incompatible with alkaline pesticides.

The acute oral $LD_{50}$ for rats is 250—65 mg/kg.

*Use.* Effective against aphids, psyllids, some flies sawflies, Woolly Aphid and red spider mites. Mainly used against fruit flies — Olive Fly and Cherry Fly, and aphids.

*Caution*
(*a*) Dangerous to bees.
(*b*) Harmful to fish, livestock, game, wild birds and animals.
(*c*) Do not use on chrysanthemums, hops or on ornamental *Prunus* spp.
(*d*) Pre-harvest interval for edible crops — 1 week.
(*e*) Pre-access interval for livestock to treated areas — 1 week.
*Formulations.* 20 and 40% e.c.; 20% w.p.; 5% granules.

### 32. Disulfoton (*Disyston, Murvin 50, Parsolin*)

*Properties.* A systemic insecticide and acaricide used mainly as a seed dressing or granules to protect seedlings from insect attack. It is metabolized in the plant to the sulphoxide and sulphone. It is a colourless oil with a characteristic odour, only slightly soluble in water, but readily soluble in most organic solvents. It is relatively stable to hydrolysis at pH below 8.0.

The acute oral $LD_{50}$ for male rats is 12.5 mg/kg.

*Use.* Effective against aphids on vegetables and fruit, and Carrot Fly, leafhoppers on rice, vegetables, cotton, some flies, some leafminers, some beetles.

*Caution*
(*a*) This is a poisonous substance (Part II) — full protective clothing should be worn.
(*b*) Dangerous to fish.
(*c*) Pre-harvest interval for edible crops — 6 weeks.

*Formulations. Disyston*, 50% impregnated on activated carbon; also 5 and 10% granules, based on Fuller's Earth (FE) or pumice (P).

### 33. Ethion (*Embathion, Nialate, Hylemox, Rhodocide*)

*Properties.* It is a non-systemic insecticide and acaricide, used mainly in combination with petroleum oils on dormant fruit as an ovicide and scalecide. It is non-phytotoxic. It is a pale-coloured liquid very slightly soluble in most organic solvents including kerosene and petroleum oils. It is slowly oxidized in air and is subject to hydrolysis by both acids and alkalis.

The acute oral $LD_{50}$ for rats is 208 mg/kg (for the pure substance) and 96 mg/kg (technical grade).

*Use.* Effective against eggs and dormant stages of pests (scales, leafhoppers, red spider mites, Heteroptera) on fruit trees. Some use against anthomyiid fly maggots on cereals and vegetables (in the soil).

*Caution.* This is a Part III poisonous substance — full protective clothing should be worn. It is phytotoxic to some varieties of apple.

*Formulations.* 25% w.p.; e.c.; 4% dust, 50% seed dressings.

### 34. Ethoate-methyl (*Fitios*)

*Properties.* It is a systemic insecticide and acaricide with contact action, particularly effective against fruit flies. The pure compound is a white crystalline solid, almost insoluble in water, but soluble in benzene, chloroform, acetone, ethanol. It is stable in aqueous solution but hydrolysed by alkali.

The acute $LD_{50}$ for male rats is 340 mg/kg; non-irritant.

*Use.* Effective especially against Olive Fly (60 g a.i./l) and fruit flies (50 g/100 l); recommended for control of aphids, and red spider mites on fruit, arable and vegetable crops at rates of 70—170 g a.i./450— 1125 l/ha.

*Caution*
(*a*) Harmful to fish, livestock, game, wild birds and animals.
(*b*) Dangerous to bees.
(*c*) Pre-harvest interval for arable crops — 1 week.
(*d*) Pre-access interval for animals to treated areas — 1 week.

*Formulations.* 20, 40% e.c.; 25% w.p.; 5% dust and 5% granules.

## 35. Fenitrothion (*Accothion, Folithion, Sumithion*)

*Properties.* It is a contact and stomach insecticide, particularly effective against rice stem borers, but has a wide spectrum of activity; also a selective acaricide but of low ovicidal activity. It is a brownish-yellow liquid, practically insoluble in water, but soluble in most organic solvents, and is hydrolysed by alkali. Of moderate persistence.

The acute oral $LD_{50}$ for rats is $250-500$ mg/kg.

*Use.* Effective against lepidopterous larvae (rice stem borers especially), aphids, whiteflies, scales, mealybugs, capsids, psyllids, some fly larvae, some beetles, locusts, thrips and sawflies.

*Caution*
(*a*) Harmful to fish, bees, livestock, game, wild birds and animals.
(*b*) Pre-harvest interval for edible crops – 2 to 3 weeks.
(*c*) Pre-access interval for livestock to treated areas – 1 week.

*Formulations.* 50% e.c.; 40 and 15% w.p.; 5, 3, and 2% dusts.

## 36. Fenthion (*Baytex, Lebaycid, Queleton,* etc.)

*Properties.* It is a contact and stomach insecticide with a useful penetrant action, which, by virtue of low volatility and stability to hydrolysis, is of high persistence. It is a colourless liquid, practically insoluble in water, but readily soluble in most organic solvents. It is stable at temperatures up to $210°C$ and is resistant to light and to alkaline hydrolysis.

The acute oral $LD_{50}$ for male rats is 215 mg/kg. It is of greater toxicity to dogs and birds, and is used for the control of weaver birds in Africa.

*Use.* Effective against fruit flies, many caterpillars, leafhoppers and plant bugs, aphids, thrips, mites sawflies, some beetles; also weaver birds.

*Caution*
(*a*) Harmful to birds and wildlife.
(*b*) Pre-harvest interval for edible crops –
7 to 42 days according to country and crop.

*Formulations.* 50, 40, 25%, w.p.; 60% fogging concentrate; 50% e.c.; 3% dust, *Queleton* for use against weaver birds.

## 37. Fonofos (*Dyfonate*)

*Properties.* It is an insecticide particularly suitable for the control of soil maggots and other soil insects. It has caused some damage to seeds when placed in their proximity. Persistence in soil is moderate – of the order of eight weeks.

It is a pale yellow liquid; practically insoluble in water, but miscible with most organic solvents such as kerosene, xylene, etc. Stable under normal conditions.

Acute oral $LD_{50}$ for male rats is $8-17$ mg/kg.

*Use.* Effective against most soil pests, such as root maggots, soil caterpillars, wireworms and other beetle larvae, crickets, symphylids.

*Caution*
(*a*) This is a very poisonous substance (Part II), and full protective clothing should be worn.
(*b*) Dangerous to fish.
(*c*) Pre-harvest interval for edible crops - 6 weeks.

*Formulations.* 5 and 10% granules.

## 38. Formothion (*Anthio, Aflix*)

*Properties.* It is a contact and systemic insecticide and acaricide effective against sap-sucking insects and mites. In plants, it is metabolized to dimethoate. In loamy soil the half-life is 14 days.

It is a yellow viscous oil or crystalline mass, slightly soluble in water, miscible with alcohols, chloroform, ether, ketones, and benzene. It is stable in non-polar solvents, but is hydrolysed by alkali and incompatible with alkaline pesticides.

The acute oral $LD_{50}$ for male rats is $375-535$ mg/kg.

*Use.* Effective against many sap-sucking insects and mites, especially aphids and red spider mites, and some fly larvae.

*Caution*
(*a*) Dangerous to bees.
(*b*) Harmful to fish and livestock.
(*c*) Pre-harvest interval for edible crops – 1 week.
(*d*) Access of animals to treated areas – 1 week.
*Formulations.* 25% w/v e.c. is the usual formulation.

### 39. Heptenophos (*Hostaquick, Ragadan*)

*Properties.* A translocatable insecticide with rapid initial action and short residual effect. It penetrates plant tissues and is quickly translocated in all directions. It is used against sap-sucking insects, and some Diptera, and ectoparasites of domesticated animals.

A pale, brown liquid, scarcely soluble in water, but quite soluble in most organic solvents.

The acute oral $LD_{50}$ for rats is 96–121 mg/kg.

*Use.* Effective against many sap-sucking insects, but especially aphids.

*Caution*
(*a*) This is a poisonous substance (Part III) – protective clothing should be worn.
(*b*) Pre-harvest interval for edible crops is 1 day.
(*c*) Dangerous to bees.
(*d*) Harmful to fish.
*Formulation.* For agricultural use only as an e.c. 500 g a.i./l.

### 40. Iodofenphos (*Nuvanol N, Elocril, Alfacron*)

*Properties.* A non-systemic, contact and stomach insecticide and acaricide, used mainly for control of stored product pests. For crop protection purposes persistence is one to two weeks.

It forms colourless crystals, almost insoluble in water, slightly soluble in benzene and acetone. It is stable in neutral media but unstable in strong acids and alkalis.

The acute oral $LD_{50}$ for rats is 2100 mg/kg.

*Use.* Effective against various pests of stored products and public hygiene, and for crop protection against a wide range of coleopterous, dipterous, and lepidopterous pests.

*Caution*
(*a*) Toxic to bees.
(*b*) Pre-harvest interval for edible crops – 7 to 14 days.
*Formulations.* 50% w.p.; 20% e.c.; and 5% powder.

### 41. Malathion (*Malathion, Malastan, Malathixo*)

*Properties.* It is a wide-spectrum non-systemic insecticide and acaricide, of brief to moderate persistence and low mammalian toxicity. It is generally non-phytotoxic, but may damage cucurbits under glasshouse conditions, and various flower species.

It is a colourless or pale brown liquid, of slight solubility in water, miscible with most organic solvents though not in petroleum oils. Hydrolysis is rapid at pH above 7.0 and below 5.0; it is incompatible with alkaline pesticides and corrosive to iron.

The acute oral $LD_{50}$ for rats is 2800 mg/kg.

*Use.* Effective against aphids, thrips, leafhoppers, spider mites, mealybugs, scales, various beetles, caterpillars and flies.

*Caution*
(*a*) Harmful to bees and fish.
(*b*) To avoid possible taint to edible crops allow 4 days from application to harvest (7 days for crops for processing).
(*c*) Pre-harvest interval – 1 day.

*Formulations.* As e.c. from 25 to 86% (many of 60%), 25 and 50% w.p.; dusts of 4%, and as atomizing concentrates (95%) for u.l.v. applications.

### 42. Mecarbam (*Murfotox, Pestan, Afos*)

*Properties.* An insecticide and acaricide with slight systemic properties, used for control of Hemiptera and Diptera. At recommended rates it persists in soil for 4–6 weeks.

It is a pale brown oil, almost insoluble in water, slightly soluble in aliphatic hydrocarbons, miscible with alcohols, aromatic hydrocarbons, ketones and esters. It is subject to hydrolysis; is compatible with all but highly alkaline pesticides; slowly attacks metals.

The acute oral $LD_{50}$ for rats is 36 mg/kg.

*Use.* Effective against scale insects and other Hemiptera, Olive Fly and other fruit flies, leafhoppers and stem flies of rice, and rootfly maggots on vegetables.

*Caution*

(*a*) Pre-harvest interval for edible crops – 2 weeks.

(*b*) Pre-access interval to treated areas – 2 weeks.

(*c*) Harmful to bees, fish, livestock and game.

*Formulations.* E.c. 68, 40%; w.p. 25%; dusts; *Murfotox Oil* 5% in petroleum oil; 5% granules.

## 43. Menazon (*Saphizon, Saphi-Col, Sayfos, Aphex*)

*Properties.* It is a systemic insecticide, used mainly against aphids. It is regarded as non-phytotoxic.

It forms colourless insoluble crystals; stable up to $35°C$; is weakly basic; and is compatible with all but strongly alkaline pesticides, but may be decomposed by the reactive surfaces of some 'inert' fillers.

The acute oral $LD_{50}$ for female rats is 1950 mg/kg.

*Use.* Most frequently used against aphids, as a seed dressing, also as a drench and a root dip. Also used for Woolly Aphid on apple.

*Caution*

(*a*) Harmful to bees and livestock.

(*b*) Pre-harvest interval for edible crops – 3 weeks.

*Formulations.* Seed dressings as 50 and 80%, 70% w.p.

## 44. Methidathion (*Supracide, Ultracide*)

*Properties.* It is a non-systemic insecticide, with some acaricidal activity, and capable of foliar penetration. Non-phytotoxic to all plants tested; rapidly metabolized and excreted by plants and animals.

It forms colourless crystals, of slight solubility in water, readily soluble in acetone, benzene, and methanol. It is stable in neutral and weakly acid media, but much less stable in alkali. It is compatible with many fungicides and acaricides.

Acute oral $LD_{50}$ for rats is 25–48 mg/kg.

*Use.* Of promise against lepidopterous larvae; foliar penetration enables it to be used against leaf-rollers. Used on a wide variety of crops against leaf-eating and sucking insects and mites, especially against scale insects.

*Caution*

(*a*) This is a poisonous substance (Part III), and protective clothing should be worn.

(*b*) Harmful to bees and livestock.

(*c*) Dangerous to fish.

(*d*) Pre-harvest interval for edible crops – 3 weeks.

(*e*) Pre-access interval for livestock to treated areas – 2 weeks.

*Formulations.* 40% e.c. and w.p.; and 20% e.c. and w.p.

## 45. Mevinphos (*Phosdrin, Menite, Phosfene*)

*Properties.* It is a contact and systemic insecticide and acaricide of short persistence. Although non-persistent, its high initial kill provides a relatively long period before build-up recurs. It is non-phytotoxic.

The technical product is a pale yellow liquid, miscible with water, alcohols, ketones, chlorinated hydrocarbons, aromatic hydrocarbons, but only slightly soluble in aliphatic hydrocarbons. Stable at ordinary temperatures, but hydrolysed in aqueous solution, and rapidly decomposed by alkalis, hence incompatible with alkaline fertilizers and pesticides. It is corrosive to cast iron, mild and some stainless steels, and brass; relatively non-corrosive to copper, nickel and aluminium; non-corrosive to glass, and many plastics, but passes slowly through thin films of polyethylene.

Acute oral $LD_{50}$ for rats is 3.7–12 mg/kg.

*Use.* Effective against sap-feeding insects (aphids, etc.) at 140–280 g/ha, mites and beetles at 210–350 g/ha, caterpillars at 280–560 g/ha; and some fly larvae. Especially useful for giving rapid kill close to harvest.

*Caution*

(*a*) This is a poisonous substance (Part II); full protective clothing should be worn.

(*b*) Dangerous to livestock, bees, fish, game and wild animals.

(*c*) Pre-harvest interval for edible crops – 3 days.

(*d*) Pre-access interval to treated areas – 1 day.

*Formulations.* Being water soluble, formulation is unnecessary, but e.c. of 5, 10, 18, 24, 48 and 50% technical are available, together with dusts and w.p.'s.

### 46. Monocrotophos (*Nuvacron, Azodrin, Monocron*)

*Properties.* A fast-acting insecticide with systemic, stomach and contact action, used against a wide range of pests on a variety of crops; persistence of 1 to 2 weeks. It has caused phytotoxicity under cool conditions to some apples, cherries, and sorghum varieties, and is incompatible with alkaline pesticides.

It is a crystalline solid, miscible with water, soluble in acetone, and ethanol, sparingly soluble in xylene but almost insoluble in kerosene and diesel oils. Unstable in low molecular weight alcohols and glycols. Corrosive to iron, steel, and brass, but does not attack glass, aluminium and stainless steel.

Acute $LD_{50}$ for rats is 13.23 mg/kg.

*Use.* Effective against a wide range of pests including mites, bugs, leaf-eating beetles, leaf miners and caterpillars.

*Caution*

(*a*) Dangerous to fish and livestock.

(*b*) Pre-harvest interval is 3–30 days according to crop and country.

*Formulations.* Water miscible concentrates contain 200/600 g a.i./l.

### 47. Naled (*Dibrom, Ortho, Bromex*)

*Properties.* A non-systemic, contact and stomach insecticide and acaricide with some fumigant action, used mainly under glass and in mushroom houses.

The technical product is a yellow liquid, insoluble in water, slightly soluble in aliphatic solvents, and readily soluble in aromatic solvents. It is stable under anhydrous conditions, but rapidly hydrolysed in water (90–100% in 48 hours at room temperature), and by alkali; stable in glass containers, but in the presence of metals and reducing agents, rapidly loses bromine and reverts to dichlorvos.

Acute $LD_{50}$ for rats is 430 mg/kg.

*Use.* Mainly against glasshouse and mushroom pests.

*Caution*

(*a*) Pre-harvest interval – 24 hours.

(*b*) It is phytotoxic to cucurbits.

*Formulations.* 4% dust and e.c. 1 kg/l.

### 48. Omethoate (*Folimat*)

*Properties.* A systemic insecticide and acaricide with a broad range of action and little phytotoxicity, except to some peach varieties.

It is a colourless oily liquid, readily soluble in water, acetone, ethanol and many hydrocarbons; insoluble in light petroleum; hydrolysed by alkali.

Acute oral $LD_{50}$ for male rats is about 50 mg/kg.

*Use.* Effective against a wide range of insects, particularly caterpillars, and Homoptera; also Orthoptera, thrips, some beetles, and phytophagous mites.

*Caution*

(*a*) Harmful to bees.

(*b*) Damaging to various varieties of peach.

(*c*) Pre-harvest interval for edible crops – 21 to 28 days.

*Formulations.* Include e.c. and granules with a range of a.i. contents.

### 49. Oxydemeton-methyl (*Metasystox-R*)

*Properties.* A systemic and contact insecticide and

acaricide, used against sap-sucking insects and mites, with a fast kill and moderate persistence.

It is a clear brown liquid, miscible with water and soluble in most organic solvents except light petroleum. It is hydrolysed by alkali.

The acute oral $LD_{50}$ for male rats is 65 mg/kg.

*Use.* Effective against aphids, and red spider mites on most crops, also leafhoppers, whiteflies, psyllids, thrips, some flies, sawflies. Only a limited effect on *Brassica* aphids.

*Caution*

(*a*) This is a poisonous substance (part III), and protective clothing should be worn.

(*b*) Harmful to bees, livestock, fish, game and wild animals.

(*c*) Pre-harvest interval for edible crops – 2 to 3 weeks.

(*d*) Pre-access interval for livestock to treated areas 2 weeks.

*Formulations.* As e.c. of various a.i. contents (25, 50%).

## 50. Oxydisulfoton (*Disyston-S*)

*Properties.* A systemic insecticide and acaricide particularly suitable for seed treatment against virus vectors.

It is a pale coloured liquid, slightly soluble in water, readily soluble in most organic solvents.

The acute oral $LD_{50}$ for rats is about 3.5 mg/kg.

*Use.* Particularly effective against sap-sucking insects and mites.

*Caution.* This is a very poisonous substance, and full protective clothing should be worn.

*Formulations.* As seed dresings of various a.i. contents; also as e.c. and granules.

## 51. Parathion (*Folidol, Bladan, Fosfex, Fosferno, Thiophos*)

*Properties.* A non-systemic, contact and stomach insecticide and acaricide, with some fumigant action.

It is non-phytotoxic except to some ornamentals, and under certain weather conditions, to pears and some apple varieties; hazardous to operators.

It is a pale yellow liquid, scarcely soluble in water, slightly soluble in petroleum oils, but miscible with most organic solvents. In alkaline solution it rapidly hydrolyses; on heating it isomerizes.

The acute oral $LD_{50}$ for male rats is 13 mg/kg; for females 3.6 mg/kg; acute dermal $LD_{50}$ respectively 21 and 6.8 mg/kg.

*Use.* Effective against most Homoptera, Diptera, springtails, mites, millipedes and some nematodes. Prolonged use may result in extensive destruction of predators and parasites.

*Caution*

(*a*) This is a very poisonous substance (Part II), and full protective clothing should be worn; easily absorbed through the skin.

(*b*) Dangerous to bees, fish, livestock, game, wild birds and animals.

(*c*) Pre-harvest interval for edible crops – 4 weeks.

(*d*) Pre-access interval for livestock to treated areas – 10 days.

*Formulations.* To w.p. and e.c. of various a.i. contents; also to dusts, smokes and aerosols.

## 52. Parathion-methyl (*Dalf, Metacide, Folidol-M, Nitrox-80*)

*Properties.* A non-systemic, contact and stomach insecticide, with some fumigant action, and a range of action similar to that of parathion but of lower mammalian toxicity. It is non-phytotoxic.

The technical product is a brown liquid of about 80% purity, scarcely soluble in water, slightly soluble in light petroleum and mineral oils, but soluble in most other organic solvents. It is hydrolysed by alkali at a faster rate than parathion, and readily isomerizes on heating. Compatible with most other pesticides.

The acute oral $LD_{50}$ for male rats is 14 mg/kg;

for female rats 24 mg/kg. It is hazardous to wild life but is of brief persistence.

*Use.* As for parathion but not effective against Acarina.

*Caution*

(*a*) This is a poisonous substance, so protective clothing should be worn.

(*b*) Hazardous to wild life.

*Formulations.* To e.c. and dusts of various a.i. contents, *Nitrox 80* is 80% solution in an aromatic petroleum solvent.

### 53. Phenisobromolate (*Acarol, Neoron*)

*Properties.* A contact acaricide with residual activity, of promise for use on many crops.

It is a crystalline solid, insoluble in water, but readily soluble in most organic solvents; stable in neutral media.

The acute oral $LD_{50}$ for rats is 5000 mg/kg.

*Use.* Of promise for use against mites on pome and stone fruits, citrus, hops, cotton, beans, cucurbits, tomatoes, strawberries and ornamentals.

*Caution*

*Formulations.* As 500 and 250 g/l e.c.

### 54. Phenthoate (*Elsan, Cidial, Papthion, Tanone*)

*Properties.* A non-systemic insecticide and acaricide with contact and stomach action. It may be phytotoxic to some peach, fig and grape varieties, and may discolour some red-skinned apple varieties.

It is a crystalline solid, almost insoluble in water, but miscible with most organic solvents.

The acute oral $LD_{50}$ for rats is 250—300 mg/kg.

*Use.* Effective against caterpillars, aphids, jassids, mites, and is also used for the protection of stored grain.

*Caution*

*Formulations.* 50% technical; 5% in mineral oil; 40% w.p.; 2% granules; 85% pure compound.

### 55. Phorate (*Thimet, Rampart, Granutox, Timet*)

*Properties.* A persistent systemic insecticide used in granular and e.c. formulations for the protection of seedlings from sap-feeding and soil insects; some fumigant action. In temperate climates effective soil persistence of 15—20 weeks is expected.

It is a clear liquid, only slightly soluble in water, but miscible with carbon tetrachloride, dioxane, xylene, and vegetable oils. It is hydrolysed by alkalis and in the presence of moisture.

The acute oral $LD_{50}$ for male rats if 3.5 mg/kg; for females 1.6 mg/kg.

*Use.* Effective against aphids, wireworms, various fly maggots (Frit, Carrot), capsids, leafhoppers, various weevils.

*Caution*

(*a*) This is a very poisonous substance (Part II), and full protective clothing should be worn.

(*b*) Dangerous to fish and livestock.

(*c*) Pre-harvest interval for edible crops – 6 weeks.

(*d*) Pre-access interval for livestock to treated areas – 6 weeks.

*Formulations.* As e.c. of various a.i. content; 5, 10 and 15% granules.

### 56. Phosalone (*Zolone, Embacide, Rubitox*)

*Properties.* A non-systemic insecticide and acaricide, used on deciduous tree fruits, field and market garden crops, against a wide range of pests. It persists on plants for about two weeks before being hydrolysed.

It forms colourless crystals; insoluble in water and light petroleum, but soluble in acetone, benzene, chloroform, ethanol, methanol, toluene and xylene. It is stable under normal storage conditions, non-corrosive, and compatible with most other pesticides.

The acute oral $LD_{50}$ for male rats is 150 mg/kg.

*Use.* Effective against a wide spectrum of pests —
caterpillars on fruit, cotton bollworms, fruit fly
maggots, aphids, psyllids, jassids, thrips, various
weevils and red spider mites.

*Caution*
(*a*) Harmful to bees, fish and livestock.
(*b*) Pre-harvest interval for edible crops — 3 weeks.
(*c*) Pre-access interval for livestock to treated areas —
4 weeks.

*Formulations.* 30, 33, 35% e.c.; 30% w.p.; 2.5 and 4%
dusts. Various e.c. formulations under heading
*Zolone DT* with DDT for use on cotton.

### 57. Phosmet (*Imidan, Appa, Prolate, Germisan*)

*Properties.* A non-systemic acaricide and insecticide,
used at concentrations safe for a variety of predators
of mites and thus useful for integrated control
programmes.

It is a white crystalline solid, with an offensive
odour, scarcely soluble in water, but more than 10%
soluble in acetone, dichloromethane and xylene.

The acute oral $LD_{50}$ for male rats is 230 mg/kg;
it is readily degraded both in laboratory animals and
in the environment.

*Use.* Used mainly against phytophagous mites; when
used as recommended should not affect mite
predators.

*Caution*

*Formulations.* 20 and 30% e.c. and 50% w.p. Storage
above 45°C may lead to decomposition.

### 58. Phosphamidon (*Dimecron, Dicron, Famfos*)

*Properties.* A systemic insecticide and acaricide,
rapidly absorbed by the plant, but only a little contact action. Non-tainting, and non-phytotoxic except
to some cherry varieties and sorghum varieties related
to Red Swazi. No fumigant action.

It is a pale yellow oil, miscible with water, and
readily soluble in most organic solvents except
saturated hydrocarbons. It is stable in neutral and
acid media but is hydrolysed by alkali. Compatible
with all but highly alkaline pesticides. It corrodes
iron, tin plate and aluminium, and is packed in polyethylene containers.

The acute oral $LD_{50}$ for rats is 28.3 mg/kg. The
half-life in plants is about 2 days.

*Use.* Effective against sap-feeding insects and leaf-
eating ones; particularly aphids, bugs, many caterpillars (but not Noctuidae), rice stem borers, thrips,
Colorado Beetle, and other beetles, grasshoppers,
sawflies, fly larvae, and phytophagous mites.

*Caution*
(*a*) This is a poisonous substance (Part III) —
protective clothing should be worn.
(*b*) Dangerous to bees.
(*c*) Harmful to livestock.
(*d*) Pre-harvest interval for edible crops — 3 weeks.
(*e*) Pre-access interval for livestock to treated areas —
2 weeks.

*Formulations. Dimecron 20*: 20 kg/100 l in iso-
propanol, the water content rigidly controlled to
delay hydrolysis; similarly for *Dimecron 50* and *100;*
50% w.p.

### 59. Phoxim (*Baythion, Valexon, Volaton*)

*Properties.* An insecticide of brief persistence and no
systemic action, with low mammalian toxicity,
effective against a broad range of insects; most useful
against soil insects and stored products pests.

It is a yellow liquid virtually insoluble in water,
slightly soluble in light petroleum, soluble in alcohols,
ketones, and aromatic hydrocarbons. Stable to water
and acid media, but unstable to alkali. Believed to be
compatible with most pesticides of a non-alkaline
nature.

The acute oral $LD_{50}$ for rats is more than
2000 mg/kg.

*Use.* Effective against a broad range of insects,

especially stored products pests and insects affecting man. Also successfully used against soil pests (dipterous maggots, rootworms and wireworms); u.l.v. applications against grasshoppers.

*Caution*

(*a*) Dangerous to bees by both contact and vapour effect.

(*b*) Harmful to fish.

*Formulations.* 50% e.c.; 5% granules; and a concentrate for u.l.v. application. Other experimental formulations are under test. *Baythion* is the trade name for use against pests of man and stored products — *Valexon* for agricultural use.

### 60. Pirimiphos-ethyl (*Primicid, Fernex, Primotec*)

*Properties.* It is a broad-spectrum insecticide, particularly against soil-inhabiting Diptera and Coleoptera. No phytotoxicity has been recorded using recommended rates; high rates of seed dressing have resulted in seedling abnormalities though. Stable for 5 days at $80^{\circ}$C. In the pure state it is a pale straw-coloured liquid; insoluble in water but miscible with most organic solvents; corrosive to iron and unprotected tin plate.

Acute oral $LD_{50}$ for rats is $140-200$ mg/kg.

*Use.* Effective particularly against soil-inhabiting Diptera and Coleoptera; effective as foliage spray at conventional rates against species of Lepidoptera, Coleoptera, Homoptera and Tetranychidae.

*Caution*

(*a*) Dangerous to bees.

(*b*) Hazardous to birds and wild animals.

*Formulations.* A 20% s.d.; 250 g/l, 500 g/l e.c.; 5 and 10% granules with 5 and 10% thiram added.

### 61. Pirimiphos-methyl (*Actellic, Blex, Actellifog*)

*Properties.* It is a fast-acting, broad-spectrum insecticide of limited persistence, with both contact and fumigant action; non-phytotoxic. It penetrates leaf tissue to the extent that an insect on one side of a leaf is killed by chemical applied to the other side, and there is also a slight systemic action.

It is a straw-coloured liquid, insoluble in water, but soluble in most organic solvents; decomposed by strong acids and alkalis; does not corrode brass, stainless steel, nylon or aluminium.

Acute oral $LD_{50}$ for female rats is about 800 mg/kg (low mammalian toxicity).

*Use.* Effective against species of Lepidoptera, Coleoptera, aphids, Tetranychidae, and many other crop pests (termites, Heteroptera, Thysanoptera, Diptera, Orthoptera).

*Caution*

(*a*) Dangerous to bees.

(*b*) Harmful to fish.

(*c*) Pre-harvest interval — 3 to 7 days.

*Formulations.* 25 and 50% e.c.; 5 and 10% granules, and 100 g/l and 500 g/l u.l.v. formulations.

### 62. Profenofos (*Curacron*)

*Properties.* A broad-spectrum, non-systemic insecticide, with both contact and stomach action, used against many different crop pests.

A pale yellow liquid, barely soluble in water, but miscible with most organic solvents. Fairly stable under neutral and slightly acidic conditions.

The acute oral $LD_{50}$ for rats is 358 mg/kg.

*Use.* Effective against many pests of cotton and vegetables, both insects and mites. Usual rates are $250-500$ g a.i./ha for sucking insects and mites, and $400-1200$ g a.i./ha for biting and chewing insects.

*Formulations.* e.c. 500 g a.i./l, 400 g a.i./l; u.l.v. 250 g a.i./l; granules 50 g a.i./kg; and mixtures with chlordimeform.

### 63. Prothoate (*Fac, Fostin, Oleofac, Telefos*)

*Properties.* An acaricide and insecticide, with systemic action, used mainly against phytophagous mites and some sap-sucking insects.

It is a colourless crystalline solid, virtually insoluble in water, but miscible with most organic solvents. It is stable in neutral, moderately acid and slightly alkaline media, but is rapidly decomposed in strong alkali.

The acute oral $LD_{50}$ for male rats is 8 mg/kg.

*Use.* Effective for protection of fruit, citrus and vegetable crops from tetranychid and some eriophyid mites, and some insects, notably aphids, Tingidae, Psyllidae, and Thysanoptera.

*Caution*

(*a*) This is a very poisonous substance, and full protective clothing should be worn.

(*b*) Dangerous to fish, livestock and game.

*Formulations.* 20% tech., 40% tech., and 3% tech.; 5% granules.

### 64. Quinomethionate (*Morestan, Erade, Forstan*)

*Properties.* A selective, non-systemic, acaricide; and a fungicide specific to powdery mildews.

It forms yellow crystals, practically insoluble in water and sparingly soluble in organic solvents. In chemical properties it is closely related to thioquinox, but is more stable to oxidation.

The acute oral $LD_{50}$ for rats is 2500–3000 mg/kg.

*Use.* Controls red spider and other phytophagous mites on a wide range of crops; and powdery mildews.

*Caution*

(*a*) Harmless to bees.

(*b*) Certain blackcurrant varieties may be damaged by sprays.

(*c*) Pre-harvest interval for edible crops – 3 to 28 days according to crops and country.

*Formulations.* 25% w.p., and also formulated as smokes (20%), and 2% dust.

### 65. Schradan (*Sytam, Pestox 3*)

*Properties.* A systemic insecticide and acaricide with little contact effect; effective against sap-feeding insects and mites; non-phytotoxic at insecticidal concentrations.

It is a brown viscous liquid, miscible with water and most organic solvents; slightly soluble in petroleum oils, and readily extracted from aqueous solution by chloroform. Stable to water and alkali, but hydrolysed under acid conditions.

The acute oral $LD_{50}$ for male rats is 9.1 mg/kg.

*Use.* Used mainly against aphids and red spider mites on a variety of crops.

*Caution*

(*a*) This is a very poisonous substance (Part II), and full protective clothing should be worn.

(*b*) Dangerous to fish, livestock, game, wild birds and animals.

(*c*) Safe to bees.

(*d*) Pre-harvest interval for edible crops – 4 to 6 weeks.

(*e*) Pre-access interval for livestock to treated areas – 4 weeks.

*Formulations.* 30% aqueous solution; also anhydrous, 75–80% or 60% with anhydrous surfactant.

### 66. TEPP (*Nifos T, Vapotone, Bladen, Fosvex*)

*Properties.* A non-systemic aphicide and acaricide of brief persistence; very high mammalian toxicity.

It is a colourless hygroscopic liquid, miscible with water and most organic solvents, but only slightly soluble in petroleum oils; is very rapidly hydrolysed by water, and is corrosive to most metals. It is rapidly metabolized in the animal body.

The acute oral $LD_{50}$ for rats is 1.2 mg/kg; the acute dermal $LD_{50}$ is 2.4 mg/kg.

*Use.* Used only against aphids and phytophagous mites.

*Caution*

(*a*) An extremely poisonous pesticide and full

protective clothing must be worn. High mammalian dermal toxicity.

(b) Dangerous to livestock, game, wild birds and animals.

*Formulations.* For agricultural purposes TEPP refers to a mixture of polyphosphates containing at least 40% tetraethyl pyrophosphate. As an aerosol, a solution in methyl chloride is used.

### 67. Terbufos (*Counter*)

*Properties.* An insecticide with strong initial, and with residual, activity against soil insects.

A pale yellow liquid, almost insoluble in water, but soluble in many organic solvents. It decomposes on prolonged heating at high temperatures, at low pH, and with strong alkalis.

The acute oral $LD_{50}$ for rats is 1.6–4.5 mg/kg, and the acute dermal $LD_{50}$ for rats is from 1.0–7.4 mg/kg.

*Use.* Effective against many different types of soil insects.

*Caution*

(a) This is a very poisonous substance, and full protective clothing must be worn.

(b) This poison can be absorbed through the skin.

(c) Generally dangerous to game and wildlife.

*Formulation.* Because of its toxicity it is only formulated as a granule, from 20–150 g a.i./kg.

### 68. Tetrachlorvinphos (*Gardona, Rabon, Ravap*)

*Properties.* A selective insecticide used against lepidopterous and dipterous pests on the aerial parts of crops; very low mammalian toxicity; some translaminar action.

It is a white crystalline solid, scarcely soluble in water, but soluble in chloroform (40% w/w), methyl chloride (40%); less so in xylene (15%) and acetone (20%). It is temperature stable, but slowly hydrolysed by water, particularly under alkaline conditions.

The acute oral $LD_{50}$ for rats is 4000–5000 g/kg.

*Use.* Effective against lepidopterous and dipterous pests of fruit, rice, vegetables, cotton and maize. With certain exceptions, it does not show high activity against Hemiptera, and because of rapid breakdown is not effective in the soil. Shows promise against pests of stored products.

*Caution*

*Formulations.* 240 g/l e.c.; 50 and 70% w.p.; and 5% granules.

### 69. Thiometon (*Ekatin, Intrathion*)

*Properties.* A systemic insecticide and acaricide suitable for control of sucking insects and mites on orchards, vineyards, hop gardens, and on beet. At 0.02% a.i. systemic effects persist for 2–3 weeks.

It is a colourless oil, fairly soluble in water, slightly soluble in light petroleum oils but soluble in most other organic solvents.

The acute oral $LD_{50}$ for rats is 120–30 mg/kg.

*Use.* Mainly used for aphid control (and also mites) on fruit, potatoes, beet, cereals, and vegetables.

*Caution*

(a) This is a poisonous substance (Part III), and protective clothing should be worn.

(b) Harmful to bees, fish, livestock, game, wild birds and animals.

(c) Pre-harvest interval for edible crops – 3 weeks.

(d) Pre-access interval for livestock to treated areas – 2 weeks.

*Formulations.* 20% e.c.; coloured blue; and a dry spray dust.

### 70. Thionazin (*Nemafos, Zinophos, Nemasol*)

*Properties.* A soil insecticide and nematicide, of relatively brief persistence.

It is a brown liquid, fairly soluble in water, miscible with most organic solvents; readily hydrolysed by alkali.

The acute oral $LD_{50}$ for rats is 12 mg/kg, with the dermal $LD_{50}$ being 11 mg/kg.

*Use.* Soil application is effective against symphylids and Cabbage Root Fly (drench); granules recommended for use on cotton, cucurbits, groundnuts, brassicas and tomato.

*Caution*

(*a*) This is a very poisonous substance (Part II), and full protective clothing should be worn.

(*b*) Dangerous to fish and livestock.

(*c*) Pre-access interval for livestock to treated areas – 8 weeks.

*Formulations.* 25 and 46% e.c.; 5 and 10% granules.

## 71. Thioquinox (*Eradex, Eraditon*)

*Properties.* A non-systemic acaricide effective against eggs; and a fungicide specific against powdery mildews.

It is a brown powder, practically insoluble in water and most organic solvents, but is slightly soluble in ethanol and acetone. Stable to temperature and light; resistant to hydrolysis, but susceptible to oxidation without any reduction of biological activity.

The acute oral $LD_{50}$ for rats is 3400 mg/kg; some dermal irritation caused to some operators.

*Use.* Specific against eggs of phytophagous mites; and the powdery mildews.

*Caution*

*Formulations.* 50% w.p.

## 72. Triazophos (*Hostathion*)

*Properties.* A broad-spectrum insecticide and acaricide with some nematicidal properties; used for either foliar or soil application. It can penetrate plant tissues generally but has no systemic action.

A pale brown liquid, with little solubility in water, but soluble in most organic solvents.

The acute oral $LD_{50}$ for rats is 82 mg/kg.

140

*Use.* As a foliar spray against aphids on fruit, at 75–125 g a.i./100 l, and on cereals at 320–600 g a.i. (40% e.c.)/ha. Incorporated into the soil, it has been used to control wireworms and some cutworms at 1–2 kg a.i./ha.

*Caution*

(*a*) This is a poisonous substance (Part III), so full protective clothing should be worn.

(b) Dangerous to bees, game and all wildlife.

*Formulations.* e.c. 400 g a.i./l; u.l.v. concentrates at 250 and 400 g a.i./l; w.p. 300 g a.i./kg; granules 20 and 50 g a.i./kg.

## 73. Trichloronate (*Agritox, Agricil, Phytosol*)

*Properties.* A persistent (in soil) non-systemic insecticide recommended for control of root maggots, wireworms, and other soil insects; acts as contact and stomach poison. Recently shown to be effective against Stem Eelworm. Foliar uses are under test. Soil applications have given up to five months residual control.

A brown liquid, practically insoluble in water, but soluble in acetone, ethanol, aromatic solvents, kerosene, and chlorinated hydrocarbons. It is hydrolysed by alkali.

The acute oral $LD_{50}$ for rats is 16–37 mg/kg. The acute dermal $LD_{50}$ for rats is 135–340 mg/kg.

*Use.* Effective against root maggots, wireworms and other soil and grassland insects. Also gives control of some nematodes. Very persistent in soil. Probably effective against cutworms, termites, and Collembola.

*Caution*

(*a*) A very poisonous substance, so full protective clothing must be worn. Very easily absorbed through mammalian skin.

(*b*) Use in some countries prohibited because of high toxicity and excessive persistence.

(*c*) Pre-harvest interval for edible crops varies – 4 weeks (Norway), 8 weeks (Holland).

(*d*) Dangerous to fish, livestock, game, wild birds and animals.

*Formulations.* 50% e.c. and granules of various a.i. content (2.5, 7.5%); also as a 20% seed dressing powder.

### 74. Trichlorphon (*Dipterex, Anthon, Chlorofos*)

*Properties.* A contact and stomach insecticide with penetrant action, recommended for use against flies, some bugs, some beetles, lepidopterous larvae, and ectoparasites of domestic animals. Its activity is attributed to its metabolic conversion to dichlorvos. Moderate persistence.

It is a white crystalline powder, quite soluble in water, insoluble in petroleum oils, poorly soluble in carbon tetrachloride and diethyl ether; soluble in benzene, ethanol, and most chlorinated hydrocarbons. It is stable at room temperature but is decomposed by water at higher temperatures (and acid media) to form dichlorvos.

The acute oral $LD_{50}$ for male rats is 630 mg/kg.

*Use.* Effective against most lepidopterous larvae, flies and fly maggots, some Homoptera, many Heteroptera, some beetles. Used for household and veterinary pests, under different trade names.

*Caution*
(*a*) Harmful to fish.
(*b*) Pre-harvest interval for edible crops — 2 to 14 days, according to the country.

*Formulations.* 50% w.p.; 50 and 80% soluble powders; 50% e.c.; 5% dust; and 2 and 5% granules.

### 75. Tricyclohexyltin hydroxide (*Plictran*)

*Properties.* An acaricide effective against the motile stages of a wide range of phytophagous mites. No phytotoxicity reported on deciduous fruit or on glasshouse crops, but some spotting on citrus fruit has been observed.

It is a white crystalline powder, insoluble in water, poorly soluble in most organic solvents, but 22% w/v soluble in chloroform. Stable to ambient temperatures, and in neutral and alkaline media. Degraded on exposure to u.v. light in thin layers.

The acute oral $LD_{50}$ for rats is 540 mg/kg.

*Use.* Effective against motile stages of phytophagous mites, but not the eggs.

*Caution*
(*a*) Virtually harmless to bees.
(*b*) Some phytotoxicity to citrus fruit reported.

*Formulations.* 50% a.i. w.p.; and 25% a.i. w.p. (available only in Europe).

### 76. Vamidothion (*Kilval, Vation, Vamidoate*)

*Properties.* A systemic insecticide and acaricide of high persistence and particular value against woolly aphid. It is metabolized in the plant to the sulphoxide which has similar biological activity but greater persistence.

It forms a colourless waxy solid, very soluble in water (4 g/ml) and most organic solvents, but almost insoluble in light petroleum and cyclohexane. Undergoes slight decomposition at room temperature; non-corrosive, and compatible with most pesticides.

The acute oral $LD_{50}$ for male rats is about 100 mg/kg. The toxicity of the sulphoxide is reduced by about half.

*Use.* Especially effective against woolly aphid; used for control of sap-feeding insects and mites on fruit, rice, cotton, and many other crops.

*Caution*
(*a*) This is a poisonous substance (Part III), and protective clothing should be worn.
(*b*) Harmful to bees, fish, livestock, game, wild birds and animals.
(*c*) Pre-harvest interval for edible crops — 4 to 6 weeks.

*Formulations.* Solution containing 40% w/v.

### Carbamates

The successful development of the organophosphates as pesticides directed attention to other compounds which act as anticholinesterases. This led to the discovery of carbaryl with its broad spectrum of activity. It appears that the action of the carbamates on both insects and mammals is an inhibition of the cholinesterase.

### 77. Aldicarb (*Temik*)

*Properties.* A systemic, soil-applied, insecticide, acaricide and nematicide; used against a wide range of leaf-eating and sap-sucking insects, stem and cyst nematodes, and certain mites. Usually applied to seed furrows, or as bands, or broadcast on to the soil; moisture is required to release the active chemical from the granules, so rainfall or irrigation should follow application. Uptake by the plant roots is rapid; protection is given for up to 80 days.

It is formed as colourless crystals, only slightly soluble in water, but soluble in most organic solvents. Generally stable, except to concentrated alkali; non-flammable, and non-corrosive.

The acute oral $LD_{50}$ for male rats is 0.9 mg/kg, and the acute dermal $LD_{50}$ for male rabbits is 5 mg/kg.

*Use.* Effective against a wide range of insects, mites and nematodes, especially aphids, whiteflies and leaf-miners. Mealybugs, scale insects, capsid bugs, spider mites and tarsonemid mites are also controlled. Both stem and cyst nematodes are killed. Reduces virus damage to many crops by controlling the insect and nematode vectors.

*Caution*
(*a*) This is a very poisonous substance (Part II), so full protective clothing must be worn. Easily absorbed through mammalian skin.
(*b*) Dangerous to fish, game and wildlife.
(*c*) Pre-harvest interval (for tomatoes in UK) – 6 weeks.

*Formulation.* Because of its toxicity and handling

142

hazards, only formulated as granules (50, 100, and 150 g a.i./kg) for soil application.

### 78. Bufencarb (*Bux*)

*Properties.* A non-persistent carbamate insecticide, effective against a wide range of soil and foliage insects. Degradation in soil is fairly rapid, with no expected seasonal accumulation following use of granules.

The technical product is a yellow solid, of low melting point; almost insoluble in water, very soluble in methanol and 1,2-dimethylbenzene, but less so in aliphatic hydrocarbons such as hexane. Stable in neutral or acid solutions, but the rate of hydrolysis increases with either a rise in temperature or increase in pH.

The acute oral $LD_{50}$ for rats is 87 mg/kg; acute dermal $LD_{50}$ for rabbits is 680 mg/kg.

*Use.* Active at rates of 0.5–2.0 kg a.i./ha. against a range of soil insects, such as corn rootworm, rice water weevil, root mealybugs; and foliage insects such as rice leafhoppers and planthoppers; and graminaceous stem borers.

*Caution*
(*a*) This is a very poisonous substance, and full protective clothing must be worn.
(*b*) This poison can be absorbed through the skin.
(*c*) Dangerous to fish.

*Formulations.* e.c. 240 g a.i./l and 360 g a.i./l; dusts 20 and 40 g a.i./kg; granules 100 g a.i./kg.

### 79. Carbaryl (*Sevin, Carbaryl 85, Murvin, Septon*)

*Properties.* A contact insecticide with slight systemic properties and broad-spectrum activity. No evidence of phytotoxicity at recommended rates. Has growth regulatory properties and may be used for fruit thinning (apples). Also used for killing earthworms in turf. Persistence more like an organochlorine compound than an organophosphate.

It is a white crystalline solid, barely soluble in

water, but soluble in most polar organic solvents (such as dimethyl sulphoxide). It is stable to light, heat and hydrolysis under normal conditions, and is non-corrosive. Compatible with most other pesticides, except those strongly alkaline, such as lime-sulphur or Bordeaux mixture, which cause hydrolysis.

The acute oral $LD_{50}$ for male rats is 850 mg/kg.

*Use.* Effective against many insect pests, especially caterpillars, midges, beetles, Orthoptera, capsids and other bugs. Generally more effective against chewing insects than sap-suckers.

*Caution*
(*a*) Dangerous to bees.
(*b*) Harmful to fish.
(*c*) Pre-harvest interval for edible crops — 1 week.
*Formulations.* 50 and 85% w.p.; 5 and 10% dusts.

## 80. Carbofuran (*Furadan, Curaterr, Yaltox*)

*Properties.* A broad-spectrum systemic insecticide, acaricide and nematicide, used against both foliage feeding insects, mites and soil pests. In plants its half-life is less than five days; in soil the half-life is 30–60 days.

It is a white odourless, crystalline solid, fairly soluble in water and acetone; non-inflammable, but unstable in alkaline media.

The acute oral $LD_{50}$ for rats is 8–14 mg/kg.

*Use.* Effective against various sap-feeding insects and mites, especially leafhoppers on rice, and also root-eating caterpillars and beetle larvae on cereals.

*Caution*
(*a*) This is a poisonous substance, and protective clothing should be worn.
(*b*) Harmful to fish.
*Formulations.* 75% w.p.; 4% flowable paste; 2, 3 and 10% granules.

## 81. Methiocarb (*Mesurol, Draza, Baysol*)

*Properties.* A non-systemic insecticide and acaricide,

with a broad spectrum of activity; persistent; a powerful molluscicide.

It is a white crystalline powder, practically insoluble in water, but soluble in most organic solvents. Hydrolysed by alkali.

The acute oral $LD_{50}$ for male rats is 100 mg/kg.

*Use.* Mainly used for control of slugs and snails, can control cutworms and various beetles.

*Caution*
(*a*) Harmful to fish.
(*b*) Pre-harvest interval for edible crops — 7 days.
(*c*) Pre-access interval for livestock to treated areas — 7 days.

*Formulations.* 50 and 70% w.p.; *Draza*, 4% granules.

## 82. Methomyl (*Lannate, Halvard, Nudrin*)

*Properties.* A new broad-spectrum, systemic and contact insecticide and acaricide. It is also effective as a nematicide.

It is a white crystalline solid, slightly soluble in water, soluble in acetone, ethanol, methanol; the aqueous solution is non-corrosive. It is stable in solid form and aqueous solution under normal conditions, but decomposes in moist soil.

The acute oral $LD_{50}$ for male rats is 17 mg/kg. Not a skin irritant.

*Use.* As a soil treatment it has given systemic control of certain insects and nematodes. It is of promise as a foliar spray for many insects such as aphids, Colorado Beetle, leaf-rollers and many other caterpillars, and red spider mites.

*Caution*
(*a*) This is a poisonous substance, and protective clothing should be worn.
(*b*) Pre-harvest interval for edible crops — about 3 weeks.

*Formulations.* 90% w.p.; 25% w.p.

143

## 83. Oxamyl (*Vydate*)

*Properties.* A carbamate insecticide and nematicide with both contact and systemic action; moderate residual effect. Will control nematodes with both soil and foliar application.

A white crystalline solid, stable in solid form and most solutions, but decomposes to innocuous materials in natural waters and soil. Aeration, sunlight, alkalinity and higher temperatures increase the rate of decomposition. Fairly soluble in water, but more soluble in ethanol, propanone, and methanol. Aqueous solutions are non-corrosive.

The acute oral $LD_{50}$ for male rats is 5.4 mg/kg.

*Use.* Effective against aphids, thrips, leaf beetles, flea beetles, leaf-miners, and mites, as foliar sprays at rates of 0.2–1.0 kg/a.i./ha; and nematodes by both foliar and soil application.

*Caution*

(*a*) A very poisonous substance (Part II) – full protective clothing must be worn.

(*b*) Dangerous to fish, game and wildlife.

*Formulations.* Water soluble liquid at 240 g a.i./l; granules 50 and 100 g a.i./kg.

## 84. Pirimicarb (*Pirimor, Aphox, Fernos*)

*Properties.* A selective insecticide of promise against Diptera and aphids, being effective against organo-phosphorous-resistant aphid strains. It is fast-acting, and has fumigant and translaminar properties – it is taken up by the roots and translocated in the xylem vessels. Non-acaricidal.

It is a colourless solid, virtually insoluble in water, but soluble in most organic solvents. Forms well-defined crystalline salts with acids; these salts are water soluble.

The acute oral $LD_{50}$ for rats is 147 mg/kg.

*Use.* Effective against organophosphorous-resistant strains of aphids, and shows promise against dipterous maggots.

144

*Caution*

*Formulations.* 50% w.p. and 5% granules.

## 85. Promecarb (*Carbamult, Minacide*)

*Properties.* A non-systemic contact insecticide, useful against coleopterous pests, Lepidoptera and Diptera.

It is a colourless crystalline solid, slightly soluble in water, soluble in acetone and ethylene dichloride; hydrolysed by alkali.

The acute oral $LD_{50}$ for rats is 70–100 mg/kg.

*Use.* Effective against Coleoptera, Lepidoptera, and fruit leaf-miners.

*Caution.* This is a poisonous substance, and protective clothing should be worn.

*Formulations.* 25% e.c.; 37.5 and 50% w.p.; 5% dust.

## 86. Propoxur (*Baygon, Blattanex, Unden, Suncide*)

*Properties.* A non-systemic insecticide, with rapid knockdown, used against Hemiptera, flies, millipedes, ants and other household and public health pests. Non-phytotoxic.

It is a white crystalline powder, only slightly soluble in water, but soluble in most organic solvents; unstable in highly alkaline media.

The acute oral $LD_{50}$ for rats is about 100 mg/kg.

*Use.* Effective against jassids, aphids, other bugs, flies, millipedes, termites, ants and other household pests, ticks and mites, mosquitoes.

*Caution*

(*a*) This is a poisonous substance, and protective clothing should be worn.

(*b*) Dangerous to bees.

*Formulations.* E.c.; w.p.; dusts; granules; baits; and pressurized sprays of different a.i. concentrations; some sprays with added dichlorvos.

### Miscellaneous compounds

These are organic fumigants and inorganic salts

which possess toxic qualities to insect pests. The gas and fumigants act through the insect respiratory system, but the inorganic salts act usually as stomach poisons. Most of these compounds are some of the earliest pesticides used in agriculture.

### 87. Aluminium phosphide (*Phostoxin*)

*Properties.* The phosphone liberated is highly insecticidal and a potent mammalian poison, used only for fumigation of stored products.

Aluminium phosphide is a yellow crystalline solid, stable when dry, but reacts with moist air liberating phosphine. Phosphine is highly toxic, and spontaneously inflammable in air. The residue is harmless.

*Use.* Fumigation of stored products and containers only.

*Caution.* This is a very potent poison — fumigation takes from 3 to 10 days and should only be undertaken by trained personnel.

*Formulations. Phostoxin* evolves a non-inflammable mixture of phosphine, ammonia and carbon dioxide; manufactured as 3 g tablets or pellets (0.6 g).

### 88. Copper acetoarsenite (*Paris Green*)

*Properties.* A stomach poison used as baits, very phytotoxic. Introduced about 1867 for the control of Colorado Beetle.

It is a green powder of low solubility in water; in the presence of water and carbon dioxide it is readily decomposed to give water-soluble and phytotoxic arsenical compounds.

Acute oral $LD_{50}$ for rats is 22 mg/kg; a violent poison to man when ingested.

*Use.* Effective only as baits because of high phytotoxicity.

*Caution.* A very poisonous substance which should be handled with great care; a persistent and accumulative poison.

*Formulations.* Most specifications require a content of at least 35% arsenious oxide, 20% cupric oxide, 10% acetic acid; and not more than 1.5% arsenious oxide in a water-soluble form.

### 89. Ethylene dibromide (*Bromofume, Dowfume–W*)

*Properties.* An insecticidal fumigant used against stored products pests and for treatment of fruit and vegetables, and for soil treatment against certain insects and nematodes; very phytotoxic.

It is a colourless liquid, insoluble in water, but soluble in ethanol, ether, and most organic solvents; stable and non-inflammable.

The acute oral $LD_{50}$ for male rats is 146 mg/kg, dermal application will cause severe burning.

*Use.* Used against stored product pests; for ripening treatment of fruit and vegetables; and for soil treatment against certain insects and nematodes. If used for soil treatment planting must be delayed until 8 days after treatment because of its phytotoxicity.

*Caution.* This is a dangerous substance and should be handled with care, and with protective clothing.

*Formulations. Dowfume W-85*, 83%; for soil use in solution in an inert solvent.

### 90. Lead arsenate (*Gypsine, Soprabel*)

*Properties.* A non-systemic stomach insecticide, with little contact action. non-phytotoxic, though the addition of components causing the production of water-soluble arsenical compounds may lead to leaf damage.

Formulated as either paste or powder — $PbHAsO_4$. It is poisonous to mammals when ingested, 10–50 mg/kg being fatal.

*Use.* Used against various caterpillars, sawfly larvae, and tipulid larvae. Does not give complete control of caterpillars, but is selective and does not harm predators of red spider mites.

*Caution*

(*a*) Dangerous to bees.

(b) Harmful to fish and livestock.
(c) Pre-harvest interval for edible crops – 6 weeks.
(d) Pre-access interval for livestock to treated areas –
6 weeks; 3 weeks in wet weather.

*Formulations.* As wettable powders or pastes.

### 91. Mercurous chloride (*Calomel, Cyclostan*)

*Properties.* A general poison, but being phytotoxic
its use is generally limited to soil application, mainly
for the control of root maggots. It is also used as a
fungicide and for the control of club-root of
brassicas. Its biological activity is attributable to its
reduction to metallic mercury.

It is a white powder, insoluble in water, but
soluble in ethanol and most organic solvents. In the
presence of water it is slowly dissociated to mercury
and mercuric chloride, a reaction hastened by alkalis.

The acute oral $LD_{50}$ for rats is 210 mg/kg.

*Use.* Moderately effective against soil-inhabiting root
maggots of vegetables. Use generally restricted also by
the high cost of the chemical.

*Caution*

*Formulations.* The pure compound is used as a seed
dressing; as a 4% dust on a non-alkaline carrier.

### 92. Methyl bromide (*Bromogas, Embafume*)

*Properties.* A general poison with high insecticidal
and some acaricidal properties, used for space fumigation and for the fumigation of plants and plant
products in stores. It is a soil fumigant, used for the
control of nematodes, fungi and weeds.

It is a colourless gas, scarcely soluble in water,
soluble in most organic solvents; stable, non-corrosive
and non-inflammable.

It is highly toxic to man – in many countries
its use is restricted to trained personnel.

*Use.* Effective against stored products pests and soil
pests.

*Caution.* A very poisonous gas – use should be
restricted to trained personnel.

*Formulations.* Packed as a liquid in glass ampoules
(up to 50 ml), in metal cans and cylinders for direct
use. Chloropicrin is sometimes added, up to 2% as a
warning gas.

### 93. Sulphur (Lime-sulphur) (*Cosan, Hexasul*)

*Properties.* A non-systemic direct and protective
fungicide and acaricide. Generally non-phytotoxic,
except to certain varieties known as 'sulphur-shy'.

It is a yellow solid, existing in allotropic forms;
practically insoluble in water, slightly soluble in
ethanol and ether. Slowly hydrolysed by water; compatible with most other pesticides, except petroleum
oils.

*Use.* Used mainly against phytophagous mites,
especially against 'big-bud' mites on blackcurrants.
Also effective against powdery mildews, and apple
scab.

*Caution*

(a) Some fruit varieties are 'sulphur-shy' – the
manufacturer's instructions on labels of products as
to susceptible varieties should be followed.

(b) To avoid possible taint, do not use on fruit for
processing.

*Formulations.* Dusts; w.p.; as finely ground 'colloidal'
suspensions or more usually as lime-sulphur.

### Natural organic compounds

Many plants contain toxic compounds, and
some are selectively poisonous to insects and have
proved valuable as insecticides. One point of value is
that these compounds do not appear to induce resistance in the insect pests. However, so far as the pyrethrins are concerned, these chemicals are so useful
that extensive research is now yielding synthetically
produced pyrethrins, some of which are more effective than their naturally occurring analogues.

## 94. Bioallethrin (*D-Trans, Esbiol*)

*Properties.* A powerful contact insecticide, by nature a synethic pyrethroid, with a rapid 'knockdown' effect. Metabolic detoxification is delayed by addition of synergists such as piperonyl butoxide. It is much more persistent than the natural pyrethins. It occurs as two very similar isomers. insoluble in water, but miscible with most organic solvents.

The acute oral $LD_{50}$ for male rats is $500-780$ mg/kg.

*Use.* Mostly used against household pests at present, especially Diptera.

*Formulations.* Mostly in combination with synergists and other insecticides in kerosene as fly sprays; also as aerosols, $0.1-0.6\%$ a.i.; and impregnated dusts.

## 94(a). Cypermethrin (*Cymbush, Ripcord*)

*Properties.* A synthetic pyrethroid, broad-spectrum in action, and effective against a wide range of insect pests by contact and stomach action; persistent; non-systemic; no recorded phytotoxicity.

A thick brown semi-solid substance, liquid at $60\,^{\circ}C$; stable; insoluble in water but soluble in most organic solvents; more stable in acid than alkaline media; some decomposition in sunlight.

Acute oral $LD_{50}$ for rats $303-4123$ mg/kg, depending on carrier used; for mice 138 mg/kg; a slight skin irritant and eye irritant.

*Use.* Effective against leaf and fruit-eating Lepidoptera and Coleoptera, as well as Hemiptera, cutworms in litter and soil, biting flies and animal ectoparasites.

*Caution*
(*a*) Dangerous to bees, and to fish.
(*b*) Slight skin and eye irritant.

*Formulations.* E.c. $25-400$ g a.i./l; u.l.v. $10-75$ g/l, etc.

## 95. Nicotine (*Nicofume*)

*Properties.* A non-persistent, non-systemic, contact insecticide with some ovicidal properties. Can be used as a fumigant in closed spaces, water-insoluble salts (the so-called 'fixed' nicotines) have been used as stomach insecticides. Nicotine is prepared from tobacco (*Nicotiana tabacum*) by steam distillation or solvent extraction.

A colourless liquid, darkening on exposure to air, miscible with water below $60\,^{\circ}C$ (forming a hydrate); miscible with ethanol, ether, and readily soluble in most organic solvents.

The acute oral $LD_{50}$ for rats is $50-60$ mg/kg.

*Use.* Effective against aphids, capsids, leaf-miners, thrips on a wide range of horticultural crops; sawflies and Woolly Aphid on apple. Also used for fumigation of glasshouses.

*Caution*
(*a*) Harmful to bees.
(*b*) Dangerous to fish, livestock, game, wild birds and animals (Part III poison).
(*c*) Pre-harvest interval for outdoor edible crops — 2 days.
(*d*) Pre-access interval for livestock to treated areas — 12 hours.

*Formulations.* Marketed as the 95% alkaloid, or as nicotine sulphate (40% alkaloids); also as $3-5\%$ dusts. For fumigation nicotine 'shreds' are burnt, or the liquid nicotine is applied to a heated metal surface.

## 95(a). Permethrin (*Ambush, Kalfil, Talcord*)

*Properties.* A synthetic pyrethroid insecticide, with contact action against a wide range of insect pests; persistent; non-systemic.

The technical product is a brown liquid, insoluble in water but soluble in most organic solvents; stable to heat, but some photochemical degradation; more stable in acid than alkaline media.

Acute oral $LD_{50}$ varies with *cis/trans* ratio, carrier used, and conditions of use; for rats 430– 4000 mg/kg.

*Use.* Effective against foliage and fruit-eating Lepidoptera, Coleoptera, Hemiptera, and others, as well as animal ectoparasites. In many uses now generally superceded by cypermethrin.

*Caution*
(*a*) Dangerous to bees.
(*b*) Extremely dangerous to fish.

*Formulations.* E.c. 100–500 g a.i./l; solution of 50 g/l; u.l.v. concentrates 50 and 100 g/l; etc.

## 96. Pyrethrins (*Pyrethrum*)

*Properties.* The pyrethrins is a general term usually including the cinerins as well as the pyrethrins extracted from flowers of *Pyrethrum cinerariaefolium*. They are powerful contact insecticides; non-systemic; causing a rapid paralysis or 'knockdown'. They are unstable to sunlight and are rapidly hydrolysed by alkalis with a loss of insecticidal properties. Their metabolic detoxification may be delayed by the addition of synergists such as piperonyl butoxide, sesamin, etc.

The acute oral $LD_{50}$ for rats is about 200 mg/kg.

*Use.* Effective against flies, and household pests, usually with added synergist.

*Caution*

*Formulations.* As dusts with added non-alkaline carrier; as aerosols the extract is dissolved in a volatile solvent such as methyl chloride or dichlorodifluoromethane, usually with added synergist.

## 97. Resmethrin (*Chryson, For-Syn, Synthrin*)

*Properties.* A synthetic pyrethroid, with powerful contact action against a wide range of insects. Plant toxicity is low. However it is not synergised to any extent by pyrethrin synergists. Toxicity to normal houseflies is about 20 times that of natural pyrethrins.

It occurs as a mixture of isomers; a colourless waxy solid; insoluble in water, but soluble in most organic solvents. More stable than pyrethrins, but is decomposed quite rapidly on exposure to air and light.

The acute oral $LD_{50}$ for rats is about 2000 mg/kg.

*Use.* Effective against a wide range of insect pests, both household and agricultural.

*Formulations.* May be formulated with or without other pyrethroids and pyrethrin synergists in aerosol concentrates; also as water-based sprays; e.c.; w.p.; and concentrate for u.l.v. application.

The mixture of isomers is usually 20–30% of cis-isomer, and 80–70% of the trans-isomer, the former known sometimes as Cismethrin, and the latter as Bioresmethrin.

## 98. Rotenone (*Derris, Cube, Rotacide*)

*Properties.* This is the name given to the main insecticidal compound of certain *Derris* spp., and *Lonchocarpus* spp.; known for many years to be effective as a fish poison and an insecticide. It is a selective non-systemic insecticide with some acaricidal properties; non-phytotoxic; but of low persistence in spray or dust residues. Insoluble in water but soluble in polar organic solvents; readily oxidized in presence of light and alkali to less insecticidal products.

The acute oral $LD_{50}$ for rats is 132–1500 mg/kg; it is very toxic to pigs.

*Use.* Effective against aphids, caterpillars, thrips, some beetles, and red spider mites.

*Caution*
(*a*) Dangerous to fish.
(*b*) Toxic to pigs.
(*c*) Pre-harvest interval for edible crops – 1 day.

*Formulations.* Usually as dusts of the ground root

with a non-alkaline carrier; dusts may be stabilized by addition of a small quantity of a strong acid such as phosphoric acid.

### Organic oils

Also known as hydrocarbon oils, they comprise the oils distilled from crude mineral oils (petroleum or mineral oils) or from coal tars (tar oils). These oils are complex chemically, of such a nature as to defy separation into individual compounds. They are all strongly phytotoxic and use is generally restricted to dormant season washes, when they are destructive to insect and mite eggs.

### 99. Petroleum oils (*Volck,* etc.)

*Properties.* Also known as mineral oils; refined grades are known as white oils. The use of kerosene as an insecticide probably dates from about the time of its introduction as an illuminant. Oils of higher distillation range came into use about 1922. They consist largely of aliphatic hydrocarbons, both saturated and unsaturated. They are produced by distillation and refinement of crude mineral oils. The ones used as pesticides generally distil above $310°C$ ($636°F$), namely 64–79%, 40–49%, and 10–25% respectively; density rarely exceeds 0.92 at $15°C$ ($60°F$). Viscosity and density vary according to the geographical area (oil-field) from which the crude oil came.

*Use.* Effective against certain insects such as mealybugs, scales and thrips, and against red spider mites; they are ovicidal. Their use is limited by their phytotoxicity; a semi-refined oil can be used as a dormant ovicide; for foliage use, a refined oil of narrow viscosity range is required. Relatively harmless to mammals.

*Caution.* Generally very phytotoxic; can be used as herbicide (especially TVO).

*Formulations.* Available either alone or in mixture with DNOC or DNOC and DDT.

### 100. Tar oils

*Properties.* Produced by distillation of tars resulting from the high temperature carbonization of coal and coke oven tars. Although used for wood preservation since 1890, the introduction of the formulated products known as tar oil washes for crop protection dates from about 1920. These oils are brown to black liquids, of density 1.05–1.11; insoluble in water but soluble in organic solvents. They consist mainly of aromatic hydrocarbons but contain 'phenols' and 'tar acids'; they are highly phytotoxic.

*Use.* Effective for the control of the eggs of many insect species, particularly of aphids, Lepidoptera, psyllids and other bugs. Because of their phytotoxicity use is restricted to the dormant season.

*Caution*
(*a*) Dangerous to fish.
(*b*) Irritating to skin, eyes, nose and mouth.
(*c*) Strongly phytotoxic; use only as dormant washes.

*Formulations.* As miscible winter washes, and stock emulsion winter washes. The most insecticidally active tar oil is creosote.

### Biological compounds

A number of polyhedral and granular viruses, and bacteria, and a few fungi, are commonly recorded as causing diseases in insect populations. On several occasions when part of a pest population has been found to be naturally infected by a pathogen it has been possible to make up a spray solution from a suspension of macerated diseased insect bodies and to spray this solution over uninfected parts of the population, often with spectacular success. Generally, the only insect groups that regularly suffer from viral or bacterial epizootics are Lepidoptera and Coleoptera.

The only biological compounds at present readily available commercially for use against insect pests are a spore suspension of *Bacillus thuringiensis*, for use against lepidopterous larvae only and *Helio-*

*this* Nuclear Polyhedrosis Virus. Various strains of *Bacillus popilliae* cause Japanese Beetle Milky Disease, and commercial formulations are now available in the USA.

## 101. *Bacillus thuringiensis* Berliner (*BTB-183, Thuricide, Biotrol-BTB*)

*Properties.* This biological compound is a suspension of spores in an inert powder, specific in action to lepidopterous larvae, on which it acts as a stomach poison only. There are no phytotoxic effects, and it is innocuous to both mammals, other animals, and insect predators and parasites. It is a stable product which has been kept at room temperature for over six years without any detectable loss in potency. However, since it is a viable biological compound it has to be protected from extremes of heat and light, and corrosive fumes. This bacterium, in addition to forming resistant spores, also produces a crystal of thermolabile endotoxin called a parasporal body – it is this parasporal body which is the insecticidal agent in any preparation of *B. thuringiensis*. The crystalline toxin is insoluble in water and is thus deposited intact upon the leaf surface, and after ingestion by the caterpillar acts as a stomach poison. It has no contact action whatsoever. Some caterpillars are killed after feeding within a few hours; other less susceptible species are not killed directly but suffer a gut paralysis which stops them feeding, and they die within a few days. A few important pest species of Lepidoptera do not appear to be susceptible to this pathogen.

*Use.* This is effective only against the larvae of Lepidoptera, and as already mentioned is not equally effective against all caterpillars.

*Caution.* There appear to be no real restrictions against the use of this insecticide at all. However, it is incompatible with certain insecticides which appreciably alter the pH.

*Formulations.* It is formulated as a wettable powder containing 25 billion viable spores per gram of product (2.5%), and 97.5% of inert ingredients.

## 102. *Heliothis* Nuclear Polyhedrosis Virus (Elcar)

*Properties.* This is a commercial formulation of the naturally occurring *Heliothis* Nuclear Polyhedrosis Virus, produced by Sandoz Inc. in California for use against *Heliothis* bollworms on cotton. Now it is being tested on other crops in various countries. There are no toxic effects against other animals or plants, but the virus is destroyed by high temperatures and lengthy exposure to bright sunlight.

*Use.* Effective against species of *Heliothis* only; the caterpillars die within a few days of eating the sprayed foliage.

*Formulation.* Used as a water-based spray, applied to the plant foliage.

Several other polyhedrosis viruses (*Spodoptera, Trichoplusia*, etc.) are now commercially available, but only in the USA.

### Insect growth regulators (IGR) and juvenile hormones (JH)

In the search for the effective, specific pesticides, which have minimal disruptive effects on the environment and do not encourage resistance build-up in the pests, insect hormones have shown great promise. They fall into two distinct categories. The juvenile hormones (JH), if applied to full-grown larvae (usually caterpillars), disturb the process of metamorphosis to the extent that the insect dies as a deformed pupa/adult. These hormones are normally species-specific, and so it is only economic to contemplate commercial production of JH for very important species of pests, such as Cotton Boll Weevil and Pink Bollworm. Insect growth regulators (IGR) often act as pesticides by interfering with cuticle formation at the time of ecdysis, killing the moulting larva. These chemicals are not as specific as JHs, and will be effective against members of the same group, e.g. caterpillars or fly maggots (larvae).

### 103. Diflubenzuron (*Dimilin*)

*Properties.* An insect growth regulator, produced in 1974, which operates by ingestion, and interference with the deposition of new cuticle at ecdysis. A species-dependent ovicidal (contact) action has been shown for *Spodoptera* as well as prevention of egg eclosion after uptake by females. No systemic action, so it is not effective against sap-sucking insects; in soil, it is rapidly degraded.

The technical product is pale brown crystals, insoluble in water and apolar solvents, but it is soluble in polar solvents.

The acute oral $LD_{50}$ for rats was more than 4640 mg/kg.

*Use.* Effective against a broad range of leaf-eating insects and some mites, at rates of 1.5–30 g a.i./100 l. Also used against fly and mosquito larvae.

*Formulation.* A water dispersible powder of 250 g a.i./kg.

### 104. Methoprene (*Altosid, Manta, Kabat*)

*Properties.* An insecticide of the insect growth regulator group, of short field persistence being rapidly degraded by micro-organisms and sunlight. No recorded damage to non-target organisms, especially predators.

A pale amber liquid, insoluble in water, but soluble in most organic solvents. Completely non-toxic to laboratory tested mammals.

*Use.* Most effective against Diptera (larvae); developed initially for destruction of mosquito larvae, now used for systemic treatment of cattle against hornfly, and being tested against various stored products pests. Treated larvae develop apparently normally into pupae, but the pupae die without giving rise to adults.

*Formulation.* Liquid 10% (8 g a.i./l.).

---

## Pesticides/pest chart

---

In the event of wanting to kill insects which are not dealt with here as major pests, and so do not have specific control recommendations, it can be borne in mind that various pesticides are generally toxic to groups of closely related insects. If no recommendation is available then the table may be of assistance in deciding which chemical to use. However, it must be stressed that such extrapolation can be dangerous, for although some chemicals are specific to certain groups of insects, many notable exceptions occur. For example, carbaryl is a pesticide generally effective against caterpillars in most parts of the world, but in Malawi, on the cotton crop, it is effective against the Red Bollworm (*Diparopsis castanea*) but has no toxic effect upon *Heliothis* caterpillars, and DDT had to be used against the latter pests, against a mixed infestation a spray mixture was employed.

## PESTS AGAINST WHICH GENERALLY EFFECTIVE

| CHEMICAL | Orthoptera | Isoptera | Homoptera | Aphids | Mealybugs | Scales | Heteroptera | Thysanoptera | Lepidoptera | Diptera | Coleoptera | Hymenoptera | Acarina | Nematoda | Miscellaneous |
|---|---|---|---|---|---|---|---|---|---|---|---|---|---|---|---|
| **Chlorinated hydrocarbons** | | | | | | | | | | | | | | | |
| 1. Aldrin | x | x | x | . | x | . | . | . | x | x | . | . | . | . | . |
| 2. BHC (HCH) | x | . | x | x | . | . | x | x | x | x | x | x | . | . | Spiders & Myriapoda |
| 3. Chlorobenzilate | . | . | . | . | . | . | . | . | . | . | . | . | x | . | . |
| 4. DDT | x | x | x | x | x | x | x | x | x | x | x | x | x | . | . |
| 5. Dicofol | . | . | . | . | . | . | . | . | . | . | . | . | x | . | . |
| 6. Dieldrin | x | x | x | x | x | x | x | x | x | x | x | x | . | . | Spiders |
| 7. Endosulfan | . | . | x | x | . | . | x | . | . | x | x | . | x | . | . |
| 8. Endrin | x | x | x | x | x | x | x | x | x | x | x | x | x | . | . |
| 9. Heptachlor | . | . | . | . | . | . | . | . | . | x | x | x | . | . | . |
| 10. Mirex | . | . | . | . | . | . | . | . | . | . | . | x | . | . | . |
| 11. Tetradifon | . | . | . | . | . | . | . | . | . | . | . | . | x | . | . |
| 12. Tetrasul | . | . | . | . | . | . | . | . | . | . | . | . | x | . | . |
| 13. TDE | . | . | . | . | . | . | x | x | x | . | x | . | . | . | . |
| 14. Amitraz | . | . | x | . | . | . | . | . | Eggs | . | . | . | x | . | . |
| **Substituted phenols** | | | | | | | | | | | | | | | |
| 15. Binapacryl | . | . | . | . | . | . | . | . | . | . | . | . | x | . | Fungi |
| 16. DNOC | x | . | x | x | x | x | x | . | x | . | . | . | x | . | . |
| 17. Pentachlorophenol | . | x | . | . | . | . | . | . | . | . | x | . | . | . | . |
| **Organophosphorous compounds** | | | | | | | | | | | | | | | |
| 18. Azinphos-methyl | x | . | x | x | x | x | x | . | x | x | x | . | x | . | . |
| 19. Azinphos-methyl with demeton-S-methyl sulphone | . | . | x | x | . | . | . | x | x | x | . | . | x | . | . |
| 20. Bromophos | . | . | x | . | . | . | x | . | . | x | x | . | x | . | . |
| 21. Bromopropylate | . | . | . | . | . | . | . | . | . | . | . | . | x | . | . |
| 22. Carbophenothion | . | . | . | x | x | x | . | . | . | x | . | . | x | . | . |

| | Orthoptera | Isoptera | Homoptera | Aphids | Mealybugs | Scales | Heteroptera | Thysanoptera | Lepidoptera | Diptera | Coleoptera | Hymenoptera | Acarina | Nematoda | Miscellaneous |
|---|---|---|---|---|---|---|---|---|---|---|---|---|---|---|---|
| 23. Chlorfenvinphos | . | . | x | . | . | x | . | . | x | x | x | . | . | . | . |
| 24. Chlorpyrifos | . | . | x | x | . | . | x | . | x | x | x | x | x | . | . |
| 25. Demephion | . | . | x | x | . | . | . | . | . | x | . | . | x | . | . |
| 26. Demeton | . | . | x | x | x | x | x | . | . | . | . | . | x | . | . |
| 27. Demeton-$S$-methyl | . | . | x | x | x | x | x | . | . | x | . | x | x | . | . |
| 28. Diazinon | . | . | x | x | . | x | x | x | . | x | . | x | x | . | . |
| 29. Dichlorvos | x | . | x | x | x | x | x | x | x | x | x | x | x | . | . |
| 30. Dimefox | . | . | x | x | . | . | . | . | . | . | . | . | x | . | . |
| 31. Dimethoate | . | . | x | x | . | x | . | x | . | x | . | x | x | . | . |
| 32. Disulfoton | . | . | x | x | . | . | . | . | . | x | x | . | x | . | . |
| 33. Ethion | . | . | x | x | . | x | . | . | . | x | . | . | x | . | . |
| 34. Ethoate-methyl | . | . | . | x | . | . | . | . | . | x | . | . | x | . | . |
| 35. Fenitrothion | x | . | x | x | x | x | x | x | x | x | x | x | . | . | . |
| 36. Fenthion | . | . | x | x | . | . | x | x | x | x | x | . | . | . | Birds |
| 37. Fonofos | x | . | . | . | . | . | . | . | x | x | x | . | . | . | . |
| 38. Formothion | . | . | x | x | . | . | . | . | . | x | . | . | x | . | . |
| 39. Heptenophos | . | . | x | x | . | . | x | . | . | x | . | . | x | . | Ectoparasites |
| 40. Iodofenphos | . | . | . | . | . | . | . | . | . | x | x | x | . | . | . |
| 41. Malathion | . | . | x | x | x | x | x | x | x | x | x | x | x | . | . |
| 42. Mecarbam | . | . | x | x | x | x | x | . | . | x | . | . | x | . | . |
| 43. Menazon | . | . | . | x | . | . | . | . | . | . | . | . | . | . | . |
| 44. Methidathion | . | . | x | x | x | x | x | . | x | . | . | . | x | . | . |
| 45. Mevinphos | . | . | x | x | . | . | . | . | x | x | x | . | x | . | . |
| 46. Monocrotophos | . | . | x | . | . | . | x | . | x | . | x | . | x | . | . |
| 47. Naled | . | . | x | . | . | . | . | . | . | x | . | . | x | . | . |
| 48. Omethoate | x | . | x | x | x | x | . | x | x | . | x | . | x | . | . |
| 49. Oxydemeton-methyl | . | . | x | x | . | . | x | . | . | x | . | x | x | . | . |
| 50. Oxydisulfoton | . | . | x | x | . | . | . | . | . | . | . | . | x | . | . |
| 51. Parathion | . | . | x | x | x | x | x | . | x | x | . | . | x | . | . |
| 52. Parathion-methyl | . | . | x | x | x | x | x | . | . | x | x | . | . | . | . |
| 53. Phenisobromolate | . | . | . | . | . | . | . | . | . | . | . | . | x | . | . |
| 54. Phenthoate | . | . | x | x | . | . | x | . | x | . | x | . | x | . | . |
| 55. Phorate | . | . | x | x | . | . | x | x | x | x | x | . | . | . | . |
| 56. Phosalone | . | . | x | x | . | . | . | x | x | x | x | . | x | . | . |

153

| | Orthoptera | Isoptera | Homoptera | Aphids | Mealybugs | Scales | Heteroptera | Thysanoptera | Lepidoptera | Diptera | Coleoptera | Hymenoptera | Acarina | Nematoda | Miscellaneous |
|---|---|---|---|---|---|---|---|---|---|---|---|---|---|---|---|
| 57. Phosmet | . | . | . | . | . | . | . | . | . | . | . | . | x | . | . |
| 58. Phosphamidon | x | . | x | x | . | x | x | x | x | x | x | . | x | . | . |
| 59. Phoxim | x | . | . | . | . | . | . | . | x | x | x | . | . | . | . |
| 60. Pirimiphos-ethyl | . | . | x | . | . | . | . | . | x | x | x | . | x | . | . |
| 61. Pirimiphos-methyl | x | . | x | x | x | x | . | x | x | x | x | x | x | . | . |
| 62. Profenofos | . | . | x | . | . | . | x | . | x | . | x | . | x | . | . |
| 63. Prothoate | . | . | x | . | . | . | x | x | . | . | . | . | x | . | . |
| 64. Quinomethionate | . | . | . | . | . | . | . | . | . | . | . | . | x | . | . |
| 65. Schradan | . | . | . | x | . | . | . | . | . | . | . | . | x | . | . |
| 66. TEPP | . | . | . | x | . | . | . | . | . | . | . | . | x | . | . |
| 67. Terbufos | . | . | . | . | . | . | . | . | . | . | x | x | . | . | Myriapoda |
| 68. Tetrachlorvinphos | . | . | . | . | . | . | . | . | x | x | . | . | . | . | . |
| 69. Thiometon | . | . | x | . | . | . | . | . | . | . | . | . | x | . | . |
| 70. Thionazin | . | . | . | . | . | . | . | . | . | x | . | . | x | . | . |
| 71. Thioquinox | . | . | . | . | . | . | . | . | . | . | . | . | x | . | . |
| 72. Triazophos | . | . | . | x | . | . | x | . | x | x | x | . | x | x | . |
| 73. Trichloronate | . | x | . | . | . | . | . | . | x | x | x | . | . | x | . |
| 74. Trichlorphon | . | . | x | . | . | . | x | . | x | x | x | . | . | . | . |
| 75. Tricyclohexyltin hydroxide | . | . | . | . | . | . | . | . | . | . | . | . | x | . | . |
| 76. Vamidothion | . | . | x | x | . | . | . | . | . | . | . | . | x | . | . |

**Carbamates**

| | Orthoptera | Isoptera | Homoptera | Aphids | Mealybugs | Scales | Heteroptera | Thysanoptera | Lepidoptera | Diptera | Coleoptera | Hymenoptera | Acarina | Nematoda | Miscellaneous |
|---|---|---|---|---|---|---|---|---|---|---|---|---|---|---|---|
| 77. Aldicarb | . | . | x | x | x | x | x | . | . | x | x | . | x | x | Nematodes |
| 78. Bufencarb | . | . | x | . | x | . | . | . | x | . | x | . | . | . | . |
| 79. Carbaryl | x | . | x | . | . | x | x | . | x | x | x | x | . | . | Earthworms |
| 80. Carbofuran | . | . | x | x | . | . | . | . | x | x | x | . | x | x | Myriapoda |
| 81. Methiocarb | . | . | . | . | . | . | . | . | x | . | x | . | x | . | Mollusca |
| 82. Methomyl | . | . | x | . | . | . | . | . | x | . | x | . | x | x | . |
| 83. Oxamyl | . | . | x | x | . | . | x | . | x | x | x | . | x | x | Nematodes |
| 84. Pirimicarb | . | . | x | . | . | . | . | . | . | x | . | . | . | . | . |
| 85. Promecarb | . | . | . | . | . | . | . | . | x | x | x | . | . | . | . |
| 86. Propoxur | . | x | x | x | . | . | x | . | . | x | x | . | x | . | Myriapoda |

**Miscellaneous compounds**

| | Orthoptera | Isoptera | Homoptera | Aphids | Mealybugs | Scales | Heteroptera | Thysanoptera | Lepidoptera | Diptera | Coleoptera | Hymenoptera | Acarina | Nematoda | Miscellaneous |
|---|---|---|---|---|---|---|---|---|---|---|---|---|---|---|---|
| 87. Aluminium phosphide | . | . | . | . | . | . | . | . | x | . | x | . | . | . | General toxicity |
| 88. Copper acetoarsenite | . | . | . | . | . | . | . | . | . | . | x | . | . | . | . |

| | Orthoptera | Isoptera | Homoptera | Aphids | Mealybugs | Scales | Heteroptera | Thysanoptera | Lepidoptera | Diptera | Coleoptera | Hymenoptera | Acarina | Nematoda | Miscellaneous |
|---|---|---|---|---|---|---|---|---|---|---|---|---|---|---|---|
| 89. Ethylene dibromide | . | . | . | . | . | . | . | . | x | . | x | . | . | . | . |
| 90. Lead arsenate | . | . | . | . | . | . | . | . | x | x | . | x | . | . | . |
| 91. Mercurous chloride | . | . | . | . | . | . | . | . | . | x | . | . | . | . | Fungi |
| 92. Methyl bromide | . | x | . | . | . | . | . | . | x | . | x | x | . | x | Nematodes & Fungi |
| 93. Sulphur (Lime-sulphur) | . | . | . | . | . | x | . | x | . | . | . | . | x | . | Fungi |
| **Natural organic compounds** | | | | | | | | | | | | | | | |
| 94. Bioallethrin | . | . | . | . | . | . | . | . | . | x | . | . | . | . | Household pests |
| 94a. Cypermethrin | . | . | x | . | . | . | x | . | x | . | x | . | . | . | Household pests |
| 95. Nicotine | . | . | . | x | . | . | x | x | . | x | . | x | . | . | . |
| 95a. Permethrin | . | . | x | . | . | . | x | . | x | . | x | . | . | . | Household pests |
| 96. Pyrethrins | . | . | x | . | . | . | x | . | x | x | . | x | x | . | . |
| 97. Resmethrin | . | . | . | . | . | . | . | . | . | x | . | . | . | . | Household pests |
| 98. Rotenone | . | . | x | . | . | . | x | x | . | . | x | x | x | . | . |
| **Organic oils** | | | | | | | | | | | | | | | |
| 99. Petroleum oils | . | . | . | . | x | x | x | . | . | . | . | . | x | . | . |
| 100. Tar oils | . | . | x | x | x | x | x | . | x | . | . | . | . | . | Eggs & Fungi |
| **Biological compounds** | | | | | | | | | | | | | | | |
| 101. *Bacillus thuringiensis* | . | . | . | . | . | . | . | . | x | . | . | . | . | . | |
| 102. Heliothis Nuclear Poly-hedrosis Virus | . | . | . | . | . | . | . | . | x | . | . | . | . | . | |
| **Insect growth regulators** | | | | | | | | | | | | | | | |
| 103. Diflubenzuron | . | . | x | . | . | . | . | . | x | x | . | . | x | . | . |
| 104. Methoprene | . | . | . | . | . | . | . | . | . | x | . | . | . | . | . |

*N.B.* It must be borne in mind that these tables only serve as *guides* to pesticide use, local advice must always be sought for details. Some chemicals I have no personal experience of using, and they are recommended for use 'on cotton', or 'against soil insects', 'against leaf-miners', etc.; this information does not indicate at all precisely against which pests they are effective. Also an insecticide generally effective against caterpillars may not kill all species, one that kills sawfly larvae may not kill ants (Hymenoptera), and one that kills leaf-hoppers may not kill psyllids (Homoptera). Also these tables do not take into account either local government restrictions on pesticide usage or local resistance problems.

# 7 *Major tropical crop pests*

## Descriptions, biology and control

The orders and families of pests are arranged according to the 9th edition of Imms *A General Textbook of Entomology* as revised by Richards & Davies (1960), and within the family the species are arranged alphabetically. The only exceptions are the families Scarabaeidae and Chrysomelidae of the Coleoptera, which are very large groups with well-defined subfamilies with important biological differences between the different subfamilies; with these two families the important pest species are separated into their respective subfamilies.

It must be stressed that all the chemical control recommendations in this chapter are tentative and intended more as a guide than as definite recommendations. This is in part because invariably chemical control recommendations are out of date by the time they are published, and also there is no generally agreed international list of recommendations for pesticide use: a chemical approved in one country may be banned in another as most countries have different approval criteria. This is now more apparent than ever, as some countries have imposed blanket bans on the use of the organochlorine compounds, whereas in others DDT, BHC, dieldrin and the like are still recommended. In this book it is attempted to point out which chemicals are, or have been, effective in killing particular pests, whether or not they are available, or approved, by any particular country. Local advice *must* be sought for details of recommended treatments for local pests, as these will take into account chemical availability, approval, local climatic and soil conditions, and any local resistance problems.

# Order **ORTHOPTERA**

Medium-sized or large insects, with the forewings thickened and modified as tegmina, although some have the wings reduced or absent; the hind legs are usually enlarged for jumping; biting mouthparts; metamorphosis is only slight; they often have specialized auditory and stridulatory organs; the female usually has a well-developed ovipositor.

## Family **Tettigoniidae**

(Long-horned grasshoppers)  A predominantly tropical group; the antennae are typically longer than the insect body; many apterous species; winged species are usually green or brown in colour and they live in herbage, particularly in bushes and trees; the eggs are not laid in pods, and the female ovipositor is frequently very long and curved; the eggs are laid in plant tissues or in the soil. A few species are carnivorous. There are about 4000 species.

## Family **Gryllidae**

(Crickets)  Many apterous species; some are arboreal (bush crickets); the antennae are very long; they are omnivorous and live in warm places; many burrow in the soil; the female has a long straight ovipositor and long unsegmented cerci; eggs are usually laid in the soil, and some species construct underground chambers for this purpose, other species lay their eggs in the pith of twigs. About 900 species have been described.

## Family **Gryllotalpidae**

(Mole crickets)  These are characterized by various adaptations to a subterranean habit; the forelegs are greatly expanded and armed with strong teeth for digging; eyes and antennae are reduced; ovipositor is vestigial; winged species are strong fliers, but apterous and brachypterous species occur; a subterranean nest is constructed for breeding purposes. There are about 50 species, some of which are pests of agricultural importance.

## Family **Acrididae**

(Short-horned grasshoppers and locusts) The antennae are short, much less than the length of the body; the locusts are essentially gregarious species capable of swarming both as nymphs and adults, whereas the grasshoppers are solitary; eggs are laid in pods in the ground; certain species in S.E. Asia are aquatic in habit. Many species are crop pests; when swarming, the locusts are devastating. About 5000 species are known.

# Homorocoryphus nitidulus vicinus Wlk.

**Common name.** Edible Grasshopper
**Family.** Tettigoniidae
**Hosts** (main). Rice, millets, maize, sorghum, and
wheat.

    (alternative). Many species of grasses.
**Damage.** Swarms of the green and brown long-horned
grasshoppers collect on cereal crops in the 'milk'
stage; the grains are eaten leaving the spikes intact.
Maize silks are completely grazed off, resulting in
poor fertilization and cobs with a poor complement
of grains.
**Pest status.** A sporadically serious pest in parts of
E. Africa on the crops referred to above.
**Life history.** Little is known of the breeding of the
Edible Grasshopper which apparently takes place on
wild grasses.

    Adults are slender long-horned grasshoppers,
about 6 cm long from head to wing tips. They may
be pale green or brown. They fly during the rainy
seasons both by day and night, and are often found
in great numbers round street lights. Any cereal crop
in the milk stage is very attractive to them, rice being
particularly favoured.
**Distribution.** Found only in E. Africa, but widely
occurring throughout Uganda, Kenya, and Tanzania.
**Control.** A bait of wheat bran with BHC dust or
aldrin w.p. added is effective. Water is added to the
mixture until it is a crumbly mash, and it is then
broadcast between the crop rows on a dry evening,
preferably on weed-free soil. When treating maize
the bait should be thrown into the crop so that it
lodges on the cobs.

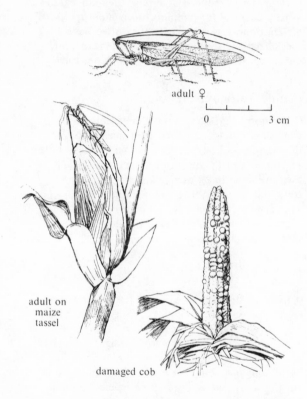

adult ♀

0         3 cm

adult on
maize
tassel

damaged cob

## Acheta spp.

**Common name.** Field Crickets
**Family.** Gryllidae
**Hosts.** General pests that live in the soil and at night attack many different herbaceous crops. Will damage tea and coffee seedlings by eating the bark.
**Damage.** Young plants generally have parts of the foliage eaten, and young woody plants may have the bark gnawed away in patches.
**Pest status.** A sporadically serious pest of many different crops, particularly at the seedling stage, only occasionally causing economic damage.
**Life history.** The elongate, banana-shaped, yellow eggs are laid in batches of up to 30 in the ground, by the female using her long ovipositor. Each female cricket lays up to about 2000 eggs. At 26 °C the eggs hatch in 10−12 days, and development of the nymphs takes 40−60 days more. Adults generally live for 2−3 months. These insects are omnivorous, and laboratory studies have shown that normal development requires a diet including other insects.

Total development takes some 50−80 days, according to climate and diet, and in the tropics there are usually four generations per year.

The adult is a dark brown or black cricket, with a body length of 2−3 cm; the cerci are long, and the female has a long ovipositor, some 15−18 mm.
**Distribution.** The two main species concerned are
*A. bimaculata* (De Geer) (Two-spotted Cricket)

found throughout Africa, southern Europe, and parts of Asia, and *A. testaceus* Wlk. found throughout S. China and S.E. Asia.
**Control.** Damage is usually not sufficiently serious to warrant control measures, but if required a bran bait mixed with BHC dust or aldrin w.p. should be effective.

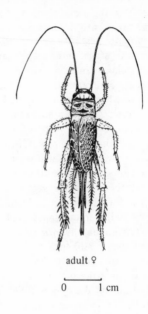

adult ♀

0      1 cm

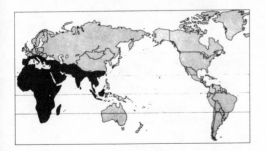

159

## Brachytrupes membranaceus (Drury)

**Common name.** Tobacco Cricket
**Family.** Gryllidae
**Hosts.** Many crops are attacked, including tobacco, tomato, tea, and cotton; seedlings are particularly vulnerable.
**Damage.** Seedlings are cut off and dragged into underground burrows, or left on the surface wilting for a few days before being taken into the burrow.
**Pest status.** A local sporadic pest of many crops, particularly at the seedling stage, and on light sandy soils where the adult crickets can easily burrow.
**Life history.** The eggs are elongate-oval, 3−4 mm long when first laid, and white, later becoming brown and expanding to 5−6 mm. They hatch after about one month.

The full-grown nymph is a fat, brown insect 4−5 cm long; there are four nymphal instars, the total nymphal period taking about eight months.

The adult is a large, fat, brown insect with a heavy square head, long thin antennae, and powerful hindlegs. The body, excluding appendages, is about 5 cm long. The adult female may live for three or four months and will lay over 300 eggs.

Newly hatched nymphs leave the burrow of the mother cricket and construct one of their own. It is gradually enlarged during the nymphal period, finally reaching a depth of 60−80 cm in sandy soil. The adult stage is reached at the start of the rainy season and eggs are laid in batches in the burrow three to four months later. Both nymphs and adults collect seedlings, leaves, and other soft vegetation, both dead and alive, and drag them into their burrow where they are stored and finally eaten. Fresh sappy material is often left on the surface for a day or two to wilt before being taken below ground.

**Distribution.** Widespread in tropical and southern Africa, down to the Transvaal.
**Control.** See *B. portentosus*.

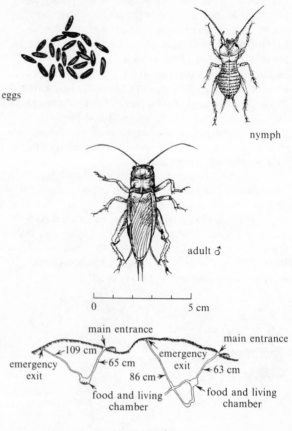

eggs

nymph

adult ♂

0          5 cm

main entrance

←109 cm

←65 cm

emergency
exit

emergency
exit

main entrance

←63 cm

86 cm

food and living
chamber

food and living
chamber

nest complex

## Brachytrupes portentosus Licht.

**Common name.** Large Brown Cricket
**Family.** Gryllidae
**Hosts.** Many field crops are attacked; seedlings are particularly vulnerable; but little host specificity is shown.
**Damage.** Typically seedlings are cut through and dragged into the underground nest where they are eaten.
**Pest status.** A sporadic pest, of polyphagous habits, feeding on many different crop plants; particularly damaging to seedlings, and more numerous on light sandy soils where burrows are easy to excavate.
**Life history.** The life history details are much the same as for *B. membranaceus* as the two species are very closely related.
**Distribution.** This genus is represented in tropical Africa by the species *B. membranaceus* and in tropical Asia by the allopatric *B. portentosus* which is also found in Papua New Guinea.
**Control.** When control is required, a bran bait is generally effective when mixed with BHC dust or aldrin w.p. Maize flour can be used but is generally not so effective as wheat bran. The moist crumbly mixture should be broadcast between the crop rows in the evening (for the pest is nocturnal) on weed-free soil.

adult     nymph

0    2 cm

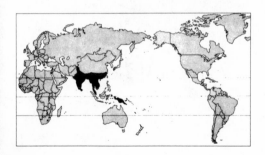

161

# Gryllotalpa africana Pal.

**Common name.** African Mole Cricket
**Family.** Gryllotalpidae
**Hosts.** A general pest attacking many herbaceous crops, especially at the seedling stage. Some shrubs (e.g. tea) are also attacked in propagating beds.
**Damage.** Heaps of soil mark the entrances to extensive burrows in the soil. Plants wilt owing to destruction of roots. Small seedlings may disappear completely during the night. Buds are eaten from sugarcane setts; potato tubers are tunnelled.
**Pest status.** A sporadically serious pest, especially at lower altitudes, and particularly in moist soil.
**Life history.** The eggs are oval, brown, and 1.5 mm long, and laid in chambers at the end of burrows 10–15 cm below the soil surface. The female mole cricket may construct three or more of these chambers and lay a total of about 100 eggs distributed between them. Eggs hatch after two or three weeks.

The first stage nymphs remain in the egg chamber and are fed by the mother. Subsequent instars live in burrows during the day and forage for food on the soil surface at night. There are 9–11 instars and the total nymphal period lasts for ten months.

The adult cricket is about 2.5 cm long, brown, and covered with short setae giving it a velvety appearance. The wings are folded and do not cover the full length of the abdomen. The forelegs are broad and curved and clearly adapted for digging. Like the nymphs the adults live in burrows or shelter under pieces of trash in the daytime and feed at night. Mating takes place about ten days after the last moult and egg-laying begins one or two weeks later. Eggs are laid in the rainy seasons, and as the total life-cycle takes about a year, there are probably two overlapping generations present in the field. Adults live for at least two or three months.
**Distribution.** Cosmopolitan in the warm regions of the Old World; comprising Africa, Mauritius, Egypt, Asia, the Far East, Australasia, and the Pacific islands (CIE map no. A293).
**Control.** As with *Brachytrupes portentosus*.

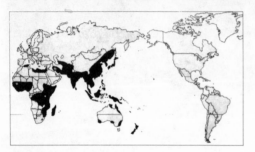

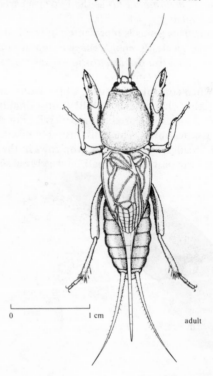

0      1 cm

adult

162

## Locusta migratoria migratorioides (R. & F.)

**Common name.** African Migratory Locust
**Family.** Acrididae
**Hosts.** A polyphagous pest but showing some preference for Gramineae, both wild and cultivated.
**Damage.** The scattered solitary forms cause negligible damage, but swarms cause complete defoliation of crops.

The last major outbreak was in 1948.

**Pest status.** A sporadically serious pest in tropical Africa, with devastating swarm damage. The outbreak area of this subspecies is in the Niger valley of W. Africa, and the invasion area covers western, central, eastern and southern Africa.

**Life history.** The eggs are laid in pods in the ground, and each female lays four or five batches. The eggs hatch in 10–25 days according to the temperature.

There are six nymphal instars.

The adults are slightly smaller than Desert Locusts, being 35–40 mm long (male) and 40–50 mm (female). The body is pale yellowish colour with fine dark lateral stripes across and along the abdomen; the elytra are translucent with many brown spots. There is no peg-like process between the bases of the forelegs, and the undersurface of the thorax is covered with fine setae ('hairy').

There is only one generation per year.

**Distribution.** Benin, Ivory Coast, Niger, Guinea, Upper Volta, Senegal, Sudan, Mauritania, Nigeria, Ghana, Togo, E. Africa, S.W. Africa, Gambia, Sierra Leone, and Zimbabwe.

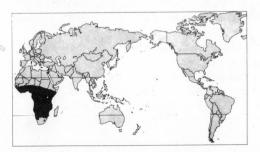

Two other important subspecies are: the Asiatic Migratory Locust (*L. m. migratoria*) in Central Asia, and the Oriental Migratory Locust (*L. m. manilensis*) in China, S.E. Asia, the Philippines and Malaysia.

**Control.** The hoppers can be killed using a bait made of wheat bran with BHC dust or aldrin w.p. added.

Spraying with carbaryl, dieldrin, or aldrin, in low- or high-volume application is effective against hoppers, and also flying swarms. DNOC is also used against flying swarms.

**Further reading**

Anti-Locust Research Centre (1966). *The Locust Handbook*, A-L Res. Centre: London.
Uvarov, B. (1966). *Grasshoppers and Locusts*, Cambridge University Press: London.

adult ♂; solitary phase

Photo: Glenda Colquhoun, Centre for Overseas Pest Research

0    1 cm

# Nomadacris septemfasciata (Serv.)

**Common name.** Red Locust
**Family.** Acrididae
**Hosts.** Preference is shown for grasses and graminaceous crops, but a great variety of crops may be attacked.
**Damage.** The leaves are eaten from the margin inwards; in a heavy attack the entire lamina is eaten away. Swarm damage can be devastating and result in complete crop defoliation.
**Pest status.** A sporadically serious pest in tropical Africa south of the Sahara. The outbreak areas are in the Rukwa rift valley of Tanzania, the Mweru marshes of Zambia, and the Chilwa plains in Malawi. The last plague began in 1930 and finally ended in 1944.
**Life history.** Eggs are laid in pods containing about 100 each, three or four pods being laid per female. The eggs are laid in the wet season (November to April), and they hatch in about 30 days without going into diapause.

There are seven stages of hoppers in the solitary phase and six in the gregarious. The hoppers are red, black and yellow, and take from 2—3 months to develop.

The adults are the largest dealt with here, the male being 50—60 mm long and the female 60—70 mm. The body is yellow-brown; prothorax with broad, yellow and red longitudinal bands. The costal and

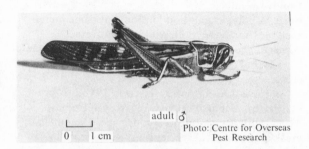

adult ♂

0    1 cm

Photo: Centre for Overseas
Pest Research

inner margin of the forewing is banded yellow, the remainder being translucent brown. The tibia of the hindlegs are reddish, and the base of the hindwing is characteristically red. There is a stout spine between the bases of the forelegs.

The adults live for about nine months, waiting until the wet season before egg laying.

There is only one generation of Red Locusts per year.
**Distribution.** Central and southern Africa, from Angola to Somalia and E. Africa and southwards to S. Africa. The breeding areas are the Lake Rukwa region of Tanzania, the Mweru marshes of Zambia, and the Chilwa plains of Malawi.
**Control.** The hoppers can be killed with the use of a bait made of wheat bran with BHC dust or aldrin w.p. added.

Spraying with carbaryl, dieldrin or aldrin, is effective against hoppers, and dieldrin, BHC or DNOC against flying swarms.

Control generally is as for Desert Locust.
**Further reading**
Anti-Locust Research Centre (1966). *The Locust Handbook*, A-L Res. Centre: London.
Uvarov, B. (1966). *Grasshoppers and Locusts*, Cambridge University Press: London.

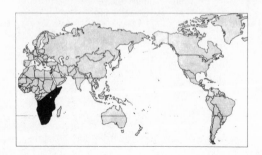

## Oxya spp.

*chinensis* Thnb.
*japonica* (Thnb.)

**Common name.** Small Rice Grasshoppers
**Family.** Acrididae
**Hosts** (main). Rice

(alternative). Many species of grasses (Gramineae), occasionally various herbaceous plants.
**Damage.** Adults and nymphs eat the foliage of the rice plants; on older plants the adults may eat the base of the rice panicle, causing it to wither and die.
**Pest status.** Of sporadic importance in different parts of their wide distribution range, only occasionally requiring control measures.
**Life history.** As with other Acrididae the eggs are laid in a pod just below the soil surface, but since these grasshoppers typically inhabit wet areas the eggs may be laid in a mass of froth (which hardens to form the protective 'pod') a few centimetres above the water level on the rice or grass foliage.

In more northern regions there is only a single generation per year, and they overwinter in the egg stage. In the warmer regions breeding appears to be more or less continuous.

Young males generally pass through six instars, whereas many females apparently have seven instars. They are alleged to swim quite well and can gain access to the rice plants in flooded fields in this way.

The adults are small, yellow and brown grasshoppers, about 3 cm in body length, with a conspicu-

ous broad, brown stripe laterally through the eyes and extending posteriorly along the tegmina. The distal end of the tibia is dark brown and conspicuous. The underparts are bright yellow, the back pale brown.

**Distribution.** The two species here included are difficult to separate taxonomically and occur throughout the region from India to north Australia and up through S.E. Asia to Japan (CIE map no. A295).

Other species are reported to occur in Africa.

**Control.** A wide range of chemicals are used against grasshopper pests, including DDT, dieldrin, carbaryl, fenthion, diazinon and phosphamidon, but local advice should be sought for use on paddy rice, partly because of the aquatic nature of the habitat and also because of resistance problems with some chemicals (e.g. diazinon).

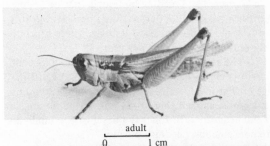

adult

0        1 cm

adult on sweet potato leaf

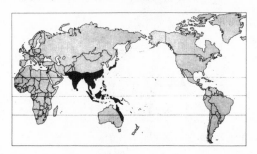

# Patanga succincta (L.)

**Common name.** Bombay Locust
**Family.** Acrididae
**Hosts** (main). Maize and other graminaceous crops.
(alternative). Wild grasses (Gramineae), and
some herbaceous crops.
**Damage.** Defoliation by both adults and nymphs; on
maize the silks and the soft grains may also be eaten.
**Pest status.** A sporadic pest throughout mainland
S.E. Asia, generally confined to higher altitudes in
this region. In S. China it is a solitary species found on
high altitude grassland with no pest status whatsoever,
but in parts of Thailand and India it behaves like a
locust and sometimes swarms in enormous numbers,
defoliating maize and other crops: in India plagues
occurred in 1835–45, 1864–6, and 1901–8.
**Life history.** The breeding biology is that of a typical
short-horned grasshopper, with egg pods laid in soft
sandy soils. In most parts of its range there is probably only one generation per year.

   The adult is a large buff and brown-coloured
grasshopper, 5–7 cm in body length, with a distinctive buff stripe dorsally on the pronotum extending
backwards, and another broad buff stripe on the side
of the pronotum. In flight the bases of the hindwings
are conspicuously reddish-purple.
**Distribution.** Recorded from S. China through mainland S.E. Asia to India, and Sri Lanka.
**Control.** The usual chemical pesticides effective
against grasshoppers will presumably give adequate
control if required (page 152).

   In 1978–80 this was a major pest in N.
Thailand, and was subject to an interesting integrated
control programme. Natural predation rates were
high, and included an egg parasite (*Scelio* sp.) and
egg predators (*Mylabris* spp.). In 1980 peasant labour
was employed to hand-catch the locusts, and 80 tons
were collected in a few days. Intercropping with
soyabean was practised, and at high temperature
the locusts left the maize to seek refuge in the shorter
legume foliage, where some were destroyed using
insecticidal sprays and others were eaten by domestic
ducks.

0          4 cm

adults

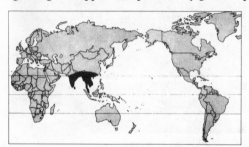

## Schistocerca gregaria (Forsk.)

**Common name.** Desert Locust
**Family.** Acrididae
**Hosts.** A very polyphagous pest, and virtually all crops are at risk so far as locusts are concerned, but there is some preference for cereals.
**Damage.** The leaves, and soft shoots, are eaten from the margin inwards leaving irregularly shaped feeding marks. Swarm damage usually results in complete defoliation of the crop.
**Pest status.** A sporadically serious pest in Africa and India; when swarms occur the damage can be devastating over a wide area. The outbreak area is very extensive and extends from W. Africa through the Sahara to Pakistan and India. Typically it can breed in any desert-type area when there is sufficient rain.
**Life history.** The eggs are laid in a hole made by the thrusting ovipositor, about 10 cm deep, in sand, embedded in a frothy mass which hardens to form a tubular egg-pod. Each egg is oval, being 1.2–1.3 mm long by 0.7–0.8 mm broad. Each egg-pod contains 70–100 eggs, and each female may lay several (4–5) pods of eggs. Egg development takes two or more weeks according to temperature.

The first larva is rather vermiform and it has to wriggle up through the egg pod to the sand surface. On reaching the surface it moults and becomes a 'hopper'. Over a period of several weeks the hoppers develop through five instars before they become adult. The hoppers exist in two distinct phases, solitary and gregarious, with differences in colour and behaviour.

The adults are large; male 40–50 mm, and female 50–60 mm, pale yellow or brownish. The elytra are greenish-yellow, translucent, with many brown spots. There are several generations per year.
**Distribution.** It occurs in Africa through the Middle East to India and Pakistan.
**Control.** The hoppers may be attacked through the use of baits (with aldrin and BHC) as well as dusting, ground spraying and aerial spraying with dieldrin, aldrin, BHC, or carbaryl, parathion-methyl or diazinon.

Against aerial swarms dieldrin, BHC, malathion and DNOC are widely employed.

Barrier spraying with persistent insecticides such as dieldrin can be very successful.

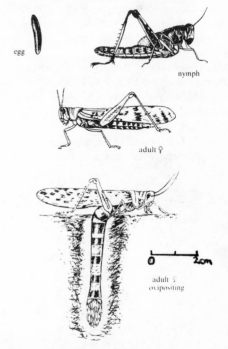

egg

nymph

adult ♀

adult ♀ ovipositing

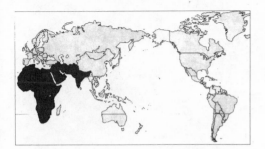

## Zonocerus spp.

*elegans* (Thun.)
*variegatus* L.

**Common name.** Elegant Grasshoppers
**Family.** Acrididae
**Hosts** (main). Many crops in the seedling stage, especially cassava, and finger millet.

(alternative). A wide range of crops and weeds, e.g. cocoa, castor, coffee, cotton, and sweet potato. Grasses and cereals are not usually attacked.
**Damage.** The leaves of seedlings are eaten, leaving ragged edges. Nymphs and adults which are both gregarious and sluggish may be found on the leaves of the crop.
**Pest status.** A sporadically severe pest of many crops, especially in the seedling stage, in parts of Africa.
**Life history.** Eggs are sausage-shaped, 6 mm long and 1.5 mm wide. They are laid in the soil in masses of froth which harden to form sponge-like packets about 2.5 cm long. Laying takes place from March to May and hatching in October and November. Each female can lay about 300 eggs.

The nymphs are typical short-horned grasshoppers about 3 cm long when full grown. They are black, the appendages ringed with yellow or white. The total nymphal period lasts for about four months; there are five instars.

The adults are handsome grasshoppers about 3.5 cm long, generally dark greenish but with much of the body boldy patterned in black, yellow and orange. Short-winged specimens which cannot fly are very common. The adult life span is about 3–4 months. They have a characteristic unpleasant smell, and are sometimes termed 'stink grasshoppers'.

There is only one generation of these grasshoppers per year.
**Distribution.** *Z. elegans* is recorded from S. Africa, Angola, Zaïre, Malawi, Mozambique, and Zimbabwe. *Z. variegatus* extends right through from W. to E. Africa.
**Control.** Dieldrin and carbaryl have been very successfully used in high- and low-volume applications.

Alternatively a bait may be used instead of a spray and then aldrin or BHC dust should be used mixed in with wheat bran. The mixture should be moistened with water to make a crumbly mash which is broadcast between the crop rows in the evening, preferably on weed-free soil.

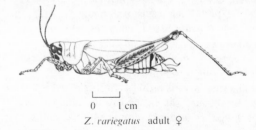

0    1 cm

*Z. variegatus*  adult ♀

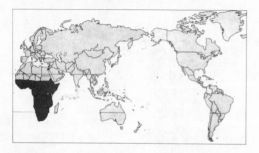

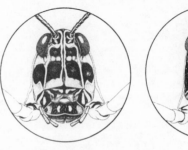

*Z. variegatus*

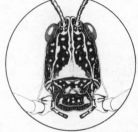

*Z. elegans*

# Order ISOPTERA

These are termites, sometimes quite erroneously called 'White Ants'. They are social insects; polymorphic, living in large communities, sometimes in elaborate nests both above and below ground, containing workers, soldiers, and reproductive forms; they have biting mouthparts; two pairs of equal-sized wings, with a fracture line near the base where the wings break off; the cerci are very small; metamorphosis is slight or absent; some genera have symbiotic bacteria in the gut which break down the cellulose (wood) eaten by the insect; others use the cellulose to cultivate fungi which are then eaten.

For further information about termites see W.V. Harris (1971).

## Family Hodotermitidae

These are wood-inhabiting and subterranean species; ocelli and fontanelle are absent; the pronotum is saddle-shaped and narrower than the head; workers are present in some genera.

## Family Rhinotermitidae

(Wet-wood Termites)  Regarded as being a primitive family, they nest in damp dead wood, and tend to be urban rather than agricultural pests; some 37 species are potentially important domestic pests attacking structural timbers and the like. The nest is typically subterranean in dead tree stumps, no mound is built. The fontanelle (opening of the frontal gland) is well developed and the soldiers use the frontal gland as a defensive organ — when alarmed they exude a drop of sticky fluid through the fontanelle. *Coptotermes* is a large genus, with 45 species found throughout the tropics, and especially well-represented in Australasia and Malaysia. Many species are pests of constructional timber, and some attack growing trees and crop plants. All members of this family have the symbiotic micro-organisms in the intestine that enable them to digest the cellulose they eat.

## Family Termitidae

(Mound-building Termites)  This is the largest family of termites; all are ground-dwelling species with a wide range of food habits and colony structure; fontanelle is present; the pronotum of workers and soldiers has a raised median anterior lobe; workers are present in all genera. Most species produce the spectacular mounds under which the colony lives; the largest mounds may be 1–2 m in height.

## Hodotermes mossambicus Hagen

**Common name.** Harvester Termite
**Family.** Hodotermitidae
**Hosts** (main). Many species of grasses.

(alternative). All species of grass appear to be attacked; also a minor pest of cotton.

**Damage.** Numerous conical earth mounds about 10 cm high scattered throughout short grass areas. During the cooler hours of the day many small grey-brown termites can be seen carrying small pieces of grass to holes in the centre of these mounds. Many bare, grassless patches in the vicinity of the mounds.

**Pest status.** A major pest of grassland below 1500 m in parts of Africa, especially during periods of drought or following overgrazing.

**Life history.** The colony consists of a number of separate hives constructed from 0.2 to more than 1 m below the soil surface, interconnected by underground passages. Each hive is a dome-shaped cavity about 0.5 m in diameter; some are full of termites in all stages of development; others are used largely for the storage of cut grass. The hives are connected to the surface by tunnels passing through the centre of conical mounds of loose earth on the soil surface. If the foraging tunnels are not in use they are sealed with mud.

Only the workers are usually seen on the soil surface, these are about 14 mm long, grey-brown, with large heads. They forage at dawn and dusk and at other times in cool weather; they cut and gather pieces of grass and wood, pieces of leaf and herbaceous twigs. The soldiers usually take up positions near the entrance of the foraging tunnels to repel intruders. The soldiers have pale bodies and large dark heads with massive jaws.

**Distribution.** S. and E. Africa.

**Control.** As a tentative recommendation, dieldrin is suggested as an overall spray on the grassland; the application of aldrin or dieldrin down the tunnels is very effective.

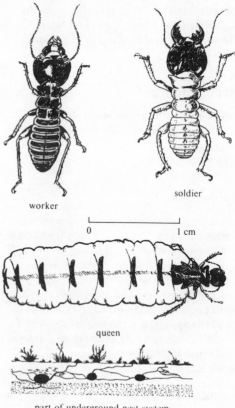

worker

soldier

0 ⊢——————⊣ 1 cm

queen

part of underground nest system

170

# Coptotermes formosanus Shiraki

**Common name.** Wet-wood Termite
**Family.** Rhinotermitidae
**Hosts.** This is essentially a polyphagous tropical forest species that usually lives in the moist stumps of dead trees, but has adapted for life under both urban and agricultural conditions. Damage is reported to have been done to forest trees, groundnuts, fruit trees, sugarcane, rice, and other food crops.
**Damage.** On woody plants the termites may construct earth-covered runways, under cover of which they eat the bark away; seedlings may be ring-barked. Roots may be eaten under the ground. Generally old and sickly, or very young, plants are the ones attacked.
**Pest status.** A pest of occasional importance agriculturally, throughout the tropical regions of the world.
**Life history.** The nest is usually in a moist dead tree stump at the edge of forested areas, but structural timber in buildings and bridges is also favoured. It is usually rather small with only a few thousand inhabitants. The wood is eventually honeycombed by a series of tunnels ramifying throughout.

The workers are small, about 4 mm in length, with soft white bodies, and pale yellow round heads with small mandibles. Soldiers are larger, but still rather small, about 5 mm long, with brown heads and slender black mandibles. At the front of the head is the fontanelle through which a white viscous fluid can be extruded when alarmed.

Winged adults swarm periodically during the early evenings of warm wet days in the summer; they are orange in colour, about 8 mm in body length, with hyaline wings of 11 mm.
**Distribution.** The genus *Coptotermes* is found throughout the tropics with 16 species, but is best represented in Australia and S.E. Asia. The species *C. formosanus* is recorded from China, Taiwan, Japan, and Hawaii.
**Control.** Dieldrin is the most effective insecticide when applied appropriately.

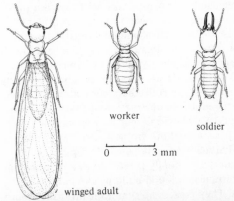

worker

soldier

0     3 mm

winged adult

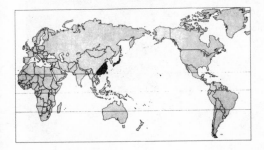

nest in tree stump

## Macrotermes spp.

**Common name.** Bark-eating Termites
**Family.** Termitidae
**Hosts.** These are polyphagous general feeders found throughout the Old World Tropics and they occasionally cause damage locally to a wide range of crops, such as, coconut, coffee, cocoa, clove, groundnuts, rice, sugarcane, fruit trees and forest trees.
**Damage.** These are the termites that build large spectacular mounds in Africa so their presence is generally obvious in an area. Typical damage is to cover tree-trunks, or plant stems, with covered runways (sometimes even sheets) composed of plant fragments, soil and saliva, and underneath the protecting cover the termites gnaw away the bark. Small plants may be completely killed, and trees may be ring-barked. Sometimes root damage subterraneously may be serious. The collected plant material is taken back to the nest for construction of fungus gardens.
**Pest status.** A widespread, but only sporadically serious, pest throughout tropical Africa and Asia, recorded attacking a wide range of crops.
**Life history.** This family is referred to as containing the subterranean, mound-building and bark-eating termites. They possess no symbiotic intestinal microorganisms and they utilize the collected plant cellulose by constructing underground fungus gardens in special chambers; the honeycombed fungus garden base is innoculated with a special species of fungus

(*Termitomyces*) and the mycelium spreads over the cellulose 'garden'. At intervals along the hyphae are found white, globular swellings known as bromatia and it is on these bromatia that the termites feed.

winged adult

nest mound (Malaysia)

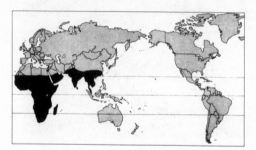

172

The young are cared for by small workers in the nest, whereas the larger workers go out and forage for food, partly through underground tunnels and partly on the soil surface and in the litter, where they are protected from predation by the large aggressive soldiers.

The mound of *M. bellicosus* in Africa may be 1–2 m in height, but in Malaysia *M. carbonarius* builds a mound no higher than 1 m, and in S. China the local *M. barneyi* has no mound at all, the nest being entirely subterranean, presumably due to the cooler climatic conditions prevailing there.

The adults are pale brown in colour, about 8–15 mm in body length, with wings 17–30 mm long, hyaline but with a yellow tinge, and they swarm periodically in vast numbers.

Each colony lasts for several years (3–8 years) but are not as long-lived as might be expected.

**Distribution.** This genus occurs as about 7–10 species throughout tropical Africa and tropical Asia, extending as far north as S. China.

**Control.** If control is required then aldrin and dieldrin, either in baits, or applied directly to the nest holes, are still the most effective pesticides.

damage to trees: eating woody tissues beneath the bark

## Odontotermes spp.

**Common name.** Bark-eating Termites
**Family.** Termitidae
**Hosts.** A polyphagous pest recorded attacking a wide range of crops, both as seedlings and as grown plants; but sugarcane, coconut and tea are attacked by several species.
**Damage.** Some damage is underground to the roots and underground stem of the plants; some is the typical bark-eating under sheets of earth/wood particles; seedlings may be completely destroyed. Tea bushes may be ring-barked.
**Pest status.** A sporadic pest throughout tropical Africa and Asia, locally important on a wide range of crops. Water-stressed or sickly plants are more often attacked than healthy plants.
**Life history.** Colonies of *Odontotermes* are generally smaller than those of *Macrotermes*, but according to W.V. Harris (1971) this genus is probably the dominant soil-dwelling termite in terms of actual numbers. The nest mounds are also generally smaller.

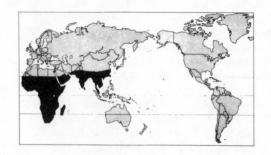

The Crater Termite (*O. badius*) is a rather different species from most of the rest and so has been dealt with separately.

As with other members of the Termitidae this is a fungus-growing species that has the typical fungus gardens in underground chambers joined by a series of narrow tunnels.

Adults are slightly smaller than *Macrotermes* spp., most being about 7–13 mm in body length with wings 15–28 mm in length; the wings of most species are quite dark brown in colour.

adult winged forms

Healthy, intact, vigorous plants are generally not damaged by most termites, but if water-stressed or sickly in any way, or physically injured, they are more likely to be attacked. Many *Odontotermes* species gain access to woody plants, such as tea, through the dead ends of pruned branch stumps from which they may invade the living tissues.

**Distribution.** W.V. Harris (1971) lists a total of 23 species of *Odontotermes* from tropical Africa throughout mainland S.E. Asia and India; all recorded as pests of various crops.

**Control.** As with other termites, if control is required then dieldrin applied down the nest entrances is still the most effective pesticide.

to bush stem

to coconut palm trunk

termite damage

# Odontotermes badius (Haviland)

**Common name.** Crater Termite
**Family.** Termitidae
**Hosts (main).** Seedlings of various crops, and grasses.
    **(alternative).** Sugarcane, and often found in the mulches used in coffee, tea, and banana plantations.
**Damage.** The nest is characterized by a series of low mounds through which a number of wide vertical shafts open to the exterior. The shafts are ringed at the top with crater-like earthen collars.
**Pest status.** Not a primary pest of growing crops except in very dry years when some seedlings may be attacked. Lawns and flower beds are badly disfigured by the nests.
**Life history.** The Crater Termite is another typical fungus-growing species. The colony is established by a pair of night-flying winged reproductives. They shed their wings and build a cell underground, which is later expanded. When the colony is well established, vertical shafts are built which one-by-one erupt to the exterior at the start of the rainy seasons. A large nest may be 2 m in diameter and be covered with 15 or more craters of various diameters. The nest contains a large number of fungus gardens.

The worker termites are pale-bodied insects about 4.5 mm long, often with large, pale brown heads. In rainy weather they can often be seen at work adding soil to the crater rims at the top of the vertical shafts. The soldiers are about 6 mm long with very large, brown heads and conspicuous black jaws.

The mature queen termite is enormously distended in the abdomen and may be 10 cm or more long. Winged reproductives are produced in very large numbers in rainy seasons and they swarm out of the colony through the vertical shafts.

The adult female has an egg-laying capacity of one egg every two seconds.
**Distribution.** Most parts of tropical Africa.
**Control.** Either aldrin or dieldrin, in solution, to be poured down each of the vertical shafts, preferably during the rainy season.

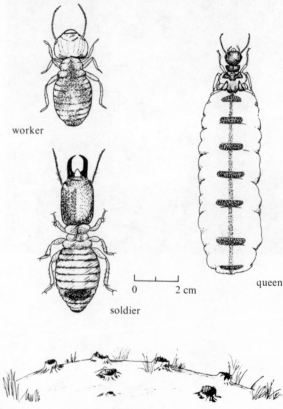

worker

soldier

queen

0    2 cm

nest site

# Pseudacanthotermes militaris (Hagen)

**Common name.** Sugarcane Termite
**Family.** Termitidae
**Hosts** (main). Sugarcane
   (alternative). Often found among fallen leaves under mango trees, where they make their characteristic 'rattling' sounds.
**Damage.** Poor germination of sugarcane setts; mature cane is encrusted with earthen tunnels; stalks are often felled when nearing maturity.
**Pest status.** A major pest of sugarcane in Kenya; elsewhere damage is only sporadic.
**Life history.** The mature colony is marked above ground level by a conical mound up to 1 m in height. If the nest is dug out, the sponge-like fungus gardens can be found. The chewed-up wood and pieces of vegetable matter are built up into the fungus gardens on which special species of fungus are cultivated; the termites then feed on the fungal mycelium and bromatia.

   The worker termites are pale-bodied insects about 4 mm long with large, brown heads. The soldiers are also pale, about 6 mm long and have very large heads with conspicuous pincer-like jaws. The queen termite has a typically enormously distended abdomen and may be 5 cm or more in length. She lives with her royal male in a special cell near the centre of the nest. Winged reproductives are produced in large numbers in the rainy seasons. Unlike most other species of termites these fly in the daytime and can often be seen swarming round the tops of tall trees.
**Distribution.** Kenya, Malawi, and Uganda.
**Control.** The planting material should be dipped in a mixture of dieldrin in water, prior to planting.

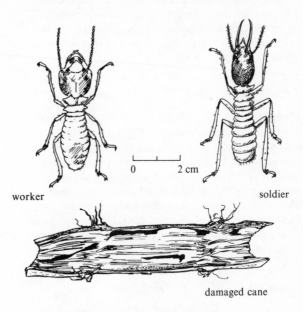

worker                                   soldier

damaged cane

nest mound

# Order **HOMOPTERA**

These are plant bugs often referred to as a separate suborder of the order Hemiptera, but here for convenience is regarded as a separate order. Two pairs of wings are usually present, but the anterior pair is uniformly thickened, and the wings are held roof-like over the body; head more or less deflexed; metamorphosis usually incomplete, but sometimes complete in males; they have piercing and sucking mouthparts, and are sap feeders.

## Family **Cercopidae**

(Froghoppers, or Spittle Bugs)  They are characterized by large stout hind tibiae bearing a terminal ring of stout setae. The nymphs of some species are subterranean, the rest are found on grasses and various herbaceous shrubs or trees; they are immersed in a froth of 'spittle' which serves to protect the soft body from desiccation and predation. The adults are capable of jumping considerable distances. Many species are pests of sugarcane in S. and Central America.

## Family **Cicadellidae (= Jassidae)**

(Leafhoppers)  A large family of bugs, second in abundance only to the Aphididae. They are slender, small insects, usually tapering posteriorly; thin tapering antennae; they jump very readily when disturbed; the hind tibiae are elongate and with a series of regularly spaced stout setae along the anterior edge. Eggs are usually laid embedded in the plant tissues. Many species are virus vectors. There are many important pest species. Some species are only found on Gramineae, but others infest trees and may be fruit pests.

For a review of the bionomics of leafhoppers see De Long (1971).

## Family **Tettigometridae**

These are superficially very like jassids, but with fulgoroid characteristics also; the antennal flagellum is segmented.

## Family **Delphacidae (= Araeopidae)**

(Planthoppers)  The characteristic feature of these insects is the large, mobile, serrulate apical spur on the hind tibiae; the antennae are very short, with a terminal arista. They are vectors of various plant viruses, mostly on Gramineae and cereal crops.

## Family **Fulgoridae**

(Lantern Bugs)  A small tropical family, not important as pests, but widespread. The anterior part of the face is drawn out into either a tapering 'snout' or a large bulbous shape rather like a peanut. They are large in size, often 5–6 cm in length.

## Family **Flattidae**

(Moth Bugs)  A small group, tropical in distribution, moth-like in appearance, gregarious in habits, with nymphs producing large quantities of wax. Many species seem to prefer hosts in the Sterculiaceae. Conspicuous insects of the tropical rain forests, but seldom serious pests.

## Family **Ricaniidae**

(Ricaniid Planthoppers)  Another group of fulgoroid planthoppers with expanded forewings, but held laterally at rest, and not moth-like. Widespread in the tropics as minor pests on many different crops and ornamentals.

## Family **Lophopidae**

A small group of tropical fulgoroids with an anterior snout-like process (being an extension of the frons) with one to three longitudinal carinae; the antennae are very short with a terminal arista.

## Family **Psyllidae**

(Jumping Plant Lice)  These bugs are about the size of aphids but resemble small cicadas; they are very active but not capable of sustained flight; the wing venation is rather simple; nymphs are typically flattened with distinctly flattened wing buds; the antennae are fairly long, simple, and segmented; the female has a short curved ovipositor; the nymphs secrete quantities of honey-dew, and sometimes waxy filaments. Some species are pests of fruit trees, others are pests of ornamental trees.

## Family **Aleyrodidae**

(Whiteflies) These are tiny, delicate, usually white, moth-like bugs, sometimes with dark or mottled wings; the whiteness is produced by a waxy dusting; the nymphs and pupae are characteristically oval-shaped and scale-like with a number of long filaments; the first nymphal stage is mobile, but after the first moult both legs and antennae are lost; honey-dew is excreted by all stages, but particularly by the nymphs.

## Family **Aphididae**

(Greenfly, Plant Lice)  The most important group of plant bugs, but more adapted for life in the temperate zones than the tropics; they have peculiar modes of development, and pronounced polymorphism; often parthenogenetic or only seasonally sexual; in the tropics most reproduction appears to be entirely parthenogenetic; the rate of reproduction may be quite phenomenal; they are found on the roots of plants as well as leaves, shoots and stems; sometimes galls are formed; all species have conspicuous cornicles or siphunculi; production of honey-dew can be quite copious, and there are often ants in attendance. They are very important vectors of plant viruses, and many species are plant pests on a wide variety of crops.

Useful reference papers on Aphididae are Eastop (1961; 1966) and Van Emden *et al.* (1969).

## Family **Pemphigidae**

(Woolly Aphids)  A small family, recently separated off from the rest of the Aphididae; they produce vast quantities of waxy filaments covering most of the body; the siphunculi are small and slit-like.

## Family **Margarodidae**

(Fluted Scales, etc.)  The females of this family have distinctly segmented bodies, usually covered with a waxy secretion, often in the form of pearl-like waxy scales which are collected as 'ground pearls' in S. Africa and the Bahamas and strung into necklaces; legs and antennae are well developed and simple eyes are present; the adult male has wings, the simple ten-segmented antennae and compound eyes. They can sometimes be mistaken for mealybugs.

## Family **Lacciferidae**

(Lac Insects)  A tropical group with highly degenerate females which are legless and have a globular body enclosed in a dense resinous cell. *Laccifer lacca* is cultivated in India and China as a source of shellac; other species are crop pests in India and S.E. Asia, mostly on species of *Citrus*.

## Family **Orthezidae**

A small group of species mainly confined to America and the Palaearctic region. Females of this family show rather distinct segmentation of the body, which is covered with white waxy plates; legs and antennae are normal; and simple eyes are present.

## Family **Pseudococcidae**

(Mealybugs) Females are usually elongate oval with a distinct segmentation and generally covered with a mealy or filamentous waxy secretion, often extended into lateral and terminal filaments; legs are well developed, antennae less so. Some species are sub-terranean; others make galls. The males are small two-winged insects with long filamentous antennae and wings with very reduced venation. They are often associated with ants, particularly the root-inhabiting species, for the honey-dew they excrete. Some species transmit viruses.

## Family **Coccidae** (= **Lecaniidae**)

(Scales, Soft Scales or Naked Scales) A large family; the females show considerable diversity of form; body segmentation is obscure, and the integument may be naked or covered with wax; the degree of development of antennae and legs varies considerably; the scale is always firmly fixed to the integument of the insect, and is not detachable. Honey-dew is usually excreted, and these insects are mostly attended by ants. There are many important pest species.

## Family **Asterolecaniidae**

(Star Scales) A small family of scales separated off from the others on rather esoteric characters, but the tiny star-shaped scale is an obvious character.

## Family **Diaspididae**

(Armoured Scales) A large family in which the adult female has lost the legs, and antennae are either lost or vestigial. The hard waxy scale is produced by the female, but is not attached to her body (as in the Coccidae); the scale may be circular, oval or very elongate, and the adult scale also comprises the scales of the nymphs. In some cases the scale is so thin as to be quite translucent. There are many important pest species. Honey-dew is normally not produced by these species.

## Tomaspis spp.
## Aeneolamia spp.

**Common name.** Sugarcane Spittlebugs
**Family.** Cercopidae
**Hosts** (main). Sugarcane.
    (alternative). Many wild grasses and sedges.
**Damage.** The feeding of adults and nymphs on the
leaves is the main source of damage – root feeding is
generally not regarded as serious. Enzymes in the
saliva cause necrosis of the plant tissues. In heavy
attacks the leaves yellow, then brown, finally wilting
and dying, and the plant becomes stunted. Necrotic
spots develop round the feeding punctures; over
one to three weeks the necrosis spreads longitudinally
to form streaks ('froghopper blight').
**Pest status.** Although cercopids are richly represented
throughout the tropics, they are common and important pests of sugarcane only in the New World.
**Life history.** Eggs are laid both in the soil and in
plant tissues; *Aeneolamia* spp. usually lay most of
their eggs in the soil and the larvae feed on the roots
of the cane. The eggs are spindle-shaped; many
species of Cercopidae have a diapausing egg stage.
Incubation takes 2–40 weeks; a moist atmosphere is
required for egg development.
    The nymphs are characterized by their pro-
duction of a frothy spittle mass, or 'cuckoo spit', in
which they are enveloped. The spittle undoubtedly
protects the soft-bodied nymph from desiccation.

Some nymphs feed on the leaves of the cane and
others largely on the roots. There are five nymphal
instars in most species. The life-cycle takes about
two months.
    The adults of many Cercopidae are distinctly
coloured – yellow, red and black, etc. *Tomaspis*
species are generally red and black, strong fliers and
jump well.
    Generally 2–4 generations per year.
**Distribution.** S. America (Brazil, Bolivia, Trinidad,
and Argentina).
    *Locris* and a few other cercopids are found in
Africa, but *Tomaspis* (4 spp.), *Aeneolamia* (14 spp.),
and *Delassor* (4 spp.), and a few smaller genera are all
confined to S. and C. America and the W. Indies.
**Control.** Dusts of γ-BHC, dieldrin, phorate, toxaphene and carbophenothion applied to the cane
stools to control the root-feeding nymphs. Adults
are sprayed or dusted by drift dusting or from the
air with γ-BHC, DDT, carbaryl and phosmet.

0    2 mm
nymph

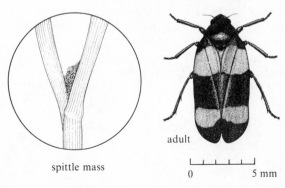

adult

spittle mass

0        5 mm

181

## Cicadulina mbila (Naudé)

(= *Balclutha mbila* Naudé)

**Common name**. Maize Leafhopper
**Family**. Cicadellidae
**Hosts** (main). Maize
     (alternative). Sugarcane, and various wild grasses.
**Damage**. Attacked plants show no signs of insect damage for this is very slight at most, but the Streak Disease symptoms are conspicuous yellow streaking against the normal green background of the leaf. Infestation of the young plant may result in its death.
**Pest status**. Actual damage by direct sap sucking is slight, but the 'active' races transmit Maize Streak Virus, which causes extensive damage to maize crops in many parts of E. and Southern Africa.
**Life history**. Eggs are laid in the plant tissue by the female, and the developmental period is 5—6 weeks in E. Africa.

The adult is a tiny leafhopper, 2—3 mm in length, with transparent wings bearing a brown longitudinal stripe. Head, thorax and abdomen are largely yellow with some dark brown markings on the dorsum. The eyes are dark brown. Adults may be found at rest on the upper surface of the young maize leaves forming the terminal cone of the plant. Field densities have been recorded as high as one leafhopper per 20 maize plants, but this is unusually

high. The leafhopper exists in two forms (biological races) – an 'active' form capable of virus transmission, and an 'inactive' form which is incapable of transmission, as shown experimentally by Storey (1932, 1961). The active form becomes infective 24 hours after feeding on a diseased plant, and will remain so for up to several months.

Three other closely related species of *Cicadulina* are also capable of transmitting this virus.
**Distribution**. E. Africa, Zimbabwe, and S. Africa. The insects are rarely seen but the incidence of Maize Streak Disease is often quite high.
**Control**. The use of resistant varieties of maize is probably the best method of control to be aimed at, but as yet insufficient plant breeding has been carried out. Otherwise, having as close a season as possible for maize growing does appear to be effective in reducing leafhopper populations. Sprays of carbaryl, dimethoate or demeton-*S*-methyl can be effective, especially if followed by roguing of diseased plants.

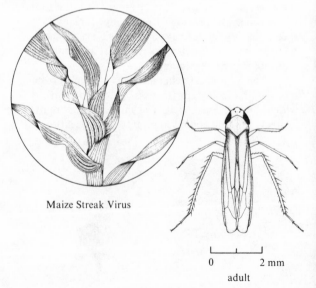

Maize Streak Virus

0     2 mm

adult

# Empoasca spp.

*fascialis* (Jacobi)
*lybica* (De Berg)

**Common name.** Cotton Jassids
**Family.** Cicadellidae (= Jassidae)
**Hosts** (main). Cotton
  (alternative). Various legumes, wild Malvaceae, groundnut, castor and many other crops.
**Damage.** The edges of leaves are down-curled and turn first yellow and then red. In severe attacks leaves may dry up and be shed. On the undersides of leaves numerous pale green bugs can be found which may move rapidly sideways when disturbed.
**Pest status.** Most of the time only a minor pest of cotton, due to the use of resistant varieties, but locally severe attacks occur from time to time.
**Life history.** The eggs are greenish, banana-shaped and about 0.8 mm long. They are embedded in one of the large leaf veins or in the leaf stalks. Hatching occurs after about 6–10 days.

There are five nymphal instars; the full-grown nymphs are yellowish-green, frog-like and about 2 mm long. Nymphs are found on the underside of large leaves during the daytime. The nymphal period lasts 14–18 days.

Adults are pale green, narrow-bodied, wedge-shaped bugs about 2.5 mm in length. The wings are semi-transparent and extend beyond the end of the body. The adult hops and flies very readily if dis-turbed, or, like the nymph, runs quickly sideways. Adult females may live 2–3 weeks or longer, and lay about 60 eggs.

**Distribution.** *E. fascialis* is recorded from 24 countries in tropical Africa (CIE map no. A250).

*E. lybica* is found in Spain, Israel, Saudi Arabia, Aden, Egypt, Ethiopia, E. Africa, Sudan, Tunisia, Libya, Mauritius, Morocco, Somalia, and S. Africa (CIE map no. A223).

**Control.** Cotton varieties with suitably hairy leaves are resistant to jassid attack. If control measures are required then the recommendations are to spray with DDT, endosulfan, dimethoate, formothion or carbaryl, etc.

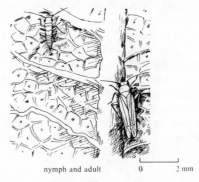

nymph and adult    0    2 mm

damaged leaf

183

## Recilia dorsalis (Mot.)

(= *Inazuma dorsalis* (Mot.))

**Common name.** Zig-zag Winged Rice Leafhopper
**Family.** Cicadellidae
**Hosts** (main). Rice
     (alternative). Not known.
**Damage.** Direct damage is done by the removal of
sap, and indirect damage by being the vector for
various viruses.
**Pest status.** One of the more important leafhoppers
which are pests of rice. This species transmits Orange
Leaf, and Rice Dwarf viruses.
**Life history.** Eggs are laid in rows within the leaf
sheath, and hatch in 7−9 days.

    There are five nymphal instars, which last
about 16−18 days. The nymphs are yellow-brown.

    The adults are 3.5−4.0 mm long, with white
forewings with pale brown bands forming the shape
of a 'W' giving the wing a zig-zagged pattern.
**Distribution.** India, Sri Lanka, Malaysia, Philippines,
China, Taiwan, and Japan.
**Control.** As with *Nephotettix*, the use of light traps
and weed removal will reduce leafhopper populations.

    Carbaryl, malathion, azinphos-methyl and
endosulfan, sprayed at weekly intervals, are successful
insecticides.

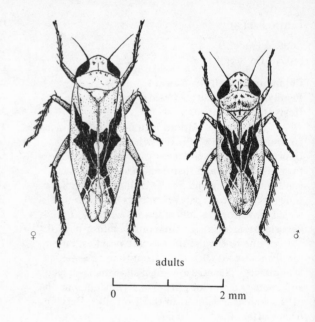

adults

0          2 mm

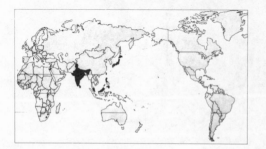

## Nephotettix nigropictus (Stål)

(= *N. apicalis* (Mot.))

**Common name.** Green Rice Leafhopper
**Family.** Cicadellidae
**Hosts** (main). Rice
    (alternative). *Panicum* spp., *Cyperus* spp.,
*Poa* spp., and other grasses.
**Damage.** Nymphs and adults cause direct damage
by sucking the sap from young leaves, and they
also transmit several plant viruses.
**Pest status.** A sporadically serious pest in many
Asiatic rice-growing areas when populations build
up rapidly. Small numbers are of little consequence
as they do negligible damage. This and two other
closely related species, *N. cincticeps* and *N. impicti-
ceps*, are vectors of the viruses causing Yellow Dwarf,
Transitory Yellowing, and other viruses.
**Life history.** Eggs are laid in the leaf sheaths, where
they hatch in about six days.

    Nymphs have a varied colour pattern on the
notum. There are five instars before they become
adult, after 16–18 days.

    Adults are 3.2–5.3 mm long, green, with black
spots on wings, and black wing tips.
**Distribution.** India, Pakistan, Burma, S.E. Asia,
Philippines, S. China, Indonesia, Papua New Guinea
and West Irian (CIE map no. A286).
**Control.** Removal of weeds from around the crop
fields can lower the pest population, for many of

these plants are alternative hosts for the leafhoppers.
The adults are greatly attracted to light, and populations can be depleted by the use of light traps.

    Suggested insecticides are carbaryl, malathion,
azinphos-methyl, and endosulfan, to be sprayed at
weekly intervals. Or diazinon as granules to be
applied to the irrigation water.

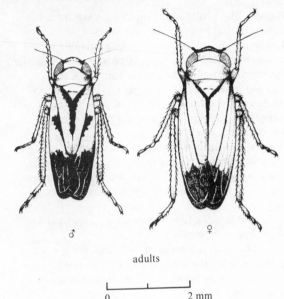

adults

0        2 mm

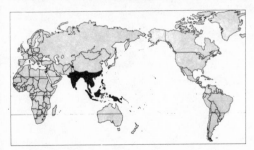

## Hilda patruelis Stål

**Common name.** Groundnut Hopper
**Family.** Tettigometridae
**Hosts** (main). Groundnut
(alternative). Various legumes, including beans, sann hemp, other *Crotalaria* spp., marigold, sunflower, cashew, etc.
**Damage.** The adults and nymphs suck sap from the stem, pegs and pods usually just below ground level. Severe damage (wilting and collapse) may be done by this bug when it occurs in large numbers. The first sign of infestation is the presence of black ants in association with the *Hilda* bugs. The ants construct chambers in the soil around the bugs and protect them from enemies. In return the bugs provide honey-dew as food for the ants.
**Pest status.** This pest is not often important, but may be locally serious. It does not transmit Rosette Virus.
**Life history.** The eggs are small, white, and elongate, and laid in batches on the stem at or below ground level, and on the pegs and pods.

The nymphs look like small versions of the adults but without wings.

The adult is a small bug 4—5 mm long, with greenish-brown markings and three lateral white patches on each forewing. Some specimens are completely green.

In Zimbabwe it is reported that one generation took about six weeks in the summer. Reproduction proceeds slowly on any overwintering plants.
**Distribution.** Africa; including Nigeria, Zaïre, Uganda, Tanzania, Zimbabwe, and Mozambique.
**Control.** Chemical treatment is not often required, but if it is then the soil may be treated with dieldrin before planting, which may kill the *Hilda* bugs and will certainly kill the ants which encourage the bugs.

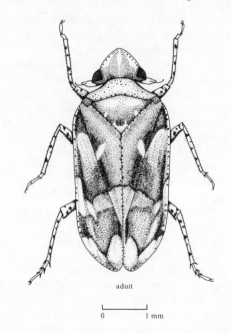

adult

0       1 mm

186

# Nilaparvata lugens (Stal)

**Common name.** Brown Rice Planthopper (BPH)
**Family.** Delphacidae
**Hosts** (main). Cultivated rice.

(alternative). Wild species of *Oryza* only.

**Damage.** Heavy infestations produce symptoms of 'hopper-burn' — leaves dry and brown after insect feeding (toxic saliva), and patches of 'burned' plants are often lodged. This insect is a vector of Grassy Stunt Virus Disease. The rice plant is most sensitive to attack at the age of 26—39 days.

**Pest status.** Ten years ago this was a minor pest of rice in parts of S.E. Asia, but its status has changed dramatically to become one of the most serious rice pests in many parts of S.E. Asia, eastern India and Japan. The reasons for this change in pest status are several, some new high-yielding rice varieties have less resistance to pest attack, some insecticides have had serious adverse effects on the natural enemies, the pest has developed resistance very quickly to many of the regularly used pesticides, and increased irrigation combined with poor drainage has resulted in much more wild rice and volunteer rice available as reservoir hosts.

**Life history.** Small white eggs are laid in batches inside the leaf sheath and on the leaf midrib; hatching requires 5—9 days. Each female lays about 200 eggs. Nymphs are brown and reach about 3 mm in length after 12—18 days, whereupon they moult into adults.

Adults may be fully winged or brachypterous (short-winged); males tend to be smaller in size than females, some 2.5 mm as opposed to 3.0 mm. The adults live for about three weeks. There are four or more generations per year.

In Japan this pest arrives annually from the China mainland, during the period mid-June to mid-July, and once on the local rice crops populations build up rapidly, there being four generations during the summer in southern Japan. However, the winter is too cold and the insects do not survive.

This pest is also unusual in that it develops biotypes in relation to insecticide resistance, which cannot be identified morphologically.

**Distribution.** From eastern India through S.E. Asia to Papua New Guinea and Australia (Queensland), to the Philippines, and to China. It occurs annually in Japan as a migrant from E. China (CIE map no. A199).

**Control.** Resistant varieties of rice are obtainable from IRRI, and various types of cultural control can be practised to reduce infestation levels. In some areas natural predation (especially by spiders and bugs) and parasitism is important. Certain pesticides should be avoided — for example, diazinon invariably causes a pest resurgence in S.E. Asia. Advice for insecticidal control should be sought locally.

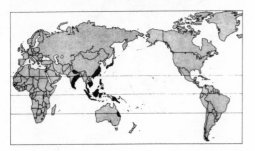

adult

0     1 mm

# Perkinsiella saccharicida (Ckll.)

**Common name.** Sugarcane Planthopper
**Family.** Delphacidae
**Hosts** (main). Sugarcane
    (alternative). A few species of grasses.
**Damage.** The nymphs and adults feed on the leaves, sucking the sap. Damage also includes laceration of tissue by the ovipositor with subsequent reddening, and desiccation of the leaves. When the insects are numerous the honey-dew excreted covers the leaves and sooty moulds are common.
**Pest status.** A pest of some importance on sugarcane; occasionally severe outbreaks have been recorded in Hawaii. It is a vector of Fiji Disease.
**Life history.** Oviposition takes place by night; each female lives for 1–2 months and will lay about 300 eggs. The eggs are laid in the midrib of the leaf, low down on the upper surface, but they may be placed in the leaf sheath, leaf blade or shoot. The egg is elongate, cylindrical, and slightly curved, about 1.0 by 0.35 mm. From 1–12 eggs may be laid in a single incision; the upper ends, which have a dome-like cap, project slightly above the leaf surface. Incubation takes 14–40 days.

    Each of the five nymphal instars lasts 4–9 days.

    Both males and females may be brachypterous or macropterous. The adult bugs regularly migrate from crop to crop. The adults rest in the leaf funnels and other places of shelter during the day.

    The whole life-cycle takes about 48–56 days; there are five or six generations per year.
**Distribution.** S. Africa, Madagascar, Malaya, Thailand, S. China, Sarawak, Java, Australia (Queensland), Hawaii, and S. America (Ecuador and Peru) (CIE map no. A150).

    There are 22 species of *Perkinsiella*, but not all are recorded from sugarcane.
**Control.** Both systemic and contact insecticides have been used to control Fiji Disease (e.g. BHC, dimethoate, dicrotophos, parathion, and dimefox).

    In Hawaii the egg predator *Tytthus mundulus* was introduced and in conjunction with other natural enemies has kept the pest under virtually complete natural control since 1923. This is one of the classic examples of very successful biological control.

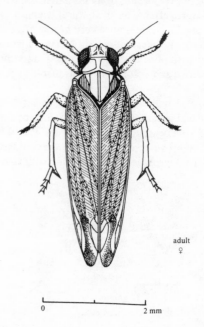

adult
♀

0          2 mm

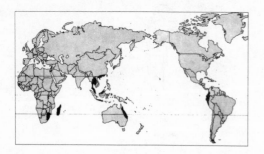

188

# Sogatella furcifera (Horv.)

**Common name.** White-backed Planthopper
**Family.** Delphacidae
**Hosts** (main). Rice

(alternative). Various species of grasses, maize and millet.

**Damage.** This species is generally found during the early growth stages of the rice crop and population build-up can be rapid. Damage is direct as a result of sap loss by the rice plants, tillering may be delayed, grain formation reduced, or the plant may even be killed. Rice Yellows cause a reddish-yellowing of the foliage (hopper-burn) and the plants become stunted.

**Pest status.** A pest of sporadic importance on rice, but it has been recorded as a virus vector of Rice Yellows and Stunt Disease which makes it a far more serious pest.

**Life history.** The eggs are laid in masses in the leaf sheath, each with a long narrow egg-cap. Hatching takes about 3–6 days.

The nymphs are pale brown, and range in size from 0.6 mm when young to 2.0 mm when fully developed. Nymphal development takes 11–12 days.

The adult is about 3–4 mm long, and is distinguished by the absence of a median transverse ridge on the vertex. The vertex is characteristic in giving the insect a long narrow face. The forewings are hyaline with dark veins, and a conspicuous dark spot in the middle of the posterior edge. The pronotum is pale yellow, and the body black. The adults live for about 18–30 days, the females living a little while longer than the males.

**Distribution.** Bangladesh, India, Sri Lanka, S.E. Asia, Malaysia, Philippines, Indonesia, China, Korea, Japan, N. Australia, and the Pacific islands (CIE map no. A200).

**Control.** Control is as for other leafhoppers (e.g. *Nephotettix*), but the other insecticides found to be effective are as follows: dimethoate, DDT, dieldrin, aldrin, endrin, phosphamidon, disulfoton, and monocrotophos.

**Further reading**

Grist, D.H. & R.J.A.W. Lever (1969). *Pests of Rice*, pp. 202–6. Longmans: London.

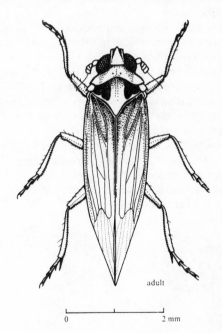

adult

0                    2 mm

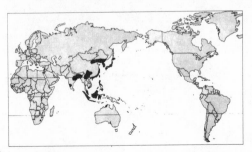

# Pyrops candelaria (L.)

**Common name.** Litchi Lantern Bug
**Family.** Fulgoridae
**Hosts** (main). Litchi and longan.
   (alternative). Occasionally seen on *Acacia confusa*.
**Damage.** Some direct damage by sap-sucking, but most fulgorids do not act as virus vectors and neither do they have toxic saliva.
**Pest status.** Generally only a minor pest, but common and very conspicuous.
**Life history.** Few details are available about the life history of this insect. Eggs are deposited into the host plant tissues via the ovipositor of the female, and nymphs are to be found developing alongside adults on the trees.

   The adults are very distinctive insects, with a mottled green coloration, long anterior 'snout' and hindwings coloured bright yellow with a terminal black band; in total body length from 40—50 mm. The adults live for several weeks, and are usually to be found sitting upright on the tree-trunk, but sometimes up in the foliage. There is usually only one generation per year in S. China.
**Distribution.** To date, only recorded from S. China.
**Control.** This insect is more of academic interest than economic, and normally does not require control measures. It is sometimes parasitized by a strange little moth larva called *Epipyrops*.

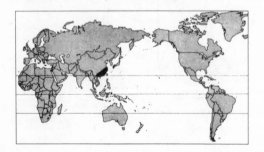

0       2 cm

## Colobesthes falcata Guér.

**Common name.** Cocoa Moth Bug
**Family.** Flattidae
**Hosts** (main). Cocoa
(alternative). Some other forest trees, usually Sterculiaceae as saplings.
**Damage.** Some damage is done directly by loss of plant sap, but when they feed on the developing pods their feeding punctures may be used as sites for invasion by pathogenic fungi and bacteria. The young larvae produce quite copious wax on the pods which has a nuisance value. The bugs are gregarious in habits and so their numbers are sometimes sufficiently large that their feeding punctures on the pods are quite damaging. Often the adult bugs seem to prefer to sit on the pod peduncles.
**Pest status.** Usually only a minor, but interesting, pest of cocoa.
**Life history.** Eggs are laid in cracks in the leaf midrib, or in twigs, in a single row of up to 100, and are covered with white waxy filaments. Young nymphs are also covered with waxy filaments, which often form long curling threads. The young nymphs tend to stay on flush leaves and shoot tips, but the older ones are usually found on the bronze leaves and young pods.

The adults are white, moth-like bugs about 15–20 mm in length, with a distinctive shape, and short legs. When disturbed they rise in a group rather like white moths.

**Distribution.** Indonesia and Malaysia.

The closely related species *Lawana candida* (F.) and *Pulastya discolorata* are also to be found on cocoa, kapok, coffee and *Sterculia* in the region from S. China down to Indonesia and Malaysia.
**Control.** Normally not required; probably the nymphs are difficult to kill with contact insecticides because of the copious wax.

adults
0      1 cm

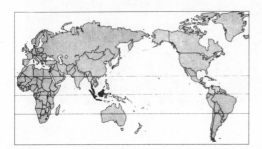

# Ricania spp.

**Common name.** Ricaniid Planthoppers
**Family.** Ricaniidae
**Hosts.** Various plants including palms, *Citrus*, cocoa, wild sugarcane, according to the species of insect.
**Damage.** Usually slight, but the insects are common in tropical regions, and often found on cultivated plants.
**Pest status.** Usually insignificant.
**Life history.** Eggs are laid in rows along the leaf midrib, from which they protrude and are quite visible on careful examination. Egg development takes from 6–9 days. The nymphs resemble psyllid nymphs in appearance and bear long waxy filaments from the end of the body; they tend to be gregarious in nature. When alarmed they can jump.

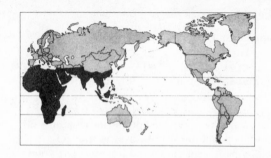

Adults have large expanded forewings which are held out sideways at rest, and give the insects a distinctive appearance; body length is about 8–10 mm. As with other Fulgoroidea the antennae are tiny and inconspicuous.

*Ricania* spp. on grasses

**Distribution.** The genus is found throughout tropical Africa and Asia, up to S. China, as a number of different species.

*R. speculum* Wlk. is blackish with white spots and found from Malaysia to China, and is a regular minor pest of oil palm.

*R. cervina* Mel. is greenish in colour and a regular minor pest on cocoa in W. Africa. A green species, to date only identified as *Ricania* sp. is of regular occurrence on *Citrus* in S. China.

Another unidentified species of *Ricania* is brown with pale bands and to be found in large numbers on various wild Gramineae in S. China.

**Control.** Not usually required.

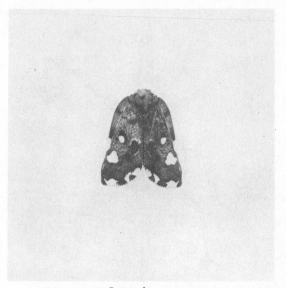

*R. speculum*

*Ricania* sp.

## Pyrilla perpusilla Wlk.

**Common name.** Indian Sugarcane Leafhopper
**Family.** Lophopidae
**Hosts** (main). Sugarcane
(alternative). Millet, wheat, maize, and other species of Gramineae, and some dicotyledons.
**Damage.** Damage consists mainly of sap removal from the plant by the feeding of the nymphs and adults.
**Pest status.** Not a serious pest, but various lophopids are of wide occurrence on crops and plants in the tropics.
**Life history.** The eggs are white, ovoid, 1.0 x 0.5 mm, and are laid almost touching on the leaf in two to four irregular rows, covered with a white waxy secretion. Each batch contains 30–50 eggs. Incubation takes 6–18 days.

After 3–4 hours the nymphs disperse; feeding is most frequent on the lower leaf surface. Honeydew is excreted in some quantities, and many ants are often in attendance; sooty moulds are also common.

There are five nymphal instars, taking a total of 5–20 weeks for completion, though eight is more usual.

The adults are not strong fliers, but they jump readily. They are 7–8 mm long, pale brown with conspicuous veins in the forewings and scattered tiny dark spots.

The complete life-cycle normally takes from 40–55 days, but in winter development is more prolonged.

There are usually 3–5 generations per year.
**Distribution.** Pakistan, India, Sri Lanka, Afghanistan, Burma, and Thailand (CIE map no. A151).

There are two species of *Pyrilla* found in India and Sri Lanka, but 15 subspecies are recognized.
**Control.** Control is seldom required.

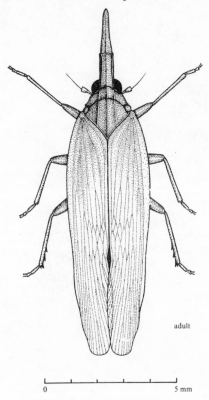

adult

0                           5 mm

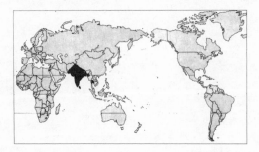

## Diaphorina citri (Kuway.)

**Common name.** Citrus Psylla
**Family.** Psyllidae
**Hosts** (main). *Citrus* species.
     (alternative). Other species of Rutaceae, mostly wild.
**Damage.** Buds and soft young shoots are attacked by the nymphs, leaves become distorted and curled; honey-dew production leads to sooty mould infestation. Badly damaged leaves die and fall, and defoliation of branches can occur. It is thought that the saliva of the nymphs is probably toxic to produce such distortion. Adults do little damage.
**Pest status.** An occasionally serious pest on *Citrus*, especially on young or newly grafted plants.
**Life history.** Eggs are laid, usually in the spring, inside the young, folded leaves in the buds, or in leaf axils. Each female may lay up to 800 eggs during her two-month life. Nymphs hatch after about five days, and after four instars become adult.
     Adults are small, about 2.5 mm in length, with pale brown wings having a broad pale stripe along the centre.
     The life-cycle takes about 20–40 days according to temperature, and there may be up to nine generations per year.
**Distribution.** From Pakistan and India through S.E. Asia to S. China, Philippines, and Indonesia; one record from Saudi Arabia; also from Mauritius and Reunion; and S. America (Brazil) (CIE map no. A335).
**Control.** Chemical control should only be applied at periods of flush growth, and is usually only required on young trees; dimethoate is usually successful, but should not be used on rough lemon trees or on non-budded rough lemon stock.

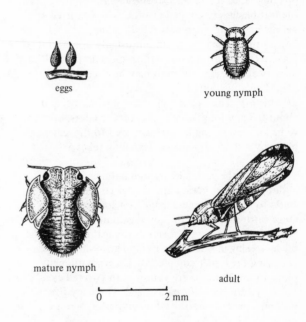

eggs

young nymph

mature nymph

adult

0     2 mm

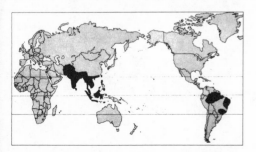

195

# Trioza erytreae (Del G.)

**Common name.** Citrus Psyllid
**Family.** Psyllidae
**Hosts** (main). *Citrus* spp.
    (alternative). Various species of wild Rutaceae.
**Damage.** The leaves are conspicuously pitted, the pits opening on to the lower leaf surface. In severe attacks the leaf blades are cupped or otherwise distorted and yellow in colour, especially when young.
**Pest status.** A very common but usually minor pest of mature *Citrus* throughout Africa; more important on nursery stocks, since growth may be checked and the plant badly disfigured.
**Life history.** The eggs are elongate pear-shape, and about 0.3 mm long; usually laid on the edges, or main veins, of very young leaves, anchored to the leaf blade by a short appendage. Hatching takes 5–6 days.

When the scale-like nymph hatches it walks about for a short period and then settles down and starts to feed on the underside of a soft young leaf. Once settled, it does not move again unless disturbed; at the feeding site a pit forms as the leaf expands, the pit increasing in size as the nymph grows but never enclosing the insect completely. Leaves with many pits curl up. Nymphs are yellow with two red eyes, but may turn brown if parasitized. There are five nymphal instars, and the whole nymphal period occupies 2–3 weeks.

The adult is aphid-like, about 2 mm long, with long transparent wings. It is green when it first emerges but later turns brown. Females may live for a month and lay 600 eggs.
**Distribution.** Tropical Africa, mainly on the eastern half; Cameroons, Zaire, Ethiopia, Sudan, E. Africa, Madagascar, Mauritius, Rwanda, Malawi, Zimbabwe, Zambia, S. Africa, and St Helena (CIE map no. A234).
**Control.** Control measures should only be applied in periods of flush growth. Treatment of mature trees is not usually economic. Young trees can be sprayed with dimethoate as a full-cover spray taking particular care to wet the flush leaves.

Dimethoate should not be used on rough lemon trees or on non-budded rough lemon stock.

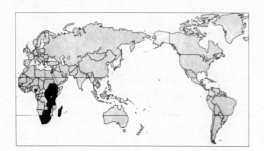

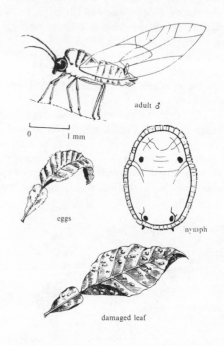

adult ♂

0    1 mm

eggs

nymph

damaged leaf

# Aleurocanthus woglumi Ashby

**Common name.** Citrus Blackfly
**Family.** Aleyrodidae
**Hosts** (main). *Citrus* spp.
    (alternative). Coffee, mango, etc.
**Damage.** Groups of shiny, black, scale-like insects on undersides of leaves. The upper sides of the leaves may have spots of sticky honey-dew or be covered with sooty mould.
**Pest status.** A serious pest in several areas on *Citrus*, and a pest of coffee in the New World.
**Life history.** The eggs are yellowish when first laid but darken to black before hatching. They are elongate oval in shape and about 0.2 mm long. One end is anchored to the leaf by a short appendage, the other tending to be raised from the surface. Batches of 30 or more eggs may be found on the undersides of leaves usually arranged in a spiral. They hatch after about 10 days.

After hatching the larva moves only a very short distance before settling down and starting to feed. There are four nymphal instars; they are all scale-like, shiny black, conspicuously spiny, and bordered by a white fringe of wax. The last instar, the so-called 'pupa', is about 1.5 mm long. The total nymphal period takes 50–100 days, according to temperature.

The adults are tiny moth-like insects, generally black, but with some white markings at the edge of the wings. The body is dusted with a bloom of grey wax. Females are about 1.2 mm long; males 0.8 mm.
**Distribution.** Tropical Asia, S.E. Asia, E. Africa, Seychelles, C. America, Ecuador and the W. Indies (CIE map no. A91).
**Control.** If new nursery stock is brought into an orchard which is still free from Blackfly the plants should be completely defoliated and dipped into a solution of dimethoate before planting.

In parts of Kenya the aphelinid wasp *Eretmoceros serius* has been introduced and has become established and gives excellent control.

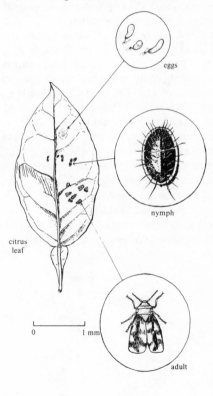

eggs

nymph

citrus leaf

0      1 mm

adult

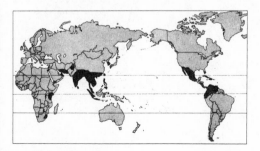

## Bemisia tabaci (Genn.)

**Common name.** Tobacco Whitefly (Cotton Whitefly)
**Family.** Aleyrodidae
**Hosts** (main). Cotton, tomato, tobacco, sweet potato and cassava.

(alternative). Many wild and cultivated plants.
**Damage.** Small white scale-like objects on the underside of the leaves. If the plant is shaken, a cloud of tiny moth-like insects flutter out but rapidly resettle.
**Pest status.** A minor pest of cotton in many parts. Attacks are common during the dry seasons, but they disappear rapidly with the onset of rain. A sporadically serious pest of tomato and tobacco. The viruses transmitted are Cassava Mosaic, Cotton Leaf-curl, Tobacco Leaf-curl, and Sweet Potato Virus B.
**Life history.** The egg is about 0.2 mm long and pear-shaped. It stands upright on the leaf, being anchored at the larger end by a tail-like appendage inserted into a stoma. Eggs are white when first laid but later turn brown. They hatch after about seven days.

When the nymphs hatch they only move a very short distance before settling down again and starting to feed. Once settled they do not move again. All the nymphal instars are greenish white, oval in outline, scale-like and somewhat spiny.

The last instar (the so-called 'pupa') is about 0.7 mm long and the red eyes of the adult can be seen through its transparent integument. The total nymphal period lasts 2–4 weeks according to temperature.

The adult is minute, about 1 mm long and emerges through a slit in the pupal skin and is covered with a white, waxy bloom. The female may lay 100 or more eggs.
**Distribution.** Cosmopolitan occurring as far north as Europe and Japan (CIE map no. A284).
**Control.** Control measures are not usually needed, but if they are then DDT, dimethoate, or pyrethrum as sprays are recommended. The spray should be directed at the undersides of the leaves as far as possible.

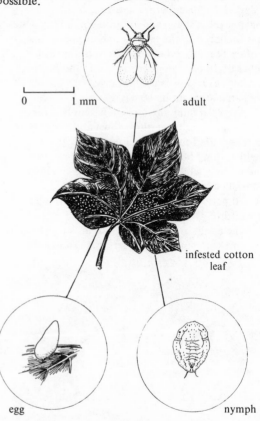

adult

infested cotton leaf

egg

nymph

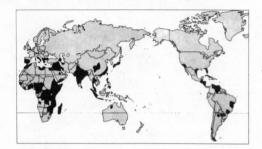

# Dialeurodes citri (Ashm.)

**Common name.** Citrus Whitefly
**Family.** Aleyrodidae
**Hosts** (main). *Citrus* spp.

(alternative). Coffee (*arabica*), *Gardenia*, *Melia*, and other ornamental trees and shrubs.
**Damage.** The direct damage is mainly the loss of sap caused by the feeding bugs, but the honey-dew excreted often leads to infestation by sooty moulds on both leaves and fruit. Thus damage is often more unsightly than real.

**Pest status.** Not a very serious pest but quite widely occurring and causing unsightly damage.
**Life history.** The eggs are tiny, pale yellow, and laid in batches on the underside of young leaves. Each female lays from 100−150 eggs. The incubation period is 8−24 days.

The young nymphs are active at first, but within a few hours they become firmly attached to the leaf. At the first moult the larvae lose their legs and antennae. There are three nymphal instars and the total nymphal period occupies some 20−30 days.

Pupation usually involves a period of quiescence and consequently may take 14−300 days.

The adults are tiny white bugs with delicate wings, about 1.2 mm long; they only live for about ten days.

The life-cycle may vary from 35 to 330 days.

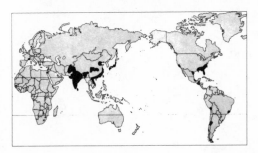

In Florida there are three generations per year.
**Distribution.** S. France, India, Sri Lanka, Pakistan, Bangladesh, China, Vietnam, Japan, USA (California, S.E. USA), Mexico, Guatemala, and S. America (Brazil, Chile, Peru, Argentina) (CIE map no. A111).

A similar species on *Citrus* with cloudy wings and black eggs is *D. citrifolii* Morg.
**Control.** Azinphos-methyl, trichlorphon and parathion-methyl will kill both nymphs and adults quite effectively.

**Further reading**

Weems, H.V. (1973). Citrus Whitefly, *Dialeurodes citri* (Ashmead) (Homoptera: Aleyrodidae). *Florida Dept. Agric., Ent. Circ. No. 128*, 2.

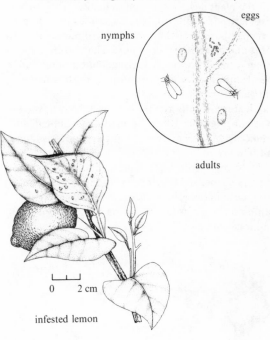

nymphs

eggs

adults

infested lemon

0    2 cm

199

## Aphis craccivora Koch

(= *A. leguminosae* Th.)
(= *A. laburni* Kalt.)

**Common name.** Groundnut Aphid
**Family.** Aphididae
**Hosts** (main). Groundnut, other Leguminosae,
   (alternative). Polyphagous on many plants.
**Damage.** Some wilting results from the sap-sucking
by the aphids in hot weather, but the most serious
damage is the transmission of Groundnut Rosette
Virus. The leaves of the infected plant typically
assume a mottled appearance with either chlorotic
or dark green spots according to the form of virus,
and the plant develops a stunted habit.
**Pest status.** In itself, this is not a serious pest, but it
is very important as the vector of Groundnut Rosette
Virus and some 13 other viruses. The virus is
brought into the crop by winged adult aphids, and is
then transmitted within the crop by both wingless
and further winged forms.
**Life history.** Adults are black or dark brown, variable
in size, being from 1.5 to 2 mm long; siphunculi and
cauda black; antennae are about two-thirds as long
as the body.

   Nymphs are wingless, dark, and fairly rounded
in body shape, and they appear in the crop soon
after germination, the adults usually have over-
wintered (or spent the dry season) on nearby leguminous plants.

The Rosette Virus is transmitted in a persistent
manner. The acquisition period is usually more than
four hours and the virus persists for more than ten
days, and through the moult. The virus is transmitted
by all stages of the insect but the nymphs are more
effective than the apterae.
**Distribution.** Virtually cosmopolitan, but records
are rather sparse in some areas; however, distribution
is expected to be continuous (CIE map no. A99).
**Control.** Cultural control can be effective through
early planting, and close spacing.

   For chemical control menazon as a seed
dressing, or dimethoate as weekly sprays, or menazon
as fortnightly sprays are recommended. The seed
dressing should give protection for about five weeks.

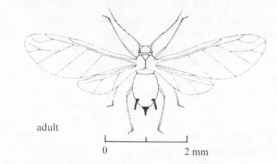

adult

0          2 mm

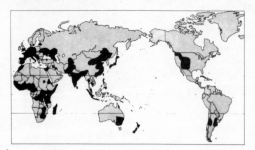

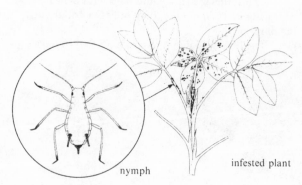

nymph          infested plant

200

# Aphis gossypii Glover

**Common name.** Cotton Aphid (Melon Aphid)
**Family.** Aphididae
**Hosts** (main). Cotton
(alternative). *Hibiscus* spp., Cucurbitaceae, many legumes, and a wide range of plants belonging to many different families; polyphagous.
**Damage.** The leaves are cupped or otherwise distorted, with clusters of soft, greenish or blackish aphids on young shoots and on the undersides of young leaves. Drops of sticky honey-dew and/or patches of sooty mould on the upper sides of leaves.
**Pest status.** Outbreaks are common on young plants in spells of dry weather which clear up rapidly with the onset of rain. Plants may be debilitated during the aphid attack but there is no evidence that the yield of seed cotton is affected. It is a greenhouse pest in Europe, especially on cucurbits. Recorded as a vector of about 44 virus diseases.
**Life history.** Only the female adults are found, which may be winged or wingless; blackish-green, small to medium-sized, about 1−2 mm long; antennae usually only about half the length of the body. Siphunculi usually black in colour and cauda not often paler; the eyes are red.

There are probably several other species going under the name of *Aphis gossypii*. The wingless females are somewhat larger, more globular, and generally paler in colour. Living young, greenish

or brownish in colour, are produced by both types of adult female. The adults may live for 2−3 weeks and produce two or more offspring each day.
**Distribution.** Completely cosmopolitan, absent only from the colder parts of Asia and Canada (CIE map no. A18).
**Control.** Control measures are not usually required on most crops. DDT tends to result in aphid outbreaks when used frequently, but when mixed with carbaryl the pest is usually kept in check.

If chemical control is required then generally sprays of carbaryl or dimethoate are recommended.

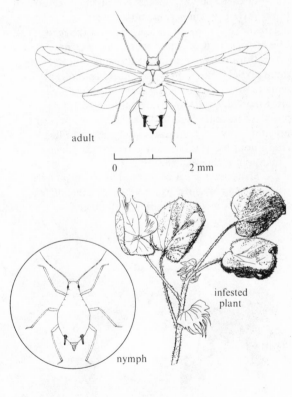

adult

0          2 mm

nymph

infested plant

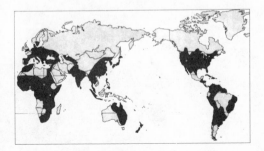

201

# Myzus persicae (Sulz.)

**Common name.** Green Peach Aphid
**Family.** Aphididae
**Hosts.** Polyphagous on many crop plants and weeds.
**Damage.** The direct damage to peach is typically distortion of young leaves and shoots (leaf-curl). On many plants these symptoms are followed by virus disease symptoms.
**Pest status.** A very important pest on many crops in many parts of the world, doing damage both by direct feeding and by virus transmission. It can transmit over 100 virus diseases of plants in about thirty different families, including beans, sugar beet, sugarcane, brassicas, potato, *Citrus* and tobacco.
**Life history.** In the tropics there is no alternation of generations between different hosts, and neither are there sexual forms — males are never found. The females breed by parthenogenesis and vivipary. Most individuals are apterous but winged forms are produced at times for the dispersal of the species. Breeding in warm countries may be more or less continuous but it is basically a temperate species and does not thrive in the tropics.

Many physiological races of *M. persicae* have been discovered, showing no morphological differences but distinct host feeding preferences.

The adult is small to medium-sized, 1.25–2.5 mm long, usually green with a darker thorax; antennae two-thirds as long as the body; siphunculi clavate, fairly long; the face viewed dorsally has a characteristic shape.
**Distribution.** Virtually cosmopolitan; northwards up to S. Scandinavia, N. China, and Canada (CIE map no. A45).
**Control.** The use of chemicals to prevent virus spread by controlling their aphid vectors is generally not successful. Thus, there is a special need for the integrated control approach, paying particular attention to predators, parasites, alternate hosts and crop manipulation.

The usual aphicides are as follows: dimethoate, malathion, demeton-*S*-methyl, menazon, demephion, disulfoton, formothion, phorate, phosphamidon, thiometon.

Frequently the use of pesticides results in destruction of predators and parasites, and so great care must be taken in their use.

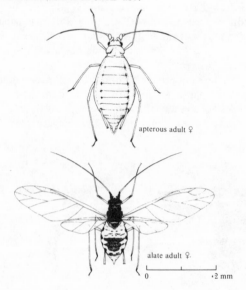

apterous adult ♀

alate adult ♀.

0                    ·2 mm

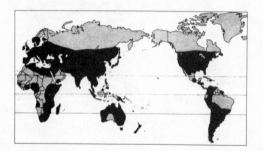

# Pentalonia nigronervosa Coq.

**Common name.** Banana Aphid
**Family.** Aphididae
**Hosts** (main). Bananas (*Musa* spp.)

(alternative). *Alpina, Heliconia, Colocasia* spp., *Costus, Zingiber, Palisota*, and tomato.

**Damage.** Direct damage is negligible but this aphid is the vector of the virus causing Bunchy Top Disease. The disease is widespread from Egypt, India, through S.E. Asia to Australia. Symptoms include dark green streaking on the leaves, midrib and petioles, progressive leaf-dwarfing, marginal chlorosis and leaf-curling. Fruit bunches are small and distorted, and the fruit is unsaleable.

**Pest status.** Important as the vector of Bunchy Top Disease, which is serious in Asiatic banana-growing areas, and three other virus diseases.

**Life history.** The adults are small to medium-sized, 1–2 mm in length, brown, with antennae as long as the body. The alate adults have a very prominent dark wing venation; the siphunculi are slightly clavate and quite long.

The aphids are found under the old leaf sheaths at the base of the pseudostem near ground level, as colonies of brown shiny wingless aphids. Ants always accompany the aphid colonies, and they are responsible for establishment of new colonies.

Winged adults are usually produced after about 7–10 apterous generations, and the winged adults migrate to new host plants.

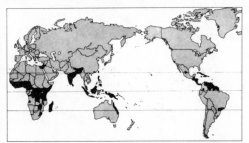

**Distribution.** Distribution is probably coexistent with banana cultivation and is more or less pantropical. (CIE map no. A242).

**Control.** Chemical treatments are generally only effective if accompanied by careful eradication of infested plants.

Suggested pesticides are parathion, phosphamidon, dicrotophos and endrin, which should be sprayed at the plant crown and pseudostem base, below soil level between the outer leaf sheath and stem, and over the surrounding soil.

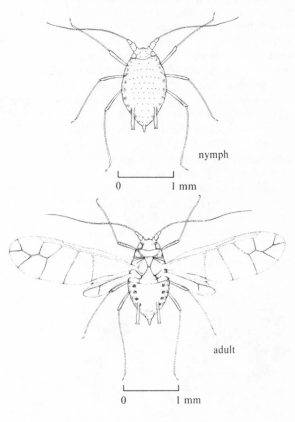

nymph

0          1 mm

adult

0          1 mm

203

## Rhopalosiphum maidis (Fitch)

(= *Aphis maidis* Fitch)

**Common name.** Maize Aphid (Corn Leaf Aphid)
**Family.** Aphididae
**Hosts** (main). Maize

(alternative). Sorghums, millets, sugarcane, wheat, barley, rice, and other Gramineae; manila hemp, tobacco, and other crops and some weeds.
**Damage.** Leaves, leaf sheath and inflorescence covered with colonies of dark green aphids, with a slight white covering. The leaves may become mottled and distorted, and new growth may be dwarfed. The inflorescence may be sufficiently damaged to become sterile. Usually a pest of young tender plants. Honey-dew production is prolific.
**Pest status.** A particularly important pest of cereals in America and parts of Europe. Mostly found on maize and sorghum, occasionally on barley, but seldom found on wheat or oats. It is a vector of many (ten) different virus diseases in cereals and other crops.
**Life history.** Adults may be winged or apterous, about 2 mm long, with characteristically short siphunculi. The cauda is pronounced with long conspicuous setae. There are dark purplish areas around the base of the siphunculi.

Reproduction is mostly or entirely partheno-genetic in most parts of the world, but males are more common in Korea, indicating a probable oriental origin for this species.

The life-cycle in the tropics takes about eight days.
**Distribution.** Almost completely cosmopolitan in distribution, throughout the tropics, subtropics and the warmer temperate regions. The northernmost records are Japan and southern Scandinavia (CIE map no. A67).
**Control.** Burning the seed crop stubbles after harvest effects a degree of cultural control.

If the plants are growing vigorously the aphids are usually kept under control by natural enemies.

Sprays of dimethoate, demephion, demeton-*S*-methyl, ethoate-methyl, formothion, or menazon are generally effective if required.

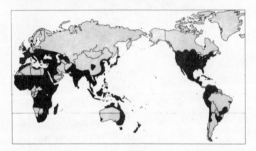

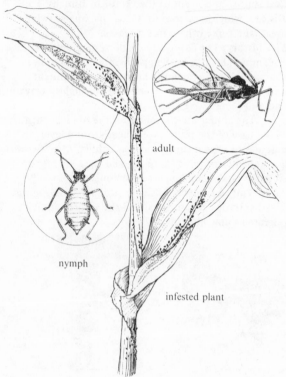

adult

nymph

infested plant

# Toxoptera aurantii (B. de F.)

**Common name.** Black Citrus Aphid
**Family.** Aphididae
**Hosts** (main). *Citrus* species
(alternative). Other Rutaceae, *Ficus* spp., tea, cocoa, coffee, and other plants; polyphagous.
**Damage.** Distortion of young leaves, with clusters of black aphids on flush growth and under young leaves. Often accompanied by sooty moulds growing on the honey-dew excreted.
**Pest status.** This aphid is universally present on *Citrus* bushes, and occasionally severe outbreaks may occur, especially in dry weather following a rainy season. It is completely polyphagous and has been recorded from 120 different host plants. It also occurs in greenhouses in temperate countries on a range of different hosts.
**Life history.** The adults are shiny black in colour, and may be winged or apterous, from 1.2—1.8 mm in body length, with relatively short antennae. Only females are found, and they produce living young, dark brown in colour, five to seven each day, up to a total of about 50 per female.

At 25°C a single generation takes as short a time as six days, but at 15°C as long as 20 days, and a similar effect is seen at higher temperatures; at temperatures above 30°C the aphid population declines sharply.

Infestations are generally attended by ants.
**Distribution.** Widely distributed throughout the warmer parts of the world, including S. Europe, Africa, Asia and Australasia, southern USA, C. and S. America (CIE map no. A131).
**Control.** Natural control measures can be encouraged by spray-banding the bushes and tree-trunks with dieldrin to discourage the attendant ants; a large number of predators and parasites have been recorded attacking this species of aphid.

nymph

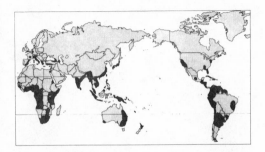

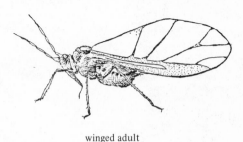

winged adult

Chemical control should be applied only at periods of flush growth at the first signs of damage. The most successful treatment is often a full-cover spray of dimethoate in water; other systemic insecticides are demeton-*S*-methyl and fenitrothion. Care must be taken to wet the flush leaves. Dimethoate should not be used on rough lemon trees or on non-budded rough lemon stock.

The usual contact aphicides can also be used as an alternative.

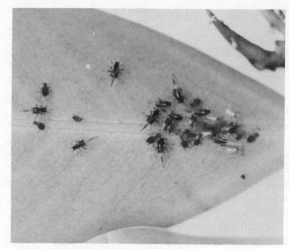

infested leaf of *Ficus microcarpa*

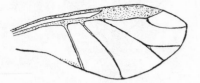

## Toxoptera citricida (Kirk.)

**Common name.** Brown Citrus Aphid (= Tropical Citrus Aphid)
**Family.** Aphididae
**Hosts** (main). *Citrus* spp.
      (alternative). Confined to members of the Rutaceae.
**Damage.** Distortion of young leaves, with clusters of dark brown aphids on flush foliage and under leaves. Sometimes sooty moulds grow on the honey-dew excreted by the aphids. Some twigs and branches show die-back symptoms as this insect is the vector of the virus causing Tristeza (die-back) Disease in Africa and S. America, and also several other virus diseases of *Citrus*.
**Pest status.** A pest species to be found on *Citrus* in most places where it is grown; occasionally serious, especially when virus diseases are transmitted. Tristeza Disease is common both in tropical Africa and S. America. Sometimes quite small aphid colonies cause serious bud drop after feeding on young flowers. Heavy infestations may regularly produce a yield loss of up to 50%.
**Life history.** Adults are dark brown or blackish in colour, either winged or apterous, about 2.0−2.8 mm in length, and in winged forms the median vein always has two branches. (*T. aurantii* has only one usually, but some specimens from S.E. Asia have two.)

Breeding is through constant viviparity and parthenogenesis, and the time required for one generation may be as short as a week, thus with four generations per month, population growth is tremendous.
**Distribution.** A more tropical and less widespread species than *T. aurantii*, it is found throughout tropical Africa, S.E. Asia, Australasia and S. America (CIE map no. A132).
**Control.** As for *T. aurantii*, but it is important to prevent the spread of alatae which are responsible for transmitting the virus diseases.

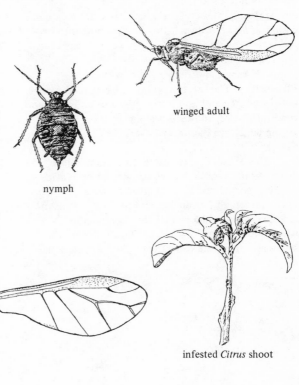

winged adult

nymph

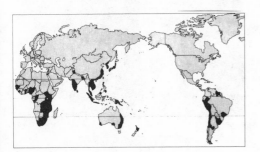

infested *Citrus* shoot

# Ceratovacuna lanigera Zhnt.

(= *Oregma*; *Cerataphis saccharivora* Mats.)

**Common name.** Sugarcane Woolly Aphid
**Family.** Pemphigidae
**Hosts** (main). Sugarcane

(alternative). Wild cane (*Saccharum spontaneum*), and *Miscanthus* spp. (Gramineae).

**Damage.** Heavy infestations on the underneath of leaves of white waxy aphids cause a loss of plant sap, and sometimes heavy sooty mould infestation occurs on the foliage. In Taiwan young plants may be killed by heavy infestations.

**Pest status.** Within its area of occurrence it is quite a common pest but does little damage and control measures are seldom required. However it has been recorded causing an estimated 20% loss of yield in China (Kwangsi and Fukien) and Taiwan.

**Life history.** The infesting population normally consists of winged and apterous, viviparous and parthenogenetic females, and each female produces 15–35 young which are copiously covered with waxy filaments.

In China it was found that the population peak was in October and November, when damage to the cane was done, and winged females, produced in the middle of November, moved over to *Miscanthus* grasses growing wild around the sugarcane fields where hibernation took place. The apterae on the sugarcane die in the early frosts. Crop infestations reappeared in June and then declined to a low level after July, presumably because of the summer temperatures.

**Distribution.** From India through mainland S.E. Asia to China and Taiwan, and the Philippines and Java.

**Control.** In the early study in China (1944–5) adequate control was achieved using sprays of nicotine, soap solution and an oil emulsion.

Normally this insect is preyed upon by many species of Coccinellidae, lacewings, lycaenid larvae and syrphid larvae, which together keep most aphid populations in check. In Taiwan (Pan, 1980) the damage to young plants by woolly aphid is sufficiently serious that pesticides have to be used, and adequate control has been achieved by foliar sprays of malathion and demeton-*S*-methyl.

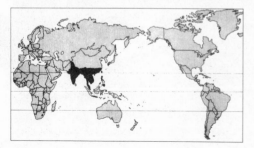

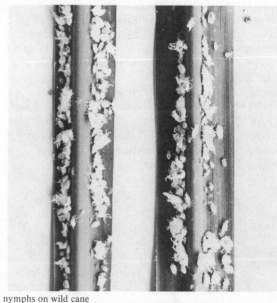

nymphs on wild cane

0       2 cm

## Pseudoregma bambusicola Tak.

**Common name.** Bamboo Woolly Aphid
**Family.** Pemphigidae
**Hosts.** Only recorded from one or two species of bamboo (*Bambusa* spp.).
**Damage.** Heavy infestations of aphids cover the stem entirely for several internodes; similarly on lateral shoots the plant is entirely obscured by a mass of fat grey aphids, which all wave their antennae frantically when excited. Sometimes the bamboo shoots succumb to this massive infestation and die. Usually though the infestation is spectacular rather than very damaging. Honey-dew excretion is copious and falls like fine rain under a heavily infested tall bamboo, and wasps, beetles, flies, and some butterflies gather to feed on the sugar. Sooty mould infestations are usually very extensive both on and around the infested bamboo stems.
**Pest status.** A striking but minor pest, although the total loss of sap with a severe infestation must be considerable.
**Life history.** No details are known, except that winged adults are seldom seen. The apterous females are large and almost globular, measuring some 4 mm in length, but presumably alatae are produced at some time in the season (probably autumn or early winter) to effect dispersal of the species to new hosts.
**Distribution.** To date only recorded from S. China, where it is very abundant.

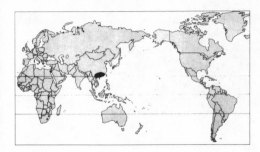

**Control.** Several large Coccinellidae are conspicuous predators and apparently eat large numbers of the aphids.

Chemical control measures are not usually needed.

nymphs on lateral shoots    0    1 cm

nymphs on main stem

209

# Icerya aegyptica (Dgl.)

**Common name.** Egyptian Fluted Scale

**Family.** Margarodidae

**Hosts.** Polyphagous in feeding habits, it has been recorded from *Citrus*, coffee, tea, guava, mulberry, wattle, date palm, grapevine, pear, jack-fruit, *Ficus* spp., and many ornamentals.

**Damage.** Usually slight, just a little loss of sap, but the insect is very conspicuous and easily noticed.

**Pest status.** Usually a minor pest, which seldom requires control measures, but it is widespread in the Old World tropics and found on many different plants.

**Life history.** Details of its life history are not available. However, the nymphs are distinguishable from those of *I. purchasi* by having a golden body colour (underneath the wax) as opposed to dark red, and by the long waxy fringes festooning the body surface. Body size is generally about 6 mm in length by 4 mm breadth. The larvae are scarcely distinguishable from *I. purchasi* morphologically.

**Distribution.** Widespread throughout the Old World tropics (CIE map no. A221).

**Control.** Usually preyed upon by many different Coccinellidae, and seldom requires chemical control.

mature ♀♀

0        1 cm

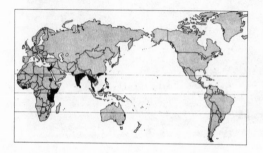

## Icerya purchasi Mask.

(= *Pericerya purchasi* Mask.)

**Common name.** Cottony Cushion Scale (Fluted Scale)
**Family.** Margarodidae
**Hosts** (main). *Citrus* spp.

(alternative). Polyphagous; attacking many other plants, especially mango and guava.
**Damage.** The leaves and twigs are infested with large, white, fluted scales, and infested leaves often turn yellow and fall prematurely. Heavily infested young shoots are killed, and in fact whole nursery trees can be killed. Copious quantities of honey-dew are excreted.
**Pest status.** A polyphagous pest, important on *Citrus*, very widely distributed throughout the world.

This pest is a native of Australia, introduced into California in 1868, and now occurring in all *Citrus*-growing areas.
**Life history.** The adult female is a distinctive insect, being quite large (about 3.5 mm), sturdy, with a brown body covered with a layer of wax. The most conspicuous part of the insect is the large, white, fluted egg-sac which is secreted by the female. The egg-sac usually contains more than 100 red, oblong eggs. The hatching period is from a few days to two months, according to climate.

The three nymphal stages are shiny, reddish insects under the wax, and they are most abundant along the midrib under the leaves. The fully grown scales are most frequently found on the twigs and shoots.

Males are seldom found, but sexual differentiation occurs during the second nymphal instar.
**Distribution.** Cosmopolitan through the warmer parts of the world; only unrecorded from a few countries (CIE map no. A51).

Called the Fluted Scale in the USA.
**Control.** This scale is usually controlled naturally by Coccinellidae which have been introduced from Australia and India into most *Citrus* growing areas.

If chemical control is required then azinphos-methyl or parathion-methyl should prove effective.

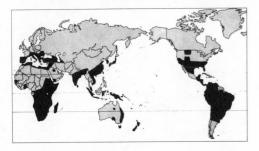

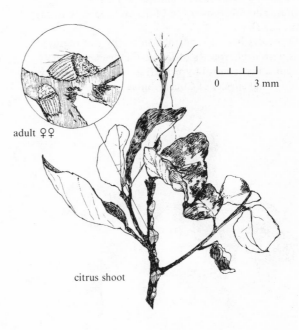

adult ♀♀

0    3 mm

citrus shoot

## Icerya seychellarum (Westw.)

**Common name.** Seychelles Fluted Scale
**Family.** Margarodidae
**Hosts.** Polyphagous; recorded attacking *Citrus*, guava, jack-fruit, mango, pear, most species of Palmae, and many other crops and ornamentals.
**Damage.** Usually slight, but a conspicuous insect producing unsightly infestations, and leaving fruits covered with waxy exudates.
**Pest status.** A very widespread and polyphagous but minor pest, which does not often require special control measures.
**Life history.** Details are not available, but it is expected to be basically similar to other species of *Icerya*.
**Distribution.** Very abundant in Madagascar, but only recorded from a few localities on mainland Africa; otherwise Seychelles, Mauritius, Sri Lanka, parts of India and S.E. Asia up to China and Japan (CIE map no. A52).
**Control.** Chemical control is normally not required; as with other species of *Icerya* numbers are usually kept under control by predation of adults and larvae of several species of Coccinellidae.

infestation on jackfruit petioles

0          1 cm

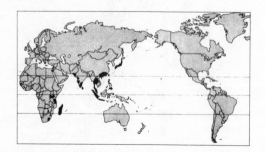

## Orthezia insignis Browne

**Common name.** Jacaranda Bug (Lantana Bug)
**Family.** Orthezidae
**Hosts** (main). Coffee, mainly *arabica*.

(alternative). *Jacaranda*, *Citrus*, *Lantana*, sweet potato, eggplant, and many other plants, especially roses.

**Damage.** The bugs are found sitting on the leaves, shoots, and fruit of the host plant. The damage consists of sap removal by the feeding bugs and is not in itself evident.

**Pest status.** A polyphagous pest on many crops and plants but seldom serious on any of them. There are some records from greenhouses in Europe and N. America. It is not well adapted to coffee and does not usually stay long on the bushes.

**Life history.** Not well known.

**Distribution.** Probably pantropical, but records from Asia are restricted to S. India, Sri Lanka, and Malaya, and there are none from Australia (CIE map no. A73).

**Control.** Not usually required.

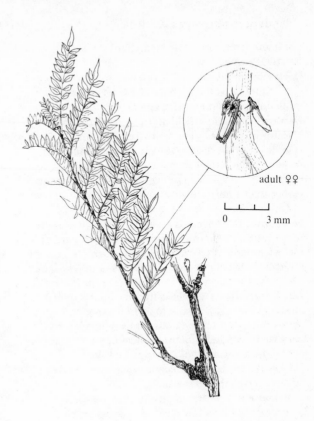

adult ♀♀

0 ——— 3 mm

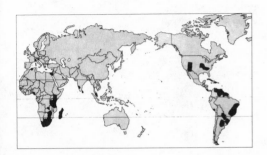

# Dysmicoccus brevipes (Ckll.)

**Common name.** Pineapple Mealybug
**Family.** Pseudococcidae
**Hosts** (main). Pineapple
(alternative). Also recorded from sugarcane, groundnut, coconut, coffee and *Pandanus*.
**Damage.** This is a particularly important pest as it is a vector of the Mealybug Wilt virus. The first symptoms of the disease usually appear in the roots which cease to grow, collapse and then rot. An apparently flourishing crop will show the symptoms earlier than a slow growing, poor crop. This is known as Quick Wilt.
**Pest status.** *D. brevipes* is a serious pest of pineapple wherever they are grown. Some varieties of pineapple are more resistant to the virus than others, the variety Cayenne being highly susceptible. It is a polyphagous pest, often found on the roots of the crops it attacks.
**Life history.** The mealybugs live in colonies underground with only a small proportion living on the leaves. The occurrence of Mealybug Wilt is largely correlated with the subterranean colony on the roots.
The aerial individuals are to be found mostly at the base of the leaves, which may have to be spread in order to make the bugs evident.
**Distribution.** Almost completely pantropical in distribution, with a few records from subtropical areas (CIE map no. A50).

**Control.** Spraying the leaves does not control the spread of the disease, as there is only a small proportion of the colony on the leaves at any time..
Control can be obtained by dipping the slips in a solution of malathion, diazinon, or parathion, and stacking the slips vertically for 24 hours to allow the insecticide to accumulate at the leaf bases, and then spraying each month at the base of the plant with parathion.

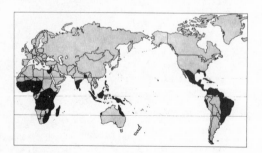

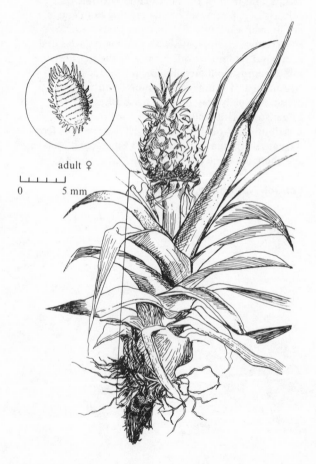

adult ♀

0      5 mm

## Ferrisia virgata (Ckll.)

(= *Ferrisiana virgata* (Ckll.)) etc.

**Common name.** Striped Mealybug
**Family.** Pseudococcidae
**Hosts** (main). Coffee
    (alternative). Cocoa, *Citrus*, cotton, jute, groundnut, beans, cassava, sugarcane, sweet potato, cashew, guava, tomato, and many other plants.
**Damage.** This insect feeds on young shoots, berries and leaves, sometimes in very large numbers. In dry weather it may move down below ground and inhabit the roots. It is generally accepted that this mealybug is favoured by dry weather; many records refer to heavy attacks following periods of prolonged drought.
**Pest status.** A polyphagous pest on many crops. Vector of Swollen Shoot disease of cocoa. A serious pest on coffee in some areas (Java and Papua New Guinea).
    There are many synonyms for this species.
**Life history.** The female lays 300—400 eggs, which hatch in a few hours, and the young nymphs move away quite rapidly. The nymphs are full grown in about six weeks.
    The adult female is a distinctive mealybug with a pair of conspicuous longitudinal submedian dark stripes, and long glassy wax threads, and a pronounced tail, and a powdery waxy secretion.
    The entire life-cycle takes about 40 days.

**Distribution.** Pantropical in distribution, but with only a few records from Australia and S. America (CIE map no. A219).
**Control.** If control is required the usual insecticides employed against mealybugs can be used. These include malathion and azinphos-methyl.
    As is usual with mealybugs it is important to make sure that the insecticide reaches the body of the insect, so it is necessary to add extra wetter to the spray solution.
**Further reading**
Le Pelley, R.H. (1968). *Pests of Coffee*, pp. 347—50. Longmans: London.

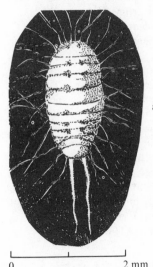

adult ♀

0            2 mm

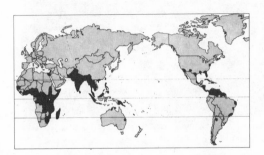

cotton leaf

# Maconellicoccus hirsutus (Green)

(= *Phenacoccus hirsutus* (Green))

**Common name.** Hibiscus Mealybug
**Family.** Pseudococcidae
**Hosts** (main). Plants in the Family Malvaceae, notably *Hibiscus* and *Gossypium* spp.

(alternative). Leguminosae, and both tropical and subtropical fruit and shade trees.
**Damage.** This mealybug is remarkable in being the only recorded species that has toxic saliva, and whose feeding results in the stunting and sometimes death of the young infested shoots.
**Pest status.** Because of the toxic saliva this can be a serious pest, as heavy infestations of young shoots will often kill the shoots completely, resulting in stunted and deformed bushes and trees. Infestations are often very heavy, and characteristic in appearance due to the shortened internodes of apical shoots, which often swell and develop a dark green colour. The leaves on these shortened shoots give the plant a 'bushy-top' appearance.
**Life history.** The adults are small (about 3 mm in length) and pink in body colour, but covered with a waxy secretion; the waxy filaments are very short and in general the insect resembles *Planococcus* in appearance.
**Distribution.** Found mostly in India and S.E. Asia, up to S. China and Taiwan, but recorded from some parts of Africa (CIE map no. A100).

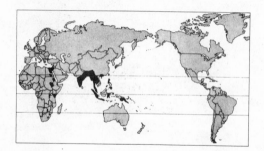

**Control.** For chemical control either malathion, diazinon, parathion, or azinphos-methyl can be used as foliar sprays, but because of the waxy body covering on the mealybugs it will probably be necessary to add extra wetter to the spray solution. Probably a second spray will be required two weeks later.

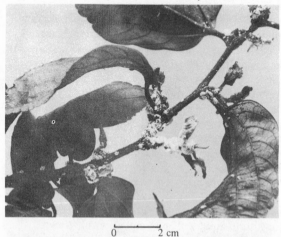

infestation of *Celtis* twig; some shoots already dead

0　　　　　2 cm

infested (dying) *Hibiscus* shoot

216

# Planococcus citri (Risso)

(= *Pseudococcus citri* (Risso))

**Common name.** Citrus Mealybug (Root Mealybug)
**Family.** Pseudococcidae
**Hosts** (main). Coffee, *Citrus*, and cocoa.
(alternative). Minor pest of cotton, and various vines.
**Damage.** The leaves wilt and turn yellow, as if affected by drought. Roots are often stunted and encased in a crust of greenish-white fungal tissue, *Polyporus* sp. If the fungus is peeled off the white mealybugs can be seen. The aerial form is found on leaves, twigs, and at the base of fruit.
**Pest status.** A minor pest of *arabica* and *robusta* coffee; occasional trees are killed, especially very young trees. Another race of *P. citri* is sometimes found on the aerial parts of the coffee trees, but very rarely causes serious damage; this race is common on *Citrus* and sometimes on cotton. Generally a polyphagous pest; occurs in greenhouses in temperate climates.
**Life history.** Very little is known about the life history of the Root Mealybug. It is probably similar to that of the Kenya Mealybug though it is known to breed more slowly. Root Mealybugs are sometimes found without the fungus. This suggests that the plant is first weakened by the feeding of the mealybug; the debilitated plant is then susceptible to fungal attack.

Citrus Mealybug is a vector of Swollen Shoot Disease of cocoa.
**Distribution.** Almost completely pantropical and also extending well into subtropical regions (CIE map no A43).

Occurs in greenhouses in temperate countries.
**Control.** Trees with green or yellow leaves can often be saved by careful treatment, though recovery is very slow. Trees with dead brown leaves are past hope and should be uprooted and replaced.

The soil under the tree should be dusted with aldrin, especially round the collar; the insecticide should be worked into the top layers of the soil. A generous layer of mulch and irrigation should also be provided round the infested trees after treatment.

When gapping up in an attacked plantation aldrin dust should be mixed with the soil in the planting hole.

Sprays of diazinon, malathion, parathion and dimethoate may be effective, especially if the malathion or parathion is mixed with white oil.
**Further reading**
Le Pelley, R.H. (1968). *Pests of Coffee*, pp. 324–30.
Longmans: London.

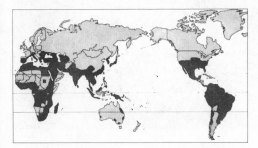

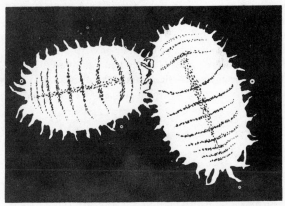

adult ♀♀

0    1 mm

# Planococcus kenyae (Le Pelley)

(= *Pseudococcus kenyae* Le Pelley)

**Common name.** Kenya Mealybug
**Family.** Pseudococcidae
**Hosts** (main). Coffee

(alternative). A large number of wild and cultivated plants, including yam, pigeon pea, passion fruit, sugarcane and sweet potato.

**Damage.** Mealy white masses of insects, especially between clusters of berries or flower buds or on sucker tips. Upper surface of leaves with spots of sticky transparent honey-dew, or covered with a crust of sooty mould growing on the honey-dew.

**Pest status.** Between 1923 and 1939 it had been a major pest of *arabica* coffee in the East Rift area of Kenya, but since the liberation of parasites from Uganda in 1938 it has been reduced to a minor pest.

**Life history.** Eggs are laid below and behind the mature female and are covered with a waxy secretion. One female may lay between 50 and 200 eggs. Females are usually fertilized by the winged males but this is not essential for fertile egg production.

The larva is flat and oval, pale brown, with six short legs; there is no wax on the body. It crawls upwards until it finds a place where a large part of its body is in contact with a surface, e.g. between the stalks of young berries or buds, or next to other mealybugs. Here it begins to feed and gradually develop the characteristic mealy wax covering. It passes through three nymphal stages before becoming adult, each stage being larger, more convex and more waxy but otherwise differing little from previous stages. It may change its position and move a short distance if conditions are becoming unfavourable, especially during the third stage. In the laboratory the female can complete her development and begin egg-laying after 36 days. The males are rarely seen and cause negligible damage. In the egg and first two stages they resemble the corresponding stages of the female. At the end of the second instar they seek out a crevice (commonly in the bark) and there turn into the so-called 'pre-pupal' stage. This is followed by another resting or 'pupal' stage from which emerges a fragile two-winged insect — the adult male.

**Distribution.** E. Africa, Nigeria, Zaïre, and Ghana (CIE map no. A384).

**Control.** Prompt stripping of unwanted sucker growth helps to reduce the number of suitable feeding sites.

The bug is best controlled indirectly; banding the stump of the tree with dieldrin keeps off the attendant ants and allows the natural enemies to clean up the infestation. The band, which may be painted on or sprayed, should be at least 15 cm wide and any bridges such as drooping primaries which would allow the ants to by-pass the band must be removed.

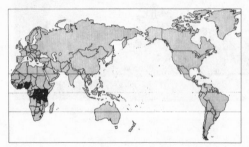

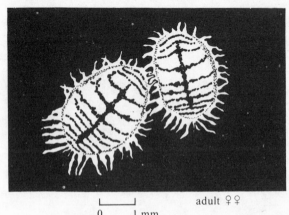

adult ♀♀

0          1 mm

## Pseudococcus adonidum (L.)

(= *P. longispinus* T.)

**Common name.** Long-tailed Mealybug
**Family.** Pseudococcidae
**Hosts** (main). *Citrus* spp.

(alternative). Coffee, cocoa, sugarcane, coconut and other palms; frequent on ornamentals and other crops; a polyphagous pest.
**Damage.** The waxy mealybugs are congregated near the tips of shoots, on the fruit, or on the leaves. By sucking the sap a heavy infestation can kill young plants. Some leaf and shoot deformation is not uncommon.
**Pest status.** Usually not a serious pest on any one crop, but very widespread and common on many crops and plants.
**Life history.** The adult female lays an egg mass of up to 100 or 200 eggs.

The young nymphs crawl away from the egg mass to find suitable feeding sites. After about 20 days the sexes become distinguishable, and the males aggregate separately, forming rough cocoons in which they become a quiescent third instar with small wing buds. After moulting again they become fourth instars with more developed wing buds. From cocoon formation to emergence of the winged males is about 10–14 days. The second instar female nymphs moult into the last immature stage. As the nymphs grow larger they gradually produce more wax. The function of the males is not known, parthenogenesis being assumed.

The adult female mealybugs are long-tailed, in that they have a pair of long filamentous posterior waxy projections (tassels). There are however several other species which have these long tassels so this character is not specific.

Ants are usually associated with the mealybugs and they feed on the honey-dew excreted by the bugs.
**Distribution.** Very widely distributed – almost cosmopolitan (CIE map no. A93).
**Control.** This pest does not often require controlling.

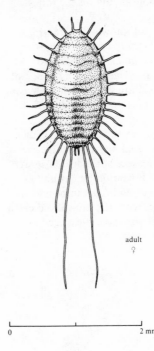

adult
♀

0                                    2 mm

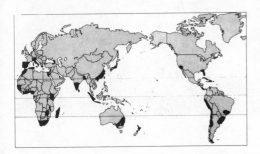

## Pseudococcus citriculus Green

**Common name.** Long-tailed Citrus Mealybug
**Family.** Pseudococcidae
**Hosts** (main). *Citrus* spp. and other Rutaceae.
     (alternative). *Hibiscus*, *Ficus* spp., various orchids and other ornamentals.
**Damage.** A polyphagous pest often found in small numbers, doing little obvious damage, but producing copious honey-dew which results in heavy sooty mould infestations on the foliage.
**Pest status.** Only a minor pest but conspicuous and widespread in S. China and usually with a disproportionate amount of sooty mould on the foliage and fruits.
**Life history.** Details are not known. The adult is a small mealybug, about 3 mm in body length, with quite long lateral waxy filaments but with two very long anal filaments of wax. Usually attended by ants. On a worldwide basis there are several other species of mealybugs with long 'tails' to be found on *Citrus* plants.
**Distribution.** To date only recorded from S. China.
**Control.** If required, the usual pesticides effective against mealybugs can be used, but repeated sprays will probably be needed. Predation by natural enemies is often quite high, and this can be encouraged by banding the trees with dieldrin-soaked cloth (or spray-banding) which will repel the attendant ants.

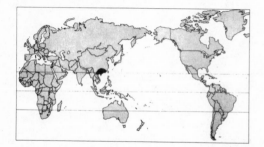

mature ♀ on fruit of mandarin orange with associated sooty mould

0       1 cm

on leaf of *Ficus elastica*

220

# Saccharicoccus sacchari (Ckll.)

**Common name.** Pink Sugarcane Mealybug
**Family.** Pseudococcidae
**Hosts** (main). Sugarcane
    (alternative). Sorghum, rice, and various grasses.
**Damage.** This mealybug is usually situated on the stem beneath the sheath but is sometimes found on the stem just below ground level, on the root crowns, on the stem buds, and underneath the leaves. The leaves often turn red at the base as a result of the insects' presence. Sooty moulds often develop in severe infestations, and ants are associated.
**Pest status.** The most important mealybug pest of sugarcane. It is often present in very large numbers, and the amount of honey-dew excreted considerable. It is probably toxicogenic; however, whether mealybugs really cause damage to sugarcane is debatable.
**Life history.** Eggs are laid under the leaf sheath; each female lays up to 1000 eggs. Hatching takes only 10–14 hours, for the eggs are retained in the genital tract of the female until development is advanced.

    First instar nymphs are quite active but generally only move from older to younger parts of the plant, or on to adjacent plants. Older nymphs are less active and move only reluctantly.

    The adult male occurs both as apterous and winged forms, but is generally rare. Parthenogenesis is the normal mode of reproduction. The adult female is pinkish and is elongate-oval to round in shape, about 7 mm long, with well-developed anal lobes; legs rather short.

    The life-cycle takes about 30 days to complete.
**Distribution.** Widely distributed throughout the tropics, but no records from W. Africa or continental S.E. Asia (CIE map no. A102).
**Control.** Cultural methods are strongly advocated, including destruction of crop residues and trash; clean cultivation; and use of uninfested cane for planting.

    Hot-water treatment can be effective.

    Dipping of planting material into fungicidal solutions with added dieldrin appears to be a promising and convenient method of control, particularly since this routine involves stripping off the sheath.

    Insecticidal application to standing cane is impracticable and usually unsuccessful.

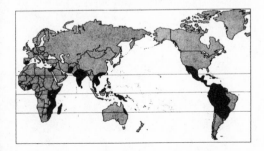

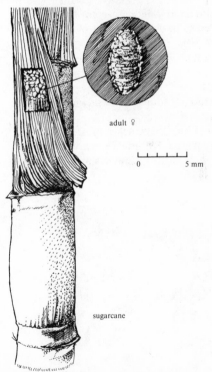

adult ♀

0     5 mm

sugarcane

## Ceroplastes rubens Mask.

**Common name.** Pink Waxy Scale
**Family.** Coccidae
**Hosts** (main). *Citrus* spp.

(alternative). Coffee, tea, mango, fig, and various other fruit trees.

**Damage.** Shoots, fruit stalks, and parts of the fruits may be covered with pink or reddish convex scales. Quantities of honey-dew are excreted and the scales are often found attended by ants. Often sooty moulds may be extensive where the honey-dew has dripped on to leaves and fruit.

**Pest status.** Not an important pest usually but widespread and frequently found on many trees. On mango the fruits may fail to develop and fall prematurely.

**Life history.** Life history details are not known, but may be expected to be similar to that of *Gascardia destructor*.

The adult female scale is covered by a pink waxy shell, often with white vertical stripes, and is 3–4 mm long.

Ants are usually found in attendance with the mature scales.

**Distribution.** E. Africa, Seychelles, India, Sri Lanka, China, Japan, Malaysia, Philippines, Solomon Isles, E. Australia, Pacific islands, New Zealand and Hawaii (CIE map no. A118).

**Control.** Control measures are seldom warranted against this pest, but if required the treatments suggested for *Gascardia destructor* should be effective.

**Further reading**

Dekle, G.W. (1971). Red Wax Scale. (*Ceroplastes rubens* Maskell) Coccidae – Homoptera. *Florida Dept. Agric., Ent. Circ. No. 115,2.*

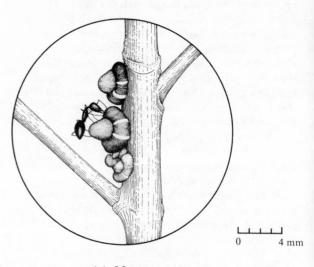

0    4 mm

adult ♀♀ with ant

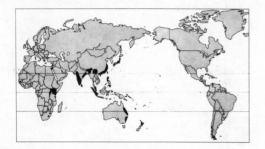

# Chloropulvinaria psidii (Mask.)

(= *Pulvinaria psidii* Mask.)

**Common name.** Guava Mealy Scale
**Family.** Coccidae
**Hosts** (main). Guava

(alternative). Coffee, tea, *Citrus* spp., mango, and many other shrubs and trees, both crop plants and ornamentals.

**Damage.** Young shoots and young leaves infested with oval green scales, sometimes causing leaf distortion and growth disturbance; often accompanied by sooty moulds.

**Pest status.** A widespread and polyphagous species to be found on many different crops in many different parts of the tropics; not often serious but often part of a pest complex that requires controlling.

**Life history.** The adult scales are shield-shaped, oval, rather convex, green in colour and about 3 mm in body length.

As with other scale insects the crawler (first instar nymph) is the active dispersive phase responsible for starting new infestations.

Eggs are laid beneath the body of the mature female in a conspicuous egg-sac, whereupon the female dies.

This scale is usually attended by ants for the honey-dew excreted.

**Distribution.** Widely distributed throughout the tropical regions of the world, but records rather sparse in some areas, and some records from temperate areas (CIE map no. A59).

**Control.** See under *Coccus viridis* (page 225).

'crawler'                    adult ♀ scale

*ex* Butani

scales on underside of guava leaf

0 ⊢—— 1 cm

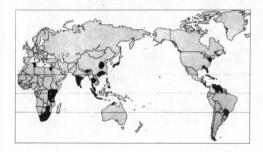

223

# Coccus alpinus De Lotto

**Common name.** Soft Green Scale
**Family.** Coccidae
**Hosts** (main). Coffee, mostly *arabica*.

(alternative). A large number of wild and cultivated plants; important on *Citrus* spp. and guava.

**Damage.** Rows of flat, oval, immobile green scale insects grouped especially along the main veins of the leaves and near the tips of green shoots. The upper surface of the leaves with honey-dew or with sooty moulds growing on the honey-dew.

**Pest status.** A common but minor pest of mature *arabica* coffee; more serious on transplanted seedlings during their first two years in the field. Another Green Scale *C. viridulus* has been found on coffee in the Nandi hills of Kenya. Common on *Citrus* above about 1300 m where it replaces the lowland *C. viridis* in E. Africa.

**Life history.** Eggs are laid below the body of the mature female scale.

When the scale hatches from the egg it is flat and oval, yellowish-green, and has six short legs. It takes up a position on a leaf or green shoot and begins to feed. It passes through three nymphal instars before becoming adult, each stage being larger and more convex than the preceding stage. Nymphs can change their position if conditions become unfavourable but the mature female is apparently fixed in position. The mature scale is 2–3 mm long.

Males have never been recorded; fertilization of the female either never occurs or else is of rare occurrence.

One generation takes less than two months.

**Distribution.** A restricted species separated from *C. viridis* only in E. Africa, where it is found generally above about 1300 mm.

**Control.** Control of this scale is best achieved indirectly by spray banding with dieldrin against the ants, as for *C. viridis*.

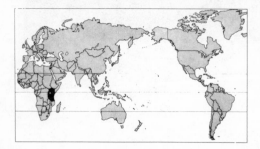

*Coccus alpinus* on *Hibiscus* stems

## Coccus viridis (Green)

**Common name.** Soft Green Scale
**Family.** Coccidae
**Hosts** (main). *Citrus* spp.

(alternative). A large number of wild and cultivated plants are attacked; coffee and guava are two important hosts.

**Damage.** Rows of flat, oval, immobile green scales especially along the main leaf veins and near the tips of green shoots. Upper surfaces of leaves have spots of sticky transparent honey-dew or covered with sooty mould growing on the honey-dew.

**Pest status.** A common, but usually minor pest of mature *Citrus*; more serious on young trees in the first two years after transplanting. *Coccus viridis* is generally found at low altitudes; above about 1300 m it is replaced by *C. alpinus* in E. Africa. Infestations are often found mixed with the Soft Brown Scale *C. hesperidum*, which has a similar life history and which can be controlled by the same sprays.

**Life history.** When the scale hatches from the egg it is flat and oval, yellowish-green, and has six short legs. It takes up a position on a leaf or green shoot and begins to feed. It passes through three nymphal instars before becoming adult, each stage being larger and more convex, but otherwise differing little from the preceding stage. Nymphs can change their position if conditions become unfavourable, but the mature female appears to be fixed in position.

Mature scales are 2−3 mm long. Eggs are laid below the body of the mature female.

One generation takes 1−2 months.

Males have never been recorded; fertilization of the female must be a rare event, if indeed it ever occurs.

**Distribution.** Cosmopolitan in the tropics with the exception of Australia (CIE map no. A305). In E. Africa it is only found up to a height of 1000−1300 m.

**Control.** Banding the tree stump with dieldrin keeps off the attendant ants and allows the natural enemies to clean up the infestation; the band should be at least 15 cm wide, and care should be taken to avoid leaving any bridges. If the trees are too small for satisfactory banding the dieldrin mixture should be sprayed on to the collar of the tree and a small area of mulch round the collar.

In severe infestations, in addition to dieldrin banding, the tree foliage should be sprayed with diazinon, malathion, or dimethoate, as a full-cover spray using as high a nozzle pressure as possible. Do not use dimethoate on rough lemon trees or non-budded rough lemon stock.

**Further reading**

Le Pelley, R.H. (1968). *Pests of Coffee*, pp. 353−64. Longmans: London.

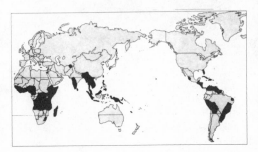

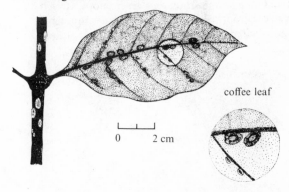

coffee leaf

0    2 cm

## Gascardia brevicauda (Hall)

(= *Ceroplastes luteolus* De Lotto)

**Common name.** White Waxy Scale
**Family.** Coccidae
**Hosts** (main). Coffee, both *arabica* and *robusta*.
(alternative). *Citrus* spp.
**Damage.** White immobile insects like blobs of cream found on green shoots and leaves. The white material is a soft wax which is easily rubbed off to reveal the shiny brown carapace of the scale.
**Pest status.** A minor pest of both *arabica* and *robusta* coffee; sporadically severe attacks occur, especially at lower altitudes. This scale, is however, a very slow feeder and enormous numbers must be present before the coffee bush is perceptibly damaged.
**Life history.** Eggs are laid under the carapace of the mature female scale which remains attached to the plant even after the eggs have hatched.

When the young nymphs hatch from the eggs they are flat and oval, purplish-brown and have six short legs. The nymphs take up positions on a leaf, usually on the upper surface and next to a main vein; here they begin to feed. A waxy plate develops on the back and a waxy fringe round the edge of the body. Later the wax covers the whole body and forms a star shape. Each scale next moves off the leaves and takes up a new position on a green shoot. Here it passes through a conical stage before assuming the more rounded form of the mature female scale.

Mature scales are about 6 mm in diameter.

Males have never been recorded.

The complete life-cycle takes about six months, the majority of scales maturing and laying their eggs during the main rainy season.
**Distribution.** Recorded only from Africa; in Angola, Kenya and Uganda.
**Control.** In a minor outbreak, badly infested branches should be cut off and left on the ground for the parasites to emerge.

Sprays have little or no effect on the older stages found on the green shoots. These stages are well protected by their waxy covering. Control measures should therefore be directed against the young stages found on the leaves. Spray to run off with an emulsion of white oil in water. The best time for spraying is usually one or two months after the end of the rainy season. Even after their death there is little change in the appearance of the young scales. Development, however, stops and there is no migration from leaves to shoots.

If ants are in attendance a dieldrin band should be applied to the trunk of the tree, at least 15 cm wide, taking care to ensure that no bridges are left.

For extra effectiveness the following insecticides can be added to the white oil: carbaryl, dimethoate, malathion, azinphos-methyl, and carbophenothion.

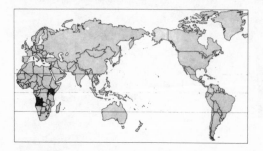

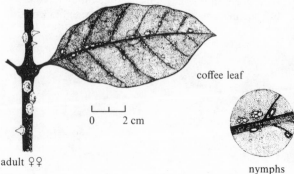

coffee leaf

0    2 cm

adult ♀♀

nymphs

## Gascardia destructor (Newst.)

(= *Ceroplastes destructor* Newst.)

**Common name.** White Waxy Scale
**Family.** Coccidae
**Hosts** (main). *Citrus* spp.

(alternative). Coffee, various fruit trees (guava, persimmon) shade trees and shrubs (gardenia).

**Damage.** The large white waxy scales encrust the twigs and leaves of the host trees; often accompanied by sooty moulds and ants feeding on the honey-dew excreted.

**Pest status.** A common pest of *Citrus* spp. and often found on coffee; only occasionally is it a serious pest.

**Life history.** Typically there is only one generation per year.

The 'crawlers' (first stage nymphs) emerge from the eggs under the female scale, and eventually settle along the veins of the leaves. After five or six weeks they crawl back on to the twigs and settle permanently in position. They start secreting their wax cover as they grow in size, and gradually mature.

After about ten months the mature females lay eggs.

If the wax covering is removed the soft reddish female scale is revealed.

**Distribution.** W., E. and southern Africa, Madagascar, Papua New Guinea, Australia, New Zealand, Florida and Mexico (CIE map no. A117).

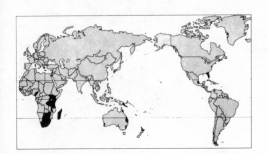

**Control.** Spraying with white oil, either alone or in combination with carbaryl, dimethoate, malathion, azinphos-methyl, or carbophenothion, while the crawlers are still on the leaves and before migration back to the twigs commences.

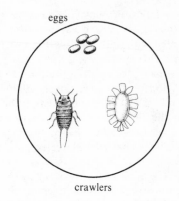

eggs

crawlers

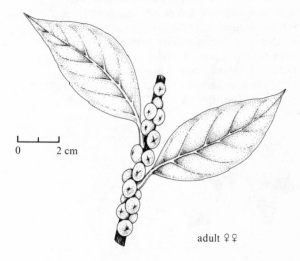

0    2 cm

adult ♀♀

227

# Parasaissetia nigra (Neitn.)

(= *Saissetia nigra* Neitn.)

**Common name.** Nigra Scale
**Family.** Coccidae
**Hosts.** A polyphagous species found on *Citrus* spp., rubber, kola nut, and other crops, as well as ornamentals such as frangipani.
**Damage.** Oval dark scales clustered on twigs, shoots and leaves, sometimes causing leaf distortion, and often associated with sooty moulds. Citrus fruits often covered with mould.
**Pest status.** Usually not a serious pest, more often part of the *Citrus* pest complex which will require overall control treatment. The infestation found on frangipani in the Seychelles was very heavy and the undersurface of almost all the leaves of several adjacent trees were covered with scales, most lying alongside veins.
**Life history.** The adult scales are oval and convex in shape, measuring some 2–3 mm in length and 1.5–2 mm in breadth. Young scales are generally paler, but the adults are dark brown or occasionally black in colour, although coloration does tend to be a variable character. The female scales reproduce parthenogenetically and lay eggs under the scale.
**Distribution.** Widely distributed throughout Africa, India, S.E. Asia up to S. China and Taiwan, Philippines, Indonesia, Australasia and New Zealand, and in California, USA.

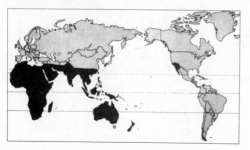

**Control.** As with other scales natural parasitism and predation usually controls population numbers most of the time, but occasionally pesticides have to be used, when the usual chemicals effective against scales may be employed. The adult scale is very difficult to kill and most spray programmes rely on killing the young nymphs, and to this effect repeated sprays at one to two week intervals are generally required.

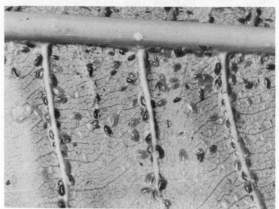

infested leaf of frangipani

0          2 cm

# Saissetia coffeae (Wlk.)

(= *S. hemisphaerica* T.-T.)

**Common name.** Helmet Scale (= Hemispherical Scale)

**Family.** Coccidae

**Hosts** (main). Coffee, both *arabica* and *robusta*.

(alternative). A wide range of alternative hosts including tea, *Citrus*, guava, mango and many other plants both wild and cultivated.

**Damage.** Immobile insects, green when young but dark brown when older, clustered on the shoots, leaves and green berries. They are often arranged in an irregular line near the edge of a leaf blade.

**Pest status.** A minor pest of *arabica* and *robusta* coffee; very occasional severe outbreaks have been recorded especially on rather unhealthy bushes. A small form of this species is found on coffee roots in the Kissi highlands of Kenya.

**Life history.** Eggs are laid beneath the carapace of the mature female scale which remains attached to the plant even after the eggs have hatched; one female may lay several hundred eggs (up to 600).

When the scale hatches from the egg it is flat and oval, greenish-brown, and has six short legs. It takes up a position on a leaf, berry or green shoot and begins to feed. It passes through three instars before becoming adult. The immature females, which can move short distances if conditions become adverse, have an H-shaped yellow mark on their body. This is diagnostic of the species. Adult females have a strongly convex helmet-shaped carapace and are dark brown; this stage is immobile. Mature scales are about 2 mm long.

Males have never been recorded; and it is presumed that reproduction is always by parthenogenesis.

One complete generation appears to take about six months in the field.

**Distribution.** Almost completely cosmopolitan; widespread through the tropics and in some subtropical areas, occurring as far north as Spain and Turkey, and California (CIE map no. A318).

**Control.** Ensure that infested trees receive optimum quantities of mulch and fertilizer. Cut off heavily infested branches and leave on the ground for the parasites to emerge.

White oil as a drenching spray is effective against the young scales, but has negligible effect on the adult females. A second spray will be required after about 3–4 weeks.

If ants are in attendance then a dieldrin band should be sprayed around the base of the trunk.

**Further reading**

Le Pelley, R.H. (1969). *Pests of Coffee*, pp. 364–7. Longmans: London.

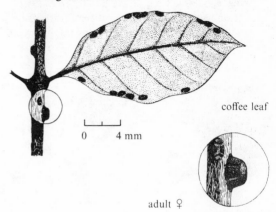

coffee leaf

0    4 mm

adult ♀

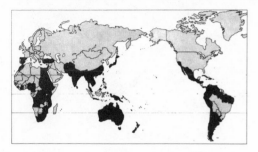

229

## Saissetia oleae (Bern.)

**Common name.** Black Scale (Olive Scale)
**Family.** Coccidae
**Hosts** (main). Olive, *Citrus* spp.

(alternative). A very wide range of plants.
**Damage.** The presence of conspicuous blackish scales on the twigs and shoots. In heavy infestations the shoots and leaves wither and fade, and there are often heavy coatings of sooty moulds; development of leaves and fruit may be impaired.

**Pest status.** A polyphagous pest recorded from many host plants, and very widely distributed throughout the world.

**Life history.** A single female lays 1000–4000 eggs (average about 2000); the egg-laying period lasts 2–4 weeks. The eggs are white, turning brown; wax flakes are deposited between the eggs to prevent them sticking together. Hatching takes 15–20 days.

The crawlers start feeding within a few hours. The nymphs tend to prefer feeding on the shoot tips and the undersides of the leaves, but the adults prefer the shoots and twigs. Nymphal development takes about 2–3 months; however, if conditions become unfavourable the nymphs may go into diapause during which they are very difficult to kill.

Winged males are only occasionally found; reproduction usually occurs without fertilization.

Under suitable climatic conditions the life-cycle takes about 3–4 months.

**Distribution.** Almost cosmopolitan in distribution, but not recorded from W. Africa; northernmost records from Japan and southern France (CIE map no. A24).

**Control.** Control recommendations are most effective when directed against the nymphal stages. The adult female scale is particularly difficult to kill with contact insecticides, but the nymphs are more susceptible.

Recommended insecticides are: azinphos-methyl, diazinon, malathion, parathion, parathion-methyl, and petroleum (white) oil.

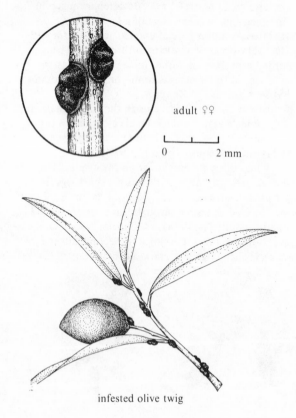

adult ♀♀

0      2 mm

infested olive twig

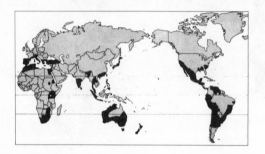

230

# Asterolecanium coffeae Newstead

**Common name.** Star Scale
**Family.** Asterolecaniidae
**Hosts** (main). Coffee, usually *arabica*.
(alternative). Jacaranda and loquat trees.
**Damage.** Green branches are sharply elbowed at the nodes with pits in the green bark on the inside of the bends. Affected nodes often bear drooping, dead leaves. Internodes beyond the elbow bends are often elongated producing whip-like branches. Numerous small red or yellow scales are usually visible in bark crevices, especially near ground level.
**Pest status.** A sporadically serious pest of *arabica* coffee grown below 1700 m.
**Life history.** When egg-laying is completed there are some 50 eggs under the carapace and they fill it completely. After hatching the crawlers leave the carapace through a small orifice at the hind end. Empty carapaces remain on the tree for many months; they are then greenish-grey and easily distinguished from those containing living females or eggs. The crawlers are flat and oval, yellow, and just visible to the unaided eye.

The immature females are reddish-brown and have coarse hair-like projections, especially at the edge of the body where they form a fringe.

Mature females are covered with a hard but transparent scale; as the eggs are laid at one end of the carapace, so the body of the female progressively shrinks to make room for them. Most eggs are laid during rainy seasons.

Development takes about six months.
**Distribution.** E. Africa, Angola, and Zaïre.
**Control.** Infested trees should be pruned severely and most of the crop stripped off.

Optimum quantities of nitrogen and mulch should be applied to infested trees. Paint as much as possible of the infested brown bark with tar oil; keep the brush very wet, allowing plenty of the mixture to soak into the crevices.

If a good kill of bark-feeding scales is obtained, experience has shown that (in the absence of road dust) parasites and predators can be relied on to kill the scales feeding on green branches. If carefully applied, only one tar oil treatment is usually required.

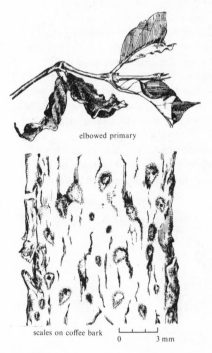

elbowed primary

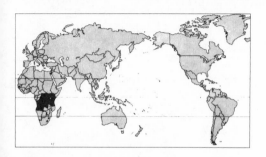

scales on coffee bark    0    3 mm

231

# Aonidiella aurantii (Maskell)

**Common name.** California Red Scale
**Family.** Diaspididae
**Hosts** (main). *Citrus* spp.

(alternative). A wide range of fruit trees, and shrubs, notably roses.

**Damage.** Infestation is indicated by the presence of numerous small circular reddish-brown scales on the trunk, branches, leaves and fruit. Severe infestations may result in branch die-back. When on a leaf the scale is often surrounded by a small pale chlorotic spot.

**Pest status.** A major pest of *Citrus* in E. Africa, and in many other *Citrus*-growing areas.

**Life history.** After copulation with the winged male, the female scale produces living young ('crawlers') at the rate of about 2–3 per day. These may shelter beneath the carapace for a short period before walking away and finding a suitable feeding site in a depression or crevice. After settling down, they do not normally move again. The developing scale moults twice before becoming an adult female. The immature males, like the females, are scale-like and reddish-brown but their bodies are elongate and they only reach about a quarter of the size of an adult female.

The adult male is a fragile two-winged insect with long filamentous antennae. The body of the female is flattened, crescent-shaped and reddish-brown, and it is covered with a circular, transparent, waxy carapace, 1.5–2.0 mm in diameter, through which the body can be seen.

**Distribution.** Cosmopolitan throughout the tropics and subtropics, but with no records from W. Africa (CIE map no. A2).

**Control.** If high-quality fruit are being produced trees should be sprayed when 25% of the fruits are infested with one or more Red Scales. Recommended sprays are either diazinon or malathion in water with added white oil, as a full-cover spray using as high a nozzle pressure as possible. A repeat spray should be made after 2–3 weeks.

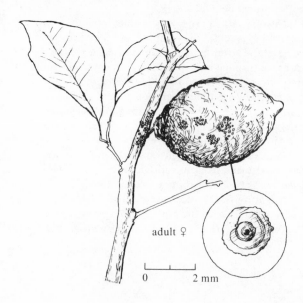

adult ♀

0    2 mm

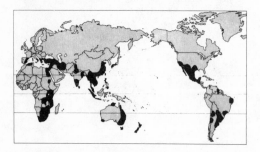

## Aonidomytilus albus (Ckll.)

**Common name.** Cassava Scale
**Family.** Diaspididae
**Hosts** (main). Cassava (*Manihot* spp.)
   (alternative). Various species of *Solanum*, and other plants.
**Damage.** The trunk and petioles are covered with mussel-shaped white scales. When young plants are attacked the leaves turn pale, wilt and fall, and root development may be impaired.
**Pest status.** Not usually a serious pest, but of some importance locally in E. Africa.
**Life history.** The females are silvery-white, mussel-shaped scales, 2.0–2.5 mm long. The brown, oval exuvium is at the anterior end of the scale. The female insect under the scale is oval, and reddish.
   The male scale is much smaller and oval, about 1.0 mm long.
**Distribution.** W. and E. Africa, Madagascar, India, Taiwan, Florida, Mexico, W. Indies, Argentina and Brazil (CIE map no. A81).
**Control.** Control measures are not usually required.

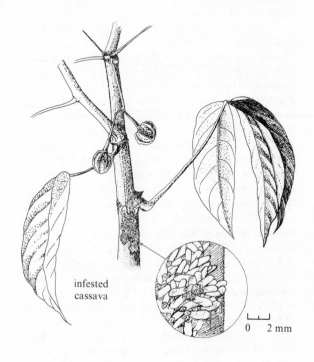

infested
cassava

0    2 mm

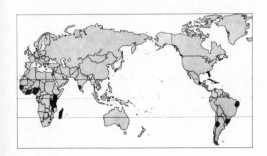

## Aspidiotus destructor Sign.

**Common name.** Coconut Scale (Transparent Scale)
**Family.** Diaspididae
**Hosts** (main). Coconut

(alternative). Other palms, mango, bananas, avocado, cocoa, *Citrus*, ginger, guava, *Artocarpus*, *Pandanus*, papaya, rubber, sugarcane, yam, etc.
**Damage.** A severe infestation forms a continuous crust over the undersurface of all leaves. The leaves first become yellow, because of sap loss and blocking of the stomata, and eventually die. The flower spikes and young nuts are also likely to be infested. Infestation is most severe in areas where rainfall is high and the palms are planted close together; neglected plantations are particularly susceptible. Infestations are usually attended by ants which feed on the honey-dew excreted by the scales; the ants usually nest in the palm crowns.
**Pest status.** One of the most serious pests of coconut, and other crops. Dispersal of this scale has been shown to be effected by both birds and bats.
**Life history.** The body of the adult female is bright yellow and nearly circular in outline, and is covered with a flimsy, semitransparent, only slightly convex scale. The scale diameter is 1.5−2.0 mm. The male scale is much smaller, oval in outline, and the insect body is reddish; on attaining maturity the male insect has a pair of wings, is motile, and leaves the scale.

The eggs are yellow, tiny, and are laid under the scale around the body of the female. Incubation takes

7−8 days. On hatching, the crawler leaves the maternal scale and takes up a position on the leaf and starts feeding. The nymph remains on this site throughout its nymphal life, and if it is a female it also stays there throughout its adult life. The male nymph moults three times, and the female twice. Larval development takes 24 days in the male, and longer in the female.

The life-cycle takes 31−35 days; and there are about ten generations per year.
**Distribution.** Pantropical; occurring up to Iran, Japan, and California, and southwards down to S. Africa and Australia (CIE map no. A218).
**Control.** The waxy scale covering the insect makes control by insecticides difficult. Successful insecticides have been parathion, malathion and dieldrin.

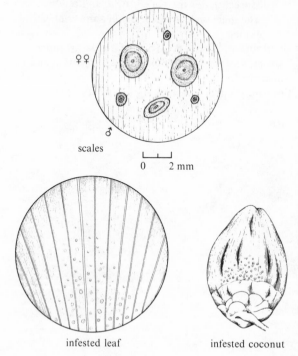

scales

0    2 mm

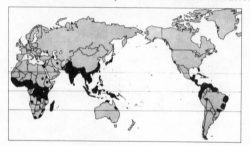

infested leaf                    infested coconut

## Aulacaspis tegalensis (Zehn.)

**Common name.** Sugarcane Scale
**Family.** Diaspididae
**Hosts** (main). Sugarcane
        (alternative). The wild grass *Erianthus* sp. in Java which is only doubtfully separable from *Saccharum*.
**Damage.** Essentially a stem-inhabiting pest, but does occur on leaves, although this may be considered as secondary and a result of crowding on the stem. Usually the bulk of the infestation is found under the leaf sheath, the looser the sheath the greater the scale population. The feeding of the scales on the leaves results in chlorotic spots which are drawn out along the length of the leaf.
**Pest status.** A serious pest of sugarcane causing appreciable loss in yield (both of canes and sugar content) and making extensive replanting necessary. The crawlers can be dispersed for considerable distances by wind or movement of vegetation by field workers and transport. Greathead (1972) found that crawlers were carried up to 1000 m on quite low wind speeds.
**Life history.** The normal post-embryonic development of Diaspididae consists of two instars in the female and four in the male; sexual dimorphism becoming apparent after the first nymphal moult.

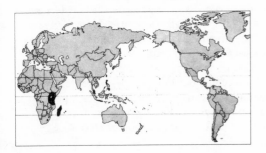

The eggs are spindle-shaped, about 250 μm by 100 μm, yellow, and covered with powdery wax, and are laid under the female's scale; each female lays 500–1000 eggs (average 750).

        The first instar (crawler) is tiny, whitish, and with two long terminal setae. After a period of wandering it selects a feeding site, inserts its stylet

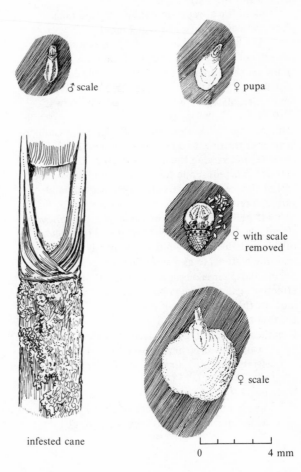

♂ scale

♀ pupa

♀ with scale removed

♀ scale

infested cane

0        4 mm

235

into the plant, then becomes inert and starts to secrete wax. The secretions do not form a definite scale during the first instar, as happens in certain other Diaspididae. Legs and antennae are lost during the first moult, and after this the sexes acquire different body forms. The second instar female assumes the pear-shaped form of the adult female, and the second instar male is more elongate with the anterior end narrowest.

Males are always present and mating takes place immediately after the final moult.

The life-cycle takes about 26–60 days according to temperature (and altitude) and in Mauritius there are eight generations per year.

**Distribution**. E. Africa, Madagascar, Mauritius, Seychelles, Malaya, Java, Philippines, and Taiwan (CIE map no. A187).

**Control**. Use of clean planting material, by washing or hot-water treatment to kill the scales and field hygiene, and varying the date of harvest, is recommended. There is scope for practical biological control of this pest using parasitic Hymenoptera and various predators.

If pesticides have to be employed the following chemicals should be effective: white oil (petroleum oil) or malathion plus white oil, either as a dip for planting material or as a spray for the setts. Various organophosphorous compounds used alone are effective against crawlers but not against eggs and most of the fixed stages of the scale.

**Further reading**

Greathead, D.J. (1972). Dispersion of the sugarcane scale *Aulacaspis tegalensis* (Zhnt.) (Hem., Diaspididae) by air currents. *Bull. ent. Res.* **61**, 547–58.

Williams, J.R. (1970). Studies on the biology, ecology and economic importance of the sugarcane scale insect *Aulacaspis tegalensis* (Zhnt.)
(Diaspididae) in Mauritius. *Bull. ent. Res.* **60**, 61–95.

Williams, J.R., J.R. Metcalfe, R.W. Mungomery & R. Mathes (eds.) (1969). *Pests of Sugar Cane*, pp. 332–5. Elsevier: London.

## Control of Armoured Scales (Homoptera; Diaspididae)

Most Diaspididae are more or less invulnerable as adults by virtue of their thick, waxy, immobile protective 'scale'. Eggs are often laid under the scale of the gravid female who gradually dies as oviposition proceeds. The young 'crawlers' emerge from the shelter of the maternal scale and find new locations on which to settle. Young scales often settle on leaves and fruits initially (presumably they are more nutritious), which of course are deciduous, and then when older move to permanent locations on twigs. They are only mobile during the first few instars, the crawler being the dispersive stage; the mature nymphs, pupae and adult females are all permanently fixed to the plant tissues by the stylet embedded into the host vascular system.

The adults are slightly susceptible to some systemic insecticides, but generally little control has been achieved in this manner. Thus the only stages vulnerable to contact pesticides are the crawlers and young nymphs. Generally, armoured scales are very difficult to kill with pesticides, but at the same time they are usually heavily parasitized by chalcidoid wasps; too often the typical result of pesticide application has been a poor kill of scales but extensive destruction of natural enemies, and often a pest population resurgence. In some situations now it is quite certain that any application of DDT to a citrus orchard will result in a scale population outbreak! Present practice is to use chemicals only as a last resort. The range of approaches to scale population control include:

(1) Cultural control

(a)     Phytosanitation – heavily infested sugarcane plants should be rogued and burned, if economically feasible; heavily infested fruit trees are usually best left for the parasites to emerge.

(b)     Clean planting material – use of clean planting material will delay any scale population build-up. Cleaning can be done with use of oils, hot water or by washing, and also by fumigation.

(2) Biological control

(a)     Natural control – allowing alternate trees, or alternate tree rows, to remain unsprayed on alternate years, to permit parasite populations to survive and to build-up.

(b)     Biological control – the supplementation of natural control by filling in gaps in the natural enemy spectrum with imported species of predators and/or parasites.

(3) Pesticides

(a)     Dips and/or winter washes with tar oils, for use on deciduous trees, against overwintering scales and eggs.

(b)     Sprays of petroleum or white oils against young nymphs, but care is required as these oils are basically phytotoxic.

(c)     Systemic insecticides, such as dimethoate; but generally the level of control is poor.

(d)     Contact insecticides, against young nymphs, such as diazinon, ethion, malathion, parathion and carbaryl; these sprays are usually more effective if mixed with white oil.

These pesticide treatments are highly destructive to the natural enemies, they show high mammalian toxicity, and may also be phytotoxic at times. Hence their use should be avoided if possible, and reliance placed on the long-term effect of natural enemies. The use of winter washes of tar oils is recommended, but is only applicable to deciduous trees in cooler regions.

## Chrysomphalus aonidum (L.)

(= *C. ficus* Ashm.)

**Common name**. Purple Scale (Florida Red Scale)
**Family**. Diaspididae
**Hosts** (main). *Citrus* spp.

(alternative). Coconut, date palm, mango, cinnamon, and a wide range of mono- and dicotyledons.

**Damage**. This scale usually occurs on leaves, but is occasionally found on green shoots and twigs. The saliva is apparently toxic and produces damage and necrosis of the host plant tissues.

**Pest status**. A serious pest of *Citrus*, and also widespread on many other crops and plants. This scale originates from the W. Indies but now is widespread throughout the tropics and subtropics, but has a preference for more humid climates. Common in greenhouses in temperate climates.

**Life history**. Eggs are laid under the scale of the female insect. The crawlers emerge within a few hours; they are easily dispersed by wind, insects and cultivation practices. The young scales are found mainly along the midribs and veins of leaves, or in depressions in the fruit. After settling the nymphs start secreting the characteristic waxy carapace or scale.

The adult female scale is purplish and circular, with a reddish-brown 'boss' or 'nipple' in the centre.

Parthenogenesis does not appear to be of significance in the life-cycle of this species.

The winged adult males are very short-lived and do not feed.

**Distribution**. Cosmopolitan in the tropics and subtropics, but no records from W. Africa. Known as the Florida Red Scale in America (CIE map no. A4).

**Control**. The usual range of insecticides recommended for use against diaspid scales is as follows: diazinon or malathion, preferably with added white oil, and carbaryl and parathion.

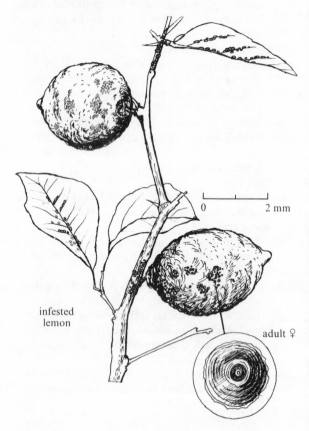

0      2 mm

infested lemon

adult ♀

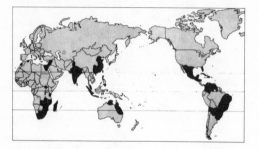

238

# Ischnaspis longirostris (Sign.)

**Common name.** Black Line Scale (Black Thread Scale)

**Family.** Diaspididae

**Hosts** (main). Coffee and coconut.

(alternative). *Citrus*, bananas, oil palm, mango, *Annona*, and other plants.

**Damage.** Leaves, shoots, and fruit can be encrusted with this scale which often occurs in very large numbers. The leaves become mottled with discoloured patches and they curl downwards. Growth of shoots can be inhibited, and yield reduced in severe cases.

**Pest status.** Not a serious pest usually, but very widespread throughout the tropics and common on many crops. Particularly harmful to coconut in the Seychelles.

**Life history.** The female scale is long and slender, black and shiny, slightly wider posteriorly. The shed skin of the first instar nymph remains conspicuously attached to the scale at the anterior end. The length of the scale is 3–4 mm. The eggs are yellow.

**Distribution.** Probably almost completely pantropical in distribution, but records at present are from Egypt, W., E. and S. Africa, Madagascar, Sri Lanka, Malaya, Java, Papua New Guinea, West Irian, N. Australia, Hawaii, and various Pacific islands, S. USA (Florida), W. Indies, and C. and S. America (CIE map no. A235).

**Control.** In the Seychelles coccinellid beetles of the genera *Chilochorus* and *Exochomus* imported from E. Africa have to some extent controlled this pest.

Insecticides are not generally required, but according to Wyniger (1968) sprays of white oil, diazinon, ethion, or parathion, are effective if applied twice at a two-week interval.

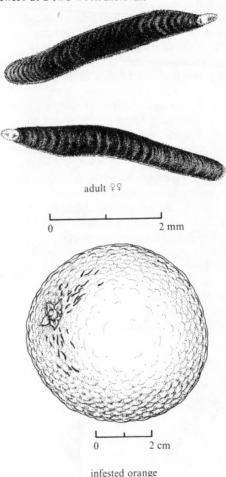

adult ♀♀

0      2 mm

infested orange

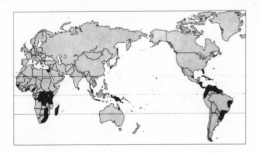

239

# Lepidosaphes beckii (Newman)

**Common name.** Citrus Mussel Scale (Purple Scale)
**Family.** Diaspididae
**Hosts** (main). *Citrus* spp., particularly severe on orange.

(alternative). All *Citrus* spp., and also on *Croton* spp. and a few other shrubs.
**Damage.** Numerous small, purplish-brown, mussel-shaped objects on the leaves, fruits and branches. Premature leaf-fall and die-back of branches may follow a severe outbreak. Very heavy infestations on branches may pass unnoticed owing to the colour of the scales.
**Pest status.** A serious pest of *Citrus* in many parts of the world. Probably introduced into E. Africa on infested nursery stock.
**Life history.** After copulation with the winged male, the female lays 50—100 eggs and they are deposited under the narrow tapering part of the carapace. The body of the female scale contracts into the broader part of the carapace as the eggs are laid down.

After the young scales ('crawlers') hatch from the eggs they walk out from under the carapace and, having selected a suitable site on leaves, fruits or stems, settle down and begin to feed. They do not move again but with each successive moult gradually change from the oval form of the crawler to the mussel shape of the adult.

The carapace of the adult female is purple-brown, and it is the shape of a mussel shell. It is about 2 mm long and 0.6 mm broad at the widest part.

Male crawlers settle down in the same way as the females and develop the same mussel shape. When about 1 mm long, growth of the carapace stops, and after a resting period the fragile, winged male emerges from under it and flies off the find a female to fertilize.

The total life-cycle probably takes 2—4 months according to temperature.
**Distribution.** Indigenous to the Orient, but now cosmopolitan throughout the tropics and subtropics (CIE map no. A49).
**Control.** The recommended insecticides are diazinon, malathion and carbaryl, to be sprayed on the trees; the first two are more effective if white oil is added to the spray.

Carbaryl is more persistent and only one spray is usually required, but two sprays of the others are necessary at an interval of 3—4 weeks.

Care should be taken not to upset the (fortuitous) natural control exerted by the specific parasite *Aphytis lepidosaphes* now present in most *Citrus*-growing areas.

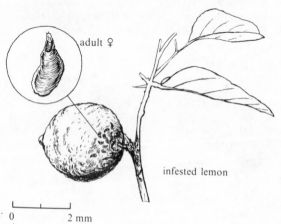

adult ♀

infested lemon

0    2 mm

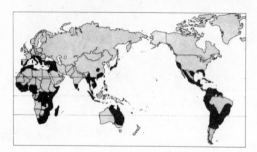

# Phenacaspis cockerelli (Cooley)

**Common name.** Mango Scale (Oleander Scale)
**Family.** Diaspididae
**Hosts** (main). Mango
  (alternative). Coconut, oil palm, and oleander.
**Damage.** Upper surface of leaves are encrusted with small rounded white scales; in heavy infestations leaves may be affected, and palm fruits are sometimes shrivelled when covered with scales.
**Pest status.** Usually a minor pest on the crops listed above; only occasionally does it appear to cause sufficient damage for concern.
**Life history.** The adult scale is rounded or slightly elongate, usually about 2–3 mm in length, and with the nymphal exuvium conspicuously dark brown at the pointed end while the rest of the scale is white.

As with many other hard scale infestations the scales are found on the upper surface of the leaves, whereas most soft scales are on the underneath of the leaves.
**Distribution.** S. and E. Africa, Madagascar, Australia, S. China and Japan; also the Seychelles and Hawaii.
**Control.** Not often required. Natural levels of parasitism are generally high. Diazinon and malathion as aqueous sprays, repeated after 2–3 weeks, would probably be effective in controlling this insect, as also would be dimethoate.

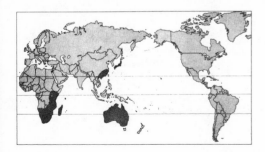

**Further reading**
Dekle, G.W. (1970). Oleander Scale (*Phenacaspis cockerelli* (Cooley)) (Homoptera: Diaspididae). *Florida Dept. Agric., Ent. Circ. No. 95*, 2.

infested oleander leaf

0   1 cm

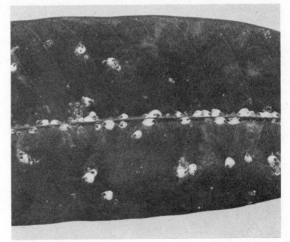

infested mango leaf

241

## Quadraspidiotus perniciosus (Comst.)

(= *Aspidiotus perniciosus* Comst.)

**Common name.** San José Scale
**Family.** Diaspididae
**Hosts** (main). Apple, peach, pear, plum, currants.

(alternative). Yam, most deciduous fruit trees and shrubs, and a wide range of trees and shrubs; more than 700 host plants are recorded).

**Damage.** Tiny circular scales, just visible to the unaided eye, can be seen on the bark of lightly infested trees. If the trees are heavily infested then the bark may be completely covered by overlapping scales. Fruits may also be attacked, especially at both stem and flower ends. The bark may crack and exude gum, and the twigs die back; infested trees often die.

**Pest status.** A polyphagous pest on many crops and plants, but probably the most serious pest of deciduous fruit trees; fruit trees are seriously debilitated or killed, and the fruit rendered unsalable.

**Life history.** This is one of the few viviparous hard scales; each female can produce 100–400 nymphs. The crawlers are minute (0.2 mm long), with well-developed legs and antennae, and yellow in colour. Overwintering occurs in the nymphal stage.

The adult female scale is flattened, nearly circular, with a raised central nipple, grey in colour, 1–2 mm in diameter, and completely covered by the circular yellow scale.

The male scale is oval, about 1.0 by 0.5 mm.

The total life-cycle takes 18–20 weeks in California, and in the warmer parts of the world there may be four or five generations per year.

**Distribution.** Widely distributed throughout the warmer parts of the world but absent from many tropical countries; recorded from S. Europe, Algeria, S. Africa, Zimbabwe, Zaïre, Iraq, Turkey, India, Pakistan, China, Korea, Japan, Australia, New Zealand, USA, W. Indies, Mexico, and much of S. America (CIE map no. A7).

**Control.** As the most important hosts are deciduous trees this scale can be best controlled by sprays during the dormant period. Recommended sprays are lime-sulphur, parathion, and petroleum oils.

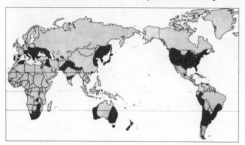

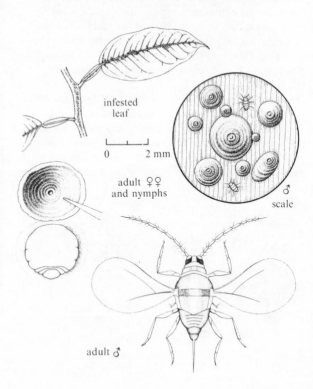

infested
leaf

0       2 mm

adult ♀♀
and nymphs

♂

scale

adult ♂

242

# Order **HETEROPTERA**

Some of these bugs are aquatic and predaceous, some are terrestrial and predators, and some are plant feeders which imbibe sap; all have piercing and sucking mouthparts. All species have toxic saliva which when injected into a plant produces necrosis and death of tissues, resulting sometimes in the death of shoots and branches. The forewings (hemelytra) have the basal two-thirds thickened and the end third is membraneous; the wings are held flat over the body.

## Family **Miridae** (= Capsidae)

(Capsids or Mirids)  Medium to small-sized bugs; usually pale in colour and delicate in structure; no ocelli. A very large family of about 5000 species. Most feed on plant sap but a few are predators. Many are important crop pests. Many species have a curious, knobbed, pin-like structure projecting from the thorax.

## Family **Lygaeidae**

Small, dark or brightly coloured forms; somewhat like smaller, softer, brighter coreids. Most are plant feeders, but a few are predatory. A large family with about 2000 species.

## Family **Pyrrhocoridae**

(Red Bugs)  Moderately-sized, brightly coloured insects without ocelli; the coloration is usually brightly contrasting. A small family; the only genus of agricultural importance is *Dysdercus*, the Cotton Stainers.

## Family **Coreidae**

Medium to large angular bugs, generally dull in colour; ocelli are present; they frequently have dilatations on the legs and antennae. The toxic saliva causes severe damage to attacked plants. More than 2000 species have been described. Many species have stink glands.

## Family **Tingidae**

(Lace Bugs)  The body and hemelytra are densely reticulate, hence the name 'Lace Bugs'. They are very characteristic and easily recognized. Ocelli are absent and the scutellum is concealed by the pronotum. About 700 species are known; they exhibit great diversity of form.

## Family **Pentatomidae**

(Shield Bugs, or Stink Bugs)  Moderate to large insects; the head has the lateral margins concealing the bases of the antennae; ocelli are almost always present; the scutellum is always large, often enormous. A large family with over 3000 species listed; most are plant feeders but some are predaceous. Most species have stink glands which emit a characteristic unpleasant odorous liquid.

# Helopeltis anacardii Miller

**Common name.** Cashew Helopeltis
**Family.** Miridae (= Capsidae)
**Hosts** (main). Cashew
    (alternative). Only sweet potato is recorded.
**Damage.** Distortion of young leaves with angular lesions along the main veins. Elongate dark lesions on green shoots, sometimes accompanied by the exudation of gum; brown sunken spots on developing apples and nuts. Bunched terminal growth follows a severe attack.
**Pest status.** A sporadically serious pest of cashew in Coast Province, Kenya.
**Life history.** Eggs are laid embedded in the soft tissues near the tips of flowering or vegetative shoots. The egg-cap bears two fine, white, thread-like processes which are visible externally.

There are five nymphal stages, the last one being yellowish in colour and about 4 mm long. Both nymphs and adults have a knobbed, hair-like projection sticking upwards from the thorax. Young nymphs feed on the undersides of young leaves. Older nymphs feed on the young shoots and developing fruits. The latter may be shed if attacked when very young. Severely damaged shoots die back and the subsequent development of numerous auxillary buds causes a 'witches broom' type of growth to appear.

The adults generally resemble the mature nymph but are more orange-brown, and are larger (males 4.5 mm; females 6.0 mm) and have transparent wings extending beyond the tip of the abdomen. Adults may live for 2–3 weeks.

After a pre-oviposition period of some 12 days the females may lay 3–4 eggs per day until they die.

The total life history, including the 12-day pre-oviposition period, takes about 48 days.
**Distribution.** E. Africa. Other species of *Helopeltis* are important pests of cashew in Africa and S.E. Asia.
**Control.** Chemical control measures must be very promptly applied to be effective. Great vigilance is needed in the July-December period or an infestation will easily be overlooked until it is too late to take effective action.

Seedling trees up to $1\frac{1}{2}$ years old: dusting with BHC is recommended, repeated if necessary at weekly intervals.

Trees $1\frac{1}{2}$-5 years old should be sprayed with a higher strength BHC solution, preferably applied with a mist blower; repeated if necessary after three weeks.
**Further reading**

Swaine, G. (1959). A preliminary note on *Helopeltis* spp. damaging cashew in Tanganyika Territory. *Bull. ent. Res.* **50**, 171–81.

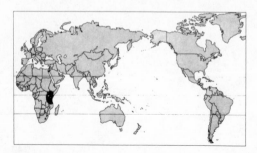

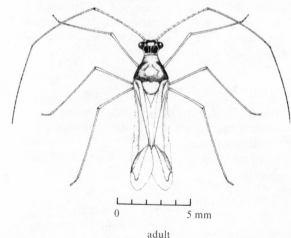

0       5 mm

adult

# Helopeltis schoutedeni Reuter

**Common name.** Cotton Helopeltis
**Family.** Miridae (= Capsidae)
**Hosts** (main). Cotton
    (alternative). A wide range of wild and cultivated plants including tea, cocoa, castor, cashew, mango, avocado, peppers, guava, and sweet potato.
**Damage.** Plants are stunted and with numerous secondary branches, and black lesions on the stems. The leaves are rolled downwards at the edge and with many brown-centred black lesions especially near the main veins. Similar crater-like lesions on large green bolls are found. If the cotton is severely attacked when young it appears as if it has been scorched by fire.
**Pest status.** A sporadic pest of cotton in various parts of Africa. Affected fields may be very severely damaged with adjacent fields almost untouched.
**Life history.** The eggs are test-tube-shaped with a rounded cap and two unequal, hair-like filaments at one end. Eggs are white and about 1.7 mm long. They are completely inserted into soft plant tissues, only the cap and filaments being visible externally. Most eggs are laid in the leaf stalk or main veins, and hatching takes place after about two weeks.
    The nymphs are slender delicate insects, yellow with pale red markings. The full-grown nymph has a body length of about 7 mm, the antennae being much longer. There are five nymphal instars all except the first having a pin-like projection sticking up from the thorax. The total nymphal period is about three weeks.
    The adult bug is 7—10 mm long with antennae nearly twice as long as the body. The antennae, head and wings are blackish. Most females have a blood-red body, and like the nymphs the adults have a pin-like projection on the thorax. After a pre-oviposition period of several days the adult female bug may live 6—10 weeks and lay 30—60 eggs.
**Distribution.** Africa; from W. through to E., and S. to Zimbabwe (CIE map no. A297).
**Control.** *Helopeltis* attacks occur very suddenly and great vigilance is necessary if they are to be effectively controlled. The pesticide usually recommended as sprays are DDT, carbaryl or phenthoate.

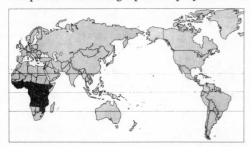

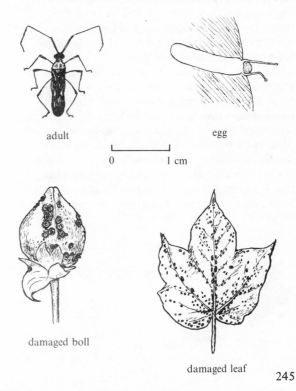

adult       egg

0      1 cm

damaged boll

damaged leaf

245

## Lamprocapsidea coffeae (China)

**Common name.** Coffee Capsid
**Family.** Miridae (= Capsidae)
**Hosts** (main). Coffee: *arabica* and *robusta*.
    (alternative). The Coffee Capsid has been recorded on a *Tricalysia* sp. but it appears to prefer coffee.
**Damage.** Flower buds blacken due to the death of the stamens and, later, the petals. The style, however, remains healthy and usually elongates. The resulting appearance is club-like, the club having a pale green shaft and a black head. The damage is due to the injection of toxic saliva by both adults and nymphs. No fruit is set by affected flowers.
**Pest status.** A sporadically serious pest of both *arabica* and *robusta* coffee in various areas. It is frequently beneficial on unshaded coffee at lower altitudes, since the pruning effect of capsid damage reduces the tendency to overbear.
**Life history.** The eggs are completely inserted into a flower bud; they cannot be detected in the field.
    The nymphs are pale green and pear-shaped. There are five instars, wing buds being visible on the fourth and fifth stages. The fifth stage is nearly 6 mm long. The nymphal period is variable depending upon the food available, but can be as little as two weeks.
    The adult bug is green and about 6 mm long. About half-way along the wings is a sharp downward bend so that both the top and posterior part of the abdomen is covered by them. The adult bug can live for at least three weeks. Both adult and nymphs feed on flower buds if they are present, but in their absence on any soft green part of the bush.
**Distribution.** Africa; Zaïre, Madagascar, and E. Africa.
**Control.** Control measures should only be applied when: developing flower buds are present; the farmer requires most of these buds to set fruit; the population of Coffee Capsid Bugs is more than an average of four to a tree.
    The following insecticides can be used: fenitrothion, fenthion, or pyrethrum as sprays in water, applying 0.5—0.9 litres per tree according to the amount of leaf present.

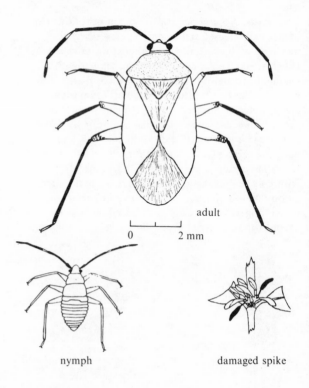

adult

0      2 mm

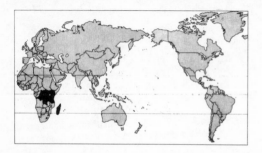

nymph

damaged spike

246

## Sahlbergella singularis Hagland

**Common name.** Cocoa Capsid
**Family.** Miridae (= Capsidae)
**Hosts** (main). Cocoa
(alternative). *Cola* spp., *Ceiba pentandra*, and *Berria* spp.

**Damage.** Both nymphs and adults feed on the pods and young shoots, and this species of mirid usually prefers mature trees. The toxic saliva injected through the feeding puncture causes a dark spot, frequently becoming infected with fungus. Each bug may make 24–36 punctures a day. After pod harvesting many bugs move to the tree canopy and feed on the young shoots; extensive damage is referred to as 'capsid blast', and severely attacked trees may die.

**Pest status.** This insect can be a serious pest in cocoa plantations, but is usually more serious to crops grown by peasant farmers. It is not a very serious pest in E. Africa, but has been so in Ghana, Nigeria, and the Ivory Coast.

**Life history.** Eggs are laid on twigs, pods or pod stalks, by being inserted into the plant tissues.

After 12–18 days the nymphs emerge, and the nymphal stage lasts some 25 days.

The adult is about 15 mm long, and speckled brown. After a week the female will start egg-laying, which persists until the end of her life some five weeks later.

**Distribution.** W. Africa (Ghana, Nigeria, Sierra Leone, Ivory Coast and Togo) through to Zaïre and Uganda (CIE map no. A22).

**Control.** A spray of lindane in water is effective, followed by a second spray of half strength after four weeks.

**Further reading**

Entwistle, I.P.F. (1972). *Pests of Cocoa*, Longmans: London.

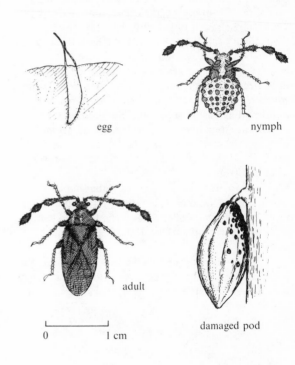

egg

nymph

adult

0        1 cm

damaged pod

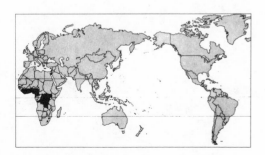

## Taylorilygus vosseleri (Popp.)

(= *Lygus vosseleri* Popp.)

**Common name.** Cotton Lygus
**Family.** Miridae (= Capsidae)
**Hosts** (main). Cotton
    (alternative). Sorghum, sesame, and various wild and cultivated legumes; recorded from plants in 14 different families.
**Damage.** Damaged plants are tall and straggly with very short side-branches; the leaves are holed, ragged, and tattered. Small flower buds turn brown and are shed. Very young bolls are shed and have black spots on the boll wall where they were punctured.
**Pest status.** A major pest of cotton in parts of E. Africa but in other areas damage may be rare and sporadic, and this pest is now not considered to be so important as was once thought. Sorghum is a most important alternative host, for large populations build up when the grain is in the milky stage, and as the grain ripens the bugs make mass migrations on to nearby cotton.
**Life history.** The egg is small, test-tube-shaped and about 0.75 mm long. It is inserted completely into the soft tissues of the plant, the flat end being level with the plant surface. Hatching after eight days.

    The full-grown nymph is about 4 mm long; the body is egg-shaped and pale green. The legs are fairly long and slender; there are five nymphal instars and the total nymphal period lasts about 17 days.

    The adult bug is 4 mm long, brown with a greenish tinge and with the hind part of the wings bent sharply down over the end of the abdomen. Adult females may live for 40 days and lay 60 eggs.
**Distribution.** Tropical Africa from W. through to E., and Madagascar.

    Other species of *Lygus* attack cotton in the USA, and also other crops in Europe and Asia.
**Control.** Recommended insecticides include DDT, trichlorphon and carbaryl.

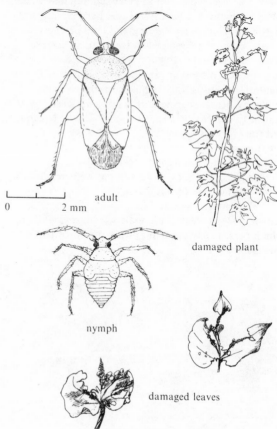

adult

0    2 mm

damaged plant

nymph

damaged leaves

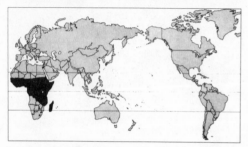

248

## Oxycarenus hyalipennis Costa

**Common name.** Cotton Seed Bug
**Family.** Lygaeidae
**Hosts** (main). Cotton
(alternative). Okra, and other Malvaceae;
*Sterculia* spp. and *Ceiba* sp.
**Damage.** The lint of the opened boll is stained, and deteriorated in quality. The seeds show brown discoloration and severe shrinking, and seed germination is severely reduced.
**Pest status.** An important pest of cotton affecting lint quality and seeds.
**Life history.** The egg is creamy, oval, about 1 by 1.2 mm, longitudinally striated and with six projections at the anterior end. Eggs are laid singly or in small groups loose amongst the seeds in the open boll. Each female lays 25–40 eggs, and incubation takes 4–10 days.

The five nymphal instars take about 2–3 weeks. All stages emit a characteristic unpleasant smell if crushed.

The adults are small, elongate bugs with pointed heads, about 4 mm long and 1.5 mm broad, dark brown or black with a red abdomen, and translucent hemelytra.

Breeding can only take place when ripe or nearly ripe seeds are available. The whole life-cycle can be completed in as little as three weeks; 3–4 generations usually take place in each crop ripening.

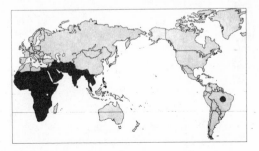

**Distribution.** Recorded from cotton throughout continental Africa, Egypt, Middle East, to India, Indo-China, and the Philippines; introduced accidentally into Brazil.

Seven other species of *Oxycarenus* are also recorded from cotton in Africa.
**Control.** Control can be effected by sprays of BHC or dusts, applied when the bugs are seen on the half-opened bolls.
**Further reading**
Pearson, E.O. (1958). *The Insect Pests of Cotton in Tropical Africa*, pp. 297–300. CIE: London.

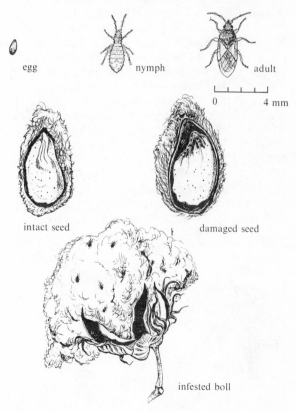

egg        nymph        adult

0        4 mm

intact seed        damaged seed

infested boll

249

## Dysdercus spp.

*cingulatus* (F.) (= *D. ornatus* Bredd.)
*fasciatus* Sign.
*nigrofasciatus* Stål
*superstitiosus* (F.)
spp. (10 +) in New World

**Common name.** Cotton Stainers (Red Cotton Bugs)
**Family.** Pyrrhocoridae
**Hosts** (main). Cotton
        (alternative). Many plants and trees including sorghum, *Hibiscus* spp. (okra, etc.), *Abutilon*, *Aznaza*, *Sterculia*, baobab and kapok trees.
**Damage.** There are conspicuous red bugs on the cotton bush, which fall to the ground if the bush is shaken. Small green bolls may abort and go brown, due to death of the seeds, but they are not shed. No damage is visible externally on large green bolls but, if the inner boll wall is examined, warty growths or water-soaked spots can be seen corresponding to patches of yellow staining on the developing lint. In very severe attacks the whole lock may be brown and shrunken.
**Pest status.** A major pest of cotton in most parts of Africa, and Asia; not important in the USA.
**Life history.** The eggs are ovoid, 1.5 × 0.9 mm, yellow when first laid but turning orange. Hatching takes about 5–8 days. They are laid in batches of

about 100 in moist soil or plant debris; moisture is essential for development and the eggs die if the soil dries out.

There are five nymphal instars. The first instar nymphs do not feed but require moisture and usually congregate near the empty egg shells. The second and

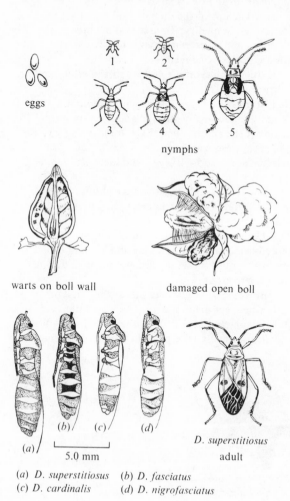

eggs

nymphs

warts on boll wall          damaged open boll

*D. superstitiosus*
adult

(a) *D. superstitiosus*     (b) *D. fasciatus*
(c) *D. cardinalis*         (d) *D. nigrofasciatus*

5.0 mm

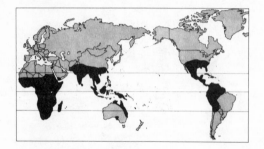

third instars feed gregariously on seeds on or near the ground. Later instars wander freely over the plant seeking suitable fruits and seeds. Nymphs are often found in large numbers on posts, tree trunks, etc. where they prefer to moult. The full-grown nymph is a bright red bug with black wing pads and is about 10−13 mm long, according to species. The total nymphal period lasts about 21−35 days.

The adult male Stainer is 12−15 mm long, according to species; the female is slightly larger. The wings are reddish and each has a black spot or bar near the middle. Stainers are able to fly strongly but usually drop to the ground and crawl if disturbed on the bush. Dispersal flights of up to 15 km have been recorded. The bugs feed by sucking sap from the seeds; the piercing and sucking proboscis is pushed through the boll wall and into the seeds. More eggs are laid by the female if she feeds on mature exposed seeds. After a pre-oviposition period of 5−14 days adult females may live a further 60 days and lay a total of 800−900 eggs.

The most serious damage done by Stainers is the injection of fungal spores of the genus *Nematospora* into the boll; the fungus grows on the lint and causes the staining.

**Distribution.** Found in the cotton-growing areas of tropical Africa, tropical Asia, Australasia, USA, C. and S. America.

*D. cingulatus* − CIE map no. A265.

*D. fasciatus* − CIE map no. A266.

*D. sidae* Montr. − CIE map no. A267, also in Australasia.

At least 10 species of *Dysdercus* on cotton in Africa, and a larger number in the New World.

**Control.** Cotton Stainers may be controlled by caging chickens in cotton plots using chicken wire; about 15 birds will keep about 0.1 ha free of Stainers. This method should *not* be combined with a chemical treatment, and is most appropriate to a small plot grown next to the homestead.

In parts of Kenya where the Mutoo tree is an important Stainer host it should be cut back each year in December or January to prevent fruiting during the cotton season.

The usual insecticides used are either carbaryl as a spray or a dust mixture of BHC and DDT.

**Further reading**

Freeman, P. (1947). A revision of the genus *Dysdercus* Boisduval (Hemiptera: Pyrrhocoridae) excluding the American species. *Trans. R. ent. Soc. Lond.* **98**, 373−424.

Pearson, E.O. (1958). *The Insect Pests of Cotton in Tropical Africa*, pp. 256−97. CIE: London.

# Acanthomia spp.

*horrida* (Germ.)
*tomentosicollis* (Stål)

**Common name.** Spiny Brown Bugs
**Family.** Coreidae
**Hosts** (main). Beans (*Phaseolus* spp.), pigeon pea, and *Dolichos labab*.

(alternative). Other pulse crops, Leguminosae, and *Solanum incanum*.

**Damage.** The external symptoms of *Acanthomia* damage on beans are dimpling of the seed coat, browning and shrivelling of the seed, and wrinkling of the seed coat. Germination ability of the seed is also impaired. These symptoms are thought to be caused by the fungus *Nematospora coryli*, rather than by the feeding bug itself.

**Pest status.** In parts of Africa these are serious pests of beans; experimental work has shown that with an infestation of only two bugs per plant the expected weight of seeds was lowered by 40–60%, the number of seeds by 25–36%, and the seed quality by 94–98%, according to the species of bug concerned. There appears to be no correlation between local infestations one year and the succeeding year.

**Life history.** Eggs are laid on the foliage of the plants and take 6–8 days to hatch under field conditions.

The different nymphal instars (five) take the following times for their development: first instar, 2–4 days; second, 3–5 days; third, 4–6 days; fourth, 4–6 days; fifth and last instar, 6–8 days. The total time for nymphal development in the field is about 28–35 days; in the laboratory this time was from 16–61 days, according to temperature (30–18°C).

The adults are small brown bugs, 7–10 mm in length, according to species.

A fungus (*Nematospora coryli*) is often associated with *Acanthomia* damage, but it is not yet certain whether the fungus is introduced by the bug itself, or whether it enters the seed after feeding via the feeding punctures.

**Distribution.** *A horrida* is found in E. Africa, Nigeria and Somalia. *A. tomentosicollis* is recorded from E. Africa, Nigeria, Guinea-Bissau and S. Africa.

**Control.** When insecticidal treatment is required a spray of DDT or endosulfan is recommended. Endosulfan is often used in *Heliothis* control.

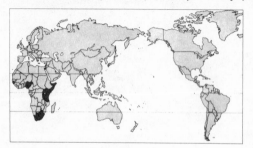

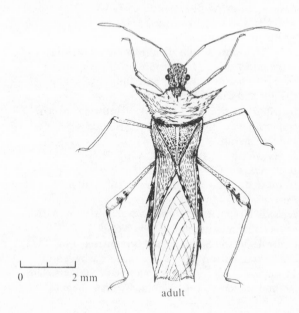

0     2 mm

adult

## Leptocorisa acuta (Thunb.)

**Common name.** Rice Seed Bug (Asian Rice Bug)
**Family.** Coreidae
**Hosts** (main). Rice
    (alternative). Various species of wild grasses.
**Damage.** The bugs usually appear in the young crop
with the early rains, and both nymphs and adults
suck sap from the developing grains at the 'milky'
stage. All soft milky grain is susceptible to attack —
the bugs suck the sap until the grain is emptied.
Before the grain is formed the bugs will feed on
succulent young shoots and leaves.
**Pest status.** Rice Bugs are very destructive in areas
where rainfall is evenly distributed throughout the
year, and also in irrigated crops. Yield losses of 10—
40% are common, and in severe infestations the entire
crop may be destroyed.
    Six other species of *Leptocorisa* are also pests
of rice in different parts of the tropics.
**Life history.** Eggs are laid in rows along the rice
leaves; they are red to black in colour, and flat in
shape. The incubation period is 5—8 days. Newly
hatched nymphs are green, but they become browner
as they grow. There are five nymphal instars, distinct
colour changes occurring after each moult. The
nymphal period lasts 17—27 days.
    The adult bugs are slender and some 15 mm in
length, greenish-brown in colour, and have been
recorded to survive up to 115 days in favourable

conditions. In the absence of rice plants the bugs
live on wild grasses.
    The complete life-cycle takes some 23—34
days, and there are several generations per year.
**Distribution.** Found in Pakistan, India, Sri Lanka,
through S.E. Asia to S. China, Philippines, Indonesia,
Papua New Guinea, West Irian and N. Australia (CIE
map no. A225).
**Control.** The same control measures that are used for
the various leafhoppers and planthoppers on rice are
generally effective against this pest.
**Further reading**
Feakin, S.D. (1976). *Pest Control in Rice*. PANS
        manual no. 3 (2nd ed.) pp. 147—9. COPR:
        London.

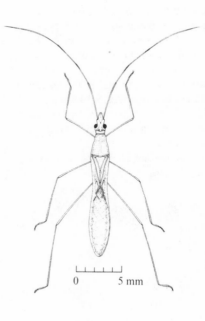

0        5 mm

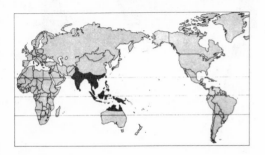

253

## Leptoglossus australis (F.)

(= *L. membranaceus* (F.))

**Common name.** Leaf-footed Plant Bug (Squash Bug)
**Family.** Coreidae
**Hosts** (main). Cucurbits
    (alternative). *Citrus* spp., groundnut, and many
legumes; sometimes found on oil palm in Malaysia,
passion fruit in Kenya; and coffee, yam, sweet
potato, cacao, and rice.
**Damage.** The young fruits show dark spots where
feeding punctures have been made. Many immature
fruits fall prematurely. The terminal shoots are fed
upon and they may wither and die off beyond the
point of attack. Similar damage is seen on *Citrus*.
**Pest status.** Not a very serious pest, but quite common
and widely occurring, and fairly polyphagous in
habits.
**Life history.** The adult is a large, brown bug about
20–25 mm long, with characteristic tibial expansions
on the hindlegs, and a pale orange stripe across the
anterior edge of the mesonotum. The antennae have
alternating black and pale orange zones.
**Distribution.** Canary Isles, W., C. and E. Africa down
to the Transvaal, Madagascar, India, Sri Lanka,
Burma, S.E. Asia, S. China, Philippines, Indonesia,
Papua New Guinea, West Irian, the Pacific Islands and
N. Australia (CIE map no. A243).

    A closely related species *L. zonatus* occurs on
*Citrus* in C. and S. America.

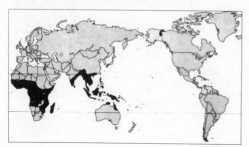

**Control.** Despite the abundance of this insect control
measures are seldom required, but both BHC and
parathion are effective against these bugs.

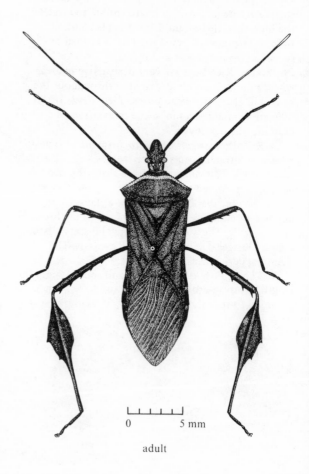

adult

# Notobitus meleagris (F.)

**Common name.** Bamboo Bug
**Family.** Coreidae
**Hosts** (main). One or two large species of bamboo.
(alternative). Not known.
**Damage.** The toxic saliva injected into the plant stem at the time of feeding causes death of plant cells and necrosis; as the plant grows so the necrotic area increases in size, usually as a longitudinal split, severely weakening the bamboo stem for any constructional purposes.
**Pest status.** Possibly a serious pest in some areas where large bamboos are grown for constructional purposes, but at present only recorded from S. China More of academic interest than directly agricultural.
**Life history.** Breeding takes place under the large leaf sheath surrounding the base of each stem internode. Under any one sheath there may be 3−15 nymphs of different ages as well as several adults. On some branching stems the developing lateral shoots may carry several bugs and sometimes the feeding of the bugs kills the young lateral shoot completely.

The adult bug is a large brown coreid, of regular body shape, and with normal legs; it is some 20−24 mm in body length.

This is a very typical coreid bug, both in appearance and habits, and resembles many other species. The saliva is apparently very toxic and feeding results in typical necrotic lesions on the stem and death of young sideshoots.
**Distribution.** To date only recorded from S. China.
**Control.** In the infestations under observation control was not regarded worthwhile as the plants were grown for ornamental purposes.

adult

0                    1 cm

necrosis of bamboo stem caused by toxic saliva of feeding bugs

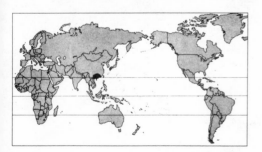

255

# Pseudotheraptus wayi Brown

(= *Therapterus* sp.)

**Common name.** Coconut Bug
**Family.** Coreidae
**Hosts** (main). Coconut

(alternative). Cashew, *Cinnamomum*, mango, guava, cocoa, and various wild legumes.

**Damage.** This bug is responsible for 'early nutfall' or 'gummosis' of coconuts in E. Africa. Young nymphs feed in the developing spadix at the base of the male flowers, or in the main stem and young branches which are still succulent. Older nymphs and adults tend to feed more on the developing nuts and female flowers. The toxic saliva of the bugs causes necrotic spots to appear. The young nuts are frequently killed (and dehisce) by the toxic saliva.

**Pest status.** A serious pest in parts of E. Africa on coconut, but actual crop loss is difficult to assess, partly because of the natural nut-fall. Over 70% of young nuts will fall 'naturally' and so many of the bug-damaged nuts would fall anyway. Two bugs per palm can cause appreciable damage. Very closely related species are found in Zaïre, Uganda and W. Africa.

**Life history.** Eggs are laid singly in the flowers or on young nuts. Each female can lay on average at least 70 eggs (probably about 100). The development time required for hatching is 8–9 days.

There are five nymphal instars taking a total of 33 days (at 24.6°C).

The adult is a medium-sized coreid bug, fawn or brown, with the membrane blackish. The pre-oviposition period in the female is nine days.

This species breeds continuously on the coconut palm and probably has nine generations per year.

**Distribution.** Found only in Africa; Kenya, Tanzania (including Zanzibar, and Pemba).

**Control.** The ant *Oecophylla longinoda* nests in palms and other trees and if present those palms are seldom inhabited by Coconut Bugs. Other predatory ants (*Anoplolepis custodiens, A. longipes,* and *Pheidole punctulata*) will remove *O. longinoda* colonies, but without preying on Coconut Bugs, and will allow the pest to severely damage palms occupied by them. The *Anoplolepis* species are largely confined to sandy soils for nesting purposes and palm trunks can be sprayed with dieldrin to stop them ascending to destroy the beneficial arboreal ant species. Dieldrin is also effective against *P. punctulata* in coffee plantations.

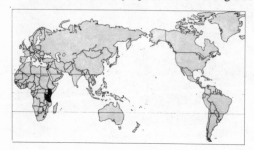

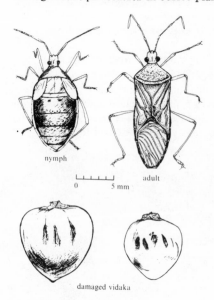

nymph

adult

0    5 mm

damaged vidaka

# Stenocoris southwoodi Ahmad

(= *Stenocoris apicalis* (Westw.))
(= *Leptocorisa apicalis* (Stål))

**Common name.** Rice Seed Bug (African Rice Bug)
**Family.** Coreidae (Alydidae)
**Hosts** (main). Rice
      (alternative). Various species of wild grasses.
**Damage.** The bugs appear in the crop usually with the early rains, and both adults and nymphs suck the sap from developing grains at the 'milky' stage; they suck the sap until the grain is emptied. Before the grain formation stage the bugs feed on the young leaves and shoots.
**Pest status.** A very destructive pest on both rain-fed and irrigated rice throughout tropical Africa. Yield losses of 10–40% are common, and in severe infestations the entire crop may be destroyed.
**Life history.** Eggs are laid in rows along the rice leaves; usually dark red to black in colour, and flat in shape; they hatch after 5–8 days. Newly hatched nymphs are green, but they become more brown with each of the five moults; total nymphal period is 17–27 days.

      The adult bugs are slender and some 15 mm in body length, greenish-brown in colour, and usually live for several months under favourable conditions. Dispersal flights are made at night occasionally during the season, and the bugs will then fly to lights.

      The complete life-cycle takes some 23–34 days, and there are several generations per year.

**Distribution.** This species is apparently confined to tropical Africa.
**Control.** The same control measures recommended for the various planthoppers and leafhoppers on rice are generally effective against this pest.

**Further reading**

Feakin, S.D. (1976) *Pest Control in Rice.* PANS manual no. 3 (2nd ed.) pp. 147–9. COPR; London.

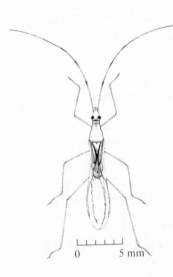

## Habrochila spp.

*placida* Horv.
*ghesquierei* Schout.

**Common name.** Coffee Lace Bugs
**Family.** Tingidae
**Hosts** (main). Coffee, especially *arabica*.
    (alternative). None recorded.
**Damage.** Yellow patches on the undersides of leaves covered with spots of shiny black liquid excreta. Severely attacked leaves turn completely yellow and then die from the edges inwards. The attack is often very localized at first, being confined to the lower leaves of a small group of coffee trees.
**Pest status.** A sporadically severe pest of *arabica* coffee over most of Kenya. Since 1956 it has extended its range into different parts of Kenya – the recent outbreaks have followed the indiscriminate use of DDT. *H. placida* apparently prefers *robusta* coffee, and *H. ghesquierei, arabica.*
**Life history.** Eggs are embedded in the undersides of leaves or in the soft tissues near the tips of green branches. Large numbers of eggs embedded near a growing point can cause checking or distortion of the terminal growth. The egg-caps, which are whitish, project slightly from the plant tissue and can just be seen with the unaided eye. Eggs hatch after 22–32 days.

    There are five nymphal instars. The newly hatched nymph is about 0.75 mm long. The fully grown nymph is about 2 mm. The nymphs feed gregariously, exclusively on the lower leaf surface. Their development is complete in 16–36 days. They are ornamented with knob-like integumental processes on the head and body. The head is darkly pigmented in the later instars.

    The adult Lace Bug is about 4 mm long. The wing carries a venation of lace-like pattern, giving the insect its common name. Dorsally, the thorax and wings carry domed outgrowths. The adults also feed on the lower surfaces of leaves but are not gregarious. Female maturation period is eight days.
**Distribution.** E. Africa, Zaïre, Burundi, and Rwanda.
**Control.** A predator (the mirid bug *Stethoconus* sp.) often keeps this pest down to low populations; spraying should only be done when the predator population is too low to keep the bugs in check.

    The recommended insecticide for Lacebug control is fenitrothion as a foliar spray; if a good cover has been achieved with the spray a high kill of Lace Bugs will result; if, however, only a poor cover was achieved a second spray will probably be required after about one month.

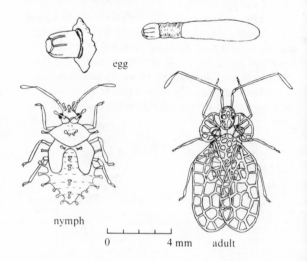

egg

nymph

0    4 mm    adult

## Stephanitis typica (Dist.)

**Common name.** Banana Lace Bug
**Family.** Tingidae
**Hosts** (main). Bananas
   (alternative). Coconut, cardamom, manila
hemp, *Alpinia* spp., and others.
**Damage.** Leaves may be dwarfed or distorted, with
yellow patches and necrotic spots; adults and nymphs
may be seen congregated on the undersurface of the
leaves.
**Pest status.** In some situations this is quite a serious
pest and control measures are required. Damage is
sometimes accentuated as the adults are usually
gregarious, and the combined effect of their feeding
and the toxic saliva may be serious.
**Life history.** Eggs are laid in the parenchymatous
tissue of the leaves.

   The adult is a small plant bug, some 5–6 mm
in length (including the closed wings), and like most
other Tingidae the wings are delicately but conspicuously veined. The thorax bears lateral expansions
extending to the sides of the head.

   One generation takes several weeks, so in hotter
climates there may be many generations per year on
perennial hosts.
**Distribution.** From India and Sri Lanka through most
of S.E. Asia, Indonesia, to Papua New Guinea, up
through the Philippines to S. China, Taiwan, Korea
and southern Japan (CIE map no. A308).

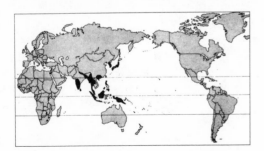

**Control.** Chemicals that have given satisfactory control are carbaryl, fenitrothion, formothion, malathion
and phosphamidon, as aqueous sprays, but care has to
be taken to ensure the wetting of the undersides of
the leaves.

nymph

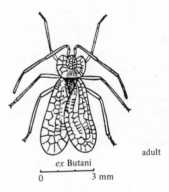

*ex* Butani

0          3 mm

adult

## Antestiopsis spp.

*orbitalis* (Westw.) group
*intricata* (Ghesq. & Carayon)

**Common name.** Antestia Bugs
**Family.** Pentatomidae
**Hosts** (main). Coffee (*arabica*).

(alternative). Antestia can live on various shrubs belonging to the same family as coffee, the Rubiaceae.

**Damage.** Blackening of the flower buds; fall of immature berries; rotting of the beans within the berries or conversion of the substance of the bean to a soft white paste ('posho beans'); multiple branching and shortening of the internodes of terminal growth.

**Pest status.** A major pest throughout Kenya and other parts of Africa, on *arabica* coffee; not a pest of *robusta* coffee.

**Life history.** The eggs are whitish and are usually laid in groups of about 12 on the underside of leaves. They hatch after about ten days.

The newly hatched nymph is about 1 mm in length. There are five nymphal instars which resemble the adults in colour but have a more rounded shape and lack functional wings. The nymphal period lasts about 3–4 months.

The body of the adult bug is shield-shaped, and generally dark brown, orange and white. Some races are much more brightly coloured than others. The body length is about 6 mm and the legs and antennae

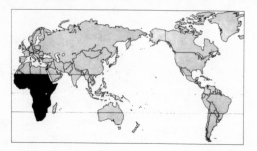

are easily visible. Adults can live for 3–4 months.

**Distribution.** Confined mainly to the Ethiopian region, but is found throughout Africa.

Greathead (1966) describes three subspecies of *A. orbitalis* and four other species on coffee *arabica* in Africa, and four species of *Antestia*; CIE maps nos. A381 & A382.

**Control.** Pruning to keep the bush open is of help in reducing bug populations for Antestia Bugs prefer dense foliage.

Spraying should be done when the average population (adults plus nymphs) is in excess of two per bush in the drier areas or one to a bush in the wetter areas. The recommended insecticides are fenitrothion, and fenthion, both as foliar sprays in water. If the infestation is heavy a second spray may be needed two weeks or more after the first.

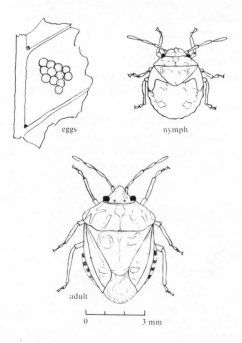

eggs

nymph

adult

0          3 mm

## Bagrada spp.

*hilaris* (Burm.)
*cruciferarum* (L.)

**Common name.** Harlequin Bugs
**Family.** Pentatomidae
**Hosts** (main). *Brassica* spp.
(alternative). Other Cruciferae; also on beet, groundnut, potato, and mallow.
**Damage.** Both adults and nymphs feed on the foliage of the crop, and the leaves wilt and dry. Young plants often die completely.
**Pest status.** A major pest of cruciferous crops in many parts of the Old World.
**Life history.** The eggs are white initially, but later turn orange; they are laid in small clusters either on the leaves or sometimes on the soil underneath. More than 100 eggs may be laid during a period of 2–3 weeks. The incubation period is 5–8 days.

There are five nymphal instars, which take 2–3 weeks for development.

The adult bug is typically shield-shaped, 5–7 mm long and 3–4 mm broad. The upper surface has a mixture of black, white and orange markings (hence Harlequin Bugs).

The whole life-cycle takes only 3–4 weeks, and there are several generations per year.
**Distribution.** *B. hilaris* is found in E. Africa, Egypt, Ethiopia, Malawi, Zimbabwe, Zaïre, Senegal, Mozambique, S. Africa, Italy, Iran, Iraq, Pakistan, India, Sri Lanka, Burma, and the USSR. (CIE map no. A417).

*B. cruciferarum* is recorded from E. Africa, India, Sri Lanka, Pakistan, S.E. Asia, and Afghanistan.
**Control.** Destruction of all cruciferous weeds will help to prevent a population build-up.

Some of the effective pesticides used to control this pest are DDT, γ-BHC, and carbaryl.

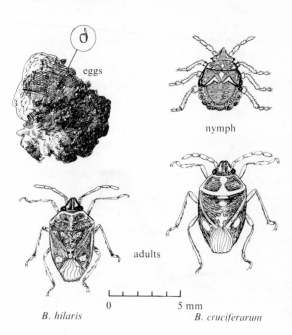

eggs

nymph

adults

*B. hilaris*          0          5 mm          *B. cruciferarum*

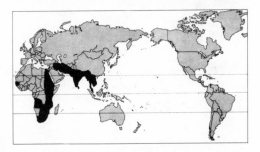

## Calidea spp.

*dregii* Germar
*bohemani* (Stål)

**Common name.** Blue Bugs
**Family.** Pentatomidae
**Hosts** (main). Cotton

(alternative). Sorghum, sunflower, castor, and many wild hosts, including *Crotalaria, Solanum, Combretum, Hibiscus,* and *Euphorbia* spp.

**Damage.** The adult bugs feed on developing seeds in unopened cotton bolls, with the result that development ceases and the boll aborts. The feeding of *Calidea* bugs results in the staining of the cotton lint and it appears that the bugs transmit the fungi of stigmatomycosis.

**Pest status.** An important pest of cotton in Tanzania. Extremely polyphagous in habits, attacking the seeds of many cultivated and wild plants.

A total of five species have been recorded from cotton in different parts of Africa.

**Life history.** *Calidea* bugs seldom breed on cotton, and usually only appear on the crop when the bolls are well formed. The life history details are taken from studies of *C. dregii* on sorghum and sunflower in Tanzania.

The eggs are spherical, 1 mm in diameter, and laid in batches of up to 40 in a closed spiral round a stalk or dried leaf, or seed head, of the host plant. They are white, turning red as they develop.

The nymphs are oval and flattened, and in colour like the adults.

Adults are strikingly coloured, with red or orange underneath, and the upper surface an iridescent blue or green, often with a bronze tinge, with a bold pattern of spots and stripes. The size range is 8–17 mm long by 4–8 mm broad.

The complete life-cycle takes from 23–56 days according to the temperature.

**Distribution.** Restricted to the Ethiopian region, including Madagascar and Arabia.

**Control.** The devastation that heavy infestations of *Calidea* can cause, and the swift and unpredictable nature of the attack, have resulted in the susceptible areas of Tanzania being avoided for cotton growing.

This pest is difficult to kill with insecticides, but the crop can be protected against combined *Calidea* and *Heliothis armigera* attack by repeated sprays of DDT (low volume) at weekly intervals for up to 12 weeks or cypermethrin.

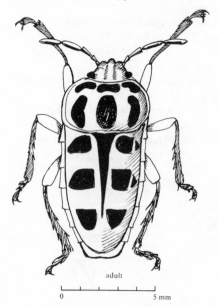

adult

0                    5 mm

## Diploxys fallax Stål

**Common name.** Rice Shield Bug
**Family.** Pentatomidae
**Hosts** (main). Rice
(alternative). The flowers of various grass species.
**Damage.** The adults feed on grass flowers before moving on to the rice crop for egg-laying. Adults and nymphs feed on the newly emerged florets before the milk stage of the grain, causing deformation or loss of grain.
**Pest status.** Occasionally a serious pest of rice in Africa.
**Life history.** About 20 eggs per female are laid in two rows on the upper leaf surface along the midrib.

The nymphs are green.

Adults are pale brown, with a pair of lateral spines on the dorsum.
**Distribution.** Africa only; in Swaziland and Madagascar.
**Control.** Cultural control can be achieved by eradication of grass flowers around the paddy fields so that a population build-up prior to the flowering of the rice crop is avoided.

Chemical control can be achieved by dusting with BHC, carbaryl, malathion, or trichlorphon.

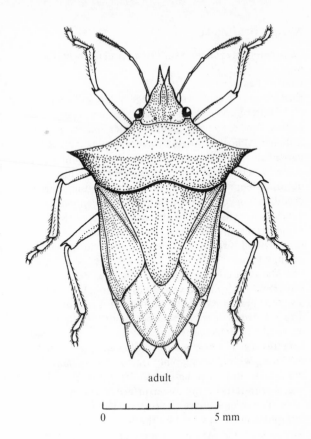

adult

0     5 mm

# Nezara viridula (L.)

**Common name.** Green Stink Bug (Green Vegetable Bug)

**Family.** Pentatomidae

**Hosts** (main). Castor, various vegetables, and cotton.

(alternative). Potato, pulses, sesame, sweet potato, *Citrus*, tomato, many legumes and cereals.

**Damage.** Almost invariably the developing fruit is the part of the plant attacked, and the feeding punctures cause local necrosis with resulting fruit spotting, deformation, or if attacked when very young, fruit-shedding. On cotton the bugs feed on green developing bolls, and it is strongly suspected that *Nezara* transmits the *Nematospora* fungus causing internal boll rot, although it is not formally proved yet in Africa.

**Pest status.** A cosmopolitan polyphagous pest, important on fruit and vegetables, although not often a serious pest in Africa. In Australia it does not attack cotton but is a serious pest with the common name of Vegetable Green Bug.

**Life history.** The eggs are barrel-shaped, 1.2 × 0.75 mm, white turning pink. Up to 300 are laid per female, usually in batches of 50–60 stuck together in rafts on the undersurface of leaves.

The nymphs have five instars. The first stage nymphs remain clustered by the egg raft and do not feed; after moulting they disperse and start feeding. They feed on sap from the softer parts of the plant, but principally from the developing seeds or fruits; absence of fruits or seeds on the host plant results in retarded or incomplete development (metamorphosis).

Development is generally slow, egg and nymphal stages occupying some eight weeks in Egypt (mean temperature 26°C), where there are believed to be three generations produced in just under nine months, after which breeding stops and the adults hibernate.

The adults are large green shieldbugs, about 15 × 8 mm, which occur in three colour varieties. The commonest is uniform apple-green above and a paler shade below, although the green colour may be sometimes replaced by a reddish-brown.

**Distribution.** Almost completely cosmopolitan in distribution from S. Europe and Japan down to Australia and S. Africa (CIE map no. A27).

**Control.** No special control measures are usually necessary in Africa. In Australia the recommended insecticides are DDT, phorate, and BHC.

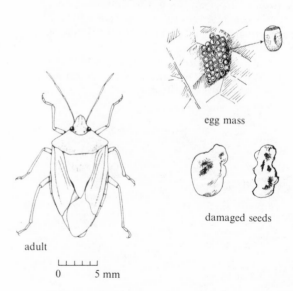

egg mass

damaged seeds

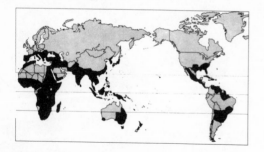

adult

0    5 mm

264

# Oebalus pugnax (F.)

**Common name.** Rice Stink Bug (Rice Seed Bug)
**Family.** Pentatomidae
**Hosts** (main). Rice
(alternative). Various species of grasses, maize and sorghums.
**Damage.** Both nymphs and adults feed on rice kernels at the milk stage, and the feeding results in the grain being infected by a fungus which enters via the bug's feeding punctures. These grains do not break during milling, and lower the grade of the rice; the condition is known as 'peckiness'. Overwintering adults emerge in early spring, and feed on developing grass seeds.
**Pest status.** A major pest of rice in America, where yields may often be reduced by half. Four other species of *Oebalus* are recorded from rice in America.
**Life history.** The eggs are bright green, turning red before hatching. They are 0.8 mm long and 0.6 mm in diameter, and are laid in batches of 10–47, on the stems, leaves, or panicles of rice plants or many grass species. They hatch in 4–8 days.

The nymphs are black with a red abdomen on hatching, but gradually pale during successive instars. There are five nymphal instars.

Adults are yellowish, 9–12 mm long and 5–6 mm broad. They have sharp shoulder spines pointing forwards. As with other pentatomids they emit a characteristic repugnant odour. Adults live for 30–40 days, but can hibernate in wood trash and 'bunch' grass.
**Distribution.** Southern USA, Cuba and the Dominican Republic.

Four other species are of importance on rice in S. America, and *O. poecilus* is important on rice in Indonesia.
**Control.** The use of light traps will deplete numbers of adults.

Chemicals applied just before rice flowering give adequate control. In the USA insecticidal control measures are advocated if numbers average ten bugs per hundred heads of rice. Recommended insecticides are DDT, BHC, aldrin, chlordane, and malathion.

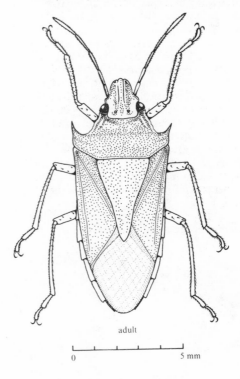

adult

0          5 mm

265

## Rhynchocoris spp.

**Common name.** Citrus Shield Bugs
**Family.** Pentatomidae
**Hosts.** *Citrus* species only to date, but probably also other Rutaceae.
**Damage.** Adults and nymphs feed on young fruit, and usually the feeding puncture becomes infested by bacteria and fungi which causes rotting and results in premature fruit-fall.
**Pest status.** Quite serious pests in the north of India, the Philippines and in parts of China.
**Life history.** Eggs (large and globular) are laid in batches of about 14, usually on the leaves, and rarely on fruit or twigs; several batches are laid per female per season. Hatching takes place after 2–9 days according to temperature. The five nymphal instars require about 30 days for development.

The adults are large shield bugs with prominent, sharp-pointed lateral expansions of the pronotum, green and brown in body colour, some 20–23 mm in body length, and with a prominent ventral keel under the thorax extending anteriorly just under the head. The pronotum is greenish, with black spots on the lateral expansions, and scutellum generally more brown; the underneath and legs are pale greenish, but the tarsi are dark, as are the distal ends of the tibiae.

There are two or three generations per year in the Philippines and in India, but only one in China; the adults hibernate in India and China over the cool winter period.
**Distribution.** *R. longirostris* occurs in the Philippines, and *R. humeralis* in India and southern China.

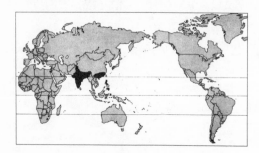

**Control.** In China it was recorded that most of the egg batches are parasitized by a small chalcid wasp and so egg mortality was high, and there was extensive predation of the bugs by spiders, mantids and ants. In Guangdong the Red Tree Ant (*Oecophylla smaragdina*) has been encouraged to nest in *Citrus* orchards since ancient times, as one of the earliest recorded examples of (fairly successful) biological control.

In the Philippines chemical control is regularly practised.

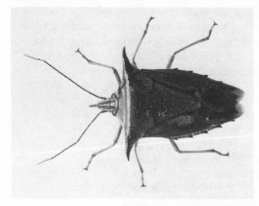

*R. longirostris*

0      1 cm

*R. humeralis*

## Scotinophara coarctata (Thunb.)

**Common name.** Black Paddy Bug
**Family.** Pentatomidae
**Hosts** (main). Rice
    (alternative). *Scirpus grossus*, *Scleria sumatrensis*, and *Hymenachne pseudointerrupta* (Gramineae).
**Damage.** Nymphs and adults feed at the base of stems, often just at water level. Infested plants are often stunted and grain fails to develop. Severe infestations, and very young, attacked plants, often die. The saliva of this bug is very toxic.
**Pest status.** A pest of rice, periodically serious, in many areas.
    Seven other species are found on rice in different regions.
**Life history.** Eggs are laid in batches of 40–50; one female laying several hundred eggs. Each egg is about 1 mm long, and is green or pink. The incubation period is 4–7 days.
    The young nymphs are brown with a green abdomen. They moult five times and become adult after 25–30 days.
    The adult bug is 8–9 mm long, brownish-black, with a few indistinct yellow spots on the thorax. The tibiae and tarsi are pink. It can live for up to seven months, and is strongly attracted to light, often appearing in large swarms.
    The life-cycle takes about 32–42 days for completion.
**Distribution.** Pakistan, India, Bangladesh, Thailand, Malaysia, Sabah, S. China and Indonesia.

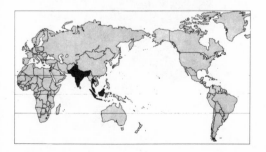

**Control.** The usual recommendation is a dust of equal parts of DDT and BHC to be applied to the plants.

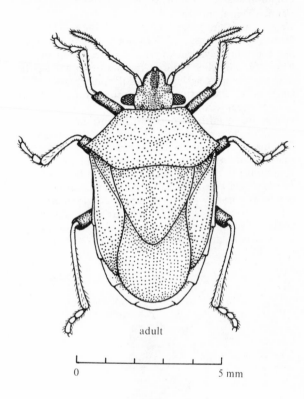

adult

0                5 mm

267

## Tesseratoma papillosa (Dru.)

**Common name.** Litchi Stink Bug
**Family.** Pentatomidae
**Hosts.** Litchi and longan only.
**Damage.** These large bugs occur in large populations and together they remove very considerable quantities of sap by their feeding; there is often damage to new shoots in the spring. The stink fluid ejected at times of fright or trauma may cause necrotic spots on young tender leaves.
**Pest status.** A fairly serious pest of both litchi and longan in the region from S. China down to Malaysia.
**Life history.** Eggs are laid in a group, stuck on the the undersurface of the leaf lamina, and there are invariably just 14 in number. Incubation takes 11 or more days according to ambient temperature, and can be as long as 34 days. Nymphal development takes up to 50 days. The nymphs are often liberally coated with a white waxy substance, both dorsally and ventrally, and adults are likewise often waxy ventrally.

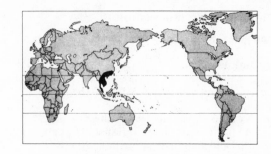

The adults are large brown bugs some 26–30 mm in body length, with fairly short antennae, to be found sitting in the tree foliage. They are strong fliers and if disturbed they fly off to another tree, usually emitting jets of yellow stink fluid as they depart. In China the adults apparently hibernate through the winter in groups in trees that have thick foliage, and in that country there is probably only the one generation per year.

nymph

adult

0          1 cm

**Distribution.** Recorded from S. China, the Indo-China peninsula and Malaya.

**Control.** In China it was recorded that predation by ants and spiders was responsible for keeping populations under some control. This is supplemented by widespread hand-picking of the adults by the orchard workmen; the most successful times for hand collection was in the autumn when the bugs congregated for hibernation, and also in the spring prior to egg-laying. In Guangdong since 1964 the egg parasite *Anastatus* sp. (Hymenoptera; Eupelmidae) has been used successfully as a means of reducing the bug populations.

nymph in longan foliage

adult in longan foliage

## Order **THYSANOPTERA**

(Thrips) Small or minute insects, with short six-segmented antennae, and asymmetrical rasping and sucking mouthparts; tarsi with one or two segments, each with a terminal protrusible vesicle. The wings are very narrow with greatly reduced venation, and long marginal setae. Metamorphosis is accompanied by one or two inactive pupal instars. Most are plant feeders but a few species are predaceous.

## Suborder **TEREBRANTIA**

### Family **Thripidae**

Most of the crop pests belong to this family within the Suborder Terebrantia; they have a saw-like ovipositor, and the apex of the abdomen is conical in the female and bluntly rounded in the male. The differences between this and the three other families tend to be rather esoteric. Many species are quite important crop pests. A striking feature of many thrips is that, whereas the adults are mostly grey, brown or black, the nymphs are typically yellow, orange or red. Thus an infested leaf may have red nymphs feeding alongside black adults. Pupation usually takes place in the soil. Damage by these thrips usually consists of scarification of the leaf lamina as a result of feeding.

## Suborder **TUBULIFERA**

No ovipositor, and tenth abdominal segment usually tubular; wings without microtrichia, and veins either absent or only one vestigial vein present.

### Family **Phlaeothripidae**

This large family contains over 300 genera, which show a great diversity of habits; some species are predaceous, some are fungus-feeders and are found in leaf litter and soil; most are phytophagous. The feeding of these thrips usually causes leaf-folding, leaf-rolling, or general leaf deformation to produce characteristic symptoms (or galls). Breeding takes place within the folded leaf, and in some species the adults show parental care and guard the eggs and nymphs. There are not many important agricultural pests within this family but a number of species are minor pests on various tree crops and woody ornamentals. *Hoplandothrips* is found on several trees where it rolls the young leaves. *Gynaikothrips* is common in the tropics; *G. ficorum* causes leaf-folding on *Ficus retusa* and *F. microcarpa*, and *G. kuwani* causes spectacular leaf distortion on *Aporusa chinensis*. *Gigantothrips elegans* is one of the largest thrips known and measures about 5 mm, and is to be found on the foliage of *Ficus* trees.

## Baliothrips biformis (Bagn.)

(= *Chloethrips oryzae* (Williams))
(= *Thrips oryzae* Williams)

**Common name.** Rice Thrips
**Family.** Thripidae
**Hosts** (main). Rice
      (alternative). Not krɔwn.
**Damage.** Essentially a pest of young rice seedlings. Nymphs and adults rasp the tissues of the leaf and suck the sap that exudes. Damaged leaves show fine yellow streaks which later join together to colour the whole leaf. Later the leaves curl longitudinally from the margin to the midline; eventually the whole plant may wither. Older plants are seldom attacked — these are plants four weeks after transplanting.
**Pest status.** Damage can be serious because of the high rate of reproduction of this pest, but only to young seedlings.
**Life history.** Eggs are laid singly into the leaf tissues of the leaf, and they hatch in about three days. They measure 0.25 × 0.1 mm.

    Nymphs are white or pale yellow, and remain in the young rolled leaves where they develop. There are usually four nymphal instars, the last being the resting 'pupal' stage. The nymphal period lasts for 10–14 days.

    The adults are minute, about 1 mm in length, dark brown, with seven-segmented antennae. At the base of the forewing is a pale spot. The tarsi end in protrusible suckers used for attachment to leaf surfaces. The adults can live for up to three weeks.

    The entire life-cycle is often not more than two weeks.
**Distribution.** India, Sri Lanka, Bangladesh, S.E. Asia, Java, Philippines, Taiwan, and Japan (CIE map no. A215).
**Control.** Removal of all infested leaves is recommended.

    Contact insecticides which have proved effective are carbaryl, DDT, BHC, malathion, diazinon. Dimethoate with its systemic action is highly effective.

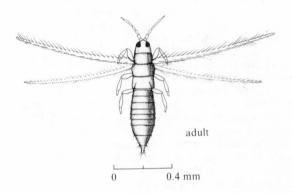

adult

0        0.4 mm

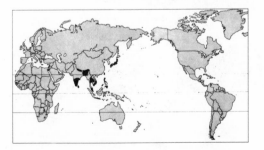

## Diarthrothrips coffeae Williams

**Common name.** Coffee Thrips
**Family.** Thripidae
**Hosts** (main). *Coffea arabica.*

(alternative). Only one wild host is definitely recorded (a *Vanguoria* sp.) but the Coffee Thrips can almost certainly breed on a large number of plants.

**Damage.** Undersides of leaves, and in severe cases the upper sides of leaves, berries and green shoots, with irregular grey or silvery patches covered by minute black spots. Death of leaves and total leaf-fall may follow a very heavy infestation.

**Pest status.** Up to about 1950, severe outbreaks occurred on *arabica* coffee in Kenya about every fourth year in the hot weather of February and March. Since then there have only been relatively isolated outbreaks and few of outstanding severity.

**Life history.** Eggs are minute kidney-shaped objects inserted into the tissues of the leaf.

There are two nymphal stages. The nymphs are cigar-shaped tiny insects, pale yellow, and just visible to the unaided eye; found on the undersides of the leaves.

At the end of the nymphal period, the nymphs drop to the ground and in an earthen cell change into pre-pupae. These then change into pupae from which the adult thrips finally emerge.

Adults crawl out of the soil, fly back into the tree and feed with the nymphs. They can be distinguished from the nymphs by their slightly larger size, their grey-brown colour and their feather-like wings.

In hot weather one generation probably takes about three weeks.

**Distribution.** E. Africa, Malawi, and Zaïre.

**Control.** Mulching reduces thrips numbers considerably and its widespread use in recent years is probably a reason for the declining importance of this pest.

Non-persistent insecticides such as fenitrothion and fenthion give effective thrips control for only a few days after spraying. DDT or dieldrin must usually be used, despite the risk of causing outbreaks of other pests.

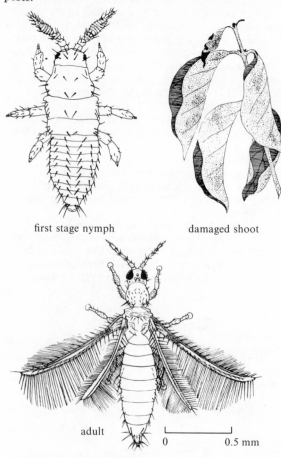

first stage nymph    damaged shoot

adult

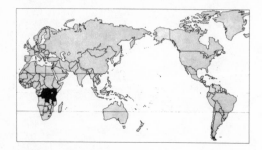

0    0.5 mm

# Frankliniella schulzei (Trybom)

**Common name.** Cotton Bud Thrips (Flower Thrips)
**Family.** Thripidae
**Hosts** (main). Groundnut, beans, cotton.

(alternative). A polyphagous pest on many crops and flowers, including coffee, sweet potato, and tomato.

**Damage.** Adults and nymphs feed in flowers and on leaves of many plants, especially legumes. They rasp the cells off the upper surface of young leaves while they are still in the bud, and these leaves become distorted. Seedling growth may be retarded by several weeks, and yield can be seriously affected. Mature plants are little affected by thrips.

**Pest status.** Various species of thrips are pests of some importance on groundnut, beans and other legumes in many parts of the world. Sometimes, although the thrips are common in the flowers, no actual damage is done. This species is a vector of Tomato Spotted Wilt virus on groundnuts; in Australia yield decreases of 90% have been recorded. This virus is widespread in groundnut growing areas but generally of low crop incidence (usually less than 5%).

**Life history.** Eggs are laid in the leaf tissue.

Nymphs are pale-coloured and wingless, and found under the curled leaves. There are three instars.

Pupation takes place in the soil.

The adults are pale brown, dark brown or black, with paler bands across the abdominal segments, and 1.0–1.5 mm long.

The life-cycle usually takes about 2–5 weeks,

so that in hot, dry conditions damage may be apparent quite suddenly.

**Distribution.** E. Africa and the Sudan.

Several other species of *Frankliniella* are found on a similar range of host plants, in Africa, Asia, and America; the genus is cosmopolitan.

**Control.** Both contact and systemic insecticides have been used successfully against thrips on groundnut, including: DDT and BHC dusts and sprays, malathion, phorate dust, and dimethoate sprays.

Ingram obtained satisfactory control with DDT and γ-BHC but no increase in crop yield.

**Further reading**

Ingram, W.R. (1969). Observations on the pest status of bean flower thrips in Uganda. *E. Afr. Agric. for. J.* **34**, 482–4.

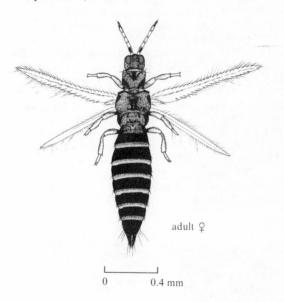

adult ♀

0        0.4 mm

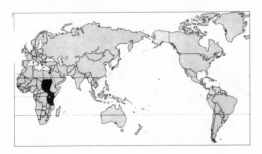

# Heliothrips haemorrhoidalis (Bouché)

**Common name.** Black Tea Thrips (Greenhouse Thrips)
**Family.** Thripidae
**Hosts** (main). Tea
 (alternative). A wide range of cultivated plants including roses, coffee, bananas, *Citrus*.
**Damage.** Silvery patches covered by black spots on the undersides of the older leaves.
**Pest status.** A polyphagous pest attacking many plants, and found in greenhouses in temperate climates. Only sporadically serious on tea in E. Africa.
**Life history.** The eggs are bean-shaped, about 0.3 mm long, and are pushed into the leaf tissue by the female, the wound being covered by a spot of excreta.

The first stage nymph is white, with red eyes, and just visible to the unaided eye. It rasps and sucks at the leaf surface causing the characteristic silvery patches by removing chlorophyll from the leaf tissue and letting air into the surface cells. The nymph carries a shining drop of greenish brown or black excreta on the tip of its upturned abdomen. It deposits these drops at intervals, causing the black spots on the leaves. The nymphs usually congregate on a damaged leaf and first change to a 'pre-pupal' stage with short wing pads; after a day or two they moult again into the 'pupal' stage which has rather longer wing pads. Both stages are yellow in colour with red eyes.

The adult thrips is dark brown or black, with whitish legs, antennae and wings. When the wings are folded they appear as an elongate T-shaped mark down the middle of the back. The Black Tea Thrips is one of the larger thrips, being about 1.5 mm long. All adults are females, which reproduce parthenogenetically. Each female lays some 25 eggs over a seven week period.

The total life-cycle takes from eight weeks at 19°C to 12 weeks at 15°C.
**Distribution.** This is a cosmopolitan species occurring in temperate countries in greenhouses; however records are sparse from Asia (CIE map no. A135).
**Control.** The usual insecticide employed against this pest is fenitrothion, as a full-cover spray directed as far as possible at the underside of the leaves.

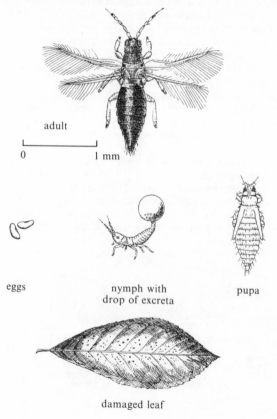

adult

0          1 mm

eggs          nymph with
              drop of excreta          pupa

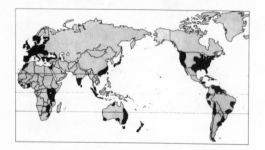

damaged leaf

274

# Hercinothrips bicinctus (Bagnall)

**Common name.** Banana Thrips
**Family.** Thripidae
**Hosts** (main). Bananas
    (alternative). None are recorded in Africa but several greenhouse crops are attacked in Europe.
**Damage.** Silvery or brown patches covered with small black spots found on the fruits. The skin of severely infested fruit may crack, and this allows secondary rots to attack the fruit.
**Pest status.** A serious pest of bananas if high-grade fruit is being produced. The damage is, however, often only a skin blemish and is of little significance on the local market.
**Life history.** Eggs are inserted into the plant tissues; a favoured site appears to be on the fruit surfaces where two young bananas are in close contact.

    Nymphs are yellowish but the abdomen may appear black and swollen due to the presence of liquid excreta. A globule of excreta is also carried at the upturned tip of the abdomen. A full-grown nymph is over 1 mm long.

    The so-called pupal stages probably occur in the soil.

    The adult is a fairly large thrips about 1.5 mm long, and is dark brown.
**Distribution.** Africa, E., W. and S. Europe, Australia, Hawaii, N. and S. America, and the W. Indies.
**Control.** The following insecticides are generally effective against this pest: dieldrin, DDT, γ-BHC, and phosphamidon.

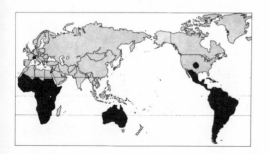

**Further reading**
Feakin, S.D. (1971a). *Pest Control in Bananas*. PANS manual no. 1, p. 105. COPR: London.

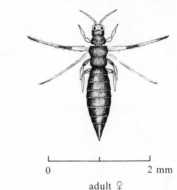

0                2 mm

adult ♀

damaged bananas

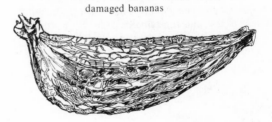

275

## Scirtothrips aurantii Fauré

**Common name.** Citrus Thrips
**Family.** Thripidae
**Hosts** (main). *Citrus* spp.

(alternative). Over thirty indigenous trees and shrubs have been recorded in S. Africa.

**Damage.** A ring of scaly, brownish tissue round the stem end of the fruit. Irregular areas of scarred tissue on other parts of the fruit. Young leaves may be damaged.

**Pest status.** A serious pest at low altitudes where an attempt to produce unblemished fruit is being made.

**Life history.** The egg is bean-shaped, very small (less than 0.2 mm long) and inserted into the soft tissues of leaves, stems and fruit. Hatching takes 1–2 weeks.

There are two nymphal stages; they are yellow to orange, cigar-shaped and just visible to the unaided eye. They feed on young fruits from petal-fall until they are about 25 mm in diameter. Most feeding takes place at the stem end, under and near the 'button'. In the absence of suitable fruits young leaves may be attacked. The nymphal period lasts 8–15 days.

When fully grown, nymphs seek out some sheltered place and then pass through two resting stages called 'pre-pupa' and 'pupa' respectively. They do not feed during these stages, but may walk a little if disturbed. The pupal stages last 1–2 weeks.

The adult thrips is reddish-orange, less than 1 mm long and like all thrips has feather-like wings. Males are rare and the females probably normally reproduce parthenogenetically. Adults may live for several weeks.

**Distribution.** Only known from Africa; Egypt, Malawi, Sudan, E. Africa, Zimbabwe, and S. Africa; most common in S. Africa (CIE map no. A137).

Another species (*S. citri*) occurs on *Citrus* in California.

**Control.** Spray the fruits towards the end of a main flowering period, when three-quarters of the petals have fallen, using a water solution of lime-sulphur. The spray should be repeated after about ten days.

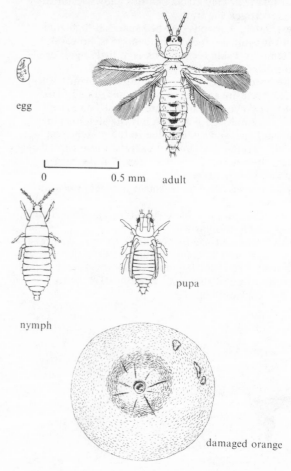

egg

0      0.5 mm      adult

nymph

pupa

damaged orange

## Selenothrips rubrocinctus (Giard)

**Common name.** Red-Banded Thrips (Cacao Thrips)
**Family.** Thripidae
**Hosts** (main). Mango
    (alternative). Avocado, pear, cashew, guava, and cacao, but usually only severe on mango.
**Damage.** The lower leaf surfaces are darkly stained, rusty in appearance, and with numerous small, shiny black spots of excreta; leaf edges are curled.
**Pest status.** A sporadically serious pest in mango nurseries; very rarely damaging to mature trees. A polyphagous pest of wide occurrence.
**Life history.** The eggs are kidney-shaped, about 0.25 mm long, and are inserted into the leaf tissue by the female thrips. Hatching takes about 12–18 days.

    The nymphal stages are yellow with a bright red band round the base of the abdomen. The full-grown second-stage nymph is about 1 mm long. Nymphs feed in company with the adults, normally on the underside of the leaf; depressions or grooves adjacent to the main veins are favoured sites. The tip of the abdomen is turned up and carries a large drop of reddish excreta. These drops are deposited at intervals on the leaf surface and dry to form shiny black spots. The total nymphal period lasts 6–10 days.

    The so-called pupal (i.e. non-feeding) stages are passed on a sheltered spot in the curl of a leaf. Both the pre-pupal and pupal stages resemble the nymphs but differ in having well-developed wing buds. The pupal stages can move but do not do so unless disturbed. After 3–6 days adults emerge from their pupal skins.

The adult thrips are dark brown and just over 1 mm long. Males are rare. Adults feed in company with the nymphs.
**Distribution.** Almost completely pantropical in distribution, but not recorded from Australia (CIE map no. A136).
**Control.** The recommended insecticide is fenitrothion as a spray directed at the undersides of the leaves.

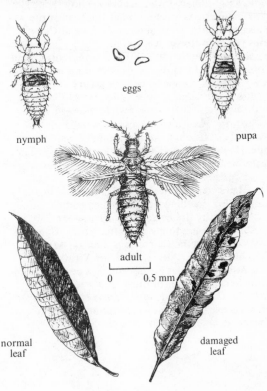

nymph    eggs    pupa

adult

0    0.5 mm

normal leaf    damaged leaf

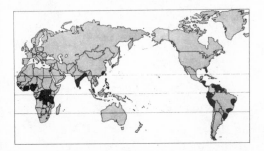

# Taeniothrips sjostedti (Trybom)

**Common name.** Bean Flower Thrips
**Family.** Thripidae
**Hosts** (main). Beans, peas, groundnut.
   (alternative). Coffee, avocado, and many other plants.
**Damage.** Both adults and nymphs are found inside the flowers of beans, other legumes, and other plants. Feeding punctures can be seen at the base of the petals and stigma. In Uganda an average of three thrips per bean flower was found.
**Pest status.** Although this thrips is commonly found in the flowers of beans and other legumes in many parts of Africa, the evidence of Ingram (1969) suggests that no real damage is done, since killing the thrips does not result in a yield increase.
**Life history.** The eggs are presumably laid in the flowers, but this observation has not actually been made. However, first and second stage nymphs can usually be found in the flower.
   Pupation occurs in the soil.
   Males were not found in Uganda, and it is assumed that breeding was parthenogenetic.
   The entire life-cycle takes 10–14 days.
**Distribution.** Found only in Malta, Africa; Gambia, Ivory Coast, Nigeria, Cameroons, Central African Republic, Zaïre, E. Africa, southern Africa.
   Another species (*T. distalis*) is a pest of groundnut flowers and leaves in India. *T. simplex* is the Gladiolus Thrips, and *T. laricivorous* is the European Larch Thrips. Other species are found on coffee either feeding on rust spores or the leaves.

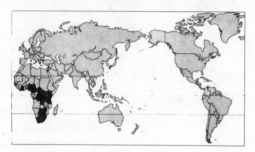

**Control.** Spraying with DDT and γ-BHC effectively controlled the thrips, but in Uganda control of the thrips did not result in a yield increase.
**Further reading**

Ingram, W.R. (1969). Observations on the pest status of bean flower thrips in Uganda. *E. Afr. Agric. for. J.* **34**, 482–4.
Zur Strassen, R. (1960). Catalogue of South African Thysanoptera. *J. ent. Soc. S. Africa,* **23**, 330.

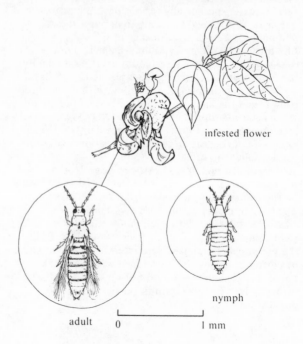

infested flower

adult     0          1 mm

nymph

# Thrips tabaci Lind.

**Common name.** Onion Thrips
**Family.** Thripidae
**Hosts** (main). Onions
    (alternative). Tobacco, tomato, pyrethrum, cotton, pineapple, peas, brassicas, beet and many other plants.
**Damage.** The leaves of attacked plants are silvered and flecked. Heavy attacks lead to the wilting of young plants. On cotton seedlings damage can be more serious causing leaf shedding. Onion leaves are often distorted, and sometimes they die; occasionally entire crops may die.
**Pest status.** A polyphagous pest on many crops; vector of virus diseases of tobacco, tomato, pineapple, and other crops.
**Life history.** Eggs are laid in notches in the epidermis of the leaves and stems of young plants. They are white, and take 4–10 days to hatch.
    Both nymphs and adults rasp the epidermis of the leaves and suck the sap that exudes. Nymphs moult twice in about five days; they are white or yellow.
    Pupation occurs in the soil, and takes 4–7 days.
    The adult is a small, yellow-brown thrips, with darker transverse bands across the thorax and abdomen, and about 1 mm long.
    One generation can take place in about three weeks. There are generally several generations per year, but there may be more (5–10) in the tropics.
**Distribution.** Almost completely cosmopolitan, but only a few records are from W. Africa; the range extends from Canada and S. Scandinavia (60°N) to S. Africa and New Zealand (CIE map no. A20).
**Control.** Dieldrin, DDT, diazinon, and malathion are all effective when required.

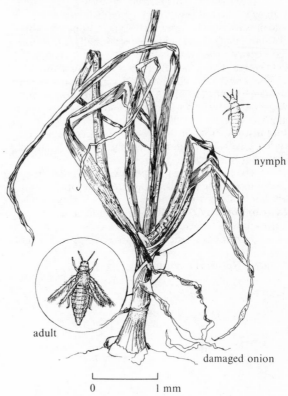

nymph

adult

damaged onion

0        1 mm

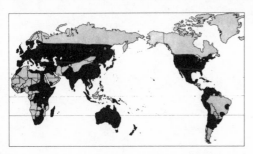

## Hoplandothrips marshalli Karny

**Common name.** Coffee Leaf-rolling Thrips
**Family.** Phlaeothripidae
**Hosts** (main). Coffee
      (alternative). Not known
**Damage.** Infested leaves curl upwards into a roll in
which the thrips live and feed. Infestations are more
common under shade, where occasionally a tree will
have many curled leaves.
**Pest status.** This is usually only a minor pest of
coffee, and mainly confined to Uganda. Many species
of thrips (20 or more) are recorded from coffee but
very few are serious pests; in fact some species are
almost certainly predators of other thrips.
**Life history.** Most of the life-cycle is spent inside the
rolled leaves, although presumably the pupal stages
are spent in the soil.
      The nymphs are pale yellow, and the adults are
dark brown. The details of the life-cycle of this pest
are not well known.
**Distribution.** E. Africa (Kenya, Uganda, and
Tanzania).
      Two other species curl coffee leaves in Tanzania
and Zaïre.
**Control.** Not usually required.

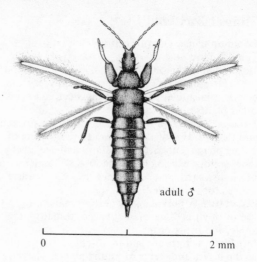

adult ♂

0               2 mm

damaged shoot

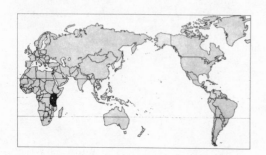

# Order **LEPIDOPTERA**

These are moths and butterflies, characterized by having two pairs of membraneous wings, usually large in size; cross-veins are few in number. The body, wings, and appendages are clothed in broad scales. Mandibles are almost always vestigial or absent; the mouthparts are generally represented by a suctorial proboscis formed by the maxillae. The larvae are caterpillars, with three pairs of thoracic legs, and usually four pairs of prolegs on the abdomen and a terminal pair of claspers.

## Family **Psychidae**

(Bagworm Moths) A small family of about 800 species, mostly (but not all) tropical, with larvae that live inside cases of silk and plant material, and degenerate females which stay inside the pupal bag; they are wingless, and in some species also legless. Several species are minor crop pests.

## Family **Gracillariidae**

(Leaf Miners) A cosmopolitan group of 1000 species; tiny moths with larvae that mine leaves of various tree crops (coffee, etc.). It includes now the species of *Leucoptera* (Coffee Leaf Miners).

## Family **Phyllocnistidae**

(Leaf Miners) The genus *Phyllocnistis* has more than 50 species; the adults are minute and very delicate; the leaf-mining larvae are unusual in being apodous.

## Family **Sesiidae** (= Aegeriidae)

(Clearwings) These are mostly small moths, characterized by the absence of scales from the greater part of both pairs of wings; the antennae are often dilated or knobbed, and the abdomen is terminated by a conspicuous fan-like tuft of scales. The forewing is usually narrow resulting from considerable reduction of the anal area. This small family (about 20 species) is characteristic of the northern hemisphere; they are diurnal fliers and superficially resemble wasps. The larvae feed in the wood of trees and bushes or in the rootstock of plants.

## Family **Yponomeutidae**

A family of about 1000 species, now including the Plutellidae. Only a few species are important crop pests, several are pests of temperate fruit trees. Some of the larvae make slight webbing, and some bore into shoots, buds and fruits.

## Family **Gelechiidae**

This family contains some 400 genera and 4000 species. The forewings are trapezoidal in shape and the wing venation is characteristic. The larvae generally feed among spun leaves or shoots.

## Family **Cossidae**

(Goat Moths; Carpenter Moths) These are large moths, of nocturnal habit, and widely distributed. The body is large and heavy, and the wings rather long and narrow. The larvae are borers in trees, but are sometimes found in herbaceous plants and reeds. The larvae are unable to digest cellulose and obtain their nutriment from the sap; they accordingly consume large quantities of wood; the tunnels they make are very extensive and larval life is long, in temperate regions being up to two years.

## Family **Metarbelidae**

(Wood-borer Moths) A small tropical family found in Africa and Asia whose larvae eat bark of trees and make tunnels into the wood. The adults are nocturnal. The genus *Indarbela* contains many species, some of which are pests on woody plants.

## Family **Limacodidae** (= Cochlidiidae)

A small family, mostly tropical in distribution; the larvae are called 'slug caterpillars' and have short, thick, fleshy bodies, a small retractile head, and tiny thoracic legs. Some bear urticating setae ('hairs') and are called 'stinging caterpillars'. Body segmentation is obscure and prolegs are absent. A number of species are pests of palms, particularly oil palms in S.E. Asia, and many species (in several genera) are pests of tea.

## Family **Tortricidae**

These are small moths, with vestigial maxillary palps, or else they are absent; they are classified by their wing venation. They have wide wings with shortish wing fringes. Most species are temperate rather than tropical in distribution. The eggs are flattened and oval, usually smooth. The larvae live concealed, usually in rolled or joined leaves, or in shoots spun together, or else in stems, fruit, buds, flower heads, seed pods, or roots. The pupa is usually found in the situation where the larvae feed, and it is protruded from the cocoon prior to the emergence of the adult. There are 1500 species described in this family. The caterpillars are often referred to in American literature as 'bud-worms'.

## Family **Pyralidae**

This family is largely tropical in distribution, and relatively scarce elsewhere. Various other families are sometimes included under the Pyralidae, otherwise they collectively constitute the Pyraloidea, a very large group. The larvae usually feed on dry (or decaying) vegetable matter; a number are stem borers in Gramineae, most are leaf-eaters. The crochets on the prolegs are either in a pair of transverse bands, or a more or less complete circle of biordinate crochets. Many of the larvae produce silk, and may roll leaves with the silk.

## Family **Nymphalidae**

(Four-footed Butterflies) This is the dominant family of butterflies and one of the largest of all the Lepidoptera, including about 5000 species. The forelegs of both sexes are reduced in size and usually folded on the thorax and functionally impotent. The antennae are slender and abruptly clubbed; the labial palps long; maxillary palps obsolete. There are several very distinct subfamilies.

## Family **Lycaenidae**

(Blues, Coppers, Hairstreaks) Small to medium-sized butterflies, well represented in most regions. The predominant colour of the upper surface of the wings is metallic blue or coppery-brown; the hindwings often have delicate 'tails'. All the legs are functional. The sexes often exhibit distinct sexual dimorphism in colour. Most of the larvae are onisciform (tapering at each end), and with broad projecting sides concealing the legs. Some of the larvae are carnivorous. Only a very few species are pests.

## Family **Pieridae**

(Whites, Yellows, etc.) A large group of butterflies, with normal legs, usually white or yellow in colour. Several species of *Pieris* are important pests of Cruciferae throughout the world. The family is equally well-represented in the tropics and in temperate regions.

## Family **Papilionidae**

(Swallowtails) A large family of tropical butterflies, but some are temperate; about 600 species are known. Most have the hindwings drawn out into conspicuous tails. Only a very few species are pests.

Some species are strikingly polymorphic; the pupae are variable in form.

## Family **Hesperiidae**

(Skippers) A very large family, widely distributed, somewhat intermediate between the butterflies and the moths. They are called Skippers because of their darting flight. The body is stout; the antennae have a gradually expanding club, often ending in a hook; the wings are proportionally smaller than in most butterflies, and are often held partly open at rest. The larvae are often concealed in the host foliage, making webs or joining leaves together with silken threads.

## Family **Drepanidae**

(Hook Tips) A small family most represented in the Indo-Malaysian region. The apex of the forewing is generally falcate. The larvae are rather slender, and without claspers on the ultimate segment, and the anal extremity is prolonged into a slender projection which is raised when at rest; the other body segments are often humped.

## Family **Geometridae**

(Loopers, or Geometers) A very large family with more than 12 000 species. The adults are typically slender-bodied with relatively large wings; not strong fliers; the wings are often held horizontally at rest. In some genera the females are apterous. The caterpillars are elongate and slender and only have one pair of prolegs (on segment six) in addition to the terminal claspers; they move in a looping manner, but at rest are cryptic and resemble a twig. Many species of larvae are defoliators of trees.

## Family **Epiplemidae**

A small group of about 550 species; inconspicuous in appearance; they commonly rest during the day with the forewings rolled up in a peculiar manner, while the hindwings are held to the sides of the body. They occur on all continents, but are best developed in Papua and thereabouts.

## Family **Saturniidae** (= Attacidae)

(Emperor Moths; Silkworm Moths) The largest and most splendid moths belong to this essentially tropical family. The larvae spin silk to make the cocoon, and are the source of some types of 'wild silk'. A few species are minor pests of tree crops in warmer climates.

## Family **Sphingidae**

(Hawk Moths) An important family of moderate-sized to very large moths; including at least 1000 species; essentially a tropical group, but some species are temperate, and some cosmopolitan. The adults are characterized by the elongate forewings with their very oblique outer margin; the proboscis is typically very long (up to 25 cm). The larvae are smooth, and the eighth abdominal segment always bears an obliquely projecting dorsal horn — in the USA the larvae are called 'hornworms'.

## Family **Noctuidae** (= Agrotidae)

The largest family of Lepidoptera, with more than 6000 species described, all bearing very similar characteristics. They are mostly medium-sized and rather dull-coloured moths; nocturnal in habit, though a few species are crepuscular; they are very attracted to light at night. There are many pest species; some of the larvae are leaf eaters, some cutworms, some armyworms, stem borers, and fruit borers, and some adults are fruit piercers. The crochets on the prolegs are all of one size, arranged in a semi-circle.

# Clania spp. etc.

**Common name.** Bagworms

**Family.** Psychidae

**Hosts.** Many bagworms show polyphagous feeding habits, and have been recorded from many different host plants. *Clania* spp. are being used as an example typifying the whole family, for field recognition purposes.

**Damage.** All bagworms do the same type of damage, that is they defoliate the host by eating the leaves, and the plant generally is festooned with the hanging cases.

**Pest status.** Bagworms generally are fairly minor pests on many different crops, with the exception of some on palms (see next page) which are very serious in Malaysia. But they do occur very regularly throughout the tropical regions of the world.

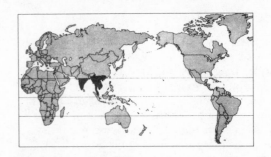

**Life history.** The male moth is winged, usually has a characteristically elongate abdomen, and flies at night to search for females. *Clania* males have a wingspan of 28–35 mm, but several other bagworms of economic importance are quite a lot smaller, with males having a wingspan of only 15 mm. The female moth is wingless and legless and never leaves the bag; after

small bagworm windowing orchid leaf

palm leaf eaten by larvae

284

mating she becomes virtually just a sac of eggs. The eggs, which number from about 200 to 3000 according to species, hatch within the bag and the young caterpillars crawl out on to the foliage of the host plant. In some species the tiny larvae spin silken threads which enable them to be lifted by air currents and carried by the wind. This is the dispersive stage. Other larvae spread from tree to tree within plantations where foliage from adjacent trees touch. The larvae soon build a tiny case out of leaf fragments in which the body is protected. Small larvae usually scrape the epidermis and make 'windows' in the leaf lamina, but as they become larger they eat the whole blade of the leaf, either from the lamina margin or make holes. As the larvae grow the case is enlarged. When feeding, the thorax and head protrude from the case, and attachment is effected by the thoracic legs holding on to the leaf. For pupation the bag is firmly attached to the leaf or twig by silken threads and dangles from the foliage. If the emerging adult is male it leaves the case by the ventral end and the pupal exuvium protrudes from the case. If it is a female then she just remains inside the case and secretes sex pheromones to guide the searching male moths to her presence.

In some large species the life-cycle takes 3—4 months, but for smaller ones generally about a month. There may be several generations per year in the warmer parts of the tropics.

**Distribution.** Bagworms are found throughout the tropical and subtropical parts of the world, but *Clania* species are only recorded from India and S.E. Asia.

**Control.** Natural parasitism is generally high amongst bagworms and so only occasionally is chemical control required. Trichlorphon has generally been effective as a foliar spray.

*Clania* larvae on *Thuja*

0           2 cm

## Mahasena corbetti Tams

**Common name.** Coconut Case Caterpillar
**Family.** Psychidae
**Hosts** (main). Coconut and oil palm.
　　(alternative). *Citrus*, kapok, derris, *Aleurites*, *Cupressus*, and others.
**Damage.** In severe infestations there may be total defoliation, not only of whole palms but occasionally of entire plantations. The leaf lamina is eaten away so that all that remains of the frond is the midrib and lateral veins.
**Pest status.** In Malaysia this has been a serious pest of oil palm for several years (B.J. Wood, 1968), but it does sporadic damage to coconut palms throughout S.E. Asia.
**Life history.** The young caterpillars scrape the leaf epidermis and make small windows, but as they grow larger they eat holes in the leaf, and the older cater-

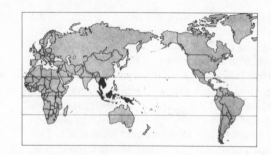

pillars eat large areas of leaf lamina and use large pieces of leaf for their cases.

　　This species lays a large number of eggs, often more than 3000 per female, which accounts for the enormous localized populations that may develop. In Sabah in 1966 some palms had 300–500 larvae on a single frond, and damage was spectacular.

bagworm

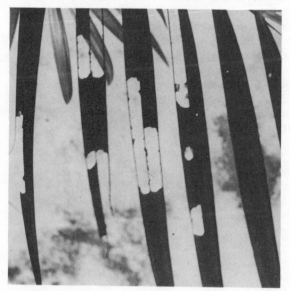

damage

286

Larvae grow to about 35 mm in body length before pupation; and the life-cycle takes about 3–4 months.

Adult males measure 25–30 mm in wingspan, and are normally winged brown-coloured moths. As with other Psychidae the female is degenerate and consists of little more than a large bag of eggs within her pupal bag.

The young caterpillars use long silken strands to aid their dispersal; they can be windborne for short distances from tree to tree; most dispersal is actually thought to be by the caterpillars walking along the ground from tree to tree.

**Distribution.** From Malaysia and Thailand, through S.E. Asia, to Papua New Guinea.

**Control.** Parasitism of the larvae by parasitic Hymenoptera and Tachinidae can be quite high, but this pest does have enormous reproductive potential and so outbreaks invariably occur. Trichlorphon and DDT have both been used successfully in the past against this pest.

## Leucoptera spp.

*caffeina* Wash.
*coma* Ghesq.
*meyricki* Ghesq.
*coffeella* (Guér.)

**Common name.** Coffee Leaf Miners
**Family.** Gracillariidae (Lyonetidae)
**Hosts** (main). Coffee, *arabica* mostly but sometimes *robusta* may be equally attacked.

(alternative). *Leucoptera* can breed on certain wild rubiaceous shrubs (*Pavetta* spp., and a *Tricalysia* sp.) in addition to other species of *Coffea*.

**Damage.** Infested plants have brown irregular blotches on the upper surface of leaves; the blotch mine is inhabited by a number of small white caterpillars. Mined leaves are usually shed prematurely.

**Pest status.** A major pest of coffee in Africa and S. America. In E. Africa, where both species occur, *L. meyricki* is dominant in unshaded coffee and *L. caffeina* in shaded coffee. All cultivated species of coffee are attacked.

**Life history.** Eggs are laid on the upper surface of the leaf; roughly oval but with a broad base on the leaf surface; they are silver when laid, turning brown just prior to hatching. Eggs of *L. meyricki* are scattered in small groups; those of *L. caffeina* are laid touching each other in a neat row along a main vein. Hatching takes place after 1–2 weeks, according to temperature.

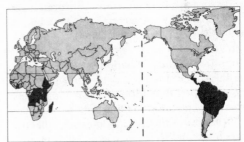

The larva is a small, white caterpillar; it bores through the floor of the egg straight into the leaf and mines just below the upper epidermis. The mines of each *L. meyricki* caterpillar are originally separate but after a few days they join up to form one large blotch mine. When a caterpillar is fully grown it cuts a semi-circular slit in the dead epidermis, comes out of the mine and lowers itself on a silken thread; it is then about 8 mm long. The total larval period is 17–35 days in the field, according to temperature.

The mature larva settles on a dead leaf on the ground or the underside of a living leaf and spins a H-shaped white cocoon about 7 mm long. In this it pupates emerging as the adult moth 1–2 weeks later.

The adults are tiny white moths about 3 mm long, and they live in the field for about two weeks. During this period the female lays about 75 eggs, mostly during the first few days after emergence.

The life-cycle takes some 4–6 weeks to complete, and in most parts breeding is continuous, there being as many as 8–9 generations per year.

**Distribution.** *L. meyricki* is the commonest African species; and this together with *L. coma* and *L. caffeina* are found only in Africa, being recorded from Ivory Coast, Angola, Zaïre, E. Africa, Ethiopia, and Madagascar (CIE map no. A316).

*L coffeella* is confined to S. and C. America, the W. Indies and Madagascar (CIE map no. A315).

**Control.** Out of a wide range of insecticides which have given varying levels of control the two most consistently successful are probably fenitrothion and fenthion as foliar sprays. A second spray may be required 2–3 weeks after the first.

Spraying should as far as possible be done when a low proportion of the population is in the cocoon stage, for these individuals will not be killed by the insecticide. The best time for spraying is about one week after the period when moths were most numerous, for then most of the insect population will consist of eggs or young larvae and a good kill is more likely to be achieved.

**Further reading**

Le Pelley, R.H. (1968). *Pests of Coffee*, pp. 217–39. Longmans: London.

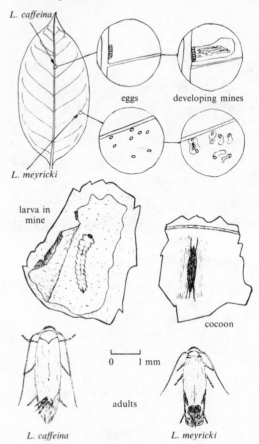

L. caffeina

eggs    developing mines

L. meyricki

larva in mine

cocoon

0    1 mm

adults

L. caffeina    L. meyricki

# Phyllocnistis citrella (Stnt.)

**Common name.** Citrus Leaf Miner
**Family.** Phyllocnistidae (formerly in Gracillariidae)
**Hosts** (main). Species of *Citrus*
(alternative). Other members of the family Rutaceae.
**Damage.** The feeding larvae make broad serpentine galleries (mines) in the leaves, leaving a distinctive dark line of faecal pellets along the centre of the tunnel. In young leaves the lamina folds over and twists, with a high degree of distortion; badly damaged leaves dry out and are clearly of little use photosynthetically.
**Pest status.** In young plants the damage can be quite serious. On older plants infestation levels may be very high occasionally, so that control measures may be required.
**Life history.** Eggs are laid singly, by the midrib on the underneath of the leaf; they are flattened, oval and white. Incubation takes 3–5 days. On hatching the young larva penetrates the epidermis and commences burrowing, eventually making a long serpentine, convoluted mine which is conspicuously silvery in colour owing to the air trapped under the epidermis. As with all leaf-mining caterpillars, the larva leaves its faecal pellets in a line down the centre of the tunnel, for it lies in the tunnel on its ventral surface. (In most leaf-mining Diptera the larva lies on its side in the tunnel and so the faecal pellets are deposited along the edges of the tunnel and are not at all conspicuous.) Larval development takes usually 16–18 days, and the mature larvae is about 3.5 mm long, and yellowish-white in colour. Pupation takes place at the edge of the leaf and the lamina margin is turned over to protect the pupa underneath.

The adult is a tiny moth 2–3 mm in body length with a wingspan of 5–8 mm, in colour greyish-white, with black eyes and four black stripes across each forewing; the hindwings are feathery.

Total life-cycle takes about three weeks, and usually there are five or more generations per year.
**Distribution.** Widely distributed throughout S.E. Asia, up to China, Korea and Japan, and across the Philippines and Indonesia, to Papua New Guinea and the northern tip of Australia. In Africa it is only recorded from the Sudan (CIE map no. A274).
**Control.** When required, it is recommended that either diazinon or phosphamidon (e.c.) be applied as a foliar spray, at weekly intervals on young plants or fortnightly on mature *Citrus*.

larval mine in leaf of mandarin orange

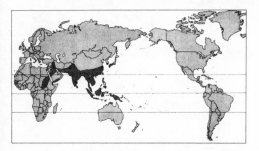

## Synanthedon dasysceles Bradley

**Common name.** Sweet Potato Clearwing
**Family.** Sesiidae
**Hosts** (main). Sweet Potato
    (alternative). Not known
**Damage.** The larvae burrow into the vines, and sometimes also into the tubers. It can be a pest of sweet potato tubers in stores. The vine base is characteristically swollen and is traversed by feeding galleries.

**Pest status.** The three closely related species of *Synanthedon* are regularly found in sweet potato in E. Africa but they are not really serious pests.

**Life history.** Eggs are laid in clusters on the stems and leaf stalks; hatching takes a few days.

The white caterpillars bore into the stems (vines) where they tunnel down to the stem base which gradually swells.

Pupation takes place in the tunnels in the vines.

The adult moths are grey-brown, without the reddish colour that the other two species possess, and with the abdomen blackish-brown with a pale central line. The male wingspan is 20–22 mm; the female 17–25 mm; the wings are hyaline, hence the common name of 'clearwing'. This species is characterized by having extensive rough scaling on the hindlegs (tibia and tarsus) which is longer in the male, and the white markings on the basal part of the hindlegs of the female.

**Distribution.** E. Africa.

Two other closely related species are also found on sweet potato in E. Africa, these being *S. leptosceles* and *S. erythromma*.

**Control.** Not really required.

**Further reading**

Bradley, J.D. (1968). Two new species of clearwing moths (Lepidoptera; Sesiidae) associated with sweet potato (*Ipomoea batatas*) in East Africa. *Bull. ent. Res.* **58**, 47–53.

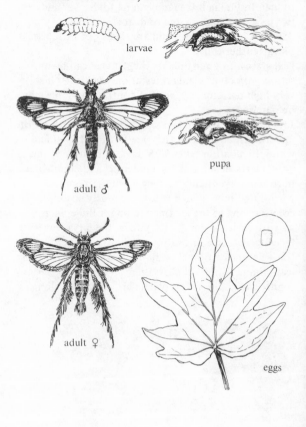

larvae

pupa

adult ♂

adult ♀

eggs

# Plutella xylostella (L.)

(= *P. maculipennis* (Curt.))

**Common name.** Diamond-back Moth
**Family.** Yponomeutidae
**Hosts** (main). Brassicae of all species.

(alternative). A wide range of wild and cultivated Crucifereae.
**Damage.** Newly hatched caterpillars crawl to the
underside of the leaf, penetrate the epidermis and
during the first instar they mine in the leaf tissue. The
three later instars feed on the underside of the leaf
making 'windows' or holes right through it.
**Pest status.** A very common and widespread pest of
cabbage, turnip, etc. Severe attacks sometimes occur,
especially in hot dry weather.
**Life history.** The tiny whitish eggs are stuck to the
upper surface of a leaf either singly or in very small
groups. Incubation takes 3−8 days.

The caterpillar is pale green, widest in the
middle of its body, and is about 12 mm long when
fully grown. Caterpillars wriggle violently if disturbed
and often drop off the leaf, remaining suspended
from it by a silk thread. The total larval period varies
from 14−28 days.

Pupation takes place inside a gauze-like silken
cocoon about 9 mm long, which is stuck to the underside of a leaf; the pupa is greenish, and the pupal
stage lasts from 5−10 days.

The adult is a small, grey moth with a wingspan
of about 15 mm. There are three pale triangular
marks along the hind margin of each forewing, and
when the wings are closed these marks form a
diamond pattern, which gives the moth its common
name. After a pre-oviposition period of 2−3 days the
female moth may live a further 14 days and lay 50−
150 eggs.
**Distribution.** Completely cosmopolitan in distribution, extending northwards up to the Arctic Circle
in Europe (CIE map no. A32).
**Control.** The recommended insecticides are DDT,
carbaryl, malathion, BTH, and pyrethrum, all as
sprays at either low- or high-volume rates.

In some parts of the world this species has
developed resistance to most of the usual insecticides.

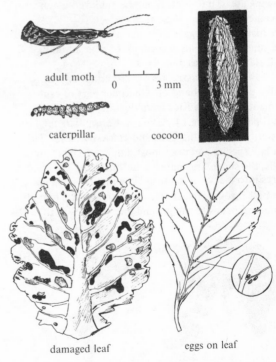

adult moth

0    3 mm

caterpillar            cocoon

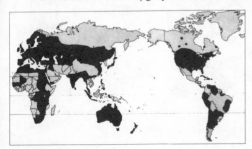

damaged leaf            eggs on leaf

## Pectinophora gossypiella (Saunders)

(= *Platyedra gossypiella* (Saund.))

**Common name.** Pink Bollworm
**Family.** Gelechiidae
**Hosts** (main). Cotton
    (alternative). *Hibiscus* and other Malvaceae, but only cotton can support a large infestation.
**Damage.** The entry hole of the caterpillar into a large green boll is almost invisible. If the boll is opened, however, the red and white caterpillar can be found, especially inside the developing seeds. Damaged bolls fail to open completely and often have secondary rots where the caterpillar has been feeding.
**Pest status.** Potentially very serious in all cotton-growing areas, but control can be achieved by enforcement of a close season.
**Life history.** The egg is oval, about 0.55 mm long; laid singly or in small groups, on or near a bud or boll. Hatching takes 4–6 days.
    Full-grown larvae are 10–12 mm long, white with a double red band on the upper part of each segment. There are four larval instars, the total larval period lasting 14–23 days. Caterpillars feed on flower buds or young bolls and cause shedding. This is a late species and does most damage to large green bolls; the small larva enters a boll near the base and bores into a developing seed. The mature larva leaves the boll through a neat circular hole about 2 mm in diameter, drops to the ground and pupates in the litter. Pupation takes place inside a loose cocoon. The pupa is brown, 7–10 mm long and about 2.5 mm broad. Pupation takes 12–14 days.

    The adult is a small brown, inconspicuous moth of wingspan 15–20 mm; nocturnal in habits. After a pre-oviposition period of about four days, it may live a further ten days and lay about 300 eggs.
**Distribution.** Completely pantropical in distribution, including the subtropical regions (CIE map no. A13).
**Control.** In some areas a mixture of DDT and endrin is recommended, or carbaryl, trichlorphon and azinphos-methyl, and sprays are applied at weekly intervals.

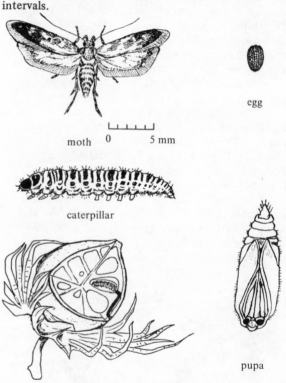

moth    0    5 mm

egg

caterpillar

pupa

section of infested boll

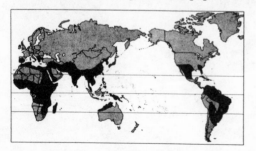

# Phthorimaea operculella (Zeller)

(= *Gnorimoschema operculella* (Zeller))

**Common name.** Potato Tuber Moth
**Family.** Gelechiidae
**Hosts** (main). Potato
      (alternative). Tobacco, tomato, eggplant, and
other Solanaceae, and *Beta vulgaris*.
**Damage.** The leaves have silver blotches caused by the
young larvae mining in the leaves. Leaf veins, petioles
and stems are tunnelled, followed by wilting of the
plants. Eventually the tubers are bored by the larger
caterpillars, and they often become infected with
fungi or bacteria.
**Pest status.** An important pest of potato in warmer
countries. Infestations arising initially in the field
and continuing during storage of the tubers. There is
a serious risk of transportation from country to
country through infested tubers.
**Life history.** The egg is minute and oval, 0.5 × 0.4 mm,
and yellow. Eggs are laid singly on the underside of
the leaf, or in tubers (usually in storage) near the eye
or on a sprout. Each female lays about 150−250 eggs.
The eggs on the leaves hatch in 3−15 days and the
first instar larvae bore into the leaf, where they make
mines. The caterpillars are pale greenish. They gradually eat their way into the leaf veins and into the
petioles, then gradually down the stem and sometimes into the tuber. The full-grown caterpillar is
9−11 mm long. The larval period lasts for 9−33 days.

Pupation takes place in a cocoon in the surface
litter or in the tuber, and takes 6−26 days.
    The adult is a small moth with narrow fringed
wings; the forewings are grey-brown with dark spots,
and the hindwings are dirty white. The wingspan is
about 15 mm. The moths are very short-lived.
    One generation takes some 3−4 weeks, and
there can be up to 12 generations per year, but
development is very dependent upon temperature.
**Distribution.** Almost completely cosmopolitan in
distribution, but with limited records from Asia and
none from W. Africa (CIE map no. A10).
**Control.** Effective insecticides are DDT, dicrotophos,
dimethoate, and parathion, all as sprays. As a preventative measure sprays should be applied every 14
days after the first mines are found in the leaves.

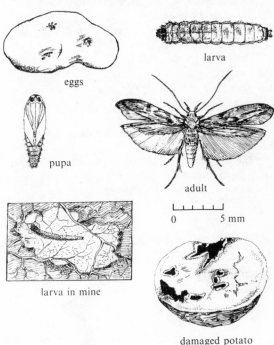

eggs

larva

pupa

adult

0      5 mm

larva in mine

damaged potato

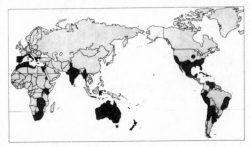

293

## Sitotroga cerealella (Ol.)

**Common name.** Angoumois Grain Moth
**Family.** Gelechiidae
**Hosts** (main). Maize and wheat, both in the field and in grain stores.

(alternative). Sorghum and other stored grains, and dried fruit.
**Damage.** Infested grains with mature larvae or pupae can be recognized by the presence of a very small window in the grain. On emergence the adult pushes its way through this small circular window and the 'trap door' is left hinged to the grain, which is characteristic of this pest.
**Pest status.** A serious pest in many parts of the world. The infestation by this moth starts in the field and may reach serious levels, before being translocated to the grain stores.
**Life history.** The eggs are ovoid and pinkish and are laid on the surface of the grain.

The larvae are elongated, dirty white, about 8 mm long. The body is covered with fine setae. The larvae bore their way into the grain and feed there; before pupation they form a channel to the surface, leaving the seed coat intact.

The pupa, which is dark brown, is enclosed in a delicate cocoon.

The adult is a small, straw-coloured moth about 7 mm long (with wings folded); the wings are 15 mm across when open and the hindwings have a long fringe. One female can lay about 100 eggs. The adults are short-lived.

The life-cycle from egg to adult moth takes about five weeks at 30°C.
**Distribution.** Cosmopolitan in the warmer parts of the world.
**Control.** Dust the maize cobs with lindane dust, and also the surface of the wheat bags; as an alternative malathion dust can be used.

Fumigation with materials such as methyl bromide should only be carried out by approved operators.

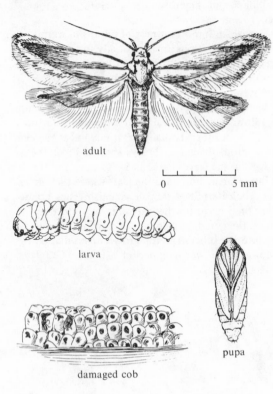

adult

0     5 mm

larva

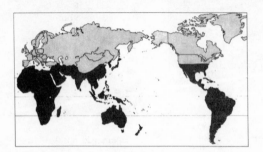

damaged cob

pupa

## Eulophonotus myrmeleon Feldr.

(= *Engyophlebus obesus* Karsh)

**Common name.** Cocoa Stem Borer
**Family.** Cossidae
**Hosts** (main). Cocoa
      (alternative). Coffee, cola, *Populus*, and
*Combretum* spp.
**Damage.** Extensive tunnels are bored in the branches
and main trunk by the larvae; sometimes even roots
are bored. Trees less than one year old are rarely
attacked. The upper parts of the tree may die.
**Pest status.** They are usually only found in small
numbers, and often are more common in plantations
frequently treated with insecticides. In such plantations infestation rates of more than 5% may occur.
However, this is not generally a serious pest in most
parts.
**Life history.** The egg period lasts about 12–13 days,
and there may be 1500 eggs laid by one female. The
eggs are laid in crevices in the bark.
      The larval period is usually at least 12 weeks,
and the pupal period another 20 days.
      The adult male is 20–28 mm across the wings,
which are almost clear and devoid of scales. The
female is 45–50 mm in wingspan, with sooty-brown
wings; the forewing having many small clear areas
without scales.
**Distribution.** W. and E. Africa only.
**Control.** Control is not often required, and is inciden-

tally difficult to achieve. The insecticides which are
sometimes used are DDT, dieldrin, endrin, and phosphamidon, and they are sprayed on the bark of the
trees at the times when egg-laying is expected.

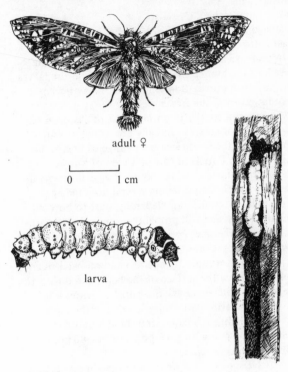

adult ♀

0      1 cm

larva

larva in stem

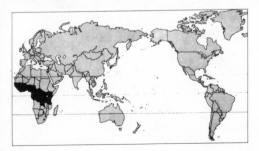

# Xyleutes capensis (Wlk.)

**Common name.** Castor Stem Borer
**Family.** Cossidae
**Hosts** (main). Castor
     (alternative). *Cassia* sp.
**Damage.** The larvae bore in branches or trunks of the castor tree causing the death of the branch, which often breaks at the level of the tunnel. Prior to the death of the branch the foliage withers and the leaves die and turn brown. Sometimes there may be sap exudation from the frass holes.
**Pest status.** A locally important pest of castor in E. Africa, but cossids generally are widely occurring throughout the world in a wide range of host trees.
**Life history.** Details of this pest are not known, but in general female Cossidae lay their eggs in cracks in the bark; each female laying several hundred eggs.

    The caterpillars immediately start to bore into the branches, and will probably spend many months as larvae; sometimes cossid larvae live for one or two years before pupation.

    Pupation typically occurs in a hollowed-out cell just below the surface of the bark, and during the process of emergence the old pupal exuvium is left with the anterior part projecting out of the bark.

    The adult is a large, stout-bodied moth about 35 mm long and with a wingspan of 70—80 mm.
**Distribution.** E. Africa.

    Six species of *Xyleutes* bore the branches and

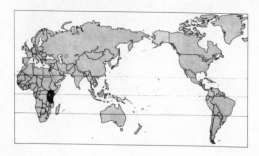

trunks of coffee in W. Africa, India, and the W. Indies.
**Control.** Control measures are seldom required on the whole, and are even less frequently successful, for once the larva is inside the branch it is quite safe from attack.

adult ♂

0       2 cm

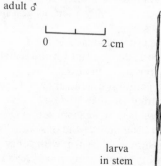
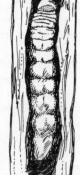

larva
in stem

## Zeuzera coffeae (Nietn.)

**Common name.** Red Coffee Borer (Red Branch Borer)
**Family.** Cossidae
**Hosts** (main). Coffee
(alternative). *Citrus*, cocoa, tea, kapok, forest trees such as teak, mahogany, sandalwood, various ornamental trees and shrubs, and wild trees.
**Damage.** The larva tunnels along branches (down the centre) and in trunks of both woody shrubs and trees, usually killing the branch distally.
**Pest status.** This polyphagous pest is widespread in the Orient, and is quite a serious pest on many different crop plants from time to time.
**Life history.** Eggs are usually laid single in crevices in the bark, and the young larvae, after hatching, bore straight into the branches. The larvae are stout-

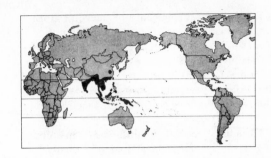

bodied and dark reddish in colour with a black head, prothorax and anal shield. They bore a cylindrical tunnel along the branch and periodically make bore holes to the surface out of which reddish brown frass (mostly faeces) is extruded. Pupation takes place in a cocoon made of silk and wood particles just under-

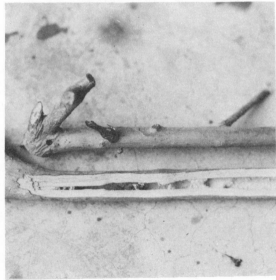

larval damage in woody stem

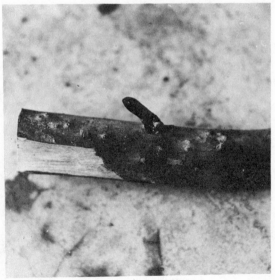

exuvium of pupa after emergence

297

neath the bark. After emergence of the adult moth the old pupal exuvium is left protruding from the emergence hole (this is characteristic of all Cossidae).

In cooler regions *Zeuzera* larvae require 2–3 years for development, but in the tropics they will develop within a year, and in some places there might even be two generations per year, but this is uncertain.

The adults are striking moths, white in colour with black spots, though the abdomen is usually grey. The female is distinctly larger with a wingspan of about 5 cm and the male about 4 cm. The adults fly to lights at night and are easily caught in light traps.

**Distribution**. From India through to S. China and throughout most of S.E. Asia, including the Philippines, West Irian and Papua New Guinea (CIE map no. A313).

**Control**. Caterpillars can be killed in their tunnels by pushing a springy wire down the hole, or by introducing an insecticide into the tunnel and then blocking the holes. Paradichlorobenzene, BHC and carbon disulphide have proved successful in this manner, with the hole being sealed with clay or putty. Chemical control by foliar sprays is generally not successful against this pest. In Sabah this was a serious pest on cocoa and most insecticidal treatments were not successful. Conway (1972*a*) reported that levels of larval parasitism were very high, and his final recommendation was to avoid the use of any contact insecticides whatsoever; the natural parasites subsequently controlled the pest population quite adequately.

It is generally recommended that infested branches be cut off and removed, but if there is a locally high rate of larval parasitism the cut branches should be left within the plantation to allow parasite survival.

0        1 cm     adult

## Indarbela spp.

**Common name.** Wood-borer Moths
**Family.** Metarbelidae
**Hosts.** A polyphagous pest found on many different species of trees and woody shrubs; recorded from *Citrus*, guava, litchi, loquat, mango, mulberry, *Ficus*, jack-fruit, *Acacia*, and many other plants.
**Damage.** The larvae bore into the trunk or branches, usually at forks or angles, to a depth of some 15–25 cm. This deep hole is the refuge of the larva during the day (and later for pupation) and at night it emerges and eats the bark of the tree in the immediate vicinity of the hole; the feeding area is covered with a web of silk and frass. Small trees are easily ring-barked by this pest and die. If enough bark is eaten away, even large trees are disturbed by the interrupted sap flow and may fail to flush.

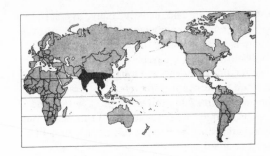

**Pest status.** This family of wood-boring moths is confined to the tropics, where several species are recorded, with different distributions on a wide range of host trees and shrubs. In most situations they are really minor pests, but are quite frequently encountered.

infested *Acacia* trunk

*Gordonia* stem ring-barked and killed by feeding caterpillars

299

**Life history.** Eggs are laid, usually singly, in cracks in the tree bark, and the young caterpillar bores into the tree. There is only one larva to each tunnel, but in heavy infestations there may be 10–30 larvae per tree. The feeding larva eats the bark and enough of the sapwood underneath to interfere with sap flow in the phloem system. Pupation takes place within the deep tunnel, and after emergence of the adult the old pupal exuvium is left projecting out of the tunnel (as with the Cossidae).

The adult moth is a creamy white colour, with brown markings on the forewings, about 2 cm in body length and 4 cm in wingspan; the body is usually rather elongate.

There is usually only one generation per year; in S. China adults emerge in the spring.

**Distribution.** At present recorded information comes from India, mainland S.E. Asia and S. China.

**Control.** Poking a wire down the tunnel will kill individual larvae in situ, or injection of DDT, BHC, endrin, carbon disulphide, paradichlorobenzene or kerosene into the tunnel, and sealing it with clay or putty, will also kill the larvae.

adult

0                 1 cm

## Latoia lepida (Cram.)

(= *Parasa lepida* Cram.)

**Common name.** Blue-striped Nettle-grub
**Family.** Limacodidae (= Cochlidiidae)
**Hosts.** No clear-cut pattern of host specificity; recorded from coconut, *Citrus,* cocoa, coffee, banana, rice, tea, mango, castor, pomegranate, rose, ornamentals.
**Damage.** The larvae defoliate the host plants, as well as having urticating body spines.
**Pest status.** Usually a minor pest on many different crops, but sporadically serious, especially in India.
**Life history.** Eggs are laid in batches of 20–30 underneath leaves, and are shiny and rather flat. Hatching occurs after 6–7 days, and larval development takes about 40 days, at which time the caterpillar measures some 25 mm. The caterpillar bears rows of fleshy protuberances each carrying a series of sharp spines; ventrally it is quite flat and fleshy and the legs are quite indistinct. The body is green in colour with blue stripes running longitudinally. The spines on the scoli are sharp and have urticating properties which makes them a nuisance when encountered in the crop by field workers. Pupation takes place in a round hard cocoon usually stuck on to the bark of the tree, and takes about three weeks; the whole life-cycle takes about three weeks in India.

    The adult moth is stout-bodied, about 15 mm in length and with a wingspan of 30–35 mm. The forewings are coloured green, with brown basal and distal portions, and hindwings pale brown.
**Distribution.** Most abundant in India and throughout S.E. Asia to S. China, Philippines, West Irian and Papua New Guinea, but also recorded from W. and S.E. Africa (CIE map no. A363).
**Control.** Spraying with BHC, and parathion have proved to be successful.

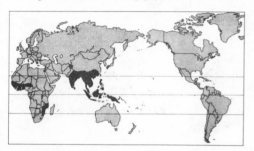

larva

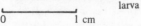

0           1 cm

adult

## Niphadolepis alianta Karsch

**Common name.** Jelly Grub
**Family.** Limacodidae (= Cochlidiidae)
**Hosts** (main). Coffee (*arabica*).

(alternative). Castor and tea are recorded as alternative hosts and these caterpillars can probably feed on a wide range of other plants.

**Damage.** The young caterpillars feed on the undersides of leaves; the feeding areas are small and circular and everything except the upper epidermis is eaten. Later this dies and falls out leaving small 'shot-holes' in the leaf. When about half-grown the larvae feed at the leaf edge, eating right through it and leaving a jagged edge. Fully hardened leaves are preferred as food.

**Pest status.** A fairly common, but unimportant, pest in eastern Africa on coffee and tea.

**Life history.** Eggs are white and scale-like; laid singly on either side of the leaf. On hatching, the shell collapses and remains as a transparent, flat, white speck on the leaf.

The larva is a slug-like caterpillar, whitish with a tinge of green, and is fully grown after 6–8 weeks when it is about 14 mm long.

The mature caterpillar spins a cocoon of white silk usually between two leaves or in the fold of a single leaf. The cocoon is 14–20 mm long, and the pupal period is from three weeks to much longer.

The adult moth is very rarely seen in the field. It is golden brown in colour with darker brown patches on the forewings. The wingspan is about 25 mm.

**Distribution.** Only found in Africa; in Malawi, Tanzania, and Kenya.

**Control.** Chemical control can be achieved with the use of either fenitrothion or fenthion sprays; the amount of spray being used is dependent upon the extent of the leaf cover.

**Further reading**

Le Pelley, R.H. (1968). *Pests of Coffee,* pp. 207–8. Longmans: London.

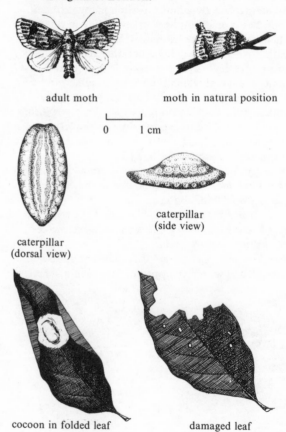

adult moth

moth in natural position

0    1 cm

caterpillar
(side view)

caterpillar
(dorsal view)

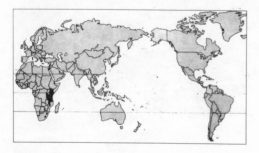

cocoon in folded leaf

damaged leaf

## Parasa vivida (Wlk.)

(= *Latoia vivida* (Wlk.))

**Common name.** Stinging Caterpillar
**Family.** Limacodidae (= Cochlidiidae)
**Hosts** (main). Coffee (*arabica*).

(alternative). Various Rubiaceae, and cocoa, groundnut, sweet potato, castor, tea and cotton.
**Damage.** The young caterpillars feed together on the underside of a leaf. They make small irregular pits, eating everything except the upper epidermis. The older caterpillars feed at the leaf edge, eating right through it and leaving a jagged edge.
**Pest status.** Usually only a minor pest of *arabica* coffee, but occasional serious outbreaks occur.
**Life history.** The eggs are greenish-yellow, and are laid in small batches, overlapping like tiles, on the underside of leaves. They hatch in about ten days.

The larva is an attractively coloured caterpillar, mainly white when young but green when older. It is covered with finger-like projections which bear stinging (urticating) hairs. The young caterpillars feed together on the underside of a leaf. The older caterpillar is solitary and feeds at the edge of the leaf. The larval period lasts about 40 days.

Pupation takes place in an oval, white cocoon which is about 14 mm long and made of tough silk. The cocoon is stuck to the branch of a tree. After spinning the cocoon, the pre-pupa often goes into a resting stage and does not actually pupate for many months. Combined pre-pupal and pupal stages last for as long as 134 days in Kenya.

The adult moth has green forewings edged with brown and yellow hindwings. The wingspan is 30 mm.
**Distribution.** This pest is confined to Africa; occurring in Ivory Coast, E. Africa, Malawi, Sierra Leone, Zimbabwe, Zaïre, Nigeria and Mozambique.
**Control.** Chemical control can be achieved with fenitrothion or fenthion as aqueous sprays, the quantity of spray depending upon the amount of leaf cover.

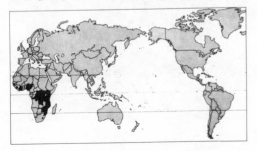

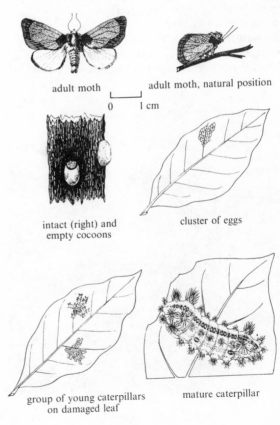

adult moth       adult moth, natural position

0       1 cm

intact (right) and
empty cocoons

cluster of eggs

group of young caterpillars
on damaged leaf

mature caterpillar

303

## Thosea sinensis (L.)

**Common name.** Slug Caterpillar
**Family.** Limacodidae (= Cochlidiidae)
**Hosts** (main). Tea and members of the Theaceae.

(alternative). *Thosea* spp. (but not including *sinensis*) are important on oil palm in Malaysia. Other species feed on Leguminosae, sorghum and millet in India.

**Damage.** As with all Limacodidae this caterpillar eats leaves and occasionally will defoliate a bush or tree, but is also a pest because it bears urticating setae on the body which are irritating to the field workers, particularly tea pickers.

**Pest status.** Usually only a minor pest, but *Thosea* occurs in S.E. Asia as at least six to eight species, some of which are of some importance as pests on tea and others on oil palm.

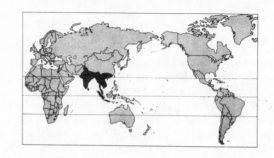

**Life history.** Details are not known for this species, but they are presumably similar to those of other members of this family.

The caterpillar is short, broad and rather flattened and bears the scoli along the lateral margins only; when mature, it is about 20−24 mm in length,

larva

0       1 cm

pupa

green in colour but with a bright yellow dorsal stripe. Pupation takes place in a subspherical, hard brown cocoon stuck on to the bark of the plant or on to a leaf surface (as illustrated). The adult of this species is a plain dark brown in colour, and with a wingspan of 3–4 cm.

**Distribution.** This species is recorded from India, mainland S.E. Asia, China, and parts of Indonesia. Within this area there are at least 6–8 other species of *Thosea* recorded from tea and oil palm, and other hosts.

**Control.** It was pointed out by Wood (1968) that on oil palm in Malaysia the outbreaks of stinging/slug caterpillars always occurred in areas with a long history of contact insecticide use. The assumption is that the pesticides upset the natural population balance by killing predators and parasites rather than the pests, which at that time only occurred in very small numbers and were not the primary targets for the contact insecticides at all.

Effective insecticides are endrin, parathion and BHC, as foliar sprays.

adult

## Cryptophlebia leucotreta (Meyrick)

(= *Argyroploce leucotreta* Meyrick)

**Common name**. False Codling Moth
**Family**. Tortricidae
**Hosts** (main). Cotton and *Citrus* spp.

(alternative). Many wild and cultivated fruits are attacked including oranges, guava, wild figs, and sodom apples; maize is also an important host.
**Damage**. Caterpillars mine in the boll wall or in the developing seeds.
**Pest status**. A serious pest of *Citrus* in Africa, especially on navel oranges, and usually a minor, although occasionally serious, pest of cotton.
**Life history**. The eggs are flat and oval in outline, whitish, and about 0.9 mm long. They are usually laid singly on large green bolls or on the surface of the fruit, but sometimes a few are laid together over-lapping like tiles. On average about eight eggs may be found on one fruit. They hatch after about 3–6 days.

The young caterpillars are whitish and spotted. Fully grown, they are about 15 mm long, pinkish with red above. The larval period lasts for 17–19 days. The young larvae wander about the fruit for some time before penetrating it. A little dark frass can be seen at the point of entry. The caterpillars normally feed on large, but not mature, bolls. The young caterpillars first mine in the walls of the bolls, but later they move into the cavity of the bolls and feed on the seeds.

Pupation takes place in the soil in a cocoon of silk and soil fragments, taking 8–12 days.

The adult is a small, brownish, night-flying moth with an average wingspan of 16 mm. The female may live for a week or more and lay 100–400 eggs.
**Distribution**. Africa, both tropical and southern temperate, from Ethiopia, Senegal, Ivory Coast, Togo, and Upper Volta down to S. Africa; Mauritius and Madagascar.
**Control**. No really successful economic chemical control measures are known. For cotton, a close season of at least two months is effective.

For *Citrus*, orchard sanitation is the only effective method of control. Infested fruit should be picked from the tree and collected from the ground at least twice a week.

Insecticides which show some success against this pest are DDT, dichlorvos, parathion, mevinphos and phosphamidon.

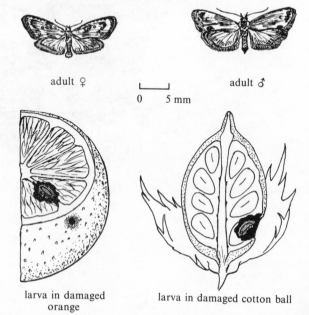

adult ♀          0   5 mm          adult ♂

larva in damaged
orange

larva in damaged cotton ball

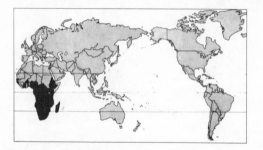

# Cydia molesta (Busck)

**Common name.** Oriental Fruit Moth
**Family.** Tortricidae
**Hosts** (main). Peach
    (alternative). Many other fruit trees.
**Damage.** The caterpillar damages both twigs and fruit. Early in the season the larvae bore into the tips of soft young shoots, causing them to wilt and die. Later as the young twigs harden and the fruit nears maturity, most of the caterpillars bore into the fruits.
**Pest status.** A serious pest of peaches in various countries, and of lesser importance on many other deciduous fruit trees.
**Life history.** The eggs are laid on the leaves and twigs.

    The caterpillars first feed on the young shoots, then enter and tunnel into the fruits causing damage very similar to that of Codling Moth. The mature caterpillars spin a cocoon in a protected place on the tree or ground, and they overwinter as mature caterpillars in the cocoon.

    The adult moths appear about the time peach trees are flowering.

    There are usually four or five generations per year.
**Distribution.** S. Europe, N. Africa (Morocco), Mauritius, Australia (New South Wales), S. China, Korea, Japan, USA, S. America (Brazil, Uruguay, Argentina) (CIE map no. A8).

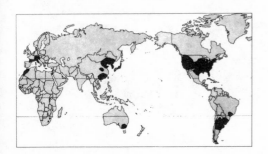

**Control.** Sprays of DDT, parathion, or the other chemicals recommended for Codling Moth should be effective; three applications are generally required for adequate control.

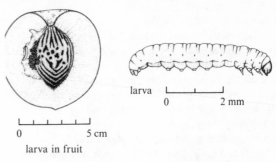

larva

0      2 mm

0      5 cm

larva in fruit

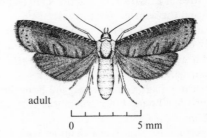

adult

0      5 mm

## Eucosma nereidopa Meyr.

(= *E. phylloseia* Meyr.)

**Common name.** Coffee Tip Borer
**Family.** Tortricidae
**Hosts** (main). Coffee (*arabica*).
      (alternative). Not known.
**Damage.** The larvae bore into terminal shoots, killing
the shoots and tips of branches. Sometimes berries
are similarly bored. The caterpillar usually bores into
the stem half-way between two nodes near the top of
a thick, succulent sucker. A slight swelling indicates
the point of entry, and the tip wilts rapidly. One
larva may destroy two or three shoot tips.
**Pest status.** A sporadically serious pest of high altitude
*arabica* coffee in E. Africa; populations are sometimes
very high and the damage extensive.
**Life history.** The eggs are flat, circular, white, and are
laid singly or in small groups in slight grooves near
the tip of a sucker or on the side of a berry. Hatching
takes about 14 days.

    The caterpillar is dark brown and bores into the
plant after a few hours. Fully grown, the caterpillar is
about 12 mm long.

    Pupation takes place in a chamber which the
caterpillar makes by enlarging a natural crevice in the
rough bark near ground level. The pupal period lasts
about 36 days.

    The adult moth is pale and dark grey and
about 8 mm long. During daylight hours they rest on
the trunks of coffee and shade trees near ground
level. The pre-oviposition period is four or more days.
**Distribution.** E. Africa.
**Control.** Control consists of killing the adult moths
before they are able to lay their eggs. When a sufficient number of moths are observed on a selection
of shade trees (ten per block of infested coffee), that
is an average of more than five moths per shade tree
before the suckers have been thinned out, or more
than one moth per shade tree after sucker thinning,
spray the shade tree trunks with DDT. Supplementary sprays will be required.

    This moth does sometimes occur in unshaded
coffee plantations, and work is still in progress to
devise adequate control measures.

damaged plant

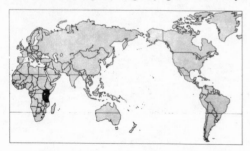

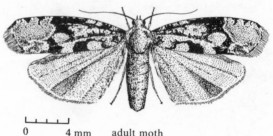

0    4 mm    adult moth

## Homona coffearia (Nietn.)

**Common name.** Tea (Coffee) Tortrix (Tea Flushworm)
**Family.** Tortricidae
**Hosts** (main). Tea
      (alternative). Coffee, *Acacia, Syzygium*, and indigo.
**Damage.** The feeding larvae roll or fold leaves longitudinally and feed within the roll, moving periodically to new sites and webbing new leaves.
**Pest status.** As a pest of tea this species is most important in Sri Lanka, but is of regular occurrence in the other regions indicated.
**Life history.** Eggs are laid in small clusters in batches of 100–150, on the flush leaves, usually on the dorsal surface. The caterpillars are slender and green with a dark dorsal shield and mature at about 20–25 mm. Pupation takes place within the folded leaf, and requires from 6–14 days, according to temperature.

      The adult moth is a typical tortricid in appearance with curved wings and prominent wingtips, brownish-yellow in colour. Wingspan is about 24–28 mm.

      In India one generation takes from 36–44 days, but elsewhere up to 60–70 days.
**Distribution.** S. India, Sri Lanka, Bangladesh, parts of S.E. Asia and up to China, and Japan. Also West Irian, Papua New Guinea and part of the Queensland coastal area (CIE map no. A330).

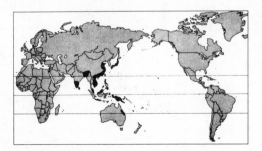

**Control.** In Sri Lanka the severe outbreaks of this pest led to consideration of biological control, and the larval parasite *Macrocentrus homonae* (Braconidae) was imported from Java where it naturally controls the pest. *Macrocentrus* has proved to be very successful and the level of parasitism on the island is high, so the tortrix is no longer regarded as a very serious pest locally. The two main reasons for the parasite's success are possibly its short breeding cycle of seven weeks as opposed to the ten weeks of the host and the fact that it is polyembryonic, and each egg may give rise to as many as 30 larvae inside the host.

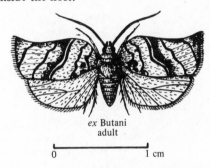

*ex* Butani
adult

0            1 cm

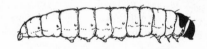

larva

rolled tea leaf

309

## Antigastra catalaunalis (Dup.)

**Common name.** Sesame Webworm
**Family.** Pyralidae
**Hosts** (main). Sesame (*Sesamum indicum*)
(alternative). No wild hosts recorded, but
two ornamentals (*Antirrhinum* and *Durante*).
**Damage.** Young leaves and shoots are webbed together
and eaten, and pods are bored by small caterpillars.
**Pest status.** A minor pest of sesame throughout
Kenya and N. Uganda, and other areas in the tropics,
with occasional serious outbreaks.
**Life history.** Eggs are oblong, 0.36 × 0.25 mm and
laid singly on young leaves or on flowers, change
from greenish-white, through yellow, grey, and
finally to red before hatching. Incubation takes
2–6 days.

   The larva is a white caterpillar when first
hatched, but later turns green with small black spots.
There are five larval instars lasting for 15–18 days
according to temperature. The mature caterpillar is
about 14 mm long. The caterpillars roll up and web
together the young leaves with silk at the top of the
shoot, and feed inside the rolled-up mass. Flowers
are also eaten, and the caterpillars may bore within
the pods.

   Pupation takes place in a silken cocoon on a
leaf or in the surface litter on the ground. The pupa is
slender, greenish-brown, and 9–10 mm long. The
pupal period varies from 4–9 days.

   The adult is an orange-brown, night-flying moth
with a wingspan of about 16 mm. After a pre-oviposition period of 2–5 days the moth may live a
further 5–6 days and lay about 20 eggs.
**Distribution.** Old World tropics and subtropics,
including S. Europe, USSR, Africa, India, Bangladesh,
Sri Lanka, and Burma.
**Control.** Planting all the sesame crops in one area in
the same rainy season so that a close season occurs
between successive crops is recommended.

   When insecticides are required effective treatment is a foliar spray of carbaryl or endosulfan.

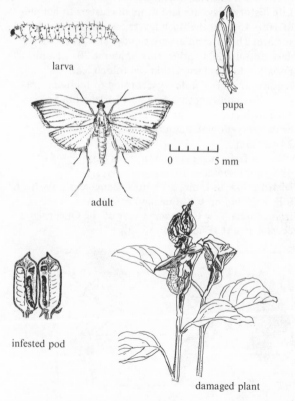

larva

pupa

0     5 mm

adult

infested pod

damaged plant

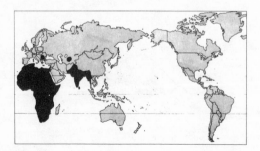

# Chilo orichalcociliella (Strand)

**Common name.** Coastal Stalk Borer
**Family.** Pyralidae
**Hosts** (main). Maize, sorghum, finger millet and sugarcane.

(alternative). Wild grasses, especially guinea grasses (*Panicum* spp.) and wild sorghums (*S. verticilliflora* and *S. versicolor*).

**Damage.** The damage is much the same as for *C. partellus*, with 'dead-hearts' in small plants, windows in the upper leaves, and caterpillars boring in the stem of older plants.

**Pest status.** This is the most important stalk borer in the coastal provinces of Kenya and Tanzania, but since 1961 *C. partellus* has become the dominant species.

**Life history.** Egg-laying starts about nine days after maize germination. The adults originate partly from standing late-planted maize and partly from wild grasses. The egg stage usually lasts about 4–6 days.

The caterpillars are almost indistinguishable from those of *C. partellus* and take some 27 days to mature. The first generation of larvae are usually not numerous and seldom do serious damage.

Pupation takes place in the stems of the host plant, and the time required for development is about six days.

The adult moths are 10–14 mm long with a wingspan of about 28 mm. The forewing is pale brown with three dark spots in the centre and a sub-terminal row of about seven small dark spots; the hindwing is pale brown.

The first generation moths lay their eggs on the same maize plants, and the second generation of larvae tends to be much larger and more damaging. Both generations take about 36 days to develop. When the second generation adult moths emerge the maize plants are generally too old to be attractive for egg-laying; these adults usually lay their eggs on wild grasses or late-planted maize.

**Distribution.** E. Africa, Nigeria, Malawi, and Madagascar.

**Control.** Destruction of all old maize plants and tall grasses by burning in the dry season before planting is advocated. Simultaneous planting of large areas of maize at the start of the rains, and the application of fertilizers to impoverished or poor soils are additional cultural methods of reducing borer populations and damage.

Only early-planted maize on fertile soil is worth spraying with insecticides; the usual insecticides employed are DDT, γ-BHC, carbaryl, etc., either as dust or spray in low volume.

See also p. 361, 'Control of cereal stalk borers'.

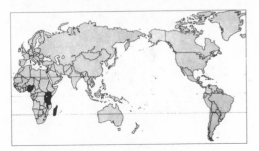

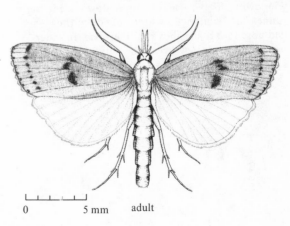

0     5 mm     adult

311

## Chilo partellus (Swinhoe)

(= *C. zonellus* (Swinhoe))

**Common name.** Spotted Stalk Borer (Pink Borer)
**Family.** Pyralidae
**Hosts** (main). Maize and sorghum, bulrush millet,
sugarcane and rice.

(alternative). Various species of wild grasses.
**Damage.** In young plants this pest causes a typical
'dead-heart'; in older plants the upper part of the
stem usually dies due to the boring of the caterpillars in the stem pith.
**Pest status.** Since 1961 this has been the dominant
pest in the coastal provinces of E. Africa. A major
pest of maize and sorghum in India and E. Africa,
not unimportant on other cereals, but actual crop
losses following attack are not easy to demonstrate.
**Life history.** The eggs are flattened, ovoid, and scale-
like, about 0.8 mm long. They are usually laid on the
underside of a leaf near the midrib, in 3—5 imbricated
rows in groups of 50—100. Hatching takes 7—10 days.

The young larvae migrate to the top of the
plant where they mine the sheaths and tunnels in the
midrib for several days, producing characteristic leaf
windowing. They then either bore down inside the
funnel, or else move down the outside of the stem
and bore into it just above an internode. In older
plants the larvae may live in the developing heads.
Larval development takes 28—35 days; the mature
caterpillar is 25 mm long, buff-coloured with four
longitudinal stripes, and a brown head capsule and
thoracic shield.

Pupation takes place in the stem in a small
chamber, and takes 7—10 days.

The adult moths are not large, being 20—30 mm
across the wings; the male is smaller and darker than
the female. The male has forewings pale brown, with
dark brown scales forming a streak along the costa;
the hindwings are a pale straw colour. The female has
much paler forewings and hindwings almost white.
The adults are short-lived.

The life-cycle takes about 29—33 days, and
there are at least six generations per year.
**Distribution.** E. Africa, Sudan, Malawi, Afghanistan,
India, Sri Lanka, Nepal, Bangladesh, Sikkim, and
Thailand (CIE map no. A184).

Essentially a pest of hot lowland areas, and is
seldom found above an altitude of 1500 m.
**Control.** Crop residues should be destroyed as should
volunteer plants. Chemical control of *Chilo* is not
generally very successful; there is a breeding programme in E. Africa breeding sorghum for resistance
to *Chilo*.

The insecticides generally used are DDT, or
endrin as dusts or sprays, but several applications are
necessary (see *C. polychrysa*).

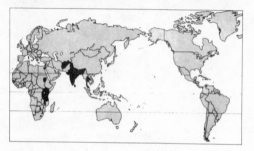

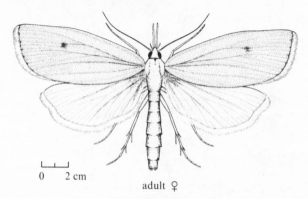

0    2 cm

adult ♀

312

# Chilo polychrysa (Meyr.)

**Common name.** Dark-headed Rice Borer
**Family.** Pyralidae
**Hosts** (main). Rice
   (alternative). Maize, sugarcane, and grasses.
**Damage.** The caterpillars bore the stem. Infested plants are liable to break at the node. Young plants show a characteristic 'dead-heart'.
**Pest status.** This is the most important rice stem borer in Malaysia, often killing the plants. The fourth instar caterpillar is the most destructive stage.
**Life history.** The eggs, which are scale-like in appearance, are laid in batches of 30–200 in rows along the undersurface of the leaves. Hatching takes 4–7 days.

The newly hatched caterpillars feed actively on the inner tissue of the leaf sheath. After a few days the caterpillars bore into the stems and feed on the stem tissue. They particularly feed at the nodes of the stem, so weakening the stems that they easily break at this point. Mature caterpillars have moulted five times, have a black head capsule and thoracic plate, and are 18–24 mm long and about 2.4 mm broad. Larval development takes about 30 (16–43) days.

Adults emerge in six days, and live for 2–5 days. The moths are 10–13 mm long with a wingspan of 17–23 mm. The forewing is uniform pale yellow with a cluster of small dark spots in the centre, and the hindwing white.

The total life-cycle takes 26–61 days; there are probably six generations per year.
**Distribution.** India, Pakistan, Bangladesh, Burma, Malaysia, Indonesia, Thailand, Vietnam, Laos, Sabah, and the Philippines.
**Control.** Clean cultivation is probably sufficient for control purposes in many areas, but in areas of intensive cultivation losses can be high. Where rice is grown under fairly natural conditions levels of parasitism are usually high and it is often preferable to avoid the use of pesticides, for they may destroy more parasites than the pests themselves.

Foliar sprays of parathion, and parathion-methyl are successful, but their mammalian toxicity is too high for safe use on a large scale. Endrin, dieldrin, trichlorphon, dichlorvos and diazinon have also been successful, but several applications at weekly intervals are required.

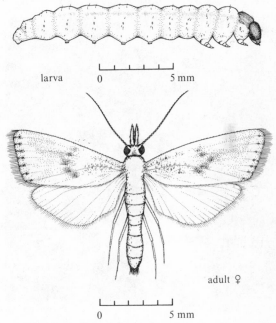

larva    0    5 mm

adult ♀

0    5 mm

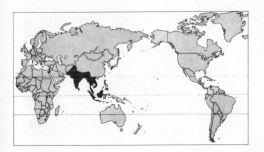

313

# Chilo sacchariphagus Bojer

(= *Proceras venosatus* (Wlk.))

**Common name.** Sugarcane Stalk Borer
**Family.** Pyralidae
**Hosts** (main). Sugarcane
     (alternative). Wild cane and other species of
Gramineae.

**Damage.** The larvae bore in the stem, making a cavity
that invariably becomes infested with bacteria, fungi
and rots. Because of the solid state of this stem each
larva usually only burrows within a single internode
and leaves a large emergence hole. Before penetrating
the stem the first instar larvae feed on the leaf-sheath
making small 'windows' in the leaves that enlarge as
the leaves expand.

**Pest status.** A serious pest of sugarcane in S.E. Asia,
Madagascar and Mauritius, but is replaced by other

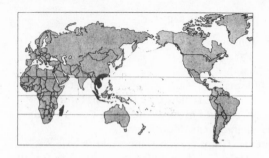

species on mainland Africa. In S.E. Asia references to
damage by '*Diatraea saccharalis*' usually refer to this
species. (*D. saccharalis* only occurs in the New
World.)

**Life history.** Details are not known, but it is expected
to be similar to that of *Diatraea saccharalis* and the
other species of *Chilo*.

emergence hole of adult moth

**Distribution.** Madagascar, Mauritius, Java, Sumatra, Malaya, Indo-China, S. China and Taiwan.

**Control.** The most effective method of controlling sugarcane borers is through the use of biological control and resistant varieties of plant. In Taiwan liber-ation of *Trichogramma australicum* resulted in stalk borer infestation decreases of 37—54%, and several varieties of cane have been shown to have definite adverse effects on borer development.

young larvae with leaf holes and frass

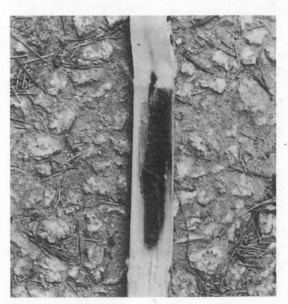

larval damage inside stem

## Chilo suppressalis (Wlk.)

(= *C. simplex*)
(= *C. oryzae*)

**Common name.** Striped Rice Stalk Borer
**Family.** Pyralidae
**Hosts** (main). Rice, maize.

(alternative). Millets, various wild species of *Oryza*, and many wild grasses.

**Damage.** Larval damage consists of boring in the stem resulting in 'dead-hearts' in the young plants, and damaged stems in older plants. One caterpillar may destroy up to ten plants.

**Pest status.** A very serious pest of rice in China and Japan especially, where crop damage of 100% has been recorded. In Japan, despite heavy pesticide applications, the rate of paddy infestation has still averaged 4–5% with an average loss of 175 kg/ha.

**Life history.** The eggs are similar to those of *C. polychrysa*, and hatch in 5–6 days.

The caterpillars have a yellowish-brown head, and have three faint dorsal, and two lateral stripes, brown in colour. After 33 days the caterpillar is fully grown and is 26 mm long and 2.5 mm broad.

The reddish-brown pupa is 11–13 mm long and 2.5 mm broad; the pupal period is six days.

The moth is very similar to *C. polychrysa* in colour but without wing spots, and is slightly larger in size; it is 13 mm long with a wingspan of 23–30 mm, although females may reach 35 mm. The adults live for 3–5 days.

The life-cycle takes 41–70 days.

**Distribution.** Found in Spain, India, Pakistan, Bangladesh, S.E. Asia, China, Korea, Japan, Philippines, Indonesia, Papua New Guinea, West Irian and N. Australia (CIE map no. A254).

**Control.** Control is as for *C. polychrysa*.

**Further reading**

Grist, D.H. & R.J.A.W. Lever (1969). *Pests of Rice*, pp. 62–70. Longmans: London.

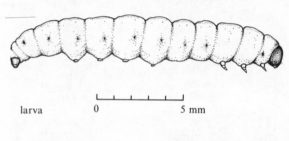

larva

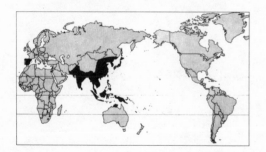

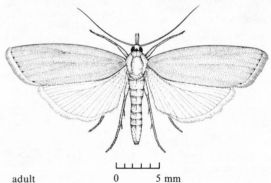

adult

# Cnaphalocrocis medinalis Gn.

**Common name**. Rice Leaf Folder (Rice Leaf Roller)
**Family**. Pyralidae
**Hosts** (main). Rice

(alternative). Grasses of various species.
**Damage**. The caterpillars infest the leaves of young plants; they fasten the edges of a leaf together and live inside the rolled leaf. The green tissues, particularly the chlorophyll, are eaten by the caterpillars and the leaf dries up. In heavy infestations the plants appear scorched, sickly and twig-like.
**Pest status**. Occasionally a rice pest of some importance. Plants are susceptible to attack up to ten weeks after transplanting.
**Life history**. The eggs are laid singly or in pairs on the young leaves; they are flat and oval, and yellow in colour. Hatching takes place after 4—7 days.

The caterpillars live inside the folded leaves for 15—25 days, and are slender and pale green.

Pupation takes place inside the rolled leaf; the pupa is dark brown, and the adult moth emerges after 6—8 days.

The adult moths often fly by day; they are small (8—10 mm long; wingspan of 12—20 mm), orange-brown with several dark, wavy lines on the wings; the outer margin of the wings is characterized by a dark terminal band.

The life-cycle generally takes 25—35 days; in some areas there are four generations per year.

**Distribution**. Madagascar, Pakistan, India, Bangladesh, Sri Lanka, S.E. Asia, China, Korea, Japan, Philippines, Indonesia, West Irian, Solomon Isles, and E. Australia (CIE map no. A212).
**Control**. Removal of grass weeds from bunds around the paddy fields helps to reduce the pest population.

Light-trapping of adults has been successful in some instances.

In severe outbreaks sprays of DDT, BHC, or dieldrin effectively control the pest. Endosulfan, fenthion, fenitrothion, and phosphamidon have also been successful as sprays, as has diflubenzuron.

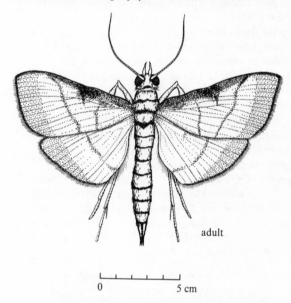

adult

0    5 cm

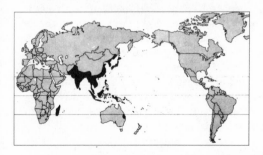

# Diatraea saccharalis (F.)

**Common name.** Sugarcane Borer
**Family.** Pyralidae
**Hosts** (main). Sugarcane
    (alternative). Maize, sorghum, rice, and many grasses.
**Damage.** The larvae bore in the internodes of the cane, and the larval excavation usually becomes infected with bacteria and/or fungi leading to rots. Usually one larvae only bores in one internode, because of the density of sugarcane stems. The emergence hole, through which the adult leaves the stem, is usually conspicuous.
**Pest status.** The major pest of sugarcane in the New World, although there are 4–5 other species of *Diatraea* important on cane in this region. Crop losses have been difficult to assess accurately because of compensatory growth, but it is thought that for every 1% of internodes bored sugar losses amount to 0.5–0.7%.
**Life history.** Eggs are usually laid within the leaf sheath (as with most stalk borers) and the young larvae feed on the epidermis of the leaves. After about a week the young larvae penetrate the stem and start to excavate the internode; larval development continues here until maturity. Mature larvae then bite an exit hole through the stem wall before they pupate, still within the stem. The larvae are creamy-white with prominent dark spots.

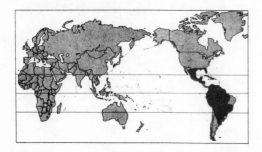

The number of larval instars varies from five to six to as many as eight, and the life-cycle takes from 35–50 days; in the tropics there may be seven generations annually, but at the northern and southern limits of its distribution four (and a partial fifth) generations are usual. In these regions larvae overwinter in a state of diapause in stubble and broken pieces of cane stalk.

The adults are small brown moths, with a wing-span of 18–29 mm; the forewings are yellow-brown with two faint oblique brown stripes, and a dark discal spot. The different species of *Diatraea* can only be reliably identified using the genitalia.
**Distribution.** Southern USA, W. Indies, C. and S. America (CIE map no. A5).
**Control.** The pest spectrum on sugarcane in most parts of the world is very large and complex, although in any one location probably not more than a dozen pests are of particular importance. In the past, pest problems often arose following prolonged use of contact insecticides, and now in most areas of intense sugarcane cultivation a careful and rational pest management programme is practised based upon the following.
Cultural practices: destruction of crop residues; careful double-cropping.
Resistant varieties: high levels of resistance are unlikely to be achieved, but a number of commercial varieties now show low levels of resistance, which helps to reduce borer damage.
Biological control: parasitic Hymenoptera are used to destroy the larvae and eggs; *Trichogramma* for the eggs, *Apanteles* and others for the larvae.
Chemical control: now generally restricted in use to the southern USA where borer generations are discrete, and only against the second and third generations if the populations are sufficiently large — monocrotophos and carbofuran as foliar sprays, and endrin or azinphos-methyl granules.

## Eldana saccharina Wlk.

**Common name.** Sugarcane Stalk Borer
**Family.** Pyralidae
**Hosts** (main). Sugarcane, and in some areas maize.
(alternative). Other cereal crops and wild grasses, including rice, cassava and *Cyperus* sp.
**Damage.** When very young plants are attacked 'dead-hearts' result, followed by tillering of the plant. Older plants or ratoon cane have internodes bored.
**Pest status.** A pest of some importance in Africa only; first recorded outbreak on sugarcane was in Tanzania in 1956, and in Uganda in 1967.
**Life history.** The eggs are oval, yellow, and laid in batches on the soil surface, although some may be laid on the leaf bases or in cracks on mature stalks. On average 200 (100–500) eggs are laid per female in batches of 10–15 (3–200 have been recorded). The female starts egg-laying the second night after emergence. The incubation period is 5–6 days at 25°C.

The larval period is 30–35 days in Uganda, and there are six larval instars. When burrowing in the stem the larvae characteristically push their faecal pellets outside. First instar larvae typically feed on the upper surface of the leaf sheath, and then later penetrate the bud and enter the stem. The larvae are mainly found in the lower parts of the stems, but in heavy infestations may be found in all parts of the stem.

Pupation takes place in the plant, in the stem or on the leaf sheath, and takes some 7–14 days (mean 10).

The adult male has a wingspan of 28–30 mm, and the female 39–40 mm. Emergence takes place at night, with mating on the following night. They have pale brown forewings, each with two small spots in the centre, and whitish hindwings with a short fringe. The adults live for 3–8 days.
**Distribution.** Africa only; Zaïre, Nigeria, Sierra Leone, Chad, Ghana, Mozambique, Zululand and E. Africa.
**Control.** Endrin as a foliar spray, applied four times at three-week intervals, gave some control of this pest in Tanzania.

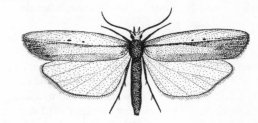

adult ♂

0          1 cm

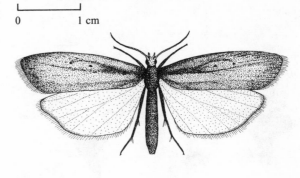

adult ♀

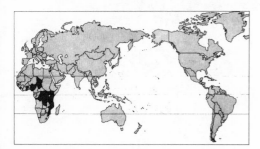

319

## Ephestia cautella (Hb.)

(= *Cadra cautella* Hb.)

**Common name.** Tropical Warehouse Moth (= Almond Moth; Dried-currant Moth)
**Family.** Pyralidae
**Hosts** (main). Maize, wheat, and other grains in store.
  (alternative). Dried fruit, beans, nuts, bananas, groundnut.
**Damage.** Webbing in the grain and on the surface of bags, with cocoons between adjacent surfaces.
**Pest status.** A very serious pest on stored produce in many parts of the world where conditions are warm, infesting a wide range of grains and foodstuffs.
**Life history.** The eggs are round and white, turning pale orange before hatching. They are laid by the female through the holes in the bags. Each female may lay 250 eggs.

The larvae are elongate caterpillars about 2 cm long, greyish-white with numerous dark setae interspersed along the body. The head is dark brown and there are two dark areas on the prothorax. The first instar larvae generally feed on the seed germ, moving about freely in the produce, and feeding until mature. They migrate to pupate in the crevices or where two bags are in contact with each other.

The pupa is pale brown and is enclosed in a cocoon.

The adult moth emerges through the cocoon and the female is responsible for the spread of infestation in the warehouses. The adult moth is greyish with rather indistinct markings on the wings. It is about 13 mm long. When at rest the wings are folded along the abdomen. The adult moths live for less than two weeks.

The life-cycle at 28°C and 70% RH is 6–8 weeks; breeding may be continuous so long as conditions are suitable.
**Distribution.** Cosmopolitan in tropical and warm temperate areas of the world.
**Control.** Store hygiene is very important in the control of this pest. After the store is thoroughly cleaned the walls should be sprayed with either a DDT + lindane mixture, or malathion solution, or lindane solution alone.

The bags and containers should be treated with malathion dust, DDT dust, or lindane dust.

Fumigation should be carried out by approved operators only, because of the hazards involved.

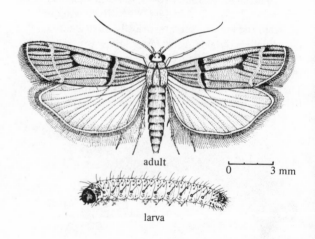

adult

0   3 mm

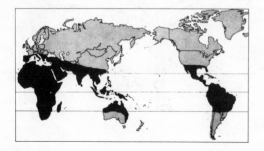

larva

# Ephestia elutella (Hub.)

**Common name.** Warehouse Moth (= Cocoa Moth)
**Family.** Pyralidae
**Hosts** (main). Dried cocoa beans
    (alternative). Dried grains, pulses, nuts, tobacco, coconut, and dried fruits.
**Damage.** This polyphagous pest attacks a very wide range of stored foodstuffs and products, and the damage done is generally eating of the foodstuffs and contamination of them with exuviae, dead bodies and frass. Damage is often serious because the infestations are so heavy. The larvae produce silk, and with heavy infestations there may be extensive webbing. In grains, the germinal (embryo) portion is usually selectively eaten.
**Pest status.** Not a common pest in the tropics, but of regular occurrence, being more abundant in the sub-tropics and temperate regions. A very important pest because of its polyphagous nature and the wide range of stored products it damages.
**Life history.** The larvae of the different species of *Ephestia* are all very similar, and all have characteristic small dark spots on each body segment; they can only be separated from each other by a microscopic study of their chaetotaxy.

The adults are small moths, about 7–8 mm in body length, with fairly distinct markings on the fore-wings. However, sometimes the adults are difficult to distinguish from *E. cautella*, and for positive identification examination of genitalia is required.

In temperate regions there is only one generation per year, but in the tropics breeding is presumably continuous and there are several generations per year. Under optimum conditions (c. 25°C and 75% RH) one generation requires only about 30 days.
**Distribution.** Cosmopolitan, but less abundant in the tropics than in the more temperate regions.
**Control.** As with most stored products pests, warehouse (godown) hygiene is most important. This is not a species likely to cause pre-harvest infestations, and so most infestations are post-harvest, usually in the stores. Fumigation can be used for short-term control, using the usual range of fumigants; for residual effects contact insecticides must be employed, such as pyrethrins, dichlorvos, and pirimiphos-methyl. Insecticides previously used were DDT, lindane (BHC), and malathion, but now the species of *Ephestia* show at least tolerance to these chemicals and in many places show high levels of resistance.

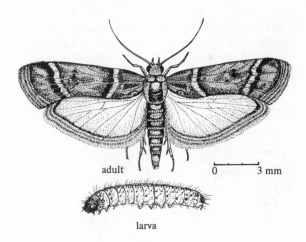

adult                    0        3 mm

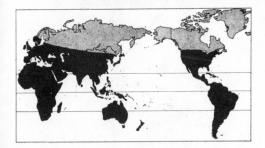

larva

## Etiella zinckenella (Treit.)

**Common name.** Pea Pod Borer (Lima Bean Pod Borer)
**Family.** Pyralidae
**Hosts** (main). Pigeon pea, lima bean.

(alternative). Green peas, cowpeas, *Dolichos labab* and other Leguminosae.

**Damage.** Early instar larvae feed inside the developing seeds, but later instars feed freely inside the pods. The partly grown caterpillar may leave the original pod and penetrate one or more fresh pods before reaching maturity.

**Pest status.** A very common pest of pigeon pea and other legume crops in many parts of the world; sometimes recorded as a serious pest.

**Life history.** The eggs are oval, shiny white, 0.6 × 0.3 mm. They are laid singly or in small groups (up to six) on immature pods. They hatch after 3–16 days, according to temperature.

The caterpillar is blue with a yellow head, and 12–17 mm long when mature. It wriggles very violently if the pod is opened and the caterpillar disturbed. After hatching from the egg, it takes about 1½ hours to select a point of penetration, spin a protective web, and bore into the pod. Cannibalism often occurs if more than two caterpillars enter the same pod. The total larval period varies from 3–5 weeks.

When the larva is fully grown it leaves the pod, drops to the ground and about 3 cm below soil level it spins a cocoon and turns into a yellowish-brown pupa 6–10 mm long. The pupal period varies from 2–4 weeks.

The adult moth is brown, with a wingspan of 24–27 mm. The female moth may live for 2–4 weeks, and lay 50–200 eggs.

**Distribution.** Almost completely pantropical in distribution, extending up into the subtropics and warmer temperate areas, but only one record from Australia (CIE map no. A105).

**Control.** Chemical control measures are not usually economic in the field. Experimental plots can be protected by spraying carbaryl, initially ten days after first flowering, with subsequent sprays every ten days thereafter.

Similar pod-boring caterpillars (e.g. *Cydia nigricana* – Pea Moth) can be controlled with fair success using tetrachlorvinphos, carbaryl, azinphos-methyl, and fenitrothion, as two sprays at high volume with a ten-day interval.

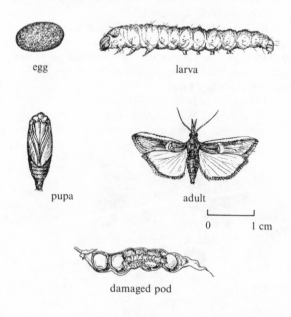

egg          larva

pupa          adult

0          1 cm

damaged pod

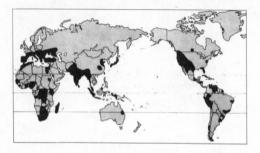

322

## Maliarpha separatella Rag.

(= *Rhinaphe vectiferella* Rag.)

**Common name.** White Rice Borer
**Family.** Pyralidae
**Hosts** (main). Rice
    (alternative). Wild rice and grasses.
**Damage.** Larval stem boring results in white heads and broken stems, although usually damage is not serious unless conditions are suitable for continuous cropping, as in parts of Madagascar. The larvae feed on the tissues inside the hollow stem of the rice plant, and can bore into the base of the stem and thence into other tillers.
**Pest status.** Only occasionally a serious pest in localities of continuous rice-cropping.
**Life history.** The eggs are laid close together in one cluster of up to 50 in number. They are stuck to the leaf by cement and as this dries it puckers the leaf so that the egg mass is enclosed inside a foliar envelope.

The larva is transparent-white with a dark brown head; as it ages it gradually turns yellowish and gets fatter. Larvae can be dispersed by wind, suspended on a silken thread. Mature caterpillars are about 18 mm long, and can go into a resting stage during a dry season. In Sierra Leone the larvae may lie dormant in the rice stubble for up to 20 weeks.

Pupation takes place in a loose cocoon in the rice stem. The pupa is brown with a red spot on the dorsal part of the fifth segment.

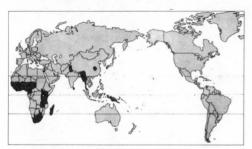

The adult male is about 15 mm long, and the female about 18 mm; wingspan is from 23−29 mm. The long, pale yellow wings overlap along the body at rest. On the paler forewings there is a marked reddish-brown line behind the costal veins. The hindwings are white with a metallic sheen.

There are usually three or four generations per year.
**Distribution.** Africa (W. and E. and S., Zambia, Malawi, and Madagascar), India, Sri Lanka, Burma, China, Papua New Guinea and West Irian (CIE map no. A271).
**Control.** Control measures are similar to those recommended for *Chilo polychrysa*.

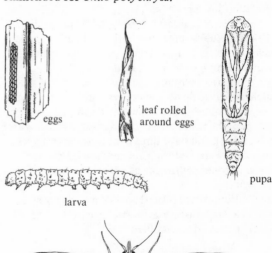

eggs

leaf rolled around eggs

pupa

larva

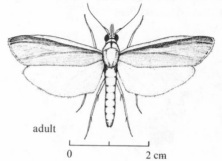

adult

0    2 cm

## Marasmia trapezalis (Gn.)

**Common name.** Maize Webworm
**Family.** Pyralidae
**Hosts** (main). Maize

(alternative). Millets, sorghum, sugarcane, rice, wheat and many wild grasses.

**Damage.** The larvae bind the two edges of the leaf together with silk to form a funnel and they feed inside by biting small pieces from the upper surface.

**Pest status.** Not usually a serious pest but infestations are quite common in some seasons, and they are quite conspicuous.

**Life history.** The eggs are laid along young leaves by the ovipositing female.

The larva is a pale greenish-yellow caterpillar, with conspicuous setae, and both head and thoracic shield reddish-brown. The fully grown caterpillar reaches a length of about 20 mm.

Pupation takes place in the rolled-up leaf to which the larvae fasten themselves with silken threads.

The adult is a small moth with 18–20 mm wingspan; the wings are greyish with shiny highlights (iridescence), and have three dark transverse stripes and a dark wide subterminal band; the hindwings have the stripes continuing and they converge on the anal point.

**Distribution.** Africe (Cameroons, Zaïre, Senegal, E. Africa, Madagascar); S.E. Asia, Australasia, Pacific islands, C. America, and Peru.

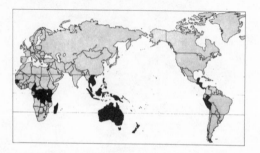

**Control.** Control measures are not usually required, but the usual caterpillar-killing insecticides, e.g. DDT, carbaryl, fenitrothion, tetrachlorvinphos, etc., should prove effective if at any time required.

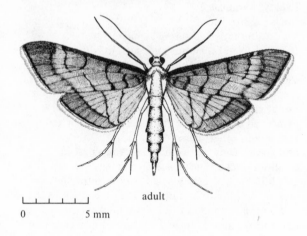

adult

0          5 mm

## Maruca testulalis (Geyer)

**Common name.** Mung Moth (Bean Pod Moth)
**Family.** Pyralidae
**Hosts** (main). Beans and peas (Leguminosae), of all species but mostly *Phaseolus* spp.

(alternative). Groundnut, castor, tobacco, rice, and *Hibiscus* spp.

**Damage.** Leaves, flowers, flower buds and pods are eaten by the caterpillars, but the more serious damage is done in the pods where the seeds are destroyed.

**Pest status.** A regular but usually minor pest of pulse crops in E. Africa and other parts of the tropics, although occasional serious outbreaks have been recorded.

**Life history.** The eggs are laid singly in the flowers or buds, or on the pods of the host plant.

The caterpillar is whitish with dark spots on each body segment, forming dorsal longitudinal rows. The mature caterpillar is about 16 mm long.

Pupation takes place in a silken cocoon in the pod, or more rarely in the soil.

The adult moth has brown forewings with three white spots, and the hindwing is greyish-white with distal brown markings; the wingspan is from 16–27 mm.

**Distribution.** Widespread in tropical and subtropical regions of the world.

**Control.** Effective chemicals include DDT, diazinon, endosulfan, pirimiphos-methyl, tetrachlorvinphos, trichlorphon, cypermethrin and permethrin, as foliar sprays.

adult

0    5 mm

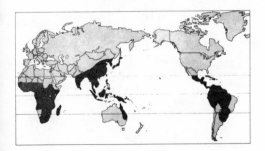

## Nacoleia octasema (Meyr.)

**Common name.** Banana Scab Moth
**Family.** Pyralidae
**Hosts** (main). Bananas

(alternative). *Pandanus*, manila hemp, maize, *Heliconia*, and nipa palm.

**Damage.** The caterpillars feed on the inflorescence of the banana as it develops, and cause a scab on the developing fruit. Further damage is caused by the frass accumulating in dark masses between the fingers and hands of the bunches. Attacks are not usually widespread, but individual bunches are severely damaged.

**Pest status.** A serious pest where it occurs, although not quite so serious in Papua New Guinea and some parts of Indonesia.

**Life history.** Eggs are laid in batches (typically about 15) on or near the flag leaf just before the bunch emerges. Each female moth can lay 80–120 eggs. The pale, greenish-white eggs hatch in about 4–6 days, and the small, transparent, yellow caterpillars crawl under the bracts of the banana inflorescence where they feed. As the larvae feed and grow they gradually turn pink. Up to 70 caterpillars may be found in a single inflorescence. The five instars take 12–21 days for completion.

Pupation takes place in a silken cocoon, constructed among the fruits, under the sheath or possibly sometimes on the ground; it takes 10–12 days.

The adult moths emerge in the evening, and only live for a few days. They vary somewhat in colour and size, wingspan being 16–26 mm (average 22 mm); colour from pale to dark brown with a series of dark spots and lines on both wings.

**Distribution.** Indonesia, Papua New Guinea, West Irian, Solomon Isles, New Caledonia, Fiji, Tonga, Samoa, and Australia (Queensland) (CIE map no. A383).

**Control.** Various attempts at biological control of the caterpillars by parasites (Tachinidae and Braconidae) are in progress at the present time.

The newly hatched caterpillars can be sprayed with DDT or a DDT + BHC mixture but these treatments are not very effective unless critically timed.

**Further reading**

Simmonds, N.W. (1970). *Bananas*, pp. 353–5. Longmans: London.

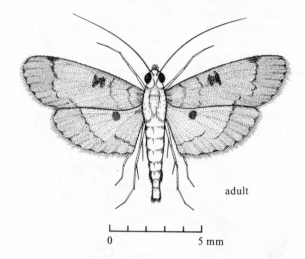

adult

0             5 mm

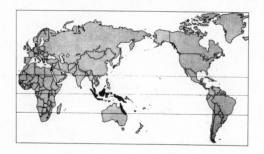

# Nymphula depunctalis Gn.

**Common name.** Rice Caseworm
**Family.** Pyralidae
**Hosts** (main). Rice
  (alternative). Various aquatic grass species.
**Damage.** The caterpillar cuts tips of leaves to make
the case in which it lives; the case is changed as the
caterpillar grows. In heavily infested crops the loss
of photosynthetic tissue can be critical and seedlings may die. Older plants generally are more
tolerant of damage, and mature plants are seldom
attacked. The larvae feed on the lower side of leaves
lying flat on the water, or on submerged leaves.
**Pest status.** A serious pest of rice seedlings in many
countries, but damage to older plants is slight.
**Life history.** The eggs are laid singly on the leaves;
they hatch in 2—6 days, and after a few days the first
instar larvae construct the first cases. One female
moth usually lays about 50 eggs.

The caterpillar is pale translucent green, with a
pale orange head. It is semi-aquatic in habits and can
withstand prolonged immersion; it has slender gills
along its sides and the case is always filled with water.
The larval stage (with four instars) lasts for 15—30
days. The fully grown caterpillar is 13—20 mm long.

Pupation takes place inside the last larval case
which is fastened to the base of the stem; it may take
place under water but more usually above water level.

The adults emerge after 4—7 days, and can live
for up to three weeks. The adults are small, delicate,
snowy-white moths with pale brown spots on the
wings; they have a wingspan of 15—25 mm.
**Distribution.** Nigeria, Ghana, Gambia, Cameroons,
Zaïre, Malawi, Mozambique, Madagascar, Mauritius,
Pakistan, India, Sri Lanka, Bangladesh, Burma,
Malaysia, Thailand, Vietnam, Indonesia, Philippines,
S. China, West Irian, Australia, and S. America
(CIE map no. A176).
**Control.** Draining the water from infested fields for
2—3 days successfully kills the caterpillars, but they
can also be killed by the addition of a kerosene film
to the water.

The more successful insecticides used are parathion, malathion, BHC and dieldrin.

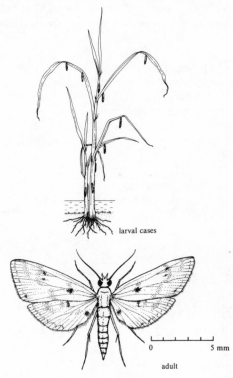

larval cases

adult

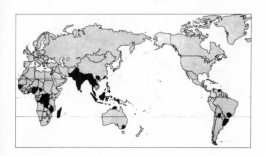

327

# Ostrinia furnacalis (Gn.)

(= *Pyrausta salientialis* (Snell.))

**Common name.** Asian Corn Borer
**Family.** Pyralidae
**Hosts** (main). Maize
    (alternative). Sorghum, millets, Indian hemp, hops, *Artemisia*, and many other plants.
**Damage.** Attacked plants may have broken stems, with tunnels up the stem, and inside the cobs. There is windowing of leaves.
**Pest status.** An important pest of maize in Asia and Australasia; with heavy infestations losses can be very serious.
**Life history.** Eggs are laid in clusters of 10–40 underneath the leaves, each female laying from 500–1500 eggs, usually about one week before the formation of the female inflorescence. Incubation takes 3–10 days.

The young larvae initially scarify the underside of the leaves, then they feed on the spike and spin the inflorescences together. Later they tunnel into ribs of the leaves, and bore into the stem. After about three or four weeks the larvae are mature.

Pupation takes place in a cocoon in the soil or in the stem.

The adult is a yellowish moth, 12–14 mm long and up to 30 mm wingspan, with dark brown terminal bands and various wavy lines. The moths are active nocturnal fliers, and may fly for up to several miles.

They live for 10–24 days.

There are one or more generations per year, according to temperature.
**Distribution.** Asia, and Australasia (CIE map no. A294).
**Control.** Cultural practices include destroying the stubble after harvest; using resistant or tolerant varieties of maize; planting late to avoid early egg-laying.

The aim of chemical control is to destroy the first instar caterpillars before they bore into the stem; recommended insecticides are trichlorphon as sprays or granules, parathion-methyl, parathion, and carbofuran.

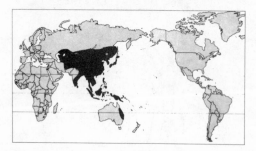

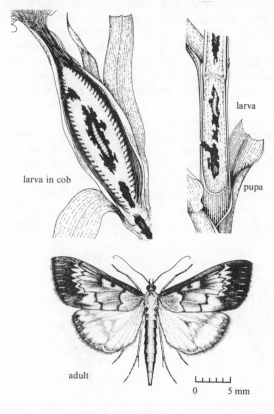

larva in cob

larva

pupa

adult

0    5 mm

## Plodia interpunctella (Hub.)

**Common name.** Indian Meal Moth
**Family.** Pyralidae
**Hosts** (main). Meals and flours, and farinaceous products.

(alternative). Dried fruits (raisins, currants, sultanas), nuts, and some pulses and cereals.

**Damage.** The direct eating of the produce, especially the germinal part of grains, is the primary damage, but the secondary damage is the contamination of foodstuffs with larvae, frass and silk webbing.

**Pest status.** An important pest of stored foodstuffs throughout the tropics and subtropics; it is regularly imported into Europe and N. America where it will survive and develop so long as temperatures are fairly high (above 18°C) and the air not too dry (RH above 40%).

**Life history.** This species does not have the dark spots on the larvae that *Ephestia* have, and so is distinct from those species.

Each female lays some 200–400 eggs which are stuck to the substrate. The eggs hatch after about 4–6 days, and the larvae live and feed in the stored foodstuff, reaching a body length of 8–10 mm after about 12–20 days. At low temperatures and low humidities larval development takes considerably longer. The total life-cycle can be as short as 24–30 days under optimum conditions, but the duration depends upon ambient temperature and humidity.

The adult moth is distinctive, with the outer half of the forewings a coppery-red separated from the creamy inner half by dark grey bands; body length is 6–7 mm and wingspan is 14–16 mm.

In Europe there may be only one or two generations per year, but in the tropics up to eight may be expected.

**Distribution.** Worldwide in the tropics and subtropics, and in some parts of the temperate regions where temperatures exceed some 20°C.

**Control.** The regular practices of fumigation and spraying contact insecticides in food stores generally keep this pest under control.

**Further reading**

Williams, G.C. (1964). The life history of the Indian Meal Moth, *Plodia interpunctella* Hbn. (Lepidoptera: Phycitidae) in a warehouse in Britain and on different foods. *Ann. appl. Biol.,* **53**, 459–75.

Tzanakakis, M.E. (1959). An ecological study of the Indian Meal Moth *Plodia interpunctella* (Hübner) with emphasis on diapause. *Hilgardia,* **29**, 205–46.

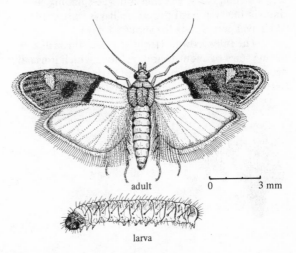

adult   0   3 mm

larva

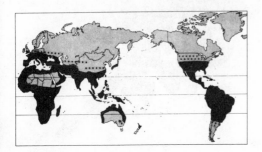

## Prophantis smaragdina (Butler)

(= *Thliptocera octoguttalis* (F. & R.))

**Common name.** Coffee Berry Moth
**Family.** Pyralidae
**Hosts** (main). Coffee (*arabica*)
    (alternative). Various woody Rubiaceae.
**Damage.** Typical symptoms are berry clusters webbed together and one or more is brown, dry and hollow. Very young berries may be grazed.
**Pest status.** A minor pest of *arabica* coffee. Frequently of benefit since it eats out a little of the crop on overbearing branches. Occasional severe attacks have occurred when the entire crop on many trees has been destroyed.
**Life history.** The eggs are scale-like, laid singly on or near green berries; they hatch in about six days.

    The larva is a reddish caterpillar about 14 mm long when fully grown. If it hatches out near a cluster of half-grown or larger berries, it bores into one of them, starting near the stalk. When one bean has been eaten, it leaves and wanders over the cluster joining the berries together with threads of silk before boring into a second berry. Feeding and web-spinning continue in this way until the caterpillar is fully grown. The larval period lasts for about 14 days.

    The fully grown caterpillar passes through a resting stage of about four days, after which it usually drops to the ground and pupates between two leaves neatly stuck together. The pupal period is very erratic, lasting from 6–42 days.

    The adult is a small, golden moth with wingspan of about 14 mm. There is a pre-oviposition period of 3–4 days. Adults live for two weeks.
**Distribution.** Africa from W. to E. and S.
**Control.** Trees should be checked at the times of flowering, and if buds or young berries are being eaten spraying with fenitrothion or fenthion should be done immediately, and repeated after 5–6 weeks.

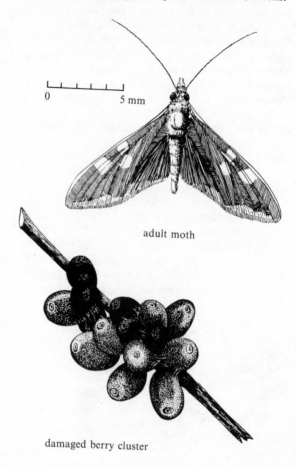

adult moth

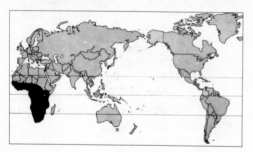

damaged berry cluster

## Sylepta derogata (F.)

**Common name.** Cotton Leaf Roller
**Family.** Pyralidae
**Hosts** (main). Cotton

(alternative). Virtually confined to the Malvales, especially *Gossypium* and *Hibiscus*.
**Damage.** The leaves are curled and drooping, as the caterpillars eat the leaf margin. The defoliation of the plants results in premature boll ripening.
**Pest status.** A serious pest usually controlled by its parasites; only rarely is chemical control warranted. Probably the most common leaf-eating caterpillar on cotton.
**Life history.** The eggs are oval, smooth, rather flattened, and pale green, and they are laid singly or in groups on either side of the leaves.

The young larvae, after an initial period of wandering, congregate within a roll of leaf which they secure by spun silken threads. Up to ten caterpillars may be found in one roll. When partly grown the caterpillars disperse and each forms a separate roll. They are very agile, and are translucent green with a black head and thoracic shield; mature at about 20 mm.

Pupation takes place in the leaf roll or in debris on the ground; the pupa is brown, 10–14 mm long.

The adult is cream-coloured with brown wavy pencillings on both wings. Wingspan is 30–40 mm.

The life-cycle generally takes 4–5 weeks.
**Distribution.** Throughout the rain-fed cotton-growing

areas of Africa, S.E. Asia, Australia and the Pacific islands (CIE map no. A397).
**Control.** Outbreaks of *S. derogata* are usually controlled naturally by its parasites. In some cases collection of the rolled leaves is recommended.

Should chemical control measures be required the following insecticides should be effective: DDT as a dust, and carbaryl as a spray, to be applied on the foliage at the first signs of damage or infestation.

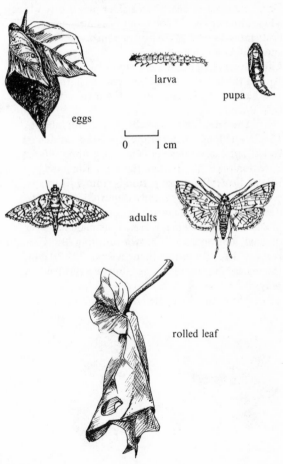

larva

pupa

eggs

0    1 cm

adults

rolled leaf

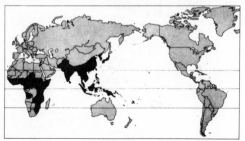

# Tryporyza incertulas (Wlk.)

(= *Schoenobius incertulas* (Wlk.))

**Common name.** Yellow Paddy Stem Borer
**Family.** Pyralidae
**Hosts** (main). Rice

(alternative). Wild rice and various wild grasses.
**Damage.** The caterpillars bore into the rice stems and hollow out the stem completely. Attacked young plants show 'dead-hearts' and older plants often break where the stem is hollowed out causing lodging.
**Pest status.** A serious pest of rice throughout India and S.E. Asia.
**Life history.** Egg masses are laid in batches of 80–150 on the leaf sheath, and covered with the brown anal hairs of the female moth. The incubation period is 4–9 days.

The caterpillars are yellow in colour, about 18–25 mm long when mature; the head capsule is black. Total larval development takes about 40 days. Pupation takes place inside the stem; the pupa is pale and soft; the pupal period is from 7–11 days.

The adults are sexually dimorphic and quite distinct; the female has one dark spot in the centre of the yellow forewing, whereas the male has a series of small dark spots on a brown forewing. Body length is about 13–16 mm and the wingspan 22–30 mm, the males being smaller. The adults do not feed and only live for 4–10 days.

The entire life-cycle takes from 35–71 days.

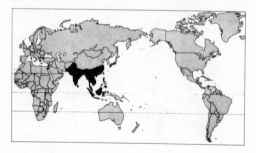

**Distribution.** Recorded from Pakistan, India, Sri Lanka, through S.E. Asia, Indonesia, Philippines, up to China and S. Japan (CIE map no. A252).
**Control.** In some areas early or late planting is recommended so as to avoid crop continuity and a build-up of the borer population. Various other types of cultural control are practised in different countries, according to the local condition. A number of rice cultivars show general resistance to stem borers.

Chemical control is as for *Chilo polychrysa*.
**Further reading**
Grist, D.H. & R.J.A.W. Lever (1969). *Pests of Rice*. pp. 93–9. Longmans: London.

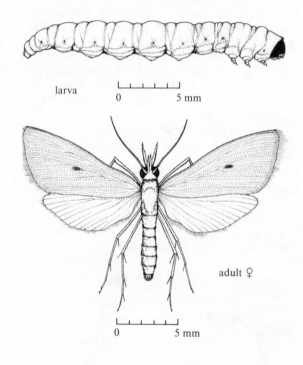

larva

| | |
|---|---|
| 0 | 5 mm |

adult ♀

| | |
|---|---|
| 0 | 5 mm |

# Tryporyza innotata (Wlk.)

(= *Scirpophaga innotata* Wlk.)

**Common name.** White Paddy Stem Borer
**Family.** Pyralidae
**Hosts** (main). Rice
    (alternative). Wild rice and various wild grasses.
**Damage.** The caterpillars bore into the rice stems and hollow out the stem internodes and nodes. Attacked young plants show typical 'dead-heart' symptoms, and older plants often break where the stem is hollowed out, and lodging results.
**Pest status.** A serious pest of rice throughout S.E. Asia.
**Life history.** Eggs are laid in batches of 80–150 on the leaf sheath and covered with a mat of brown hairs taken from the anal region of the female moth. Incubation takes 4–9 days.

    The caterpillars are white in colour with a black head capsule, and are 18–25 mm long when fully grown. Larval development takes from 19–31 days.

    Pupation takes place inside the rice stem, and the pupa is pale and soft; the pupal period takes 7–11 days.

    The adult of this species does not have any dark spots on the forewings at all, in either sex; body length is 13–16 mm and wingspan 22–30 mm, the male being smaller.

    The entire life-cycle takes 30–51 days and there are usually several generations per year.

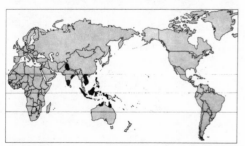

**Distribution.** Pakistan, India, Vietnam, Indonesia, Philippines, West Irian and N. Australia (CIE map no. A253).
**Control.** Several resistant varieties of rice are available, and a number of different cultural practices are used to lower borer populations.

    Chemical control is as for *Chilo polychrysa*.
**Further reading**
Grist, D.H. & R.J.A.W. Lever (1969). *Pests of Rice*. pp. 99–102. Longmans: London.

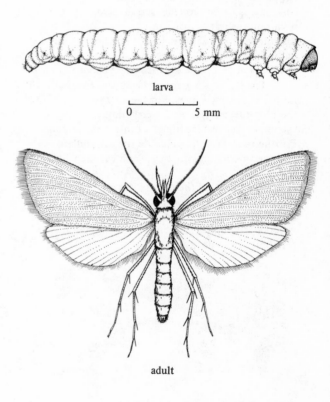

larva

0         5 mm

adult

## Acraea acerata Hew.

**Common name.** Sweet Potato Butterfly
**Family.** Nymphalidae
**Hosts** (main). Sweet Potato
     (alternative). Other species of *Ipomoea*
(Convolvulaceae).
**Damage.** The caterpillars feed on the leaves of sweet potato, there often being several per leaf. Heavy attacks can result in complete defoliation of the vines often several times in succession, over wide areas. Defoliation in young plants causes crop retardation and large reductions in yield.
**Pest status.** A common and serious pest of sweet potato in eastern Africa.
**Life history.** The eggs are laid in batches of 100–150 on both surfaces of the leaves, and hatching takes about seven days. The eggs are pale yellow.

    The larvae are greenish-black and covered with fleshy, branching spines. For the first two weeks of their life the caterpillars are gregarious, feeding on the upper leaf surface under a protective webbing. For the final week of larval life the caterpillars become solitary and nocturnal and eat the whole leaf lamina.

    For pupation the caterpillars crawl away from the crop and climb up any convenient support; here in a vertical position, often several metres from the ground, the pupa is formed. The pupal period lasts for about seven days.

    The adult butterfly is a pretty little nymphalid with orange and black wings and conspicuously knobbed antennae; wingspan is 30–40 mm.
**Distribution.** E. Africa and Zaïre only.
**Control.** Chemical control can usually be achieved by the use of DDT sprays.

adult

0     1 cm

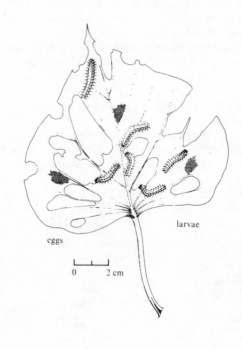

larvae

eggs

0     2 cm

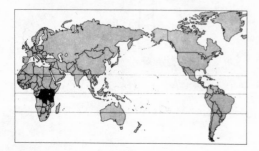

# Chilades lajus (Stoll)

**Common name.** Lime Blue Butterfly
**Family.** Lycaenidae
**Hosts** (main). *Citrus* species
   (alternative). Other members of the Rutaceae, especially *Atalantia*, *Fortunella*, and *Zanthoxylum* spp.
**Damage.** The caterpillar feeds on the flush leaves, and is usually attended by ants, but damage is seldom serious.
**Pest status.** A widespread pest of *Citrus* in tropical Asia, but usually only of minor pest status.
**Life history.** Eggs are laid singly on to the flush growth, where they are stuck firmly to the leaf lamina.

The larvae are pale green, with faint dark markings running longitudinally along the body; at maturity they are about 20 mm in body length; They are attended by ants but are not dependent upon them.

Pupation takes place on the leaf surface, and the small rounded pupa measures about 14 mm.

The adults show sexual dimorphism in that the male is blue-coloured, but the female is brownish with blue infuscation on the inner parts of the wings; both are greyish-coloured underneath with a series of dark markings; there are no 'tails' on this species. Body length is about 8–10 mm and wingspan 20–24 mm; there are distinct wet-season and dry-season forms.

There are several generations per year.
**Distribution.** Sri Lanka, India, Malaya (North), Philippines, and S.E. Asia up to S. China.
**Control.** Usually not required.

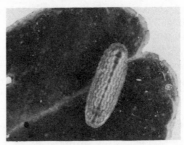

larva

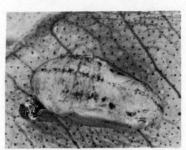

pupa

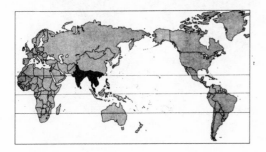

adult ♂    0    1 cm    adult ♀

335

# Lampides boeticus (L.)

**Common name.** Pea Blue Butterfly (Long-tailed Blue)
**Family.** Lycaenidae
**Hosts** (main). Various pulse crops
  (alternative). Cultivated and wild species of
Leguminosae.
**Damage.** The larva feeds inside the pods, eating the
young seeds; on small pods the larvae makes a hole
in the pod wall in order to reach the seeds.
**Pest status.** A widespread pest of legumes, commonly
encountered, but not often a serious pest.
**Life history.** Eggs are laid singly on the shoots, on or
near the young flowers, and the young caterpillar
feeds first inside the flower and then on the young
pod. A crop with large pods will have the larvae inside
the pods eating the seeds, but in crops with small
pods the caterpillars eat holes in the sides of the pods

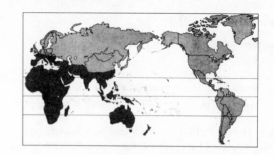

in order to reach the young seeds. The larva is green
in colour and blends well into the general colour of
the foliage. Pupation takes place on the plant foliage.

The adults show distinct sexual dimorphism in
that the upper surface of the male's wings are bright
blue (with a dark distal fringe) but the female is

adult ♂

0                    1 cm

adult ♀

predominantly brown with a blue infusation near the body; both sexes have small 'tails' on the trailing edge of the hindwings — one tail per wing, and each about 2–3 mm in length. The female illustrated had lost her 'tails'. On the hindwing by the 'tail' are two small eyespots which show on both sides of the wing. As with many other blues, these butterflies sit at rest in a 'head-down' position so that the 'tails' simulate antennae and the eyespots simulate eyes, thus creating the impression that the insect is facing in the opposite direction, and presumably misleading would-be predators.

**Distribution.** S. and C. Europe, Africa, through India and S.E. Asia up to S. China, throughout Australasia, New Zealand and Hawaii (USA). Not yet recorded from the American mainland.

**Control.** Generally not required as the populations of this species seldom reach economic injury level.

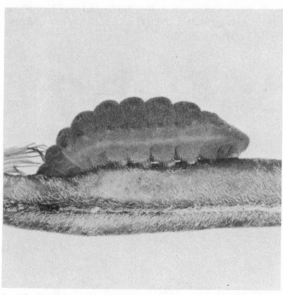

larva feeding in a pod

Photos: G. Johnston

pupa

# Virachola bimaculata (Hew.)

**Common name.** Coffee Berry Butterfly
**Family.** Lycaenidae
**Hosts** (main). Coffee (*arabica*)
(alternative). None recorded.
**Damage.** Single holes are bored into the sides or ends of large green berries. Later the berries turn brown and the edges of the holes bend up to form a distinct rim. Many affected berries are shed. The damaged berry is hollowed out and is clean inside, since the caterpillar pushes its excreta out through the entrance hole. Both beans are usually eaten.
**Pest status.** Normally only a very minor pest of *arabica* coffee but occasional severe outbreaks have occurred — especially at higher altitudes. It is only recorded from *arabica* coffee. The E. and W. Rift forms of the butterfly in Kenya have slightly different markings and may be different races.
**Life history.** The eggs are spiny, and are laid singly, stuck to the sides of berries.

The larva is a green caterpillar with brown markings which grows to a length of about 20 mm before pupating. It probably eats out several berries before reaching maturity.

Pupation usually takes place in a chamber hollowed out in dead wood. The rotting surface where an old vertical stem has been cut off is a favourite site. The pupa is brown, rather squat in appearance and about 14 mm long. The pupal period is about one month.

The male butterfly has wings which are grey below and nearly black above with bold orange-red markings. Its wingspan is a little more than 20 mm. The female has wings which are generally grey on both surfaces; its wingspan is more than 25 mm. Both sexes have delicate, tail-like prolongations of the hindwings.
**Distribution.** E. Africa and Sierra Leone.
**Control.** Attacks are rarely reported until too late for effective action, so no insecticidal trials have been performed.

Since the attack is normally confined to a small area it is suggested that DDT as a foliar spray or dust could be effective.

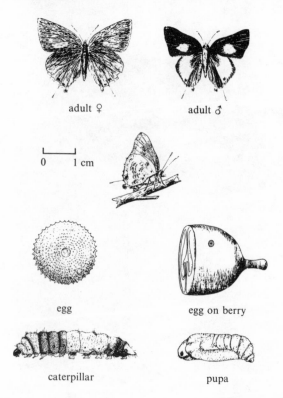

adult ♀          adult ♂

0     1 cm

egg          egg on berry

caterpillar          pupa

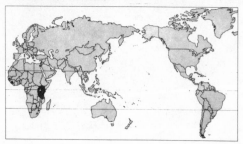

## Pieris canidia (L.)

(= *Artogeia canidia* (L.))

**Common name.** Small White Butterfly
**Family.** Pieridae
**Hosts** (main). *Brassica* species.

(alternative). Other members of the Cruciferae, and a few other plants, such as nasturtium.

**Damage.** The larvae feed singly, usually deep in the cabbage heart, making holes in the leaves, with frass accumulation; so much damage may be overlooked by casual inspection.

**Pest status.** A serious pest of *Brassica* crops in S.E.Asia and India, often occurring as a mixed infestation with other species of *Pieris* (e.g. *P. brassicae* and *rapae* in N. India); a small amount of damage is important as it is usually in the heart of the plant.

**Life history.** Eggs are laid singly on the host leaves (as opposed to the clusters laid by the larger *P. brassicae*), and the solitary caterpillars prefer to feed in the cabbage heart where they are offered more protection. The caterpillar is green with many tiny

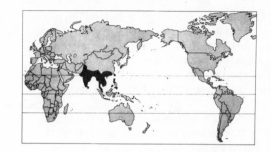

spots and fine setae so that it has a velvety appearance. There is a narrow yellow line dorsally along the back. Mature larvae are about 25 mm in length.

Pupation takes place in an upright pose with a band of silk around the thoracic region holding the chrysalis in position; the cremaster at the end of the abdomen rests upon a small pad of woven silk. The larvae often crawl off the host plant on to a firm substrate and can be found pupating on walls, fences and trees.

larva

pupa

The adults are typical white butterflies, showing distinct sexual dimorphism, the male does not have the two black spots on the upper surface of the forewings, although they are present on the undersides of the wings. Also the marginal black spots on the hindwings are larger in the female. Wingspan is about 5–6 cm.

Breeding is continuous in the warmer regions and there may be eight generations per year; each generation taking as little as 20 days. In more northern regions overwintering occurs in the pupal stage and the winter generation may last for three months.

**Distribution.** India, S.E. Asia, to S. China, Taiwan and the Philippines.

**Control.** Natural levels of predation and parasitism are usually high in most countries, but if chemical control is required then DDT, trichlorphon, or mevinphos, either as dusts or sprays, can be used. Care must be taken to apply the insecticide to the heart region of the plant, and extra wetter may need to be added to sprays to overcome the waxy nature of the plant cuticle.

0   1 cm   adult ♀

## Papilio demodocus Esp.

**Common name.** Orange Dog
**Family.** Papilionidae
**Hosts** (main). *Citrus* spp.
    (alternative). Other species of Rutaceae.
**Damage.** The caterpillars defoliate the trees; all stages feed at the edge of either flush or hard leaves.
**Pest status.** This pest is universally present on all species of *Citrus* in the Old World. It is usually only a minor pest of mature trees but severe attacks are quite common in nurseries and on small trees.
**Life history.** The eggs are pure white or have black bands or patches; spherical, and just over 1 mm in diameter; they are laid singly on flush leaves and hatch after about four days.

    The caterpillar has five instars; the first three are dark brown with white markings and resemble bird droppings; the fourth and fifth instars are pale green caterpillars with black, brown and grey markings. If the caterpillars are disturbed they shoot out a pink Y-shaped organ from just behind the head. Fully grown caterpillars are 5 cm or more long. The larval period lasts about 30 days.

    Pupation takes place on a small branch; the posterior end of the pupa touches the branch but the anterior end is about 10 mm away and connected to it by strands of silk. The pupa varies in colour from yellowish-green to brown and is about 3 cm long. The pupal stage lasts about 14 days.

    The adult is a handsome swallowtail butterfly often seen feeding on the nectar of various flowers. The general colour is dark brown with numerous pale yellow markings.
**Distribution.** Africa.
**Control.** Hand collection of the caterpillars is often effective on small trees. Otherwise, chemical control can be achieved using foliar sprays, applied to run-off, of malathion, fenthion or fenitrothion.

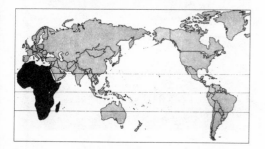

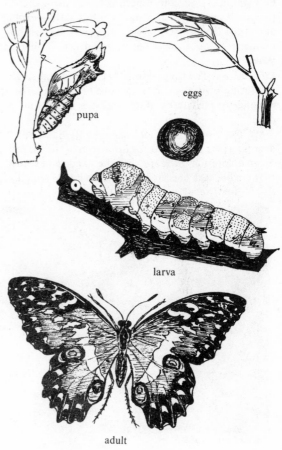

pupa

eggs

larva

adult

341

## Papilio demoleus L.

**Common name.** Lemon Butterfly (Citrus Swallowtail)
**Family.** Papilionidae
**Hosts** (main). *Citrus* species
      (alternative). Other members of the Rutaceae; *Zizyphus*, and bael fruit.
**Damage.** The larvae eat leaves, especially flush growth, and young plants may be defoliated.
**Pest status.** A widespread pest of *Citrus* throughout tropical Asia and Australasia, together with the other species of *Papilio* (see next page). A serious pest of young trees, but usually only a minor pest of mature trees.
**Life history.** The eggs are pale yellow and laid singly on flush growth, usually at the tips of the leaves or at the edge of the lamina. Hatching takes about four days.

      The young larvae look remarkably like bird droppings for the first three instars, then the bright green coloration develops with the bands and spots that enable each species of *Papilio* to be identified at

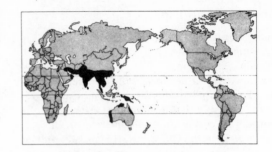

this stage. Full-grown caterpillars measure up to 5 cm at the end of about 30 days growth.

      Pupation takes place on a branch or twig, in an upright position, with the typical band of silk around the thorax to retain the chrysalis in that position. Pupation takes about 14 days.

      The adult is a beautiful, tail-less swallowtail butterfly with a wingspan of 9—10 cm.

      Usually 4—5 generations per year, maybe more in the hotter tropics. This is the Asian version of the African *Papilio demodocus* and to the non-taxonomist

egg on *Citrus* leaf tip

young larva

the two species appear the same; apparently several subspecies of both can be recognized in different parts of their range.

**Distribution.** Saudi Arabia, through Pakistan, India, S.E. Asia up to S. China and Taiwan, and parts of Australia, Papua New Guinea and West Irian (CIE map no. A396).

**Control.** See *P. demodocus*.

mature larva

pupa

adult

0           2 cm

## Papilio spp.

*helenus* L.
*memnon* L. } etc.
*polytes* L.

**Common name.** Citrus Swallowtails
**Family.** Papilionidae
**Hosts** (main). *Citrus* spp.
    (alternative). Other members of the Rutaceae, and a few other plants.
**Damage.** The larvae eat leaves, especially flush growth, and despite being solitary they are often present in numbers large enough to defoliate young trees.
**Pest status.** *Papilio* spp. occur on *Citrus* wherever it is grown throughout the world, and their feeding damage is always noticeable, but seldom serious unless on young trees or in nurseries.
**Life history.** As with the previous two species, eggs are laid singly on the flush growth, but a laying

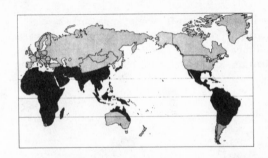

female may well lay several dozen eggs scattered over one tree. The eggs hatch after a few days into small caterpillars that closely resemble bird droppings as they sit conspicuously on the leaf lamina. In the last two instars the larvae develop their individual specific body markings on a green background. There is some variation in these body markings, but experts claim

*Papilio polytes*

defoliated *Citrus* plant

0      2 cm

to be able to recognize the different species of *Papilio* by the bands and spots on the body.

The pupae are indistinguishable generally, although there is some colour variation through different shades of green and brown.

The adults are, of course, quite different in shape and coloration and also in size to some extent. On a pantropical basis there are some 10–15 species of *Papilio* to be found feeding on *Citrus* trees and bushes, and other Rutaceae; the African, Asian, Australasian and American species generally differ from each other. Within these broad distribution zones there are some minor differences; for example,

there are some species in India that do not occur in S. China.

**Distribution.** These 10–15 species of *Papilio* are completely pantropical in distribution, although separated into species groups according to major geographical regions, as indicated above.

**Control.** Hand-picking is still recommended in many countries for removal of larvae in nurseries and from young trees.

For chemical control foliar sprays of malathion, fenthion or fenitrothion are recommended, applied to run-off.

*Papilio memnon*

0      2 cm

## Erionota spp.

*thrax* (L.)
*torus* Evans

**Common name.** Banana Skippers
**Family.** Hesperiidae
**Hosts** (main). Banana
        (alternative). Other members of the Musaceae;
oil palm.
**Damage.** The larvae cut the leaf lamina and make
individual rolls of leaf material in which they live, and
grow, and eventually pupate.
**Pest status.** Usually a widespread but minor pest in
S.E. Asia, but occasionally very heavy infestations
occur and result in crop defoliation.
**Life history.** Eggs are laid in small clusters stuck on to
the leaf lamina. The larvae is a typical skipper in
having a soft white body with vague segmentation,
narrow neck and a round black head capsule; the
body is covered with a fine waxy material.
Pupation takes place within a large leaf roll.

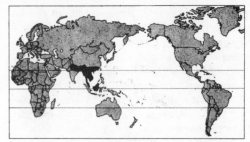

*E. torus*

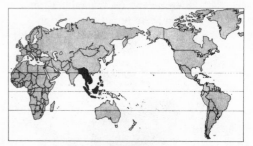

*E. thrax*

eggs

larva

346

The adult is a large brown skipper with yellow marks on the forewings, and, in life, bright shining eyes like small rubies; wingspan is about 7 cm.

**Distribution.** The two species are sympatric in Malaysia, S.E. Asia and Borneo; but *E. thrax* also occurs in Java, Sumatra, and the Philippines, and *E. torus* also occurs in N. India and S. China.

**Control.** Generally not required in most situations, which is fortunate as it would be difficult to reach the insect larvae with conventional sprays; in emergencies, however, the usual insecticides effective against caterpillars could be used, with extra wetter added because of the wax on their bodies.

pupa

0          2 cm          adult

347

# Telicota augias (L.)

**Common name.** Rice Skipper
**Family.** Hesperiidae
**Hosts** (main). Rice
    (alternative). Sugarcane and bamboo.
**Damage.** Damage is done by the caterpillars feeding on the leaves, from the margin inwards towards the midrib, which is usually left intact, and also by tying the leaf edges together to form a tube or roll. The caterpillars live inside this roll. Damage is more severe on young transplanted seedlings which may fail to recover.
**Pest status.** Only occasionally a pest of any real importance, and then usually only on young transplanted seedlings.

    Several other species of *Telicota* and several other genera of skippers are minor pests of rice in S., E. and S.E. Asia, and they are all quite similar to *T. augias* in appearance.
**Life history.** The eggs are laid singly on the leaves; they are pale yellow, and hatch in three days.

    The caterpillars are pale green with a dark head and dark spot on the anal flap, and nocturnal in habit. They are fully grown in 20–25 days, after reaching a length of 40 mm.

    Pupation occurs in the leaf tube; the pupa is pale yellow-green, and takes 8–10 days for development to be completed.

    The adult skippers are orange-brown with brown wing markings, and have the characteristic large head and clubbed antennae with recurved tips, typical of the Hesperiidae.
**Distribution.** India, Bangladesh, Malaysia, Java, Philippines, Papua New Guinea, West Irian and Australia.
**Control.** Control is similar to that for the rice leaf-rollers as they are similar in habits. Contact poisons such as dieldrin, carbaryl, and BHC as foliar sprays have given very satisfactory control.

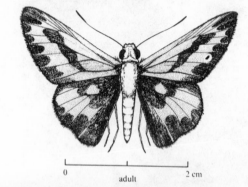

0            adult         2 cm

larval leaf roll

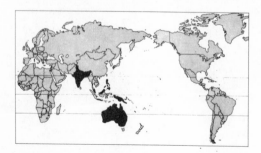

348

## Epicampoptera spp.

*marantica* Tams
*andersoni* Tams

**Common name.** Tailed Caterpillars
**Family.** Drepanidae
**Hosts** (main). Coffee: *arabica* and *robusta*.
    (alternative). None recorded.
**Damage.** The young caterpillars feed on the under surface of the leaf about half-way between the midrib and the edge. The upper surface of the leaf is left intact. The older caterpillars feed at the edge of the leaf sometimes eating everything except the midrib. If the tree is completely defoliated they will eat the berries and the green bark.
**Pest status.** *E. andersoni* is normally a minor pest of *arabica* coffee but severe outbreaks sometimes occur. Generally, *robusta* coffee is attacked by the species *E. marantica*. Three species are known from coffee in Africa.
**Life history.** The eggs are small, oval, usually laid singly; incubation takes 8–9 days.

The larva is an easily recognized caterpillar, having a humped appearance due to the large thorax and the abdomen tapering to a thin tail. The colour is variable but the caterpillars are usually green when young, becoming purple or velvety brown when older. Fully grown, they are about 55 mm long. The caterpillar stage lasts for about four weeks.

The fully grown caterpillar rolls a leaf up into

a cone, closes the open end with a pad of silk and pupates inside. The pupal period lasts about two weeks.

The adult moth is silvery grey to dark brown or mottled black in colour. The wingspan is about 40 mm. In the daytime it rests on the underside of leaves. Egg-laying occurs at night; each female moth lives for about six days and lays more than 500 eggs.
**Distribution.** Most parts of Africa.
**Control.** Chemical control can be achieved by foliar sprays of either fenitrothion or fenthion.

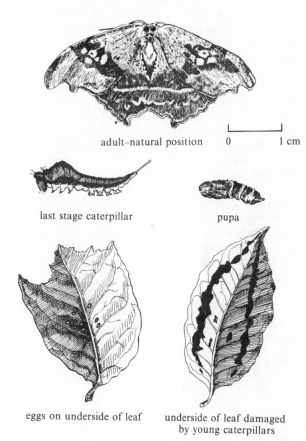

adult–natural position     0     1 cm

last stage caterpillar      pupa

eggs on underside of leaf     underside of leaf damaged
                                     by young caterpillars

349

## Ascotis selenaria reciprocaris (Wlk.)

**Common name.** Giant Looper
**Family.** Geometridae
**Hosts** (main). Coffee
   (alternative). Many other plants — 78 host species are recorded in S. Africa.

**Damage.** Young caterpillars chew pits in the upper leaf surface; older caterpillars chew circular holes right through the leaf. The oldest caterpillars eat at the margin of the leaf leaving a jagged edge. They also feed on the other green parts of the plant, flower buds, young berries, terminal shoots and the bark of sucker stems. If the terminal point is damaged, there may be fan-branching and growth distortions.

**Pest status.** This has been a minor pest of *robusta* coffee in Uganda for many years and has become a major pest of *arabica* coffee in Kenya recently, after very frequent use of parathion in coffee plantations — the caterpillars are not particularly susceptible to parathion but the predators and parasites are.

**Life history.** The eggs are pale blue-green, oval, about 0.7 mm long. They are laid in crevices in the bark. Hatching takes about 7—10 days.

There are usually five larval instars, taking about 25—42 days. The caterpillar is a typical geometrid (looper) having only one pair of prolegs (in addition to the terminal claspers), and they move in the typical looping motion. The colour varies from pale grey to dark brown. Fully grown, the caterpillar is about 5 cm.

Pupation takes place in the soil; the pupa is shiny brown, and 18—20 mm long; pupation takes 14—21 days.

The adult is a night-flying moth, with a wingspan of 5 cm; colour is variable but is commonly pale grey with many dark grey markings. Egg-laying starts 1—2 days after emergence, and each female may lay about 2000 eggs; the females usually live for about 14 days.

**Distribution.** S. and E. Africa only.

**Control.** Chemical control can be effected using foliar sprays of methomyl.

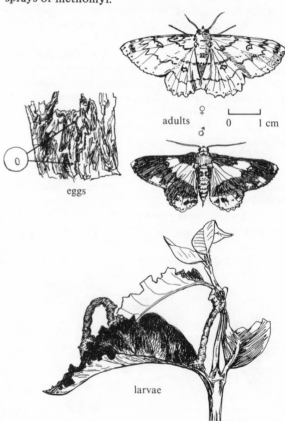

adults ♀ ♂    0    1 cm

eggs

larvae

# Leucoplema dohertyi (Warr.)

**Common name.** Coffee Leaf Skeletonizer
**Family.** Epiplemidae
**Hosts** (main). Coffee, all species.
     (alternative). None recorded.
**Damage.** The larvae feed on the undersides of leaves, usually near the midrib. Everything except the veins and upper epidermis is eaten, leaving irregular lace-like patches in the leaf.
**Pest status.** This pest attacks all cultivated species of coffee. It is usually a minor pest but severe outbreaks sometimes occur, especially in nurseries.
**Life history.** Eggs are laid singly or in small groups mostly on the underside of the leaf. Patches of old Skeletonizer damage are a favoured site. They are yellow-green, dome-shaped, and about 0.5 mm in diameter. Hatching takes about seven days.

The larva is a grey or white caterpillar with many pimple-like projections on its body. The larval period is about three weeks. On the day prior to pupation the caterpillar turns red; it is then about 10 mm long.

The mature caterpillar lowers itself on a silken thread and pupates in the ground. The pupal period lasts for about three weeks.

The adult is a grey and brown moth with a wingspan of about 14 mm. It is found resting on leaves during the day with the hindwings drawn back alongside the body and the narrow forewings held at right angles to the body.

**Distribution.** E. Africa, Zaïre, Ghana, Angola.
**Control.** Chemical control of this pest can be achieved with either fenitrothion or fenthion, as foliar sprays, the amount of spray used varying with the amount of leaf on the tree.

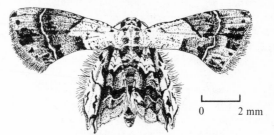

adult, natural position

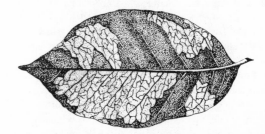

damaged leaf

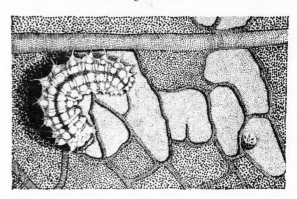

caterpillar and egg on patch of damage

351

# Attacus atlas (L.)

**Common name.** Atlas Moth
**Family.** Saturniidae
**Hosts** (main). Cinnamon

(alternative) Cashew, coffee, guava, mango, tea, and other Lauraceae including camphor; also *Sapium* and *Schefflera* trees in the forests.

**Damage.** The larvae are leaf-eaters and because of their large size a small population may defoliate small trees.

**Pest status.** A serious pest of cinnamon in parts of Malaysia; elsewhere generally only a minor pest.

**Life history.** Eggs are laid in batches on the leaves of the host plant, and the young caterpillars soon disperse generally over the plant. There are five larval instars and the mature caterpillars are very large (up to 10 cm in length), white and coated with a flaky white wax. There are dark lateral tubercles along the body and several rows of long white tubercles dorsally along the abdomen.

Pupation takes place within a large silken cocoon spun between two leaves.

The adult is probably the largest known moth and females may measure up to 25 cm in wingspan. They are completely nocturnal and will never be seen in flight during the day. The females of this family are well known for the copious quantities of sex pheromone produced by newly emerged virgins, and their 'calling' is so successful that males come from considerable distances (1–2 km).

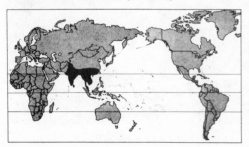

There is usually only one generation per year, but there might be two in Malaysia.

**Distribution.** The genus occurs throughout the Old World tropics, but this species is found through India, S.E. Asia and up to China.

**Control.** Usually control measures are not required; the larvae are so large and conspicuous that on small trees hand-picking would be feasible.

larva

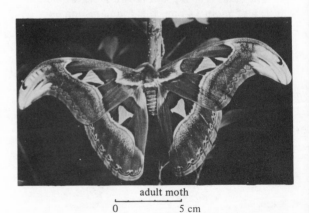

adult moth

0                    5 cm

# Acherontia atropos (L.)

**Common name.** Death's Head Hawk Moth
**Family.** Sphingidae
**Hosts** (main). Potato, tomato, eggplant, tobacco.
    (alternative). Wild members of the Solanaceae; olive, sesame and some other plants.
**Damage.** The larvae eat leaves, and although solitary they are very large caterpillars so a small number can cause defoliation of the host plant. However, frequently there is only one larva per plant.
**Pest status.** Generally only a minor pest.
**Life history.** Eggs, often coloured reddish or blue, are laid singly on the foliage of the host plant — usually only one per plant.

The larva is white initially, with a yellow head and black 'horn', but when larger the colour is predominantly yellow, green or grey with diagonal blue stripes. The characteristic 'horn' on the last body segment, which gives all sphingid larvae the collective name of 'hornworms' in the USA, is the same colour as the body and is distinctly hooked dorsally at the tip. Mature larvae measure up to 15 cm in body length and larval development usually takes several months.

As with all Sphingidae pupation takes place in the soil in an earthen cocoon. In the more northern parts of its range it overwinters in the pupal stage and is univoltine. The length of the pupa is about 80 mm.

The adult is a large stout-bodied moth of striking appearance with long, dark forewings and short, yellow hindwings with black barring. The dorsum bears the skull-like marking from which it derives its common name. The abdomen is banded yellow and black. Body length is 50–55 mm and wingspan 8–12 cm. The natural resting position is shown in the top illustration. As with most other Sphingidae the adults are nocturnal in habits. This species is remarkable in that the adult moth can make a high-pitched chirping noise and is reputed to enter bee hives to steal honey.
**Distribution.** Europe, the whole of Africa, Middle East, India, S.E. Asia, China and Japan.
**Control.** Generally not needed.

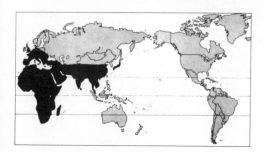

adult in natural resting position

0      3 cm

adult in pinned or flying position

353

## Agrius convolvuli (L.)

(= *Herse convolvuli* (L.))

**Common name.** Sweet Potato Moth (Convolvulus Hawk Moth)
**Family.** Sphingidae
**Hosts** (main). Sweet Potato
    (alternative). Other Convolvulaceae, and some Leguminoseae, for example *Phaseolus* spp. and *Glycine*; also sunflower.
**Damage.** The larvae defoliate the plant.
**Pest status.** A very widespread species, common in Africa, but not usually a serious pest for the damage to the sweet potato plants is not extensive and the plants can tolerate some defoliation. It has been very serious in parts of India on beans.

It is known as the Convolvulus Hawk Moth in Europe.
**Life history.** The eggs are small for a sphingid, sub-spherical, 1 mm diameter, laid singly on any part of the food plant.

There are five larval instars, each with a conspicuous posterior 'horn' which gives the family its common name in the USA of 'hornworms'. The colour is variable, usually either greenish or brownish. Fifth instar caterpillars are large — up to 95 mm long and 14 mm broad. The larval period lasts 3—4 weeks.

Pupation takes place in the soil, some 8—10 cm down; pupal duration is variable, it may be 17—26 days or as long as 4—6 months according to the climate.

Adults are large grey hawk moths with black lines on the wings and broad incomplete pink bands on the abdomen. Wingspan is 80—120 mm. The adults are crepuscular in habits and feed on flowers, particularly those with a long tubular calyx (e.g. *Hibiscus, Ipomoea, Begonia,* etc.).
**Distribution.** Europe, most of Africa; Iran, India, Bangladesh, Burma, Malaysia, Indonesia, Australia, New Zealand, Papua New Guinea, West Irian, S. China and the Pacific islands.
**Control.** Should insecticidal treatment be required then dusts or sprays of TDE, or sprays of parathion, applied when the first signs of damage are evident, should be effective.

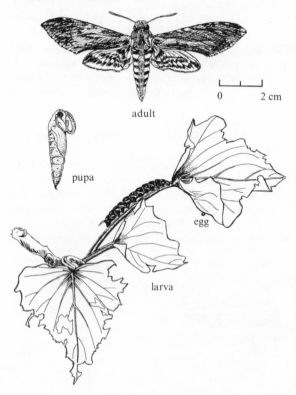

adult

pupa

egg

larva

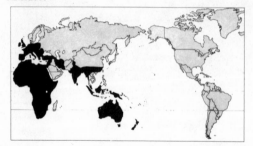

354

# Cephonodes hylas (Wall.)

**Common name.** Coffee Hawk Moth (Bee Hawk Moth)
**Family.** Sphingidae
**Hosts** (main). Coffee
    (alternative). *Gardenia*, and some other (but not all) members of the Rubiaceae.
**Damage.** The larvae eat leaves and in heavy infestations may actually defoliate the host tree (bush).
**Pest status.** A widespread species, frequently encountered, and usually only a minor pest; but in Malaya at the turn of the century it was a very serious pest.
**Life history.** Eggs are laid singly on the coffee leaves, and each female may lay about 90 eggs. Hatching occurs after three days.

The caterpillars are green laterally, with conspicuous red spiracles, a dorso-lateral stripe of white separates the green flanks from the blue-coloured back, and when mature they measure about 5–6 cm in length. Larval development takes 20–22 days.

Pupation takes place in the soil inside an earthen cocoon, at a depth of about 5 cm, or in the leaf litter at the base of the tree. Pupal development in Malaya takes some 12–14 days, but in China and Japan the pupa overwinters.

In the tropics there are several generations per year, but further north there are probably only one or two generations.

The adult is a smallish hawk moth with a wingspan of 5–6 cm and characteristic hyaline wings; it is one of the few diurnal species. The abdomen has a median red band on two segments followed by two yellowish segments. The species is quite distinctive.
**Distribution.** From W. to E. Africa, Madagascar, India and Sri Lanka, through Malaysia and S.E. Asia up to China and southern Japan; also Papua New Guinea and Queensland (Australia).
**Control.** At the present time control measures are normally not required for this species on coffee, but DDT or carbaryl should be effective.

adult moth

0     1 cm

mature larva
Photo: F. Bascombe

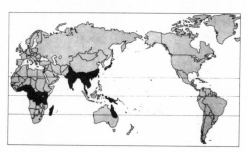

355

## Hyles lineata (F.)

(= *Celerio lineata* F.)

**Common name.** Striped Hawk Moth
**Family.** Sphingidae
**Hosts** (main). Grapevine

(alternative). Cotton, olive, buckwheat, *Prunus* spp., sweet potato and others.

**Damage.** The caterpillars are leaf-eaters and cause a certain amount of defoliation, but since they are solitary damage is usually only slight, although one larva can eat many leaves.

**Pest status.** Generally only a minor pest, but widespread in distribution and quite frequently encountered in the field.

**Life history.** Eggs are laid singly on the plant foliage; they are green and about 1 mm in diameter.

The larvae are greyish or greenish, with a brown-coloured horn, and finally reach a size of about 7–8 cm, after about 25–35 days.

Pupation takes place in the soil inside an earthen cocoon, either in the soil or sometimes in the leaf litter.

The adult is basically brown above but more orange-rufous ventrally, with the hindwings conspicuously red and a red stripe laterally along the body, and a white stripe dorso-laterally along the thorax and head. The wingspan is 5–6 cm.

In warm climates one generation takes 30–40 days, but in Israel the winter generation requires

seven months, or more. Southern Europe is generally too cold for this species to develop successfully, and most populations originate from spring migrants from North Africa which even reach as far north as southern England.

**Distribution.** S. Europe, N. Africa, E. and S. Africa, Middle East, India, China, Japan, parts of Australia, most of the USA, and parts of C. and S. America (CIE map no. A312). Within this range there are several recognized subspecies.

**Control.** Usually not required.

adult

0    2 cm

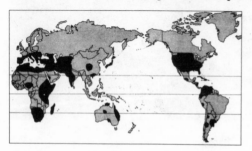

## Agrotis ipsilon (Hfn.)

(= *Euxoa ypsilon* Rott.)
(= *Scotia ipsilon* Rott.)

**Common name.** Black (Greasy) Cutworm
**Family.** Noctuidae
**Hosts.** A polyphagous cutworm attacking the seedlings of most crops, in particular cotton, rice, potato, tobacco, cereals, and crucifers.
**Damage.** The young larvae feed on the leaves of many crops; the older caterpillars feed at the base of crop plants or on the roots or stems underground. Seedlings are typically cut through at ground level; one caterpillar may destroy a number of seedlings in this manner in a single night.
**Pest status.** A cosmopolitan pest of sporadic importance on many crops in different parts of the world. It can cause severe damage on rice in S.E. Asia and in Australasia. On other crops the occasional severe infestation usually results in devastating damage.
**Life history.** The eggs are white, globular, and ribbed; 0.5 mm in diameter; and hatch in 2–9 days. Each female may lay as many as 1800 eggs.

The larvae are brownish above with a broad pale grey band along the mid-line, and with grey-green sides with lateral blackish stripes. The head capsule is brownish-black with two white spots. The general appearance of the caterpillar is blackish. The mature caterpillar is 25–35 mm long; larval development takes 28–34 days. In temperate countries some larvae overwinter as such, and pupate in the late spring. The first two instars feed on the foliage of the plant, the third instar becoming non-gregarious, in fact often cannibalistic, and adopting cutworm habits.

The pupa is dark brown, 20 mm long, with a posterior spine; pupation takes 10–30 days, according to temperature.

The adults are large, dark noctuids with wingspan of 40–50 mm, with a grey body, grey forewings with dark brownish-black markings; the hindwings are almost white basally but with a dark terminal fringe; paler in the males.

The life-cycle from egg to adult takes 32 days at $30^{\circ}$C, 41 days at $26^{\circ}$C, and 67 days at $20^{\circ}$C.
**Distribution.** Almost completely cosmopolitan, from northern Europe, Canada, Japan, down to New Zealand, S. Africa, and S. America. It has not been recorded to date from a few areas in the tropics (e.g. S. India, N.E. South America) (CIE map no. A261).
**Control.** See following section on the control of cutworms.

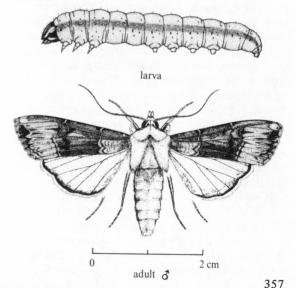

larva

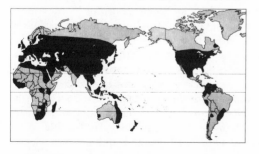

0       2 cm

adult ♂

357

# Control of Cutworms
## (Lepidoptera; Noctuidae)

Cutworms are the caterpillars of various Noctuidae, belonging mostly to the genera *Agrotis* and *Euxoa*; the group (and some species) is completely cosmopolitan in distribution. They are generally more important in temperate rather than tropical situations. (The genera *Agrotis* and *Euxoa* are very closely related and separated only by minor esoteric taxonomic characters, and some species have been placed in both genera at different times.)

The larvae are nocturnal in habits and spend the day hiding in the litter or in the soil sometimes to a depth of up to 10 cm. Whilst subterranean they will feed on plant roots and tubers, boring a wide shallow hole (somewhat like slug damage) in potatoes (see p. 76) and other root crops. Thick-stemmed vegetables, such as lettuce and brassicas may have the stem below ground completely hollowed out; the attacked plant first wilts and then dies. At night the larvae customarily come to the soil surface and feed on plant stems at about ground level. Damage to seedlings and close-planted crops (carrots, lettuce, celery, red beet, some brassicas) is particularly serious, and root crops such as potato, turnip and parsnip are often severely damaged. Typical damage is for the cutworm to move along the row of seedlings cutting each one through the stem at ground level. Root damage is generally most serious in light soils where the soft-bodied caterpillars can burrow more easily.

The female moths lay many eggs (1000–3000), in a series of batches, usually on leaves and stems of weeds or crop plants, or sometimes on litter or plant debris. The first instar larvae feed on the leaves of the host plants, and when larger they descend to the soil and adopt typical cutworm habits. There are typically six (sometimes five) larval instars.

Some species are migratory, and most are characterized by extreme population fluctuations, being abundant in some years and scarce in others.

Suggested control methods of population control are as follows.

(1) Cultural methods

(a)  Weed destruction — these plants are often preferred sites for oviposition, and food for the first instar larvae.

(b)  Hand-collection of larvae — often more suitable for gardens and small-holdings

(c)  Flooding of the infested field may be feasible for some crops.

(d)  Deep ploughing will bring larvae and pupae to soil surface for exposure to predators and sun.

(2) Chemical methods

(a)  High-volume sprays (at least 1000 l/ha) of insecticides.

| | |
|---|---|
| DDT (25%) | 4.2 l/ha |
| endrin (w.p.) | 3.75 l/ha |
| chlorpyrifos (48% e.c.) | 2.5 l/ha |
| triazophos (40% e.c.) | 2 l/ha |

The spray should be directed along the plant rows, aiming at run-off to the soil below. Timing should ideally be aimed at the young caterpillars whilst still feeding on the plant leaves or on the soil surface.

(b)  Soil application of bromophos (w.p.) or chlorpyrifos granules.

(c)  Baits of moist bran mixed with DDT, γ-BHC, or endrin may be effective against older caterpillars.

Generally cutworms are extremely difficult to control, especially in 'boom' years, for by the time infestations are apparent the susceptible stages of the larvae are often past, and damage may be already quite serious. The sporadic nature of cutworm population outbreaks makes preventative treatments rather futile in most areas. Finally the soil-dwelling larvae, often under dense and continuous crop foliage, make targets difficult to 'hit' with insecticides, and in many areas such high-volume spraying is not feasible because of water shortages and equipment restrictions.

# Agrotis segetum (D. & S.)

(= *Euxoa segetum* Schiff.)

**Common name.** Common Cutworm (Turnip Moth)
**Family.** Noctuidae
**Hosts.** A polyphagous cutworm attacking the seedlings of many crops, and many vegetables and root crops.
**Damage.** Seedlings are cut through the stems just above ground level; one caterpillar may cut through a number of seedlings in one night. Root crops may be deeply gnawed, often at levels well below the soil surface. The bark of young coffee trees may be scarred.
**Pest status.** A cosmopolitan pest of sporadic importance on a wide range of crops; occasionally the damage by this pest may be devastating.
**Life history.** The eggs are laid on the stems of weeds or crop plants, or on the soil, and they hatch in 10–14 days. Each female may lay up to 1000 eggs.

The early instars generally remain on the foliage of the host plants for a week or two, then as the caterpillars develop they gradually move down into the soil and assume the cutworm habits. They usually remain in the soil during the day and come to the surface to feed at night. Sometimes the older caterpillars may remain in the soil all the time feeding on the roots of carrots, other root crops and potatoes. The clay-coloured caterpillar is about 30–40 mm long at maturity with faint dark lines along the sides of the body.

The pupa is smooth shiny brown with two spines at the rear; 20–22 mm long. Pupation takes place in the soil.

The adult moth is 30–40 mm across the wings; the forewing is grey-brown with a dark brown or black kidney-shaped marking; the hindwings are almost white in the male, but darker in the female.

There is only one generation per year in Europe but there may be four per year in Africa.
**Distribution.** Cosmopolitan, with the exception of the American continent; occurring in all of Africa, Europe, USSR, the Asian mainland, Sri Lanka, Taiwan, Japan, and Indonesia.
**Control.** As for *A. ipsilon*.

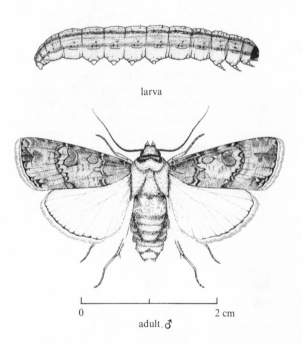

larva

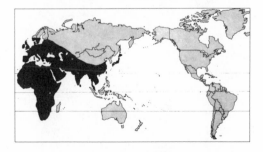

0                     2 cm

adult ♂

## Busseola fusca (Fuller)

**Common name.** Maize Stalk Borer
**Family.** Noctuidae
**Hosts** (main). Maize and sorghum.
 (alternative). Young caterpillars can be found in many species of grass and cereals, but only those with thick stems can support the larvae to maturity.
**Damage.** Young plants have holes and 'windows' in the leaves, and small dark caterpillars may be seen in the funnel. In severe attacks the central leaves die. In older plants the first generation caterpillars bore in the main stem and later some of the second generation caterpillars may be found boring in the cobs.
**Pest status.** A major pest of maize and sorghum in E. Africa and other parts of tropical Africa, in areas with an altitude greater than about 700 m.
**Life history.** The globular eggs are about 1 mm in diameter, and are laid under a leaf sheath in a long column stretching up the stem. They are white when first laid but darken with age. Hatching takes place after about ten days.
 The larva is a buff or pinkish caterpillar with more or less distinct black spots along the body; and the full-grown size is about 40 mm long. On hatching the first instar larvae are blackish; they crawl up the plant into the funnel where they eat the leaf tissues, leaving only the epidermal layer on the supper surface (i.e. 'windows'). After some time they either move to another plant or bore down the funnel into the centre

of the stalk where they feed until fully grown. The mature caterpillar cuts a hole in the side of the stem before pupating within the tunnel. The total larval period is usually 35 days or more. There are usually two generations of Stalk Borer before the crop ripens. In the second generation some eggs may be laid on the cobs, where the caterpillars also feed but move into the stem when fully grown. The mature caterpillar of the second generation often goes into a diapause which will be broken at the onset of the

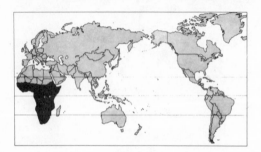

360

next rainy season when it will prepare a pupal chamber in the stem and pupate.

The pupa is brown and about 2.5 mm long; the pupal stage lasting ten days or more, according to temperature.

The adult is a brown night-flying moth with a wingspan of about 35 mm. It emerges through the hole in the stem prepared by the mature caterpillar. There is a pre-oviposition period of 2–3 days.

**Distribution.** A widespread pest in the maize-growing areas of tropical and subtropical Africa, from S. of the Sahara down to S. Africa.

**Control.** See following section on the control of cereal stalk borers.

# Control of Cereal Stalk Borers (Pyralidae & Noctuidae)

As with the other borers (except fruit flies) the eggs are laid on the plant foliage and the young larvae have to find their way to the stem. But these caterpillars are slightly different from other borers in that the first instars usually feed on the leaves for a while before entering the stem (see illustration of feeding damage on page 67). The site of oviposition is usually under the leaf sheath. The young larvae are thus vulnerable to contact insecticides while they are feeding inside the funnel of the cereal seedling.

Experience has shown that cereal stem borers are often best attacked by an IPM programme, as demonstrated by the successes on sugarcane in the USA (Hensley, 1980), and maize in New Zealand (page 82). The different ways in which stalk borers can be controlled are enumerated below.

(1) Cultural control

(a)     Resistant varieties – some varieties with tight or extensive leaf sheaths are not favoured for oviposition; other varieties have an increased silica content in their tissues and feeding larvae usually die.

(b)     Simultaneous sowings over a large area prevent population build-up.

(c)     Destruction of crop residues – important for killing of pupae left in old stems and tall stubble.

(d)     Destruction of thick-stemmed grass weeds which would act as alternative hosts.

(e)     Close season of at least two months to prevent population continuity.

(2) Biological control

(a)     *Trichogramma* spp. (Chalcidoidea) as egg parasites.

(b)     *Apanteles* spp. (Braconidae) as larval parasites.

(3) Insecticidal control

(a)     Dusts ⎫ Contact insecticides applied down the
(b)     Sprays ⎬ funnel of young plants to kill the
          ⎭ emerging and feeding first instar larvae.

(c)     Granules – applied either by foliar lodging, to the soil, or to the water for paddy rice; usually systemic in action.

(d)     Systemics applied as sprays.

The list of effective chemicals being used against cereal stalk borers is extensive, and includes DDT, γ-BHC, endrin, azinphos-methyl, diazinon, endosulfan, fenthion, fenitrothion, monocrotophos, phorate, phosphamidon, tetrachlorvinphos, triazophos, carbaryl and carbofuran, either as powders, sprays or granules.

Resistance is a problem in some areas, and also certain species are less sensitive to some chemicals, so that local advice should be sought for control of any particular stalk borer.

# Diparopsis spp.

*castanea* (Hamps.)
*watersi* (Roths.)

**Common name.** Red Bollworms (Red Bollworm;
Sudan Bollworm)
**Family.** Noctuidae
**Hosts** (main). Cotton
(alternative). This pest is oligophagous (almost
monophagous) in being restricted to *Gossypium* and
the two closely related genera *Cienfuegosia* and
*Gossypioides*, although odd records from other
Malvaceae have been made.
**Damage.** The young larvae bore into flower buds
which are then eaten out and later dehisce. Usually
a few buds are eaten before the larva penetrates a
boll and remains inside eating out the contents.

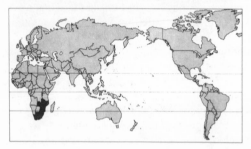

D. castanea

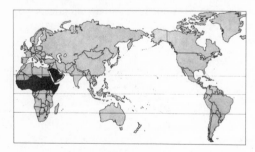

D. watersi

**Pest status.** These species are major pests of cotton
in Africa; *D. castanea* being found S. of the Equator,
and *D. watersi* N. of the Equator. There are two other
species of *Diparopsis* attacking cotton in Africa.
**Life history.** The eggs are subspherical, 0.5–0.7 mm
in diameter, bluish in colour, with both vertical and
horizontal ribbing; usually laid singly on young leaves
or stems, taking 4–10 days to hatch. Each female
may lay up to 500 eggs.

The newly hatched larva is pale, with head, pro-
thoracic plate, anal plate, and legs black. The characteristic red markings appear in the second instar, and
consist of a median dorsal mark flanked with an
oblique one on either side, and with a broad lateral
mark above each spiracle. The mature larva is
25–30 mm long. Larval development takes 11–23
days. The young larvae soon start boring into buds

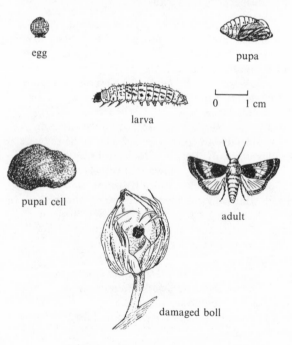

egg

pupa

larva

0 1 cm

pupal cell

adult

damaged boll

and young bolls; each larva may consume six or more flower buds, or alternatively it may spend most of its life within one cotton boll.

Pupation takes place in an earthen cell in the soil at depths of up to 15 cm, and lasts some 2—3 weeks, unless diapause is involved when the period of quiescence may be as long as 35 weeks.

The adult moths are stout-bodied with wing-span of about 25—35 mm. The abdomen and hind-wings are silvery cream, the latter slightly infuscate at the margins which are fringed. The forewing colour is quite variable but the commonest pattern is with the central area reddish, basal and distal bands a shade darker, and the penultimate band grey-brown. However, both yellowish and greenish-pink forms occur. The two species are difficult to separate on the grounds of forewing coloration.

The moths are sexually mature upon emergence, and mating and oviposition may occur on the night of emergence.

**Distribution.** *D. castanea* occurs in S.E. Africa, in the Transvaal, Natal, Swaziland, Mozambique, Zimbabwe, Malawi, and Zambia.

*D. watersi* is found in the Sudan, and from Somalia across to Senegal, all areas being north of the Equator in Africa (including Sierra Leone, Guinea, Ivory Coast, Ghana, Upper Volta, Mali, Benin, Nigeria, Cameroons, Chad, Central African Republic and Ethiopia). It is also found in Arabia.

This genus does not occur in E. Africa or Zaïre, except for records in the southernmost tip of Tanzania.

**Control.** Control in E. Africa is by legislative means through the maintenance of a cotton-free zone in S. Tanzania, which has to date effectively prevented the spread of this pest from Zambia and Malawi into E. Africa.

The recommended insecticides for the control of Red Bollworm are sprays of carbaryl applied weekly, starting when the first flower buds appear, the results generally improving as the number of

sprays is increased from four to eight.

DDT, BHC/DDT, DDT/toxaphene, ethion, and monocrotophos were also recommended by Wyniger (1968).

**Further reading**

Pearson, E.O. (1958). *The Insect Pests of Cotton in Tropical Africa*, pp. 96—141. CIE: London.

## Control of Bollworms on Cotton

The caterpillars that bore developing cotton bolls belong to several different families (i.e. Gelechiidae, Tortricidae), but most are Noctuidae. Eggs are laid on the young foliage (terminal shoots, leaves and squares) and the first instar larvae have to search for the young bolls or buds. A single larva may destroy several buds or may spend its entire larval period inside a single boll. As with other borers once the larva is inside the boll it is safe from contact insecticides, and generally it is not worthwhile trying to control bollworms with systemics, for once the larva is in situ the developing lint is already damaged. Thus, when using contact insecticides, the timing of application is absolutely vital for control of cotton bollworms.

Scouting for eggs in the crop, or field-sampling of foliage for eggs, is generally recommended twice weekly at the appropriate time, and insecticide sprays applied when the first eggs are found. Sprays should be applied to the crop foliage at weekly intervals so long as the egg-laying period persists. The first eggs are generally found at the time of the first flower buds, which is some six weeks after germination.

The insecticides used at present are mainly carbaryl, DDT, cypermethrin and endosulfan, but for American Bollworm usually DDT and endosulfan are recommended, as carbaryl was only reported effective against it in areas of low rainfall. For mixed bollworm infestations, a spray mixture is generally advised although cypermethrin is reported to be effective.

# Earias spp.

*biplaga* Wlk.
*insulana* (Boisd.)
*vittella* (F.) (= *E. fabia* Stoll)

**Common name.** Spiny Bollworms

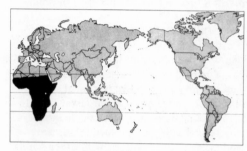

*E. biplaga*

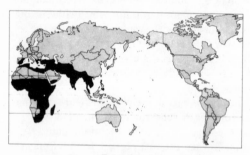

*E. insulana*

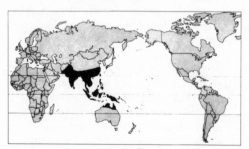

*E. vittella*

**Family.** Noctuidae
**Hosts** (main). Cotton, okra.

(alternative). Species of *Hibiscus* and *Abutilon* and other Malvaceae; also cocoa and a few members of the Tiliaceae and Sterculiaceae.
**Damage.** Terminal shoots of young cotton plants are bored, causing death of the tip and subsequent development of sideshoots and branches. Flower buds and young bolls are shed after being bored by the caterpillars; large bolls are bored but are not shed.
**Pest status.** Present in many Old World cotton-growing areas most seasons and sometimes very severe attacks occur.

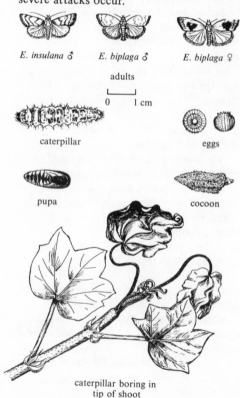

*E. insulana* ♂    *E. biplaga* ♂    *E. biplaga* ♀

adults

0    1 cm

caterpillar

eggs

pupa

cocoon

caterpillar boring in
tip of shoot

364

A total of seven species of *Earias* are found on cotton in different parts of the Old World.

**Life history.** The eggs are blue, subspherical, about 0.5 mm in diameter; the shell is ribbed and alternate ribs project above the egg forming a crown. Eggs of *E. insulana* have longer projections than those of *E. biplaga*. Oviposition occurs anywhere on the plant; on young plants they are usually found singly on young shoots; on older plants they are usually on the stalks or bracteoles of flower buds or young bolls. Hatching takes 3–4 days.

The larva is a stout, spindle-shaped caterpillar which, when fully grown, is 15–18 mm long. Most segments have two pairs of fleshy, finger-like tubercles which give the caterpillar its common name. The colour is variable but is usually pale brown tinged with green or grey and with yellowish spots. *E. insulana* larvae are usually paler in colour than *E. biplaga*. There are five larval instars, the larval period taking 12–18 days. In young plants the caterpillars bore in the soft terminal shoots causing death of the growing point. Older larvae feed on flower buds (squares) and green bolls of various ages. The bracteoles of damaged flower buds open out, causing the condition known as 'flared squares'. The entrance hole of the caterpillar in a bud or boll is neat and circular and may be blocked with frass.

The mature caterpillar spins its cocoon on the plant or among the plant debris on the soil surface and pupates inside it. The pupa is brown with rounded ends and is about 13 mm long; the pupal stage lasts 7–12 days.

The adult *E. insulana* is a small moth with green or yellowish-green wings, pale hindwings, and a wing-span of 20–22 mm. The adult *E. biplaga* male usually has yellow forewings with a brown edge and brown markings. The female has greenish forewings which have a brown edge and a brown patch in the centre of the wing.

After a pre-oviposition period of 3–4 days the female moth may live a further 40 days, and lay 300–600 eggs.

**Distribution.** The genus *Earias* is confined to the Old World including Australasia; *E. insulana* covers most of Africa, to the Mediterranean and S. Europe, Middle East, to India and S.E. Asia including the Philippines (CIE map no. A251).

*E. biplaga* is confined to Africa south of the Sahara.

*E. vittella* is the common Oriental species, found from India and China to N. Australia (CIE map no. A282).

**Control.** To control Spiny Bollworms effectively it is necessary to apply the sprays while the caterpillars are still small. The recommended insecticides are carbaryl and a mixture of DDT + BHC dusts.

**Further reading**

Pearson, E.O. (1958). *The Insect Pests of Cotton in Tropical Africa*, pp. 74–95. CIE: London.

## Heliothis armigera (Hb.)

(= *Heliothis obsoleta*)
(= *Helicoverpa armigera* Hb.)

**Common name.** American Bollworm
**Family.** Noctuidae
**Hosts** (main). Cotton, beans, maize and sorghum.
    (alternative). Tobacco, tomato, many legumes, some vegetables, and other plants.
**Damage.** Clean circular holes are bored in flower buds, cotton bolls of all sizes and some fruits.
**Pest status.** A sporadically very serious pest of cotton and beans in many parts of the Old World; completely polyphagous, and a minor pest on many cultivated fruits (in the botanical sense).
**Life history.** Eggs are spherical, 0.5 mm in diameter, yellow turning brown; hatching takes 2–4 days. Each female moth may lay 1000 or more eggs.

    The larva is stout, greenish or brown, but variable in coloration. Body bears long dark and pale bands; fully grown larvae are 40 mm long. There are six larval instars and the total period lasts 14–24 days, but 51 days at 17°C.

    Pupation takes place in the soil; the shiny brown pupa is about 16 mm long; pupal development takes 10–14 days in the tropics.

    The adult is a stout-bodied, brown nocturnal moth of wingspan about 40 mm.

    The complete life-cycle can be as short as about 28 days in the tropics, and there are several generations each season; in Europe there are probably only two generations per year, the second over-wintering as pupae in diapause.

**Distribution.** Widespread throughout the tropics, subtropics and warmer temperate regions of the Old World, extending as far north as Germany and Japan (CIE map no. A15).

**Control.** Maize during the tasselling period, and flowering legumes, are attractive to this species and will divert egg-laying females from the cotton; little damage is done to the maize and beans because of the high larval mortality on these crops.

    For chemical control, apply insecticides when caterpillars are small; the usual insecticides are DDT, BHC (or a mixture of the two), endosulfan, carbaryl, and now the polyhedral virus is available.

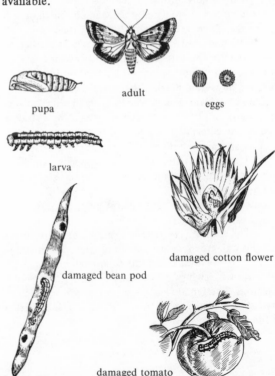

adult

eggs

pupa

larva

damaged cotton flower

damaged bean pod

damaged tomato

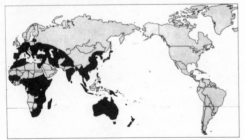

## Heliothis zea (Boddie)

**Common name.** Cotton Bollworm; Corn Earworm; Tomato Fruitworm.

**Family.** Noctuidae

**Hosts** (main). Cotton, maize, beans, tobacco, tomato.

(alternative). Okra, sorghum, cabbage, other legumes, cucurbits, sunflower, capsicums, strawberry, clovers and many other crops and wild hosts.

**Damage.** Caterpillars generally feed on the fruiting points of the host; on maize they eat the tassels and the young soft grains at the top of the cob. Often secondary rots develop in the insect feeding sites.

**Pest status.** A major pest in the New World, especially on cotton, because of its polyphagous feeding habits, wide distribution and large size of the larvae.

**Life history.** The spherical eggs are laid singly, stuck to the plant, and one female has been recorded laying as many as 2500 eggs.

Larvae are variable in colour and have several alternating pale and dark lines along the body, and after five or six instars they attain a length of 45 mm. Larval development takes from 28–60 days.

Pupation takes place in the soil, as with most Noctuidae, and the dark brown pupa measures about 20 mm.

The adults are brown moths with a pale hind-wing bordered by a dark band, about 40–44 mm in wingspan.

There are usually 2–3 generations per year,

according to location.

**Distribution.** The New World from Canada down to Argentina, including the W. Indies, and also Hawaii (CIE map no. A239).

**Control.** Various aspects of cultural control are practised against this pest, including the use of resistant varieties of crop plants, destruction of crop residues, uniform and early planting.

Natural control is important in many areas. For chemical control see *H. armigera*.

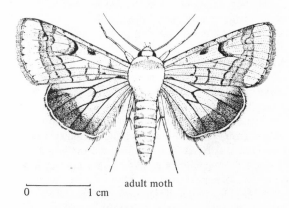

adult moth

0        1 cm

mature larva on cabbage heart

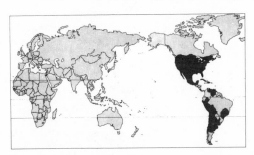

367

## Mythimna loreyi (Dup.)

(= *Cirphis loreyi* (Dup.))
(= *Leucania loreyi* (Dup.))

**Common name.** Rice Armyworm
**Family.** Noctuidae
**Hosts** (main). Maize, rice.
(alternative). Sugarcane, sorghum, wheat, etc.;
essentially polyphagous.
**Damage.** Leaves are skeletonized by young larvae, and
later the older caterpillars become gregarious and feed
voraciously, eating entire leaves and sometimes the
whole plant, usually during the night.
**Pest status.** An important pest on a number of different graminaceous crops; sporadically serious, when
entire crops may be destroyed.
**Life history.** Eggs are laid in batches of up to 100
between the leaf sheath and the stem, and hatch in
about five days.

The caterpillars are quite variable in colour, but
usually have several distinctive longitudinal stripes.
There are usually six larval instars, and the larger
caterpillars are generally gregarious. Mature size is
some 35–40 mm.

Life history details are probably similar to
those of *M. separata*. The adult moth is pale brown,
with a small spot in the middle of the forewings and
whitish hindwings; wingspan is from 35–50 mm.

There are usually several generations per year.

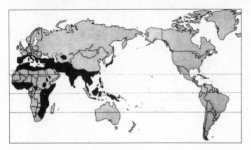

**Distribution.** Recorded from the Mediterranean
region, W. and E. Africa, Middle East, India through
to China and Japan, extending down to Papua New
Guinea (CIE map no. A275); a closely related but
different species occurs in Australia.
**Control.** See under *M. unipuncta*.

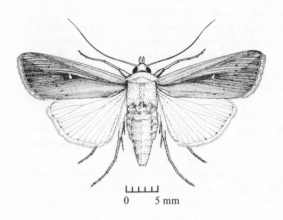

⌙⌙⌙⌙⌙
0       5 mm

368

## Mythimna separata (Wlk.)

(= *Leucania separata* Wlk.)
(= *Pseudaletia separata* (Wlk.))

**Common name.** Rice Ear-cutting Caterpillar (Paddy Armyworm)
**Family.** Noctuidae
**Hosts** (main). Rice
 (alternative). Sugarcane, sorghum, maize, millets, wheat, oats, barley, rye, legumes, *Brassica*, tobacco and various wild grasses and sedges.
**Damage.** The young larvae eat the leaves of the host plant, and generally defoliate, but the sixth instar caterpillars may also climb the peduncle and cut off the rice panicles. Large caterpillars are often gregarious and may occur in vast numbers.
**Pest status.** A serious pest of rice throughout S.E. Asia, and also on some other crops, especially when the larvae aggregate in large numbers, when their nocturnal feeding activities may completely defoliate crops.
**Life history.** Eggs are laid in batches of about 100 inside the rolled leaves or between the leaf sheath and stem; they are subspherical (about 0.5 mm), greenish-white turning yellow, and take about five days (4–13) to hatch.
 Larvae are variable in colour, from green to pinkish, but have four longitudinal black stripes laterally, with a white mid-dorsal stripe in young stages, darkening later. Mature length is 35–40 mm.

There are six larval instars usually, and the last instar larva eats about 80% of the total food consumed as a larva; large caterpillars are voracious in appetite. Larval development takes about 18 days (14–22); large larvae are usually gregarious, occasionally vast swarms of larvae are found and crop defoliation may be widespread.
 The pupa is dark brown, 15–19 mm long, and is formed in an oval cocoon about 4 cm deep in the soil; pupation takes about 18 days (7–29).
 Adult moths are brownish in colour with some faint marks on the forewings; wingspan is from 35–50 mm. Males live for about three days and females for about seven; both sexes may feed on honey-dew excreted by various Homoptera.
 The life-cycle takes about 30 days in the tropics, but varies from 25–64 days; there are often five generations per year.
**Distribution.** Pakistan, India, S.E. Asia, Papua New Guinea, West Irian, Australia (Queensland), New Zealand, China, Korea, and Japan (CIE map no. A230).
**Control.** See *M. unipuncta*.

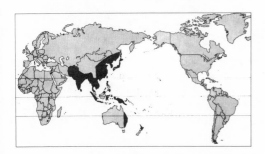

369

## Mythimna unipuncta (Haw.)

(= *Cirphis*, *Leucania* & *Pseudaletia unipuncta* (Haw.))

**Common name.** Rice Armyworm (American Armyworm)
**Family.** Noctuidae
**Hosts** (main). Rice
(alternative). Other cereals and forage crops, including maize, sugarcane, flax, wheat, rye, barley, oats, buckwheat and Jerusalem artichoke.
**Damage.** Defoliation by the feeding larvae; when gregarious swarms occur then complete crop destruction may follow.
**Pest status.** A sporadic pest, but occasionally serious, especially so when swarming occurs.
**Life history.** Very similar to *M. separata* in all stages, both morphologically and in life history details. Identification usually requires a taxonomic expert.

A series of closely related species are to be found on rice and other crops throughout the warmer parts of the world, but their identification is difficult.
**Distribution.** S. Europe, Mediterranean region, W. Africa, Somalia, Iran, Israel, USSR, USA, Hawaii, C. and S. America (CIE map no. A231).
**Control.** For all the species of *Mythimna* cultural methods such as ploughing or burning of stubble, flooding infested fields, removal of grass and alternative hosts from around the fields, all help to reduce the pest populations.

Some species are regularly parasitized by braconid wasps, ichneumons, and Tachinidae, in the larval stages, which are also susceptible to polyhedrosis virus and *Bacillus thuringiensis*.

As with other sporadic pests, chemical control is difficult to time because the pest population is usually large by the time it is noticed. Dust and spray with contact insecticides such as DDT, BHC, endrin, parathion, dichlorvos, trichlorphon or fenitrothion. In some areas the use of the organochlorine compounds is no longer approved, but the latter chemicals should prove effective.

Baits are sometimes used against the swarming caterpillars.

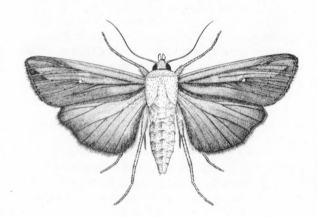

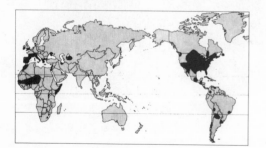

# Othreis fullonia (Cl.)

(= *Ophideres fullonica* L.)

**Common name.** Fruit-piercing Moth
**Family.** Noctuidae
**Hosts** (main). *Citrus*, mango, papaya, guava.

(alternative). Banana, tomato, grapes, and some wild fruits.

**Damage.** The adult moths pierce the ripening fruits to obtain sap. The proboscis is short and stout with a barbed tip. The damaged fruits usually develop secondary rots and fall prematurely.

**Pest status.** Only occasionally do fruit-piercing moths cause any appreciable damage, then it is usually *Achaea* spp. in parts of W. Africa; *O. fullonia* is more of academic interest than economic.

**Life history.** No details are available.

There are several genera of Noctuidae involved as fruit-piercing moths, including *Achaea* (Africa and India), *Serrodes*, *Anomis*, and, in S.E. Asia, *Calpe*, *Othreis* and *Ophiusa* spp.; there are also a few other less well-known genera. The larvae of *Achaea* spp. are defoliators of castor plants, and one species in India is known as the Caster Semi-looper.

Two species from S.E. Asia are illustrated here: *O. fullonia* has brown, leaf-like forewings and bright yellow hindwings — it has a very stout proboscis and apparently feeds mostly on *Citrus; Ophiusa tirhaca* has forewings which are basically greenish with brown markings, and yellow hindwings with a dark band —

its proboscis is less stout and it feeds on softer fruit (Banzinger, *pers. comm.*).

**Distribution.** *Othreis fullonia* is recorded from parts of Africa, through India, S.E. Asia up to China, Korea and Japan, and through the Philippines to Papua New Guinea and N.E. Australia (CIE map no. A377).

**Control.** Fortunately, control is seldom required, for it would be most difficult, partly because the larvae do not develop in the crops that are attacked by the adults, and partly because of the nature of the adult damage; fleeting feeding visits by the adult moths could only be combated by coating all ripening fruit with a potent contact insecticide.

0      1 cm          *Ophiusa tirhaca*

0      1 cm          *Othreis fullonia*

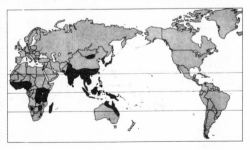

## Sesamia calamistis Hamps.

**Common name.** Pink Stalk Borer
**Family.** Noctuidae
**Hosts** (main). Maize, sorghum, finger millet, rice, and sugarcane.

(alternative). Various species of wild grasses.
**Damage.** The larvae bore in the stem of the various graminaceous crops, weakening the stem mechanically, and reducing the crop yield. Early damage results in cereal 'dead-hearts' with the destruction of the central shoot, although tillering may compensate somewhat for this damage.
**Pest status.** A pest of sporadical importance on a wide range of graminaceous crops. Three other species also occur in E. Africa.
**Life history.** Eggs are laid on the leaf sheath in groups of up to 40. They hatch a week later and the larvae immediately start boring into the stem. The larval period is 6–10 weeks. The mature caterpillar is about 30 mm long and 3.5 mm broad, with a brown head and buff body with pink dorsal markings.

The pupal period lasts about ten days.

The adult moths are pale buff with darker markings on the forewings; the male is smaller (22–30 mm wingspan) than the female (24–36 mm), and the hindwings are white.

The total life-cycle takes from 30 days for completion, according to climatic conditions.

**Distribution.** Most of tropical Africa (CIE map no. A414).

Several other species of *Sesamia* also occur widely in Africa on the same range of host plants.
**Control.** Cultural control measures such as weeding, crop hygiene, removal of alternative hosts in the vicinity of the crop, do help to lower the pest populations.

The chemical control measures recommended are the same as recommended on page 361.
**Further reading**
Williams, J.R. *et al.* (1969). *Pests of Sugar Cane*, pp. 207–23. Elsevier: London and New York.

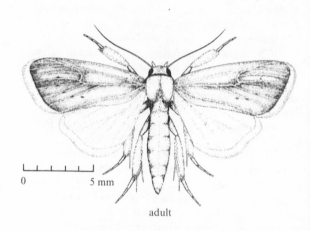

0        5 mm

adult

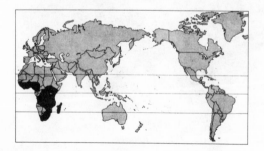

## Sesamia inferens (Wlk.)

**Common name.** Purple Stem Borer
**Family.** Noctuidae
**Hosts** (main). Rice, and sugarcane.

(alternative). Maize, sorghum, wheat, other cereals, *Eleusine coracana*, and many other grasses.
**Damage.** Small plants typically show 'dead-hearts', and older plants have extensive parts of the stem hollowed out, with a consequent physical weakening of the stem, and a reduction of crop yield.
**Pest status.** One of the major pests of rice and sugarcane; a polyphagous pest, and of importance on several other cereals in the tropics. Sugarcane is not a preferred host for oviposition — rice and grasses being preferred for this.
**Life history.** The eggs are bead-like, and laid in rows within the leaf-sheath; some 30–100 eggs per batch. Incubation takes about seven days.

The caterpillar is purple-pink dorsally and white ventrally; the head capsule is orange-red. After about 36 days the mature caterpillar is up to 35 mm long and 3 mm broad.

The pupa is dark brown with a purple tinge in the head region, and is about 18 mm by 4 mm. Pupation takes about ten days.

The adult moth is fawn-coloured with dark brown streaks on the forewings and white hindwings. The body length is 14–17 mm and wingspan up to 33 mm. The adults survive in the field for 4–6 days.

The total life-cycle takes 46–83 days.
**Distribution.** Pakistan, India, Bangladesh, Sri Lanka, S.E. Asia, China, Korea, Japan, Philippines, Indonesia, Papua New Guinea, West Irian and the Solomon Isles (CIE map no. A237).
**Control.** Control measures are as on page 361.
**Further reading**
Williams, J.R. *et al.* (1969). *Pests of Sugar Cane*, pp. 208–23. Elsevier: London and New York.

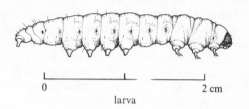

larva

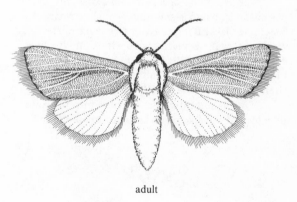

adult

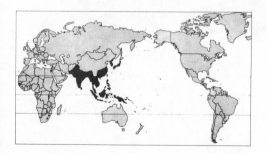

## Spodoptera exempta (Wlk.)

(= *Laphygma exempta* (Wlk.))

**Common name.** African Armyworm
**Family.** Noctuidae
**Hosts** (main). Grasses, maize, rice, and sorghum.
(alternative). Many cereals and wild grasses.
**Damage.** Leaves of cereals are holed or eaten down to the midrib. Blackish velvety caterpillars are present which drop to the ground if disturbed.
**Pest status.** A major pest in outbreak years; outbreaks often follow rain in the hot season; a second outbreak generation may follow the first, though not necessarily in the same district. In non-outbreak years they are cryptically coloured and non-gregarious.
**Life history.** Eggs are laid in masses of one or more layers on the leaves, 10–300 or more. The egg mass is covered with hairs from the female. The eggs are white turning dark brown; hatching takes 2–5 days.

The caterpillar occurs in two forms; the gregarious outbreak phase is greyish-green when small, becoming blackish in the latter two instars. The fully grown caterpillar is black above with thin blue lines down the middle of the back; on each side of the black area are several greenish-yellow lines and a midlateral black line. The larval period lasts 14–32 days.

The mature caterpillar burrows into the soil to pupate. The pupa is brown or black in colour and about 17 mm long. It is enclosed in a delicate cocoon of soil particles held together by silk. The pupal period lasts 7–21 days.

The adult is a grey-brown night-flying moth with pale hindwings and a conspicuous kidney-shaped whitish mark on the forewings; the wingspan is about 28 mm. After a pre-oviposition period of 2–4 days the adult female may live a further seven days and lay 400 or more eggs.

**Distribution.** Africa, Madagascar, India, Sri Lanka, Burma, Malaya, Sumatra, Java, Philippines, Celebes, Hawaii, Borneo, Papua New Guinea, West Irian and Australia (CIE map no. A53 revised).

**Control.** There is no great advantage in killing mature caterpillars ready for pupation, as the second generation rarely occurs in the same area. The following insecticides are generally effective for armyworm control when applied as sprays or dusts: DDT, endosulfan, malathion, trichlorphon, and carbaryl.

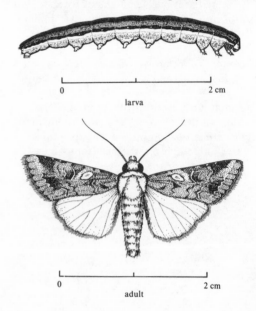

larva

0         2 cm

adult

0         2 cm

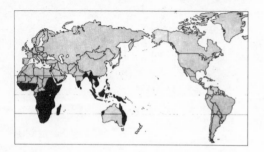

## Control of Armyworms (Lepidoptera; Noctuidae).

Armyworms are the caterpillars of various Noctuidae, mostly *Spodoptera* spp., which under certain conditions of high population density behave gregariously; swarms will march from field to field devastatingly defoliating entire crops. Most species are migratory as adults (within Africa, and parts of Asia and the Americas) and are renowned for their spectacular population fluctuations which can lead to a severe outbreak year resulting in catastrophic damage. Egg batches are laid close together, and in a severe year the high density of caterpillars results in a rapid demolition of the local supply of food plants. When this happens the caterpillars may 'march' off over the ground like an 'army' to fresh feeding locations, feeding as they go. With large populations the ground may be literally covered with the gregarious 'marching' band of caterpillars, in many respects similar to a locust swarm.

The plants attacked are mostly Gramineae (cereals and grasses), but the genus *Spodoptera* (not all of which act as armyworms) is recorded feeding on plants from 40 different families, containing at least 87 species of economic importance (Brown & Dewhurst, 1975). In Africa *Spodoptera exempta* (African Armyworm) is of such importance that, following the severe outbreak in East Africa in 1961, there was established in 1962 an Armyworm Research Unit at EAAFRO (Kenya) in collaboration with COPR (London). It was to monitor field populations, migrations and flight behaviour in an attempt to predict population outbreaks and immigrations, and to co-ordinate co-operative control measures (E.S. Brown, 1972). This basic research is still being pursued by a unit at ICIPE, Nairobi, Kenya.

The term 'armyworm' is essentially behavioural (and physiological) and is not strictly taxonomic. Some species of *Spodoptera* do not act as armyworms, and species belonging to other genera of Noctuidae show swarming tendencies and may be classed as armyworms, including *Achaea* spp., *Cerapteryx* (especially *C. graminis*) and *Mythimna unipunctata*.

Some species are very versatile behaviourally and may behave quite differently under different circumstances. For example, *Spodoptera litura/littoralis* in small numbers feeds on the leaves and foliage of many plants, but will sometimes act as a cutworm and at other times will swarm gregariously and act as a typical armyworm. Common names for this species complex include Cotton Leafworm (Africa), Rice Cutworm (S.E. Asia), Fall Armyworm (China), Climbing Cutworm (Mediterranean) and Cluster Caterpillar (Australia); the diversity of common names for these two almost identical species (formerly regarded as one species) gives an indication of their behavioural diversity. Thus, it can be rather difficult to categorize particular species of pests according to the nature of the damage done, and care should be taken when using these terms.

Control of armyworms is often a large-scale or even international collaborative venture, and usually consists now of warnings based on light-trap or pheromone-trap catches, or prognostications based on prevailing meteorological conditions in certain areas, followed by u.l.v. drift spraying from light aircraft or helicopters or ground-based, hand-held sprayers. Effective insecticides used, to date, include DDT, endosulfan, fenitrothion, malathion, tetrachlorvinphos and trichlorphon. Dieldrin and BHC are also effective but seldom used because of environmental hazards associated with their widespread employment. Choice of chemical used depends in part upon the plants (crops) to be sprayed, and also local government restrictions (DDT may be banned).

## Spodoptera exigua (Hb.)

**Common name.** Lesser Armyworm (Beet Armyworm)
**Family.** Noctuidae
**Hosts** (main). Rice; upland rice being most commonly attacked.

(alternative). Cotton, sugar beet, lucerne, tobacco, tomato, groundnut.

**Damage.** The caterpillars are gregarious, moving in swarms, and destroying the young leaves and stems of the rice plants. Young seedlings can be completely destroyed, but older plants often recover after an attack and may tiller vigorously.

**Pest status.** A sporadic pest of importance mainly on upland rice; young plants are often completely destroyed. Called Beet Armyworm in Europe.

**Life history.** Eggs are laid on the leaves of the rice plant, and hatch after 2–4 days.

The larvae are blackish and there are six instars. The maximum size is 37–50 mm. Larval development takes 10–12 days.

Pupation takes place in the soil and lasts for about six days.

The adult is a small brown moth of wingspan up to 25 mm, and lives for 8–10 days. They do not as a rule fly far, and generally lay their eggs close to their place of emergence.

The total life-cycle takes about 21 days.

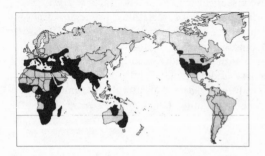

**Distribution.** Africa, S.E. Asia, C. and S. Europe, Middle East, Australia, southern USA, Madagascar, India, S. China, Philippines, Java, West Irian and Canada (CIE map no. A302).

**Control.** Cultural methods of control such as ploughing and burning of crop stubble, flooding infested fields, and removal of weeds all help to lower the pest populations.

Dusting and spraying with contact insecticides such as DDT, BHC, endrin, and parathion have been very effective.

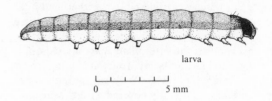

larva

0        5 mm

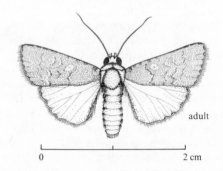

adult

0        2 cm

## Spodoptera littoralis (Boisd.)

*(= Prodenia litura* (F.) *auctt.)*

**Common name.** Was called 'Cotton Leafworm'
**Family.** Noctuidae
**Hosts.** A polyphagous pest attacking cotton, rice, tobacco, tomato, maize, castor, *Citrus*, mulberry, Cruciferae, legumes, many other vegetables, grasses, and ornamentals.
**Damage.** This caterpillar is essentially a leaf-eater, but does occasionally behave like a cutworm. Heavy infestations result in severe defoliation, but these are not of frequent occurrence. The young larvae are gregarious but they disperse as they become older.
**Pest status.** Not often a serious pest on any one crop but very frequently of minor importance on very many crops. Also called Mediterranean Climbing Cutworm.
**Life history.** The eggs are spherical, 0.3 mm in diameter, and laid on the undersides of leaves in batches of 100–300 and covered with hair-scales; one female lays from 1500–2000 eggs. Hatching takes 2–6 days, but can take up to 26 days in cooler regions.

The newly hatched larvae are gregarious, but later they disperse. Development through six instars takes 2–4 weeks. The caterpillars are pale green at first, becoming brown with dark markings, with yellow lateral and dorsal stripes. The length of the mature caterpillar is 35–50 mm.

Pupation takes place in the soil in an earthen cell, just beneath the surface; the pupa is dark red, 15–20 mm long. Pupation takes 6–11 days.

The adult has a whitish body with red tinges; the forewings are yellow-brown with varied white bands; the hindwings are whitish.

In the wet tropics breeding is virtually continuous with up to eight generations per year, the life-cycle taking 24–35 days.
**Distribution.** Africa, and the Mediterranean region, the Near East, and Madagascar (CIE map no. A232).

This species has for many years been inseparable from *S. litura* (the names being regarded as synonyms); the adults are only distinguishable by their genitalia; the larvae are very variable in colouration and cannot be definitely separated.
**Control.** See control of armyworms, page 375.

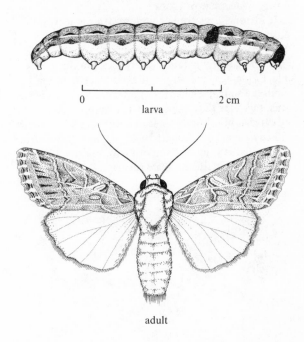

larva

adult

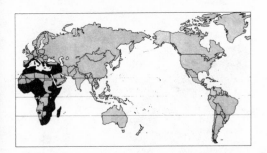

## Spodoptera litura (F.)

(= *Prodenia litura* (F.) *auctt.*)

**Common name.** Fall Armyworm (Cluster Caterpillar; Rice Cutworm)

**Family.** Noctuidae

**Hosts** (main). A polyphagous pest of major status on cotton, rice, tomato, and tobacco.

(alternative). *Citrus*, cocoa, sweet potato, rubber, groundnut, castor, legumes, millets, sorghum, maize, and many vegetables.

**Damage.** As with *S. littoralis* this caterpillar is basically a leaf-eater, but does rarely act like a cutworm with crop seedlings. Heavy infestations can seriously defoliate a crop, but this is not a common happening.

**Pest status.** Not very frequently a serious pest on any one particular crop but of very regular occurrence on a very wide range of crops.

**Life history.** Eggs are laid underneath the leaves, in clusters of 200—300, and covered with hair scales. They hatch in 3—4 days.

The newly hatched caterpillars are tiny, blackish-green, and with a distinct black band on the first abdominal segment. For a while they are gregarious, but later they disperse. The caterpillars are nocturnal in habits and become fully grown in about 20 days, reaching a length of 40—50 mm. The mature caterpillar is stout and smooth with scattered short setae, dull greyish and blackish-green with yellow dorsal and lateral stripes. The lateral yellow stripe is bordered dorsally with a series of semi-lunar black marks. The head capsule is black.

Pupation takes place in the soil in an earthen cell, and the adult emerges after 6—7 days.

The whole life-cycle takes about 30 days, and in wet tropics there may be as many as eight generations.

**Distribution.** South and eastern Old World tropics, including Pakistan, India, Bangladesh, Sri Lanka, S.E. Asia, China, Korea, Japan, Philippines, Indonesia, Australasia, Pacific islands, Hawaii and Fiji (CIE map no. A61).

Only recently separated from *S. littoralis* by the genitalia of the adult moths — the larvae are not really separable. The two species are quite allopatric in distribution however.

**Control.** See control of armyworms, page 375.

larva         pupa

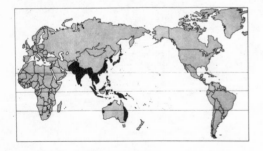

0      2 cm      adult

# Spodoptera mauritia (Boisd.)

**Common name.** Paddy Armyworm
**Family.** Noctuidae
**Hosts** (main). Rice
    (alternative). Maize, sugarcane, Cruciferae, and other species of Gramineae.
**Damage.** The small caterpillars eat the leaves of rice seedlings, at first nibbling the surface. Later as the caterpillars grow they become voracious and can destroy whole crops in a short period of time before moving on to other fields. The caterpillars are nocturnal and feed at night. Some older plants are attacked and damaged but this pest is really only a threat to seedlings.
**Pest status.** A sporadically serious pest of rice seedlings. This is sometimes called the 'Paddy Swarming Caterpillar'. In severe attacks whole crops are destroyed.
**Life history.** The spherical eggs are laid in clusters of 100–300 at the tips of upright leaves, and are covered with setae from the female body; hatching takes 3–9 days.

    The newly hatched caterpillars are green, about 2 mm long, and difficult to see on the foliage. As they grow they become more brown, and after 15–24 days they are fully grown, and some 35–40 mm long. The head capsule is dark with a pale forked line; there are three lateral lines along the body with dark segmental marks above.

Pupation occurs in the soil; the pupa is dark brown and has two slender apical spines. Pupation takes 7–14 days.

    The adult is a grey-brown moth, with 30–40 mm wingspan, and 15–20 mm long. The forewings are marked with several dark lines and a conspicuous black spot; the hindwings are whitish-brown with a thin dark margin.

    The entire life-cycle takes 37–40 days typically.
**Distribution.** Madagascar, Tanzania, Uganda, Pakistan, India, Bangladesh, Sri Lanka, S.E. Asia, S. China, Philippines, Indonesia, Australia, Pacific islands, Hawaii and Fiji (CIE map no. A162 revised).
**Control.** See control of armyworms, page 375.

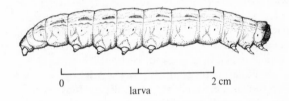

0            2 cm

larva

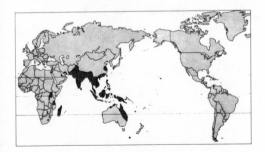

adult

## Anomis flava (F.)

(= *Cosmophila flava* (F.))

**Common name.** Cotton Semi-looper
**Family.** Noctuidae (Plusiinae)
**Hosts** (main). Cotton

(alternative). Other Malvaceae, especially *Hibiscus, Abutilon,* and *Sida* spp., and *Althaea rosea* (hollyhock); also tomato and okra.

**Damage.** The caterpillar, which is a semi-looper (subfamily Plusiinae), eats the leaves of the cotton. Sometimes attacks are heavy and the plants may be completely defoliated.

**Pest status.** A pest of sporadic importance on cotton in the countries listed.

**Life history.** The eggs are laid singly on the leaves on which the larvae feed.

The larva is a semi-looper, and has only three pairs of prolegs. The body colour is pale yellowish-green with five fine lines longitudinally on the dorsal surface. When fully grown it is about 30 mm long.

Pupation takes place in the soil debris, or in a flap of leaf, between bract and boll, in a loose cocoon.

The adult is an attractive small moth with reddish-brown forewings traversed by two darker zigzagged bands. The hindwings are pale brown; wing-span is about 30 mm.

The total life-cycle takes 4–6 weeks. In China it is reported that the pupa overwinters.

**Distribution.** Found throughout the cotton-growing areas of Africa, Asia, and Australasia. A related species *Cosmophila erosa* (Hb.) occurs in the New World (CIE map no. A379).

**Control.** Recommended insecticides are diazinon, parathion and carbaryl, as foliar sprays to be applied when caterpillars are seen on the leaves of the crop.

**Further reading**

Pearson, E.O. (1958). *The Insect Pests of Cotton in Tropical Africa*, p. 70. CIE: London.

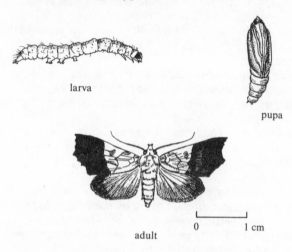

larva

pupa

adult

0    1 cm

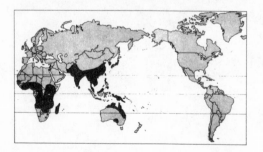

## Trichoplusia ni (Hb.)

(= *Autographa brassicae* Riley)
(= *Plusia ni* Hb.)

**Common name.** Cabbage Semi-looper
**Family.** Noctuidae (Plusiinae)
**Hosts** (main). *Brassica* spp. and other Cruciferae.

(alternative). Cotton, legumes, Solanaceae, sweet potato, some cucurbits, and many others.

**Damage.** The larvae eat irregular holes in the leaf lamina, and in cabbage they tend to attack the heart, eating a great deal and contaminating with frass.

**Pest status.** A serious pest on *Brassica* spp. throughout most parts of the world, though its pest status and hosts attacked tend to vary.

**Life history.** Eggs are laid singly on the underside of the host plant leaves, and hatch after 2–3 days.

Larvae are basically green in colour, with a thin, white, lateral line, and two white lines along the middle of the back. There are two pairs of prolegs so the caterpillars walk with a 'looping' action characteristic of the Plusiinae. Larval development takes some 30–35 days, usually five instars.

Pupation takes place within a silken cocoon, usually in the leaf litter or crop debris, and development takes about 15 days under optimum conditions.

The adult is a dark brownish moth with two small white markings on the forewing, sometimes resembling a figure '8', wingspan is about 35 mm. Adults live for about three weeks.

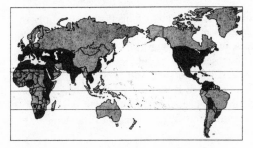

There is no evidence for diapause, in temperate areas the caterpillars continue to be active at low temperatures; generally flight ceases at about 16°C and larval development at about 12°C.

In warm regions there may be five generations per year, or more as a result of continuous breeding.

**Distribution.** Very widely distributed throughout the tropics and subtropics, with the exception of Australasia, and extending up into the warmer parts of S. Canada and Europe (CIE map no. A328) where populations are reinforced by annual migrations.

**Control.** Many natural predators and parasites are recorded for this species, and field mortality may sometimes be high. Nuclear Polyhedrosis Virus and *Bacillus thuringiensis* are effective against this pest.

Ultra-violet light traps, used with sex pheromone, have controlled populations in some localities in the USA, but this is more a diagnostic tool than control measure.

Chemical control has been difficult to achieve; resistance is established to many insecticides including DDT, carbaryl, parathion, methomyl, and others.

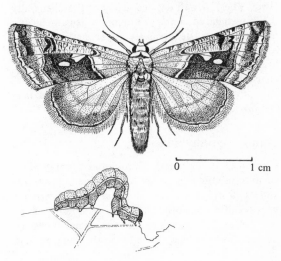

0     1 cm

## Order **DIPTERA**

These are the true flies, with a single pair of membraneous wings, the hindwings modified into halteres. Metamorphosis is complete, the larvae are eruciform and apodous, usually with the head reduced and retracted. Many are crop pests, but more are pests of medical and veterinary importance.

### Family **Cecidomyiidae**

(Gall Midges) These are minute delicate flies, with long, moniliform antennae bearing conspicuous whorls of setae; wings with reduced venation. Most of the larvae live in plants and are phytophagous. The phytophagous species may be found in any part of the plant body — roots, stem, leaves, buds, flowers, or fruit, and sometimes galls are formed. Both larvae and adults are often red, orange, or yellow in colour.

For information on gall midges on cultivated plants see the eight volumes by Barnes (1946–59).

### Family **Tephritidae** (= Trypetidae)

(Fruit Flies) A well defined group of flies, generally easily recognized, being of moderate size and having mottled wings. The larvae are phytophagous, often living inside fruits, but some are found in flower heads, and inside stems and leaves. The major pest genera are *Dacus* and *Ceratitis*.

For a review of fruit fly ecology see Bateman (1972), also see Drew, Hooper & Bateman (1978).

### Family **Agromyzidae**

(Dipterous Leaf Miners) An ill-defined family of small to minute flies whose larvae mine in the leaves and stems of many plants. Some species are virtually host-specific while others attack a wide range of plants. Pupation occurs either in the plant or else in the soil. Some species are endoparasitic on Coccidae.

There is an important monograph on this family by Spencer (1973).

### Family **Diopsidae**

(Stalk-eyed Flies) This peculiar group of flies is largely confined to Africa and S.E. Asia. The larvae are miners in the leaves of Gramineae or else saprozoic.

For publications of Diopsidae see Shillito (1971, etc.).

### Family **Ephydridae**

(Shore Flies) These are black or darkly coloured flies inhabiting damp, marshy places, and are closely related to the Drosophilidae. One genus (*Hydrellia*) is a pest of graminaceous crops, particularly important on paddy rice, and also various pond weeds (*Potamogeton* spp.).

### Family **Muscidae**

(House Flies, etc.) A large group of flies of small to medium size usually resembling the House Fly, only some of which are crop pests — these being mainly shoot flies. Most of the Muscidae are medical, veterinary or household pests. Some authorities lump the Anthomyiidae in with the Muscidae as a subfamily.

### Family **Anthomyiidae**

(Root Flies; Root Maggots) This is a small family of phytophagous species, most of which are temperate crop pests, but some occur in the tropics. The larvae feed on sown seeds, roots, and in the stems of many crops, especially vegetables and cereals, and are collectively referred to as root maggots. The Anthomyiidae are separated from the Muscidae by having the anal vein extended right to the wing margin.

## Asphondylia sesami Felt

**Common name.** Sesame Gall Midge
**Family.** Cecidomyiidae
**Hosts** (main). Sesame (*Sesamum indicum*)
 (alternative). A wild species, *S. angustifolium*,
has been recorded in Uganda.
**Damage.** Flower buds or capsules are galled by the
larvae, and then become twisted and stunted.
**Pest status.** Usually only a minor pest but occasionally high infestations occur with resulting consider-
able crop losses.
**Life history.** Details of the life history are not well
known.
 Eggs are laid along the veins of the terminal
leaves.
 The larvae are white, and pupation occurs in
the galls on the capsules.
 The adult is a large (5 mm long) red-bodied
midge.
 Generally plants with green capsules appear
to be more susceptible to attack by this pest than do
ones with black capsules.
**Distribution.** Only E. Africa and S. India.
**Control.** No control measures are recorded to date.
**Further reading**

Barnes, H.F. (1949). *Gall Midges of Economic
 Importance*, vol. VI. *Gall Midges of Miscel-
 laneous Crops*, pp. 146–7. Crosby Lockwood:
 London.

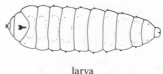

eggs

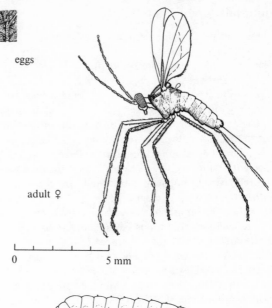

adult ♀

0           5 mm

larva

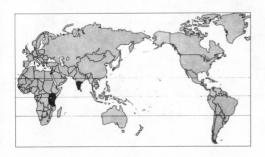

# Contarinia sorghicola (Coq.)

**Common name.** Sorghum Midge

**Family.** Cecidomyiidae

**Hosts** (main). *Sorghum*; cultivated and wild species.

(alternative). *Andropogon gayanus* in Nigeria.

**Damage.** The larvae feed on the developing seeds, often only one larva per spikelet, but this pest density is sufficient to cause complete loss of grain. In the USA in high infestations there may be 8–10 larvae on the same seed developing to maturity. The grain head is flattened with tiny shrunken seeds, and the orange-coloured larvae or pupae may be seen in the head to confirm the infestation diagnosis.

**Pest status.** Of common occurrence in Africa; sometimes infestations are serious. Some indigenous varieties of *Sorghum* show resistance to attack by *Contarinia*. In the USA millions of dollars are lost annually as a result of Sorghum Midge attack.

**Life history.** The eggs are laid while the spikelet is in bloom over a period of some days; some 20–130 eggs being laid per female. Usually the egg is placed near the spikelet tip. After 2–4 days the eggs hatch.

The larvae move down into the ovary and lie there, feeding on the nutrients which would normally nourish the embryo. The fully grown larvae are dark orange. Larval development takes 9–11 days.

The pupae may be either naked or in cocoons according to the weather conditions. Pupation takes 2–6 days if the pupae are naked. Cocooned pupae may hibernate or aestivate.

The adults are stout-bodied, about 2 mm long, and the females have a dark orange abdomen.

The total life-cycle takes 19–25 days.

**Distribution.** Scattered throughout the tropics and subtropics; in tropical Africa, from W. through to E.; Java, Australia, southern USA, W. Indies, and S. America (Venezuela) (CIE map no. A72).

**Control.** Many indigenous African species of *Sorghum* are naturally resistant to Midge attack in that they are less favoured by the female for oviposition. Control measures have generally relied upon the growing of these resistant varieties whenever possible.

Chemical control is difficult in practice; frequent sprays of DDT or carbaryl are fairly effective but chemical residues in the crop become excessive.

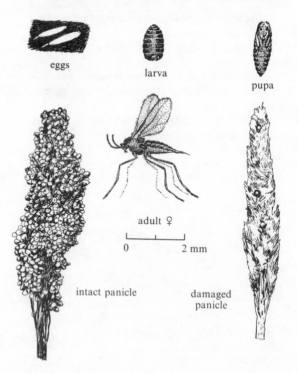

eggs    larva    pupa

adult ♀

0    2 mm

intact panicle    damaged panicle

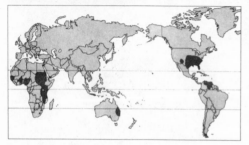

## Orseolia oryzae (W.–M.)

(= *Pachydiplosis oryzae* (W.–M.))

**Common name.** Rice Stem Gall Midge
**Family.** Cecidomyiidae
**Hosts** (main). Rice

(alternative). Wild species of *Oryza*, and grasses.
**Damage.** The severity of damage is related to the time of attack. The larvae move down between the leaf sheaths until they reach the apical bud or one of the lateral buds. There they lacerate the tissues of the bud and feed until pupation. The feeding causes formation of a gall called a 'silver' or 'onion' shoot.
**Pest status.** In some areas this is a very serious pest causing crop losses of 30–50% with some regularity and occasionally losses of 100%.
**Life history.** Fertilized females start egg-laying within a few hours of emergence. They lay 100–300 eggs each. The eggs are elongate, tubular, and white, pink or red in colour. Incubation takes 3–4 days, or more.

The larvae are 1 mm long on hatching, with a pointed anterior end, and a pale colour. They eventually grow to 3 mm long and become red.

Pupation takes place at the base of the gall, and the pupa is 2.0–2.5 mm long and 0.6–0.8 mm broad, pink initially becoming red. Before the adult emerges the pupa makes a hole in the top of the gall with its spines and projects halfway; the skin splits and the adult emerges. Pupation takes 2–8 days.

The adult is a delicate little midge, 3.5 mm long, brown in colour, and with long strong legs.

The whole life-cycle takes 9–26 days on rice, and slightly less on grasses. After one or two generations on grasses the midge generally moves to rice.
**Distribution.** W. Africa, Sudan, Pakistan, India, Bangladesh, Sri Lanka, S.E. Asia, S. China, Java, and West Irian (CIE map no. A171).
**Control.** Careful timing of planting can avoid damage by this pest. Once past the tillering stage the plant is not suitable as a host. Considerable build-up of midge populations can occur on grasses near the rice crop.

The success of chemical control is very much dependent on accurate spray-timing to coincide with the emergence of each brood of midges. Several insecticides have been recommended in different countries, with somewhat differing results; these include phosphamidon, parathion, carbaryl, phorate, diazinon, dimethoate, BHC, dieldrin and endrin. Several sprays are necessary during the vulnerable period, that is 20–45 days after transplanting.

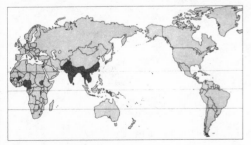

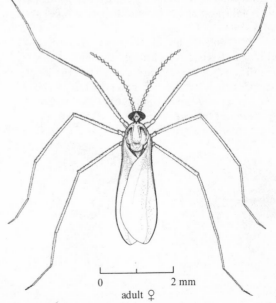

0        2 mm
adult ♀

# Ceratitis capitata (Wied.)

**Common name.** Mediterranean Fruit Fly (Medfly)
**Family.** Tephritidae
**Hosts** (main). Peach and *Citrus* fruits.

(alternative). Coffee berries, cocoa, *Ficus*, mango, guava, *Prunus* spp., *Solanum* spp., etc.
**Damage.** The larvae bore into the fruits, often accompanied by fungi and bacteria which rot the fruit. Severely attacked fruits fall prematurely.
**Pest status.** A serious pest of many deciduous and subtropical fruits.
**Life history.** The eggs are laid in groups with the female's protrusible ovipositor into the pulp of the fruit just under the skin. Incubation takes 2–3 days.

The whitish maggots bore through the pulp of the fruit; up to 10–12 maggots per fruit, but more than 100 have been recorded. The three larval instars take 10–14 days for development.

Pupation takes place in the soil under the tree, in an elongate, brown puparium. By the time the maggot comes to leave the rotted fruit the fruit has usually fallen on to the ground. Pupation takes about 14 days.

The adult is a bright, decorative fly, with red and blue iridescent eyes, brown-headed, 5–6 mm long, thorax black with white and yellow markings. The abdomen is yellow with two grey transverse bands. The male fly has characteristic triangular expansions at the end of the antennal arista. The female fly becomes sexually mature 4–5 days after emergence, and the first oviposition occurs after eight days. The adults require sugary foods at this time – with food they live for 5–6 months.

There may be 8–10 generations per year.
**Distribution.** S. Europe, Near East, Africa, S.W. Australia, Hawaii, C. and S. America (CIE map no. A1).
**Control.** See following section on control.

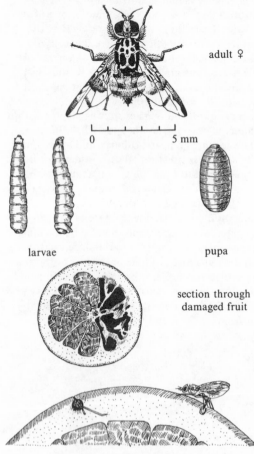

adult ♀

0       5 mm

larvae       pupa

section through damaged fruit

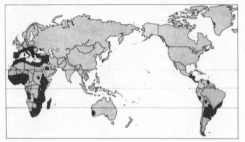

Egg-laying female and egg cavity with larval tunnel

386

# Control of Fruit Flies (Tephritidae)

It appears that fruit fly control in most parts of the world now falls into three different categories, as suggested by Drew, Hooper & Bateman (1978).

(1) Control — defined as local procedures aimed at the protection of individual orchards; these procedures usually have little effect on the breeding population of flies in the general area.

(a) Collection and destruction of all infested fruits.

(b) The maggots cannot easily be destroyed for they are inside the fruits and therefore inaccessible, but some success is claimed for the systemic insecticide fenthion.

(c) Use of protein bait sprays, and sex attractants.

(2) Suppression — methods aimed at the control of an entire breeding population, for example in a large valley, part of a State, island, etc., the object being to reduce the level of fruit infestation below the economic threshold. Such wide-scale suppression programmes are appropriate only in areas where the fruit flies are likely to recur each year, and therefore they must normally be repeated in successive seasons.

(a) Protein bait spot sprays — this technique consists of mixing 20 g of protein 'solids' with 10 g malathion per litre of solution, and squirting a 'spot' of this viscous liquid on the crop foliage of about 100 ml in volume. If the 'spots' are dispersed over the foliage in sufficient density, with adequate uniformity over the entire area occupied by the breeding population, then all the winged adults will be killed. Field observations with *D. tryoni* showed that trivial movements of mature females usually daily exceeded 10 m; thus a grid of 'spots' at about 15 m intervals should be adequate to ensure that at least one spot is within the expected daily wandering range of each female fly. For long-term suppression regular application is required, generally weekly sprays at the height of the fly season, and fortnightly at other times, ceasing during the overwintering period.

(3) Eradication — methods aimed at the killing of all individuals in a breeding population; usually undertaken only in areas where the subject species is not endemic and where it has been accidentally introduced. The three methods employed are as follows.

(a) Protein hydrolysate bait sprays — these can be used at short notice and are generally effective if the outbreak is fairly confined, for example the Medfly invasion of Florida in 1956, which involved more than $3000 \, km^2$.

(b) Male annihilation — by using traps baited with female sex pheromones or sex attractants. Generally this is a tool for population monitoring rather than actual population control, but it has been used successfully a few times to eradicate a fly population. Methyl eugenol has a strong attraction for males of various *Dacus* species, and Cu-lure is very effective against *D. cucurbitae* and *D. tryoni*. The sex-lure traps contain suitable insecticides so that entering flies are killed. Generally these traps used together with protein bait sprays are very effective.

(c) Sterile insect release method (SIRM) — following the success of this method against Screw-worm Fly in USA and Curacao in 1958–9 it was developed for use against fruit flies in the USA and elsewhere. Success was achieved in several cases; *D. dorsalis* was eradicated from Guam in 1963, and *D. cucurbitae* from Rota in 1962–3.

Because of the discontinuous distribution of some species of fruit fly, and their enormous potential as fruit pests, several species are subject to quarantine legislation in different countries or states (USA, Australia), involving restriction of importation of fruit likely to contain the larvae. The main species involved are *Ceratitis capitata*, *Dacus dorsalis* and *D. tryoni*.

# Ceratitis coffeae (Bezzi)

(= *Trirhithrum coffeae* (Bezzi))

**Common name.** Coffee Fruit Fly
**Family.** Tephritidae
**Hosts** (main). Coffee fruit, both *arabica* and *robusta*.
   (alternative). Other *Coffea* spp., rarely in
*Citrus*, cocoa, mango, *Solanum* spp., and carambola.
**Damage.** The maggots bore into the mucilage of the
ripe berries, or the fruit.
**Pest status.** No real damage is done to *robusta* coffee
in E. Africa, with any effect on quality. However, if
picking is irregular fruit-fly-induced shedding can be
responsible for considerable crop loss.
**Life history.** Eggs are laid under the skin of the berry,
in groups. Hatching usually takes place within two
days.

   The maggots feed on the fruit mucilage, several
larvae can develop to maturity in one fruit. The larval
stage usually lasts about 21 days. The mature larvae
drop out from the berry into the surface litter under
the tree.

   Pupation takes place in the litter under the tree,
and the pupal period lasts for about ten days.

   The adult is a small dark fruit fly with the
typical mottled wings shown by this genus.
**Distribution.** E. Africa only.
**Control.** So far pesticide treatment has not been
effective against this pest, and in point of fact

dieldrin and trichlorphon appeared to reduce the
proportion of larvae parasitized in trials carried out in
Uganda.
**Further reading**
Le Pelley, R. H. (1968). *Pests of Coffee*, pp. 247–8.
   Longmans: London.

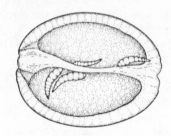

larvae in berry

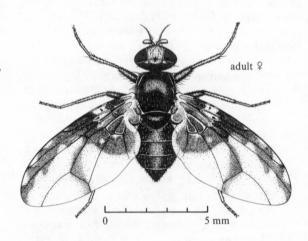

adult ♀

0                    5 mm

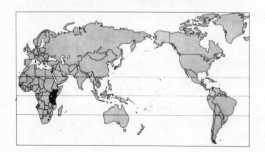

# Ceratitis cosyra (Wlk.)

(= *Pardalaspis cosyra* Wlk.)

**Common name.** Mango Fruit Fly
**Family.** Tephritidae
**Hosts** (main). Mango
     (alternative). Peach, *Warburgia, Acokanthera*, and *Cordyla* spp.
**Damage.** The fruits show oviposition punctures with dark stains (rotting) around them. The pulp is heavily mined and the mines contain many small white maggots. The prematurely ripening fruits fall off the tree.
**Pest status.** Not often a serious pest.
**Life history.** The biology of this pest is similar to that of *C. capitata*.

     The female flies pierce the ripening fruit and insert the eggs into the puncture.

     The maggots feed on the pulp, making it worthless as a crop.

     Pupation takes place either inside the fruit or underground.

     The adult is a small fly, which holds its wings partly extended at rest, and is about 4–5 mm long, and 10 mm wingspan.

     There are probably only two generations per year.
**Distribution.** Africa only, from E. and S. Africa, Zimbabwe, Cameroons, and Zanzibar.

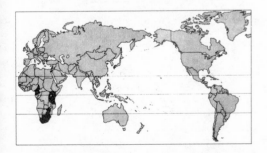

**Control.** For control measures see *Ceratitis capitata* (p. 387).

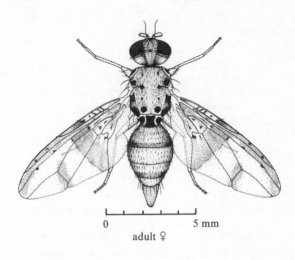

adult ♀

## Ceratitis rosa Karsch

(= *Pterandrus rosa* Karsh)

**Common name.** Natal Fruit Fly
**Family.** Tephritidae
**Hosts** (main). Peach, and *Citrus* spp.

(alternative). Many deciduous and subtropical fruits in S. and E. Africa.
**Damage.** Fruits show 'stings' caused by the female ovipositor, surrounded by soft dark patches; insect attack is often followed by secondary disease infection. The fruit pulp becomes soft and rotten and contains many small, white maggots. Infested fruits usually fall prematurely.
**Pest status.** An important pest of fruit in tropical and subtropical Africa, south of the Sahara. This fruit fly is able to breed in harder and greener fruit, and also has more wild hosts than *C. capitata*.
**Life history.** The eggs are tiny, creamy-white, and elliptical, 0.9 mm long. They are laid in batches in cavities beneath the rind (skin) of the fruit. Several punctures ('stings') may be made without ovipositing and these holes are entry points for fungi and bacteria. Incubation takes 2–4 days.

The maggots are white and legless, tapering to a narrow anterior point; typical of most fly maggots. The three instars take 10–14 days, according to the fruit and the season. The full-grown maggots are 8 mm long, and they leave the fallen fruit to pupate in the soil some 5–20 cm below the surface.

The puparium is elongate and reddish-brown, and the pupal period lasts for 11–12 days.

The adult is a small fly, with a wingspan of about 10–12 mm, with golden bars and markings on otherwise hyaline wings, and 4–6 mm long.

The complete life-cycle takes 21–28 days, and there may be 6–12 generations per year, according to the climate.
**Distribution.** Only Africa; recorded from Angola, E. Africa, Mauritius, Mozambique, Nigeria, Malawi, Zimbabwe, S. Africa, and Zanzibar (CIE map no. A153).
**Control.** For control measures see *Ceratitis capitata* (p. 387).

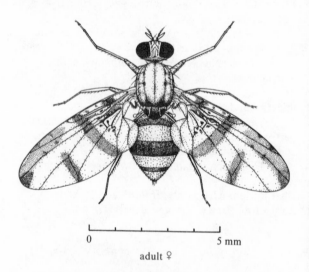

0          5 mm

adult ♀

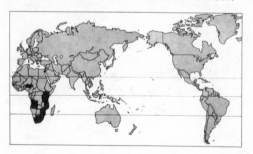

## Dacus cucurbitae Coq.

(= *Strumeta cucurbitae* Coq.)

**Common name.** Melon Fly
**Family.** Tephritidae
**Hosts** (main). Melon
    (alternative). Other cucurbits, both cultivated and wild; and also recorded from cotton, *Citrus*, sunflower, and lettuce.
**Damage.** The larvae tunnel in the fruit, contaminating them with frass and providing entry points for fungi and bacteria which cause the fruit to rot. Young fruit can be destroyed in a few days; older fruit show less obvious symptoms, but on cutting open they are found to contain a mass of maggots in the pulp.
**Pest status.** A very important pest of cucurbits in Africa, India, and Hawaii, rendering these crops quite uncommercial in many areas. The distribution of this pest in India is largely determined by moisture; the population expands when rainfall is adequate and contracts during dry periods.
**Life history.** Eggs are laid in groups under the skin of young fruit by means of the quite sharp ovipositor of the female.

    The larvae are typical dipterous maggots, 10–12 mm long when fully grown, and they bore in the pulp of the fruit.

    Pupation takes place in the soil, but occasionally in the fruit, and it takes about ten days. The puparium is elongate, oval, brown, and 6–8 mm long. In drier areas the pupa may enter diapause.

    The adult is a large, brown fly, 8–10 mm long, including ovipositor, with a wingspan of 12–15 mm. The eyes and head are dark brown. Wings are hyaline with a dark brown costal stripe extending right up to the tip of the wing; there are a few small infuscate areas in the wings. There are three bright yellow stripes on the dorsum, and the scutellum is yellow. The adults feed on nectar, bird faeces, plant sap, and juices from tissues of damaged or decaying fruit.

    The life-cycle takes about 3–4 weeks, and many generations can occur in one year.
**Distribution.** E. Africa, Mauritius, Pakistan, India, Bangladesh, Sri Lanka, Burma, Malaysia, Indonesia, Thailand, Sarawak, Philippines, Taiwan, China, S. Japan, Ryukyu Isles, Hawaii, and N. Australia (CIE map no. A64).

    Several other species of *Dacus* are important fruit pests; *D. oleae* – the Olive Fly; *D. ciliata* – the Lesser Melon Fly; *D. dorsalis* – the Oriental Fruit Fly; *D. tryoni* – the Queensland Fruit Fly; and *D. zonatus* – the Peach Fruit Fly of India.
**Control.** Control recommendations are as for *Ceratitis capitata* (p. 387).

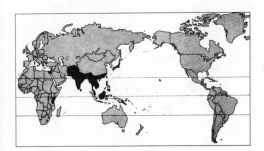

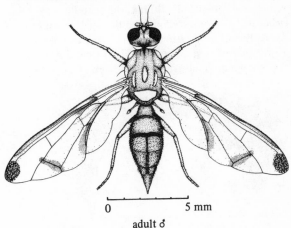

0     5 mm

adult ♂

# Dacus dorsalis (Hend.)

(= *Dacus ferrugineus*)
(= *Strumeta dorsalis* Hend.)

**Common name.** Oriental Fruit Fly
**Family.** Tephritidae
**Hosts** (main). Guava, mango, *Citrus*, banana, avocado, papaya, etc.

(alternative). Peach, passion fruit, coffee, melons, pineapple, jackfruit, strawberry; in Hawaii 173 species of plants in 112 genera were recorded.

**Damage.** Females oviposit through the skin of the fruits and sap may ooze from the punctures. The maggots feed inside the fruits, and their infestation may be associated with fungal and bacterial rots.

**Pest status.** A serious pest of all fleshy fruits and vegetables in the general S.E. Asia region, and Hawaii.

**Life history.** The female fly uses her ovipositor to deposit eggs about 5 mm beneath the surface of ripening fruits. The batches of eggs hatch in about two days, and larval development can be as short as seven days before the mature maggots drop out of the fruits to pupate in the soil. Pupal development takes about ten days.

The adult flies are dark brown with bright yellow markings on the thorax, the scutellum is either white or pale yellow, and the hyaline wings have a line of infuscation along the leading edge and down the anal veins. These markings are quite specific, as is shown in the illustrations of Drew, Hooper &

Bateman (1978). Adult wingspan is about 15 mm, with body length about 8 mm. The female maturation period is 5–7 days. The entire life-cycle takes only about 25 days in the tropics, where there may be many generations per year, but in cooler regions development is much slower. Temperature limits are 14 °C for larval development and 21 °C for adult.

**Distribution.** From Pakistan and India through S.E. Asia to N. Australia, and to China, Taiwan and the Ryukyu Islands and Hawaii (CIE map no. A109).

**Control.** Bagging of fruits is practised in S.E. Asia to deter ovipositing female flies. Other methods of control include destruction of unmarketable fruits.

Pheromone traps are used for fruit fly monitoring purposes in many orchards. Natural parasitism levels are often quite high, and this pest has, to a great extent, been controlled in Hawaii by the parasite *Opius* (Braconidae).

Chemical control is clearly difficult, and is generally a suppressive procedure aimed at the female flies, usually with a protein bait spray incorporating malathion (or naled), and a chemical attractant.

Some countries have quarantine regulations aimed specifically at this pest.

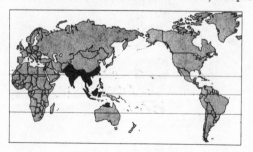

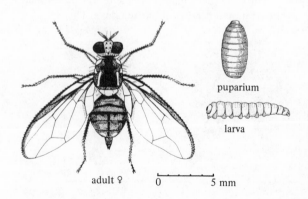

puparium

larva

adult ♀    0    5 mm

# Dacus oleae (Gmel.)

**Common name.** Olive Fruit Fly
**Family.** Tephritidae
**Hosts** (main). Olive, both cultivated and wild.
   (alternative). None recorded.
**Damage.** The fruits fall prematurely, and are mottled with a hollowed interior inhabited by a white maggot. Early damage shows as slightly sunken brown necrotic spots. The stone of the fruit is not damaged. Yield can be reduced by as much as 80–90%. The oil from attacked fruit is inferior and has an unpleasant flavour.
**Pest status.** The most serious pest of olive in the Mediterranean region.
**Life history.** The female fly lays a single egg on the young olives (about the size of a pea) – if several maggots are found in one fruit then they have come from eggs laid by different females. The egg is deposited under the skin of the fruit; hatching takes 2–3 days.

The maggots are 1–6 mm long, according to their age; larval development takes 10–15 days, according to temperature.

During the summer pupation takes place in the fruit, but the last generation maggots pupate in the soil under the tree, where they overwinter, at a depth of 5–10 cm.

The adult is a small, dark brown fly, about 5 mm long, with hyaline wings having a small dark terminal spot; the female has a prominent ovipositor.

The complete life-cycle in the summer takes about four weeks and there are typically three or four generations per year.
**Distribution.** The Mediterranean region, Canary Isles, Pakistan, Caucasus, Egypt, Eritrean region of Ethiopia, and S. Africa (CIE map no. A74).
**Control.** Owing to the site of oviposition being under the skin of the fruit, only systemic or translocatory insecticides will be effective against the larvae. Parathion-methyl is the generally recommended insecticide for use against this pest.

In Greece the practice is to use bait sprays with 4–12% protein, in both aerial and ground application, against the adult flies; most trees receive 2–3 treatments per season; a high level of control is regularly achieved.

Considerable work is in progress on various aspects of biological control for this pest in several European countries.

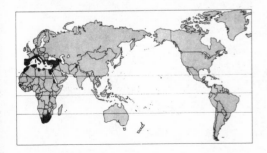

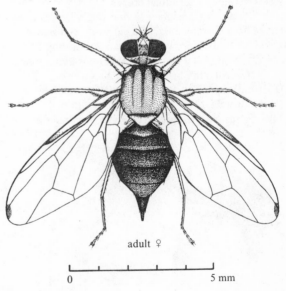

adult ♀

0            5 mm

## Ophiomyia phaseoli (Tryon)

(= *Melanagromyza phaseoli* (Tryon))

**Common name.** Bean Fly
**Family.** Agromyzidae
**Hosts** (main). Beans of various species, including *Phaseolus, Vicia, Glycine* spp.

(alternative). A wide range of leguminous crops.
**Damage.** Attacked plants are yellow and stunted; often many are dead. Stems just above soil level are thickened and usually cracked. Attacks on older plants are confined to the leaf petioles.
**Pest status.** A major pest of beans in Old World tropics.
**Life history.** The slender, white eggs, 1 mm long, are laid singly in holes made on the upper surface of young leaves, near the petiole end of the leaf.

The larva is a small, white maggot which bores down inside the stem where it feeds just above ground level. The leaves often turn yellow, and the stems develop longitudinal cracks.

Pupation takes place in the stem where the larvae have been feeding. The barrel-shaped pupae are black or dark brown and about 3 mm long.

The adult is a tiny black fly about 2 mm long.

The total life history takes 2—3 weeks.
**Distribution.** Africa, Pakistan, India, Bangladesh, Sri Lanka, Burma, Malaysia, China, Philippines, Taiwan, Java, Papua New Guinea, West Irian, Australasia, Samoa, Fiji, Caroline and Mariana Isles (CIE map no. A130). Four other species are important pests of grain legumes throughout parts of S.E. Asia.
**Control.** Successive, overlapping crops of beans should be avoided. Crop residues should be destroyed and volunteer plants removed.

Chemical control can be easily and cheaply achieved by seed dressings of aldrin or dieldrin.

Insecticides effective as sprays are mono-crotophos, omethoate, oxamyl, triazophos, dimethoate and some synthetic pyrethroids; generally they are applied as two pre-flowering sprays, two and 12 days after crop emergence.

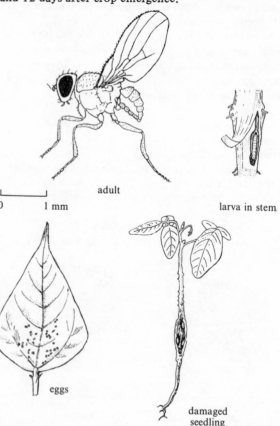

adult

larva in stem

0    1 mm

eggs

damaged seedling

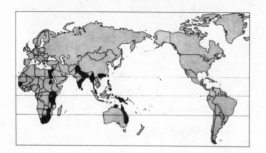

394

## Tropicomyia theae (Cotes)

(= *Melanagromyza theae* Bigot)

**Common name.** Tea Leaf Miner
**Family.** Agromyzidae
**Hosts** (main). Tea
      (alternative). None recorded.
**Damage.** The larvae make tunnel mines in the leaves of the host plant, leading to leaf distortion and sometimes destruction by fungal attack.
**Pest status.** Of regular occurrence within its distribution range, on tea foliage, but usually only a minor pest.
**Life history.** Eggs are laid singly underneath the upper epidermis of the leaf; either one or two eggs per leaf. The second and third leaves from the bud seem to be preferred. The larva makes an irregular silvery tunnel mine in the upper part of the leaf. After about 11 days pupation takes place within the end of the mine, with the anterior spiracles projecting through the epidermis; pupation takes about 15 days.

      The adult is a small black fly about 3 mm in body length, which resembles many other adult Agromyzidae and is not easily identified.
**Distribution.** India, Sri Lanka and the Seychelles. Spencer (1973) established this new genus to accommodate a group of tiny leaf-miners occurring widely throughout the Old World tropics and reaching Australia and Japan, on tea and some other plants.

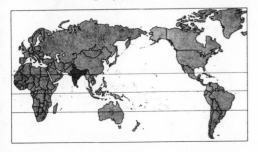

Four other species also mine tea leaves; *T. atomella* (Mall.) in India, Sri Lanka, Thailand and Taiwan; *T. flacourtiae* (Seg.) in S., E., and W. Africa, and Madagascar (also in coffee, *Citrus*, cotton, and many wild hosts); *T. polyphyta* (Klein.) in Australia, on many different crops and wild hosts; and *T. styricola* (Sasakawa) in Japan, on several different hosts.
**Control.** Normally not required; in most places larval and pupal parasitism by parasitic wasps keeps the populations down to a fairly low level. Field observations reveal that levels of parasitism are often as high as 70%.
**Further reading**

Spencer, K. A. (1973). *Agromyzidae (Diptera) of Economic Importance*, W. Junk: The Hague.

leaves of tea with larval mines

## Diopsis thoracica Westw.

**Common name.** Stalk-eyed Fly (Stalk-eyed Borer)
**Family.** Diopsidae
**Hosts** (main). Rice, and *Sorghum*.
    (alternative). Probably occurs on wild grasses
also but this is not known for certain.
**Damage.** The maggot feeds on the central shoot of
the young rice plant, causing a typical 'dead-heart'.
Later generations of larvae feed on the flower head
before it emerges.
**Pest status.** A serious pest of rice in Swaziland and
North Cameroon, and Sierra Leone, but generally
this pest is more of academic interest than economic.
**Life history.** The eggs are 1.7 × 0.4 mm, white, boat-
shaped, with a characteristic anterior projection.
Each female lays about 20 eggs over a 10-day period.
The eggs are laid singly on the upper surface of young
rice leaves, usually on the subterminal leaf, fixed to
the leaf by cement which prevents their being washed
off in heavy rain.

    The larva on emergence moves down inside the
leaf sheath and feeds on the central shoot above the
meristem. Later larvae feed on the flower head before
it emerges. The mature larvae are 18 × 3 mm, white
with terminal yellow markings, and with very small
heads. Larval development takes 25–33 days.

    The pupae are red with brown dorsal bands,
fat, and almost triangular in section because of their
compression inside the rice stem. The pupal period is
10–12 days.

    The adults are typical diopsid flies with charac-
teristic eyes and antennae borne on the ends of long
lateral stalks. *D. thoracica* has a red abdomen and two
long posterior spines on the thorax.
**Distribution.** Africa only; from Somalia and E.
through Zaïre to Cameroons and W. Africa, down to
S. Africa.

    There are many species of *Diopsis* in Africa, at
least five of which have been associated with rice, and
several species in the Oriental region.
**Control.** The economic importance of this pest is
not clearly established.

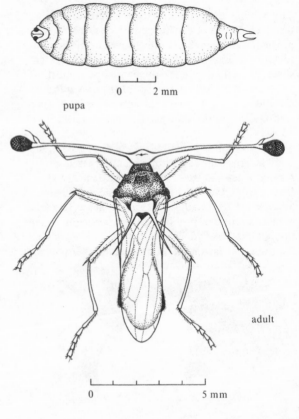

pupa

adult

## Hydrellia griseola Fall.

**Common name.** Rice Whorl Maggot (Rice Leaf Miner)
**Family.** Ephydridae
**Hosts** (main). Rice
(alternative). Wheat, barley, oats, and many
species of grasses and sedges.
**Damage.** The maggots bore into the leaves and feed
on the mesophyll tissue, and make mines; linear
initially later they coalesce and become a blotch
mine. The attacked leaves shrivel and lie flat on the
surface of the water. Heavily attacked plants may
die.
**Pest status.** An important pest of rice in California,
and other areas; losses can be heavy. Other species
occur on *Potamogeton* pond-weeds.
**Life history.** The eggs are laid singly on the leaf
blades close to the water surface. They are banana-
shaped, with irregular sculpturing; 0.6 × 0.16 mm.
Each female lays 50–100 eggs. Hatching takes
3–5 days.

The maggots on hatching immediately bore
into the leaf. Larval development takes 7–10 days
or longer (up to 40) in cooler climates.

Pupation takes place in the leaf, and the brown
oval puparium is easily visible inside the transparent
mine in the leaf. Pupation takes from 5–40 days,
according to temperature.

The adults look like small, grey houseflies with
long legs, and with a conspicuous pale, shining, frontal
lunule. Wingspan is 2.5–3.2 mm, the males being
smaller. The females start egg-laying 3–4 days after
emergence, and can live for 3–4 months.

The number of generations per year varies with
the climate, from eleven in California to eight in
N. Japan.
**Distribution.** Europe, N. Africa, Egypt, Malaysia,
Korea, Japan, USA and S. America.

Another species occurs on rice in Japan –
*H. sasakii* – and in Europe *H. griseola* larvae mine the
leaves of cereals (wheat, barley, etc.).
**Control.** Best control is achieved by a combination of
water management and insecticide use. The crop
should be started in shallow water gradually becoming
deeper to allow the plants to emerge quickly and
develop robustly.

Sprays of dieldrin or heptachlor are effective in
killing both adults and the mining maggots.

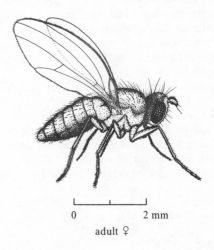

L____l____l
0          2 mm

adult ♀

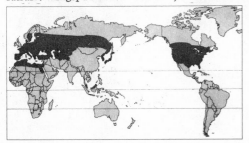

## Atherigona oryzae Mall.

**Common name.** Rice Stem Fly (Corn Seedling Maggot)
**Family.** Muscidae
**Hosts** (main). Rice
  (alternative). Maize, sorghum, wheat, barley, rye, and various grasses.
**Damage.** This pest only attacks upland paddy (i.e. non-irrigated rice). The maggots bore into the stem and feed on the basal part of the youngest leaf causing a typical dipterous 'dead-heart'.
**Pest status.** A pest of upland rice of some importance locally, in various different areas.
**Life history.** Eggs are laid on the upper and lower surfaces of rice seedling leaves; they are relatively large, being 1.5 mm in length, elongate, and white.

  The tiny larvae move down the leaves, pass between the leaf sheath and stem, and bore into the stem of the seedling, at about the level of the growing point. They feed on the developing shoot causing the youngest leaf to die, which then turns brown and withers.

  Pupation takes place in the soil, and the larva pupates within a pale brown puparium.

  The adult looks like a small grey housefly, about 3 mm long, with a yellow spotted abdomen and yellow legs. The head has a distinctive angular shape with deep-set antennae.

  The total life-cycle takes 15–32 days.

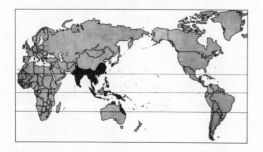

**Distribution.** India, Bangladesh, Sri Lanka, Malaysia, Indonesia, Philippines, China, Japan, Papua New Guinea, West Irian and Australia. (CIE map no. A411).

  *Atherigona exigua* Stein is the Rice Seedling Fly of S.E. Asia, and several other species of *Atherigona* are recorded from rice.
**Control.** Inundation of nurseries at intervals can be effective, but is not really feasible. Early sowing can be advantageous in some areas.

  Seed dressings using aldrin, dieldrin, and endrin, have been quite effective. Cover sprays of parathion and endrin, applied at the first sign of damage, have given good results.

  Certain varieties of rice are apparently not attacked by this pest.

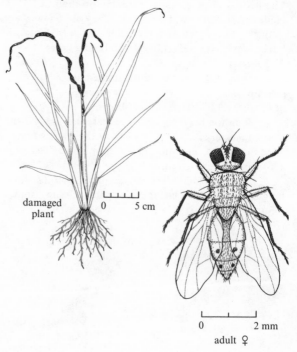

damaged
plant

0    5 cm

0    2 mm

adult ♀

## Atherigona soccata Rond.

**Common name.** Sorghum Shoot Fly
**Family.** Muscidae
**Hosts** (main). *Sorghum* spp.

(alternative). Maize, finger millet, bulrush millet, rice, wheat, and the grasses *Andropogon sorghum, A. s. saccharatum, Cynodon dactylon, Elusine* spp., and *Panicum* spp.

**Damage.** The maggot feeds on the growing point of the shoot of the seedling causing a typical dipterous 'dead-heart'. Attack usually results in tillering, and in severe attacks the tillers in turn may be attacked. The damage by this fly is indistinguishable from that by other Muscidae, Chloropidae and Oscinellidae.

**Pest status.** The most serious shoot fly pest of *Sorghum* seedlings in many parts of the Old World tropics, but many other species of shoot flies do occur. Deeming (1971) recorded 50 species of *Atherigona* from cereal crops in N. Nigeria, and described 23 as new species.

**Life history.** Eggs are laid on the underside of the leaves of seedlings which are 7–8 days old, or on young tillers. Often a single egg is laid per leaf, but up to three have been recorded. The eggs are white, elongate, 0.8 × 0.2 mm, with a raised, flattened, longitudinal ridge. Hatching takes 2–3 days.

The young larvae crawl down inside the sheath and then bore horizontally into the base of the young shoot, killing the growing point and the youngest leaf which eventually turns brown and withers. The third instar larva is white to yellowish, about 10 mm long and 1.3 mm broad, with the anterior spiracle of a rosette type with 8–10 digitations. Larval development takes 7–12 days.

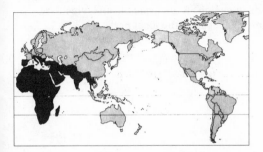

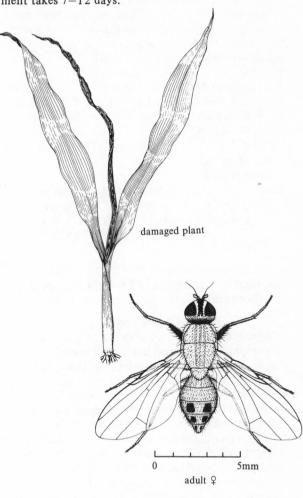

damaged plant

0      5mm

adult ♀

399

Pupation takes place in the base of the necrotic shoot, or rarely in the soil, and takes about seven days. Under unfavourable conditions the pupae may aestivate.

The adult is rather like a small housefly in appearance, 4–5 mm long. The female has head and thorax pale grey, and abdomen yellowish with paired brown patches; the male is blacker. Under the warmest conditions the life-cycle only takes 17 days, but this may be 21 days in cooler weather.

**Distribution.** Old World tropics from the Canary Isles to Central Asia; W., E. and S. Africa, Sudan, Ethiopia, Zaïre, Madagascar, Mauritius; and from N. Italy to India, Burma, and Thailand (CIE map no. A311).

**Control.** See following section on the control of cereal shoot flies.

**Further reading**

Clearwater, J. R. (1981). Practical identification of the females of five species of *Atherigona* Rondani (Diptera, Muscidae) in Kenya. *Tropical Pest Management* (formerly PANS), **27**, 303–12.

# Control of Cereal Shoot Flies (Muscidae & Anthomyiidae)

The eggs are laid either on the young leaves of the seedling or on the soil at the plant base, and the first instar maggots crawl inside the seedling leaf-sheath and bore horizontally into the shoot, killing the growing point and causing the characteristic 'dead-heart'. Once inside the shoot the larva is relatively safe from contact insecticides and can only be attacked using systemics. The period of pest vulnerability is clearly the first instar larva as it leaves the eggshell and crawls inside the leaf sheath. Chemical application should be made only when really necessary, and then with care for natural egg mortality is normally high (about 70%), due mostly to predation.

The various ways in which cereal shoot flies can be combated are as follows.

(1) Early sowing – only the young seedling is vulnerable to attack, so by sowing early it is often possible to have the period of crop vulnerability over by the time the flies emerge.

(2) Resistant crop varieties – many cereals show some degree of resistance to shoot fly larvae; certain rice varieties show a high level of resistance.

(3) Insecticides

(a)  Seed dressings may be used in areas at high risk, sometimes followed by cover spray later; the usual insecticides employed have been dieldrin, γ-BHC, heptachlor, carbofenothion, etc.

(b)  Granules – chlorfenvinphos, endosulfan, disulfoton, and phorate were especially formulated as granules for such use, and other chemicals are being formulated this way now; they are systemic, sometimes slightly fumigant, and persist in soil as granules for several weeks. They are usually applied by the 'bow-wave' technique at drilling.

(c)  Cover sprays – usually applied post-emergence along the rows, and the chemicals used for the various shoot flies in the past have included dieldrin, endrin, carbofuran, carbofenothion, dimethoate, formothion, ethoate methyl and parathion.

The different shoot flies respond slightly differently to these chemicals and local advice should be sought as to which chemicals are available and which are recommended against the local pests.

# Delia arambourgi (Seguy)

(= *Hylemya arambourgi* Seguy)

**Common name.** Barley Fly
**Family.** Anthomyiidae
**Hosts** (main). Barley
 (alternative). Maize, wheat, bulrush millet, and some grasses.

**Damage.** The feeding larva eats the stem of the central shoot, causing a typical shoot borer 'dead-heart'. The central shoot dies, and turns brown, and may be easily pulled out of the plant. One larva may destroy three or four shoots.

**Pest status.** An important pest of barley in Africa. Heavy infestations occur; if at times of drought, there may be a complete loss of the crop.

**Life history.** Eggs are usually laid on the soil within 2 cm of the plant, though occasionally recorded on the tips of young leaves. Hatching takes 3–4 days.

 The young larvae make their way over the soil surface, climb the plant to just above the first leaf sheaths, and bore down through the tissue to the growing point. This results in the death of the central shoot, producing a 'dead-heart'. The larva remains in this shoot during the first two instars. Soon after the second moult the larva quits that shoot, and attacks either another on the same plant, if tillering has commenced, or else another plant. The new shoot is penetrated by eating through the leaf sheath, and again the central shoot is destroyed. The larva moves to a third or even fourth shoot before reaching maturity some 12 days after hatching.

 Pupation takes place in the soil, amongst the plant roots; the pupal stage takes some seven days.

 The adult is a medium-sized fly about 7–8 mm long, and looking rather like a small housefly; the female has a pointed abdomen and is grey; the male is blackish and has a rounded abdomen apex.

**Distribution.** Africa: Nigeria, Zimbabwe, Kenya, Uganda, Sudan, Ethiopia, S. Africa, and the S. Arabian peninsula.

**Control.** The most successful results have been obtained using seed dressings of dieldrin and heptachlor.

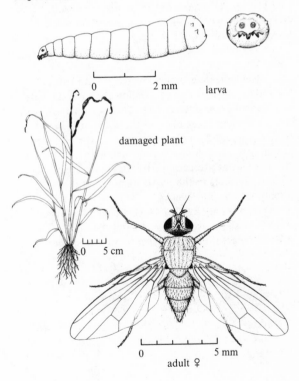

larva

damaged plant

adult ♀

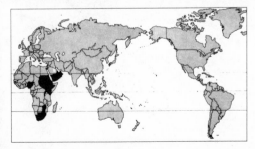

401

## Delia platura (Meig.)

(= *Hylemya platura* Meig.)
(= *H. cilicrura* (Rond.))
(= *Chortophila cilicrura* Rond.)

**Common name.** Bean Seed Fly (Corn Seed Maggot)
**Family.** Anthomyiidae
**Hosts** (main). Sown seeds of beans and maize.

    (alternative). Also onions, tobacco, marrow, cucumber, lettuce, peas, and crucifers.
**Damage.** The maggots bore the cotyledons of sown seeds or into stems and petioles of young seedlings.
**Pest status.** A serious pest of beans in many areas, and of maize in Europe and the USA; locally important.
**Life history.** The eggs are laid on disturbed soil. They are elongate, white, with a reticulate pattern. Each female lays about 100 eggs, a few at a time, over a period of 3–4 weeks. Hatching takes 2–4 days.

    The larvae are typical maggots, with three instars, taking, 3, 3 and 6–10 days for each stage of development respectively (12–16 days in total).

    Pupation takes place in the soil a little way from the plant, 2–4 cm under the soil surface. The puparia are 5 mm long, dark brown, with a posterior circlet of stout projections. In temperate areas over-wintering occurs in the pupal stage, otherwise pupation takes 2–3 weeks.

    Adults may live for 4–10 weeks; the female has a greyish pointed abdomen and the male a rounded blackish one. A definite pre-oviposition period occurs in the female from one to several weeks.

    There are 2–5 generations per year, according to climatic conditions; the life-cycle may be completed in 4–5 weeks under warm conditions.
**Distribution.** Almost completely cosmopolitan, occurring from the Arctic Circle down to S. Africa, New Zealand, Tasmania and Argentina, but not recorded from the north-eastern part of S. America, W. Africa, India or the Malaysia/Indonesia peninsula area (CIE map no. A141).
**Control.** Seed dressings of dieldrin were very successful, but in most areas resistance to dieldrin has become established. In these cases ethion, diazinon and pirimiphos-ethyl and trichloronate have proved to be effective. However, care must be taken with diazinon, for higher dose rates show marked phyto-toxicity; and because of toxicity, the use of trichloronate in some countries is not approved.

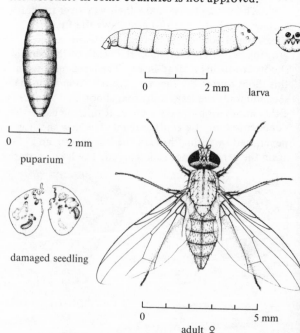

puparium

larva

damaged seedling

adult ♀

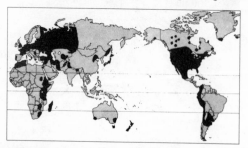

402

# Order **HYMENOPTERA**

An enormous group of about 100 000 species, comprising the ants, bees, wasps, chalcids, ichneumonids, etc. They all possess two pairs of membraneous wings, the hindwings smaller and interlocking with the forewings by small hooks. The mouthparts are typically for biting, but sometimes for sucking. The abdomen is typically basally constricted and its first segment fused with the metathorax. An ovipositor is always present, and modified for sawing, piercing or stinging. Metamorphosis is complete; the larva is generally apodous with a more or less well-defined head.

## Family **Tenthredinidae**

(Sawflies)  This is the largest family of sawflies containing about 4000 species. The eggs are usually laid in young shoots or leaves, and the saw of the ovipositor modified according to its role. The larvae are typified by having on the abdomen six pairs of prolegs in addition to the terminal claspers. The larvae are either exposed on the leaf surface, sometimes gregarious, or else they make galls, or live inside the stems. A few species have slug-like larvae. Pupation takes place in a silken cocoon in the soil.

## Family **Formicidae**

(Ants)  A large family containing all the described species of ants — some 3500 species in total. They are usually social insects with well-defined castes, including workers, soldiers, and royal forms. Some ants do cause direct damage to crops but others are most important as a nuisance to the plantation workers by their biting and stings. Some ants are also of considerable importance in their attendance on various homopterous pests such as aphids, mealybugs and scale insects — generally their presence is thought to keep the bugs free from predation and parasitism. Some of the aggressive arboreal species (*Oecophylla* spp.) that may attack field workers are also predaceous and will kill some species of pests in trees and palms. But the arboreal ants may in turn be killed or driven away by certain more aggressive ground-nesting species that forage in trees (*Anoplolepis* and *Pheidole* spp.).

## Family **Vespidae**

(Wasps)  A large family of social insects, mostly tropical in distribution, but some are temperate. Adults and larvae are basically predaceous, and so the adult wasps are part of the local natural predator complex and important in the natural control of pest species. However, the adults also are attracted to sugar, and can be found eating honey-dew; they also attack ripe and ripening fruits for the sweet sap and may do considerable damage in some orchards. Some species, such as *Polistes*, tend to nest in bushes and small trees, close to the ground and are very aggressive if disturbed and fiercely attack any disturbers.

## Athalia spp.

**Common name.** Cabbage Sawfly
**Family.** Tenthredinidae
**Hosts** (main). Brassicas of all species
    (alternative). Other members of the Cruciferae.
**Damage.** Leaves are eaten by the larvae, often leaving only the midrib. Blackish larvae are present on the plant and they fall to the ground if the plants are shaken.
**Pest status.** A sporadically serious pest of all cruciferous crops; turnip, chinese cabbage, kale and crambe are particularly susceptible to attack.
**Life history.** Eggs are laid singly in small pockets cut in the leaf by the female sawfly.

The larva closely resembles a lepidopterous caterpillar, the important difference is that these sawfly larvae have six pairs of prolegs on the abdomen instead of the four found in caterpillars. The full-grown larva is about 2.5 cm long, and is oily black or green; black specimens often have yellow spots. The head is shiny black and that part of the body just behind the head is often slightly swollen, giving the larva a humped appearance.

The full-grown larva burrows into the soil and spins a tough silk cocoon to which particles of soil adhere. The yellowish pupa forms within the cocoon.

The adult remains in the cocoon for a while before it emerges and pushes its way through the soil to the surface. It is about 1.5 cm long and has a dark

head and thorax with a bright yellow abdomen. The adults may often be seen flying about slowly, just above the crop.

**Distribution.** Great Britain, and Europe from Spain through to Siberia and Japan, down to Asia Minor, N. Africa, E. and S. Africa, and S. America.

**Control.** Destruction of wild crucifers on the headlands will help to keep the pest population down.

The following pesticides have been found to be effective as foliar sprays: DDT, carbaryl and pyrethrum.

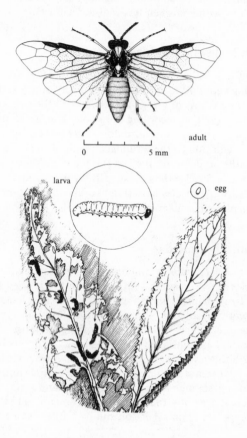

adult

larva

egg

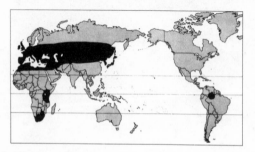

## Atta spp.

**Common name.** Leaf-cutting Ants (Fungus, or Parasol Ants)
**Family.** Formicidae (Attini)
**Hosts** (main). *Citrus*, cocoa, coffee, maize.
    (alternative). Cotton, cassava, mango, beans, sweet potato, groundnut, banana, pineapple, rice, wheat, etc., and many forest trees (especially teak and *Pinus*) and many grasses and wild plants.
**Damage.** They are generally polyphagous feeders, but certain species show host-specificity and some prefer monocotyledonous plants whilst others prefer dicotyledonous ones. The foraging adult ants cut pieces of living leaf foliage and carry them back to the nest for construction of fungus gardens. Severe attacks result in total defoliation, and many trees die.
**Pest status.** Damage in the neotropics was estimated by Cramer (1967) to be US $1000 million annually, due to defoliation by species of *Atta* and *Acromyrmex*. There are 18 species of *Atta*, collectively regarded as serious pests in the region between Texas, USA, and northern Argentina, because of their defoliation of cultivated plants.

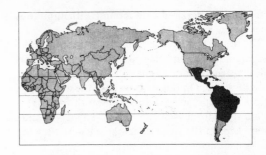

**Life history.** The tribe Attini are fungus-growing ants. The largest individuals belong to the genera *Atta* and *Acromyrmex* and are known as leaf-cutter ants because they use pieces of leaf tissue for the construction of their underground fungus gardens. The ants are colonial, living in large underground nests. *Atta* spp. are apparently forest dwellers for most nests are located in the forest areas, and they are most serious as pests in cultivated land near forest edges. The nest is generally large, measuring up to 10–15 m in diameter, 4 m in depth, and containing up to 2 million ants. Queens have been recorded living up to 20 years, and so many nests may be long-lived. The nest area is

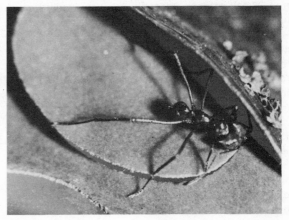

*A. cephalotes* workers cutting pieces of leaf

Photos: J. M. Cherrett

evident as a slightly raised bare mound with numerous entrance holes.

The nest colony consists of many small underground interlinked chambers which contain the fungus gardens constructed out of chewed leaf fragments mixed with saliva. On this organic base the ant fungus mycelium develops, and the ants feed on the staphylae which are specialized swollen hyphae. The larvae and pupae are kept in the fungus chambers where they are nursed by the smallest worker ants.

The workers (sterile females) occur as several distinct size castes; the smallest remain in the nest for nursing and cultivation duties, and the largest go out to forage for food under the protection of the large aggressive soldiers.

The ants forage communally using pheromone trails and cut pieces of living leaf material; they then carry each piece back to the nest, where it is chopped up into small pieces and stored in special chambers. *Atta* spp. generally forage for distances up to 150 m from the nest.

Sexual adults are produced once annually, and the nuptial/dispersal flights occur usually in the middle of the rainy season; dispersal has been recorded up to 10 km. *Atta* spp. generally produce a small number of large queens; after founding a new colony sexuals are not produced until after three years.

**Distribution.** These ants are confined to the New World tropics. The 18 species of *Atta* occur as a group throughout 23 countries from Texas (USA), Mexico, W. Indies, C. America and S. America down to Uruguay and northern Argentina. Each country usually has just a few species of *Atta*, but Brazil has a total of nine species. The most important species is *Atta cephalotes* (L.), found in 17 countries.

**Control.** Because of the feeding habits of these ants, and their colonial life-style, destruction of the nest population is generally aimed at rather than any direct plant protection. Chemical control is of two basic types; poison baits or organochlorine compounds injected into the nest tunnel system.

Baits are either Mirex in a pellet formulation, or else a matrix (base) of organic material with aldrin or chlordane as the toxicant. The matrix may be citrus pulp (meal) or other meals, and may have vegetable oils added. Recent work is testing the use of trail pheromones added to the bait to make it more attractive.

The organochlorine compounds used are generally aldrin or chlordane, either blown down the nest entrances as a dust, or poured down as an emulsion, usually in water. Some baits (such as Mirex 450 pellets) which are successful in dry weather tend to disintegrate during rainy weather, unless made water-repellant by additives. Baits applied just before the annual nuptial flight are most effective because at that time the population of ants is largest and conrol reduces the spread of new colonies.

**Further reading**

Cherrett, J. M. & D. J. Peregrine (1976). A review of the status of leaf-cutting ants and their control. *Ann. appl. Biol.*, **84**, 128–33.

Weber, N. A. (1972). Gardening ants: the Attines. *Mem. Amer. Phil. Soc.*, **92**, 1–146.

nest site of *A. vollenuxideri* in Paraguay

Photo: J. M. Cherrett

## Acromyrmex spp.

**Common name.** Leaf-cutting Ants
**Family.** Formicidae (Attini)
**Hosts** (main). *Citrus*, cocoa, coffee, maize.

(alternative). Cotton, cassava, mango, beans, sweet potato, groundnut, banana, pineapple, rice, wheat, many grasses and many wild plants.

**Damage.** These are general feeders that show some host-specificity as some species prefer monocotyledonous plants whereas others prefer dicotyledonous ones. The foraging worker ants cut pieces of living leaf material and carry them back to the nest for construction of fungus gardens. Severe attacks result in total defoliation and the damaged plants usually die.

**Pest status.** *Acromyrmex* spp. are regarded collectively as serious pests in the tropical Americas from southern USA to Uruguay and Argentina (25 countries in all).

**Life history.** These ants are also social and live in underground nests, but the nests are small in comparison to those of *Atta* spp., seldom measuring more than a metre in diameter. The life history details and feeding habits are much the same as the species of *Atta* except that the colony is smaller, shorter lived, and generally more adapted for life in disturbed ground. Colonies are usually found in urban areas or cultivated ground, and less frequently in forest. Foraging generally only occurs up to about 30 m from the nest, but nest density in agricultural areas is usually from 3–60 nests per hectare. *Acromyrmex* species typically produce a large number of small queens, and dispersal flights are seldom more than about 2 km; queens have been recorded living for seven years, and the nuptial/dispersal flights generally take place at the start of the first rains.

**Distribution.** These ants only occur in the neotropics. The 23 species of *Acromyrmex* are found from southern USA, Mexico, the W. Indies, C. and S. America down to Uruguay and Argentina, in a total of 25 countries. Most countries have only one or a few species of *Acromyrmex*, but Brazil has 18 species, Bolivia has 10, and Argentina has 17 species.

The most important species is *Acromyrmex octospinosus* (Reich.) which occurs in 20 different countries.

**Control.** For control see *Atta* spp.
**Further reading**
See *Atta* spp.

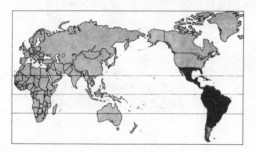

## Macromischoides aculeatus (Mayr)

(= *Tetramorium aculeatum*)

**Common name**. Biting Ant
**Family**. Formicidae
**Hosts** (main). Coffee bushes, especially *robusta*.
      (alternative). Many wild bushes.
**Damage**. Small papery nests are constructed between
the leaves; when the nest is disturbed by movement
of the leaves the ants rush forth with great activity.
**Pest status**. This is an indirect pest only occasionally
important. No direct damage is done to the plant,
neither do the ants seem to encourage infestations by
scales or mealybugs. But workers in the plantation
may be severely bitten at times, even to the extent
where they refuse to carry on picking or pruning. The
wound is a compound one consisting of a bite by the
mandibles followed by formic acid injection into it,
making it most irritating.
**Life history**. The adult is a small brown ant, about
5 mm long, found in numbers of up to several hundred in the small nests made between leaves on the
coffee bushes. Often additional plant material may be
incorporated into the nest.

The adults are characterized by having two
sharp, posteriorly pointing spines at the back of the
thorax.

The immature stages are confined solely to the
nest and are of no direct importance as pests.
**Distribution**. Africa only; Zaïre, Uganda, and
Tanzania.

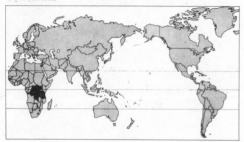

**Control**. Sprays of malathion and dieldrin, as two
foliar applications at a 14-day interval, gave almost
70% control of this pest in Uganda.

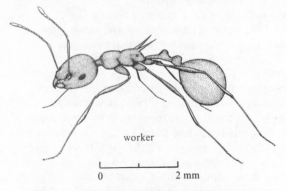

worker

0          2 mm

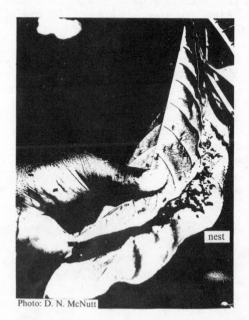

nest

Photo: D. N. McNutt

408

## Messor barbarus L.

**Common name.** Harvester Ant
**Family.** Formicidae
**Hosts** (main). Grasses of many species
(alternative). They also damage cereals and rob seed beds.
**Damage.** These ants remove grass plants, and cereal plants, from around their nests so the vicinity of the nest is quite bare. Through the use of tracks 5–8 cm wide the ants may forage at distances of 30–50 m or more from their nest, causing areas of severe defoliation.
**Pest status.** Harvester Ant occurs throughout the drier parts of E. Africa and in places can be serious pests of grassland. Ten or 12 nests commonly occur per hectare and it is not uncommon to find a loss of grazing of the magnitude of 10–20% in some areas.
**Life history.** The adults are reddish-brown ants with the characteristic narrow petiole (waist) connecting thorax and gaster. The nests are readily recognized by the bare circular areas on the ground surface from which vegetation has been removed. These areas are often slightly depressed and contain up to six slightly raised conical mounds in which are the entrances to the underground nests. There are no ventilator shafts to these nests, such as are produced in termite nests. Seeds are stored underground, often in considerable quantities, and so starving the colony out by a band of insecticide is not often successful.

**Distribution.** E. Africa.
**Control.** One successful method used has been to use a bait of aldrin in maize posho; the bait was placed in and around the nest entrances and was carried down into the nests by the workers. Some ants probably died as a result of ingesting the poison and others probably because of the fumigant action.

It is recommended that treated nests be inspected after a month and if required, to repeat the treatment.

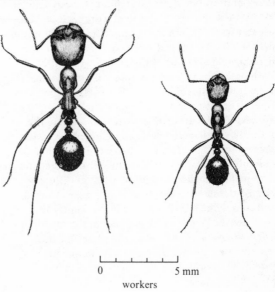

0          5 mm

workers

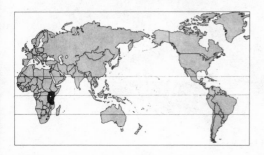

## Oecophylla spp.

*longinoda* (Latr.)
*smaragdina* (F.)

**Common name.** Red Tree Ants (Tailor Ants)
**Family.** Formicidae
**Hosts.** Any plantation or orchard crop.
**Damage.** Nests are constructed by sewing leaves together using silken threads so that the nest is a mixture of silk, living and dead leaves. These ants are almost entirely aerial, and respond to vibration of the nest by furious activity and field workers attacked.
**Pest status.** An indirect pest only occasionally important when it attacks field workers. However, it does encourage infestations of Coccoidea which it attends. On the other hand, this ant does kill arboreal insects in the trees it inhabits and so has a definite beneficial effect; it is basically a carnivorous ant.
**Life history.** An arboreal ant, seldom seen on the ground, living in aerial nests constructed between leaves amongst the twigs of the tree. The main nest is usually large, up to 20 cm diameter or more, and there will be smaller, subsidiary nests in different parts of the same tree or in adjacent trees. The silk used to sew the leaves together is apparently produced by the larvae which are held in the mandibles of the worker ants and used rather like shuttles. There may be several thousand ants in one nest system. They are adept at using telephone wires and cables as a means of aerial translocation. Most nests last for a year or two in the same location.
**Distribution.** *O. smaragdina* occurs throughout S.E. Asia, from India to S. China and Australia. The closely related *O. longinoda* (Latr.) is found throughout tropical Africa.
**Control.** In some situations the predation of crop pests by these ants outweighs their pest value, but occasionally they have to be controlled. Sometimes the nest branch can be cut and the entire nest removed, otherwise sprays of dieldrin, or malathion, at a two week interval should be effective.

worker

0      1 cm

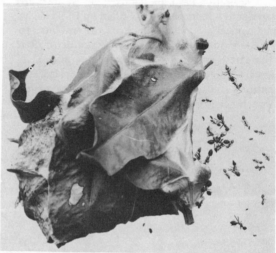

nest with soldiers and workers

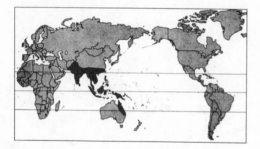

## Solenopsis geminata (F.)

**Common name.** Fire Ant
**Family.** Formicidae
**Hosts** (main). *Citrus* spp., avocado.

(alternative). Coffee, cocoa, and other fruit trees. Seeds and seedlings, notably tobacco, can be destroyed.

**Damage.** The fruit trees are damaged by girdling, where the ants bite through the bark. Branches, shoots, buds, flowers, and fruit can be injured by small gnawing marks. The ants lick the exuding sap. The ants are very aggressive and their bite is painful.

**Pest status.** A minor pest of various fruit trees, but of considerable importance as a hazard and deterrent to field workers in areas where it is abundant.

**Life history.** The ants live in nest burrows in the soil around the base of the tree trunk and make earth galleries on the tree where the bark is gnawed off. They also attend various aphid species on the trees concerned.

The adults are dark reddish-brown ants; the winged females measure about 5 mm, the small-headed workers about 3 mm, and the large-headed workers (soldiers) 5–6 mm long.

**Distribution.** W. Africa, Mauritius, India, Bangladesh, Sri Lanka, S.E. Asia, parts of China, Philippines, Indonesia, Papua New Guinea, West Irian, N. Australia, Hawaii, Samoa, and various Pacific islands, S. USA,

C. America, W. Indies and the northern half of S. America (CIE map no. A95).

A closely related species is *S. saevissima* (F. Smith) the 'Imported Fire Ant' of the S. USA and S. America.

**Control.** Control can be achieved by the application of contact insecticides in powder or spray form to both nests and entrances, and the foliage of the trees. The insecticides recommended are dieldrin, γ-BHC, and diazinon.

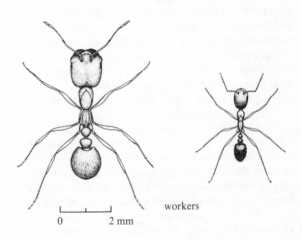

workers

0          2 mm

## Polistes spp.

**Common name.** Paper Wasps

**Family.** Vespidae (Polistinae)

**Hosts and Damage.** These wasps make nests in bushes and small trees, and if disturbed will attack and sting with vigour. The author was once stung 38 times by two wasps from an accidentally disturbed nest in the Seychelles. The adults are attracted to ripe fruits and will damage them by feeding.

**Pest status.** An occasional indirect pest when they nest in the foliage of plantation or orchard crops. The adults are also beneficial in that they are carnivorous and prey on many insects that are crop pests. The photograph illustrates a *Polistes* wasp eating the remains of a *Papilio* pupa on a *Citrus* bush.

**Life history.** A social wasp that occurs as several dozen species throughout the tropics and subtropics; most species bear a strong resemblance to each other, and only differ in minor characters.

The nest is open and small, with only 100–300 cells, and hangs free in vegetation (usually bushes or small trees) from a median pedicel. Some species have the nest flat and plate-like, suspended horizontally, whereas others (as illustrated) have the nest hanging laterally and twisted, with the pedicel at the top edge.

As with all social wasp nests, the colony is annual, and any one nest only lasts for one season, which is inevitably less than one year.

**Distribution.** The 20–30 species that occur world-wide are found throughout the tropics and subtropics, and into southern Europe and Canada.

**Control.** Since these are an integral part of the local insect predator community, making a contribution to the natural control of various pests, it would be preferable for their nests to be left alone. However, there are occasions when a particular nest has to be destroyed because of its location; then a contact insecticide with a rapid knockdown effect is required. Domestic aerosols based upon synthetic pyrethroids are quite effective.

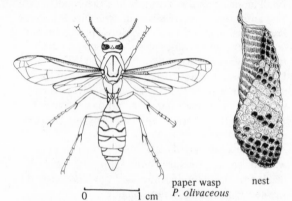

paper wasp     nest
*P. olivaceous*

0      1 cm

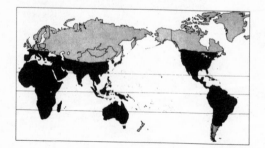

paper wasp eating *Papilio* pupa on *Citrus* bush

412

## Vespa/Vespula spp.

**Common name.** Common Wasps
**Family.** Vespidae
**Hosts and Damage.** These social wasps build nests in the ground, in hollow trees, in buildings and in some cases in a large subspherical papery structure hanging free from a pedicel in the foliage of trees. Nests in, or in the vicinity of, plantation crops or orchards may be disturbed by field workers who then will probably be attacked by the wasps and stung. The adults are attracted by ripe fruits (as are *Polistes*) and they will damage ripe fruits in order to feed on the sugary sap; the fruits usually attacked include papaya, mango, guava, grape, peach, apple, pear, plum, strawberry, and sometimes *Citrus*.
**Pest status.** An indirect pest when they nest in plantation crops or near human dwellings, having a pain-

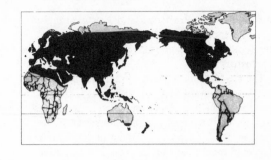

ful sting and aggressive nature. Also a beneficial insect in that they are predators and will undoubtedly be feeding upon some insects that are crop pests. In all parts of the world adult wasps are attracted to ripe fruits and will damage soft-skinned fruits that are left on the tree after ripening.

adult

nest in tree

**Life history.** The genera *Vespa* and *Vespula* include some 10–20 species on a worldwide basis, most of which resemble each other quite closely in morphology and life history.

The nest is usually large, sometimes built underground, sometimes in a hollow tree, often in buildings and dwelling places, and also hanging free from a tree. The free nest is either spherical or elongate, measuring from 20–30 cm in diameter and 20–40 cm in length, and hangs from a single thick pedicel attached to a branch. The nest is constructed from chewed tree bark and wood fragments mixed with saliva and has a papery appearance. The entrance holes are at the base, and internally the nest consists of several layers of cells, or combs, each cell being the home for a single larva. The larvae are fed on chewed insect remains and sometimes honey-dew or sugar obtained from ripe fruits. The entire larval and pupal development takes place within the cells of the nest combs. The largest nests will contain several thousand individuals.

The colony only lives for one season, even in the tropics, and at the end of the season (usually autumn) young queens emerge and fly off to hibernate until the following spring when they start to build a nest of their own. The old nest disintegrates in the winter storms.

**Distribution.** The 10–20 species of *Vespa* and *Vespula* are collectively completely cosmopolitan, and one or two species are very widely distributed.

**Control.** Should control be essential a contact insecticide with rapid knockdown effect should be used, but protective clothing is essential for the spray operator.

0        1 cm

## Order **COLEOPTERA**

These are the beetles; characterized by having fore-wings which are thickened and hardened into elytra, and which are used to protect the delicate, folded hindwings. They range in size, from minute to gigantic insects, and this is the largest order in the animal kingdom, numbering some 220 000 species. The group exhibits great diversity of form and habits.

## Family **Scarabeidae**

(Chafers or White Grubs) A very large family of over 19 000 species, which falls into several distinct sub-families. The larvae of all are very similar in general appearance, and differ mainly in size. They are fleshy grubs with a swollen abdomen, usually adopting a C-shaped position, a well-developed head capsule and large jaws, and thoracic legs. This shape is known as 'scarabeiform'. The larvae live in the soil and rotting vegetation, and eat plant roots; many are important pests known as Chafer Grubs and White Grubs.

## Subfamily **Cetoninae**

(Rose Chafers) These beetles have a flattened body, and the elytra are incurved level with the hind legs; in flight the elytra are raised slightly and the wings protrude through these emarginations. Diurnal in habits; weak mouthparts so usually feed on pollen, nectar and liquids (over-ripe fruits). Adults usually not pests (*Protaetia, Cetonia*, etc.).

## Subfamily **Coprinae** (= Scarabaeinae)

(Dung Beetles) These black, rounded beetles are not pests, and neither are their larvae, but are common in pasture lands.

## Subfamily **Dynastinae**

(Elephant and Rhinoceros Beetles) Large, black beetles usually; typically tropical rain forest species; the adults do some leaf-eating but are usually not pests, except for *Oryctes* spp. The spectacular *Xylotrupes* belongs to this group. Nocturnal in habits.

## Subfamily **Melolonthinae**

(Cockchafers) Dull brown beetles with fat, rounded bodies; nocturnal; elytra held vertically in flight; claws of hind legs of equal size and not movable. Larvae are often serious soil pests eating roots; adults have strong mouthparts and eat leaves, and sometimes young fruits. Genera include *Melolontha, Serica, Holotrichia, Leucopholis*, and *Dermolepida*.

## Subfamily **Rutelinae**

(Flower Beetles and June Beetles) The body is smooth, oval, shiny, and sometimes brightly metallic; the hind legs have thickened tibiae and long movable claws of unequal size. They are nocturnal and will fly to lights at night. Adults have well-developed mouthparts and may defoliate crops and ornamentals, eating both leaves and flowers. Common genera that are pests include *Adoretus, Anomala* and *Popillia*.

## Family **Buprestidae**

(Jewel Beetles and Flat-headed Borers) Essentially a tropical group with larvae that bore tree trunks and branches; a few are temperate. The adults are oval, often brightly coloured with metallic sheen. The larvae are legless and have an expanded flattened prothorax; they usually bore in the sapwood under the bark but also make deep tunnels. Pests of timber and trees. *Agrilus* has about 700 spp. and some are pests of fruit and ornamental trees; *Chrysobothris* has about 300 spp. worldwide.

## Family **Dermestidae**

(Carpet and Hide Beetles) A small family (700 spp.) of small beetles that infest animal cadavers and dried animal products, furs, carpets, etc. *Trogoderma* is a very important pest of stored cereals and is the only phytophagous member of the family.

## Family **Anobiidae**

(Timber Beetles) These are pests of timber and stored products – some 1100 species have been described. The larvae tunnel in the timber, or eat the foodstuffs. Pupation usually takes place in the wood. The head of the adult beetle is deflexed under the anterior edge of the prothorax.

## Family **Ptinidae**

(Spider Beetles) A small group (about 700 spp.) of stored products pests; mostly tropical; body shape is rounded and the legs rather long, which gives them a characteristic shape.

## Family **Bostrychidae**

(Black Borers) A family of cylindrically shaped beetles, making cylindrical burrows in felled timber, dried wood, or occasionally standing trees. The head is bent down and hidden by the projecting prothorax. Most of the damage is done by the boring adults, and breeding takes place in the tunnels made by the adults.

## Family **Lymexylidae**

A small tropical family (40 spp.); adults have a soft elongate body (especially *Atractocerus*) some with reduced elytra; larvae bore living trees and palms. Some temperate species.

## Family **Nitidulidae**

A large family (2200 spp.) of small beetles of varied habits and habit; on flowers, fungi of decaying material. Some stored products pests.

## Family **Silvanidae**

A small family with clubbed antennae, found mostly on plants or on plant material in storehouses. Only a couple of species are pests, and these are pests of stored products.

## Family **Coccinellidae**

(Ladybird Beetles) A large family of 5000 species, mostly brightly coloured, of medium size and convex shape. The vast majority are carnivorous, being predators of aphids, coccids, and other soft-bodied insects, and are of great importance in the natural control of many pests. One genus (and recently a second has been split off) is of considerable importance in being phytophagous and a pest on several different crops – this is *Epilachna*.

## Family **Tenebrionidae**

A large family of some 10 000 species, with larvae of striking similarity but adults exhibiting a wide diversity of form. Many are ground beetles rather like the Carabidae. Some are stored products pests, and a few species are damaging to small trees and bushes by ring-barking them.

## Family **Meloidae**

(Blister Beetles; Oil Beetles) An interesting group of beetles, numbering 2000 species, with soft bodies, long legs, a deflexed head on a narrow neck. The larvae exhibit hypermetamorphosis and are predators upon the egg-pods of grasshoppers and locusts. The beetles contain a substance known as cantharidin, which is the blistering material possessed by these beetles.

## Family **Cerambycidae**

(Longhorn Beetles; Longicorn Beetles) A large family of 15 000 species, of elaborate form, and attractive coloration, often of very large size. The larvae bore in timber, but a few are confined to the roots or pith of herbaceous plants.

## Family **Bruchidae** (= Lariidae)

Over 1000 species are known and most pests are found boring in the seeds of leguminous plants and crops, both in the field and in storage. Many species are widely distributed as a result of being extensively transported with foodstuffs. In addition to Leguminosae many species are found in the fruits and seeds of Umbelliferae and Convolvulaceae.

## Family **Chrysomelidae**

(Leaf Beetles) A large family with more than 26 000 spp. and many are important crop pests. Great diversity of form and habits, and the family is divided into distinct subfamilies.

## Subfamily **Cassidinae**

(Tortoise Beetles) Adults are small beetles with an expanded dorsal skeleton, often brightly coloured; larvae are softly spiny and carry old exuviae posteriorly. Both adults and larvae eat holes in leaves; many species attack sweet potato.

## Subfamily **Chrysomelinae**

(Leaf Beetles) Most of the species belong to this group; both adults and larvae are leaf-eaters; convex in shape; the notorious Colorado Beetle belongs here.

## Subfamily **Criocerinae**

Adults rather elongate, but larvae short and stubby, sometimes covered with their own excreta. *Oulema* and *Crioceris* are important pests.

## Subfamily **Eumolpinae**

Small, rounded beetles; the adults eat holes in leaves, and the larvae generally live in the soil. *Colaspis* spp. larvae are pests of rice and other crops in S. USA.

## Subfamily **Galerucinae**

Also beetles convex in shape, adults and larvae usually together eating leaves. Generally considerable diversity of forms and habits within the group. Many are pests of crops and growing plants.

## Subfamily **Halticinae**

(Flea Beetles) Adults are tiny, with swollen hind femora, and jump readily. Many are serious crop pests. Larval habits vary considerably; some live in the soil, some with the adults on the leaves, some are leaf-miners and some make galls in plant stems.

## Subfamily **Hispinae**

Adults are small and spiny; tropical in distribution; adults scrape the leaf lamina; larvae mine inside the leaves, mostly on Gramineae.

417

## Family **Apionidae**

(Weevils) Adults have clubbed antennae, normally not geniculate (elbowed). Larvae are typically weevils (i.e. Curculionoidea) being legless but with large mandibles; larvae live in the soil or roots, or in tubers; 1000 spp.; *Apion* larvae develop in seeds, stems or roots. *Cylas* is the Sweet Potato Weevil.

## Family **Brenthidae**

(Weevils) Antennae of adults neither geniculate nor clavate (clubbed); often narrow and elongate in body; in tropical forests.

## Family **Curculionidae**

(Weevils proper) A very large group, of more than 60 000 spp.; the adults have antennae geniculate and clavate. The snout may be elongate (rostrum) and narrow, or else short and broad; sometimes referred to as a 'nose' but the mouth is terminal. The elongate rostrum in many species functions as a boring instrument for egg-placing into the host plant tissues. The larvae show great similarity of form, but diversity of habits; some bore timber, some live in plant galls (all parts of the plant body are attacked), most live in the soil where the larvae eat roots. Some species lack wings; some lack males.

## Family **Scolytidae**

(Bark and Ambrosia Beetles) Sometimes regarded as a subfamily of weevils, but are treated here as a separate family. The adults bore tunnels under the bark of trees and construct extensive breeding galleries. The Ambrosia Beetles are species which carry symbiotic fungi and cultivate the fungus in the breeding gallery for the larvae to eat. The larvae do not eat the wood of the host. Mostly pests of timber, but some species bore in thin branches and twigs and are pests of tea, coffee and other small trees and bushes. Many species are really secondary pests, but some are clearly primary pests.

418

# Heteronychus spp.

*arator* (F.)
*consimilis* Kolbe
*licas* (Klug)

**Common name.** Black Maize Beetle (Sugarcane Beetle) Black Wheat Beetle
**Family.** Scarabaeidae (Dynastinae)
**Hosts** (main). Maize, and wheat.
(alternative). Sugarcane, other cereals, yam, tobacco, vegetables and various wild plants.
**Damage.** The main damage is that done by the adult beetles to the young shoots just below ground level, the stems being eaten through. One adult beetle may destroy several seedlings in a row. Losses of young plants can be extensive. Some adult damage consists only of a hole in the side of the stem, with the resulting 'dead-heart'. The larval damage is less significant and consists of the roots being eaten away; the symptoms of larval attack are general wilting and yellowing of the leaves.
**Pest status.** These beetles are important pests of cereals and sugarcane in many parts of Africa, and *H. licas* is of some importance as a pest of yam.
**Life history.** Eggs are laid in moist soil at the base of the plants.

The larvae are soft bodied and fleshy, curved in a C-shape, with well-developed thoracic legs, like typical chafer grubs. The mature larvae are about 35 mm long. They feed below ground on the roots of the host plants, and they often destroy sugarcane stools and maize seedlings.

The adult is a black, rounded scarab beetle between 15 and 20 mm long.

There is usually only one generation per year.
**Distribution.** *H. arator* occurs in southern Africa, Madagascar, Australia and New Zealand (CIE map no. A163).

*H. licas* is found throughout tropical Africa, in Nigeria, Zaïre, Swaziland, E. Africa, Zimbabwe and S. Africa.

*H. consimilis* is apparently confined to E. Africa.
**Control.** The usual insecticidal treatment recommended is the use of seed dressings of aldrin or dieldrin. DDT, γ-BHC and chlordane are also recommended. Dieldrin sprays can be applied to the soil before planting, and also the spraying of young sugarcane stools is recommended when one or more adult beetles are found in 20 damaged stools.

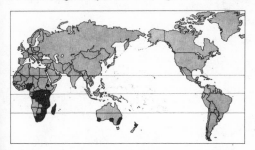

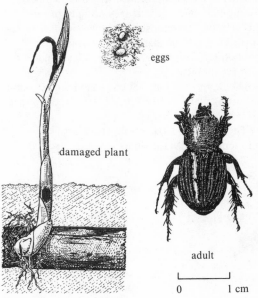

eggs

damaged plant

adult

0      1 cm

# Oryctes boas (F.)

**Common name.** African Rhinoceros Beetle
**Family.** Scarabaeidae (Dynastinae)
**Hosts** (main). Coconut palm
  (alternative). Oil palm, date palm, and other Palmae.
**Damage.** The adult beetles feed on the growing point of the palm, eventually producing V-shaped cuts through the leaflets of mature palm leaves. Severely attacked palms will die and remain standing but leafless.
**Pest status.** A major pest of palms in tropical Africa. It may occur in areas where palms are not found in any number, such as Uganda.
**Life hsitory.** Eggs are laid in rotting vegetation, especially in the trunks of rotting palms; they are white and oval, and about 3.5 mm long when freshly laid, later expanding to about 4 mm. Hatching takes place after 10—12 days.

  The full-grown larva is a soft, white, wrinkled grub some 6 cm or more in length, usually found curled up in its characteristic C-shaped position in the moist rotten vegetable matter on which it feeds. There are three larval instars; the total larval period lasts for two months.

  The fat, brown pupa, about 4 cm long, is found in the rotting plant matter with the larvae; the pupal period is about three weeks.

  The adult is a large, black, shiny beetle about 4 cm in length. It has the rhinoceros-type frontal horn which is well developed in the male, but short in the female. Adults rest during the day but fly strongly at night, and feed upon the 'cabbage' of the palm (the large terminal bud). With slight damage the leaves later unfold to show the characteristic V-shaped cuts, but if the actual growing point is eaten then the palm dies. The female beetle may live for 3—4 months and lay more than 50 eggs.
**Distribution.** This species is confined to tropical Africa and Madagascar (CIE map no. A298).
**Control.** See *Oryctes rhinoceros.*

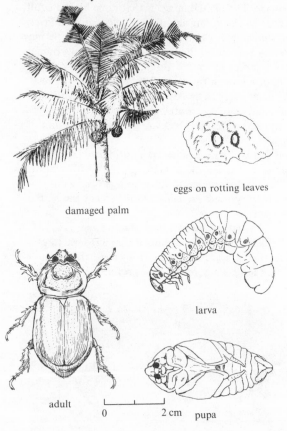

eggs on rotting leaves

damaged palm

larva

adult

pupa

0  2 cm

420

## Oryctes monoceros (Oliver)

**Common name.** African Rhinoceros Beetle
**Family.** Scarabaeidae (Dynastinae)
**Hosts** (main). Coconut palm
    (alternative). Oil palm, date palm, and other Palmae. For the adults only.
**Damage.** The feeding adults attack the growing point of the palm, and the eaten leaves eventually expand and produce the characteristic V-shaped cuts, as illustrated. The larvae and pupae are to be found in rotting vegetable matter and in soil.
**Pest status.** A major pest of palms in tropical Africa and its associated islands.
**Life history.** The life history details are similar to those of *O. boas*, on the previous page.
**Distribution.** *O. monoceros* is also found in tropical Africa, including Madagascar, Mauritius and the Seychelles (CIE map no. A188).
**Control.** See *Oryctes rhinoceros*.

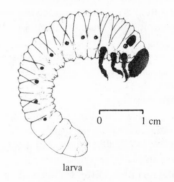

larva

damage

*Oryctes monoceros*

421

## Oryctes rhinoceros (L.)

**Common name.** Asiatic Rhinoceros Beetle
**Family.** Scarabaeidae (Dynastinae)
**Hosts** (main). Coconut, oil and date palms.

(alternative). Wild species of Palmae, sugarcane, banana, sisal, pineapple, papaya, and other plants.
**Damage.** The adult beetles feed on the growing point of the palm, and their feeding on the young leaves eventually produces the typical V-shaped cuts seen on mature leaves (page 000). Heavy attacks regularly cause a serious reduction in crop yield. Damaged palms often become secondarily infected by fungal rots.
**Pest status.** A serious pest of cultivated palms throughout S.E. Asia and the Pacific region. Heavy yield losses have been reported.
**Life history.** The white, oval eggs are laid in rotting plant material, especially dead palm trunks, compost heaps and rubbish dumps. Hatching takes 10−12 days usually. The larvae grow to a size of 6−8 cm in length, after 2−4 months, according to temperature.

Within the pupal chamber in the soil or rotting plant material the mature larva turns first into a pre-pupa, before becoming a pupa proper; pupation takes, in total, 3−4 weeks, but the adult beetle may not emerge from the cocoon immediately, and the maturation period may be spent within the pupal cocoon in the soil.

The adult beetle is large and stout, ranging in size from 3−5 cm long; shiny and black above but a reddish brown ventrally. Adult longevity may be six months, during which each female may lay 90−100 eggs.

Under optimum conditions one generation may only take 4−5 months; there is usually either one or two generations per year, according to locality.
**Distribution.** This is the common Asiatic species, being recorded from Pakistan and India, through S.E. Asia to Hainan and Taiwan, Philippines, Indonesia, Fiji, Samoa, etc. (CIE map no. A54).

According to Schmütterer (in Kranz, Schmütterer & Koch, 1979) there are altogether 42 species of *Oryctes* known, most of which attack palms, and many of which are similar in appearance.
**Control.** A single beetle is capable of flying long distances and attacking many different palms during its long adult life; it is essential therefore to organize an integrated programme simultaneously over a large area.

The recommended methods of cultural control are as follows.

(a)  Plant all palms at the same time in close regular spacing so that a continuous canopy of foliage develops. Isolated palms, and palms of irregular heights, are particularly liable to be attacked.

(b)  All dead palms should be cut down to short stumps and burned.

(c)  All rubbish and rotting vegetation should be dried and burned.

Chemical control can be effected with sprays of BHC, DDT, dieldrin, diazinon, or carbaryl.

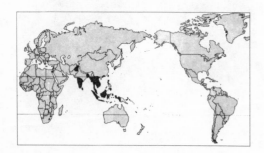

# Prionoryctes caniculus Arr.

**Common name.** Yam Beetle
**Family.** Scarabaeidae (Dynastinae)
**Hosts** (main). Yam tubers.

(alternative). Roots of other plants such as bananas, coffee, grasses etc. in marshy areas.
**Damage.** Holes are bored in the tubers by both larvae and adults but mostly by the feeding adults, the feeding lesions generally being hemispherical and 1–2 cm in diameter. The adults do the damage on their 'feeding migration' from the swampy areas in the forests.
**Pest status.** This and the other dynastine beetles are important pests of yam in W. Africa, particularly in Nigeria. Yield losses can be high and marketability of tubers reduced.
**Life history.** Eggs are laid in moist soil early in the dry season. The polyphagous young larvae initially feed on organic debris, and later feed on roots. At this stage the larvae are usually in swampy areas where yams are not often available.

After pupation in these areas the adults emerge early in the rainy season; usually a storm bringing at least 1–5 cm of rain is required to stimulate emergence of most of the adults. After emergence the adults make their migratory flight to the feeding areas where the yams grow. At this stage the beetles are sexually immature, and this migration is referred to as the 'feeding migration'. On the arrival in the yam fields, the beetles burrow in the soil around the base of the yam plants and here they feed on the tubers, making holes and tunnels. At the end of the rainy season the adults fly back to the breeding grounds in the swamps or river flood plains.

There are three larval instars, a quiescent pre-pupal stage, and the pupal stage. The larvae are white or grey, with a pale brown head capsule. The relative lengths of the different developmental stages are as follows: 18–21 days; 17–23 days; 65–78 days; prepupal stage 6–9 days; and pupal stage 17–20 days.

The total developmental period is recorded from 138–171 days in the laboratory.
**Distribution.** Tropical Africa only.
**Control.** The use of organochlorine contact insecticides applied to the yam setts or the soil has been very successful. The recommended insecticides were aldrin, dieldrin, chlordane, γ-BHC. The best results have been obtained by dusting the planting setts with aldrin and γ-BHC.

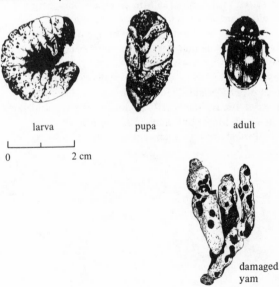

larva    pupa    adult

0    2 cm

damaged yam

# Cochliotis melolonthoides (Gerst.)

**Common name.** Sugarcane Whitegrub
**Family.** Scarabaeidae (Melolonthinae)
**Hosts** (main). Sugarcane roots.

(alternative). Roots of many other plants and many trees.

**Damage.** The attacked crop appears as if affected by drought, with the yellowing of leaves and drooping of the spindle. Later the leaves may die, and in severe cases the stool is deprived of its roots and will fall because of its own weight. Occasionally the base of the stool may be eaten.

**Pest status.** This has been a serious pest in Tanzania where it has reduced potential yields of 110–195 tonnes of plant per hectare to less than 24 tonnes. 49 000 or more grubs per hectare cause serious damage; 19 000–49 000 cause patches of yellowed cane in the crop.

**Life histroy.** The eggs are subspherical, whitish, and deposited in the soil at depths to ensure moisture for development. They hatch after 10–25 days.

Three larval instars occur; during development the head capsule increases in size only at ecdysis but the soft flexible body does increase all the time. The larvae (grubs or whitegrubs) are difficult to distinguish from other Scarabaeidae. The first instar lasts about 30 days, the second a little more, and the third instar lasts much longer and may be six months or more. The third-stage larva feeds voraciously for 3–4

months and then burrows deeper into the soil (0.5–1.0 m) and becomes inactive.

After a while pupation takes place, in an earthen cell, and it takes about 30 days. The young adult remains in the cell while the cuticle hardens.

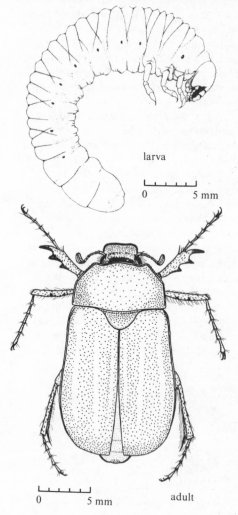

larva

0   5 mm

0   5 mm   adult

In Tanzania, eggs and young are abundant from December onwards, third-stage larvae are most abundant in the period June—August, after which they pupate. Adults emerge in October and November. The life-cycle takes about one year to complete. Adults emerge in the evening, after rain, and during the day they hide in the soil. They are large, fat, brown beetles of typical scarab appearance, 25—28 mm long, with a faint pale speckling of white scales over the body.

**Distribution.** Only recorded to date from Tanzania.

It is, however, a typical chafer grub or white grub, and is representative of the family which is of considerable importance in all parts of the world on a wide range of crops. For example, Box (1953) listed 80 species of Melolonthinae, 40 of Rutelinae and 65 of Dynastinae (all Scarabaeidae) associated with sugarcane. As reported by F. Wilson (1971), in the grasslands of Australia there are times when there is a total weight of chafer grubs in the soil greater than that of the sheep grazing on the surface.

**Control.** See following section on the control of chafer grubs.

**Further reading**

Jepson, W. F. (1956). The biology and control of the sugarcane chafer beetles in Tanganyika. *Bull. ent. Res.* **47**, 377—97.

Williams, J. R. *et al.* (1969). *Pest of Sugar Cane*, pp. 237—58. Elsevier: London.

## Control of Chafer Grubs (Scarabaeidae)

Chafer (White) Grubs live their larval and pupal lives in the soil; the larvae eat plant roots, mostly at depths of 2—10 cm. They are most serious as pests of pastures, quality grassland (e.g. golf courses), sugarcane and some cereals, especially maize. The life-cycle is lengthy, and most species have only one generation per year.

They are generally quite difficult pests to kill with chemicals, mostly because of the problem of actually getting the pesticide into contact with the insect; some pesticides are rapidly degraded on contact with soil. To date, the only really successful insecticides against chafer grubs have been the organochlorines.

For turf protection the chemicals are generally used as a curative programme, but for sugarcane preventative measures are most effective, and the standard recommendation was to apply dieldrin, aldrin, γ-BHC or heptachlor as sprays along the furrows before the cane was planted, then to make further application along the top of the row in the first ratoons, and finally to cover with soil. In the USA, dieldrin and chlordane were used with great success in the 1950s and early 1960s; but, by 1969, resistance was first noted (12 years after first use) to these chemicals and now in many parts of the USA the grubs are resistant to most organochlorines.

In parts of the tropics, where dieldrin has not been so extensively used, it will still give good control of chafer grubs.

A recent organophosphate compound being tested for control of chafer grubs on golf course turf in the USA is isofenphos; in 1980 it gave good results at 2.2 g a.i./ha, and it is expected to be registered for use on corn and turf-grass in the USA in 1981.

In some areas light trapping (u.v.) for adults has proved successful in long-term population control.

## *Schizonycha* spp.

**Common name.** Chafer Grubs
**Family.** Scarabaeidae (Melolonthinae)
**Hosts.** Wheat, maize, sugarcane, sorghum, groundnut, sunflower, yam, and many other crops and weeds.
**Damage.** The larvae are soil dwellers and damage the roots of many crop plants, and the pods of groundnut. Freshly turned soil, especially with a high organic content, is reputed to be attractive to the adult beetles. The adults do a certain amount of foliage damage but this is generally not important.
**Pest status.** A fairly important pest found on many crops in Africa. As a group the polyphagous chafer grubs are found throughout the world and the total damage to crops by this group must be tremendous.
**Life history.** Eggs are laid in the soil, especially where there is a high organic content.

The larvae are white, fleshy, with a curved C-shaped body and swollen abdomen, and well-developed thoracic legs. The larval stage lasts for about six months.

Pupation takes place in an earthen cell in the soil, and lasts for about 14 days.

The adult is a shiny brown cockchafer, about 15 mm long, and nocturnal in habits.

The life-cycle takes about seven months, and there is only one generation per year.
**Distribution.** Africa only; E. Africa, Sudan, Egypt.

**Control.** Deep ploughing will kill a number of larvae and pupae both directly and by exposure to sunlight and predators. Large numbers of adults can be caught in light traps.

Seed dressings or soil treatments with aldrin, dieldrin, heptachlor, or γ-BHC are generally effective.

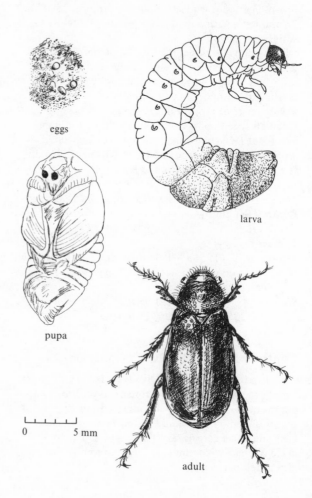

eggs

larva

pupa

0      5 mm

adult

426

## Anomala cupripes (Hope)

**Common name.** Green Flower Beetle
**Family.** Scarabaeidae (Rutelinae)
**Hosts and Damage.** The genera *Anomala, Adoretus*
and *Popillia* are collectively known as Flower Beetles
(subfamily Rutelinae), and the adults damage the
foliage (leaves and inflorescences) and sometimes
fruits, of a wide range of crop plants and ornamentals
throughout most parts of the world, but more abun-
dantly in warmer climates. The larvae are typical
Chafer Grubs or White Grubs, along with the other
members of the Scaraboidea, and they live in the soil
(or dead trees) and eat the roots of many different
plants. As a group they are probably most damaging
to pastures.
**Pest status.** Individually any one species is probably
not particularly important (except for *Popillia
japonica* in N. America), but, collectively they are
quite abundant in most parts of the world, and
commonly encountered on crops and ornamentals,
and in soils, where the different species do much the
same sort of damage.
**Life history.** Eggs are laid in the soil, and the
developing larvae live underground for many weeks or
months, where they feed on fine roots, eating them
with their biting and chewing mouthparts. Larval
development in soil is generally rather slow, and these
larvae probably require several weeks or months for
their development, according to their geographical

larva

0                    1 cm

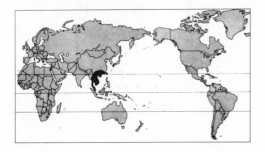

adult

427

location. The larvae are referred to as 'scarabaeiform', being oligopodous and eucephalous, with well-developed thoracic legs and head capsule bearing strong mandibles, but the distal end of the body is generally swollen and the whole body is held in a C-shaped position, when not crawling. Mature larvae measure some 20–60 mm in length, according to species; *A. cupripes* is about 35 mm long.

Pupation takes place in the soil in an earthen cocoon; in cooler climates the pupa overwinters in a state of diapause. There are usually one or two generations per year.

The adult is a smooth, shiny beetle, a green colour dorsally and bright coppery underneath, and about 25 mm in body length. The tibia on the hind legs are rather thickened, and there are long movable claws of unequal size on the tarsi. The adults are nocturnal and fly to lights at night. They can be found on vegetation during the day, but most activity is at night when they eat the leaves and flowers, and sometimes the fruits, of the host plants.

**Distribution.** This species of *Anomala* is confined to S.E. Asia (mainland) but other species occur in India and tropical Africa, and China. The group of similar genera belonging to the Rutelinae are cosmopolitan throughout the warmer parts of the world.

**Control.** When control of the soil-inhabiting larvae has been required in the past, dieldrin and chlordane gave good results, but in some parts of the world the chafer grubs are now totally resistant to most organochlorines; isofenphos is giving promising results in the USA.

damaged cocoa leaves (Malaysia)

## Agrilus spp.

**Common name.** Citrus Jewel Beetles; Citrus Bark Borers

**Family.** Buprestidae

**Hosts** (main). *Citrus* spp.

(alternative). Other members of the Rutaceae.

**Damage.** The larvae bore flat winding galleries just beneath the trunk of the tree; often sap exudes from the cracked bark, but there are usually no frass holes; heavy infestations generally kill the tree by ring-barking.

**Pest status.** An important pest of *Citrus* in the Philippines (*A. occipitalis*) where damage is sometimes serious; somewhat less important in China (*A. auriventris*). The genus is however widespread and is a pest of several ornamental trees in the USA and Europe, as well as chestnut, pear and rose, and individuals are sometimes found in domestic situations when adults emerge from wood used in furniture.

**Life history.** Eggs are laid in cracks in the tree bark, and the young larvae bore into the bark, where eventually they make long winding galleries (tunnels) that are flat in section. Larvae measure from 15–25 mm according to species, and are legless, with a reduced head and laterally expanded prothorax, which gives them their common name of Flat-Headed Borers (F. Buprestidae).

Pupation takes place within the gallery, just under the surface of the bark. Buprestid larvae gener-ally do not make frass holes as they burrow (as do Cerambycidae) but they pack the frass tightly in the gallery behind them as they tunnel.

The adults are small, greenish or bronze beetles, 15–20 mm in length; they may live for several weeks and feed on the tree foliage.

**Distribution.** *Agrilus occipitalis* is recorded from the Philippines; *A. auriventris* on *Citrus* in China. *A. acutus* is the Jute Stem Buprestid of India, also on kenaf and roselle; and other species occur in S.E. Asia and Papua New Guinea. A couple of species occur in Europe, on oak and other forest trees, and six species in the USA on pear, poplar, birch, chestnut, raspberry and blackberry canes. On a worldwide basis, 700 spp. of *Agrilus* are known.

**Control.** When control has been required, the systemic insecticide dimethoate as a foliar spray has been effective. Good husbandry will reduce the likelihood of attack by this pest, as it has been noticed that trees under physiological (or water) stress are most heavily infested.

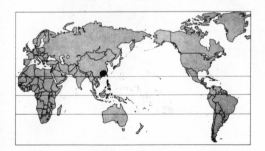

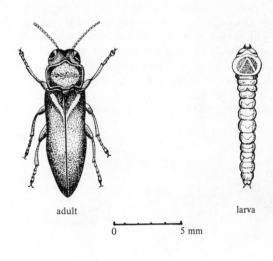

adult

0        5 mm

larva

429

# Trogoderma granarium Everts.

**Common name.** Khapra Beetle
**Family.** Dermestidae
**Hosts** (main). Cereals and groundnut.

(alternative). Pulses, spices, and various cereal and pulse products (cakes).

**Damage.** The larvae bore in the stored cereal grains and pulses, usually hollowing out the grain. Development is rapid in the hot humid tropics and very large populations may build up quickly. The pest is fairly polyphagous and can survive in facultative diapause for a year or longer in the absence of food.

**Pest status.** Although virtually cosmopolitan, this pest is absent from E. Africa and S. Africa, and these countries have legislation to prevent its importation. A very serious pest of stored grains and pulses in most of the warmer parts of the world. It is the only member of the Dermestidae that is phytophagous.

**Life history.** Eggs are laid in the stored produce, and the larvae are regarded as primary pests of stored grains in that they can damage intact grains and seeds. The larvae often develop at different rates, some coming to maturity in two weeks and others taking months or even more than a year. In the absence of food the larvae can go into a state of facultative diapause for many months which ensures their survival at times when the store is empty; at these times the larvae congregate in large numbers in crevices and cracks in the buildings or storage containers.

The adults are small dark beetles 3–4 mm in body length; they are wingless, neither do they feed; they live for about 14 days. Thus, this species is spread almost entirely through the agency of man, and quarantine regulations against it may be easily enforced.

Under optimum conditions the life-cycle can be completed in about three weeks (37°C and 25% RH).

**Distribution.** Almost cosmopolitan in the warmer parts of the world, but as yet kept out of the countries of E. and S. Africa; it prefers a hot dry climate.

**Control.** A difficult pest to control in that many of the usual contact insecticides used against Coleoptera have little effect; the effect of its being able to practice facultative diapause results in infestations reappearing after a long storage period. Fumigation with toxic materials such as methyl bromide are most likely to give the best control results, but pirimiphos-methyl is reputed to be effective.

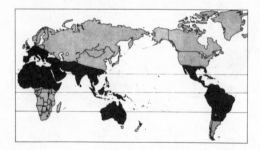

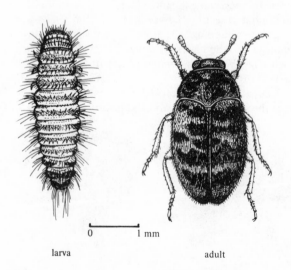

0        1 mm

larva                    adult

## Lasioderma serricorne (F.)

**Common name.** Cigarette Beetle (Tobacco Beetle)
**Family.** Anobiidae
**Hosts** (main). Tobacco (stored leaf and cigarettes).

(alternative). Cocoa beans, groundnut, peas and beans, many stored grains, flours and foodstuffs.
**Damage.** The larvae can attack undamaged cereal grains and pulse seeds, and often show preference for the germ of the seed for feeding. In packaged cigarettes holes are made in the packets by larvae and adults.
**Pest status.** This can be a very serious pest in many stores, owing to the wide range of foodstuffs attacked. High populations can build up very quickly and cause considerable damage. The adults fly readily at dusk, but most dispersal is effected by trade.
**Life history.** This beetle breeds anywhere at temperatures above $19°C$ and above a RH of 20–30%, but $30–35°C$ and 60–80% RH are optima.

Eggs are laid soon after the emergence of the females, and require 6–10 days to hatch.

The newly hatched larvae are very active, negatively phototaxic, and capable of penetrating tiny holes in search of food. In closely packed produce such as meal, the infestation remains peripheral. Older larvae are scarabaeiform and less active. The fourth larval instar stops feeding and builds a cell on some firm foundation for pupation. Larval development on good foodstuffs takes 17–30 days.

Pupation takes 3–10 days, and the pre-emergence maturation period of the adult is also 3–10 days.

The adults are small brown beetles 3–4 mm long with the typical deflexed head of this family. The adults drink but do not feed, and cause most of their damage by making emergence holes when they bite their way out of their cocoons; they live for 2–6 days.

The total developmental cycle takes about 26–50 days, according to climate and the quality of the food, tobacco being a poor food (at 50 days).
**Distribution.** Cosmopolitan in the warmer parts of the world, but found only in heated stores in colder regions; distribution is essentially controlled by temperature and humidity (above $19°C$ and 30% RH).
**Control.** This pest can be killed by exposure to low temperatures; below $18°C$ development ceases.

Chemical control in foodstuffs can be effected by most fumigants, and also such contact and stomach poisons as $\gamma$-BHC, which also have some fumigant action.

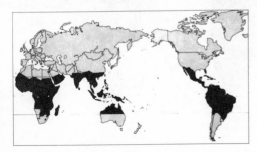

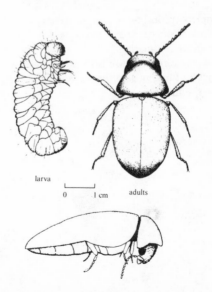

larva

0    1 cm    adults

431

## Apate spp.

*monachus* Boh.
*indistincta* Murray

**Common name.** Black Borers
**Family.** Bostrychidae
**Hosts** (main). Coffee species.
   (alternative). Not known, but must be wild trees of various species.
**Damage.** The beetle makes a clean-cut, circular, fairly straight tunnel about 6 mm in diameter obliquely upwards in the main stem. Sawdust-like fragments drop to the ground whenever the beetle is actively boring.
**Pest status.** A minor pest of coffee, attacking all cultivated species, although usually only a few trees in a plantation are attacked.
**Life history.** Egg, larval, and pupal stages have not been recorded from coffee. They probably occur in the dead branches of shade and other trees.

   The adult beetle is black and nearly 20 mm long. It is rather square at the front end; the head is not visible from above, being deflexed under the thorax.
**Distribution.** Tropical Africa, N. Africa, S. Africa, Madagascar; Sardinia, Corsica, Spain, Syria, Israel; W. Indies, and tropical S. America.
**Control.** The usual recommendation is to spear the beetle in its tunnel by pushing a springy wire (e.g. a bicycle spoke) up the hole!

Alternatively, a plug of cotton wool can be soaked in dieldrin liquid and pushed up the tunnel.

   It is advisable to clear away the sawdust-like frass from the base of the tree when control measures are applied; if they fail to kill some of the beetles then fresh frass will be seen on the ground again after a few days.

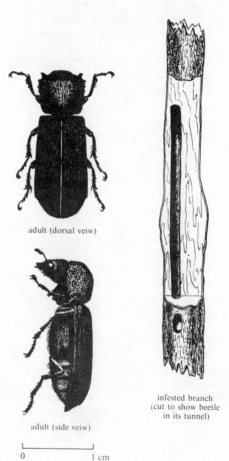

adult (dorsal veiw)

adult (side veiw)

infested branch
(cut to show beetle
in its tunnel)

0                    1 cm

# Rhizopertha dominica (F.)

**Common name.** Lesser Grain Borer
**Family.** Bostrychidae
**Hosts** (main). Stored cereals.

(alternative). Other stored foodstuffs including cassava, cereal products, flours etc.

**Damage.** Both larvae and adults feed on the grains, usually from the outside, and in a rather haphazard manner. The adults are quite long-lived. They are both primary pests and can attack rice grains (paddy rice) more readily than *Sitophilus*.

**Pest status.** A serious pest of stored grains throughout the warmer parts of the world. During World War I wheat from Australia was heavily infested with *R. dominica* which then became widely established throughout the USA and other countries, causing serious post-harvest losses in various cereals. A particularly important pest in Australasia.

**Life history.** Eggs are laid amongst the cereal grains. The larvae are scarabaeiform and have legs, so are considerably more mobile than weevil larvae. The larvae eat into the grains in a haphazard manner, usually from the outside. Pupation usually takes place within the eaten grain.

The adult is a tiny, dark beetle 2–3 mm in body length, and its cylindrical body shape, deflexed head and round prothorax with conspicuous sculpturing makes it easy to recognize. The other members of the family are timber borers, and it is the adult

stage that does the actual boring, not the larva. The adults are long-lived and feed quite voraciously.

At 34°C the life-cycle takes about four weeks.

**Distribution.** Originally described from S. America, it is now cosmopolitan in all the warmer parts of the world. It has not established itself in the cooler parts of western Europe.

**Control.** The usual practices employed against stored products pests, particularly fumigation is effective against this pest. The threshold temperature for development is about 18°C. Pirimiphos-methyl is reported to be particularly effective against this pest; BHC, carbaryl and the pyrethroids are also used successfully.

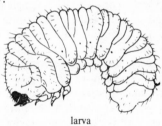

larva

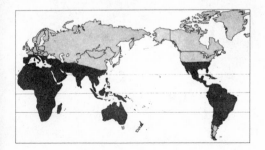

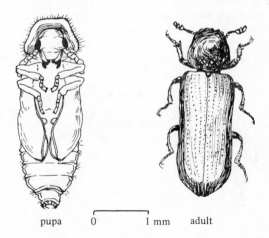

pupa  0  1 mm  adult

433

## Melittomma insulare Fairm.

**Common name.** Coconut Palm Borer
**Family.** Lymexylidae
**Hosts** (main). Coconut palm
    (alternative). Oil palm, and various wild Palmae.
**Damage.** The beetle larvae bore in the bole and trunk of the palm, causing general impairment of growth, and some palms break just over ground-level, at the rot-cavity.
**Pest status.** A serious pest with a very restricted distribution, confined virtually to the coconut palm as host.
**Life history.** Eggs are laid at the base of the trunk in clusters of up to 100 or more; each female probably lays about 250 eggs on average. Each egg is white, sausage-shaped and about 1.2 mm by 0.3 mm in size. The larvae are creamy white, rather thin and with a strange 'tail-piece', a hoof-shaped, heavily chitinized structure with the posterior face concave and studded with about 18 small pits. The tail-piece is used to ram the borings towards the mouth of the tunnel where they appear as small pellets resembling sawdust. Mature larvae are about 20 mm in length and 3 mm in body thickness. The larvae are also odd in that they apparently do not eat the woody material of the palm trunk (as do most timber borers) but they chew pieces to extract the sap and then push the chewed remains with their legs to the back of the tunnel. The duration of the larval stage is about one year, or more under unfavourable conditions.

Pupation takes place in the larval tunnel, after the larva turns round to face the exterior, and takes some 10–12 days.

The adult is a slender, dark brown beetle, 10–15 mm in body length, with large eyes and serrate antennae. Males are usually smaller than females. Males have a conspicuous branched appendage on the maxillary palps, which is characteristic of the family Lymexylidae. The adults are short-lived, and survive for only 2–6 days.

There is usually only one generation per year, and there does not seem to be any seasonal life-cycle.
**Distribution.** Only recorded from Madagascar and the Seychelles (CIE map no. A152).
**Control.** Various cultural practices can reduce the level of infestation; for example, the cutting of steps

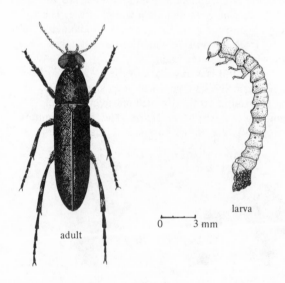

adult

larva

0    3 mm

434

in the trunk encourages infestation, the burning of fallen palms will kill the larvae and pupae inside the trunk, careful cultivation to promote good growth will reduce attack levels, etc.

For chemical control the most successful was to scrape out the damaged tissues and then to apply the fumigant paradichlorobenzene as crystals to the damaged trunk base and to earth-up the bole of the trunk.

**Further reading**

Brown, E. S. (1954). The biology of the coconut pest *Melittomma insulare* (Col., Lymexylonidae), and its control in the Seychelles. *Bull. ent. Res.*, **45**, (1), 1–66.

damage to bole of coconut palm in Seychelles

# Carpophilus hemipterus (L.)

**Common name.** Dried Fruit Beetle
**Family.** Nitidulidae
**Hosts** (main). Primary pest on dried fruits, especially currants, raisins and figs.

(alternative). In the field, these are found as secondary pests in cotton bolls, maize cobs, and on different types of fruit. In stores they also attack copra, cocoa beans, and groundnuts.
**Damage.** Both adults and larvae feed on the dried fruits, causing direct damage and also soiling the produce with their frass.
**Pest status.** A serious pest of dried fruits, and some other produce, in the warmer parts of the world; not serious as a crop pest.
**Life history.** Life history details are not available, but it is known that eggs are laid amongst the dried fruits and that both larvae and adults feed on the fruits. The larvae are campodeiform in shape, white or yellow in colour, and bear two small pairs of horns at the end of the abdomen.

The adults are small, flat beetles, 3–4 mm in body length, have characteristic yellow patches on the elytra, and the terminal two body segments are left uncovered by the short elytra.

There are three other species to be found in food stores all over the world, the others being *C. dimidiatus* (F.), *C. obsoletus* Erich., and *C. ligneus* Murray; all four species are difficult to separate.
**Distribution.** Cosmopolitan, and extending up into northern parts of Europe and N. America.
**Control.** The usual recommendations for stored products pests are effective against this pest. The normal practice is for general warehouse fumigation, at intervals of time, which effectively controls all the pests that are likely to be present.

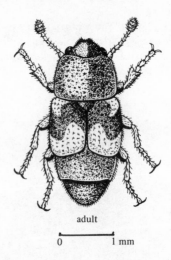

adult

0          1 mm

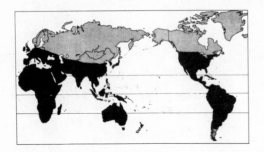

# Oryzaephilus spp.

*surinamensis* (L.)
*mercator* (Fauvel)

**Common name.** Saw-toothed Grain Beetle;
Merchant Grain Beetle
**Family.** Silvanidae
**Hosts** (main). Stored grains.

(alternative). Other plant and animal stored products.
**Damage.** *Oryzaephilus* beetles are general feeders, and usually secondary on stored products, following the more destructive primary pests such as Grain Weevils and pyralid moths. Their actual diet consists of fragments of animal and plant debris.
*O. surinamensis* is more frequently found on cereal products, and *O. mercator* on oil-seed products.
**Pest status.** Not particularly important, since they are both secondary pests, but these two species are of very common occurrence and are widespread in distribution. Because of their small size they can easily hide in small crevices and are easily transported.
**Life history.** The entire life-cycle takes place in the stored produce. Egg mortality is high below 20°C; 30°C appears to be about optimum for development; egg development takes 4–12 days.

The larvae are free-living however, and only spend part of their time within the grain, but when there they prefer to feed on the germ. Higher humidities are preferred (60–90% RH) for larval development, which takes about 12–20 days.

The pupal period is 5–15 days.

The adults are small, narrow, flattened beetles, some 2.5–3.5 mm in length, with 11-segmented antennae, six large lateral 'teeth' on either side of the prothorax and three low longitudinal ridges. The pre-oviposition period is 3–6 days in the female.

*O. surinamensis* is distinguished from *O. mercator* by the shape of the head behind the eyes – the temple being drawn out into a rounded point in *mercator*, but flat and nearly equal to the vertical eye diameter in *surinamensis* – and in the male genitalia.
**Distribution.** Both species are virtually cosmopolitan in distribution in food stores and godowns.
**Control.** These species cannot breed at temperatures of less than 19°C, so in temperate countries infestations can easily be controlled by cooling the grain. *O. mercator* is the more temperature-sensitive.

For chemical control methyl bromide fumigation is particularly effective against the adult beetles. Aluminium phosphide, and a mixture of ethylene dichloride and carbon tetrachloride, are both commonly used on farms.

Insecticides recommended as sprays include malathion, fenitrothion, and γ-BHC.

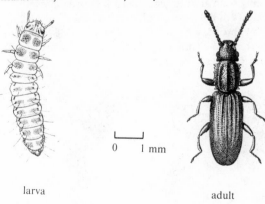

0      1 mm

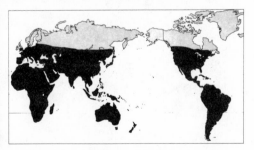

larva

adult

437

## Epilachna spp.

**Common name.** Epilachna Beetles (*E. varivestis* is known as the Mexican Bean Beetle)
**Family.** Coccinellidae
**Hosts** (main). Cucurbits, and solanaceae; in USA, beans (*Phaseolus* spp.).

(alternative). Maize, sorghum, finger millet, rice, wheat, cotton, sesame, lettuce; soybean and cowpea in N. America; and solanaceous weeds.
**Damage.** Both adults and larvae feed on the leaves and fruits of cucurbits and other crop plants. The leaves are eaten between the veins, sometimes being completely stripped to the midrib. Stems are often gnawed and holes are eaten in the fruits.
**Pest status.** A quite serious pest of many crops in Asia and Africa, and the Mexican Bean Beetle is a serious pest of various legumes in N. America.
**Life history.** The eggs are pale yellow, elongate-oval, with comb-like hexagonal sculpturing, and are 0.5 mm long. The eggs are laid in clusters, usually on the underside of the leaves and placed vertically. Each female lays on average 12 clusters, each with 22 eggs (up to 50). Incubation takes 4–5 days.

The larvae are pale yellow, covered with delicate spines when first hatched. The young larvae start feeding soon after hatching, making rows of small windows in the leaves. Fully grown larvae are dark yellow, broad, with a dark head, and strong branched spines, and 6–7 mm long. Larval development takes about 16 days.

Pupation takes place on the leaves of the host plant, and the pupa is dark yellow.

The adult beetles are oval, 6–8 mm long, reddish to brownish-yellow, but colour is variable. Each elytron is marked with a series of black spots. The adults look like typical 'Ladybirds' but have the distinction of being the only phytophagous representatives of this family; they are strong fliers.

The whole life-cycle takes about 35 days, and in Africa there are five generations per year.
**Distribution.** There are many species of *Epilachna* known, but the most important are:
*E. chrysomelina* (F.) – Europe, Asia and Africa (CIE map no. A409).
*E. similis* (Thnb.)
*E. fulvosignata* Reiche  } – Africa
*E. sparsa* (Hbst.) – S.E. Asia
*E. spp.* – India, S.E. Asia, USA
*E. varivestis* Muls. – Mexican Bean Beetle (CIE map no. A46).
**Control.** Sprays of dieldrin, carbaryl, methomyl, parathion-methyl, toxaphene or malathion are said to be effective against these pests.

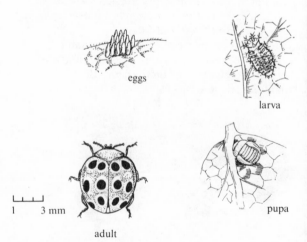
eggs

larva

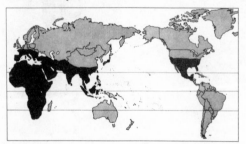

1    3 mm

adult

pupa

# Gonocephalum simplex (F.)

**Common name.** Dusty Brown Beetle
**Family.** Tenebrionidae
**Hosts** (main). Coffee
(alternative). Many wild and cultivated plants are attacked; the species is a well-known pest of cereals.
**Damage.** Patches of young brown bark are chewed away from coffee stems or branches. Green berries are found on the ground with their stalks chewed off.
**Pest status.** A minor sporadic pest of coffee, especially young bushes; sometimes attacks cereals.
**Life history.** The eggs are presumably laid in the soil.

The larvae, which are called 'false wireworms', are found in the soil. They eat many kinds of seeds and may do slight damage to coffee roots.

Pupation takes place in the soil.

The adult beetle is a dusty brownish-black; it is about 8 mm long, oval in outline, and flattened. The hard forewings (elytra) have longitudinal ridges. The beetles live in the mulch or the upper layers of the soil during the day. They climb up the coffee bush to feed at night, feeding principally on bark that has turned brown. Branches or stems may be completely ring-barked; more often irregular patches are chewed off, and the stalks of large green cherries are also cut through.
**Distribution.** Africa, S. of the Sahara.

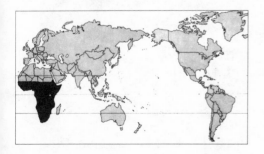

**Control.** If coffee is to be planted in an area heavily infested with Dusty Brown Beetles it is recommended that aldrin dust should be mixed with the soil of each planting hole.

Recommended insecticidal treatment for established trees is a dieldrin spray applied round the base of each tree.
**Further reading**
Le Pelley, R. H. (1968). *Pests of Coffee*, pp. 104–5.
Longmans: London.

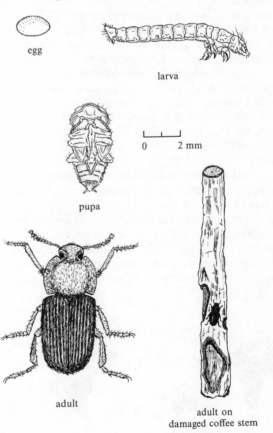

egg

larva

pupa

0    2 mm

adult

adult on
damaged coffee stem

439

## Tribolium castaneum (Herbst)

**Common name.** Red Flour Beetle
**Family.** Tenebrionidae
**Hosts** (main). Maize, wheat and other stored grains.
(alternative). Many types of stored foodstuffs.
**Damage.** Infestation is apparent by the appearance of adults on the surface of the grain; there is extensive damage to previously holed or broken grains, or grain damaged by other pests. Damage is done by both larvae and adults.
**Pest status.** A serious secondary pest throughout the warmer parts of the world in food stores.
**Life history.** The eggs are small, cylindrical, and white. The female lays the eggs scattered in the produce.

The larvae are yellowish-white, about 6 mm long when fully grown. The head is pale brown, and the last segment of the abdomen has two upturned dark pointed structures. The larvae live and develop inside the grain till pupation. The damage to the stored grains is done by the larvae of the Red Flour Beetle.

The pupa is yellowish-white, later becoming brown, the dorsum hairy, and the tip of the abdomen having two spine-like processes.

The adult is rather flat, oblong, reddish-brown in colour, and about 3–4 mm long. Each female is capable of laying about 400–500 eggs and the adults are long-lived, under some circumstances living for a year or more. The life period from egg to adult is 35 days at 30°C. Adults fly in large numbers in the late afternoon.
**Distribution.** This pest is cosmopolitan in warmer countries.
**Control.** The shelled grain should be thoroughly admixed with BHC dust or powder, if locally permitted.

Fumigation should only be carried out by approved operators.

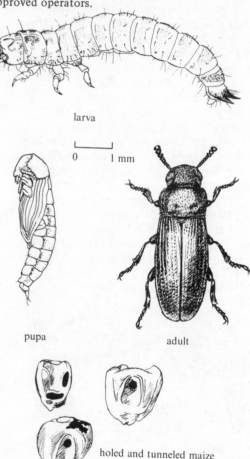

larva

0        1 mm

pupa                    adult

holed and tunneled maize

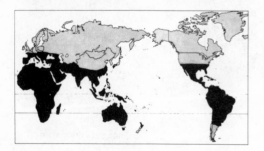

## Coryna spp.

**Common name.** Pollen Beetles
**Family.** Meloidae
**Hosts.** Flowers of pulse crops, cotton, and many flowering plants.
**Damage.** The adults eat the pollen out of the open flowers, often destroying the anthers in the process. The larvae are not pests.
**Pest status.** A widespread and common pest, but not economically serious, found on many flowering crops. It can be of importance on research stations by interfering with cotton boll-setting in breeding material, and similarly on pulse crops.
**Life history.** As with *Mylabris* spp., the eggs are laid in the soil, developing initially into very active and mobile triungulin larvae which seek out eggs of Orthoptera. Older larvae become eruciform, sluggish, with a large body and reduced legs.

Pupation is in the soil.

Adults are elongate, 10–16 mm long, with a black hairy head and thorax, club-shaped antennae with a yellowish club, smooth flexible elytra with three transverse yellow and black stripes. The coloration of the distal segments of the antennae is an important specific character in this genus.
**Distribution.** Tropical Africa.
**Control.** For control see *Mylabris* spp. (p. 443).

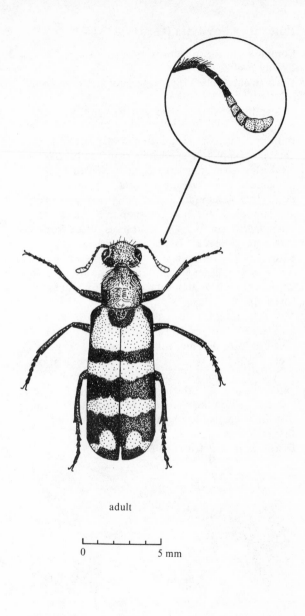

adult

0          5 mm

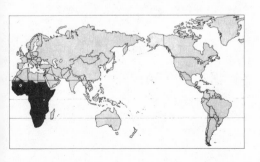

# Epicauta albovittata (Gestro)

**Common name.** Striped Blister Beetle
**Family.** Meloidae
**Hosts** (main). Pulse crops, especially groundnut and soybean.

(alternative). Tomato, potato, eggplant, capsicums.
**Damage.** This pest often occurs in very large numbers and may completely defoliate a crop; the beetles feed on the leaves, making large irregular-shaped holes in the lamina. The larvae are not pests.
**Pest status.** A sporadic pest of pulse crops, occasionally important, but quite common in E. Africa, and other species are found in other parts of the tropics.
**Life history.** Eggs are laid in holes in the soil, in clusters of 100 or more.

As with other Meloidae the larvae are predators on the egg-pods of grasshoppers. The triungulins hatch after 5–8 days.

The adult is a black blister beetle 13–20 mm long, with large eyes, and a pale whitish stripe running down the dorsum from the head and thorax and thence round the edges of the elytra. There is also a whitish stripe along the centre of each elytron extending nearly to the posterior apex, and transverse white stripes on the abdomen. The stripes are bands of white scale-like setae, and there are white setae on the legs.
**Distribution.** E. Africa and Somalia.

The closely related species *E. vittata* is a serious pest of beans in S. America, *E. limbatipennis* can do severe damage to finger millet in E. Africa, and *E. aethiops* is an important pest of vegetables and fodder crops in the Sudan. Other species occur in S.E. Asia and N. America.
**Control.** Chemical control is difficult because the pest has a high level of natural resistance to insecticides such as DDT and $\gamma$-BHC, but high dose levels of dieldrin and parathion usually give a good kill.

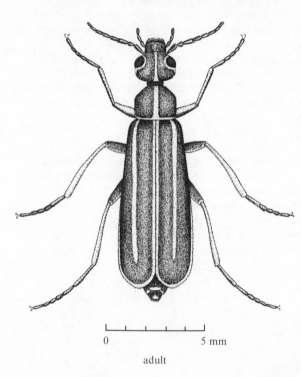

0      5 mm

adult

# Mylabris spp.

**Common name.** Blister Beetles (Flower Beetles)
**Family.** Meloidae
**Hosts** (main). Pulse crops
(alternative). Cotton, and many flowers and ornamentals (e.g. *Hibiscus*).
**Damage.** The flowers are eaten by the adult beetles, causing a loss of pods in leguminous crops and conspicuous damage to various flowering ornamentals.
**Pest status.** A widespread and common, but not commercially serious, pest of many flowering crops.
**Life history.** Eggs are laid in the soil in batches, and hatch into very active triungulin larvae which feed on egg pods of Orthoptera. The later larval stages are often sluggish with a large body and reduced legs. An abundance of meloid beetles has often been noted following locust invasions.

Pupation takes place in the soil.

The adults are large beetles, 25–35 mm long, with a bright conspicuous red (or yellow) and black patterned coloration. They are rather sluggish in behaviour but are strong fliers. If handled the adults exude an acrid yellow fluid containing cantharidin, the effect of which upon the skin accounts for the common name of 'blister beetle'.
**Distribution.** Africa, India, Bangladesh, Sri Lanka, and S.E. Asia.
**Control.** The adult beetles are difficult to control because of their mobility; also the damage done is usually only slight and so chemical control is seldom warranted commercially, but it is expected that DDT sprays might be effective.

**Further reading**

Greathead, D. J. (1963). A review of the insect enemies of Acridoidea (Orthoptera). *Trans. Roy. ent. Soc. Lond.* **114**, 437–517 (456–9).

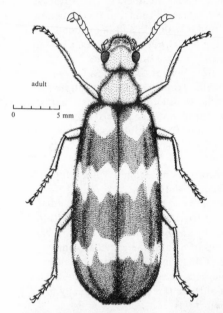

adult

0    5 mm

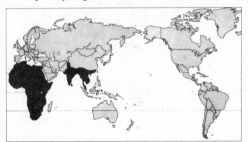

*M. cinchorii* on flowers of pigeon pea

# Anoplophora chinensis (Forst.)

(= *Melanauster chinensis* Forst.)

**Common name.** Citrus Longhorn Beetle
**Family.** Cerambycidae
**Hosts** (main). *Citrus* spp.
  (alternative). Other members of the Rutaceae.
**Damage.** The feeding larvae tunnel in the branches and trunks just under the bark, and occasionally they bore deep into the heartwood. Small trees, and branches of larger trees, are often killed.
**Pest status.** A serious pest of *Citrus* in S. China.
**Life history.** Details are not known, but can be presumed to be similar to other Cerambycidae. Larval development will probably take at least a year.

The adult is a distinctive longhorn beetle, with long antennae and striking black and white body coloration.
**Distribution.** Recorded only from S. China, Indo-China and Malaysia to date.
**Control.** Heavily infested branches should be cut and burned to destroy the larvae and pupae within. Spraying the trunks and branches with dieldrin has in the past proved to be of some value. In some cases a mixture of dieldrin and kerosene has been injected into the frass holes to kill the boring larvae. A combination of these three methods usually gives adequate control.

After any chemical treatment has been applied the old frass under the infested tree should be removed so that any new frass to be expelled will be immediately obvious, and the chemical treatment can be repeated. Sometimes it is recommended that a marker dye (methylene blue for example) be added to the dieldrin solution so that it will indicate the extent of the solution penetration within the larval tunnel system.

young *Citrus* stem killed by feeding larvae

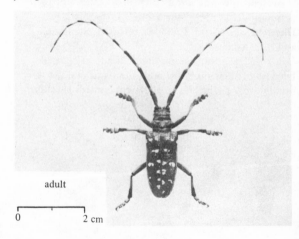

adult

0          2 cm

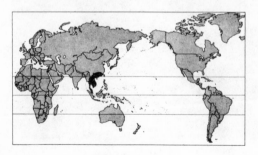

444

## Anthores leuconotus Pasc.

**Common name.** White Coffee Borer
**Family.** Cerambycidae
**Hosts** (main). Coffee, particularly *arabica*.
(alternative). Various wild Rubiaceae (shrubs).
**Damage.** Attack is indicated by a yellowing of the foliage and eventual death of the trees. Wood shavings extruded by the larvae from their burrows in the bark are diagnostic, as are the round exit-holes of the adult beetles in the trunks of the trees.
**Pest status.** A serious pest of *arabica* coffee below about 1700 m in Africa.
**Life history.** Eggs are inserted beneath the bark of the tree usually within 0.5 m of the ground. They require three weeks to hatch.

The young larvae bore just under the bark of the tree downwards from the point of insertion of the eggs. In these early stages the most serious damage, in the form of ring-barking, is done. Complete ring-barking does not invariably occur. The larvae continue downwards towards the ground, under the bark, and usually penetrate the wood of the tree at the junction of a lateral root with the stem of the tree. The later instars bore in the wood cylinder. There are thought to be seven larval instars, and the larval stages last about 20 months.

The full-grown larva excavates a large chamber within the trunk in which pupation takes place; the duration of the pupal stage varies between 2–4 months.

Adult beetles are about 30 mm long; they are greyish with a dark head and thorax and dark markings near the end of the wing cases. At the start of the rains they emerge from the tree trunk by cutting circular holes to the exterior, which are about 8 mm in diameter. The beetles do little damage and feed only on the bark of the branches. A single female beetle has been known to lay 23 eggs.
**Distribution.** The southern half of Africa only (CIE map no. A196).
**Control.** The recommended insecticide is dieldrin with added methylene blue dye as a marker. The mixture should be applied to the trunks of the trees from ground level to a height of about 0.5 m. Application can be by spray lance or brush. The best time is just before the onset of the rains. The spray should be repeated one year later, and after that every second year. The adult beetles are killed when they touch sprayed bark during oviposition or emergence.

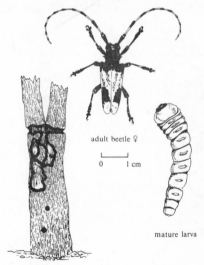

adult beetle ♀

0    1 cm

mature larva

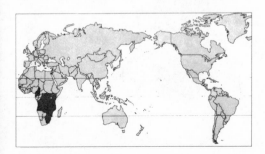

tree stump
showing ring barking and exit holes

445

## Apriona germarii Hope

**Common name.** Jackfruit Longhorn Beetle
**Family.** Cerambycidae
**Hosts** (main). Jackfruit, fig, mulberry.
    (alternative). Apple, peach, and some other trees, especially wild *Ficus* spp.
**Damage.** The larvae tunnel along the branches and in the tree trunk just under the bark, and sometimes into the heartwood. Frass expulsion holes are made at intervals along the main gallery, so frass can be seen externally, and often dark red sap exudes from these holes making wet patches on the tree trunk; some butterflies and wasps are attracted to the oozing sap. Heavily attacked trees often die.
**Pest status.** A regular pest of several different tree crops throughout S.E. Asia, but not often serious; frequently stressed trees appear to be selected as hosts.
**Life history.** Eggs are laid on the trunk or branches of the tree, usually in crevices in the bark. The eggs

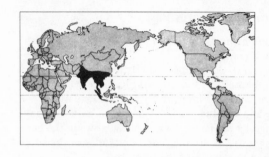

hatch in about 7–10 days, and the larvae start burrowing under the bark in the sapwood. The length of the larval life is probably two years in most localities, although in the hotter parts of S.E. Asia development could be completed in one year. Frass expulsion holes are made at intervals, and these provide conspicuous symptoms of attack. Frequently dark red sap oozes from some of the holes making a wet patch on the bark. Pupation takes place at the end of the tunnel

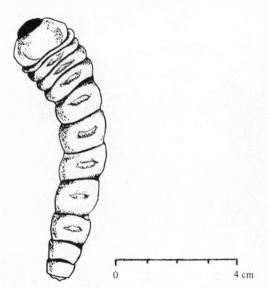

0              4 cm

adult

446

which is blocked by a mass of chewed wood fragments.

Adults emerge in the spring and may be found mating on the tree trunk a few days after emergence. The adults are large, dark grey beetles, some 4—6 cm in body length, with conspicuous dark (blackish) tubercles at the bases of the elytra. As with other longhorns the males are smaller and have longer antennae than the females. Both adults feed a little and eat the bark of the tree in patches. A closely related species in India (*A. cinerae* Chevr. — Apple Stem Borer) also eats shoots and young leaves, which is unusual for a longhorn beetle.

The life-cycle is probably a two-year one in most parts of its range.

**Distribution.** This species is known from India through S.E. Asia to S. China. In India there are several other closely related species of *Apriona*, but the full range of this, and the other closely related species, is not yet known.

**Control.** Not often required, but the methods recommended for *Anoplophora chinensis* should suffice.

larva in branch of *Ficus carica*

# Batocera rubus (L.)

**Common name.** White-spotted Longhorn Beetle
**Family.** Cerambycidae
**Hosts** (main). Fig, mango, jackfruit.
  (alternative). Various other tree species.
**Damage.** The larvae burrow through the sapwood
under the bark of the trees, either on the trunk or on
the main branches. Frass expulsion holes are made at
intervals and sometimes sap oozes out of the holes,
making obvious symptoms. On damaged branches
the foliage may die and fruitset will be impaired.
**Pest status.** A widespread pest, together with its
larger relative (*B. rufomaculata*), are frequently
encountered in fruit trees and various ornamentals
but seldom a serious pest.

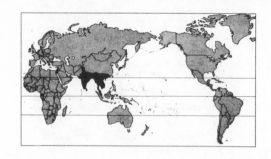

**Life history.** Eggs are laid singly in the bark of the
tree, up to 200 in total, into small cuts made by the
female's mandibles. The young larvae bore straight
into the wood, feeding in the vascular tissues of the

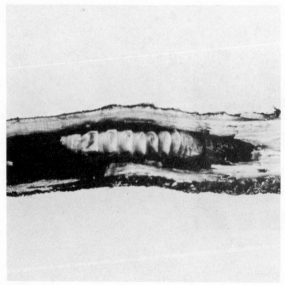

larva in situ

emergence hole of adult

448

tree, eventually making long irregular tunnels in the sapwood. In multiple infestations the flow of sap and water is interrupted, which has obvious adverse effects on the tree. Pupation takes place within the larval tunnel system, usually just under the bark.

The adult is a greyish-brown beetle, about 3–4 cm in body length, with a series of conspicuous white spots on the elytra. The adults live for several weeks and feed on the bark of the tree to some extent.

The complete life-cycle could be as short as one year, but might be two years.

**Distribution.** From India through S.E. Asia to S. China.

**Control.** If required, then the control methods given for *Anoplophora chinensis* (p. 444) should be employed.

adult

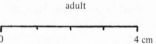

0                                              4 cm

# Batocera rufomaculata (De Geer)

**Common name.** Red-spotted Longhorn Beetle
**Family.** Cerambycidae
**Hosts** (main). Fig, mango, guava, jackfruit, pomegranate, apple, rubber and walnut.

(alternative). In India recorded from more than 30 different host plants.

**Damage.** The larvae tunnel through the sapwood and, because of their size, they make large tunnels, which interfere with sap flow and affect foliage and fruit production. Heavily attacked trees may die.

**Pest status.** A serious pest of Edible Fig, and of regular occurrence on the other fruit trees mentioned.

**Life history.** The female beetle cuts the tree bark and lays eggs singly into these cuts; laying a total of up to 200 eggs. On hatching the larvae start to tunnel into the sapwood of the trunk or branches. Larval development takes probably two years. As a very large species, the larval tunnel, measuring 2–3 cm in width, is correspondingly large and very damaging to the tree.

The adult beetles emerge in the spring (in S. China), they are especially large and measure up to 70 mm in body length, with very long antennae. In life the spots on the elytra and the prothorax are bright red (hence the name) but after death they fade to white or pale yellow.

It is thought that the life-cycle takes two years to complete.

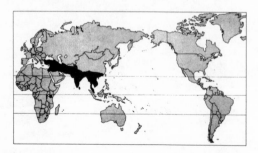

**Distribution.** This species is recorded from S. China and S.E. Asia, through India to the eastern Mediterranean. Wyniger (1962) records it from E. Africa, but the specimen illustrated looks rather like *Apriona*!

**Control.** As suggested for *Anoplophora chinensis*. (p. 444).

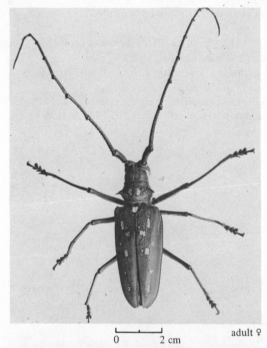

0    2 cm      adult ♀

larva in sapwood tunnel of *Sapium discolor*

## Chlorophorus annularis (F.)

**Common name.** Bamboo Longhorn Beetle
**Family.** Cerambycidae
**Hosts** (main). Bamboos
     (alternative). Sugarcane.
**Damage.** The tunnelling larvae bore through the internode walls and through the nodes of bamboo stems,
and generally inside the stem of sugarcane.
**Pest status.** A common Oriental species of longhorn
to be found regularly in various species of bamboo,
and occasionally in sugarcane, but not of particular
importance.
**Life history.** Details are not known at present.

     The adult beetle is quite small and slender,
some 10–16 mm in length, with yellow and black
markings; the distal ends of the elytra are deeply
emarginate, ending in two spines. This species regularly emerges in domestic premises from bamboo

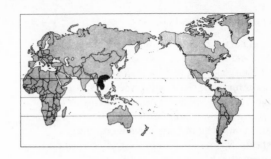

stems made into furniture. Adults have emerged from
pieces of dried bamboo that have been stored for
longer than a year, so it probably has a two-year life-
cycle.
**Distribution.** An Oriental species from S. China and
Indo-China.
**Control.** Normally not required.

larval damage

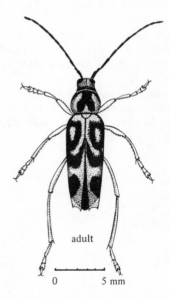

adult

0      5 mm

## Dirphya nigricornis (Ol.)

**Common name.** Yellow-headed Stem Borer
**Family.** Cerambycidae
**Hosts** (main). Coffee, mainly *arabica*.
　(alternative). Other wild woody Rubiaceae.
**Damage.** Attacked plants have wilted tips to the primaries; a series of holes down one side of a branch or main stem; broken branches are common.
**Pest status.** Normally a minor pest of *arabica* coffee, but severe attacks have occurred locally.
**Life history.** Eggs are laid singly near the tip of a branch under a small flap of green bark.

　The larva is red or brown when young. On hatching it bores into the green shoot, causing it to wilt. Later it bores down the primary towards the main stem, making a flute-like series of holes to the outside. Through these holes it throws out its sawdust-like frass. Burrowing continues right down the main stem. The mature larva is yellow or orange and about 50 mm long. The duration of the larval period is about ten months.

　The pupa is about 30 mm long. It is found in a cell excavated by the mature larva usually near ground level. The pupal period is about two months.

　The adult is a slender beetle about 25 mm long with long black antennae. The body is generally brown but the head, thorax and about the first quarter of the wing cases are orange or yellow. Most of the beetles emerge during the rainy seasons.

**Distribution.** Only E. Africa.
**Control.** Wilted primaries should be cut off and destroyed, and old heads should be burned.

　For chemical control it is suggested that the lowest frass hole should be enlarged and (using a pen filler or oil can) a mixture of dieldrin and kerosene (1:100) or water, squirted in.

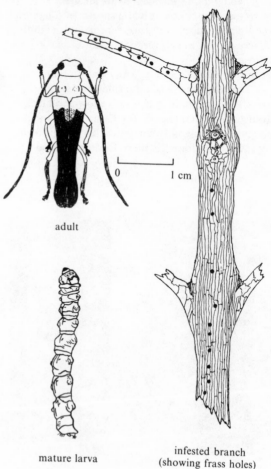

adult

mature larva

infested branch
(showing frass holes)

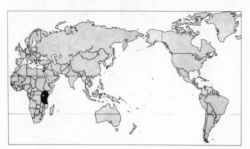

## Paranaleptes reticulata (Thoms.)

**Common name**. Cashew Stem Girdler
**Family**. Cerambycidae
**Hosts** (main). Cashew
    (alternative). All wild members of the family
Bombacaceae (e.g. Baobab) are probably Stem
Girdler hosts. *Hibiscus*, kapok, *Bougainvillea*, cotton,
*Acacia*, *Citrus*, and *Ceiba pentandra* are also attacked.
**Damage**. Branches from 3−8 cm in diameter are completely girdled by the adult beetles with a **V**-section
cut. Only a narrow, central pillar round the pith zone
is left, which eventually breaks off. The distal part of
the girdled branch is usually much marked with
impressions of the adult beetle's jaws.
**Pest status**. A common, but usually minor pest of
cashew in Coast Province, Kenya. However, neglected
plantations may be severely damaged.
**Life history**. Eggs are elongate, about 5 mm long, and
are laid singly in transverse slits made in the bark or
the girdled branch at points above the girdle.
    The larva, which is yellow, mines in the dead
wood of the girdled branch. It reaches a length of
45 mm, when fully grown.
    Pupation takes place in the dead wood in a
chamber prepared by the larva.
    The adult is a typical longhorn beetle with a
body length of 25−35 mm and with antennae longer
than the insect body. The head and thorax are very
dark brown. The wing cases are orange with large
polygonal black blotches giving them a reticulated
appearance.
    The total life-cycle takes one year. Adults are
on the wing and girdling and egg-laying taking place
in the period from May to October.
**Distribution**. Kenya and Tanzania only.
**Control**. Once a year in November or December all
girdled branches should be collected up and burned.
Only the dead or dying part of the branch above the
girdle need be collected.
**Further reading**

Duffey, E. A. J. (1957). *African Timber Beetles*,
    p. 225. British Museum (NH): London.

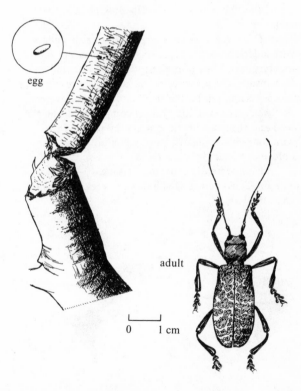

egg

adult

0    1 cm

# Acanthoscelides obtectus (Say)

**Common name.** Bean Bruchid
**Family.** Bruchidae
**Hosts** (main). Beans of various species.
(alternative). Other pulse crops.
**Damage.** The infested seeds having mature larvae or pupae inside can be recognized by the presence of a small window; emergence holes are about 2 mm in diameter.
**Pest status.** A serious pest of beans in many parts of the world, more particularly in tropical countries.
**Life history.** The infestation by this pest often starts in the field.

The eggs, which are dirty white and pointed, are laid by the female on the ripening pods in the crop.

The tiny larvae are dirty white or pale yellow, with a dark brown head, strong mandibles and rudimentary legs. They bore their way into the seed and feed inside. The presence of mature larvae or pupae can be recognized by the small circular windows on the bean seeds. The life-cycle is completed inside the seed and the adult beetle emerges by pushing the window, which falls off, leaving behind a neat round hole about 2 mm in diameter.

Each female is capable of laying 40—60 eggs. The life-cycle period is about 4—6 weeks at 28 °C and 70% RH.

**Distribution.** Widely distributed in Europe, Africa, New Zealand, USA, C. and S. America.
**Control.** For chemical control the beans should be thoroughly mixed with γ-BHC dust or pyrethrins.

Fumigation should be carried out by approved operators only.

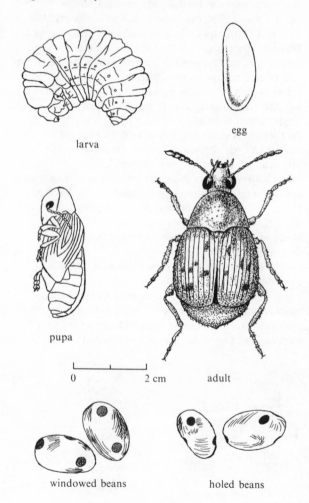

larva

egg

pupa

0        2 cm        adult

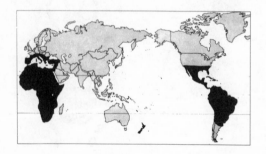

windowed beans          holed beans

454

## Callosobruchus spp.

*chinensis* (L.)
*maculatus* (F.)

**Common name.** Cowpea Bruchids
**Family.** Bruchidae
**Hosts** (main). Cowpea
    (alternative). Soybean, and other pulses.
**Damage.** The larvae bore into the pea or bean. Infestations usually originate from farm stores but the adult beetles can fly for up to about half a mile. The infested pods are then harvested and taken into the farm stores where further development takes place.
**Pest status.** These are important pests of pulse crops in Africa and Asia both on field crops and in stores.
**Life history.** Eggs are laid, stuck on to the outside of the pods, by the female beetle; each female laying up to 90 eggs. If the pods have dehisced, eggs are laid directly onto the seeds. Hatching takes about six days.

The larvae spend their entire life within the pea or bean. On hatching, the larva is scarabaeiform. The larval period is about 20 days.

Pupation takes place in a chamber just under the testa of the seed, this being known as the 'window' stage; pupation takes about seven days to complete.

The adults are small brownish beetles, with characteristically emarginated eyes. Distinctive sexual dimorphism is shown in the antennae.

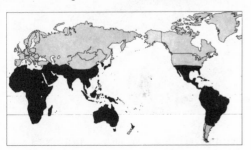

*C. maculatus* is a more elongate species with the posterior part of the abdomen not covered by the elytra, and it is more definitely spotted.

The whole life-cycle takes about 4—5 weeks, and about six or seven generations are usual.
**Distribution.** Cosmopolitan throughout most of the tropics and subtropics.

Several other species of *Callosobruchus* attack pulse crops in the tropics.
**Control.** Cultural control can be effective in growing vulnerable crops at least half a mile from farm crop stores which are the primary source of infestation.

Fumigation with methyl bromide in the stores is very effective.

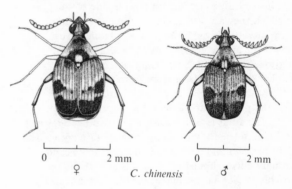

♀    *C. chinensis*    ♂

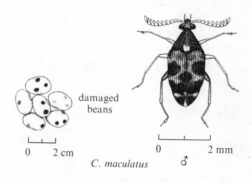

damaged beans

*C. maculatus*    ♂

455

## Caryedon serratus (Ol.)

(= *C. gonagra* (F.))

**Common name.** Groundnut Borer; Seed Beetle
**Family.** Bruchidae
**Hosts** (main). Groundnuts
      (alternative). Other legumes.

**Damage.** The larvae bore into the kernels, and a single larva makes a large hole in the cotyledons; the emerging adult beetle makes a large hole in the pod. Pods are attacked both in the field and in post-harvest storage.

**Pest status.** A serious primary pest of groundnuts in West Africa.

**Life history.** Eggs are laid on the outside of the pod, to which they are stuck. On hatching the young larva bores directly through the pod wall from the egg. It then feeds on the cotyledons of the kernel until mature. At this stage there is no sign of damage visible externally. The mature larvae either pupate within the pod or else emerge by boring a large hole and pupate outside in a thin papery cocoon.

The adult is an oval-shaped brown beetle, some 4–7 mm in length, with quite long, serrate antennae, stout hind femora with a row of distal spines and strongly curved hind tibiae. This is one of the Bruchidae that has normal eyes and there are also small blackish spots on the elytra.

The life-cycle takes about 42 days under optimum conditions (30 °C and 70% RH).

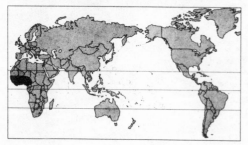

**Distribution.** This pest is common in W. Africa; it also occurs in E. Africa but has not been recorded attacking groundnut there.

**Control.** Clearly a difficult pest to control with the larvae completely inside the protective pods, but repeated fumigation will destroy infestations.

At present there is no information regarding control of field infestations; however, it is thought that they are seldom at all serious.

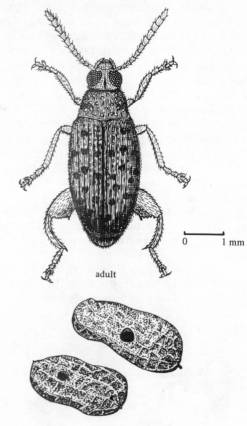

adult

attacked groundnuts showing adult emergence holes

456

## Aspidomorpha spp.

**Common name.** Tortoise Beetles
**Family.** Chrysomelidae (Cassidinae)
**Hosts** (main). Sweet Potato
(alternative). Other Convolvulaceae (e.g. morning glory), coffee, beet, potato, and various flowers.
**Damage.** Large round holes are eaten in the leaves, by both adults and larvae; occasionally attacks are sufficiently severe to completely skeletonize the leaves.
**Pest status.** Seldom a serious pest, but one which is very widely distributed and often very common and the damage is quite conspicuous.
**Life history.** Eggs are laid on the leaves, usually singly and usually on the underside. Hatching takes about ten days.

The larvae are oval, with a fringe of spines along the margin, and a forked tail carried held up over the back, usually with all the previous cast skins (exuviae) adhering.

The pupa is less spiny than the larva, and is fixed inert to the leaf.

The adult is oval and shield-like, hence the common name of 'tortoise beetle', 6—8 mm long, with broad and flat elytra, often with a beautiful golden iridescence.

There are several generations per year in the tropics, but generally only one in temperate countries.

**Distribution.** Africa, S. China, S.E. Asia, Papua New Guinea, West Irian and the W. Indies.

There are more than 12 species of *Aspidomorpha* in Africa and a dozen in Asia, mostly recorded from sweet potato, and this host also has more than five other genera of Cassidinae in Africa and Asia.
**Control.** Usually this pest does not warrant control, but DDT sprays would be effective if control should be required.

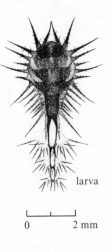

larva

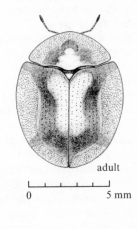

adult

0          5 mm

*A. cincta*

0     2 mm

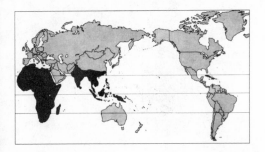

## Oulema oryzae (Kuw.)

(= *Lema oryzae* Kuw.)

**Common name.** Rice Leaf Beetle
**Family.** Chrysomelidae (Criocerinae)
**Hosts** (main). Rice
    (alternative). Many species of grass.
**Damage.** Damage consists of the removal of longitudinal strips of the upper leaf epidermis, giving the
leaf a bleached appearance. Excessive damage can
result in death of the plant. Damage to floral growth
occurs, and can be particularly serious when at the
heading stage.
**Pest status.** A very serious pest of paddy rice in
northern Japan, China and Korea, both adults and
larvae causing severe leaf damage with regular losses
in crop yield of 20–30%.
**Life history.** Eggs are cylindrical, rounded, 0.8 by
0.3 mm, and they are laid in groups on the upper leaf
surface. The oviposition period lasts for about 15
days, and the incubation period 5–11 days.

    The larva is squat, with brown nodules on a
yellow base, covered with a crust of its own excre-
ment as camouflage, about 4.7 mm long. Larval
development takes 13–19 days.

    The whitish papery cocoon is usually found on
the leaf, but in upland rice fields they may be underground; its size is usually about 4.0–4.5 mm long.

    The adult is a small leaf beetle about 4–5 mm
long, with shiny black elytra, conspicuously punctate,
and a reddish-brown thorax and head. The legs are
pale. The adult beetle feeds for only about three
months, and spends the rest of the year hibernating
in vegetable debris.
**Distribution.** Japan, China, Ryukyu Isles, Taiwan,
Manchuria, Korea, and eastern Siberia.

    *O. bilineata* is a serious pest of tobacco in
S. Africa.

    *O. melanopus* is the Cereal Leaf Beetle, wide-
spread in Europe, western and central Asia, and a
serious cereal pest in N.E. USA. It is a vector of a
virus on cocksfoot grass in Europe (CIE map no.
A260).
**Control.** Applications of DDT, BHC, and phosphamidon have been successful in controlling this
pest.

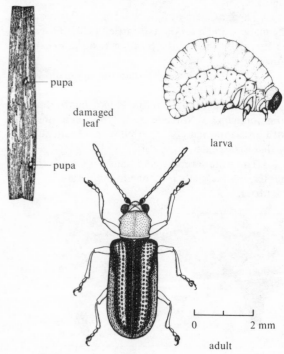

pupa

damaged
leaf

pupa

larva

0    2 mm

adult

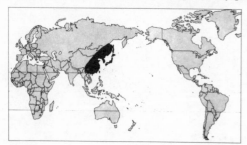

458

## Colaspis hypochlora Lefèvre

**Common name.** Banana Fruit-scarring Beetle
**Family.** Chrysomelidae (Eumolpinae)
**Hosts** (main). Bananas
    (alternative). A wide range of weeds and grasses.
**Damage.** The adult beetle feeds on the young unfurled leaves and stems of banana plants, and also eats the skin of young fruit, making scars which spoil the fruit and make it unsalable, and allowing the entry of pathogens.
**Pest status.** A pest of some past importance in C. and S. American banana-growing areas.
**Life history.** Eggs are laid in the soil around the banana roots, or in holes gnawed in the roots, singly or in groups of 5–45. Each female can lay several hundred eggs. The incubation period is 6–9 days.

    The larvae remain in the soil feeding on the roots of grasses, often to a depth of 25 cm. Larval development takes 20–22 days.

    Pupation takes place in the soil, and lasts for 7–10 days.

    The adult beetles upon emergence feed on various weeds, as well as the young leaves and fruit of bananas. They are small beetles 5–6 mm long, nocturnal in habit, and they can fly strongly; their normal life span is probably several months. The adult beetles gnaw the banana roots before laying eggs in the soil around the roots.

**Distribution.** Mexico, Costa Rica, Panama, Nicaragua, Honduras, Guyana, Guatemala, and Colombia.
**Control.** Clean cultivation, mainly being the removal of grass weeds from plantations, will help to reduce populations often enough to avoid the use of insecticides.

    Should chemical control really be necessary then endrin, aldrin, and dieldrin have been found to be effective.
**Further reading**
Simmonds, N. W. (1970). *Bananas*, 2nd ed.
    pp. 355–8. Longmans: London.

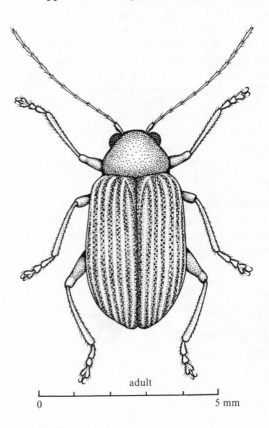

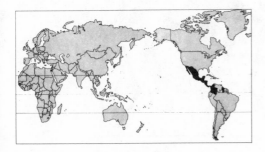

adult

0                        5 mm

## Diabrotica undecimpunctata Mann.

**Common name.** Spotted Cucumber Beetle (adult);
Southern Corn Rootworm (larva)
**Family.** Chrysomelidae (Galerucinae)
**Hosts** (main). Groundnut, cucurbits
(alternative). Maize
**Damage.** The larvae bore into the underground stem,
and cotyledons of young plants, and bore small
round holes in developing groundnut pods and eat
the kernels within.

The adult beetles are often contaminated with
bacteria which cause Bacterial Wilt of cucurbits.
**Pest status.** The larva only is a pest of groundnut; in
severe attacks 80% of the pods may be bored.

The adult is a serious pest of young cucurbits
in the USA. Other closely related species attack
legumes in southern USA and S. America.
**Life history.** The eggs are laid in the soil.

The larvae are wrinkled, yellowish-white, and
have a brown head capsule. They grow to 10–18 mm.
The total time required for larval development is
about 30 days.

The adult beetle is yellowish-green, about 6 mm
long, with a black head and 12 conspicuous black
spots on the elytra.

In the USA the life-cycle takes some 50 days,
and there are generally three generations per year.
**Distribution.** Confined to N. America, from Canada
to Mexico.

Other closely related species are found in
S. America.
**Control.** In areas at risk from this pest it is advisable
to avoid winter crops such as lucerne which encourage the pest in the spring.

Unnecessary and excessive use of insecticides
should be avoided, as this pest has developed resistance to the chlorinated hydrocarbons in the USA. If
pesticides are really necessary then diazinon and
phorate granules are effective. Parathion granules are
also effective, but this chemical has very high mammalian toxicity.

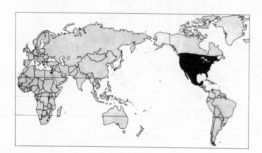

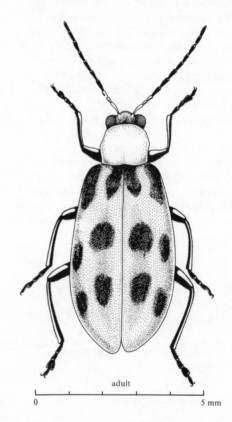

adult

0                                    5 mm

460

## Megalognatha rufiventris Baly

**Common name.** Maize Tassel Beetle
**Family.** Chrysomelidae (Galerucinae)
**Hosts** (main). Maize
     (alternative). Many other plants
**Damage.** On maize this pest congregates on the cobs
and they devour the silks, and occasionally they
penetrate the sheaths and destroy the seed embryos.
Damage is done solely by the adult beetles.
**Pest status.** This pest periodically occurs in large
numbers in E. Africa and causes severe damage to
many crops when this happens.
**Life history.** Details of the life history of this beetle
are not known.

     The adult is a small black beetle with a pale
reddish-brown abdomen, the tip of which protudes
below the end of the elytra.
**Distribution.** E. Africa only.
**Control.** Control is usually achieved easily by dusting
the crop with DDT.

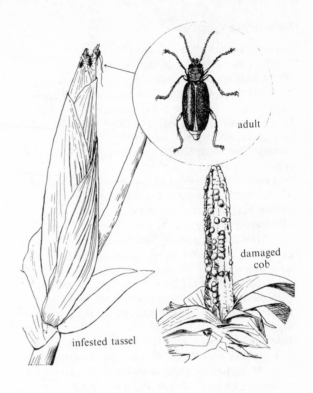

adult

damaged
cob

infested tassel

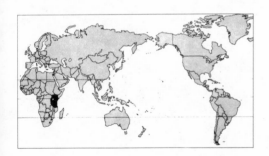

## Monolepta signata

**Common name.** White-spotted Leaf Beetle
**Family.** Chrysomelidae (Galerucinae)
**Hosts.** This is a regular minor pest on a number of different crops, including Cucurbitaceae and grapevine; closely related species in Africa and Australia feed on maize, *Citrus*, various stone fruits, cotton, groundnut, mango and cashew.
**Damage.** The adult beetles eat holes in the leaf lamina.
**Pest status.** A common and widespread pest, especially at generic level, on many different hosts, but usually only important as part of a pest complex.
**Life history.** No details are known for this pest.

The adult is a small, dark beetle, about 2—3 mm in body length, with long antennae and two pairs of large pale spots on the elytra.

The other recorded pest species are similarly patterned, but *M. australis* (Red-shouldered Leaf Beetle) is basically yellow with red spots at the elytra bases.
**Distribution.** This species is found from India, through S.E. Asia, to S. China and the Philippines.

*M. australis* (Jac.) is found in Australia; *M. dahlmanni* (Jac.) and *M. duplicata* (Sahlb.) in tropical Africa.
**Control.** Usually not required.

adult

0 — 2 mm

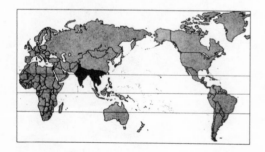

462

## Ootheca mutabilis (Sahlb.)

**Common name.** Brown Leaf Beetle
**Family.** Chrysomelidae (Galerucinae)
**Hosts** (main). Groundnut, sesame, beans, cowpea, and other pulses.
    (alternative). Coffee, cocoa, and cotton.
**Damage.** Young leaves are attacked and eaten by the adult beetles. Damage is typical of that done by leaf-beetles. It is reported as being a vector of Cowpea Yellow Mosaic virus, and in Nigeria it sometimes occurs in considerable swarms, when damage can be extensive.
**Pest status.** Not a serious pest usually, but of quite frequent occurrence on many leguminous crops.
**Life history.** The biology of this pest is not known in detail.

Eggs are laid in the soil.

The larvae feed on the roots of various plants.

Pupation takes place in the soil, during the dry season.

The adults emerge with the early rains, and attack the seedling plants. The adult is an oval, shiny, convex-shaped beetle, about 6–8 mm long, yellowish-red or brownish, with a black head and legs.
**Distribution.** Nigeria, and E. Africa.

Several other closely related species are found mostly on leguminous crops.
**Control.** Control measures are not usually advocated.

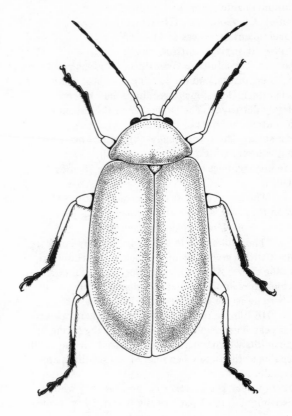

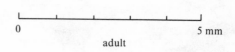

0                    5 mm

adult

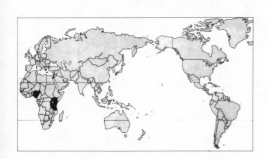

## Phyllotreta spp.

**Common name.** Cabbage Flea Beetles
**Family.** Chrysomelidae (Halticinae)
**Hosts** (main). Brassicas and Cruciferae.
  (alternative). Cotton, cereals.
**Damage.** The adults feed on the cotyledons and leaves of young plants, and the feeding produces a shot-hole effect. Occasionally seedlings may be completely destroyed. The larvae live in the soil and feed upon the roots of the host plants.
**Pest status.** These are usually only minor pests but they are very widespread and common.
**Life history.** Eggs are laid in the soil by the host plant.

The larvae of most species of *Phyllotreta* feed upon the roots, and do little damage.

Pupation takes place in the soil.

The adults vary in colour from shiny black, to black with a green sheen, to black with yellow stripes on the elytra. All species have very stout femora with which they jump in a flea-like manner. The adults hibernate in the soil litter or in hedgerows.

In Europe there are two or three generations per year. The main damage is done in the spring by the adults which emerge from hibernation and resume feeding at the time when many crop seedlings are available.

**Distribution.** The genus *Phyllotreta* is very widely distributed in most parts of the world; 12 + species

of *Phyllotreta* are pests on cruciferous crops; one on cereals, and one on cotton.

Four of the most common species are:
*P. cruciferae* (Goeze) – in Europe, Asia, Middle East, Egypt, USSR, and N. America; *P. nemorum* (L.) – Europe, Asia, USSR, Korea and S.E. Australia; *P. cheiranthei* Weise – Egypt, E. Africa, Sudan and Sri Lanka; and *P. striolata* – S. China.

Twenty or more genera of flea beetles are recorded from a wide range of crops throughout the tropical parts of the world.

**Control.** If control is required, a seed dressing of BHC, or treatment with DDT, BHC or derris dust, is generally effective.

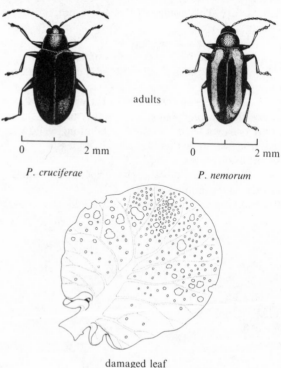

adults

| 0        2 mm | 0        2 mm |

*P. cruciferae*          *P. nemorum*

damaged leaf

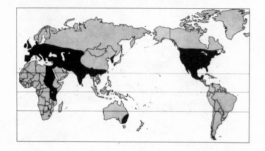

464

## Prodagricomela nigricollis Chen

**Common name.** Citrus Flea Beetle; Citrus Leaf-miner
**Family.** Chrysomelidae (Halticinae)
**Hosts** (main). *Citrus* spp.

(alternative). Some other species of Rutaceae.
**Damage.** This is a most unusual member of the Halticinae in that both adult and larvae damage the host plant. The adult eats windows in the leaves, from the underneath, and in heavy attacks can skeletonize the whole leaf; the larvae mines in the leaf making a broad tunnel mine with a conspicuous line of faecal pellets along the centre. Heavy attacks result in considerable leaf loss.
**Pest status.** A common pest of *Citrus* in S. China, but not classed as serious; more of academic interest because of the unusual life-cycle.

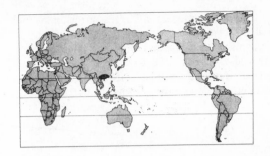

**Life history.** Egg-laying starts in early April, lasting for about a month, and by the end of April the first larval mines are to be found. The larva is orange-yellow in colour with an elongate body and reduced legs, and when full grown is about 5 mm in length.

adult on damaged *Citrus* leaf

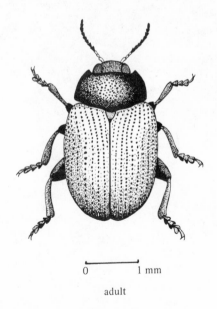

0          1 mm

adult

465

Pupation takes place in the soil or leaf litter under the infested tree. Adults emerge in June in S. China whereupon they aestivate and hibernate underground until the following March/April.

There is only the one generation per year.

**Distribution.** To date, only recorded from S. China.

**Control.** In China the hand-picking of infested leaves to control the eggs and larvae has been practised, together with clean cultivation to control the pupae. In Hong Kong chemical control is practised against this pest.

larval mine in *Citrus* leaf with larva in situ

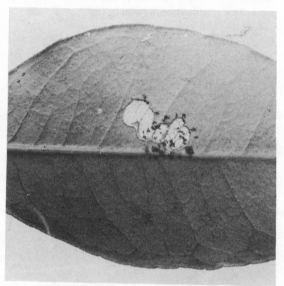

adult damage to underside of *Citrus* leaf

# Dicladispa armigera (Ol.)

**Common name.** Paddy Hispid
**Family.** Chrysomelidae (Hispinae)
**Hosts** (main). Rice
　　　(alternative). Various grass species.
**Damage.** Young rice plants are attacked by both adults and larvae; the adults feed on the green part of the leaf, leaving only the epidermal membranes — the feeding damage showing as characteristic white streaks along the long axis of the leaf. The larvae mine in the leaves between the epidermal membranes, producing elongate white patches. Damage starts from the leaf tip and extends back towards the leaf base; attacked leaves wither and die.
**Pest status.** A serious pest of rice, particularly in Bangladesh, but also important in other parts of S.E. Asia.
**Life history.** Eggs are laid singly, embedded in the lower epidermis of the rice leaves, near the leaf tip. Hatching takes 4—5 days. Each female lays about 55 eggs.

　　The larvae are minute, pale yellow, depressed, and about 2.5 mm long. Immediately after hatching the larvae burrow into the leaf, where they feed for 7—12 days before pupation.

　　The pupae are brown, depressed, exarate, and lie in the larval tunnel. Pupation takes 4—5 days.

　　The adults are small, shiny, blue-black beetles, about 5.5 mm long with spines on the thorax and elytra.

**Distribution.** Pakistan, India, Bangladesh, Sri Lanka, Burma, Malaysia, Sumatra, Java, Cambodia, Thailand, Laos, Vietnam, S. China, and West Irian (CIE map no. A228).
**Control.** Removal of grass weeds from around the paddy fields will lower the pest population, as also will the cutting off of the tips of the rice leaves at the start of an infestation.

　　Pesticides which have been effective are dieldrin, endrin, phosphamidon, demeton-S-methyl, BHC, DDT, and diazinon, as foliar sprays, and carbofuran, thiodemeton, phorate and disulfoton as granules.

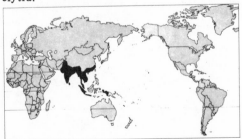

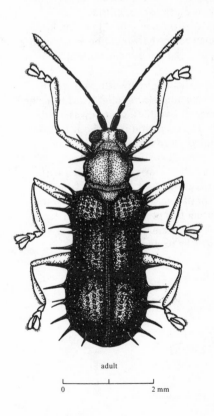

adult

0　　　　　　　　　2 mm

467

# Trichispa sericea (Guérin)

**Common name.** Rice Hispid
**Family.** Chrysomelidae (Hispinae)
**Hosts** (main). Rice
    (alternative). None recorded, but probably species of wild grasses.
**Damage.** Attacked plants have irregular pale brown patches and narrow whitish streaks on the leaves; the pale brown patches are the larval mines, and the whitish streaks are the feeding scars produced by the adults.
**Pest status.** An important pest in rice nurseries, only sporadically serious on transplanted rice.
**Life history.** Eggs are laid singly in slits in the leaf made by the adult beetle, the wound being covered by a spot of excreta. Hatching takes place after 3–4 days.

    The larva is a slender, yellowish grub which when fully grown is about 6 mm long. It feeds inside the leaf, the mine being visible externally as a pale brown blotch. The larval period lasts about ten days.

    Pupation takes place within the mine, the pupal stage lasting about six days.

    The adult is a dark grey beetle covered with upright spines. It is about 3–4 mm long. Adult females live for about two weeks and may lay more than 100 eggs during this period. Adults feed externally on the leaves, the damage being visible as narrow whitish streaks parallel to the veins.

**Distribution.** Only recorded from Africa; Angola, Senegal, Mali, Cameroons, Nigeria, Togo, Ivory Coast, Zaïre, Swaziland, Sudan, Ethiopia, Burundi, Rwanda, Uganda, Kenya, Tanzania including Zanzibar, S. Africa, and Madagascar (CIE map no. A257).
**Control.** Sprays of BHC or DDT are recommended in either high- or low-volume according to the spraying machinery available; BHC is to be preferred as it kills both larvae and adults.
**Further reading**
Grist, D. H. & R. J. A. W. Lever (1969). *Pests of Rice*, pp. 226–7. Longmans: London.

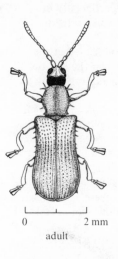

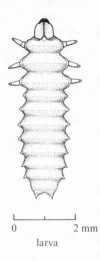

adult          larva

## Cylas formicarius (F.)

**Common name.** Sweet Potato Weevil
**Family.** Apionidae (Curculionoidea)
**Hosts** (main). Sweet potato
(alternative). Some other species of *Ipomoea*
(not all).
**Damage.** The larvae bore into the tubers and stems,
where they feed, and eventually pupate. The tunnel
systems are usually infected with fungi and bacteria
causing extensive rotting of the tubers. Adults are
also found in the tunnel systems and on the leaves of
the plant, on which they feed.
**Pest status.** A serious pest of sweet potato causing
extensive damage to field crops throughout the
tropical parts of the world; the damage to the tubers
continues during storage, which makes this an even
more important pest.
**Life history.** Eggs are laid singly in hollows in the
stem, or else inserted directly into the tubers. Hatching requires about one week.

The larvae are white, curved, and apodous, and
they tunnel inside both stems and tubers, for about
2–4 weeks, making tunnels some 3 mm in diameter.
Pupation takes place within the larval tunnels and
requires about one week.

The adult is a small, black weevil with brown
thorax and legs, about 6–8 mm in length. Sexual
dimorphism is apparent in the antennae, the males

having a long antennal club. Adults are long-lived,
active, and fly quite readily.

The life-cycle takes some 6–7 weeks usually;
and there are several generations each year.
**Distribution.** Recorded as common from tropical
Africa, India, S.E. Asia, Australasia, Hawaii, S. USA,
W. Indies and S. America (CIE map no. A278).
**Control.** See under *Cylas puncticollis*.

larva     adult ♂♂

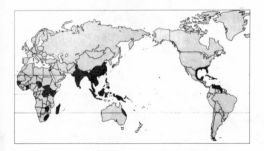

damaged tuber

469

## Cylas puncticollis Boh.

**Common name.** African Sweet Potato Weevil
**Family.** Apionidae (Curculionoidea)
**Hosts** (main). Sweet potato
(alternative). Some other species of *Ipomoea*,
and also maize.
**Damage.** The larvae bore in the stems and tubers,
which eventually develop extensive rotting patches,
and the adults sometimes eat the leaves.
**Pest status.** A serious pest of sweet potato in tropical
Africa, partly because of the direct damage done to
the tubers, but also because of the associated rots
which mean that damage continues during storage
after harvest.
**Life history.** See under *C. formicarius.*

The adults are slightly different from the previous species in that they are entirely black in colour.
In both E. and W. Africa many infestations of sweet
potato tubers contain both species of *Cylas.*
**Distribution.** This species is confined to tropical
Africa, mostly E. and W. Africa (CIE map no. A279).
**Control.** Continuous cropping of sweet potato can
keep the weevil populations very high and so crop
rotation is recommended. Varietal resistance to weevil
infestation has been both claimed by some workers
and denied by others as being only very short-lived,
and so at present its possible value is dubious.
Destruction of infested crop material and crop
residues will also help to lower pest populations.

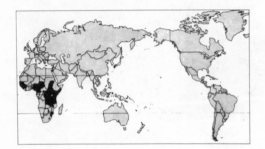

In W. Africa a fungus (*Beauvaria* sp.) has been
found to attack the adult weevils during the rainy
season.

Foliar sprays of insecticides have generally been
of little use, but the dipping of planting slips in a
DDT solution gave good control in Uganda. DDT
and BHC dusts used at planting have also given good
control. Dipping and dusting has to be done with care
for if the chemical is too strong phytotoxicity may
result.

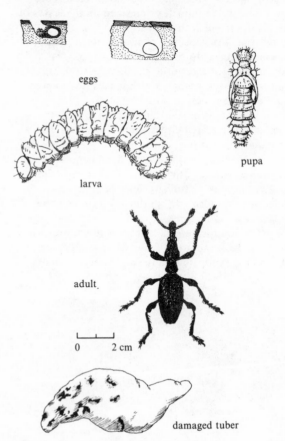

eggs

larva

pupa

adult

0    2 cm

damaged tuber

# Araeocerus fasciculatus (De Geer)

**Common name.** Coffee Bean Weevil; Nutmeg Weevil
**Family.** Brenthidae (Curculionoidea)
**Hosts** (main). Nutmeg

   (alternative). Coffee beans, cocoa beans, seeds of various types, both in the field and in storage.
**Pest status.** Most serious as a stored products pest, but field infestations are common; in Hong Kong field infestations of nasturtium seeds are heavy.
**Life history.** Eggs are laid singly on the ripening, or fully ripe, seeds, and the white, legless larvae burrow into the seeds, each larva usually spending its imma-ture life inside the same seed. Pupation takes place with the seed.

   The adult is a small brown beetle about 3 mm in body length; it looks rather like a bruchid in appearance but has distinctively clubbed antennae. Adults fly strongly, and in Hong Kong are serious domestic pests where they can be seen flying into flats and houses during the day at certain times of the year.
**Distribution.** Cosmopolitan throughout the warmer parts of the world; it is found quite regularly in ware-houses in the UK but generally fails to survive the winter.
**Control.** Generally not required, as field infestations are usually light, and warehouse infestations are gen-erally controlled by the regular fumigations.

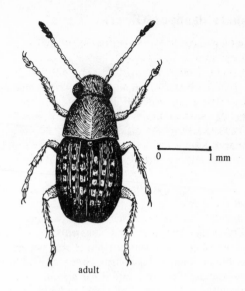

0      1 mm

adult

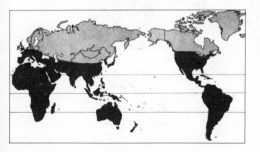

## Alcidodes dentipes (Oliver)

**Common name.** Striped Sweet Potato Weevil
**Family.** Curculionidae
**Hosts** (main). Sweet Potato, and groundnut.

(alternative). Cotton, and other woody legumes.
**Damage.** The adult weevil girdles the stem of the plants just above ground level; the plants then wilt and die. The larvae bore inside the stem, making galls.
**Pest status.** Sometimes a serious pest of sweet potato, but usually only a minor pest of groundnut and cotton.

*Alcidodes gossypii* is the Cotton Stem-girdling Weevil.
**Life history.** The larvae are believed to feed inside the pith of the stem, forming visible galls.

The adult weevil is of moderate size, being about 14 mm long, and is conspicuously striped longitudinally along the elytra, and it has pronounced spines on the inner edge of all tibiae and also on the fore-femora.
**Distribution.** Tropical Africa.
**Control.** Not usually required.

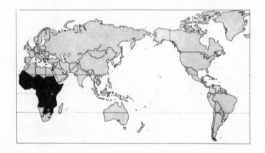

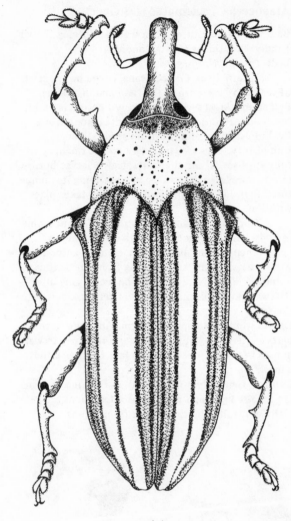

adult

0                                    5 mm

## Anthonomus grandis Boh.

**Common name.** Cotton Boll Weevil
**Family.** Curculionidae
**Hosts** (main). Cotton
     (alternative). Wild species of *Gossypium*, *Abutilon* spp., *Hibiscus* spp., and *Thurbaria thespesioides*.
**Damage.** The larvae bore into the bolls and squares; the squares turn yellow and die. Most punctured squares and small bolls are shed. Large, punctured bolls are not shed but the locus on which the larva feeds fails to develop properly, and the lint becomes cut, stained and decayed.
**Pest status.** A serious pest of the cotton-growing areas of the USA and Mexico where crop losses can be high.
**Life history.** Eggs are laid singly in deep punctures within the squares or bolls. Incubation requires 3–5 days.

     The larvae feed in the squares and bolls for 7–14 days, and then they pupate. Pupation takes 3–5 days and the adults cut their way out of the squares.

     The adult weevils spend the winter in soil litter and trash. In the spring they return to the cotton fields, and stay there until the first frosts. The adults feed on the squares and flowers, or bolls, and after 3–4 days the females start egg-laying.

     The life-cycle takes only about 21 days, and there may be seven generations per year.

**Distribution.** Southern USA, Mexico, C. America, W. Indies, and S. America (Venezuela and Colombia) (CIE map no. A12).
**Control.** Cultural practices which help to reduce boll weevil populations include: using good fertile land with sufficient fertilizer, the growing of early maturing varieties, early planting with close spacing, frequent cultivation, picking the crop early and cleanly, and finally the destruction of crop residues.

     Chemical sprays recommended are: BHC, toxaphene, aldrin, and dieldrin, but care has to be taken with their use or else other pests may become more troublesome.

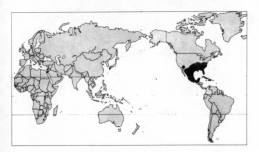

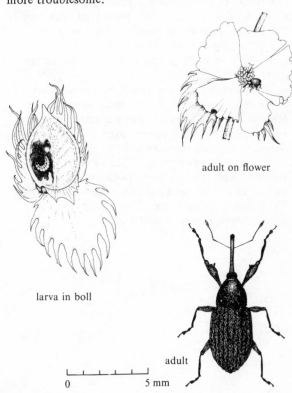

adult on flower

larva in boll

adult

0          5 mm

473

# Aperitmetus brunneus (Hust.)

**Common name.** Tea Root Weevil
**Family.** Curculionidae
**Hosts** (main). Tea
    (alternative). Coffee, beans, *Brassica* spp.
**Damage.** The larvae feed on the tap root, gnawing channels the length of the root, causing wilting, stunting, and eventual death of the young plant. Stems of young plants may be ring-barked at ground level by the larvae. Adult weevils feed on the foliage and chew irregularly shaped holes through the surface and also chew the leaf edges.
**Pest status.** In Kenya this is a serious pest in tea nurseries, where losses of 30—50% are not uncommon.
**Life history.** Life history details are not available.

    The larvae of this weevil feed on the roots, particularly of seedlings where they feed on the tap root.

    The adult is a distinctive black weevil about 7—9 mm long, which feeds on the leaves of tea and other plants. It looks rather like a *Systates* weevil, but has pale grey scales on its body.
**Distribution.** Only recorded from Kenya.

    Twenty-seven species of weevils have been recorded feeding on the roots and leaves of tea bushes, in nurseries and recently established gardens. The other main pests are *Nematocerus* spp., and *Eutypotrachelus meyeri* Kolbe.

**Control.** Foliar sprays of DDT or dieldrin are generally effective against the adult weevils.

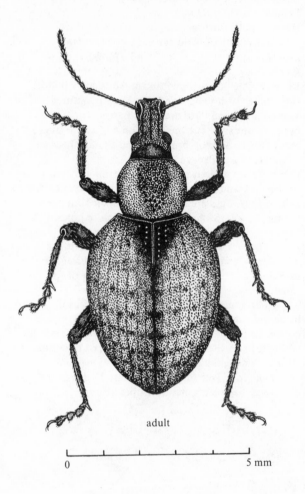

adult

0                                5 mm

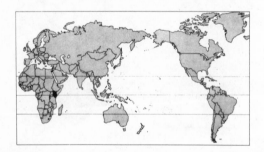

## Cosmopolites sordidus (Germ.)

**Common name.** Banana Weevil
**Family.** Curculionidae
**Hosts** (main). Bananas (*Musa* spp.).
    (alternative). Recorded from cocoa stems.
**Damage.** The larva bores irregular tunnels in the rhizome and pseudostem at ground level. The tissue at the edge of the tunnels turns brown and rots. If the stem is small, the banana variety susceptible or the infestation very heavy, the plant will die.
**Pest status.** A major pest of bananas throughout the tropics, and in some areas it is still spreading.
**Life history.** The eggs are laid singly in small pits made in the pseudostem near ground level by the female weevil; they are elongate-oval, white, and about 2–3 mm long. Hatching takes 5–8 days.

    The larva is a white, legless grub with a brown head capsule. The larva period occupies 14–21 days.

    Pupation takes place in holes bored by the larvae; the pupal period lasts 5–7 days. The pupa is white and about 12 mm long.

    The newly emerged beetle is brown, turning almost black after a few days. Its normal food is dead or dying banana plants. It does not usually fly and may live for up to two years. Each female may lay 10–50 or more eggs. They are nocturnal in habits.
**Distribution.** Pantropical but with some areas not inhabited (CIE map no. A41).

    *C. minutus* occurs in the Pacific region.

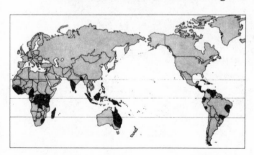

**Control.** Cultural methods are the most important and in many areas will be sufficient to keep the population level down. These methods include the use of clean suckers only; old stems should be cut off at ground level and the cut rhizome covered with impacted soil; old stems should be cut into strips and used for mulch; and good weed control.

    The most successful insecticides for use against this pest are aldrin and dieldrin, applied as a dust around the bases of the pseudostems, and applied to the cut surfaces of the rhizomes before they are covered with soil. Planting suckers which are suspected of being infested should be dipped into dieldrin solution.

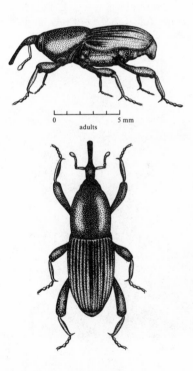

0          5 mm
adults

## Cyrtotrachelus longimanus

**Common name.** Bamboo Weevil
**Family.** Curculionidae
**Hosts.** Various larger species of bamboos.
**Damage.** The larva develops in the apical shoot, which is entirely eaten away, thus terminating the growth of that stem and initiating lateral bud development, causing terminal branching. The feeding/oviposition site is marked by a ragged hole on the stem.
**Pest status.** A serious pest of bamboo grown for constructional purposes in southern China.
**Life history.** The female weevil feeds by biting into the apical shoot of the bamboo and after feeding she lays a single egg into the feeding site. The egg develops into a white, legless larva, with a brown head capsule and well-developed jaws — a typical weevil larva. The feeding larva destroys the apical shoot entirely, and

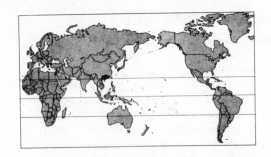

after about four weeks, and attaining a body length of 4—5 cm, it drops out of the stem and pupates in the soil. During the feeding process the larva is protected and shielded by the stiff leaf-scales that normally protect the growing point.

damage

larva

476

The pupa overwinters inside an earthen cocoon in the soil, and usually the whole pupal period lasts for about ten months.

In S. China the adults are generally only seen in July and feeding larvae found in August, although it is thought that occasionally there might be a second infestation period in early October. Usually the damaged bamboos have a second period of growth in November which to some extent compensates for the damage done in July/August.

**Distribution.** To date only recorded from southern China.

**Control.** No details are known as to whether control is attempted in China.

adult ♀

0      1 cm

## Diocalandra frumenti (F.)

(= *D. stigmaticollis* Gyll.)

**Common name.** Four-spotted Coconut Weevil
**Family.** Curculionidae
**Hosts** (main). Coconut palm

(alternative). Date, oil, and nipa palms; also sorghum.

**Damage.** The larvae attack all parts of the palm, especially roots, leaves and fruit stalks, and cause premature fruit-fall. The leaf bases are bored from the trunk out to the leaflets. In some areas the trunk is also bored, at all heights.

**Pest status.** The precise status of these weevils as pests is open to dispute; some entomologists believe that the damage is primary and results in appreciable crop losses, but others maintain that this damage is purely secondary.

**Life history.** Eggs are laid in crevices at the base of the adventitious roots, at the foot of the trunk, or in the flowers, in the petiole, or at the base of the peduncle. Incubation takes 4–9 days.

The larvae bore into the tissues and cause gum to exude from the opening of the gallery. Larval development takes 8–10 weeks.

Pupation takes place within the larval gallery, taking some 10–12 days, but no cocoon is made.

The adults are small weevils, about 6–8 mm in length, shiny blackish, with four large reddish spots on the elytra. Coloration varies somewhat and is not a reliable taxonomic character – the adults are only distinguishable to an expert. Sexual dimorphism is evident, as with many other weevils, by the shape of the posterior apex, and the length and thickness of the rostrum ('snout'), the male rostrum being shorter, thicker, and more curved.

The life-cycle takes 10–12 weeks.

**Distribution.** Recorded from Tanzania, Somalia, Seychelles, Madagascar, S. India, Sri Lanka, Bangladesh, Burma, Malaysia, Thailand, Indonesia, Philippines, Papua New Guniea, West Irian, N. Australia, Solomon Isles, Samoa, Caroline and Mariana Isles (CIE map no. A249).

**Control.** See *D. taitense*.

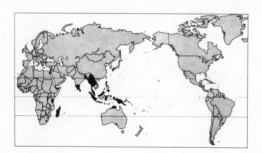

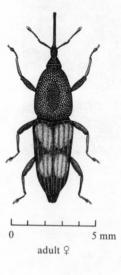

adult ♀

adult ♂

## Diocalandra taitense (Guer.)

**Common name.** Tahiti Coconut Weevil
**Family.** Curculionidae
**Hosts** (main). Coconut palm
    (alternative). None recorded.
**Damage.** The larvae attack all parts of the palm, sometimes the trunk (at all heights), but mostly the roots, leaves and fruit stalks. The leaf bases are bored from the trunk out to the leaflets. The end result of attack is usually premature fruit-fall.
**Pest status.** It is not clearly established yet whether this pest is primary or secondary, and so its pest status is uncertain.
**Life history.** Eggs are laid in crevices at the base of the roots, at the bole of the trunk, in the flowers, in the petioles or at the base of the peduncles. Incubation takes 4–9 days.

The larvae bore through the tissues of the palm and cause gum to exude from the gallery opening. Larval development takes 8–10 weeks.

Pupation takes place within the gallery, taking some 10–12 days; no cocoon is made in this species.

Adults are small weevils, about 6–8 mm in length, shiny blackish, with four large reddish spots on the elytra, and to the non-expert indistinguishable from those of *D. frumenti* (see previous page). As with the previous species, sexual dimorphism is apparent.

The life-cycle takes 10–12 weeks.
**Distribution.** Madagascar (introduced 1943), Papua New Guinea, Hawaii and most of the Pacific Islands (CIE map no. A248).
**Control.** Cultural methods include the avoidance of knife slash marks on the trunks, and earthing-up the base of the trunk to cover the adventitious roots.

A braconid parasite (*Spathius apicalis* Westw.) reportedly destroys up to 40% of the weevil larvae, and the predatory beetle *Plasius javanus* Eric. (Histeridae) is important.

The application of tar to the roots and the base of the trunk, and spraying with dieldrin, is said to be successful.
**Further reading**
Lever, R. J. A. W. (1969). *Pests of the Coconut Palm*, pp. 121–4. F.A.O.: Rome.

## Graphognathus spp.

**Common name.** White-fringed Weevils (Beetles)
**Family.** Curculionidae
**Hosts** (main). Groundnut

(alternative). Maize, cotton, various legumes and vegetables, ornamentals and wild grasses.
**Damage.** Both adults and larvae are pests but larval damage is the more important. The adults feed on the leaves, and do sometimes defoliate the plants, but the larvae in the soil do serious damage by eating the plant roots.
**Pest status.** There are at least four closely related species, and the larvae are quite serious soil pests feeding on the roots of various crop plants.
**Life history.** Parthenogenetic females lay their eggs on the soil 5–15 days after emergence. Each female lays about 1500 eggs. Incubation takes 11–30 days.

The newly hatched larvae live in the soil, where they feed on the roots of many plants and over-winter. The larvae may already be in the soil when the crops are planted.

Pupation takes place in the soil.

The adults do not disperse very far, so heavy infestations can build up locally unless a suitable crop rotation is practised, or insecticides used. The adult is a very blunt-nosed weevil, about 12 mm long, with conspicuous white edges to the lateral edges of the elytra.

There is only one generation per year.

**Distribution.** Limited to the S.E. of the USA, S. America (Chile, Peru, Argentina, Brazil, Uruguay), Australia (New South Wales, Victoria), New Zealand (N. Island), and S. Africa (Cape Province).

The most important species is *G. leucoloma* (Boh.), and *G. imitator* and *G. striatus* have been regarded as synonyms; the other main species is *G. peregrinus.*

The distribution of *G. leucoloma* has been mapped by the CIE (map no. A179).
**Control.** Larval populations can be assessed by taking soil samples. The numbers of larvae and pupae can be reduced by harrowing or discing.

Soil treatment with DDT is generally effective against the larvae.

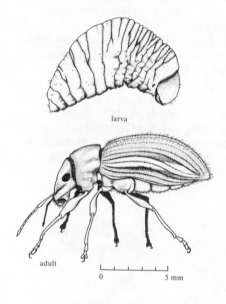

larva

adult

0       5 mm

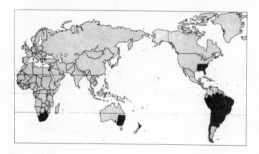

# Hypomeces squamosus (F.)

**Common name.** Gold-dust Weevil
**Family.** Curculionidae
**Hosts** (main). *Citrus* spp., and sweet potato.
  (alternative). Other species of *Ipomea*, and various other plants.
**Damage.** The adults eat notches out of the leaf margin.
**Pest status.** A common and widespread pest in S.E. Asia on several different crops, but seldom serious.
**Life history.** So far as is known the larvae live in the soil and feed on plant roots; pupation takes place in the soil.

The adult is a broad-nosed weevil of about 10–15 mm body length, greyish in colour and with the body surface covered with a fine golden-green 'dust' (the dust-like appearance is due to the body of the weevil being covered with tiny flattened circular iridescent setae).

In different parts of the world there are different broad-nosed weevils that inflict almost identical damage on the foliage of their host-plants by eating notches out of the edge of the leaf lamina; in all cases the damage is done by the adult weevils and in most cases the larvae live in the soil and inflict no noticeable damage; these weevils include *Hypomeces* in S.E. Asia, *Myllocerus* spp. in India, *Systates* spp.

in Africa, *Sitona* spp. in Europe and N. America, etc.
**Distribution.** This species is confined to S.E. Asia and the Philippines, but very similar species occur as mentioned above.
**Control.** Generally control is not required.

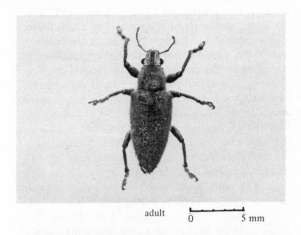

adult    0 ⊢——————⊣ 5 mm

adults on leaves of grapefruit

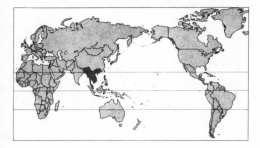

481

## Lissorhoptrus oryzophilus Kusch.

(= *L. simplex auctt.*)

**Common name.** Rice Water Weevil
**Family.** Curculionidae
**Hosts** (main). Rice
    (alternative). Various grasses and sedges.
**Damage.** Rice is attacked by both adults, feeding on the leaves, and larvae on the roots. The adults leave long scars on the leaf where the surface layers have been eaten. Larval damage is the more important, for the larvae eat the roots resulting in delayed crop maturity, stunting of plants and loss of grain yield.
**Pest status.** A serious pest of rice in the rice-growing areas of the USA, and also in Alberta, Canada.
**Life history.** Eggs are laid under water on the basal half of the submerged part of the leaf sheath, and occasionally on the roots. Each female may lay up to 35 eggs. The eggs hatch in 7–10 days.

The white larvae, with reduced legs, and brown head capsule, feed on the roots of the rice plant, and can apparently move as much as 15 cm through the soil. The larvae have paired dorsal hooks modified (spiracles) on the abdominal segments used for obtaining air from the plant roots. There are four larval instars, which take about 50 days to complete.

Pupation takes place in an earthen cocoon attached to the roots, and takes about 21 days.

The adults are small, broad-nosed weevils; when newly emerged they fly at night to adjacent fields of young rice. The over-wintering period is spent by the adult in grass or rice stubbles.

The entire life-cycle takes about 78 days in California: there are two generations per year there.
**Distribution.** Eastern and southern states of the USA, and California; and Alberta, Canada (CIE map no. A270). Now established in S. Japan (Kyushu).

Several other species of *Lissorhoptrus* are found attacking rice in S. America.
**Control.** Draining and drying of the rice fields is effective as a control measure but is generally not economically feasible.

Insecticides used as sprays, granules or seed dressings have been effective; the most successful method being aldrin as a seed dressing.

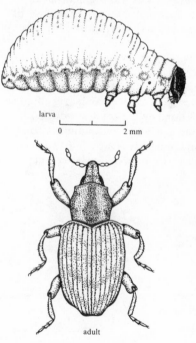

larva
0    2 mm

adult

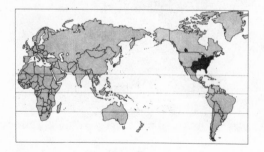

482

## Mecocorynus loripes Chevr.

**Common name.** Cashew Weevil
**Family.** Curculionidae
**Hosts** (main). Cashew trees.
(alternative). *Afzelia* sp. (Caesalpiniaceae).
**Damage.** The larva bores in the sapwood of the tree.
Brown-black gummy frass is seen on the trunk and
main branches; in severe attacks the tree may die.
**Pest status.** This is usually only a minor pest of
cashew in Coast Province, Kenya, but neglected
plantations are liable to be severely attacked.
**Life history.** Eggs are laid singly in small holes by the
female weevil in the bark of the trunk or branch.

The larva is a legless grub, whitish in colour
with a brown head. It bores through the bark and
moves downwards feeding on the sapwood of the
tree. At intervals it makes frass-ejection holes to the
exterior. Heavily infested trees become ringed by
damaged sapwood and eventually die. When the
larva becomes full-grown it constructs a pupal
chamber about 2 cm below the bark. The tunnel
from the chamber to the exterior is stuffed with
wood fragments before pupation.

The adult is a dark grey-brown weevil about
2 cm long, and of a knobbled appearance. It has fully
developed wings but is not known to fly.

The complete life-cycle takes about six months.
**Distribution.** Kenya, Mozambique, and Tanzania
only.

**Control.** Very severely infested trees should be
destroyed, in the following manner. All adult weevils
should be collected and destroyed; the tree should be
felled and de-barked to expose all the larval galleries;
all larvae and pupae should be killed; and after not
more than two months the tree should be burned.

Lightly infested trees can be treated by killing
all evident adults, cutting off bark to expose the
larval and pupal galleries, and then removing and
killing the larvae and pupae found. This treatment
should be repeated every month for a further six
months if required.

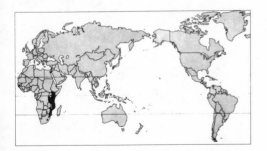

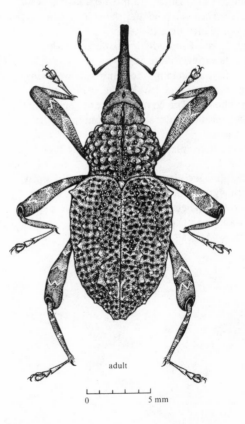

adult

0          5 mm

## Nematocerus spp.

**Common name.** Shiny Cereal Weevils
**Family.** Curculionidae
**Hosts** (main). Maize, barley, wheat, and other cereals.
(alternative). Coffee, tea, beans, and many other crops and plants.
**Damage.** The adult weevils feed on the leaves, making characteristic notch-like damage to the leaf margin. In severe attacks there can be almost complete defoliation. The larvae live in the soil and eat the roots, the underground stem, and germinating seeds. Seedlings are preferred as host plants, and most damage is done to the stem between the seed and the soil surface, although older plants are attacked.
**Pest status.** A pest of sporadic importance in Kenya, and found on a variety of crops.
**Life history.** Eggs are laid in a fold in the leaves. On hatching the young larvae drop to the ground and burrow into the soil. The larvae are white, legless, and with a brown head capsule; about 12 mm long when mature.

Pupation takes place in an earthen cell in the soil.

The adult weevil is 6–12 mm long, and is shiny with smooth elytra which are fused together, so the adults cannot fly. When they emerge from the pupal cells they walk on to the nearest host plant.
**Distribution.** E. Africa.

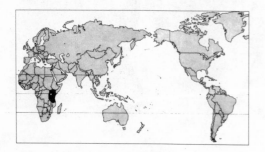

**Control.** Adults can be controlled by DDT dusts or sprays, but there are no insecticidal recommendations available for the larvae.

In E. Africa emphasis has been placed on various aspects of cultural control, such as the avoidance of double cropping, early planting, and application of fertilizer to marginal soils.

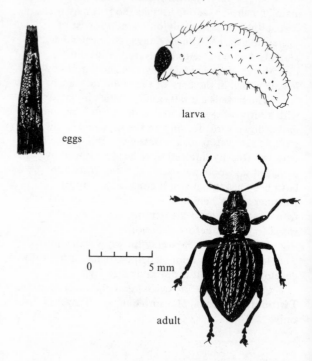

eggs

larva

0    5 mm

adult

## Odoiporus longicollis (Oliv.)

**Common name.** Banana Stem Weevil
**Family.** Curculionidae
**Hosts** (main). Bananas (*Musa* spp.).
    (alternative). None recorded.
**Damage.** The larvae bore in the pseudostem and the peduncles of the fruit, and the tunnels often become infected with rots. In heavily attacked plants the pseudostem is severely weakened and easily breaks.
**Pest status.** A serious pest of bananas in S.E. Asia and India.
**Life history.** Eggs are laid singly in small cuts in the pseudostem and the young larvae bore straight into it. The burrowing larvae make long tunnels through the pseudostem and in some cases right up into the

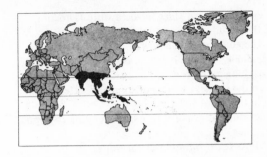

fruit bunch peduncles. In India and S.E. Asia some banana plants are attacked simultaneously by both species of banana weevil, so that both rhizome and pseudostem are tunnelled.

damage

485

Pupation takes place inside a fibrous cocoon inside the pseudostem, and takes about a week.

The adult is a small weevil, brown after emergence gradually turning black, about 10 mm in length, with an elongate narrow snout. It is long-lived and feeds on the tissues of the banana plant, often being found in the larval galleries.

**Distribution.** India, and S.E. Asia extending to Papua New Guinea.

**Control.** As for *Cosmopolites sordidus*.

larva and pupa

adult

0                    1 cm

# Rhynchophorus ferrugineus (Oliv.)

(= *R. schach* Oliv.)

**Common name.** Asiatic (Red) Palm Weevil; (Red Stripe Weevil)
**Family.** Curculionidae
**Hosts** (main). Coconut and oil palm.
     (alternative). Date, sago, and other species of Palmae.
**Damage.** The feeding larvae bore the crown of the palm and may destroy it. Initially the outer leaves turn chlorotic and die; this gradually spreads to the innermost leaves. Later the trunk becomes tunnelled and weakened, and may break.
**Pest status.** A fairly serious pest of coconut and oil palm throughout S.E. Asia.
**Life history.** Eggs are laid in the crown of the palm, often in holes made by other insects (*Oryctes* spp. etc.), or by man; the females may actively search for cut petioles as oviposition sites. Each female may lay 200–500 eggs. Hatching takes place after about three days.

    The larvae are yellowish-white, legless (apodous), rather oval in shape, with a reddish-brown head capsule; at maturity they are about 5–6 cm long. They penetrate the crown initially, and later to most parts of the upper trunk, making tunnels of up to 1 m in length. They are voracious feeders, and the damaged tissues soon turn necrotic and decay,

resulting in a characteristic unpleasant odour. As the galleries become more extensive the trunk weakens and in a storm the tree may easily be decapitated. The larval period lasts 2–4 months, but has been recorded as only 24 days when feeding on the nutritious palm 'cabbage'.

    Pupation takes place in a cocoon (80 × 35 mm) of fragments under the bark, the actual emergence hole being blocked with a fibrous plug. The pupal stage lasts for 14–28 days.

    The young adult stays in the cocoon for 8–14 days before emerging and starting to feed. The adult beetle is a large (but smaller than the other species of *Rhynchophorus*) reddish-brown weevil some 32–34 mm in length (full range 25–50 mm), and this species has either spots or a stripe of red on the thorax. The variation in size and colour pattern has led to some taxonomic confusion, and the *R. schach* of Wood (1968) is thought to be a synonym and colour variation of *R. ferrugineus*.
**Distribution.** Pakistan, India, Sri Lanka, S.E. Asia, to China, Taiwan, and the Solomon Isles (CIE map no. A258).
**Control.** See under *R. palmarum*.

0      2 cm
♀

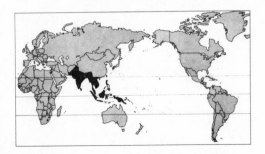

# Rhynchophorus palmarum (L.)

**Common name.** South American Palm Weevil
**Family.** Curculionidae
**Hosts** (main). Coconut and oil palm.

(alternative). Date, sago and other palms; also sugarcane.

**Damage.** The larvae burrow in the crown of the palm, feeding on the young tissues, and sometimes destroy the growing point, when the palm will die. The leaves turn chlorotic and die, and the trunk becomes tunnelled and weakened, and may break in a storm.

**Pest status.** A serious pest of palms in C. and S. America.

**Life history.** For details see *R. ferrugineus.*

The adult weevil is larger than the previous species being some 43–54 mm in length, on average, and is usually darker brown in colour, sometimes almost black. The usual weevil sexual dimorphism is clearly apparent in this genus.

**Distribution.** Recorded from Mexico, C. America, W. Indies, and the northern half of S. America (CIE map no. A259).

**Control.** Many cultural control methods can be applied against this pest in order to reduce the population size, such as elimination of breeding sites by restriction of physical injury to the palms, control of *Oryctes* etc., destruction of infested palms, trapping of adults, etc.

Recommended insecticides are aldrin and dieldrin sprayed on the crowns and trunks of the palms. Or, alternatively, demeton-*S*-methyl, para-dichlorobenzene, oxydemeton-methyl, or carbaryl, can be injected into the infested galleries, or the liquids can be injected into the trunk a little way above the infested region.

**Further reading**

Lever, R. J. A. W. (1969). *Pests of the Coconut Palm*, pp. 113–17. FAO: Rome.

Wood, B. J. (1968). *Pests of Oil Palms in Malaysia and their Control*, pp. 144–6. Inc. Soc. Planters: Kuala Lumpur.

0      2 cm

♀

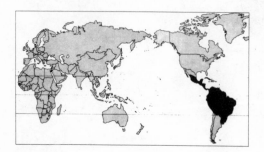

## Rhynchophorus phoenicis (F.)

**Common name.** African Palm Weevil
**Family.** Curculionidae
**Hosts** (main). Coconut and oil palm.
     (alternative). Date, sago, and other species of
Palmae.
**Damage.** The larvae bore in the crown and feed on
the shoot and young leaves, sometimes destroying
the growing point, and killing the palm. Leaves turn
chlorotic and die. The trunk eventually becomes
tunnelled and may break during a storm.
**Pest status.** A serious pest of palms in tropical Africa.
**Life history.** For details see *R. ferrugineus.*

     The adult is a large species, measuring some
40—55 mm in length, reddish-brown in colour, with
two reddish bands on the thorax.
**Distribution.** This species is confined to tropical
Africa, and has been recorded from Ivory Coast,
Sierra Leone, Nigeria, Angola, Ghana, Zaïre, and
E. Africa.

     There are actually many different species of
*Rhynchophorus* to be found, many in S.E. Asia and
Africa, and many attacking species of Palmae, so field
identifications can only be tentative (Wattanapongsiri,
1966).
**Control.** See under *R. palmarum.*

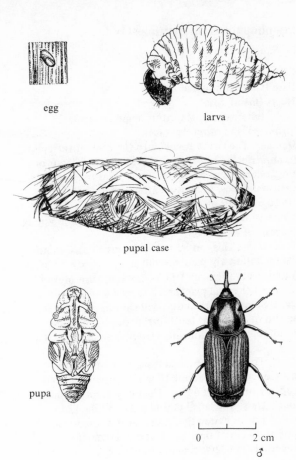

egg

larva

pupal case

pupa

0     2 cm

♂

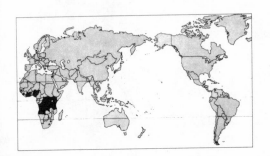

## Scyphophorus interstitialis Gyll.

(= *S. acupunctatus* Gyll.)

**Common name.** Sisal Weevil
**Family.** Curculionidae
**Hosts** (main). Sisal

    (alternative). Mauritius hemp (*Furcraea gigantea*), and other Agavaceae.
**Damage.** The larvae tunnel into the bole of the spike in nurseries, and in field sisal grey patches occur on the undersides of the lower leaves, and the plant eventually dies. In a severe attack there may be up to 60% loss of plants in nurseries. Adult damage consists of groups of feeding punctures on the young leaves.
**Pest status.** This can be a serious pest of sisal in nurseries, and newly planted field sisal.
**Life history.** Eggs are laid 2–6 at a time in mushy tissue of the sisal plant; each female lays 25–50 eggs. Sometimes the adult will eat out a small cavity in the spike so that local rotting occurs, making a suitable oviposition site. Hatching requires 3–5 days.

    There are five larval instars, and the fully developed larva is about 18 mm long, with a head capsule breadth of 4 mm. The body is soft, crinkled, and legless. The larval period is 21–53 days.

    For pupation the larva makes a rough cocoon out of pieces of fibre and leaf debris cemented together. The larva remains inside the cocoon for a few days before pupating, passing first through a pre-pupal stage of 3–10 days. The pupal stage lasts for 7–23 days, typically 12–16.

    The adults are small, dull black weevils, varying in length (9–15 mm). The adults tend to remain in the area of their origin and generally dispersal is slow. They are long-lived, and have been kept in captivity for ten months.

    The total life-cycle takes 50–90 days; and there may be four or five generations per year.
**Distribution.** Kenya, Tanzania, Sumatra, Java, Hawaii, Mexico, southern USA (Arizona, Texas, California, and Colorado), W. Indies, and northern S. America (Colombia, and Venezuela) (CIE map no. A66).
**Control.** The rotting boles should be treated with dieldrin. Adults feeding at the base of the spike can be killed with aldrin, and this insecticide should also be used to protect the young field sisal.

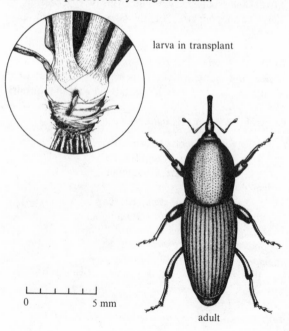

larva in transplant

0           5 mm

adult

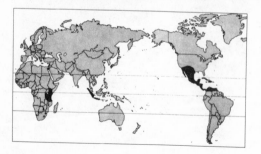

490

# Sitophilus oryzae (L.)

(= *Calandra oryzae* L.)

**Common name.** Rice Weevil
**Family.** Curculionidae
**Hosts** (main). Rice
     (alternative). Maize and other cereals in storage.
**Damage.** The developing larva lives and feeds inside the grain hollowing it out in the process. In rice (the preferred host) the entire grain is usually destroyed by the time the adult emerges.
**Pest status.** A very serious major (primary) pest of stored rice and other cereals in the warmer parts of the world.
**Life history.** Identical to *S. zeamais* so far as is known, but preferably taking place in rice.
**Distribution.** The two species, *Sitophilus oryzae* and *S. zeamais*, are virtually cosmopolitan throughout the warmer parts of the world, extending as far north as Japan and southern Europe. They have been recorded from Europe (Spain, Italy, Turkey); Africa (Angola, Arabia, S. Africa, E. Africa, Ethiopia, Ghana, Madagascar, Madeira, Mauritius, Morocco, Mozambique, Nigeria); Asia (India, Tibet, Malaysia, Molucca Isles, Borneo, Japan); Australasia (Australia – New South Wales, Queensland, W. Australia; New Guinea, New Zealand, Pacific Islands); USA (Texas, Florida); Central America (Costa Rica, Mexico, W. Indies); S. America (Argentina, Brazil, Guyana, Honduras, Chile, Ecuador, Guatemala, Nicaragua, Venezuela).

In Europe they are replaced by the temperate Palaearctic species *S. granarius* which is distinguished by the punctate sculpturing on the prothorax and elytra, and by the fact that it is wingless and hence cannot fly. For map, see *S. zeamais*.
**Control.** Infested buildings should be thoroughly cleaned and sprayed with BHC or malathion, and any parts of the building which cannot be reached with sprays should be fumigated with the use of DDT/

γ-BHC smoke generators. Infested grain can be mixed with malathion w.p. which is generally successful in achieving control.

Fumigation of infested grain with methyl bromide, or an ethylene dichloride and carbon tetrachloride mixture is effective but should only be carried out by approved operators because of the toxicity hazards.
**Further reading**
Halstead, D. G. H. (1964). The separation of *Sitophilus oryzae* (L.) and *S. zeamais* Motschulsky (Col., Curculionidae), with a summary of their distribution. *Ent. mon. Mag.* **99**, 72–4.

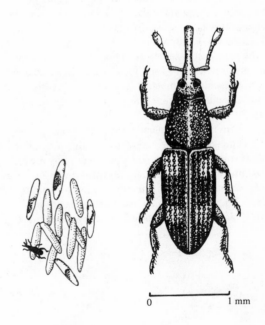

0              1 mm

## Sitophilus zeamais Motsch.

(= *Calandra zeamais* Motsch.)

**Common name.** Maize Weevil
**Family.** Curculionidae
**Hosts** (main). Maize
     (alternative). Sorghum, rice, and other cereals
in storage.
**Damage.** A thin tunnel is bored by the larva from the
surface to the inside of the grain. Circular exit holes
on the surface of the grain kernel are characteristic.
**Pest status.** This species is a very serious major
(primary) pest of stored grain throughout the warmer
parts of the world. Infestation often starts in the
field, and is later carried into the grain stores.
**Life history.** Eggs are white and oval. The female lays
the eggs inside the grain by chewing a minute hole in
which each egg is deposited, followed by the sealing
of the hole with a secretion. These eggs hatch into
tiny grubs which stay and feed inside the grain and
are responsible for most of the damage. Mature
larvae are plump, legless and white, about 4 mm
long.

    Pupation takes place inside the grain.

    The adult beetle emerges by biting a circular
hole through outer layers of the grain. They are
small brown weevils, virtually indistinguishable from
each other, about 3.5–4.0 mm long with rostrum and
thorax large and conspicuous. The elytra are uniformly dark brown. Each female is capable of laying

300–400 eggs, and the adults live for up to five
months and are capable fliers.

    The life-cycle is about five weeks at 30°C and
70% RH; optimum conditions for development are
27–31°C and more than 60% RH; below 17°C
development ceases. *S. zeamais* has only recently
been separated off from *oryzae* and the two species
are only distinguishable by their genitalia.
**Distribution.** See *S. oryzae*.
**Control.** See *S. oryzae*.

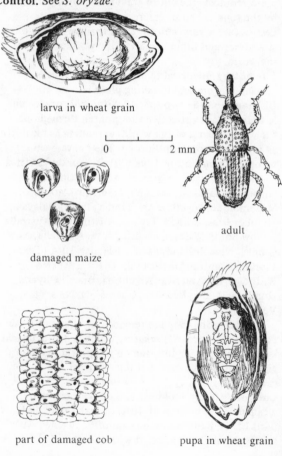

larva in wheat grain

0      2 mm

damaged maize

adult

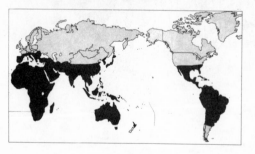

part of damaged cob

pupa in wheat grain

492

## Sternochetus mangiferae (F.)

(= *Cryptorynchus mangiferae* F.)
(= *Acryptorynchus mangiferae* F.)

**Common name.** Mango Seed Weevil (Mango Stone Weevil)
**Family.** Curculionidae
**Hosts** (main). Mango
(alternative). None recorded.
**Damage.** There are no external symptoms of attack by this pest. Infested fruits usually fall.
**Pest status.** The effect of the weevil on yield appears in most years to be quite small. On certain varieties of mango many seeds may be destroyed without the edible part of the fruit being affected.
**Life history.** The female weevil makes small cuts in the skin of young fruits and inserts a single egg through each cut. The cut normally heals over completely and becomes invisible to the unaided eye.

The larva is a white, legless grub with a brown head. On hatching from the egg it bores through the pulp of the fruit and into the developing seed where it feeds until mature.

Pupation takes place in the seed within the stone of the fruit.

Adult is dark brown with paler patches, some 6—9 mm long. It is a small, stout weevil with a reduced head, and the body covered with scales.

The total life-cycle takes 40—50 days.

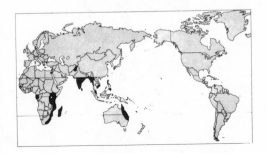

**Distribution.** Africa (Gabon, E. and S. Africa, Madagascar, Mauritius); Pakistan, India, Bangladesh, Sri Lanka, Burma, Malaya, S. Vietnam, Philippines, E. Australia, New Caledonia, and Hawaii (CIE map no. A180).

A closely related species *S. frigidus* F. occurs in S.E. Asia, but it inhabits the pulp of the fruit and not the seed, hence its name of Mango Weevil.
**Control.** All fallen fruits should be collected and destroyed twice a week. Before planting a mango stone the husk should be cut off to avoid damaging the embryo; if the weevil larva is feeding upon the cotyledons it should be removed, and the seed planted immediately.

Chemical control measures have not proved to be practical against this pest.

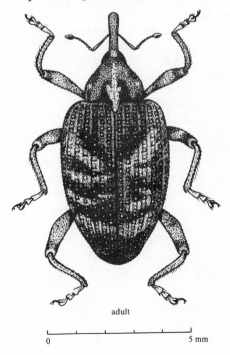

adult

0                                              5 mm

493

# Systates pollinosus Gerst.

**Common name.** Systates Weevil
**Family.** Curculionidae
**Hosts (main).** *Citrus* and coffee species.
    (alternative). Similar leaf damage is seen on a wide range of cultivated and wild plants.
**Damage.** The edges of attacked leaves have characteristic fjord-like indentations, where the adult weevils have eaten away the lamina.
**Pest status.** A minor pest of all cultivated *Citrus* and coffee species. Young plants are particularly liable to be attacked and often receive a severe setback.
    Several other closely related species of *Systates* are equally important as pests.
**Life history.** Egg, larval and pupal stages have not been recorded from either *Citrus* or coffee; they presumably all occur in the soil.
    The adult is a black weevil about 12 mm long with a swollen, rounded abdomen, and long, thin, elbowed antennae. It cannot fly, and is rarely seen during the daytime. At night it feeds on the edges of leaves producing the characteristic feeding damage. Daylight hours are usually spent in the mulch or loose soil near the *Citrus* or coffee trees.
**Distribution.** E. Africa.
**Control.** A persistent contact or stomach-acting insecticide must be used both as a foliar spray and on the mulch. The usual insecticides recommended are DDT and dieldrin. For *Citrus* control measures are

only required on nursery stock and trees in the first year or two after transplanting.

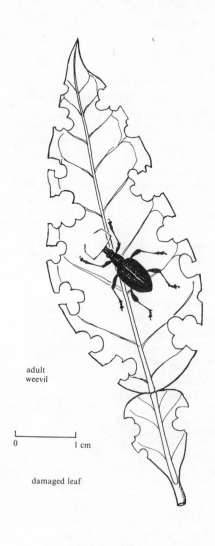

adult weevil

0         1 cm

damaged leaf

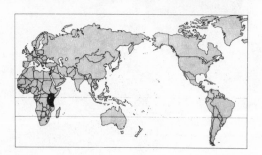

494

# Hypothenemus hampei (Ferr.)

(= *Stephanoderes hampei* Ferr.)

**Common name.** Coffee Berry Borer
**Family.** Scolytidae
**Hosts** (main). *Coffea* spp.

(alternative). Various Rubiaceae and Leguminoseae, including *Phaseolus*, and *Hibiscus* spp.

**Damage.** One or more round holes can be seen near the apex of large green or ripe berries. The damaged beans, which have a distinctive blue-green staining, contain up to 20 larvae of different sizes.

**Pest status.** A serious pest of *robusta* and low-altitude *arabica* coffee in many countries.

**Life history.** Eggs are laid in batches of 8–12 in chambers cut in the hardened maturing coffee bean. Each female lays 30–60 eggs over a period of 3–7 weeks. The eggs hatch in 8–9 days.

The larvae are legless, white with brown heads. They feed by tunnelling in the tissues of the beans. The male larva develops through two instars in 15 days, and the female through three instars in 19 days.

The naked pupal stage is passed in 7–8 days in the larval galleries.

The adult female beetle is about 2.5 mm long, and the male about 1.6 mm. Females are more numerous (sex ratio about 10:1) and fly from tree to tree to oviposit. The males are flightless and remain in the berry, fertilizing females of the same brood.

Infestations are carried over between peaks of fruiting by breeding in over-ripe berries left on the tree or fallen to the ground.

**Distribution.** Recorded from tropical Africa from W. through to E., Sri Lanka, S.E. Asia, Indonesia, Papua New Guinea, New Caledonia, Caroline, Society and Mariana Isles, and S. America (Brazil, Colombia, and Surinam) (CIE map no. A170).

**Control.** Heavy shade, from either shade trees or inadequately pruned coffee causes conditions unsuitable for the natural enemies of the borer and should be removed. All over-ripe or dried berries should be removed and destroyed. Old crop remains should be stripped completely. These cultural measures, efficiently applied, should be sufficient to control Berry Borer. Dieldrin or BHC foliar sprays should only be regarded as a supplement to the cultural measures.

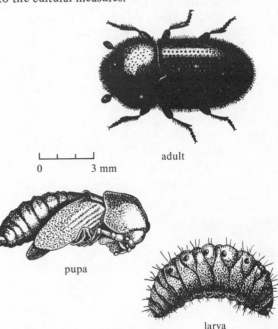

0        3 mm        adult

pupa

larva

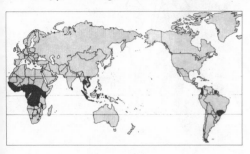

495

# Xyleborus fornicatus (Eichh.)

**Common name.** Tea Shot-hole Borer
**Family.** Scolytidae
**Hosts** (main). Tea
     (alternative). Cocoa, avocado, *Citrus*, castor, rubber, cinchona, etc.

**Damage.** The adult beetles bore in woody stems generally about 1–2 cm in thickness, and they excavate a system of tunnels which become infected with fungus; this is one of the ambrosia beetles and so the larvae feed on the gallery fungus rather than on the woody tissues of the host. Infestation of stems and branches usually leads to wilting and often the branch breaks at the infestation site; pathogenic fungi are often carried by ambrosia beetles (for example, Dutch Elm Disease in Europe). Fungal attack usually discolours the wood in a very characteristic manner.

**Pest status.** A primary pest of tea, cocoa, etc., for this species usually attacks healthy trees (many Scolytidae are really secondary pests in that they infest sickly or moribund trees), and because of the fungal infections associated the host tree may be killed.

**Life history.** The adult beetles bore galleries in the sapwood under the bark of the host and lay the eggs in the galleries. The fungal spores for the ambrosia are carried in special pockets in the beetles' bodies. The larvae feed on the ambrosia fungus in the gallery. Pupation takes place within the gallery.

     Male beetles are flightless and usually there is a 1:10 sex ratio, the numerous females being fertilized by the few males in the breeding gallery before dispersal. The females make dispersal flights to new trees during daylight. The adult beetles are small (4–5 mm long), black in colour, cylindrical in shape, and the male usually much smaller than the female, and, as mentioned, flightless.

     The life-cycle takes 30–35 days, including the preparation of the gallery for oviposition, and there are some 30–50 offspring per gallery (per female).

**Distribution.** This species is found from Madagascar, India, S.E. Asia, including Papua New Guinea (CIE map no. A319).

     Four other species are of particular importance agriculturally, and the adults are virtually indistinguishable from *X. fornicatus*. They are:
*X. perforans* (Woll) – Coconut Shot-hole Borer; pantropical (CIE map no. A320)
*X. ferrugineus* (F.) – many hosts; pantropical (CIE map no. A277)
*Xylosandrus compactus* (Eichh.) – Black Twig Borer; widespread (CIE map no. A244)
*X. morigerus* (Bldf.) – Brown Coffee Borer; S.E. Asia (CIE map no. A292).

**Control.** Cultural methods have met with little success, and in the past sprays of dieldrin with added sticker, sometimes with Bordeaux mixture, have given adequate levels of control.

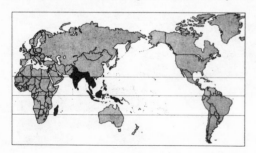

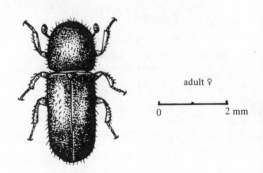

adult ♀

0          2 mm

# Class **ARACHNIDA**

## Order **ACARINA**

These arthropods belong to a separate class from the Insecta called the Arachnida. They are characterized by having four pairs of legs in the adult stage. The body is divided into proterosoma (gnathosoma and propodosoma) and hysterosoma, instead of head, thorax and abdomen. The first stage in the life-cycle is a six-legged larva which on moulting becomes an eight-legged nymph.

## Family **Tetranychidae**

(Spider Mites) Sometimes referred to as the Red Spider Mites because many of the species are bright red in colour; others are green or orange. They are all plant feeders and resemble small, reddish spiders. The eggs are relatively very large, often red, globular and scattered singly over the leaf surface. They are of moderate size (for mites), being about 0.8 mm long, with a soft integument without skeletal plates, and there is a pair of eyes on either side of the propodosoma. There are many important pest species both in the tropics and temperate regions. Attacked plants are often covered with an extensive fine webbing.

## Family **Tenuipalpidae** (= Phytoptipalpidae)

Very small, reddish mites (0.2–0.3 mm long), which feed on plants. A suture separating the propodosoma from the hysterosoma may or may not be present. Body form varies somewhat within this group. The adults usually have four pairs of legs but some genera only have three pairs.

## Family **Tarsonemidae**

Tarsonemids have a segmented body and an anterior dorsal shield which lacks a roof-like projection. They are placed in the suborder Trombidiformes, in part because of their anatomical degeneration. Heteromorphic males occur in this family. They have a simplified life-cycle in which the nymphal stage is quiescent; the active feeding larva transfers through the quiescent nymphal stage into a mature adult. There are several important polyphagous pest species.

## Family **Eriophyidae**

(Gall Mites) These mites are also in the suborder Trombidiformes; they are minute in size, rather worm-like, and possess only two pairs of anterior legs. The propodosoma is shield-like and has distinctive specific patterns; the hysterosoma is elongated and annulate. As a group these mites are either free-living on plants or gall makers; amongst the galls they produce are the characteristic 'erinia' on the under-surface of leaves. These moss-like outgrowths are induced by the mites which then inhabit them.

# Eutetranychus orientalis (Klein)

**Common name.** Oriental Mite
**Family.** Tetranychidae
**Hosts** (main). *Citrus* spp.

(alternative). A wide range of alternative hosts including castor, cotton, *Ficus*, *Morus*, cucurbits, frangipani, *Croton*, and *Euphorbia*.
**Damage.** Upper surfaces of fully expanded leaves turn a yellow or red-brown.
**Pest status.** A sporadically serious pest of *Citrus*, especially at lower altitudes.
**Life history.** Eggs, which are subspherical and pale brown, are usually laid along the main veins on the upper surface of the leaves. They are just visible to the unaided eye, being a diameter of 0.14 mm.

The six-legged larva is brownish.

After a resting phase the nymph emerges from the larval skin. The female mite passes through a second resting stage and second nymphal stage before becoming adult. The male has only one nymphal stage. Nymphs have eight legs and are similar in general form to the adult.

The adult female is round-bodied, brownish and almost 0.5 mm long. It emerges from the second nymphal skin after a third resting period. The male has a more slender and elongate body and is smaller and redder than the female. The females have dorsal striae of the propodosoma more or less parallel, and the lateral setae are moderately slender and spatulate.

All active stages feed and moult on the upper surfaces of fully expanded leaves. The cast skins remain stuck to the leaf surface and appear as small white irregular specks.

The total life-cycle probably takes between one and three weeks, according to temperature.
**Distribution.** Africa (Egypt, Senegal, Aden, E. and S. Africa, Sudan); Cyprus, Jordan, Lebanon, Afghanistan, Israel, Iran, India, Bangladesh and Taiwan.
**Control.** The recommended control measures are foliar sprays of dicofol or white oil.

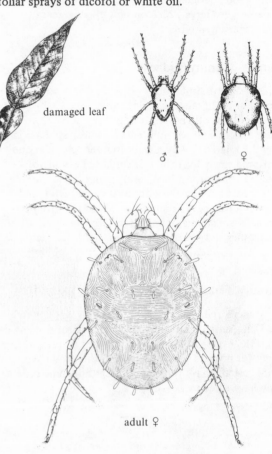

damaged leaf

♂          ♀

adult ♀

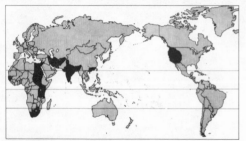

## Oligonychus coffeae (Nietner)

**Common name.** Red Coffee Mite (Red Tea Mite)
**Family.** Tetranychidae
**Hosts** (main). Tea, and coffee.

(alternative). A wide range of trees and shrubs including castor, and *Grevillea*.

**Damage.** The upper surface of fully hardened leaves turn a yellowish-brown, rusty or purple colour. If the tea bush is drought-stressed, flush leaves may also be attacked.

**Pest status.** An occasional pest of tea in many areas; attacks are usually confined to a few bushes. This pest is sometimes called the Red Tea Mite.

**Life history.** Eggs are laid singly on the upper leaf surfaces, often near a main vein; they are just visible to the unaided eye. They are nearly spherical but have a fine filament projecting on the upper side. They are bright red, changing to orange just before hatching. They hatch after 8–12 days.

The larva is six-legged, almost spherical, orange, and slightly larger than the egg.

There are two nymphal stages; the protonymph and the deutonymph. They are more oval than the larva and have four pairs of legs. The front part of the body is red, the posterior half reddish-brown or purple. The total period spent in the larval and nymphal stages is 9–12 days.

Adults are little less than 0.5 mm, and are similar in coloration to the nymph. Female mites, which are usually more numerous than the males, usually lay 4–6 eggs per day for 2–3 weeks, starting immediately after the final moult.

All active stages feed together on the upper surfaces of the leaves, and the cast skins of the larval and nymphal stages remain stuck on to the leaf and may be seen with the unaided eye as irregular white spots.

This species is scarcely separable from *Tetranychus cinnabarinus* without the use of microscopic taxonomic characters.

**Distribution.** Widely scattered records have been made from Africa (Egypt, Ethiopia, E. Africa, Malawi, Zaïre and S. Africa), USSR (Transcaucasia), Asia (S. India, Sri Lanka, Burma, Indo-China, Java, Sumatra, and Taiwan), Australia (Brisbane), USA (Florida), C. America (Costa Rica), and S. America (Colombia, and Ecuador) (CIE map no. A165).

**Control.** The usual recommendation is to spray the foliage with either dicofol or dimethoate. The other acaricides referred to for control of *T. cinnabarinus* would probably be equally effective against this pest.

**Further reading**

Le Pelley, R. H. (1968). *Pests of Coffee*, pp. 398–9. Longmans: London.

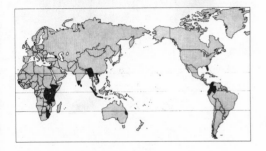

## Panonychus citri (McG.)

**Common name.** Citrus Red Spider Mite
**Family.** Tetranychidae
**Hosts** (main). *Citrus* spp.
    (alternative). Broad-leaved evergreen ornamentals; peach and various deciduous fruits.
**Damage.** Symptoms of attack are silvering of leaves, with severely damaged leaves developing dead areas – the dead leaves sometimes remain on the tree for a long time. Die-back of young twigs may occur and fruits on affected branches may drop.
**Pest status.** A major pest of *Citrus* in California and Florida especially; damage is most evident on the leaves, and is of little consequence to mature fruit, but severe infestation weakens the trees and may be followed by a poor crop in the next season.
**Life history.** The egg is minute, bright red, spherical, stalked, and with threads radiating to the leaf surface, which is characteristic of the species. The egg period lasts 6–20 days, according to season.
    Both nymphs and adults are dark red, rounded, and have characteristic white setae arising from tubercles on the body. Adult males are about 0.5 mm and females 0.6 mm. The nymphal period lasts 14 days under optimum conditions. A temperature of 25°C and RH of 60–70% favours breeding. Dispersal is mainly by wind, and transportation of infested nursery stock.

**Distribution.** S.W. Europe, Tunisia, Iran, Turkey, S. Africa, India, Sri Lanka, Vietnam, S. China, Japan, Australia (Queensland), New Zealand, Hawaii, S. USA (California, Florida, etc.), W. Indies, and S. America (Brazil, Argentina, Chile, Peru, and Colombia) (CIE map no. A192).
**Control.** Control is similar to that for *T. cinnabarinus* (p. 502).

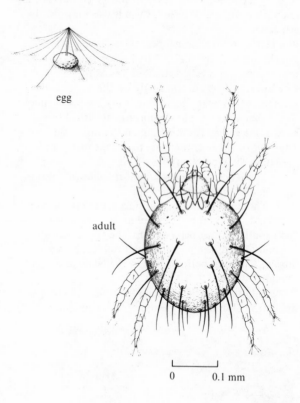

egg

adult

0        0.1 mm

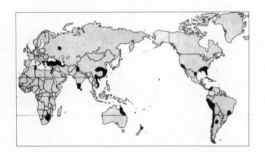

# Tetranychus cinnabarinus (Boisd.)

(= *T. telarius* (L.))

**Common name.** Carmine Spider Mite (Tropical Red Spider Mite)
**Family.** Tetranychidae
**Hosts** (main). Cotton

(alternative). A very wide range of wild and cultivated plants are attacked by this pest.
**Damage.** Clusters of yellow spots are visible on the upper side of the leaf especially between the main veins near the leaf stalk. Later the affected areas spread, the leaf reddens and finally withers and is shed. Red or greenish mites just visible to the unaided eye can be seen on the underside of the leaf.
**Pest status.** An occasional pest of cotton in many areas. A single bush or a small patch of bushes are often severely attacked leaving the rest of the field undamaged.
**Life history.** The eggs are spherical, whitish, about 0.1 mm in diameter. They are laid singly on the underside of leaves stuck to the leaf surface or on the strands of silken web spun by the adult mites. They hatch after 4—7 days.

The larva is six-legged, pinkish, and slightly larger than the egg. The larval stage lasts 3—5 days.

There are two nymphal stages, the protonymph and the deutonymph. They have four pairs of legs and are greenish or reddish. The total nymphal period lasts 6—10 days.

The adult females are oval, red or greenish, and 0.4—0.5 mm long. The males are slightly smaller. Fine strands of silk are spun by the adults which form an open web above the leaf surface. The adult female may live for three weeks and lay 200 eggs.

All active stages feed together on the lower sides of the leaves between the main veins. Yellow spots appear where a group have been feeding together.
**Distribution.** Africa, Middle East, Pakistan, India, Sri Lanka, S.E. Asia, Australasia, S. USA, Japan, C. and S. America; generally distributed within the tropics and subtropics.

The Linnean name *T. telarius* has now been sunk taxonomically and replaced by two species, *T. cinnabarinus* the pantropical species which has a red-coloured summer female, and *T. urticae* the temperate species with a green-coloured summer female. The winter females (green) of both species

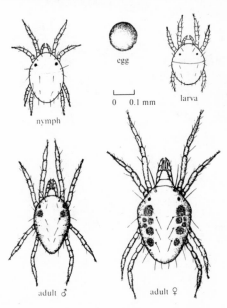

egg

larva

nymph

0    0.1 mm

adult ♂    adult ♀

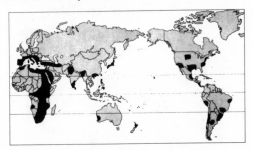

501

are indistinguishable, even by the taxonomic experts, and the precise status of these two 'species' is not yet really clear, but both names are accepted for use. Most distribution records are not clearly attributable to either species.

**Control.** A predaceous mite *Phytoseilus riegeli* (Phytoseidae) has been used in glasshouses in Europe for the control of *T. urticae* with considerable success. This predator has been used in field control of *T. cinnabarinus* in areas of both Kenya and Uganda.

Chemical control measures are not usually required; but, if very heavy infestations are found early in the season, foliar sprays of dimethoate are recommended. Other pesticides generally effective against this pest are demeton-*S*-methyl, derris, dichlorvos, dicofol, formothion, malathion, oxy-demeton-methyl, petroleum oil, phosphamidon, quinomethionate, tetradifon and vamidothion, according to the crop concerned and the conditions at the time of spraying.

**Further reading**

Jeppson, L.R., Keifer, H.H. & E.W. Baker (1975). *Mites Injurious to Economic Plants,* pp. 221–222. University of California Press: Berkeley.

# Brevipalpus phoenicis (Geijskes)

**Common name.** Red Crevice Tea Mite
**Family.** Tenuipalpidae (= Phytoptipalpidae)
**Hosts (main).** Tea, and *Citrus*.

(alternative). Coffee, rubber, *Phoenix* spp., *Grevillea* shade trees, and other trees.
**Damage.** Corky areas are to be found on the undersides of leaves, especially between the main veins at the petiole end of the leaves; the leaves may then dry up and will be prematurely shed. Numerous tiny red mites can be seen in the bark crevices of the new wood.
**Pest status.** A sporadic pest of tea in many parts of the world; a few bushes may be very heavily attacked, leaving the majority almost free from attack. There are a number of records from glasshouses in Europe.
**Life history.** The eggs are oval, about 0.1 mm long and bright red. They are stuck firmly to the undersides of leaves or in crevices in young bark. They hatch after about three weeks.

The larva is a six-legged, scarlet creature which grows to a length of 0.15 mm.

After a resting stage of a few days a protonymph emerges from the larval skin. After feeding for about a week there is a second resting stage, from which emerges the deutonymph. Both nymphal stages are flat-bodied, oval in outline and scarlet. They both have four pairs of legs.

After a further resting period the adult mite emerges from the skin of the deutonymph. Adults resemble the nymphs but are somewhat larger, reaching a length of 0.3 mm. Each female mite may lay an average of one egg per day for a period of 7–8 weeks.

All active stages feed on the undersides of leaves, especially between the main veins at the petiole end. Leaves of all ages are attacked.

The total life-cycle takes about six weeks.
**Distribution.** Possibly a cosmopolitan species throughout the warmer parts of the world, but the records to date are very scattered. (CIE map no. A106).
**Control.** Recommended control practice is to spot-spray the affected bushes with either dicofol or chlorobenzilate.

In severe infestations the spray should be repeated 2–3 weeks later.

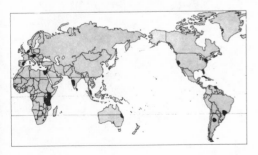

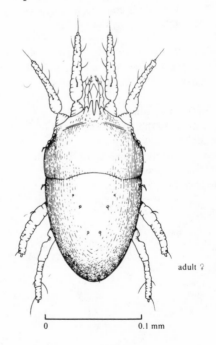

adult ♀

0          0.1 mm

## Polyphagotarsonemus latus (Banks)

(= *Hemitarsonemus latus* (Banks))

**Common name.** Yellow Tea Mite (Broad Mite)
**Family.** Tarsonemidae
**Hosts** (main). Tea, and cotton.

    (alternative). Coffee, jute, tomato, potato, castor, beans, peppers, avocado, *Citrus* and mango.
**Damage.** The blades of flush leaves are cupped or otherwise distorted, with corky brown areas between the main veins on the underside of the leaf. The corky areas are often bounded by two brown lines parallel to the main vein.
**Pest status.** A sporadically serious pest in tea nurseries. Generally a polyphagous minor pest.
**Life history.** Eggs are laid singly on the undersides of flush leaves. They are oval in outline but flattened on the lower side. The upper side is covered with five or six rows of white tubercles. The eggs are 0.7 mm long and hatch in 2–3 days.

    The larvae are minute, white, and pear-shaped. They usually remain feeding near the egg shells from which they have emerged. The larval stage lasts 2–3 days.

    The larva turns into a quiescent pupal stage (nymph) which is stuck to the underside of the leaf. This stage also lasts 2–3 days.

    Female pupae are usually picked up by adult males and carried on to leaves which have newly opened from the bud. Male pupae are not usually moved but when the adult male mites emerge they migrate to younger leaves. The female pupa is carried cross-wise at the tip of the male abdomen forming a T-shape. Adults are yellow and about 1.5 mm long. A typical female mite lives for about ten days, laying 2–4 eggs per day.

    All active stages are found on the underside of flush leaves. A favourite site is in the groove between the two halves of the leaf lamina before it has unfurled. It is this feeding which causes the parallel brown lines when the blade is fully expanded.
**Distribution.** Virtually cosmopolitan, but records are sparse in some areas (CIE map no. A191).
**Control.** Reduction of overhead shade in tea nurseries appears to reduce tea mite attacks.

    Recommended chemical control measures are a foliar spray of dicofol; only the flush leaves need be sprayed. In very severe attacks a repeat spray may be needed one week after the first.

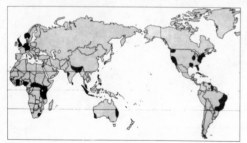

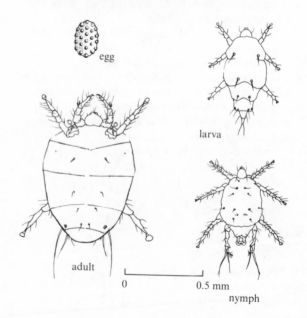

egg

larva

adult

0     0.5 mm

nymph

## Aceria sheldoni (Ewing)

**Common name.** Citrus Bud Mite
**Family.** Eriophyidae
**Hosts** (main). *Citrus* spp., especially lemon.
    (alternative). None recorded — it appears to be confined to *Citrus*.
**Damage.** Twigs are bunched and twisted, blossoms misshapen, and fruits often assume grotesque shapes.
**Pest status.** A sporadically serious pest of all *Citrus* spp. in various parts of the world. Especially serious on grapefruit and lemon trees.
**Life history.** The eggs are extremely small, whitish, spherical objects (about 0.04 mm).

    The larva is a minute triangular shape, about 0.1 mm long.

    After a quiescent period the nymph emerges from the larval skin. It is about 0.13 mm long, cylindrical, but tapering at the posterior end. Only two pairs of legs are present, and these at the anterior end.

    After a second quiescent period the adult emerges. It generally resembles the nymph but is yellowish or pinkish and about 0.18 mm long.

    All stages of mite are found in protected places between the leaves or scales or developing buds. The total life-cycle takes 1—3 weeks, according to temperature.
**Distribution.** Recorded from the Mediterranean (Spain, Cyprus, Israel, Turkey, Italy, Sicily, Greece), Africa (Algeria, Zaïre, Kenya, Libya, S. Africa, Zimbabwe, Tunisia, Uganda), Java, Australia (Queensland, New South Wales), Hawaii, USA (Florida, California), S. America (Brazil and Argentina) (CIE map no. A127).
**Control.** The usual control recommendation is to spray infested trees at periods of blossom or flush growth with either lime-sulphur or chlorobenzilate, as a full-cover spray using as high a nozzle pressure as possible.

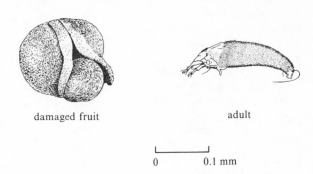

damaged fruit                          adult

0           0.1 mm

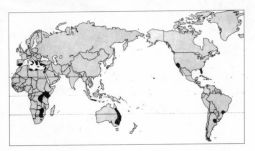

## Phyllocoptruta oleivora (Ashmead)

**Common name.** Citrus Rust Mite
**Family.** Eriophyidae
**Hosts** (main). *Citrus* spp.

(alternative). Apparently confined to *Citrus* as a host, but *Fortunella* is recently recorded.

**Damage.** Lemon fruits become a silver colour; oranges and grapefruits russet-coloured. The skins of injured fruit are thicker than usual and the fruits are smaller. Leaves and young shoots may also be damaged.

**Pest status.** A common and locally serious *Citrus* pest in many countries.

**Life history.** Eggs are minute, spherical, whitish, and laid in depressions on the fruit or leaves. They hatch after 3—7 days.

The larva is very small, yellowish, and has a worm-like tapering cylindrical shape with two pairs of short legs at the anterior end. Larval stage lasts 2—4 days.

After a quiescent period the nymph emerges from the larval skin. The nymph is similar to the larva but more yellow and slightly larger. The nymphal stage also lasts 2—4 days.

After a second quiescent period the adult emerges from the nymphal skin. It generally resembles the nymph but is somewhat darker. Adults are about 0.1 mm long. Males have not been recorded. The female lives for about two weeks, during which time she lays 20—30 eggs.

**Distribution.** Almost cosmopolitan throughout the warmer parts of the world; recorded from Europe (Italy, Malta, Yugoslavia), Near East (Cyprus, Iran, Israel, Gaza Strip, Jordan, Lebanon, Turkey, Syria), USSR (Georgia, Krasnodar Krai), Africa (Angola, Zaïre, Kenya, Madagascar, Malawi, Mauritius, Mozambique, Nigeria, Senegal, S. Africa, Tanzania, Uganda, Zambia), Bangladesh, India, China, Japan, Philippines, Taiwan, S. Vietnam, Australia (Queensland, New South Wales, Western Australia), Fiji, Hawaii, Cook Isles, USA (Florida, California, Alabama, Louisiana, Mississippi, Texas), C. America (Mexico, Guatemala), W. Indies, and S. America (Argentina, Brazil, Colombia, Ecuador, Peru, Uruguay, Venezuela).

**Control.** Fruits should be examined regularly with a good hand lens from blossom shed onwards. They are usually damaged when about 20—30 mm in diameter. The recommended sprays are either lime-sulphur or chlorobenzilate, as sprays to run-off taking particular care to wet the fruits. The fruits should be examined after 4—5 days with a hand lens and if living mites are present the spray should be repeated.

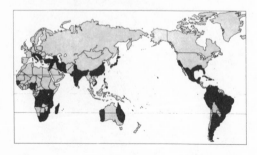

damaged fruit

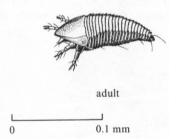

adult

0    0.1 mm

# 8 Tropical crops and their pest spectra

## CROPS INCLUDED

Almond
Apple
Apricot
Avocado
Bamboo
Bananas
Beans and grams
Betel palm
Betel-pepper
Brassicas
Breadfruit
Capsicums
Cardamom
Cashew
Cassava
Castor
Chickpea
Cinchona
Cinnamon
Citrus
Clove
Cocoa
Coconut
Cocoyam
Coffee
Cotton
Cowpea
Cucurbits
Custard apple
Date palm
Deccan hemp
Eggplant
Fig
Ginger

Grapevine
Grass
Groundnut
Guava
Hemp
Hyacinth bean
Jackfruit
Jujube
Jute
Kapok
Kola
Lentil
Lettuce
Litchi
Longan
Loquat
Macadamia
Maize
Mango
Manila hemp
Millets
Mulberry
Nutmeg
Oil palm
Okra
Olive
Onions
Opium poppy
Papaya
Passion fruit
Pea
Peach
Pecan
Pepper

Pigeon pea
Pineapple
Pistachio
Plum
Pomegranate
Potato
Pyrethrum
Quince
Rambutan
Rice
Rose apple
Roselle
Rubber
Safflower
Sann hemp
Sapodilla
Sesame
Sisal
Sorghum
Soybean
Sugarcane
Sunflower
Sweet potato
Tamarind
Taro
Tea
Tobacco
Tomato
Turmeric
Vanilla
Walnut
Watercress
Wheat (incl. barley and oats)
Yam

Pests of seedlings and general
  pests
Pests of stored products

# ALMOND (*Prunus amygdalus* — Rosacea)

A native of eastern Mediterranean region; occurs as two distinct varieties, var. *dulcis* is the Sweet Almond, and var. *amara* is the Bitter Almond. The tree is small and closely resembles the near-relative peach. The trees are often cultivated as ornamentals for their convenient size and delicate blossoms. The nuts are very popular, and this is probably the nut sold in largest quantities in the world. The seed is eaten green, though most frequently roasted or salted, and also made into paste for cake making. There are many cultivars with quite different shell thicknesses and seed flavour. The bitter almond contains a bitter glucoside (amygdalin) which readily breaks down into cyanic acid and so prevents its use as food. However, it is grown in southern Europe as a source of the oil of bitter almond, which is used after the cyanic acid is extracted, for flavouring. The bitter almond trees are also used as stock for grafting sweet almonds on to. Sweet almonds are grown throughout southern Europe, S. Africa, Australia and California.

## MAJOR PESTS

| | | | | |
|---|---|---|---|---|
| *Pterochloroides persicae* Chol. | Peach Aphid | Aphididae | Mediterranean, C. Asia | Infest foliage |
| *Cimbex quadrimaculatus* Mull. | Sawfly | Cimbicidae | Italy, E. Med. | Larvae defoliate |
| *Eurytoma amygdali* End. | Almond Stone Wasp | Eurytomidae | Bulgaria, Greece, Iran, Turkey, Afghanistan, S. Russia | Larvae bore in kernel of fruit |
| *Cossus cossus* L. | Goat Moth | Cossidae | Europe, E. Med., C. Asia | Larvae bore branches |
| *Ephestia cautella* (Hb.) | Almond Moth | Pyralidae | Cosmopolitan in warmer countries | Larvae feed on kernel |
| *Paramyelois transitella* (Wlk.) | Navel Orangeworm | Pyralidae | USA (California) | Larvae bore fruits |
| *Odinodiplosis amygdali* (Anag.) | Almond Gall Midge | Cecidomyiidae | Greece, Lebanon | Larvae destroy flowers & fruits |
| *Capnodis* spp. | Flat-headed Borers | Buprestidae | W. Palaearctic | Larvae bore branches |
| *Aceria phloeocoptes* (Nal.) | Almond Bud Mite | Eriophyidae | Europe, C. Asia | Gall buds |

## MINOR PESTS

| | | | | |
|---|---|---|---|---|
| *Myzus persicae* (Sulz.) | Green Peach Aphid | Aphididae | India | Infest foliage |
| *Hyalopterus pruni* Koch. | Mealy Plum Aphid | Aphididae | Palaearctic | Infest foliage |
| *Brachycaudus helichrysi* Kalt. | Peach Leaf-curl Aphid | Aphididae | India | Cause leaf-curl |
| *Didesmococcus onifasciatus* | Soft Scale | Coccidae | Lebanon, S. Russia, Afghanistan | Infest foliage |
| *Pseudaulacaspis pentagona* (Targ.) | White Scale | Diaspididae | India | Infest foliage |
| *Monosteira lobuliferia* Reut. | Lace Bug | Tingidae | Lebanon, Libya, Syria | Sap-sucker; toxic saliva |

| | | | | |
|---|---|---|---|---|
| *Anarsia lineatella* Zell. | Peach Twig Borer | Gelechiidae | USA (California) | Larvae bore shoots |
| *Cydia funebrana* (Treit.) | Red Plum Maggot | Tortricidae | Europe, Asia | Larvae bore fruits |
| *Cydia molesta* (Busck) | Oriental Fruit Moth | Tortricidae | USA (California) | Larvae bore fruits |
| *Malacosoma indica* Wlk. | Tent Caterpillar | Lasiocampidae | India | Larvae defoliate |
| *Polyphylla fullo* L. | Chafer Beetle | Scarabaeidae | E. Med., Greece, S. Russia | Adults defoliate |
| *Mimastra cyanura* Hope | Leaf Beetle | Chrysomelidae | India | Adults eat leaves |
| *Cerambyx clux* Fald. | Longhorn Beetle | Cerambycidae | E. Med., Greece | Larvae bore trunk |
| *Sphenoptera lafertei* Thom. | Stem Borer | Buprestidae | India | Larvae bore trunk |
| *Ruguloscolytus amygdali* Guen. | Almond Bark Beetle | Scolytidae | Med. | Adults bore trunk |
| *Ruguloscolytus* spp. | Bark Beetles | Scolytidae | Med. | Adults bore bark |
| *Myllocerus* spp. | Grey Weevils | Curculionidae | India | Adults eat leaves |
| *Otiorhynchus cribricollis* Gylh. | Apple Weevil | Curculionidae | Med., Australia, W. USA | Adults eat leaves; larvae eat roots |
| *Tanymecus cribricollis* Gylh. | Southern Grey Weevil | Curculionidae | E. Europe | Adults eat leaves |
| *Bryobia rubrioculus* Schent. | Brown Mite | Tetranychidae | E. Med., Iran, Afghanistan | Scarify leaves |

# APPLE (*Pyrus malus* – Rosaceae)

Apple is essentially a temperate fruit, native to E. Europe and W. Asia, and has been cultivated for more than 3000 years. Some 6500 horticultural forms are known. The crop is grown commercially in Britain, Europe, Canada, N. USA, Australia, New Zealand and in some sub-tropical areas such as S. Africa, the Mediterranean region and in the cooler parts of India. There is a recent trend generally for crop diversification in most countries, and this has led to the introduction of apples as a crop to parts of Indonesia, Philippines and Kenya, in areas of high altitude where the climate is more sub-tropical.

## MAJOR PESTS

| | | | | |
|---|---|---|---|---|
| *Psylla mali* Sch. | Apple Sucker | Psyllidae | Europe, India | Nymphs destroy flower trusses |
| *Rhopalosiphum insertum* (Wlk.) | Apple-grass Aphid | Aphididae | Europe | Infest leaves |
| *Dysaphis mali* (Ferr.) | Rosy Apple Aphid | Aphididae | Europe | Infest foliage |
| *Aphis pomi* (Deg.) | Green Apple Aphid | Aphididae | Europe, W. Asia, N. America | Infest leaves |
| *Dysaphis devecta* (Wlk.) | Rosy Leaf-curling Aphid | Aphididae | Europe | Infest leaves |
| *Eriosoma lanigerum* (Ham.) | Woolly Apple Aphid | Pemphigidae | Cosmopolitan (esp. Iran) | Infest foliage |
| *Quadraspidiotus perniciosus* (Comst.) | San José Scale | Diaspididae | Cosmopolitan | Infest foliage & fruit; very damaging |
| *Lygocoris pabulinus* (L.) | Common Green Capsid | Miridae | Europe | Sap-sucker; toxic saliva |
| *Cydia pomonella* (L.) | Codling Moth | Tortricidae | Cosmopolitan | Larvae bore fruits |
| *Archips podana* (Scop.) | Fruit Tree Tortrix | Tortricidae | Europe | Larvae feed on flowers & buds |
| *Operophtera brumata* (L.) | Winter Moth | Geometridae | Europe | Larvae defoliate |
| *Hoplocampa testudinea* (Klug.) | Apple Sawfly | Tenthredinidae | Europe, N. America | Larvae damage fruitlets |
| *Panonychus ulmi* (Koch) | Fruit Tree Red Spider Mite | Tetranychidae | Europe | Scarify leaves |

## MINOR PESTS

| | | | | |
|---|---|---|---|---|
| *Empoasca* spp. | Green Leafhoppers | Cicadellidae | Europe, N., C. & S. America | Infest foliage |
| *Typhlocyba* spp. | Fruit Tree Leafhoppers | Cicadellidae | Europe, Australia, NZ | Infest foliage |
| *Aphis gossypii* Glov. | Cotton Aphid | Aphididae | India | Infest foliage |
| *Lachnus krishnii* Glov. | Pear Aphid | Aphididae | India | Infest foliage |
| *Hemiberlesia* spp. | 'Mealybugs' | Margarodidae | India | Infest foliage |
| *Icerya purchasi* Mask. | Cottony Cushion Scale | Margarodidae | India | Infest foliage |

510

| *Parthenolecanium persicae* (F.) | Peach Scale | Coccidae | Cosmopolitan | Infest foliage |
|---|---|---|---|---|
| *Lepidosaphes ulmi* (L.) | Oystershell Scale | Diaspididae | Cosmopolitan | Infest twigs |
| *Aspidiotus nerii* Bche. | Oleander Scale | Diaspididae | Europe, Africa, Australasia, N., C. & S. America | Infest foliage |
| *Parlatoria oleae* (Colv.) | Olive Scale | Diaspididae | Europe, India, N. & S. America | Infest foliage |
| *Pseudaulacaspis pentagona* T. | White Scale | Diaspididae | India | Infest foliage |
| *Quadraspidiotus ostraeformis* (Curt.) | Oystershell Scale | Diaspididae | Europe | Infest foliage |
| *Helopeltis antonii* S. | Capsid Bug | Miridae | India | Sap-sucker; toxic saliva; scar fruitlets, may kill shoots |
| *Lygus viridanus* M. | Capsid Bug | Miridae | India | |
| *Plesiocoris rugicollis* Fall. | Apple Capsid | Miridae | Europe | |
| *Thrips flavus* Schm. | Flower Thrips | Thripidae | India | Infest flowers |
| *Thrips hawaiiensis* (Morg.) | Thrips | Thripidae | India | Infest foliage |
| *Ametastegia glabrata* (Fall.) | Dock Sawfly | Tenthredinidae | Europe | Larvae defoliate |
| *Vespula/Vespa* spp. | Common Wasps | Vespidae | Cosmopolitan | Adults puncture ripe fruits |
| *Dasyneura mali* (Kieff.) | Apple Leaf Midge | Cecidomyiidae | Europe | Larvae roll leaves |
| *Thomasiniana oculiperda* (Rubs.) | Red Bud Borer | Cecidomyiidae | Europe | Larvae bore buds of young stock |
| *Dacus* spp. | Fruit Flies | Tephritidae | India | Larvae bore fruits |
| *Rhagoletis pomonella* (Walsh) | Apple Fruit Fly | Tephritidae | USA | Larvae bore fruits |
| *Synanthedon myopiformis* (Bork.) | Apple Clearwing Moth | Sesiidae | Europe | Larvae bore under bark |
| *Alsophila aescularia* (Schiff.) | March Moth | Geometridae | Europe | Larvae defoliate |
| *Erannis defoliaria* (Clerck) | Mottled Umber Moth | Geometridae | Europe | Larvae defoliate |
| *Zeuzera* spp. | Leopard Moths | Cossidae | Europe, Med., India, Japan, USA | Larvae bore branches |
| *Yponomeuta* spp. | Small Ermine Moths | Yponomeutidae | Europe | Larvae defoliate |
| *Adoxophyes orana* (F.v.R.) | Summer Fruit Tortrix | Tortricidae | Europe, China | Larvae eat fruit surface |
| *Acroclita naevana* Hbn. | Tortrix Moth | Tortricidae | Europe, India | Larvae eat leaves |
| *Cacaecia oporana* L. | Fruit Tree Tortrix | Tortricidae | Europe | Larvae damage fruits |
| *Cydia funebrana* (Treit.) | Red Plum Maggot | Tortricidae | Europe, Asia | Larvae bore fruits |
| *Epiphyas postvittana* (Wlk.) | Light Brown Apple Moth | Tortricidae | Australia, Hawaii, NZ | Larvae damage fruits |
| *Pammene rhediella* (Clerck) | Fruitlet Mining Tortrix | Tortricidae | Europe | Larvae eat leaves |
| *Spilonota ocellana* Schiff. | Bud Moth | Tortricidae | Europe, India | Larvae bore buds |
| *Orthosia* spp. | Leaf-eating Caterpillars | Noctuidae | Europe | Larvae eat leaves |
| *Malacosoma indica* Wlk. | Tent Caterpillar | Lasiocampidae | India | Larvae defoliate |
| *Arctias selene* Hbn. | Moon Moth | Saturniidae | India | Larvae defoliate |

| | | | | |
|---|---|---|---|---|
| *Smerinthus ocellata* (L.) | Eyed Hawk Moth | Sphingidae | Europe, USA | Larvae defoliate |
| *Euproctis* spp. | Tussock Moths | Lymantriidae | Europe, India | Larvae defoliate |
| *Orygia* spp. | Tussock Moths | Lymantriidae | Europe, Australia | Larvae defoliate |
| *Merista* spp. | Leaf Beetles | Chrysomelidae | India | Adults eat leaves |
| *Eubrachis indica* Baly | Apple Shoot Beetle | Chrysomelidae | India | Shoot bored |
| *Brahmina* spp. | Cockchafers | Scarabaeidae | India ⎫ | Adults damage young |
| *Melolontha* spp. | Cockchafers | Scarabaeidae | Europe, India ⎬ | fruits & flowers, |
| *Serica* spp. | Cockchafers | Scarabaeidae | Europe, India ⎭ | and eat pieces of leaves; larvae in soil |
| *Xylotrupes gideon* L. | Unicorn Beetle | Scarabaeidae | India | Adults damage foliage |
| *Adoretus* spp. | Flower Beetles | Scarabaeidae | India ⎫ | Adults eat leaves & |
| *Anomala* spp. | Flower Beetles | Scarabaeidae | India ⎬ | flowers |
| *Popillia* spp. | Flower Beetles | Scarabaeidae | India ⎭ | |
| *Sphenoptera lafertei* Thom. | Jewel Beetle | Buprestidae | India | Larvae bore trunk |
| *Apriona cinerea* Chev. | Longhorn Beetle | Cerambycidae | India | Larvae bore trunk |
| *Betula* spp. | Longhorn Beetles | Cerambycidae | India | Larvae bore trunk |
| *Dorysthenus hugelii* Redt. | Apple Root Borer | Cerambycidae | India | Larvae bore trunk |
| *Lophosternus hugelii* Redt. | Longhorn Beetle | Cerambycidae | India | Larvae bore trunk |
| *Anthonomus pomorum* (L.) | Apple Blossom Weevil | Curculionidae | Europe | Larvae eat flowers |
| *Caenorhinus aequatus* (L.) | Apple Fruit Rhynchites | Curculionidae | Europe | Adults hole fruitlets |
| *Rhynchites coeruleus* (Deg.) | Apple Twig Cutter | Curculionidae | Europe | Adults cut off shoots |
| *Phyllobius* spp. | Leaf Weevils | Curculionidae | Europe | Adults eat leaves |
| *Myllocerus* spp. | Grey Weevils | Curculionidae | India | Adults eat leaves |
| *Scolytus rugulosus* (Muller) | Fruit Bark Borer | Scolytidae | Eurasia, N. & S. America | Adults bore bark |
| *Aculus schlechtendali* (Nal.) | Leaf & Bud Mite | Eriophyidae | Europe, Canada, USA | Blister leaves |
| *Eriophyes pyri* (Pgst.) | Pear Leaf Blister Mite | Eriophyidae | Europe, S. Africa, Australia, NZ, N. & S. America | Blister leaves |
| *Bryobia* spp. | Brown Mites | Tetranychidae | Europe | Scarify leaves |
| *Tetranychus urticae* (Koch) | Temperate Red Spider Mite | Tetranychidae | Europe, Australia | Scarify leaves |

# APRICOT (*Prunus armeniaca* — Rosaceae)

A native of Asia, this small tree (6–10 m tall) has long been cultivated in China, India, Egypt, and Iran, and is now grown in Europe, parts of Africa, and the warmer parts of the New World. This plant is susceptible to frost, so is grown in warm temperate regions and the sub-tropics. The fruit is like a peach but the stone is smooth. Apricots are used as table fruits in the areas where they are grown, and are also dried, frozen, canned, candied, and made into a paste; an oil is extracted from the seeds.

## MAJOR PESTS

| | | | | |
|---|---|---|---|---|
| *Quadraspidiotus perniciosus* (Comst.) | San José Scale | Diaspididae | India | Infest foliage |
| *Anarsia lineatella* Zell. | Peach Twig Borer | Gelechiidae | India | Larvae bore buds |

## MINOR PESTS

| | | | | |
|---|---|---|---|---|
| *Hyalopterus pruni* (Geoff.) | Mealy Plum Aphid | Aphididae | India | Infest foliage |
| *Brachycaudus helichrysi* Kalt. | Peach Leaf-curl Aphid | Aphididae | India | Infest foliage |
| *Myzus persicae* (Sulz.) | Green Peach Aphid | Aphididae | India | Infest foliage |
| *Eulecanium corylil* L. | Soft Scale | Coccidae | India | Infest foliage |
| *Parthenolecanium corni* (Bch.) | Plum Scale | Coccidae | Europe, W. Asia | Infest foliage |
| *Drosicha mangiferae* (Green) | Mango Giant Mealybug | Margarodidae | India | Infest foliage |
| *Eucosma ocellana* Schiff. | Bud Caterpillar | Tortricidae | India | Larvae bore buds |
| *Cacoecia sarcostega* Meyr. | — | Tortricidae | India | Larvae feed on buds |
| *Cydia funebrana* (Treit.) | Red Plum Maggot | Tortricidae | Europe, Asia | Larvae bore fruits |
| *Archips* spp. | Fruit Tree Tortrixes | Tortricidae | India | Larvae feed on buds |
| *Malacosoma indica* Wlk. | Tent Caterpillar | Lasiocampidae | India | Larvae defoliate |
| *Lymantria obfuscata* Wlk. | Tussock Moth | Lymantriidae | India | Larvae defoliate |
| *Dacus dorsalis* Hend. | Oriental Fruit Fly | Tephritidae | India | Larvae inside fruits |
| *Brahmina* spp. | Cockchafers | Scarabaeidae | India | Adults eat leaves |
| *Melolontha* spp. | Cockchafers | Scarabaeidae | India | Adults eat leaves |
| *Serica* spp. | Cockchafers | Scarabaeidae | India | Adults eat leaves |
| *Anomala* spp. | Flower Beetles | Scarabaeidae | India | Adults eat leaves |
| *Adoretus* spp. | Flower Beetles | Scarabaeidae | India | Adults eat leaves |
| *Sphenoptera lafertei* Thom. | Stem Borer | Buprestidae | India | Larvae bore trunk |
| *Aeolesthes holosericea* F. | Apple Stem Borer | Cerambycidae | India | Larvae bore trunk |
| *Dorysthenes hugeli* Redt. | — | Cerambycidae | India | Larvae bore trunk |
| *Lophosternus hugelii* Redt. | Stem Borer | Cerambycidae | India | Larvae bore trunk |
| *Mimastra cyanura* Hope | Almond Beetle | Chrysomelidae | India | Adults defoliate |
| *Amblyrrhinus poricollis* Boh. | Plum Weevil | Curculionidae | India | Adults eat leaves |
| *Myllocerus* spp. | Grey Weevils | Curculionidae | India | Adults eat leaves |

# AVOCADO (*Persea americana* — Lauraceae)

Avocado originated from C. America, was introduced into the W. Indies in the mid-17th century, and spread into Asia by the mid-19th century. Now it is grown in most tropical and sub-tropical countries. The main commercial crop-producing countries are S. Africa, Hawaii, Austrialia, USA. C. and S. America. It occurs as three ecological races of a single species; the West Indian race produces the best fruit, but hybrids are now cultivated. It will grow in a variety of soils but cannot stand waterlogging, and is liable to wind damage. The best fruit are grown in the lowland hot tropics. The fruit is a large fleshy single-seeded berry, pyriform in shape, green in colour, with a high oil content rich in vitamins A, B and E. The tree is evergreen and grows to a height of about 20 m at the most. (See also D. Smith, 1973).

## MAJOR PESTS

| | | | | |
|---|---|---|---|---|
| *Aleurocanthus woglumi* Ashby | Citrus Blackfly | Aleyrodidae | Pantropical | Infest leaves |
| *Amblypelta* spp. | Fruit-spotting Bugs | Coreidae | E. Australia | Necrotic spots on fruit |
| *Dacus ferrugineus* (F.) | Fruit Fly | Tephritidae | Malaysia | Larvae in fruits |
| *Dacus* spp. | Fruit Flies | Tephritidae | S.E. Asia, Australia | Larvae in fruits |
| *Solenopsis geminata* (F.) | Fire Ant | Formicidae | India, S.E. Asia | Sting workers |
| *Cryptophlebia leucotreta* (Meyr.) | False Codling Moth | Tortricidae | W. Africa | Larvae bore fruits |

## MINOR PESTS

| | | | | |
|---|---|---|---|---|
| *Aleurotuberculatus psidii* (Singh) | Guava Whitefly | Aleyrodidae | Malaysia | Infest leaves |
| *Aphis gossypii* Glov. | Cotton Aphid | Aphididae | Cosmopolitan (Philippines) | Infest foliage |
| *Ceroplastes* spp. | Waxy Scales | Coccidae | E. Australia | Encrust foliage |
| *Nipaecoccus nipae* (Mask.) | Nipa Mealybug | Pseudococcidae | Africa, India, C. America | Encrust foliage |
| *Ferrisia virgata* (Ckll.) | Striped Mealybug | Pseudococcidae | S.E. Asia | Infest stems |
| *Pseudococcus* spp. | Mealybugs | Pseudococcidae | S.E. Asia | Encrust stems |
| *Gascardia destructor* (Newst.) | White Waxy Scale | Margarodidae | E. Australia | Encrust twigs |
| *Aspidiotus destructor* Sign. | Coconut Scale | Diaspididae | Pantropical | On fruits & twigs |
| *Pseudaonidia trilobitiformis* (Green) | Trilobite Scale | Diaspididae | Pantropical | On fruits & twigs |
| *Hemiberlesia lataniae* (Sign.) | Latania Scale | Diaspididae | E. Australia | On fruits & twigs |
| *Colgar* sp. | Moth Bug | Flattidae | Papua NG | Sapsucker |
| *Terentius nubifasciatus* | Treehopper | Membracidae | Papua NG | Sapsucker |
| *Euricania villica* | Ricanid Planthopper | Ricaniidae | Papua NG | Sapsucker |
| *Tarundia glaucesenus* | Ricanid Planthopper | Ricaniidae | Papua NG | Sapsucker |
| *Helopeltis* spp. | Capsid Bugs | Miridae | Africa, S.E. Asia, India | Attack shoots; toxic saliva |
| *Selenothrips rubrocinctus* (Giard) | Red-banded Thrips | Thripidae | S.E. Asia, India | Infest leaves |

| | | | | |
|---|---|---|---|---|
| *Dacus dorsalis* (Hend.) | Oriental Fruit Fly | Tephritidae | Philippines | Larvae in fruits |
| *Clania gigantea* (Dndg.) | Giant Bagworm | Psychidae | S.E. Asia | Larvae defoliate |
| *Oiketicus elongatus* Saund. | Large Bagworm | Psychidae | E. Australia | Larvae defoliate |
| *Attacus atlas* (L.) | Atlas Moth | Saturniidae | Malaysia, Philippines | Larvae defoliate |
| *Zeuzera coffeae* Nietn. | Red Coffee Borer | Cossidae | Philippines | Larvae bore branches |
| *Papilio* spp. | Swallowtails | Papilionidae | Philippines | Larvae defoliate |
| *Lophodes miserana* (Wlk.) | Brown Looper | Geometridae | E. Australia | Larvae defoliate |
| *Anomala* spp. | White Grubs | Scarabaeidae | S.E. Asia | Larvae eat roots |
| *Leucopholis irrorata* (Chevr.) | White Grub | Scarabaeidae | Philippines | Larvae eat roots |
| *Niphonoclea* spp. | Twig Borers | Cerambycidae | Philippines | Larvae bore twigs |
| *Monolepta australis* (Jac.) | Red-shouldered Leaf Beetle | Chrysomelidae | E. Australia | Destroy foliage |
| *Platypus bicornis* (Schedl.) | Pin-hole Borer | Platypodidae | Malaysia | Adults bore bark |
| *Carpophilus marginellus* (Mots.) | Sap Beetle | Nitidulidae | Malaysia | Larvae bore fruit |
| *Xylosandrus compactus* (Eichh.) | Black Twig Borer | Scolytidae | India, S.E. Asia | Adults bore twigs |
| *Xyleborus morstatti* Hagdn. | Shot-hole Borer | Scolytidae | India | Adults bore twigs |
| *Xyleborus fornicatus* Eichh. | Tea Shot-hole Borer | Scolytidae | India, S.E. Asia | Adults bore twigs |
| *Polyphagotarsonemus latus* (Banks) | Yellow Tea Mite | Tarsonemidae | Pantropical | Damage leaves |

515

# BAMBOO (*Bambusa vulgaris*, etc. — Gramineae)

About 45 genera, and more than 200 species, of woody perennials belong to the Bambuseae, generally regarded as the most primitive tribe within the Gramineae. They usually grow gregariously in clumps, the largest species being 20–30 m in height. As a group they are found from sea-level to the snow line in the tropics, sub-tropics and warmer temperate regions of the world. The greatest number of species are found in the Indo-Malaysian region, extending through China to Korea and Japan. Africa only has a few species. The important species number less than ten, and most of the cultivated species are now unknown in the wild state. The uses of bamboo are many, for example in the building of houses, bridges, for scaffolding, making paper, furniture, weaving, the young shoot is eaten in China as a vegetable and the fruits provide an edible grain, although typically each bamboo clump only flowers once and then dies.

## MAJOR PESTS

| | | | | |
|---|---|---|---|---|
| *Dinoderus* spp. | Bamboo Borers | Bostrychidae | India, China, Philippines | Adults bore stems |
| *Cyrtotrachelus longimanus* | Bamboo Weevil | Curculionidae | S. China | Larva destroys shoot |

## MINOR PESTS

| | | | | |
|---|---|---|---|---|
| *Ceracris kiangsu* Tsai | Bamboo Locust | Acrididae | S. China ⎫ | Adults and nymphs |
| *Locusta migratoria manilensis* (Meyen) | Mig. Locust | Acrididae | Philippines ⎰ | defoliate |
| *Macrotermes* spp. | Termites | Termitidae | Philippines | Workers eat roots |
| *Pseudoregma bambusicola* Tak. | Bamboo Woolly Aphid | Pemphigidae | S. China, Philippines | Encrust shoots |
| *Purohita fuscovenosa* Muir | Planthopper | Delphacidae | S. China | Feed on stems |
| *Saccharicoccus* sp. | Sugarcane Mealybug | Pseudococcidae | S. China, Philippines | Under leaf sheaths |
| *Asterolecanium bambusae* (Boisd.) | Bamboo Star Scale | Asterolecanidae | Philippines, USA | Encrust stem |
| *Notobitus meleagris* (F.) | Bamboo Bug | Coreidae | S. China | Toxic saliva scars stems |
| *Xylocopa iridipennis* Lep. | Bamboo Carpenter Bee | Xylocopidae | S. China | Nest in dead internodes |
| *Arge* spp. | Sawflies | Argidae | S. China | Larvae defoliate |
| *Atrachea vulagris* | Shoot Borer | – | China | Larvae bore shoots |
| *Agrotis segetum* (Schiff.) | Common Cutworm | Noctuidae | China | Larvae are cutworms |
| *Erionota thrax* L. | Banana Skipper | Hesperiidae | Philippines | Larvae roll leaves |
| *Telicota augias* (L.) | Rice Skipper | Hesperiidae | Philippines | Larvae roll leaves |
| *Bostrychopsis parallela* (Lesne) | Bamboo Borer | Bostrychidae | Philippines | Adult bores stems |
| *Dactylispa* spp. | Bamboo Hispids | Chrysomelidae | Philippines | Larvae mine leaves |
| *Chlorophorus annularis* (F.) | Bamboo Longhorn | Cerambycidae | Philippines, S. China | Larvae bore stems |
| *Xylotrechus nauticus* | Longhorn | Cerambycidae | S. China | Larvae bore stems |

# BANANAS (*Musa sapientum* varieties — Musaceae)

Also known as plantain; there are many varieties, some with high sugar content eaten for dessert, some with high starch for cooking or beer-brewing. They occur wild in the area from India to New Guinea, but are now cultivated throughout the tropics. It is essentially a tropical crop, growing best on well-drained fertile soils; it cannot tolerate frost, and is very susceptible to wind damage. It is basically a giant herb with the pseudostem formed by the overlapping leaf bases. The fruit is formed as 'fingers' on a series of successional 'hands' along the flower stalk. The main production areas are Ecuador, C. America, W. Indies, W. Africa, Cameroons and the Pacific Islands.

## MAJOR PESTS

| | | | | |
|---|---|---|---|---|
| *Pentalonia nigronervosa* Coq. | Banana Aphid | Aphididae | Pantropical | Under leaf sheaths |
| *Hercinothrips bicinctus* Bagn. | Banana Thrips | Thripidae | E. Africa, Australia | Infest flowers |
| *Nacoleia octasema* (Meyr.) | Banana Scab Moth | Pyralidae | Indonesia | Larvae scab fruits |
| *Erionota* spp. | Banana Skippers | Hesperiidae | India, S.E. Asia, China | Larvae roll leaves |
| *Cosmopolites sordidus* (Germ.) | Banana Weevil | Curculionidae | Pantropical | Larvae bore rhizome |
| *Odoiporus longicollis* (Oliv.) | Banana Stem Weevil | Curculionidae | S.E. Asia | Larvae bore stem |
| *Colaspis hypochlora* Lefevre | Banana Fruit-scarring Beetle | Chrysomelidae | C. & S. America | Larvae scar fruits |

## MINOR PESTS

| | | | | |
|---|---|---|---|---|
| *Hieroglyphus banian* (F.) | Large Rice Grasshopper | Acrididae | Thailand, Laos ⎫ | Adults & nymphs eat leaves |
| *Sexava* spp. | Longhorned Grasshoppers | Tettigoniidae | Papua NG ⎭ | |
| *Patanga succincta* (L.) | Bombay Locust | Acrididae | Thailand, Laos | Adults & nymphs eat leaves |
| *Aphis gossypii* Glov. | Melon/Cotton Aphid | Aphididae | Pantropical | Sap suckers |
| *Aleurocanthus woglumi* Ashby | Citrus Blackfly | Aleyrodidae | Pantropical | Under leaves |
| *Pseudococcus comstocki* Kuw. | Banana Mealybug | Pseudococcidae | S.E. Asia | Encrust foliage |
| *Dysmicoccus brevipes* (Ckll.) | Pineapple Mealybug | Pseudococcidae | S.E. Asia | Encrust foliage |
| *Aspidiotus destructor* Sign. | Coconut Scale | Diaspididae | Pantropical | Encrust foliage |
| *Ischnaspis longirostris* (Sign.) | Black Line Scale | Diaspididae | Pantropical | Encrust foliage |
| *Aonidiella aurantii* (Mask.) | California Red Scale | Diaspididae | India | Encrust foliage |
| *Aspidiotus* spp. | Armoured Scales | Diaspididae | Old World tropics ⎫ | Infest fruit & foliage |
| *Chrysomphalus aonidum* (L.) | Purple Scale | Diaspididae | C. & S. America ⎭ | |
| *Pocillocarda mitrata* Gerst. | Leafhopper | Cicadellidae | E. Africa | Infests leaves |
| *Stephanitis typica* (Dist.) | Banana Lace Bug | Tingidae | India, S.E. Asia, Korea, Japan | Sap-sucker; toxic saliva |

| *Chaetanaphothrips signipennis* (Bag.) | Banana Rust Thrips | Thripidae | India, Australia, C. America | Infest flowers & fruit |
|---|---|---|---|---|
| *Hercinothrips femoralis* (Reut.) | Banded Greenhouse Thrips | Thripidae | Cosmopolitan | Infest foliage |
| *Heliothrips haemorrhoidalis* (Bouché) | Black Tea Thrips | Thripidae | S.E. Asia | Leaves & fruit silvered |
| *Thrips florum* Schmutz | Banana Flower Thrips | Thripidae | Australia | Infest flowers |
| *Tiracola plagiata* (Wlk.) | Banana Fruit Caterpillar | Noctuidae | India, Australia, S.E. Asia | Larvae damage fruit |
| *Spodoptera litura* (F.) | Rice Cutworm | Noctuidae | India, S.E. Asia | Larvae eat leaves |
| *Castniomera humbolti* (Boisd.) | Banana Stem Borer | Castniidae | C. & S. America | Larvae bore pseudostem |
| *Platynota rostrana* (Wlk.) | – | Tortricidae | C. & S. America | Larvae damage fruit |
| *Antichloris viridis* (Druce) | Leaf-cutting Caterpillar | Syntomidae | C. & S. America | Larvae defoliate |
| *Ecpantheria icasia* (Cramer) | Tiger Moth | Arctiidae | C. & S. Amercia | Larvae eat fruit peel |
| *Othreis fullonia* (Cl.) | Fruit-piercing Moth | Noctuidae | Old World tropics | Adults pierce fruit |
| *Latoia lepida* (Cram.) | Blue-striped Nettlegrub | Limacodidae | India | Larvae defoliate |
| *Dacus curvipennis* Frogg. | Banana Fruit Fly | Tephritidae | Fiji | Larvae in fruits |
| *Dacus* spp. | Fruit Flies | Tephritidae | India, S.E. Asia, Australasia, Pacific | Larvae in fruits |
| *Trigona* spp. | Fruit-scarring Bees | Apidae | C. & S. America | Adults scar fruits |
| *Prionoryctes caniculus* Arr. | Yam Beetle | Scarabaeidae | Africa | Larvae damage roots |
| *Colaspis* spp. | Banana Fruit-scarring Beetles | Chrysomelidae | India, C. & S. America | Larvae scar fruits |
| *Nodostoma* spp. | Leaf Beetles | Chrysomelidae | India | Adults damage fruits |
| *Sphaeroderma varipennis* Jacoby | Leaf Beetle | Chrysomelidae | Thailand | Adults defoliate |
| *Sphaeroderma* spp. | Leaf Beetles | Chrysomelidae | Laos | Adults defoliate |
| *Metamasius hemipterus* L. | West Indian Cane Weevil | Curculionidae | W. Africa, W. Indies, S. America | Scarify flowers |
| *Tomnoschoita nigroplagiata* Qued. | – | Curculionidae | E. Africa, Zaïre | Larvae bore stem |
| *Philicoptus waltoni* (Boh.) | – | Curculionidae | Philippines | Adults damage fruit |
| *Araeocerus fasciculatus* (De. G.) | Coffee Bean Weevil | Brenthidae | Ghana, Thailand, Bermuda | Adults attack fruits |
| *Doticus palmaris* Pascoe | – | Anthribidae | Australia (Queensland) | Adults attack fruits |
| *Tetranychus* spp. | Red Spider Mites | Tetranychidae | Pantropical | Scarify foliage |

More than 470 species of insects and mites recorded by Ostmark (1974) as major and minor pests of bananas.

# BEANS AND GRAMS (*Phaseolus* spp. — Leguminosae)

(*P. vulgaris* — Common, Garden, Field, French, Haricot, or Kidney Bean)
(*P. lunatus* — Lima Bean          *P. coccineus* — Scarlet Runner Bean)
(*P. aureus* — Mung Bean, Green Gram   *P. munga* — Black Gram)
(*P. calcareus* — Rice Bean        *P. angularis* — Adzuki Bean)

The species listed above are the usual ones referred to as beans and grams, but a few other obscure species are known. The plant is an annual climber, which can be made to assume a bushy habit in some species. The common bean is a native of S. America, as is also the lima bean (Peru and Brazil); scarlet runner is from C. America; mung bean, black gram and rice bean are ancient legumes from India; adzuki bean is of great importance in China and Japan. Growth require-ments vary somewhat as some species are more temperate than others. The immature pods are cooked green as a vegetable, and the seeds may be dried and processed; the whole plant may be used as forage. As grain legumes these plants are an important source of plant protein in many parts of the world. Generally the different species of *Phaseolus* show a similar pest spectrum.

## MAJOR PESTS

| | | | | |
|---|---|---|---|---|
| *Aphis fabae* Scop. | Black Bean Aphid | Aphididae | Cosmopolitan (not Australia) | Infest foliage |
| *Acanthomia* spp. | Spiny Brown Bugs | Coreidae | Africa | Sap-suckers; toxic saliva |
| *Lamprosema* spp. | Bean Leaf Rollers | Pyralidae | India, S.E. Asia | Larvae roll leaves |
| *Maruca testulalis* (Geyer) | Mung Moth | Pyralidae | Pantropical | Larvae bore pods |
| *Heliothis armigera* (Hb.) | American Bollworm | Noctuidae | Cosmopolitan in Old World | Larvae bore pods |
| *Taeniothrips sjostedti* (Tryb.) | Bean Flower Thrips | Thripidae | Africa | Infest flowers |
| *Ophiomyia phaseolii* (Tryon) | Bean Fly | Agromyzidae | Europe, Africa, India, S.E. Asia, Australasia | Larvae bore inside swollen stem |
| *Delia platura* (Meign.) | Bean Seed Fly | Anthomyiidae | Cosmopolitan | Larvae bore sown seed |
| *Coryna* spp. | Pollen Beetles | Meloidae | Africa | Adults eat pollen |
| *Mylabris* spp. | Banded Blister Beetles | Meloidae | Africa, India, S.E. Asia | Adults eat flowers |
| *Epicauta* spp. | Black Blister Beetles | Meloidae | Africa, Asia, S. America | Adults eat flowers |
| *Epilachna varivestis* Muls. | Mexican Bean Beetle | Coccinellidae | USA, Mexico | Adults & larvae defoliate |
| *Acanthoscelides obtectus* (Say) | Bean Bruchid | Bruchidae | Cosmopolitan | Attack ripe seeds |
| *Callosobruchus* spp. | Cowpea Bruchids | Bruchidae | Cosmopolitan in warmer regions | Attack ripe seeds |
| *Colaspis brunnea* (F.) | Grape Colaspis | Chrysomelidae | Southern USA | Larvae feed on roots |
| *Ootheca mutabilis* (Sahlb.) | Brown Leaf Beetle | Chrysomelidae | E. Africa, Nigeria | Adults eat leaves |
| *Apion* spp. | Apion Weevils | Apionidae | Cosmopolitan | Infest flowers |
| *Tetranychus* spp. | Red Spider Mites | Tetranychidae | Cosmopolitan | Scarify foliage |

519

## MINOR PESTS

| | | | | |
|---|---|---|---|---|
| *Empoasca* spp. | Green Leafhoppers | Cicadellidae | Cosmopolitan | Infest foliage |
| *Aphis craccivora* Koch | Groundnut Aphid | Aphididae | Cosmopolitan | Infest foliage |
| *Macrosiphum* spp. | – | Aphididae | Africa, S.E. Asia | Infest foliage |
| *Coccus* spp. | Soft Scales | Coccidae | Cosmopolitan | Infest foliage |
| *Ferrisia virgata* (Ckll.) | Striped Mealybug | Pseudococcidae | Pantropical | Infest foliage |
| *Amblypelta* spp. | Coreid Bugs | Coreidae | S.E. Asia | Sap-sucker; toxic saliva |
| *Anaplocnemis horrida* Germ. | Coreid Bug | Coreidae | Africa | Sap-sucker; toxic saliva |
| *Leptoglossus australis* (F.) | Leaf-footed Plant Bug | Coreidae | Africa, India, S.E. Asia, Australasia | Sap-sucker; toxic saliva |
| *Riptortus pedestris* F. | Coreid Bug | Coreidae | India | Sap-sucker; toxic saliva |
| *Lygus* spp. | Capsid Bugs | Miridae | Europe, Africa | Sap-suckers; toxic saliva |
| *Calocoris norvegicus* (Gmel.) | Potato Capsid | Miridae | Europe | Sap-sucker; toxic saliva |
| *Halticus tibialis* | Capsid Bug | Miridae | S.E. Asia | Sap-sucker; toxic saliva |
| *Lygocoris pabulinus* (L.) | Common Green Capsid | Miridae | Europe | Sap-sucker; toxic saliva |
| *Coptosoma cribraria* F. | – | Pentatomidae | India | Sap-sucker; toxic saliva |
| *Nezara viridula* (L.) | Green Stink Bug | Pentatomidae | Cosmopolitan | Sap-sucker; toxic saliva |
| *Taeniothrips cinctipennis* (Bagn.) | Thrips | Thripidae | S.E. Asia | Infest flowers |
| *Thrips tabaci* Lind. | Onion Thrips | Thripidae | Cosmopolitan | Infest foliage |
| *Thrips palmi* | Palm Thrips | Thripidae | S.E. Asia, China | Infest foliage |
| *Kakothrips robustus* (Uzel) | Pea Thrips | Thripidae | Europe | Infest flowers |
| *Homona coffearia* (Niet.) | Tea Tortrix | Tortricidae | S.E. Asia | Larvae roll leaves |
| *Eucosma melanaula* (Meyr.) | Leaf Roller | Tortricidae | India | Larvae roll leaves |
| *Cnephasia* spp. | Leaf Rollers | Tortricidae | Europe | Larvae roll leaves |
| *Lampides boeticus* L. | Pea Blue Butterfly | Lycaenidae | India | Larvae bore pods |
| *Agrius convolvuli* (L.) | Convolvulus Hawk Moth | Sphingidae | India | Larvae defoliate |
| *Amsacta* spp. | Red Hairy Caterpillar | Arctiidae | Africa, India | Larvae defoliate |
| *Diacrisia obliqua* Wlk. | Tiger Moth | Arctiidae | India | Larvae defoliate |
| *Anticarsia gemmatalis* (Hb.) | Velvetbean Caterpillar | Noctuidae | USA, S. America | Larvae bore pods |
| *Anticarsia irrorata* B. | Green Leaf Caterpillar | Noctuidae | India | Larvae defoliate |
| *Spodoptera litura* (F.) | Rice Cutworm | Noctuidae | Asia, India, Australasia | Larvae defoliate |

| | | | | |
|---|---|---|---|---|
| *Spodoptera littoralis* (Boisd.) | Cotton Leafworm | Noctuidae | Africa | Larvae defoliate |
| *Achaea* spp. | Semi-loopers | Noctuidae | Africa, S.E. Asia | Larvae defoliate |
| *Plusia* spp. | Semi-loopers | Noctuidae | S.E. Asia, India | Larvae defoliate |
| *Ophiomyia* spp. | Bean Flies | Agromyzidae | S.E. Asia | Larvae bore stem |
| *Phytomyza horticola* Gour. | Pea Leaf Miner | Agromyzidae | Cosmopolitan in Old World | Larvae mine leaves |
| *Delia trichodactyla* Rond. | Bean Seed Fly | Anthomyiidae | Europe | Larvae bore sown seed |
| *Tipula* spp. | Leatherjackets | Tipulidae | Europe | Larvae damage seedlings |
| *Epilachna* spp. | Epilachna Beetles | Coccinellidae | S.E. Asia | Adults & larvae defoliate |
| *Henosepilachna capensis* (Thnb.) | Epilachna Beetle | Coccinellidae | S. Africa | Adults & larvae defoliate |
| *Monolepta elegantula* (Boh.) | Leaf Beetle | Chrysomelidae | Malaysia | Adults defoliate |
| *Plagiodera inclusa* Stål | Leaf Beetle | Chrysomelidae | India | Adults defoliate |
| *Alcidodes* spp. | Striped Weevils | Curculionidae | Africa, India | Adults girdle stems |
| *Aperitmetus brunneus* (Hust.) | Tea Root Weevil | Curculionidae | E. Africa | Larvae eat roots; adults eat leaves |
| *Graphognathus* spp | White-fringed Weevils | Curculionidae | Australia, USA, S. America | Larvae eat roots; adults eat leaves |
| *Nematocerus* spp. | Nematocerus Weevils | Curculionidae | E. Africa | Adults eat leaves |
| *Oribius* spp. | Leaf Weevils | Curculionidae | Papua NG | Adults eat leaves |
| *Sitona* spp. | Pea and Bean Weevils | Curculionidae | Europe | Adults eat leaves; larvae in root nodules |

# BETEL PALM (*Areca catechu* — Palmae)

Betel 'nuts' are the seeds of a palm native to Malaysia, but now widely grown throughout the tropics wherever the habit of betel chewing is practised, that is from Africa to India, S.E. Asia, and throughout the Pacific region. On a world basis it outrivals chewing gum as a masticatory. The hard dried endosperm of the seeds (both ripe and unripe), misnamed 'nuts', may be chewed alone, but the usual practice is to wrap pieces of nut in a leaf of betel-pepper on which a dab of slaked lime has been added. The quid is chewed slowly, causing continuous salivation, and the whole mouth becomes stained bright red. The practice of betel-chewing appears to be quite addictive, but is thought to be harmless. The plant is a tall slender monoecious palm, living for some 60–100 years. It flourishes in wet maritime climates in the tropics, and is grown at altitudes from sea-level up to 900 m.

## MAJOR PESTS

## MINOR PESTS

| | | | | |
|---|---|---|---|---|
| *Cerataphis variabilis* Hrl. | Coconut Aphid | Aphididae | India | Infest foliage |
| *Icerya aegyptica* (Dgl.) | Egyptian Fluted Scale | Margarodidae | India | Infest foliage |
| *Chrysomphalus aonidum* (L.) | Purple Scale | Diaspididae | India | Encrust foliage |
| *Pinnaspis* spp. | Armoured Scales | Diaspididae | India | Encrust foliage |
| *Coccus hesperidum* L. | – | Coccidae | India | Encrust foliage |
| *Dysmicoccus brevipes* (Ckll.) | Pineapple Mealybug | Pseudococcidae | India | Encrust foliage |
| *Carvalhoia arecae* Mill. | Areca Bug | Miridae | India | Sap-sucker; toxic saliva |
| *Rhipiphorothrips cruentatus* H. | Leaf Thrips | Thripidae | India | Infest foliage |
| *Thrips hawaiiensis* (Morg.) | Flower Thrips | Thripidae | India | Infest flowers |
| *Contheyla rotunda* | Slug Caterpillar | Limacodidae | India | Larvae defoliate |
| *Leucopholis lepidophora* Bl. | Cockchafer | Scarabaeidae | India | Adults eat leaves; larvae in soil eat roots |
| *Promecotheca cummingii* Baly | Coconut Hispid | Chrysomelidae | S.E. Asia | Larvae mine leaves |
| *Araeocerus fasciculatus* (De. G.) | Coffee Bean Weevil | Brenthidae | India | Larvae feed inside 'nuts' |
| *Coccotrypes carpophagus* Horn. | Stored Arecanut Beetle | Scolytidae | India | Adults bore stored 'nuts' |

# BETEL-PEPPER (*Piper betle* — Piperaceae)

This tall woody vine, native to Malaya, is cultivated for its leaves which are chewed together with betel nut as a masticatory. The habit is of great antiquity, and it is thought that more than 400 million people chew the betel pan, in an area from E. Africa, through India and S.E. Asia, to most of the Pacific region. A number of different cultivars are recognized in India. The pungency of the leaves is due to certain volatile oils (phenols), and the leaves are rich in vitamins B and C. In its wild state the plant grows in the tropical rain forests of Malaysia, but it may be grown under irrigation in drier areas providing the soil is fertile.

## MAJOR PESTS

## MINOR PESTS

| | | | | |
|---|---|---|---|---|
| *Odontotermes obesus* Ramb. | Scavenging Termite | Termitidae | India | Workers collect plant material |
| *Aleurocanthus rugosa* Singh | Whitefly | Aleyrodidae | India | Infest foliage |
| *Dialurodes pallida* Singh | Whitefly | Aleyrodidae | India | Infest foliage |
| *Aphis gossypii* Glov. | Cotton Aphid | Aphididae | India | Infest foliage |
| *Ferrisia virgata* (Ckll.) | Striped Mealybug | Pseudococcidae | India | Infest foliage |
| *Planococcus citri* (Risso) | Root Mealybug | Pseudococcidae | India | Infest roots |
| *Lepidosaphes cornutus* Ramk. | Betelvine Scale | Diaspididae | India | Infest foliage |
| *Pachypeltis politus* Dist. | Betelvine Bug | Miridae | India | Sap-sucker; toxic saliva |
| *Cyclopelta siccifolia* Westw. | Black Bug | Pentatomidae | India | Sap-sucker; toxic saliva |

# BRASSICAS (*Brassica* spp. – Cruciferae)
## (Cabbage, Kale, Cauliflower, Mustards, Broccoli, Turnip, Brussels Sprout, Rape)

An agriculturally diverse group of crops of European origin, and of great antiquity. They are cultivated from the Arctic to the sub-tropics, and at higher altitudes in the tropics. Certain species and varieties are more adapted to the tropics than others. As a group together with Radish (*Raphanus sativus*), they seem to have a similar spectrum of pests in most regions.

## MAJOR PESTS

| | | | | |
|---|---|---|---|---|
| *Brevicoryne brassicae* (L.) | Cabbage Aphid | Aphididae | Cosmopolitan in cooler regions | Infest foliage; virus vector |
| *Lipaphis erysimi* (Kalt.) | Turnip Aphid | Aphididae | Cosmopolitan in warmer regions | Infest foliage; virus vector |
| *Aleyrodes brassicae* (L.) | Brassica Whitefly | Aleyrodidae | Europe | Infest foliage |
| *Bagrada* spp. | Harlequin Bugs | Pentatomidae | Africa, India, S.E. Asia | Sap-suckers |
| *Plutella xylostella* (L.) | Diamond-back Moth | Yponomeutidae | Cosmopolitan | Larvae hole leaves |
| *Pieris canidia* (Sparr.) | Small White Butterfly | Pieridae | S.E. Asia, India | Larvae defoliate |
| *Pieris rapae* (L.) | Small White Butterfly | Pieridae | Europe, Asia | Larvae defoliate |
| *Agrotis ipsilon* (Hfn.) | Black Cutworm | Noctuidae | Cosmopolitan | Larvae are cutworms |
| *Agrotis segetum* (D. & S.) | Common Cutworm | Noctuidae | Cosmopolitan in Old World | Larvae are cutworms |
| *Delia brassicae* (Bouché) | Cabbage Root Fly | Anthomyiidae | Europe, N. America | Larvae eat roots |
| *Athalia* spp. | Cabbage Sawflies | Tenthredinidae | Cosmopolitan | Larvae defoliate |
| *Aulacophora similis* (Ol.) | Red Melon Beetle | Chrysomelidae | Thailand | Adults eat leaves |
| *Phyllotreta* spp. | Cabbage Flea Beetles | Chrysomelidae | Cosmopolitan | Adults hole leaves |
| *Leucopholis irrorata* (Chevr.) | White Grub | Scarabaeidae | Philippines | Larvae eat roots |

## MINOR PESTS

| | | | | |
|---|---|---|---|---|
| *Acheta testaceus* (Wlk.) | Field Cricket | Gryllidae | S.E. Asia | Roots & seedlings eaten |
| *Gryllotalpa africana* (Pal.) | African Mole Cricket | Gryllotalpidae | Old World | Roots & seedlings eaten |
| *Myzus persicae* (Sulz.) | Peach–Potato Aphid | Aphididae | Cosmopolitan | Infest shoots & leaves |
| *Aphis gossypii* Glov. | Cotton Aphid | Aphididae | S.E. Asia | Infest shoots & leaves |
| *Eurydema pulchrum* (Westw.) | Harlequin Bug | Pentatomidae | Laos, Philippines | Sap-sucker |
| *Nezara viridula* (L.) | Green Stink Bug | Pentatomidae | Pantropical | Sap-sucker; toxic saliva |

| | | | | |
|---|---|---|---|---|
| *Dysdercus* spp. | Cotton Stainers | Pyrrhocoridae | Philippines | Sap-suckers |
| *Bemisia inconspicua* (Quaint.) | Whitefly | Aleyrodidae | Philippines | Infest leaves |
| *Thrips angusticeps* Uzel | Cabbage Thrips | Thripidae | Europe | Infest foliage |
| *Thrips tabaci* Lind. | Onion Thrips | Thripidae | Cosmopolitan | Infest foliage |
| *Hepialus* spp. | Swift Moths | Hepialidae | Europe | Larvae eat roots |
| *Pieris brassicae* (L.) | Large White Butterfly | Pieridae | Europe, India, N. America | Larvae eat leaves |
| *Mythimna separata* (Wlk.) | Rice Ear-cutting Caterpillar | Noctuidae | S.E. Asia | Larvae defoliate |
| *Mamestra brassicae* (L.) | Cabbage Moth | Noctuidae | Europe | Larvae defoliate |
| *Trichoplusia ni* (Hubner) | Cabbage Semi-looper | Noctuidae | S.E. Asia, USA | Larvae defoliate |
| *Chrysodeixis chalcites* (Esp.) | Cabbage Semi-looper | Noctuidae | Africa, Med., India, S.E. Asia Australasia, Japan | Larvae defoliate |
| *Plusia orichalcea* (Hub.) | – | Noctuidae | S.E. Asia | Larvae defoliate |
| *Spodoptera exigua* (Hub.) | Beet Armyworm | Noctuidae | Thailand | Larvae defoliate |
| *Spodoptera littoralis* (Boisd.) | Cotton Leafworm | Noctuidae | Africa, Med. | Larvae defoliate |
| *Spodoptera litura* (F.) | Rice Cutworm | Noctuidae | India, S.E. Asia | Larvae defoliate |
| *Spodoptera exempta* (Wlk.) | African Armyworm | Noctuidae | Papua NG | Larvae defoliate |
| *Spodoptera mauritia* (Boisd.) | Paddy Armyworm | Noctuidae | S.E. Asia | Larvae defoliate |
| *Heliothis armigera* (Hub.) | American Bollworm | Noctuidae | Old World | Larvae defoliate |
| *Xestia c-nigrum* (L.) | Spotted Cutworm | Noctuidae | Europe, Asia, N. America | Larvae defoliate |
| *Crocidolomia binotalis* (Zell.) | Cabbage Cluster-caterpillar | Pyralidae | Africa, India, S.E. Asia, Australia | Larvae defoliate |
| *Hellula phidilealis* Wlk. | Cabbage Webworm | Pyralidae | Sierra Leone, C. & S. America | Larvae defoliate |
| *Hellula undalis* (F.) | Oriental Cabbage Webworm | Pyralidae | N. & W. Africa, Near East, S.E. Asia, Australia, NZ | Larvae defoliate |
| *Evergestis* spp. | Cabbageworms | Pyralidae | Europe, USA | Larvae defoliate |
| *Hymenia recurvalis* | Cabbageworms | Pyralidae | Papua NG | Larvae defoliate |
| *Tipula* spp. | Leatherjackets | Tipulidae | Europe | Larvae eat roots |
| *Phytomyza horticola* Goureau | Pea Leaf Miner | Agromyzidae | Cosmopolitan in Old World | Larvae mine leaves |
| *Phytomyza rufipes* Meig. | Cabbage Leaf Miner | Agromyzidae | Europe | Larvae mine leaves |
| *Phytomyza* spp. | Leaf Miners | Agromyzidae | Asia | Larvae mine leaves |
| *Liriomyza brassicae* (Riley) | Cabbage Leaf Miner | Agromyzidae | Cosmopolitan | Larvae mine leaves |
| *Contarinia nasturtii* (Kieff.) | Swede Midge | Cecidomyiidae | Europe | Larvae gall shoot |
| *Delia platura* (Meign.) | Bean Seed Fly | Anthomyiidae | Cosmopolitan | Larvae destroy roots |
| *Delia floralis* (Fall.) | Turnip Root Fly | Anthomyiidae | Europe | Larvae bore stem |
| *Meligethes aeneus* Fab. | Blossom Beetle | Nitidulidae | Europe | Adults infest flowers |
| *Phaedon* spp. | Mustard Beetles | Chrysomelidae | Europe | Adults infest flowers |
| *Psylliodes chrysocephala* (L.) | Cabbage Stem Flea Beetle | Chrysomelidae | Europe | Larvae gall stems |
| *Adoretus* spp. | White Grubs | Scarabaeidae | S.E. Asia | Larvae eat roots |
| *Anomala* spp. | White Grubs | Scarabaeidae | S.E. Asia | Larvae eat roots |
| *Ceutorhynchus quadridens* (Panz.) | Cabbage Stem Weevil | Curculionidae | Europe | Larvae gall stem |

| | | | | |
|---|---|---|---|---|
| *Ceutorhynchus assimilis* (Payk.) | Cabbage Seed Weevil | Curculionidae | Europe | Larvae gall seeds |
| *Ceutorhynchus pleurostigma* (March.) | Turnip Gall Weevil | Curculionidae | Europe | Larvae gall root |
| *Aperitmetus brunneus* (Hust.) | Tea Root Weevil | Curculionidae | E. Africa | Larvae eat roots |

# BREADFRUIT (*Artocarpus altilis* – Moraceae)

A native of Polynesia, this is one of the most important food fruits in the world. It is now widespread in the tropics and particularly abundant in Polynesia as a staple food. Cultivation has been practised since antiquity and more than 100 varieties are known. A handsome tree, 15–20 m in height, bears prickly fruits the size of a melon in the leaf axils; technically the fruit is a syncarp. It is usually grown only for local consumption. The fruit is mostly starch in composition.

## MAJOR PESTS

| | | | | |
|---|---|---|---|---|
| *Ferrisia virgata* (Ckll.) | Striped Mealybug | Pseudococcidae | Philippines | Encrust leaves |
| *Pseudococcus* spp. | Mealybugs | Pseudococcidae | Philippines | Encrust leaves |

## MINOR PESTS

| | | | | |
|---|---|---|---|---|
| *Drosicha townsendi* (Ckll.) | Giant Mealybug | Margarodidae | Philippines | Infest leaves |
| *Icerya seychellarum* (Westw.) | Seychelles Fluted Scale | Margarodidae | Philippines | Infest leaves |
| *Chrysomphalum aonidum* (L.) | Florida Red Scale | Diaspididae | Philippines | Infest leaves |
| *Aspidiotus destructor* Sing. | Coconut Scale | Diaspididae | S.E. Asia | Infest leaves |
| *Aonidiella aurantii* (Mask.) | California Red Scale | Diaspididae | Philippines | Infest leaves |
| *Pulvinaria psidii* Mask. | Green Shield Scale | Coccidae | Philippines | Infest leaves |
| *Leptocorisa acuta* (Thumb.) | Rice Bug | Coreidae | S.E. Asia | Sap-sucker; toxic saliva |
| *Margaronia caesalis* (Wlk.) | – | Pyralidae | Malaysia | Larvae eat leaves |
| *Zeuzera coffeae* Nietn. | Red Coffee Borer | Cossidae | Philippines | Larvae bore branches |
| *Dacus umbrosus* (F.) | Fruit Fly | Tephretidae | Philippines | Larvae bore fruit |
| *Batocera rubus* (L.) | White-spotted Longhorn | Cerambycidae | Philippines | Larvae bore branches |
| *Xyleborus perforans* (Woll.) | Coconut Shot-hole Borer | Scolytidae | Pantropical | Adults bore branches |

# CAPSICUMS (*Capsicum* spp. – Solanaceae)
## (= Sweet Peppers and Chilli)

The centre of origin is uncertain, but probably was Peru; they spread throughout the New World very early and are now grown widely throughout the tropics and sub-tropics, and under glass (or polythene) in temperate regions. They can be grown from sea-level to 2000 m or more in the tropics, preferably with a rainfall of 60–120 cm per annum. They are sensitive to frost, water-logging, and too much rain. In habit the plant is a very variable herb, erect, many-branched, and is grown as an annual. The main areas of production are India, Thailand, Indonesia, Japan, Mexico, Uganda, Kenya, Nigeria, and Sudan. Sweet peppers are large and green, turning red as they ripen, and are used in salads or cooked as vegetables. Chillies are small, pungent, and bright red, used in curries or dried to make cayenne pepper and paprika. (See also Butani, 1976c).

## MAJOR PESTS

| | | | | |
|---|---|---|---|---|
| *Aphis gossypii* Glover | Melon/Cotton Aphid | Aphididae | Cosmopolitan | Infest leaves & stems |
| *Myzus persicae* (Sulz.) | Peach–Potato Aphid | Aphididae | Cosmopolitan | Infest leaves & stems |
| *Epicauta albovittata* (Gestro) | Striped Blister Beetle | Meloidae | E. Africa, Somalia | Adults defoliate |
| *Epilachna* spp. | Epilachna Beetles | Coccinellidae | S.E. Asia | Leaves eaten |
| *Heliothis armigera* (Hb.) | American Bollworm | Noctuidae | S.E. Asia, India | Larvae bore fruits |
| *Polyphagotarsonemus latus* (Banks) | Yellow Tea Mite | Tarsonemidae | Cosmopolitan | Leaves scarified |
| *Tetranychus cinnabarinus* (Boisd.) | Tropical Red Spider Mite | Tetranychidae | Pantropical | Leaves scarified |

## MINOR PESTS

| | | | | |
|---|---|---|---|---|
| *Brachytrupes portentosus* Licht. | Large Brown Cricket | Gryllidae | India | Leaves attacked |
| *Gryllotalpa africana* (Pal.) | African Mole Cricket | Gryllotalpidae | S.E. Asia | Eat roots |
| *Hodotermes mossambicus* Hag. | Harvester Termite | Hodotermitidae | S. & E. Africa | Destroy plant |
| *Empoasca* spp. | Green Leafhoppers | Cicadellidae | S.E. Asia | Sap-suckers on foliage |
| *Ferrisia virgata* (Ckll.) | Striped Mealybug | Pseudococcidae | S.E. Asia, India | Encrust foliage |
| *Saissetia coffeae* (Wlk.) | Helmet Scale | Coccidae | India | Infest foliage |
| *Aspidiotus desctructor* Sign. | Coconut Scale | Diaspididae | Pantropical | Infest foliage |
| *Bemisia tabaci* (Genn.) | Cotton Whitefly | Aleyrodidae | S.E. Asia, India | Infest foliage |
| *Cyrtopeltis tenuis* Reuter | Tomato Mirid | Miridae | Philippines | Sap-sucker; toxic saliva |
| *Helopeltis theobromae* (Miller) | Cocoa Capsid | Miridae | Malaysia | Sap-sucker; toxic saliva |
| *Helopeltis westwoodi* White | Capsid Bug | Miridae | Africa | Sap-sucker; toxic saliva |
| *Helopeltis schoutedeni* Reuter | Cotton Jassid | Miridae | Africa | Sap-sucker; toxic saliva |
| *Acanthocoris* spp. | Coreid Bugs | Coreidae | Philippines | Sap-suckers; toxic saliva |

| | | | | |
|---|---|---|---|---|
| *Thrips tabaci* Lind. | Onion Thrips | Thripidae | Cosmopolitan | Infest foliage |
| *Scirtothrips dorsalis* Hood | Chilli Thrips | Thripidae | India, Sri Lanka, Thailand | Infest foliage |
| *Agrotis ipsilon* (Hfn.) | Black Cutworm | Noctuidae | S.E. Asia, India | Larvae are cutworms |
| *Spodoptera exigua* (Hub.) | Beet Armyworm | Noctuidae | India | Larvae bore fruits |
| *Spodoptera litura* (F.) | Rice Cutworm | Noctuidae | Malaysia, Laos, Philippines, India | Larvae bore fruits |
| *Spodoptera littoralis* (Boisd.) | Cotton Leafworm | Noctuidae | Africa | Larvae defoliate |
| *Tiracola plagiata* (Wlk.) | Banana Fruit Caterpillar | Noctuidae | S.E. Asia | Larvae eat leaves or bore fruits |
| *Mythimna separata* (Wlk.) | Rice Ear-cutting Caterpillar | Noctuidae | Philippines | Larvae eat leaves or bore fruits |
| *Mythimna* spp. | Rice Cutworms | Noctuidae | S.E. Asia ⎫ | Larvae eat leaves or bore fruits |
| *Plusia* spp. | Semi-loopers | Noctuidae | India ⎬ | |
| *Agrius convolvuli* (L.) | Convolvulus Hawk | Sphingidae | S.E. Asia | Larvae defoliate |
| *Dacus* spp. | Fruit Flies | Tephretidae | S.E. Asia | Larvae inside fruit |
| *Asphondylia capsici* Barnes | Capsicum Gall Midge | Cecidomyiidae | Med. | Larvae gall fruit |
| *Anomala* spp. | White Grubs | Scarabaeidae | Philippines, India | Larvae eat roots |
| *Leucopholis irrorata* (Chevr.) | White Grub | Scarabaeidae | Philippines | Larvae eat roots |
| *Psylliodes* spp. | Flea Beetles | Chrysomelidae | Philippines | Adults hole leaves |
| *Orthaulaca similis* Oliv. | Leaf Beetle | Chrysomelidae | Philippines | Adults eat leaves |
| *Monolepta signata* Oliv. | Leaf Beetle | Chrysomelidae | India | Adults eat leaves |
| *Tarsonemus translucens* Green | Leaf Mite | Tarsonemidae | India | Scarify foliage |
| *Calacarus carinatus* (Green) | Purple Mite | Eriophyidae | India, Japan, S.E. Asia, S. USA, Australia | Distort foliage |

# CARDAMOM (*Elettaria cardamomum* – Zingiberaceae)

A native of India, but now grown in other tropical countries, especially C. America. It is a perennial herb 2–4 m in height, with white flowers and thin triangular fruits (capsules) containing the small pale seeds. The seeds are used in curries, pickles, cakes and other culinary purposes, as a masticatory in India, and for medicinal purposes.

## MAJOR PESTS

| | | | | |
|---|---|---|---|---|
| *Ragwelellus horvathi* Popp. | Cardamom Capsid | Miridae | Papua NG | Sap-sucker |
| *Sciothrips cardamomi* (Ramk.) | Cardamom Thrips | Thripidae | India, Papua NG | Infest foliage |
| *Lenodera vittata* Wlk. | Leaf Caterpillar | Lasiocampidae | India | Larvae eat leaves |
| *Eupterote* spp. | Hairy Caterpillars | Bombycidae | India | Larvae eat leaves |

## MINOR PESTS

| | | | | |
|---|---|---|---|---|
| *Orthacris* sp. | Leaf Grasshopper | Acrididae | India | Eat young leaves |
| *Pentalonia nigronervosa* Coq. | Banana Aphid | Aphididae | India | Infest leaf bases |
| *Aphrophora nuwarana* Dist. | Spittle Bug | Cercopidae | India | Suck sap |
| *Tettigoniella ferruginea* | Leafhopper | Cicadellidae | India | Infest young leaves |
| *Diaspis* spp. | Armoured Scales | Diaspididae | India | Encrust foliage |
| *Mittilaspis* spp. | Soft Scales | Coccidae | India | Encrust berries |
| *Stephanitis typica* Dist. | Banana Lace Bug | Tingidae | India, S.E. Asia, Papua NG | Sap-sucker; toxic saliva |
| *Riptortus pedestris* F. | Coreid Bug | Coreidae | India | Sap-sucker |
| *Leewania maculans* P. & S. | Thrips | Thripidae | India | Infest foliage |
| *Dichocrocis punctiferalis* (Guen.) | Castor Capsule Borer | Pyralidae | India | Larvae bore shoots & capsules |
| *Hilarographa caminodes* Meyr. | – | Plutellidae | India | Larvae eat leaves |
| *Acanthopsyche bispar* Wlk. | Bagworm | Psychidae | India | Larvae eat leaves |
| *Homona* sp. | Tortrix Moth | Tortricidae | India | Larvae roll leaves |
| *Eumelia rosalia* Cram. | Looper Caterpillar | Geometridae | India | Larvae defoliate |
| *Attacus atlas* L. | Atlas Moth | Saturniidae | India | Larvae defoliate |
| *Plesioneura alysos* M. | Black Skipper | Hesperiidae | India | Larvae fold leaves |
| *Lampides elpis* Godt. | Blue Butterfly | Lycaenidae | India | Larvae eat flowers & pods |
| *Alphaea biguttata* Wlk. | Tiger Moth | Arctiidae | India | Larvae eat leaves |
| *Euproctis lutifacia* Hamp. | Tussock Moth | Lymantriidae | India | Larvae defoliate |
| *Hallomyia cardamomi* Nayer | Root Gall Midge | Cecidomyiidae | India | Larvae gall root |
| *Formosina flavipes* M. | Shoot Fly | Chloropidae | India | Larvae bore shoots |
| *Oulema* sp. | Leaf Beetle | Chrysomelidae | India | Larvae mine leaves; adults eat strips |
| *Prodioctes haematicus* Chev. | Rhizome Weevil | Curculionidae | India, Sri Lanka | Larvae in rhizome |

# CASHEW (*Anacardium occidental* – Anacardiaceae)

Cashew originated in C. and S. America and the W. Indies, and was widely distributed by early Portuguese and Spanish adventurers. It was first brought to India from Brazil in the 16th century, and also reached Malaya and the E. African coast at about the same time. It is now naturalized in many tropical countries, particularly in coastal areas. The spreading evergreen tree, up to 12 m in height, is hardy and drought-resistant, and can be grown under varied conditions of climate and soil from sea-level to 1500 m with 40–350 cm of rain. It is easily damaged by frost. The fruit is a kidney-shaped nut partly embedded in a large fleshy pedicel (Cashew Apple). The main production areas are the coastal strips of S. India, Mozambique and Tanzania.

## MAJOR PESTS

| | | | | |
|---|---|---|---|---|
| *Toxoptera odinae* van d.G. | Mango Aphid | Aphididae | India | Infest foliage |
| *Aleurocanthus woglumi* Ashby | Citrus Blackfly | Aleyrodidae | Pantropical | Infest leaves |
| *Helopeltis* spp. | Helopeltis Bugs | Miridae | S.E. Asia, Africa | Toxic saliva |
| *Helopeltis anacardii* Miller | Cashew Helopeltis | Miridae | India, E. Africa | Toxic saliva |
| *Indarbela tetraonis* (Moore) | Bark Borer | Metarbelidae | India | Larvae eat bark |
| *Macalla moncusalis* Wlk. | Shoot Webber | Pyralidae | India | Larvae web & eat young leaves |
| *Paranaleptes reticulata* (Thom.) | Cashew Stem Girdler | Cerambycidae | E. Africa | Larva girdles stem |
| *Rhytidodera simulans* (White) | Cashew Longhorn | Cerambycidae | Malaysia | Larvae bore trunk |
| *Plocoederus ferrugineus* L. | Longhorn Beetle | Cerambycidae | India | Larvae bore trunk |
| *Mecocorynus loripes* Chevr. | Cashew Weevil | Curculionidae | E. Africa, Mozambique | Larvae bore under bark |

## MINOR PESTS

| | | | | |
|---|---|---|---|---|
| *Aphis craccivora* Koch | Groundnut Aphid | Aphididae | Cosmopolitan | Infest leaves |
| *Ferrisia virgata* (Ckll.) | Striped Mealybug | Pseudococcidae | Pantropical | Infest twigs |
| *Egropa malayensis* (Dist.) | Treehopper | Membracidae | Malaysia | Infest twigs |
| *Pseudotheraptus wayi* Brown | Coconut Bug | Coreidae | E. Africa | Toxic saliva |
| *Selenothrips rubrocinctus* (Giard) | Red-banded Thrips | Thripidae | Philippines, India | Infest leaves |
| *Attacus atlas* (L.) | Atlas Moth | Saturniidae | Malaysia | Larvae defoliate |
| *Orthaga incarusalis* (Wlk.) | – | Pyralidae | Malaysia, India | Larvae web leaves |
| *Euproctis* spp. | Tussock Moths | Lymantriidae | Philippines | Larvae defoliate |
| *Orgyia australis* | Tussock Moth | Lymantriidae | Philippines | Larvae defoliate |
| *Xylotrupes gideon* (L.) | Unicorn Beetle | Scarabaeidae | Malaysia | Adults defoliate |
| *Niphonoclea* spp. | Twig Borers | Cerambycidae | Philippines | Larvae bore twigs |
| *Plocoederus* spp. | Longhorn Beetles | Cerambycidae | India | Larvae bore trunk |
| *Sthenias grisator* F. | Stem Girdler | Cerambycidae | India | Larvae girdle stem |
| *Myllocerus* spp. | Grey Weevils | Curculionidae | India | Adults eat leaves |

# CASSAVA (*Manihot esculenta* — Euphorbiaceae)
## (= Manioc; Tapioca; Yuca)

Cassava is unknown in the wild state; it probably originated from either S. Mexico or Brazil. It is a lowland tropical crop and can be grown under a variety of conditions, but favours a light sandy soil. It is a short-lived shrub, 1–5 m in height, with latex in all parts. The tubers develop as swellings on adventitious roots close to the stem, 5–10 per plant. The core is rich in starch (20–30%), also calcium and vitamin C.

Cyanic acid is present in the tubers and has to be destroyed before the tubers are eaten. More is grown in Africa than elsewhere, and here it is for local consumption. Cassava is exported from Indonesia, Malaysia, Madagascar and Brazil. Propagation is by stem cuttings. (See also Bellotti & van Schoonhoven, 1978).

## MAJOR PESTS

| | | | | |
|---|---|---|---|---|
| *Zonocerus elegans* (Thun.) | Elegant Grasshopper | Acrididae | Africa | Defoliate |
| *Zonocerus variegatus* L. | Variegated Grasshopper | Acrididae | Africa | Defoliate |
| *Bemisia tabaci* (Genn.) | Tobacco Whitefly | Aleyrodidae | Pantropical | Infest foliage; virus vector |
| *Aonidomytilus albus* (Ckll.) | Cassava Scale | Diaspididae | Africa, India, Florida, C. & S. America | Encrust stems |
| *Erinnyis ello* (L.) | Cassava Hornworm | Sphingidae | USA, C. & S. America | Larvae defoliate |

## MINOR PESTS

| | | | | |
|---|---|---|---|---|
| *Brachytrupes portentosus* (Lich.) | Large Brown Cricket | Gryllidae | Papua NG | Eat roots & cuttings |
| *Coptotermes* spp. | Subterranean Termites | Rhinotermitidae | Madagascar | Eat cuttings |
| *Aleurotrachelus* sp. | Whitefly | Aleyrodidae | USA, S. America | Infest foliage; virus vectors |
| *Bemisia* spp. | Whiteflies | Aleyrodidae | Africa, S. America | |
| *Trialeurodes variabilis* | Whitefly | Aleyrodidae | USA, S. America | |
| *Ferrisia virgata* (Ckll.) | Striped Mealybug | Pseudococcidae | Pantropical | Infest leaves |
| *Phenacoccus gossypii* T. & C. | Mexican Mealybug | Pseudococcidae | S. America, Hawaii | Infest leaves |
| *Planococcus citri* (Risso) | Root Mealybug | Pseudococcidae | Pantropical | Infest leaves & roots |
| *Coccus viridis* (Green) | Soft Green Scale | Coccidae | Madagascar, Pacific Isl. | Infest leaves |
| *Pseudaulacaspis pentagona* (T.-T.) | White Peach Scale | Coccidae | Pacific Isl. | Encrust stems |
| *Saissetia coffeae* (Wlk.) | Helmet Scale | Coccidae | Cosmopolitan | Encrust stems |
| *Saissetia nigra* (Nietn.) | Nigra Scale | Coccidae | Malaysia, Pacific Isl. | Encrust stems |
| *Vatiga manihotae* | Cassava Lacebug | Tingidae | S. America | |
| *Amblypelta* spp. | Leaf-footed Bugs | Coreidae | Solomon Isl., Papua NG | Sap-suckers; toxic saliva; tatter young leaves |
| *Dasynus manihotis* | Cassava Bug | Coreidae | Papua NG | |
| *Helopeltis bergrothi* Reut. | Mosquito Bug | Miridae | Africa | |
| *Leptoglossus australis* (F.) | Squash Bug | Coreidae | S.E. Asia | |
| *Nezara viridula* (L.) | Green Stink Bug | Pentatomidae | S.E. Asia | |

| | | | | |
|---|---|---|---|---|
| *Scirtothrips manihoti* Bondar | Cassava Thrips | Thripidae | C. & S. America | Infest foliage |
| *Selenothrips rubrocinctus* (Giard) | Red-banded Thrips | Thripidae | Pacific Isl. | Infest foliage |
| *Atta* spp. | Leaf-cutter Ants | Formicidae | S. America | Adults defoliate |
| *Acromyrmex* spp. | Leaf-cutter Ants | Formicidae | S. America | Adults defoliate |
| *Zeuzera coffeae* Niet. | Red Coffee Borer | Cossidae | Papua NG, Solomon Isl. | Larvae bore stems |
| *Eldana saccharina* Wlk. | Sugarcane Stalk Borer | Pyralidae | Africa | Larvae bore stems |
| *Agrotis ipsilon* (Wlk.) | Black Cutworm | Noctuidae | S.E. Asia, S. America | Larvae are cutworms |
| *Spodoptera littoralis* (Boisd.) | Cotton Leafworm | Noctuidae | Madagascar | Larvae are cutworms |
| *Tiracola plagiata* (Wlk.) | Banana Fruit Caterpillar | Noctuidae | S.E. Asia, Pacific Isl., Australasia | Larvae defoliate |
| *Dasychira horsfieldi* (Saund.) | Tussock Moth | Lymantriidae | Malaysia | Larvae defoliate |
| *Jatrophobia brasiliensis* Rubs. | Gall Midge | Cecidomyiidae | USA, S. America | Larvae gall young leaves |
| *Anastrepha* spp. | Fruit Flies | Tephritidae | S. America | Larvae bore fruits |
| *Lonchaea* spp. | Shoot Flies | Lonchaeidae | S. America | Larvae bore shoots |
| *Silba pendula* | Shoot Fly | Lonchaeidae | S. America | Larvae bore shoots |
| *Atherigona* spp. | Shoot Fly | Muscidae | Japan, Pacific Isl. | Larvae bore shoots |
| *Anomala* spp. | White Grubs | Scarabaeidae | S.E. Asia | Larvae eat roots |
| *Phyllophaga* spp. | White Grubs | Scarabaeidae | S. America | Larvae eat roots |
| *Lepidiota stigma* F. | White Grubs | Scarabaeidae | Thailand, Papua NG | Larvae eat roots |
| *Leucopholis rorida* F. | White Grub | Scarabaeidae | Indonesia, Papua NG | Larvae eat roots |
| *Lagochirus* spp. | Longhorn Beetles | Cerambycidae | Indonesia | Larvae bore stems |
| *Coelosternus* spp. | Stem Weevils | Curculionidae | S. America | Larvae bore stems |
| *Apirocaulus cornutus* Pasc. | Leaf-eating Weevil | Curculionidae | Papua NG | Adults eat leaves |
| *Hypomeces squamosus* F. | Gold-dust Weevil | Curculionidae | Thailand, Laos | Adults eat leaves |
| *Sepiomus* sp. | Weevil | Curculionidae | Thailand | Adults eat leaves |
| *Mononychellus tanajoa* (Bondar) | Spider Mite | Tetranychidae | Uganda, S. America | Infest foliage & scarify leaves |
| *Oligonychus* spp. | Spider Mites | Tetranychidae | Australia, Pacific Isl., S. America | |
| *Tetranychus cinnabarinus* (Boisd.) | Carmine Mite | Tetranychidae | Pantropical | |

533

# CASTOR (*Ricinus communis* – Euphorbiaceae)

Castor grows wild in N. and E. Africa, Yemen, Near and Middle East, was established very early in Egypt, and from there was taken to India and China. Now it is cultivated in many tropical and sub-tropical countries. It needs a warm climate and 180 frost-free days. It is grown under a wide range of conditions of altitude and rainfall, but is killed by frost and waterlogged soil, or temperatures for any length of time over 30°C. It is an annual herb 1–7 m in height; sometimes a short-lived perennial. The main production areas are Brazil, India, Thailand, USA, Ecuador, S. Africa, Ethiopia, and Tanzania. The seeds contain copious endosperm with 40–55% oil content; the oil is used for medicinal purposes, and industrially in lubricants, soaps, inks, dyes, and for tanning. Propagation is by seed.

## MAJOR PESTS

| | | | | |
|---|---|---|---|---|
| *Helopeltis schoutedeni* Reut. | Cotton Helopeltis | Miridae | Africa | Sap-sucker; toxic saliva |
| *Helopeltis* spp. | Helopeltis Bugs | Miridae | S.E. Asia, Africa | Sap-suckers with toxic saliva |
| *Nezara viridula* (L.) | Green Stink Bug | Pentatomidae | Cosmopolitan � | Sap-suckers with toxic saliva |
| *Calidea* spp. | Blue Bugs | Pentatomidae | Africa | |
| *Dichocrocis punctiferalis* Gn. | Castor Capsule Borer | Pyralidae | India | Larvae bore capsules & shoots |
| *Achaea janata* (L.) | Castor Semi-looper (Fruit-piercing Moth) | Noctuidae | Malaysia, India, Laos, Philippines | Larvae defoliate |
| *Cryptophlebia leucotreta* (Meyr.) | False Codling Moth | Tortricidae | Africa | Larvae bore fruits |
| *Xyleutes capensis* (Wlk.) | Castor Stem Borer | Cossidae | E. Africa | Larva bores stem |

## MINOR PESTS

| | | | | |
|---|---|---|---|---|
| *Zonocerus variegatus* L. | Variegated Grasshopper | Acrididae | Africa | Defoliate |
| *Empoasca formosana* Paoli | Castor Jassid | Cicadellidae | S.E. Asia ⎫ | Sap-suckers, infest leaves; toxic saliva |
| *Empoasca fascialis* (Jacobi) | Cotton Jassid | Cicadellidae | Pantropical ⎬ | |
| *Empoasca flavescens* (F.) | Leaf-hopper | Cicadellidae | Malaysia ⎭ | |
| *Ptyelus grossus* F. | Spittlebug | Cercopidae | E. Africa | Infest foliage |
| *Ferrisia virgata* (Ckll.) | Striped Mealybug | Pseudococcidae | Philippines | Infest stems |
| *Euristylus capensis* | Capsid Bug | Miridae | Mozambique ⎫ | Sap-sucker; toxic saliva; feed on fruits |
| *Taylorilygus ricini* | Castor Capsid | Miridae | Mozambique ⎭ | |
| *Retithrips syriacus* (Mayer) | Castor Thrips | Thripidae | India, Africa | Infest foliage |
| *Maruca testulalis* (Geyer) | Maruca Moth | Pyralidae | S.E. Asia | Larvae bore fruits |
| *Dichocrocis punctiferalis* (Guen.) | Castor Capsule Borer | Pyralidae | India, S.E. Asia, Japan | Larvae bore capsules |
| *Spodoptera litura* (F.) | Rice Cutworm | Noctuidae | S.E. Asia | Larvae defoliate |
| *Spodoptera littoralis* (Boisd.) | Cotton Leafworm | Noctuidae | Africa | Larvae defoliate |
| *Heliothis armigera* (Hb.) | American Bollworm | Noctuidae | Philippines | Larvae eat leaves |

| | | | | |
|---|---|---|---|---|
| *Tiracola plagiata* (Wlk.) | Banana Fruit Caterpillar | Noctuidae | S.E. Asia | Larvae attack fruits |
| *Attacus ricini* | Castor Silkworm | Saturniidae | Philippines | Larvae defoliate |
| *Samia cynthia* | Lesser Atlas Moth | Saturniidae | Taiwan | Larvae defoliate |
| *Euproctis varians* (Wlk.) | Tussock Moth | Lymantriidae | Philippines | Larvae defoliate |
| *Euproctis producta* Wlk. | Tussock Moth | Lymantriidae | Africa | Larvae defoliate |
| *Orgyia australis* | Tussock Moth | Lymantriidae | Philippines | Larvae defoliate |
| *Dasychira mendosa* Hb. | Tussock Moth | Lymantriidae | Philippines, Laos | Larvae defoliate |
| *Niphadolepis alianta* Karsh | Jelly Grub | Limacodidae | Africa | Larvae defoliate |
| *Parasa vivida* (Wlk.) | Stinging Caterpillar | Limacodidae | Africa, India | Larvae defoliate |
| *Latoia lepida* (Wlk.) | Blue-striped Nettlegrub | Limacodidae | S.E. Asia, India | Larvae defoliate |
| *Stomophastis conflua* | Leaf Miner | Gracillariidae | Mozambique | Larvae mine leaves |
| *Asphondylia ricini* Mani· | Castor Gall Midge | Cecidomyiidae | India, Mozambique | Larvae gall fruits |
| *Dihammus vastator* Newm. | Longhorn Beetle | Cerambycidae | Philippines | Larvae bore stems |
| *Hypomeces squamosus* (F.) | Gold-dust Weevil | Curculionidae | Laos | Adults eat leaves |
| *Xyleborus fornicatus* Eichh. | Tea Shot-hole Borer | Scolytidae | India, S.E. Asia | Adults bore stems |
| *Polyphagotarsonemus latus* (Banks) | Yellow Tea Mite | Tarsonemidae | S. Africa, S.E. Asia | Scarify foliage |
| *Oligonychus coffeae* (Niet.) | Red Coffee Mite | Tetranychidae | Pantropical | Scarify foliage |
| *Eutetranychus orientalis* (Klein) | Oriental Mite | Tetranychidae | Africa, India | Scarify foliage |
| *Tetranychus cinnabarinus* (Boisd.) | Tropical Red Spider Mite | Tetranychidae | Pantropical | Scarify foliage |
| *Tetranychus* spp. | Red Spider Mites | Tetranychidae | Cosmopolitan | Scarify foliage |

# CHICKPEA (*Cicer arietinum* − Leguminosae)
## ( = Gram; Garbanzo Bean)

Also called Gram, it is thought to have originated in western Asia, and has been the most important pulse crop of India since early times. Now cultivated widely throughout the tropics. A drought-resistant crop requiring a cool dry climate. It is grown in India as a winter crop; generally unsuccessful in the hot wet tropics. A spreading, branched annual herb, up to 0.5 m high, hairy, with short pods containing 1−2 seeds only. Seeds vary in colour from white to red to black. Areas of maximum cultivation are India and the Middle East; the crop is grown largely for local consumption; India grows over 8 million hectares annually, producing 4−5 million tonnes. The foliage is, however, toxic and is not used for fodder.

## MAJOR PESTS

| | | | | |
|---|---|---|---|---|
| *Tricentrus bicolor* Dist. | Tree-hopper | Membracidae | India | Infest stems; suck sap |
| *Agrotis ipsilon* (Hfn.) | Black Cutworm | Noctuidae | India | Larvae are cutworms |
| *Heliothis armigera* (Hub.) | American Bollworm | Noctuidae | India | Larvae eat shoots; bore pods |
| *Melanagromyza obtusa* (Mall.) | Bean Pod Fly | Agromyzidae | India | Larvae bore young seeds in pod |
| *Tanymecus indicus* Fst. | Surface Weevil | Curculionidae | India | Adults cut stems of seedlings |

## MINOR PESTS

| | | | | |
|---|---|---|---|---|
| *Atractomorpha crenulata* F. | Grasshopper | Acrididae | India | Adults & nymphs eat foliage |
| *Bemisia tabaci* Genn. | Cotton Whitefly | Aleyrodidae | India | Infest leaves |
| *Ferrisia virgata* (Ckll.) | Striped Mealybug | Pseudococcidae | India | Infest foliage |
| *Lamprosema indicata* F. | Webworm | Pyralidae | India | Larvae web leaves |
| *Agrotis segetum* Schiff. | Common Cutworm | Noctuidae | India | Larvae are cutworms |
| *Agrotis* spp. | Cutworms | Noctuidae | India | Larvae are cutworms |
| *Spodoptera exigua* (Hb.) | Lesser Armyworm | Noctuidae | India | Larvae defoliate |
| *Spodoptera litura* (F.) | Rice Cutworm | Noctuidae | India | Larvae defoliate |
| *Plusia orichalcea* (F.) | − | Noctuidae | India | Larvae defoliate |
| *Mythimna loreyi* (Dup.) | Rice Armyworm | Noctuidae | India | Larvae defoliate |
| *Mythimna separata* (Wlk.) | Rice Ear-cutting Caterpillar | Noctuidae | India | Larvae defoliate |
| *Lampides boeticus* L. | Pea Blue Butterfly | Lycaenidae | India | Larvae inside pods |

# CINCHONA (*Cinchona* spp. — Rubiaceae)
## (= Quinine)

This is a small evergreen tree, or shrub, native to the Andes in S. America, and from the bark is extracted quinine, the specific remedy for the cure of malaria. The use of cinchona bark was known to the S. American Indians, and later to the Jesuits who were responsible for its introduction to the rest of the world and its widespread cultivation. It is now cultivated in India, Java, and E. Africa, in addition to S. America. Four or five species are grown for quinine production, but two are probably only varieties.

## MAJOR PESTS

## MINOR PESTS

| | | | | |
|---|---|---|---|---|
| *Helopeltis antonii* Sign. | Tea Mosquito Bug | Miridae | India, S.E. Asia | Sap-suckers; toxic saliva |
| *Helopeltis bergrothi* Reut. | Mosquito Bug | Miridae | Africa | |
| *Pachypeltis humeralis* (W.) | Mirid Bug | Miridae | India | |
| *Margaronia marginata* Hmps. | – | Pyralidae | Malaysia | Larvae eat leaves |
| *Belippa laleana* M. | Slug Caterpillar | Limacodidae | India | Larvae defoliate |
| *Popillia chlorion* N. | Flower Beetle | Scarabaeidae | India | Adults defoliate |
| *Holotrichia repetita* S. | Cockchafer | Scarabaeidae | India | Adults eat leaves |
| *Rhizotrogus rufus* A. | Cockchafer | Scarabaeidae | India | Adults eat leaves |
| *Serica nilgiriensis* S. | Cockchafer | Scarabaeidae | India | Adults eat leaves |
| *Sympiezomias decipiens* M. | Leaf Weevil | Curculionidae | India | Adults eat leaves |
| *Xyleborus fornicatus* Eichh. | Tea Shot-hole Borer | Scolytidae | India, S.E. Asia, Madagascar | Adults bore stems and twigs |

# CINNAMON (*Cinnamomum zeylandicum* — Lauraceae)

Cinnamon is extracted from the young bark of a small evergreen tree with dark coriaceous aromatic leaves. Native to Sri Lanka, it is now cultivated in India, Burma, Malaya, Seychelles, and also the W. Indies and parts of S. America.

The leaves and roots also produce oil but of different quality and generally inferior to the essential oil from the bark. A popular spice for food flavouring, and widely used in medicine.

## MAJOR PESTS

| | | | | |
|---|---|---|---|---|
| *Pauropsylla depressa* Crawford | Fig-leaf Psyllid | Psyllidae | India | Young shoots galled |
| *Attacus atlas* (L.) | Atlas Moth | Saturniidae | Malaysia | Larvae defoliate |

## MINOR PESTS

| | | | | |
|---|---|---|---|---|
| *Vinsonia stellifera* Westw. | Star Scale | Coccidae | Seychelles | Infest leaves |
| *Phyllocnistis chrysophthalina* | Cinnamon Leaf Miner | Phyllocnistidae | India | Larvae mine leaves |
| *Homona coffearia* (Niet.) | Tea Tortrix | Tortricidae | Malaysia | Larvae roll leaves |
| *Papilio* sp. | Swallowtail Butterfly | Papilionidae | India | Larvae eat leaves |
| *Dasychira mendosa* (Hubn.) | Jute Hairy Caterpillar | Lymantriidae | India, Malaysia | Larvae defoliate |

# CITRUS (*Citrus* spp. – Rutaceae)
## (Orange, Lemon, Lime, Mandarin, Tangerine, Grapefruit, Pomelo)

The cultivated species of *Citrus* are native to S.E. Asia, where they originated in the drier monsoon areas, but they are now grown throughout the tropics and sub-tropics, often under irrigation. They are thorny, aromatic shrubs or small trees with leathery evergreen leaves. The white or purple flowers are usually very fragrant. They are cultivated from about 45° N to 35° S, between sea-level and 1000 m, and are susceptible to frost unless the tree is dormant. Generally they require 100 cm of rain or else irrigation; they do not grow well in the very humid tropics. The main areas of production are in sub-tropical regions, and are S. USA, the Mediterranean region, S. Africa, C. America, Australia, China, and Japan. Although still a young industry in many places, it is now an exceedingly valuable one. (See also Butani, 1979*b*.)

## *MAJOR PESTS*

| | | | | |
|---|---|---|---|---|
| *Trioza erytreae* (Del G.) | Citrus Psyllid | Psyllidae | E. & S. Africa | Nymphs pit leaves |
| *Aleurocanthus woglumi* Ashby | Citrus Blackfly | Aleyrodidae | Pantropical | Infest leaves |
| *Dialeurodes citri* (Ashm.) | Citrus Whitefly | Aleyrodidae | India, S.E. Asia | Infest leaves |
| *Toxoptera aurantii* (B. de F.) | Black Citrus Aphid | Aphididae | Cosmopolitan | Infest foliage |
| *Toxoptera citricida* (Kirk.) | Brown Citrus Aphid | Aphididae | Pantropical | Infest foliage |
| *Planococcus citri* (Ricco) | Root Mealybug | Pseudococcidae | Pantropical | Infest foliage |
| *Pseudococcus adonidum* (L.) | Long-tailed Mealybug | Pseudococcidae | Pantropical | Infest foliage |
| *Pseudococcus citriculus* Green | Citrus Mealybug | Pseudococcidae | S.E. Asia | Infest foliage |
| *Lepidosaphes beckii* (Neumann) | Mussel Scale | Diaspididae | Cosmopolitan | Infest foliage |
| *Aonidiella aurantii* (Maskell) | California Red Scale | Diaspididae | Cosmopolitan | Infest foliage |
| *Chrysomphalus aonidum* (L.) | Purple Scale | Diaspididae | Cosmopolitan | Infest foliage |
| *Saissetia oleae* (Bern.) | Black Scale | Coccidae | Cosmopolitan | Infest foliage |
| *Coccus viridis* (Green) | Soft Green Scale | Coccidae | Pantropical | Infest foliage |
| *Coccus alpinus* De Lotto | Soft Green Scale | Coccidae | E. Africa | Infest foliage |
| *Ceroplastes rubens* Mask | Pink Waxy Scale | Coccidae | Old World tropics | Infest foliage |
| *Gascardia destructor* (Newst.) | White Waxy Scale | Coccidae | Africa, Australasia, Florida, Mexico | Infest foliage |
| *Icerya purchasi* Mask. | Cottony Cushion Scale | Margarodidae | Pantropical | Infest foliage |
| *Rhynchocoris* spp. | Citrus Green Bugs | Pentatomidae | Africa, Asia, China | Suck sap; toxic saliva |
| *Heliothrips haemorrhoidalis* (Bouché) | Black Tea Thrips | Thripidae | S.E. Asia | Scarify leaves |
| *Scirtothrips aurantii* Fauré | Citrus Thrips | Thripidae | Africa | Scar fruits |
| *Papilio demoleus* L. | Lemon Butterfly | Papilionidae | India, S.E. Asia, China | Larvae defoliate |
| *Papilio demodocus* Esp. | Orange Dog | Papilionidae | Africa | Larvae defoliate |
| *Papilio polytes* L. | Common Mormon | Papilionidae | S.E. Asia | Larvae defoliate |
| *Othreis fullonia* Cl. | Fruit-piercing Moth | Noctuidae | India, S.E. Asia, Australasia, China, Africa | Adults pierce fruits |
| *Ceratitis capitata* (Wied.) | Medfly | Tephritidae | Cosmopolitan | Larvae inside fruit |
| *Ceratitis rosa* Karsch | Natal Fruit Fly | Tephritidae | Africa | Larvae inside fruit |
| *Dacus dorsalis* (Hend.) | Oriental Fruit Fly | Tephritidae | S.E. Asia | Larvae in fruits |

| *Solenopsis geminata* (F.) | Fire Ant | Formicidae | India, S.E. Asia, Africa, C. & S. America | Sting workers |
|---|---|---|---|---|
| *Leucopholis irrorata* (Chevr.) | White Grub | Scarabaeidae | Philippines | Larvae eat roots |
| *Systates pollinosus* Gerst. | Systates Weevil | Curculionidae | E. Africa | Adults eat leaves |
| *Agrilus* spp. | Citrus Bark Borers | Buprestidae | Philippines, India, China | Larvae bore bark |
| *Anoplophora chinensis* (Forst.) | Citrus Longhorn | Cerambycidae | China, Taiwan | Larvae bore trunk |
| *Tetranychus cinnabarinus* (Boisd.) | Tropical Red Spider Mite | Tetranychidae | Pantropical | Scarify leaves & fruit |
| *Panonychus citri* (Mc G.) | Citrus Red Spider Mite | Tetranychidae | S. Africa, Asia, Australia, USA, S. America | Scarify leaves & fruit |
| *Eutetranychus orientalis* (Klein) | Oriental Mite | Tetranychidae | India, S.E. Asia, Africa, Taiwan | Scarify leaves & fruit |
| *Aceria sheldoni* (Ew.) | Citrus Bud Mite | Eriophyidae | Italy, Africa, USA, Australia | Feed on leaves and buds |
| *Phyllocoptruta oleivora* (Ashm.) | Citrus Rust Mite | Eriophyidae | Cosmopolitan | Feed on leaves and buds |
| *Brevipalpus phoenicia* (Geijskes) | Red Crevice Tea Mite | Tenuipalpidae | Pantropical | Feed on leaves and buds |

## MINOR PESTS

| *Brachytrupes* spp. | Large Brown Crickets | Gryllidae | Africa, Asia | Nursery pests |
|---|---|---|---|---|
| *Nasutitermes* spp. | Termites | Termitidae | S. America | Nursery pests |
| *Diaphorina citri* Kew. | Citrus Psyllid | Psyllidae | India, S.E. Asia, China, Japan | Vector of Greening Virus |
| *Aleurocanthus spiniferus* Quaint | Orange Spiny Whitefly | Aleyrodidae | India, S.E. Asia, China, Japan | Infest leaves |
| *Aphis gossypii* Glov. | Cotton Aphid | Aphididae | Philippines, Africa | Infest foliage |
| *Aphis spiraecola* Patch | Spiraea Aphid | Aphididae | Cosmopolitan | Infest foliage |
| *Aphis tavaresi* (Kirk.) | – | Aphididae | Malaysia | Infest foliage |
| *Myzus persicae* (Sulz.) | Green Peach Aphid | Aphididae | Philippines | Infest foliage |
| *Coccus hesperidum* L. | Brown Scale | Coccidae | Cosmopolitan | Infest foliage |
| *Gascardia brevicauda* (Hall) | White Waxy Scale | Coccidae | E. Africa | Encrust foliage |
| *Chloropulvinaria psidii* (Mask.) | Guava Scale | Coccidae | Pantropical | Infest foliage |
| *Ceroplastes sinensis* Del G. | Chinese Waxy Scale | Coccidae | Widespread | Encrust foliage |
| *Saissetia coffeae* (Wlk.) | Helmet Scale | Coccidae | Cosmopolitan | Infest foliage |
| *Pulvinaria* spp. | Soft Scales | Coccidae | S.E. Asia | Infest foliage |
| *Orthezia insignis* Browne | Jacaranda Bug | Ortheziidae | Africa, India, Malaysia, N., C. & S. America | Infest foliage |
| *Drosicha stebbingii* Stebb. | Giant Mealybug | Margarodidae | India, Pakistan | Infest foliage |

| | | | | |
|---|---|---|---|---|
| *Icerya seychellarum* (Westw.) | Seychelles Fluted Scale | Margarodidae | S.E. Asia | Infest foliage |
| *Ferrisia virgata* (Ckll.) | Striped Mealybug | Pseudococcidae | Pantropical | Infest foliage |
| *Pseudococcus maritimus* (Ehrh.) | Grape Mealybug | Pseudococcidae | Widespread | Infest foliage |
| *Ricania speculum* (Wlk.) | Black Planthopper | Ricaniidae | S.E. Asia | Infest twigs |
| *Ricania* sp. | Green Planthopper | Ricaniidae | S.E. Asia | Infest twigs |
| *Tricentrus* spp. | Treehoppers | Membracidae | Philippines, S. China | Infest twigs |
| *Dictyophara* sp. | – | Dictyopharidae | S. China | Infest foliage |
| *Lepidosaphes gloverii* (Packard) | Glover Scale | Diaspididae | S. America | Infest foliage |
| *Aonidiella orientalis* (Newst.) | Oriental Yellow Scale | Diaspididae | Pantropical | Infest foliage |
| *Aspidiotus destructor* Sign. | Coconut Scale | Diaspididae | Pantropical | Infest foliage |
| *Pseudaonidia trilobitiformis* (Green) | Trilobite Scale | Diaspididae | Pantropical | Infest foliage |
| *Pinnaspis aspidistrae* (Sign.) | Fern Scale | Diaspididae | Philippines | Infest foliage |
| *Pinnaspis* spp. | – | Diaspididae | Africa | Infest foliage |
| *Parlatoria pergandii* Comst. | Chaff Scale | Diaspididae | Pantropical | Infest foliage |
| *Selanaspidus articulatus* | – | Diaspididae | S. America | Infest foliage |
| *Unaspis citri* (Cmst.) | Citrus Snow Scale | Diaspididae | Pantropical | Infest foliage |
| *Ischnaspis longirostris* (Sign.) | Black Line Scale | Diaspididae | Pantropical | Infest foliage |
| *Distantiella theobroma* (Dist.) | Cocoa Capsid | Miridae | W. Africa | Sap-sucker; toxic saliva |
| *Helopeltis collaris* Stal | Capsid | Miridae | Philippines | Sap-sucker; toxic saliva |
| *Leptoglossus* spp. | Leaf-footed Plant Bugs | Coreidae | Pantropical | Sap-suckers; toxic saliva |
| *Nezara viridula* L. | Green Stink Bug | Pentatomidae | Pantropical | Sap-sucker; toxic saliva |
| *Scirtothrips citri* (Moulten) | Citrus Thrips | Thripidae | India, USA (California) | Scar fruits |
| *Papilio* spp. (10 + spp.) | Citrus Butterflies | Papilionidae | S.E. Asia, India, Africa, Americas | Larvae defoliate |
| *Prays citri* Mill | Citrus Flower Moth | Yponomeutidae | S. Europe, India, Malaysia, Philippines, Australasia | Larvae eat flowers |
| *Prays endocarpa* Meyr. | Citrus Rind Borer | Yponomeutidae | India, Indonesia | Larvae bore rind |
| *Phyllocnistis citrella* Stnt. | Citrus Leaf Miner | Phyllocnistidae | N.E. Africa, India, China, Japan, S.E. Asia | Larvae mine leaves |
| *Indarbela* spp. | Wood-borer Moths | Metarbelidae | India | Larvae eat bark |
| *Paramyelois transitella* (Wlk.) | Navel Orangeworm | Pyralidae | USA (California) | Larvae bore fruits |
| *Chilades lajus* (Stoll) | Citrus Blue | Lycaenidae | S.E. Asia | Larvae defoliate |
| *Metanastra hyrtaca* (Cramer) | Grisly Citrus Caterpillar | Lasiocampidae | S.E. Asia | Larvae defoliate |
| *Spodoptera* spp. | Cotton Leafworms | Noctuidae | Old World tropics | Larvae defoliate |
| *Othreis* spp. | Fruit Piercing Moths | Noctuidae | India, S.E. Asia | Adults pierce fruit |

| *Achaea* spp. | Fruit Piercing Moths | Noctuidae | Africa, Asia | Adults pierce fruit |
|---|---|---|---|---|
| *Tiracola plagiata* (Wlk.) | Banana Fruit Caterpillar | Noctuidae | S.E. Asia | Larvae destroy fruit |
| *Euproctis similis* (Fuess.) | Tussock Moth | Lymantriidae | Europe, Asia | Larvae defoliate |
| *Zaprionus multistriata* | Citrus Fruit Fly | Drosophilidae | S.E. Asia, India | Larvae in fruit pulp |
| *Dasyneura citri* G. & P. | Citrus Blossom Midge | Cecidomyiidae | India | Larvae in flowers; distort fruits |
| *Anastrepha ludens* (Lw.) | Mexican Fruit Fly | Tephritidae | Mexico, C. America | Larvae in fruit |
| *Dacus cucurbitae* Coq. | Melon Fly | Tephritidae | Africa, Asia, Australasia | Larvae in fruit |
| *Dacus* spp. | Fruit Flies | Tephritidae | India | Larvae in fruit |
| *Paradalaspis quinaria* Bez. | Rhodesian Fruit Fly | Tephritidae | S. & N.W. Africa | Larvae in fruit |
| *Tetradacus tsuneonis* (Miyake) | Japanese Orange Fruit Fly | Tephritidae | Japan, China | Larvae in fruit |
| *Vespa/Vespula* spp. | Common Wasps | Vespidae | Cosmopolitan | Adults pierce ripe fruits |
| *Atta* spp. | Leaf-cutting Ants | Formicidae | C. & S. America | Adults cut leaves |
| *Oecophylla* spp. | Red Tree Ants | Formicidae | Africa, S.E. Asia, Australasia | Nest in foliage; attack workers |
| *Anomala* spp. | Flower Beetles | Scarabaeidae | India, S.E. Asia ⎫ | Adults eat leaves; larvae eat roots in soil; nursery pests |
| *Schizonycha* spp. | Cockchafers | Scarabaeidae | India ⎭ | |
| *Apate monachus* F. | Black Borer | Bostrychidae | Africa, W. Indies | Adults bore stems |
| *Chrysochroa fulminans* (F.) | Jewel Beetle | Buprestidae | Philippines | Larvae bore branches |
| *Prodagricomela nigricollis* | Citrus Flea Beetle | Chrysomelidae | S. China | Larvae mine leaf; adults skeletonize |
| *Argopistes* spp. | Citrus Flea Beetles | Chrysomelidae | S. China | Larvae mine leaves |
| *Monochamus* spp. | Citrus Longhorns | Cerambycidae | India, S.E. Asia | Larvae bore branches |
| *Niphonoclea capito* (Newm.) | Mango Twig Borer | Cerambycidae | Philippines | Larvae bore twigs |
| *Pachnaeus litus* | Weevil | Curculionidae | Cuba | Adults eat leaves |
| *Hypomeces squamosus* (F.) | Gold-dust Weevil | Curculionidae | S.E. Asia | Adults eat leaves |
| *Brevipalpus californicus* (Banks) | — | Tenuipalpidae | Cosmopolitan | Scarify foliage |
| *Polyphagotarsonemus latus* (Banks) | Yellow Mite | Tarsonemidae | Pantropical | Scarify leaves |

# CLOVE (*Eugenia caryophyllus* — Myrtaceae)
## (= *Syzygium aromaticum*)

One of the earliest and most important of the spices; native to the Molucca Islands, but today it is grown in many tropical countries. Cloves are the unopened flower buds of a small evergreen tree of symmetrical shape. They are very aromatic and have widespread uses, both whole and ground, as a culinary spice, for flavouring, and in medicine. The essential oil has many medicinal and industrial uses. Zanzibar produces 90% of the total world output, and the other producing countries are Indonesia, Mauritius and the W. Indies.

*MAJOR PESTS*

| | | | | |
|---|---|---|---|---|
| *Saissetia eugeniae* | Clove Scale | Coccidae | Zanzibar | Infest foliage |
| *Xyleborus dedevigranulatus* (Schedl.) | — | Scolytidae | Malaysia | Adults bore stems |

*MINOR PESTS*

| | | | | |
|---|---|---|---|---|
| *Macrotermes bellicosus* (Smeath) | War-like Termite | Termitidae | Zanzibar | Destroy seedlings |
| *Aleurotuberculatus eugeniae* (Corb.) | Clove Whitefly | Aleyrodidae | Malaysia | Infest foliage |
| *Lecanium* spp. | Soft Scales | Coccidae | S. India | Encrust foliage |
| *Saissetia coffeae* (Wlk.) | Helmet Scale | Coccidae | Malaysia | Encrust foliage |
| *Saissetia nigra* (Nietn.) | Nigra Scale | Coccidae | India | Encrust foliage |
| *Dasychira horsfieldi* (Saund.) | Tussock Moth | Lymantriidae | Malaysia | Larvae defoliate |
| *Oecophylla smaragdina* F. | Red Tree Ant | Formicidae | Zanzibar | Nest in foliage; attack workers |
| *Chelidonium brevicorne* Schwarz | Longhorn Beetle | Cerambycidae | Malaya | Larvae bore trunk |

# COCOA (*Theobroma cacao* – Sterculiaceae)

Cocoa originated in the evergree forest of the Brazilian Andes, and has been cultivated there since early times. It was spread through the New World in the 16th century, and in the 18th the Spanish took it to S.E. Asia and W. Africa. It is a small tree of the lower forest strata, essentially tropical; most are grown within 10° of the equator. The beans are borne inside pods on the trunks and branches. The main production areas are Ghana, Nigeria, S. & C. America. W. Indies, New Guinea and Samoa. (See also Entwistle, 1972.)

## MAJOR PESTS

| | | | | |
|---|---|---|---|---|
| *Toxoptera aurantii* (B. de F.) | Black Citrus Aphid | Aphididae | Cosmopolitan | Infest foliage |
| *Planococcus citri* (Risso) | Root Mealybug | Pseudococcidae | Cosmopolitan | Infest foliage |
| *Coccus viridis* (Green) | Soft Green Scale | Coccidae | Pantropical | Infest foliage |
| *Helopeltis schoutedeni* Reuter | Cotton Helopeltis | Miridae | Africa | Sap-sucker; toxic saliva |
| *Sahlbergella singularis* Haglund | Cocoa Capsid | Miridae | W. & E. Africa | Sap-sucker; toxic saliva; spot pods |
| *Heliothrips haemorrhoidalis* (Bouché) | Black Tea Thrips | Thripidae | Cosmopolitan | Scarify foliage |
| *Eulophonotus myrmelon* Feldr. | Cocoa Stem Borer | Cossidae | W. & E. Africa | Larvae bore stems |
| *Zeuzera coffeae* Nietn. | Red Coffee Borer | Cossidae | Philippines, Malaysia, Indonesia | Larvae bore stems |

## MINOR PESTS

| | | | | |
|---|---|---|---|---|
| *Zonocerus variegatus* L. | Variegated Grasshopper | Acrididae | Africa | Defoliate |
| *Macrotermes bellicosus* (Smeath.) | War-like Termite | Termitidae | W. Africa | Workers eat bark |
| *Empoasca fascialis* (Jacobi) | Cotton Jassid | Cicadellidae | W. Africa | Infest foliage |
| *Aphis gossypii* Glov. | Cotton Aphid | Aphididae | Philippines | Infest foliage |
| *Mesohomotoma tessmanni* Aulm. | Cocoa Psylla | Psyllidae | W. Africa | Nymphs in shoots |
| *Ferrisia virgata* (Ckll.) | Striped Mealybug | Pseudococcidae | Pantropical ⎫ | Infest foliage; virus vectors |
| *Planococcoides njalensis* (Laing) | Cocoa Mealybug | Pseudococcidae | Africa ⎬ | |
| *Pseudococcus adonidum* (L.) | Long-tailed Mealybug | Pseudococcidae | Pantropical ⎭ | |
| *Pseudococcus hispidus* (Morr.) | Mealybug | Pseudococcidae | Malaysia | Infest foliage |
| *Ricania speculum* Wlk. | Black Planthopper | Ricaniidae | Philippines | Infest foliage |
| *Lawana candida* (F.) | Moth Bug | Flattidae | Philippines ⎫ | Infest foliage |
| *Colobesthes falcata* Guér. | Cocoa Moth Bug | Flattidae | Malaysia ⎬ | |
| *Pulastya discolorata* | Moth Bug | Flattidae | S.E. Asia ⎭ | |
| *Stictococcus sjostedti* Ckll. | – | Diaspididae | W. Africa | Infest foliage |
| *Pseudaonidia trilobitiformis* (Green) | Trilobite Scale | Diaspididae | Pantropical | Infest foliage |
| *Helopeltis theobromae* (Miller) | Cocoa Helopeltis | Miridae | Malaysia, Indonesia, Philippines ⎫ | Sap-sucker; toxic saliva |
| *Helopeltis bergrothi* Rent. | Cocoa Mosquito Bug | Miridae | Tropical Africa ⎭ | |

| *Distantiella theobroma* (Dist.) | Cocoa Capsid | Miridae | W. Africa | Sap-suckers; toxic |
| *Bryocoropsis laticollis* | Capsid Bug | Miridae | W. Africa | saliva causes pod spotting |
| *Pseudotheraptus wayi* Brown | Coconut Bug | Coreidae | E. Africa | Sap-sucker; toxic saliva |
| *Leptoglossus australis* (F.) | Leaf-footed Plant Bug | Coreidae | Old World tropics | Sap-sucker; toxic saliva |
| *Nezara viridula* (L.) | Green Stink Bug | Pentatomidae | Cosmopolitan (Philippines) | Sap-sucker; toxic saliva |
| *Bathycoelia thalassina* | – | Pentatomidae | W. Africa | Sap-sucker; toxic saliva |
| *Selenothrips rubrocinctus* (Giard) | Red-banded Thrips | Thripidae | W. Africa, S.E. Asia, C. & S. America | Scarify foliage |
| *Ceratitis capitata* (Wied.) | Medfly | Tephritidae | Africa, Australia, C. & S. America | Larvae in pods |
| *Pardalaspis punctata* Wied. | – | Tephritidae | Africa | Larvae inside pods |
| *Parasa vivida* (Wlk.) | Stinging Caterpillar | Limacodidae | W. & E. Africa | Larvae defoliate |
| *Latoia lepida* (Cram.) | Blue-striped Nettlegrub | Limacodidae | Philippines | Larvae defoliate |
| *Acrocercops cramerella* Snell. | Cocoa Pod Borer | Gracillariidae | Africa, S.E. Asia | Larvae bore pods |
| *Characoma stictograpta* Hmps. | Pod Husk Borer | Noctuidae | W. Africa | Larvae bore pods |
| *Earias biplaga* Wlk. | Spiny Bollworm | Noctuidae | Africa | Larvae bore pods |
| *Tiracola plagiata* (Wlk.) | Banana Fruit Caterpillar | Noctuidae | S.E. Asia | Larvae damage pods |
| *Spodoptera littoralis* (Boisd.) | Cotton Leafworm | Noctuidae | Africa | Larvae defoliate |
| *Spodoptera litura* (F.) | Rice Cutworm | Noctuidae | Asia, Australasia | Larvae defoliate |
| *Achaea janata* (L.) | Castor Semi-looper | Noctuidae | S.E. Asia | Larvae defoliate |
| *Thecla* spp. | Hairstreaks | Lycaenidae | S. America | Larvae eat leaves |
| *Laspeyresia toocosma* | – | Tortricidae | W. Africa | Larvae eat leaves |
| *Homona coffearia* Niet.) | Tea Tortrix | Tortricidae | Philippines | Larvae roll leaves |
| *Kotochalia junodi* (Heyl.) | Bagworm | Psychidae | Africa | Larvae defoliate |
| *Cryptothelea* spp. | Bagworms | Psychidae | Philippines | Larvae defoliate |
| *Anomala* spp. | White Grubs | Scarabaeidae | Philippines, Malaysia | Adults eat edges of leaves; larvae eat |
| *Apogonia cribricollis* (Burm.) | White Grub | Scarabaeidae | Malaysia | roots in soil |
| *Leucopholis* spp. | White Grubs | Scarabaeidae | Philippines | |
| *Ootheca mutabilis* (Sahlb.) | Brown Leaf Beetle | Chrysomelidae | Africa | Adults eat leaves |
| *Xylosandrus compactus* (Eichh.) | Black Twig Borer | Scolytidae | Africa, India, S.E. Asia | Adults bore twigs |
| *Xyleborus ferrugineus* (F.) | Black Twig Borer | Scolytidae | Africa, S.E. Asia, N., C. & S. America | Adults bore twigs |
| *Xyleborus similis* (Ferr.) | Black Twig Borer | Scolytidae | Malaysia | Adults bore twigs |
| *Xyleborus fornicatus* Eichh. | Tea Shot-hole Borer | Scolytidae | India, S.E. Asia | Adults bore twigs |
| *Chrysochroa* spp. | Jewel Beetles | Buprestidae | Philippines | Larvae bore stems |

| | | | | |
|---|---|---|---|---|
| *Niphonoclea* spp. | Twig Borers | Cerambycidae | Philippines | Larvae bore stems |
| *Mallodon downesi* F. | Stem Borer | Cerambycidae | Africa | Larvae bore stems |
| *Steirastoma breve* Guby | Cocoa Longhorn | Cerambycidae | C. & S. America | Larvae bore stems |
| *Systates* spp. | Systates Weevils | Curculionidae | Africa | Adults eat leaves |

# COCONUT (*Cocos nucifera* – Palmae)

Owing to the normal method of seed dispersal being by sea, the centre of origin of the coconut is uncertain; it has been abundant in the Old World and the Americas since early times, and is now typical of tropical coasts. It is confined to the tropics and is only successful if grown in the lowlands just above sea-level. The trees are tall, being up to 30 m in height, with a slender, often curved, trunk. Fruit-bearing starts after six years. The endosperm of the nut is dried to make copra from which oil is extracted. Propagation is from fruits planted in nurseries. The main production areas are the Philippines, Indonesia, India, New Guinea and the Pacific Islands. (See also Lever, 1969.)

## MAJOR PESTS

| | | | | |
|---|---|---|---|---|
| *Pseudotheraptus wayi* Brown | Coconut Bug | Coreidae | E. Africa | Sap-sucker; toxic saliva; cause nutfall |
| *Aspidiotus destructor* Sign. | Coconut Scale | Diaspididae | Pantropical ⎫ | Encrust leaves & fruits |
| *Ischnaspis longirostris* (Sign.) | Black Line Scale | Diaspididae | Pantropical ⎭ | |
| *Artona catoxantha* (Hmps.) | Coconut Leaf Skeletonizers | Zygaenidae | S.E. Asia | Larvae skeletonize leaves |
| *Oryctes boas* (F.) | Rhinoceros Beetle | Scarabaeidae | Africa ⎫ | Adults eat crown; larvae live in rotting vegetation |
| *Oryctes rhinoceros* L. | Rhinoceros Beetle | Scarabaeidae | India, S.E. Asia, Pacific | |
| *Oryctes monoceros* (Ol.) | Rhinoceros Beetle | Scarabaeidae | Africa ⎭ | |
| *Scapanes australis* Arrow | Taro Beetle | Scarabaeidae | Moluccas | Adults bore shoot and stem |
| *Diocalandra taitense* (Gucr.) | Tahiti Coconut Weevil | Curculionidae | New Guinea, Solomon Isl., Madagascar | Larvae bore leaves & fruit stalks |
| *Diocalandra frumenti* (F.) | Four-spotted Coconut Weevil | Curculionidae | S.E. Asia, India, Africa, Australasia | Larvae bore leaves & fruit stalks |
| *Rhynchophorus ferrugineus* (Oliv.) | Asiatic Palm Weevil | Curculionidae | S.E. Asia, India, Australia | Larvae bore into crown and trunk |
| *Rhynchophorus phoenicis* (F.) | African Palm Weevil | Curculionidae | Africa | Larvae bore into crown and trunk |
| *Rhynchophorus palmarum* (L.) | South American Palm Weevil | Curculionidae | C. & S. America | Larvae bore into crown & trunk |

## MINOR PESTS

| | | | | |
|---|---|---|---|---|
| *Sexava* spp. | Longhorned Grasshoppers | Tettigoniidae | New Guinea, Celebes | Defoliation by adults & nymphs |
| *Locusta migratoria* spp. | Migratory Locusts | Acrididae | Africa, India, Asia | Defoliation by adults & nymphs |
| *Tropidacris* spp. | – | Acrididae | C. & S. America | Defoliate |
| *Aularches miliaris* L. | Spotted Grasshopper | Acrididae | Thailand | Defoliation by adults & nymphs |

| | | | | |
|---|---|---|---|---|
| *Odontotermes* spp. | Scavenging Termites | Termitidae | India, Mozambique | Nest on trunks & in ground |
| *Macrotermes bellicosus* | War-like Termite | Termitidae | E. Africa | Damage trunk |
| *Nasutitermes* spp. | – | Termitidae | S.E. Asia | Nest on trunks & in ground |
| *Coptotermes* spp. | Moist-wood Termites | Rhinotermitidae | S.E. Asia | Nest on trunks & in ground |
| *Microcerotermes* spp. | Live-wood Termites | Termitidae | S.E. Asia, Pacific | Nest on trunks & in ground |
| *Graeffea crouani* Le Guillou | Stick Insect | Phasmidae | Polynesia, Melanesia | Defoliate |
| *Aleurodicus destructor* Quaint | Coconut Whitefly | Aleyrodidae | Indonesia, Malaysia, Philippines | Infest foliage |
| *Cerataphis variabilis* H.R.L. | – | Aphididae | E. Africa, Hawaii, Pacific, C. & S. America | Infest foliage |
| *Pinnaspis buxi* (Bch.) | – | Diaspididae | Pantropical | |
| *Aonidiella orientalis* (Newst.) | Oriental Yellow Scale | Diaspididae | Pantropical | |
| *Chrysomphalus aonidum* (L.) | Purple Scale | Diaspididae | Pantropical | Infest foliage & fruits |
| *Chrysomphalus dictyospermi* (Morg.) | Spanish Red Scale | Diaspididae | Pantropical | |
| *Pseudaonidia trilobitiformis* (Green) | Trilobite Scale | Diaspididae | Pantropical | |
| *Hemiberlesia palmae* (Ckll.) | Palm Scale | Coccidae | Pantropical | Infest foliage & fruits |
| *Vinsonia stellifera* Westw. | Star Scale | Coccidae | E. Africa, Seychelles, India, S. America | Encrust leaves |
| *Pseudococcus adonidum* (L.) | Long-tailed Mealybug | Pseudococcidae | Pantropical | Infest foliage & fruits |
| *Dysmicoccus brevipes* (Ckll.) | Pineapple Mealybug | Pseudococcidae | Pantropical | Infest foliage & fruits |
| *Icerya seychellarum* (West.) | Seychelles Fluted Scale | Margarodidae | Philippines | Infest foliage & fruits |
| *Amblypelta cocophaga* China | Coconut Bug | Coreidae | Solomon Isl., Fiji | Sap-sucker; toxic saliva |
| *Axiagastus campbelli* Dist. | Shield Bug | Pentatomidae | Papua NG, Solomon Isl. | |
| *Stephanitis typica* (Dist.) | Lace Bug | Tingidae | S.E. Asia, India, Korea, Japan | Sap-sucker; toxic saliva |
| *Acritocera negligens* | Spathe Borer | Cossidae | Fiji | Larvae bore spathe |
| *Erionota thrax* L. | Banana Skipper | Hesperiidae | Thailand, Philippines | Larvae cut & roll leaves |
| *Gangara thyrsis* (F.) | – | Hesperiidae | India, S.E. Asia | Larvae roll leaves |
| *Hidari irava* Moore | Leafbinder | Hesperiidae | S.E. Asia | Larvae bind leaves |
| *Lotongus calathus* How | Leafbinder | Hesperiidae | Thailand | Larvae bind leaves |
| *Telicota palmarum* Moore | Coconut Skipper | Hesperiidae | India, S.E. Asia | Larvae bind leaves |
| *Batrachedra arenosella* Wlk. | – | Cosmopterygidae | India, S.E. Asia, Australasia, Zaïre, Guiana | Larvae damage flowers |
| *Nephantis serinopa* Meyr. | Black-headed Caterpillar | Xyloryctidae | India, Burma | Larvae defoliate |

| | | | | |
|---|---|---|---|---|
| *Tirabatha* spp. | Flower-eating Caterpillars | Pyralidae | Thailand, Pacific, Australasia | Larvae feed on flowers |
| *Setora nitens* (Wlk.) | Nettlegrub | Limacodidae | S.E. Asia | Larvae defoliate |
| *Latoia lepida* (Cram.) | Nettlegrub | Limacodidae | S.E. Asia | Larvae defoliate |
| *Thosea sinensis* Wlk. | Nettle Caterpillar | Limacodidae | Thailand, Laos | Larvae defoliate |
| *Amathusia phidippus* L. | Palm Butterfly | Nymphalidae | Malaysia | Larvae defoliate |
| *Castnia* spp. | Giant Stalk Borers | Castniidae | C. & S. America | Larvae bore trunk |
| *Mahasena corbetti* Tams | Bagworm | Psychidae | S.E. Asia | Larvae eat leaves |
| *Dasychira horsfieldi* Saund. | Tussock Moth | Lymantriidae | Thailand | Larvae eat leaves |
| *Oecophylla smaragdina* Wlk. | Red Tree Ant | Formicidae | S.E. Asia | Attack workers |
| *Plesispa reichei* Chapuis | Coconut Hispid | Chrysomelidae | S.E. Asia | Larvae mine leaves |
| *Promecotheca* spp. | Coconut Leaf Miners | Chrysomelidae | S.E. Asia, Pacific | Larvae mine leaves |
| *Coelaenomenodera elaeidis* Wlk. | Oil Palm Leaf Miner | Chrysomelidae | W. & C. Africa | Larvae mine leaves; adults attack crown |
| *Brontispa* spp. | Coconut Leaf Beetles | Chrysomelidae | S.E. Asia, Pacific | Seedlings damaged by adults & larvae |
| *Scapanes* spp. | – | Scarabaeidae | Papua NG, Solomon Isl. | Adults eat foliage |
| *Xylotrupes gideon* (L.) | Unicorn Beetle | Scarabaeidae | S.E. Asia | Adults eat foliage |
| *Papuana laevipennis* Arrow | Taro Beetle | Scarabaeidae | Moluccas | Adults eat foliage |
| *Xyleborus ferrugineus* (F.) | Shot-hole Borer | Scolytidae | Pantropical | Adults bore trunk |
| *Xyleborus perforans* (Woll.) | Coconut Shot-hole Borer | Scolytidae | Pantropical | Adults bore trunk |
| *Melittomma insulare* Fairm. | – | Lymexylidae | Madagascar, Seychelles | Larvae bore trunk base |
| *Rhynchophorus vulneratus* (Panzer) | Asiatic Palm Weevil | Curculionidae | Thailand | Larvae bore crown |
| *Rhina afzelii* Fhs. | – | Curculionidae | Africa, Madagascar | Larvae bore trunk |
| *Rhina barbirostris* F. | Bearded Weevil | Curculionidae | Mexico, Trinidad, S. America | Larvae bore trunk |
| *Raoiella indica* Hirst. | Date Palm Scarlet Mite | Tenuipalpidae | India, Egypt, Mauritius | Infest foliage |
| *Aceria guerreronis* Keifer | – | Eriophyidae | Colombia | Infest foliage |

Lepesme (1947) listed 751 insect species recorded on *Cocos*, of which 22% are specific to coconut.

# COCOYAM (*Xanthosoma safittifolium* — Araceae)
## (= Tannia)

This is a native of the New World, cultivated there in tropical rain forest regions in pre-Columbian times, and only quite recently has been spread throughout the wet tropics; for example it was introduced into Ghana in 1841. Because of its resemblance to *Colocasia* it was called 'Cocoyam', and is sometimes distinguished by the name of 'New Cocoyam', but it is sometimes called 'Taro' also. It is a typical aroid in that it is a robust herb growing out from underground corms, in height up to 2 m or more, but the leaves are sagittate and sharply pointed, which distinguishes the plant from *Colocasia*. The food base consists of starch grains in the corm, but the grains are distinctly larger than those in Taro.

## MAJOR PESTS

## MINOR PESTS

| | | | | |
|---|---|---|---|---|
| *Pentalonia nigronervosa* Coq. | Banana Aphid | Aphididae | Pantropical | Infest foliage |
| *Ligyrus ebenus* (De G.) | — | Scarabaeidae | W. Indies | Larvae damage tubers |
| *Araeocerus fasciculatus* (De Geer) | Coffee Bean Weevil | Brenthidae | USA | Attack tubers in stores mostly |

# COFFEE (*Coffea arabica* & *C. robusta* — Rubiaceae)

*Arabica* coffee originated in Ethiopia in forests at about 1500 to 2000 m, and was early taken to Arabia. It was introduced into Java in the late 17th century, to India and Ceylon at about the same time, and to E. Africa in the late 19th century. The crop is now widely grown throughout the tropics.

*Robusta* coffee grows wild in African equatorial forests, and is also now widely distributed throughout the tropics. This species is more successful at lower altitudes than is *arabica*, and is the more important species in Asia and tropical Africa. Both are evergreen shrubs or small trees, growing to 5 m if unpruned, bearing continuous clusters of berries along the smaller branches, crimson when ripe. The main production areas are Brazil, Java, Kenya, W. Indies, and several other S. American countries. (See also Le Pelley, 1968.)

## MAJOR PESTS

| | | | | |
|---|---|---|---|---|
| *Planococcus citri* (Risso) | Root Mealybug | Pseudococcidae | Pantropical | Encrust roots mainly |
| *Planococcus kenyae* (Le Pelley) | Kenya Mealybug | Pseudococcidae | E. & W. Africa | Encrust foliage |
| *Ferrisia virgata* (Ckll.) | Striped Mealybug | Pseudococcidae | Pantropical | Encrust foliage |
| *Pseudococcus* sp. | Root Mealybug | Pseudococcidae | S. America | Encrust roots |
| *Asterolecanium coffeae* Newstead | Star Scale | Asterolecanidae | E. Africa, Zaïre | Encrust branches |
| *Gascardia brevicauda* (Hall) | White Waxy Scale | Coccidae | E. Africa | Encrust foliage |
| *Coccus alpinus* De Lotto | Soft Green Scale | Coccidae | E. Africa | Infest foliage |
| *Coccus viridis* (Green) | Soft Green Scale | Coccidae | Pantropical | Infest foliage |
| *Saissetia coffeae* (Wlk.) | Helmet Scale | Coccidae | Cosmopolitan | Infest foliage |
| *Orthezia insignis* Browne | Jacaranda Bug | Orthezidae | Africa, S. America | Infest foliage |
| *Antestiopsis* spp. | Antestia Bugs | Pentatomidae | Africa | Sap-suckers; toxic saliva |
| *Lamprocapsidea coffeae* (China) | Coffee Capsid | Miridae | E. Africa, Zaïre | Sap-suckers; toxic saliva |
| *Habrochila ghesquierei* Schout. | Coffee Lace Bug | Tingidae | E. Africa, Zaïre ⎫ | Sap-suckers; toxic saliva |
| *Habrochila placida* Hory. | Coffee Lace Bug | Tingidae | E. Africa, Zaïre ⎭ | |
| *Hoplandothrips marshalli* Karny | Coffee Leaf-rolling Thrips | Phlaeothripidae | Uganda | Cause leaf-rolling |
| *Diarthrothrips coffeae* Williams | Coffee Thrips | Thripidae | Africa | Scarify foliage |
| *Parasa vivida* (Wlk.) | Stinging Caterpillar | Limacodidae | Africa | Larvae defoliate |
| *Latoia lepida* (Cram.) | Blue-striped Nettlegrub | Limacodidae | Philippines | Larvae defoliate |
| *Thosea sinensis* (Wlk.) | Slug Caterpillar | Limacodidae | Philippines | Larvae defoliate |
| *Niphadolepis alianta* Karsch | Jelly Grub | Limacodidae | E. Africa, Malawi | Larvae defoliate |
| *Eucosma nereidopa* Meyr. | Coffee Tip Borer | Tortricidae | Kenya | Larvae bore shoots |
| *Leucoptera* spp. | Coffee Leaf Miners | Gracillariidae | Africa, S. America | Larvae mine leaves |
| *Prophantis smaragdina* Butler | Coffee Berry Moth | Pyralidae | Africa | Larvae web berries |
| *Virachola bimaculata* (Hew.) | Coffee Berry Butterfly | Lycaenidae | W. & E. Africa | Larvae bore berries |
| *Epicampoptera marantica* Tams | Tailed Caterpillar | Drepanidae | Tanzania, Zaïre | Larvae defoliate |
| *Epicampoptera andersoni* Tams | Tailed Caterpillar | Drepanidae | Kenya, Uganda, Nigeria | Larvae defoliate |
| *Ascotis selenaria* (Wlk.) | Coffee Looper | Geometridae | E. & S. Africa | Larvae defoliate |
| *Leucoplema dohertyi* (Warr.) | Leaf Skeletonizer | Epiplemidae | E. Africa, Zaïre | Larvae skeletonize leaves |

| Ceratitis coffeae (Bezzi) | Coffee Fruit Fly | Tephritidae | E. & W. Africa, Zaïre | Larvae inside fruits |
|---|---|---|---|---|
| Macromischoides aculeatus Mayr | Biting Ant | Formicidae | Uganda, Zaïre, Tanzania | Nest in foliage; attack workers |
| Gonocephalum simplex (F.) | Dusty Brown Beetle | Tenebrionidae | Africa | Adults eat bark |
| Anthores leuconotus (Ol.) | White Coffee Borer | Cerambycidae | E. Africa | Larvae bore stems |
| Dirphya nigricornis (Ol.) | Yellow-headed Borer | Cerambycidae | E. Africa | Larvae bore stems |
| Hypothenemus hampei (Ferr.) | Coffee Berry Borer | Scolytidae | Africa, S.E. Asia, S. America | Adults bore berries |
| Apate monachus Boh. | Black Borer | Bostrychidae ⎫ | Africa, Med., | Adults bore |
| Apate indistincta Murray | Black Borer | Bostrychidae ⎬ | C. & S. America | branches |
| Systates pollinosus Gerst. | Systates Weevil | Curculionidae | E. Africa | Adults eat leaves |
| Oligonychus coffeae (Nietn.) | Red Coffee Mite | Tetranychidae | Pantropical | Scarify foliage |

## MINOR PESTS

| Zonocerus variegatus L. | Variegated Grasshopper | Acrididae | Africa | Defoliate |
|---|---|---|---|---|
| Bothrogonia sp. | Leafhopper | Cicadellidae | Philippines | Infest foliage |
| Aleurothrix floccosus (Mask.) | – | Aleyrodidae | S. America | Infest foliage |
| Dialeurodes citri (Ashm.) | Citrus Whitefly | Aleyrodidae | S.E. Asia, S. America | Infest foliage |
| Aleurocanthus woglumi Ashby | Citrus Blackfly | Aleyrodidae | S. America | Infest foliage |
| Toxoptera aurantii (B. de F.) | Black Citrus Aphid | Aphididae | Cosmopolitan | Infest foliage |
| Dysmicoccus brevipes (Ckll.) | Pineapple Mealybug | Pseudococcidae | Pantropical | Infest foliage |
| Geococcus coffeae Green | Coffee Root Mealybug | Pseudococcidae | Pantropical | Infest roots |
| Pseudococcus adonidum (L.) | Long-tailed Mealybug | Pseudococcidae | Pantropical | Infest foliage |
| Cerococcus spp. | Soft Scales | Coccidae | Africa, Philippines, S. America | Infest foliage |
| Saissetia nigra (Nietn.) | Nigra Scale | Coccidae | Philippines | Infest foliage |
| Saissetia oleae (Bern.) | Black Scale | Coccidae | S. America | Infest foliage |
| Gascardia destructor (Newst.) | White Waxy Scale | Coccidae | Africa, Papua NG | Infest foliage |
| Ceroplastes rubens Mask. | Pink Waxy Scale | Coccidae | E. Africa, Asia | Infest foliage |
| Pseudaonidia trilobitiformis (Green) | Trilobite Scale | Diaspididae | Pantropical | Encrust foliage |
| Ischnaspis longirostris (Sign.) | Black Line Scale | Diaspididae | Africa, S.E. Asia, C. & S. America | Encrust foliage |
| Selanaspidus articulatus (Morg.) | West Indian Red Scale | Diaspididae | Pantropical | Encrust foliage |
| Lawana candida F. | Coffee Moth Bug | Flattidae | Indonesia, Java, Vietnam, Philippines | Infest foliage |
| Ricania speculum (Wlk.) | Black Planthopper | Ricaniidae | Philippines | Infest foliage |
| Leptoglossus australis F. | Leaf-footed Plant Bug | Coreidae | Africa, Asia | Sap-sucker; toxic saliva |

| | | | | |
|---|---|---|---|---|
| *Anoplocnemis curvipes* F. | Coreid Bug | Coreidae | Africa | Sap-sucker; toxic saliva |
| *Frankliniella schulzei* (Trybom) | Cotton Flower Thrips | Thripidae | E. Africa, Sudan | Infest flowers |
| *Heliothrips haemorrhoidalis* (Bouche) | Black Tea Thrips | Thripidae | Pantropical | Infest foliage |
| *Taeniothrips sjostedti* (Trybom) | Bean Flower Thrips | Thripidae | Africa | Infest flowers |
| *Selenothrips rubrocinctus* (Giard) | Red-banded Thrips | Thripidae | Philippines | Infest foliage |
| *Melanagromyza coffeae* (Koning.) | Coffee Leaf Miner | Agromyzidae | E. Africa, India, Java, Papua NG | Larvae mine leaves |
| *Ceratitis capitata* (Wied.) | Medfly | Tephritidae | Africa, Hawaii, S. America | Larvae in berries |
| *Ceratitis rosa* (Ksh.) | Natal Fruit Fly | Tephritidae | E. Africa | Larvae in berries |
| *Dacus dorsalis* Hend. | Oriental Fruit Fly | Tephritidae | Philippines | Larvae in berries |
| *Atta* spp. | Leaf-cutting Ants | Formicidae | S. & C. America | Adults defoliate |
| *Xyleutes* spp. | Stem Borers | Cossidae | Africa, S. USA, C. America, India | Larvae bore stems |
| *Zeuzera coffeae* Nietn. | Red Coffee Borer | Cossidae | Indonesia, Thailand, Malaysia, Philippines | Larvae bore stems |
| *Eulophonotus myrmeleon* Feldr. | Cocoa Stem Borer | Cossidae | E. & W. Africa | Larvae bore stems |
| *Epigynopteryx coffeae* Prout | Coffee Looper | Geometridae | Kenya | Larvae defoliate |
| *Cryptophlebia leucotreta* (Meyr.) | False Codling Moth | Tortricidae | Africa | Larvae bore berries |
| *Tortrix dinota* Meyr. | Coffee Tortrix | Tortricidae | E. Africa | Larvae roll leaves |
| *Archips occidentalis* Wals. | Coffee Tortrix | Tortricidae | E. Africa | Larvae roll leaves |
| *Homona coffearia* Nietn. | Tea Tortrix | Tortricidae | Philippines | Larvae roll leaves |
| *Tiracola plagiata* (Wlk.) | Banana Fruit Caterpillar | Noctuidae | S.E. Asia | Larvae damage fruits |
| *Attacus atlas* (L.) | Atlas Moth | Saturniidae | Philippines | Larvae defoliate |
| *Cephonodes hylas* L. | Coffee Hawk Moth | Sphingidae | Africa, India, S.E. Asia, Japan, Australasia | Larvae defoliate |
| *Dasychira mendosa* (Hb.) | Tussock Moth | Lymantriidae | Philippines | Larvae defoliate |
| *Orgyia australis* (Wlk.) | Tussock Moth | Lymantriidae | Philippines | Larvae defoliate |
| *Prionoryctes caniculus* Arr. | Yam Beetle | Scarabaeidae | Africa | Larvae damage roots |
| *Pachnoda sinuata* (F.) | Rose Beetle | Scarabaeidae | Kenya | Adults eat leaves |
| *Anomala* spp. | White Grubs | Scarabaeidae | Philippines ⎫ | Adults eat leaves; |
| *Leucopholis* spp. | White Grubs | Scarabaeidae | Philippines ⎭ | larvae in soil eat roots |
| *Aspidomorpha* spp. | Tortoise Beetles | Chrysomelidae | Africa | Adults & larvae eat leaves |
| *Ootheca mutabilis* (Sahlb.) | Brown Leaf Beetle | Chrysomelidae | Africa | Adults & larvae eat leaves |
| *Bixadus sierricola* White | Coffee Stem Borer | Cerambycidae | Africa | Larvae bore stems |
| *Xylotrechus quadripes* Chevr. | Coffee Stem Borer | Cerambycidae | Philippines | Larvae bore stems |
| *Aperitmetus brunneus* (Hust.) | Tea Root Weevil | Curculionidae | Kenya | Larvae eat roots |
| *Araeocerus fasciculatus* (Deg.) | Coffee Bean Weevil | Brenthidae | Pantropical | Attack berries on tree and in store |
| *Xylosandrus compactus* (Eichh.) | Black Twig Borer | Scolytidae | Africa, India, S.E. Asia | Adults bore branches |

553

| | | | | |
|---|---|---|---|---|
| *Xylosandrus morigerus* (Bland.) | Black Twig Borer | Scolytidae | S.E. Asia | Adults bore branches |
| *Brevipalpus phoenicis* (Geijskes) | Red Crevice Tea Mite | Tenuipalpidae | E. Africa, India, Mexico, Brazil | Scarify foliage |
| *Polyphagotarsonemus latus* (Banks) | Yellow Tea Mite | Tarsonemidae | Pantropical | Scarify foliage |

# COTTON (*Gossypium* spp. – Malvaceae)

Wild species are found in many parts of the tropics and sub-tropics. Commercial crops are grown now in the New World between 37° N and 32° S, and in the Old World between 47° N and 30° S. It cannot be grown successfully in India above 1000 m and in Africa above 2000 m; it needs 200 frost-free days, the optimum temperature for growth is 32 °C, and the crop must have full sunshine (no shade); it can be grown on a variety of soils, but cannot tolerate very heavy rainfall; when grown 'rain-fed' the average rainfall needed is 100–150 cm, but in some areas it is grown under irrigation. The wild species are xerophytic. In habit they are annual sub-shrubs, 1–1.5 m high. The lint is used to make processed cotton; the seeds contain 18–24% edible oil, and the residual cake is rich in protein and used for cattle food. The main production areas are USA, China, USSR, Egypt, Mexico, Uganda, Nigeria, Tanzania, Sudan, India and the W. Indies. (See also Pearson, 1958, and Butani, 1975*a*.)

## MAJOR PESTS

| | | | | |
|---|---|---|---|---|
| *Aphis gossypii* Glov. | Cotton Aphid | Aphididae | Cosmopolitan | Infest foliage |
| *Empoasca fascialis* Jacobi | Cotton Jassid | Cicadellidae | Africa | Infest foliage; virus vector |
| *Empoasca lybica* (De Berg) | Cotton Jassid | Cicadellidae | Africa, Spain, Israel | Infest foliage; virus vector |
| *Bemisia tabaci* (Genn.) | Tobacco Whitefly | Aleyrodidae | Africa, India | Infest foliage; virus vector |
| *Calidea dregii* Germ. | Blue Bug | Pentatomidae | Africa ⎫ | Sap-suckers; toxic saliva |
| *Calidea bohemani* (Stal) | Blue Bug | Pentatomidae | Africa ⎬ | |
| *Nezara viridula* (L.) | Green Stink Bug | Pentatomidae | Cosmopolitan | Sap-sucker; toxic saliva |
| *Oxycarenus hyalipennis* Costa | Cotton Seed Bug | Lygaeidae | Africa, India, S.E. Asia, Brazil | Suck sap from seeds |
| *Taylorilygus vosseleri* (Popp.) | Cotton Lygus | Miridae | Africa | Cause leaf tattering |
| *Helopeltis schoutedeni* Reuter | Cotton Helopeltis | Miridae | Africa | Sap-sucker; toxic saliva |
| *Dysdercus* spp. | Cotton Stainers | Pyrrhocoridae | Africa, India | Sap-suckers; toxic saliva |
| *Frankliniella schulzei* (Trybom) | Cotton Flower Thrips | Thripidae | E. Africa, Sudan | Infest flowers |
| *Sylepta derogata* (F.) | Cotton Leaf Roller | Pyralidae | Africa, India, Australasia, China | Larvae roll leaves |
| *Pectinophora gossypiella* (Saund.) | Pink Bollworm | Gelechiidae | Pantropical | Larvae bore bolls |
| *Cryptophlebia leucotreta* (Meyr.) | False Codling Moth | Tortricidae | Africa | Larvae bore bolls |
| *Heliothis armigera* (Hub.) | American Bollworm | Noctuidae | Cosmopolitan in Old World | Larvae bore bolls |
| *Heliothis zea* Boddie | Cotton Bollworm | Noctuidae | USA | Larvae bore bolls |
| *Earias biplaga* Wlk. | Spiny Bollworm | Noctuidae | Africa | Larvae bore bolls |
| *Earias insulana* (Boisd.) | Spiny Bollworm | Noctuidae | Africa, India, S.E. Asia | Larvae bore bolls |
| *Spodoptera littoralis* (Boisd.) | Cotton Leafworm | Noctuidae | Africa, Near East | Larvae defoliate |
| *Spodoptera litura* (F.) | Rice Cutworm | Noctuidae | India, S.E. Asia, Australasia | Larvae defoliate |

| *Anomis flava* (F.) | Cotton Semi-looper | Noctuidae | Africa, Asia, Australasia | Larvae eat leaves and buds |
|---|---|---|---|---|
| *Diparopsis castanea* Hmps. | Red Bollworm | Noctuidae | Southern Africa | Larvae bore bolls |
| *Diparopsis watersi* (Roths.) | Sudan Red Bollworm | Noctuidae | W. & N.E. Africa | Larvae bore bolls |
| *Anthonomus grandis* Boh. | Cotton Boll Weevil | Curculionidae | S. USA | Larvae in bolls |
| *Pempheres affinis* Faust. | Cotton Stem Weevil | Curculionidae | India, Thailand, Philippines | Larvae bore stems |
| *Tetranychus cinnabarinus* (Boisd.) | Tropical Red Spider Mite | Tetranychidae | Pantropical | Scarify leaves |
| *Polyphagotarsonemus latus* (Banks) | Yellow Tea Mite | Tarsonemidae | Cosmopolitan | Scarify leaves |

## MINOR PESTS

| *Zonocerus variegatus* L. | Variegated Grasshopper | Acrididae | Africa | Defoliate |
|---|---|---|---|---|
| *Austracris guttulosa* Wlk. | Spur-throated Locust | Acrididae | Australia | Defoliate |
| *Brachytrupes membranaceus* (Drury) | Tobacco Cricket | Gryllidae | Africa | Destroy seedlings; & eat roots |
| *Gryllus* spp. | Field Crickets | Gryllidae | India | Destroy seedlings |
| *Hodotermes mossambicus* (Hagen) | Harvester Termite | Hodotermitidae | E. & S. Africa | Damage foliage |
| *Microcerotermes parvus* Haviland | – | Termitidae | E. Africa | Damage foliage |
| *Odontotermes obesus* (Ram.) | Scavenging Termite | Termitidae | India | Damage foliage & seedlings |
| *Empoasca devastans* Dist. | Green Leafhopper | Cicadellidae | India, Thailand, Philippines | Sap-sucker; infests leaves |
| *Amrasca* spp. | Cotton Jassids | Cicadellidae | India, S.E. Asia | Sap-sucker; infests leaves |
| *Planococcus citri* (Risso) | Root Mealybug | Pseudococcidae | Pantropical | Infest roots & foliage |
| *Ferrisia virgata* (Ckll.) | Striped Mealybug | Pseudococcidae | Pantropical | Infest foliage |
| *Saissetia coffeae* (Wlk.) | Helmet Scale | Coccidae | Philippines | Infest foliage |
| *Ricania speculum* (Wlk.) | Black Planthopper | Ricaniidae | Philippines | Infest foliage |
| *Lygus oblineatus* (Say) | Tarnished Plant Bug | Miridae | N. America ⎫ | Sap-sucker; toxic saliva |
| *Creontiades pallidus* Ramb. | Cotton Mirid | Miridae | Africa, India ⎭ | |
| *Helopeltis* spp. | Helopeltis Bugs | Miridae | Philippines ⎫ | Sap-suckers; toxic saliva |
| *Bagrada hilaris* (Burm.) | Harlequin Bug | Pentatomidae | Africa, India ⎭ | |
| *Thrips tabaci* Lind. | Onion Thrips | Thripidae | Cosmopolitan | Scarify foliage |
| *Hercinothrips femoralis* Reut. | Banded Green-house Thrips | Thripidae | Cosmopolitan | Scarify foliage |
| *Scirtothrips dorsalis* Hood | Flower Thrips | Thripidae | India | Infest flowers |
| *Ayyardia chaetophora* Karny | Cotton Thrips | Thripidae | Thailand | Infest flowers |
| *Zeuzera coffeae* Nietn. | Red Coffee Borer | Cossidae | Philippines | Larvae bore stems |
| *Acrocercops bifasciata* Wlsm. | Cotton Leaf Miner | Gracillariidae | Africa | Larvae mine leaves |
| *Parasa vivida* (Wlk.) | Stinging Caterpillar | Limacodidae | E. & W. Africa | Larvae defoliate |
| *Alabama argillacea* (Hb.) | Alabama Leafworm | Noctuidae | N., C & S. America | Larvae defoliate |

| | | | | |
|---|---|---|---|---|
| *Trichoplusia ni* (Hb.) | Cabbage Semi-looper | Noctuidae | USA, Mexico | Larvae defoliate |
| *Chrysodeixis chalcites* (Es.) | Cabbage Semi-looper | Noctuidae | Old World | Larvae defoliate |
| *Spodoptera exigua* (Hb.) | Lesser Armyworm | Noctuidae | Europe, Africa, India, Japan, USA | Larvae defoliate |
| *Earias vittella* Stoll | Spotted (Spiny) Bollworm | Noctuidae | India | Larvae bore bolls |
| *Agrotis ipsilon* (Hfn.) | Black Cutworm | Noctuidae | Cosmopolitan | Larvae are cutworms |
| *Sacodes pyralis* Dyar. | Sugarcane Stalk Borer | Noctuidae | S. America | Larvae bore bolls |
| *Tarache* spp. | Cotton Semi-loopers | Noctuidae | India | Larvae defoliate |
| *Xestia c-nigrum* (L) | Spotted Cutworm | Noctuidae | Europe, Asia, N. America | Larvae are cutworms |
| *Hyles lineata* (F.) | Silver-striped Hawk Moth | Sphingidae | Pantropical | Larvae defoliate |
| *Bucculatrix thurberiella* Busck | Cotton Leaf Perforator | Zygaenidae | USA, C. America | Larvae hole leaves |
| *Diacrisia* spp. | Tiger Moths | Arctiidae | Africa, India, S.E. Asia | Larvae defoliate |
| *Estigmene acrea* Drury | Salt Marsh Caterpillar | Arctiidae | USA | Larvae defoliate |
| *Amsacta* spp. | Tiger Moths | Arctiidae | India | Larvae defoliate |
| *Euproctis* spp. | Tussock Moths | Lymantriidae | E. Africa, India | Larvae defoliate |
| *Contarinia gossypii* Felt | Cotton Gall Midge | Cecidomyiidae | India, USA, W. Indies | Larvae infest buds |
| *Dasyneura gossypii* Felt | Cotton Flower Bud Midge | Cecidomyiidae | India | Larvae infest flower buds |
| *Dacus* spp. | Fruit Flies | Tephritidae | Africa, India, Australasia, Philippines | Larvae inside bolls |
| *Podagrica puncticollis* Weise | Cotton Flea Beetle | Chrysomelidae | Africa | Adults hole leaves |
| *Phyllotreta* sp. | Flea Beetle | Chrysomelidae | Africa | Adults hole leaves |
| *Ootheca mutabilis* (Sahlb.) | Brown Leaf Beetle | Chrysomelidae | Africa | Adults eat leaves |
| *Psallus seriatus* (Reut.) | Cotton Fleahopper | Chrysomelidae | USA | Adults hole leaves |
| *Apate* spp. | Black Borers | Bostrychidae | Africa, Asia, C. & S. America | Adults bore stems |
| *Coryna* spp. | Pollen Beetles | Meloidae | Africa | Adults eat pollen |
| *Mylabris* spp. | Banded Blister Beetles | Meloidae | Africa, India, S.E. Asia | Adults eat flowers |
| *Eriesthis vulpina* Brum. | – | Scarabaeidae | Africa ⎫ | Larvae in soil attack roots; adults eat leaves |
| *Pachnoda sinuata* (F.) | Flower Beetle | Scarabaeidae | Africa ⎭ | |
| *Epilachna* spp. | Epilachna Beetles | Coccinellidae | Africa, Middle East | Adults & larvae eat leaves |

| | | | | |
|---|---|---|---|---|
| *Sphenoptera gossypii* Banks | Cotton Jewel Beetle | Buprestidae | India | Larvae bore stems |
| *Graphognathus* spp. | White-fringed Weevils | Curculionidae | Africa | Larvae eat roots; adults eat leaves |
| *Hypomeces squamosus* (F.) | Gold-dust Weevil | Curculionidae | Thailand, China | Adults eat leaves |
| *Myllocerus* spp. | Grey Weevils | Curculionidae | India | Adults eat leaves |
| *Amorphoidea lata* Mot. | Cotton Boll Weevil | Curculionidae | Philippines | Larvae in bolls |
| *Amorphoidea* sp. | Blossom Weevil | Curculionidae | Thailand | Adults eat flowers |
| *Alcidodes gossypii* (Hust.) | Striped Cotton Weevil | Curculionidae | Africa | Adult eats bark; girdling the stem |
| *Apion soleatum* Wagn. | Apion Weevil | Apionidae | E. Africa | Infest flowers |
| *Aceria gossypii* (Banks) | Cotton Gall Mite | Eriophyidae | India | Gall leaves |
| *Eutetranychus orientalis* (Klein) | Oriental Mite | Tetranychidae | Africa, India | Scarify foliage |

# COWPEA ( *Vigna sinensis* — Leguminosae)

Despite its common name this plant is more closely related to beans than to peas. A vigorous bushy annual with cylindrical pendant pods. A crop of great antiquity, thought to be native to either C. Africa or C. America, but cultivated throughout S.E. Asia for more than 2000 years. Now widely grown throughout the tropics, mainly as a forage crop, cover crop, or as green manure; the seeds are fed to cattle and poultry. An important crop in India, China, and southern USA. It is susceptible to frost and heavy rainfall, and grown only in warm humid areas in light soils. The Oriental Longbean is *Vigna sesquipedalis* and is grown as a vegetable, and has a similar pest spectrum.

## MAJOR PESTS

| | | | | |
|---|---|---|---|---|
| *Aphis craccivora* (Koch) | Groundnut Aphid | Aphididae | Malaysia, Laos, India | Infest foliage |
| *Lampides boeticus* (L.) | Pea Blue Butterfly | Lycaenidae | Thailand, India | Larvae bore pods |
| *Ophiomyia phaseoli* (Coq.) | Bean Fly | Agromyzidae | Malaysia, India | Larvae mine stems |
| *Callosobruchus chinensis* (L.) | Oriental Cowpea Bruchid | Bruchidae | Thailand, Laos, | Attack ripe seeds in pods |

## MINOR PESTS

| | | | | |
|---|---|---|---|---|
| *Colemania sphenaroides* Bol. | Deccan Wingless Grasshopper | Acrididae | India | Adults & nymphs defoliate |
| *Empoasca* spp. | Green Leafhoppers | Cicadellidae | India | Infest foliage & suck sap |
| *Bemisia tabaci* (Genn.) | Tobacco Whitefly | Aleyrodidae | India | Infest foliage |
| *Anchon pilosum* W. | Treehopper | Membracidae | India | Infest stems |
| *Creontiades pallidifer* Wlk. | Capsid Bug | Miridae | India | Sap-sucker; toxic saliva |
| *Chauliops fallax* Scott | Cowpea Pod Bug | Lygaeidae | India | Sap-sucker; toxic saliva |
| *Acanthocoris scabrator* (F.) | Coreid Bug | Coreidae | Malaysia | Sap-sucker; toxic saliva |
| *Anoplocnemis phasiana* (F.) | Coreid Bug | Coreidae | India | Sap-sucker; toxic saliva |
| *Riptortus* spp. | Coreid Bugs | Coreidae | India | Sap-suckers; toxic saliva |
| *Caliothrips indicus* (Bagn.) | Thrips | Thripidae | India | Infest flowers |
| *Taeniothrips* spp. | Flower Thrips | Thripidae | India | Infest flowers |
| *Acrocercops* spp. | Leaf Miners | Gracillariidae | India | Larvae mine leaves |
| *Cydia tricentra* (Meyr.) | – | Tortricidae | India | Larvae eat foliage |
| *Thosea aperiens* (Wlk.) | Slug Caterpillar | Limacodidae | India | Larvae eat leaves |
| *Etiella zinckenella* (Triet.) | Pea Pod Borer | Pyralidae | India | Larvae bore pods |
| *Maruca testulalis* (Geyer) | Maruca Moth | Pyralidae | Malaysia, Laos, China | Larvae bore pods |
| *Lamprosema* spp. | Bean Leaf Rollers | Pyralidae | Malaysia, India | Larvae roll leaves |

| | | | | |
|---|---|---|---|---|
| *Nacoleia vulgaris* Guen. | – | Pyralidae | India | Larvae eat foliage & pods |
| *Amata passalis* F. | Wasp-moth | Amatidae | India | Larvae eat leaves |
| *Euchrysops cnejus* | Blue Butterfly | Lycaenidae | India | Larvae bore pods |
| *Amsacta* spp. | Tiger Moths | Arctiidae | India | Larvae defoliate |
| *Agrotis ipsilon* (Hfn.) | Black Cutworm | Noctuidae | India | Larvae are cutworms |
| *Anticarsia irrorata* B. | Green Leaf Caterpillar | Noctuidae | India | Larvae eats leaves |
| *Heliothis armigera* (Hbn.) | American Bollworm | Noctuidae | India | Larvae eat young shoots & bore pods |
| *Spodoptera exigua* (Hbn.) | Lesser Armyworm | Noctuidae | India | Larvae eat leaves |
| *Spodoptera litura* (F.) | Rice Armyworm | Noctuidae | India, Thailand | Larvae defoliate |
| *Euproctis fraterna* (Moore) | Tussock Moth | Lymantriidae | India | Larvae defoliate |
| *Dasychira mendosa* (Hbn.) | Tussock Moth | Lymantriidae | Malaysia | Larvae defoliate |
| *Melanagromya obtusa* M. | Bean Pod Fly | Agromyzidae | India | Larvae bore young seeds in pod |
| *Anomala benghalensis* Blanchard | Flower Beetle | Scarabaeidae | India ⎫ | Adults attack foliage; |
| *Holotrichia* spp. | Cockchafers | Scarabaeidae | India ⎭ | larvae eat roots in soil |
| *Mylabris phalerata* (Pall.) | Large Yellow-banded Blister Beetle | Meloidae | Laos | Adults eat leaves & flowers |
| *Mylabris pustulata* (Thunb.) | Banded Blister Beetle | Meloidae | India | Adults eat flowers |
| *Epilachna* spp. | Epilachna Beetles | Coccinellidae | India | Adults & larvae eat leaves |
| *Oulema* spp. | Leaf Beetles | Chrysomelidae | India | Adults & larvae eat leaves |
| *Phyllotreta sinuata* (Steph.) | Flea Beetle | Chrysomelidae | Malaysia | Adults hole leaves |
| *Callosobruchus maculatus* (F.) | Spotted Cowpea Bruchid | Bruchidae | Thailand | Attack ripe seeds in open pods |
| *Obera brevis* S. | Stem Borer | Cerambycidae | India | Larvae bore stems |

# CUCURBITS (Marrow, Pumpkin, Melon, Watermelon, Squash, (Cucumber, Loofah, etc — Cucurbitaceae)

Important cultivated species belong to nine separate genera within this family. Agriculturally the crops are really very diverse, but are biologically similar and have very similar pest spectra. Different species are native to different parts of the tropics (e.g. Watermelon to Africa; Marrow to the New World; Loofah in Asia). They are tendril-climbing or prostrate annuals with soft stems, table fruits, others as vegetables, some form gourds, and others loofahs. In temperate countries most cucurbits are cultivated under glass or polythene. (See also Butani & Varma, 1977.)

## MAJOR PESTS

| | | | | |
|---|---|---|---|---|
| *Trialeurodes vaporarorium* (Westw.) | Glasshouse Whitefly | Aleyrodidae | Europe | Infest foliage (greenhouse pest) |
| *Leptoglossus australis* (F.) | Leaf-footed Plant Bug | Coreidae | Africa, India, S.E. Asia, Australasia | Suck sap; toxic saliva |
| *Dacus cucurbitae* Coq. | Melon Fly | Tephritidae | E. Africa, India, S.E. Asia, Australasia | Larvae in fruits |
| *Dacus dorsalis* Hend. | Oriental Fruit Fly | Tephritidae | S.E. Asia | Larvae in fruits |
| *Dacus* spp. | Fruit Flies | Tephritidae | S.E. Asia, India, Australasia | Larvae in fruits |
| *Epilachna* spp. | Epilachna Beetles | Coccinellidae | S.E. Asia | Adults & larvae eat leaves |
| *Diabrotica undecimpunctata* Mann. | Spotted Cucumber Beetle | Chrysomelidae | N. America | Defoliate |
| *Raphidopalpa foveicollis* (Lucas) | Red Pumpkin Beetle | Chrysomelidae | Med., India, Burma, Australia | Defoliate |
| *Aulacophora* spp. | Cucumber Beetles | Chrysomelidae | S.E. Asia | Adults defoliate |
| *Tetranychus cinnabarinus* (Boisd.) | Carmine Spider Mite | Tetranychidae | Cosmopolitan | Scarify leaves & web foliage |
| *Tetranychus* spp. | Red Spider Mites | Tetranychidae | S.E. Asia | |
| *Tetranychus urticae* (Koch) | Temperate Red Spider Mite | Tetranychidae | Europe, Asia, N. America | (greenhouse pest) |

## MINOR PESTS

| | | | | |
|---|---|---|---|---|
| *Onychiurus* spp. | Springtails | Onychiuridae | Europe | Damage seedlings |
| *Mecopoda elongata* (L.) | Leaf Grasshopper | Tettigoniidae | Philippines | Eat leaves |
| *Empoasca* spp. | Green Leafhoppers | Cicadellidae | S.E. Asia | |
| *Bothrogonia ferruginea* | Large Brown Leafhopper | Cicadellidae | S.E. Asia | Infest leaves; sap-suckers; virus vectors |
| *Aphis gossypii* Glov. | Melon/Cotton Aphid | Aphididae | Cosmopolitan | |
| *Myzus persicae* (Sulz.) | Peach–Potato Aphid | Aphididae | Cosmopolitan | |

| | | | | |
|---|---|---|---|---|
| *Ferrisia virgata* (Ckll.) | Striped Mealybug | Pseudococcidae | S.E. Asia | Infest foliage |
| *Piezosternum calidum* F. | Shield Bug | Pentatomidae | Africa | |
| *Cyclopelta obscura* (L. & S.) | Shield Bug | Pentatomidae | Philippines | |
| *Coridius janus* F. | Cucurbit Stink Bug | Pentatomidae | India | Sap-suckers; toxic saliva |
| *Aspangopus brunneus* Thunb. | Cucurbit Stink Bug | Pentatomidae | India | |
| *Anoplocnemis phasiana* F. | Coreid Bug | Coreidae | Philippines | |
| *Thrips tabaci* Lind. | Onion Thrips | Thripidae | Europe | Infest foliage |
| *Thrips fuscipennis* Hal. | Rose Thrips | Thripidae | Europe | Infest foliage |
| *Adoxophyes privatana* (Wlk.) | Tortrix Moth | Tortricidae | Malaysia | Larvae roll leaves |
| *Sphenarches caffer* Zell. | Plume Moth | Pterophoridae | Africa, Asia, W. Indies | Larvae hole leaves |
| *Palpita indica* (Saund.) | Leaf-roller | Pyralidae | S. China | Larvae roll leaves |
| *Heliothis armigera* Hb. | American Bollworm | Noctuidae | Laos | Larvae bore fruit |
| *Plusia* spp. | Semi-loopers | Noctuidae | India, S.E. Asia | Larvae defoliate |
| *Agrotis segetum* (D. & S.) | Common Cutworm | Noctuidae | Philippines | Larvae are cutworms |
| *Spodoptera litura* (F.) | Rice Cutworm | Noctuidae | Philippines | Larvae are cutworms |
| *Sciara* spp. | Sciarid Flies | Sciaridae | Europe | Larvae damage roots (greenhouse pest) |
| *Phytomyza horticola* Gour. | Pea Leaf Miner | Agromyzidae | Cosmopolitan in Old World | Larvae mine leaves |
| *Lasioptera falcata* F. | Gall Midge | Cecidomyiidae | India | Larvae gall stems |
| *Mylabris* spp. | Banded Blister Beetles | Meloidae | Africa, India, S.E. Asia | Adults eat flowers |
| *Ceratia frontalis* | Leaf Beetle | Chrysomelidae | Thailand, Philippines | Adults defoliate |
| *Copa kunowi* Weise | Brown Flower Beetle | Chrysomelidae | Africa | Adults eat flowers |
| *Raphidopalpa similis* | Leaf Beetle | Chrysomelidae | Thailand | Adults defoliate |
| *Phyllotreta crucifera* (Goeze) | Cabbage Flea Beetle | Chrysomelidae | Africa, India | Adults hole leaves |
| *Monolepta bifasciata* (Hornst.) | Leaf Beetle | Chrysomelidae | Malaysia | Adults defoliate |
| *Diabrotica* spp. | Cucumber Beetles | Chrysomelidae | S. USA | Adults & larvae damage plants |
| *Anomala* spp. | White Grubs | Scarabaeidae | Philippines | Larvae eat roots |
| *Leucopholis irrorata* (Chevr.) | White Grub | Scarabaeidae | Philippines | Larvae eat roots |
| *Apomecyna* spp. | Vine Borers | Cerambycidae | Philippines, India | Larvae bore vines |
| *Tyrophagus dimidiatus* | Mushroom Mite | Tyroglyphidae | Europe | Damage foliage (greenhouse pest) |
| *Eutetranychus orientalis* (Klein) | Oriental Mite | Tetranychidae | Africa, India | Scarify foliage |

# CUSTARD APPLE (*Annona squamosa* – Annonaceae)
## (= Sugar Apple; Sweetsop)

A native to the W. Indies and S. America, but now grown widely throughout the tropics for its edible fruit. The fruit is large (7–10 cm diameter), heart-shaped, green in colour, covered with rounded fleshy tubercles which represent the loosely joined carpels. The pulp is granular, white, sweet, and rather like custard. It is used mainly as a dessert fruit. The plant is a woody shrub, seldom grown commercially but commonly in gardens or near houses. The ripe fruits are soft and perishable and difficult to transport, and are mostly consumed locally.

*Annona muricata* is the Soursop of C. America; *A. reticulata* is Bullock's Heart; and *A. montana* is the Mountain Soursop. (See also Butani, 1976*b*.)

## MAJOR PESTS

| | | | | |
|---|---|---|---|---|
| *Planococcus* spp. | Mealybugs | Pseudococcidae | S.E. Asia, India | Infest fruits |
| *Ferrisia virgata* (Ckll.) | Striped Mealybug | Pseudococcidae | India | Infest fruits |
| *Anonaepestis bengalella* (Rag.) | Fruit Borer | Pyralidae | India | Larvae bore fruit |

## MINOR PESTS

| | | | | |
|---|---|---|---|---|
| *Dialeuropora decempunctata* (Q. & B.) | Whitefly | Aleyrodidae | India | Infest foliage |
| *Ceroplastes floridensis* Comst. | Soft Scale | Coccidae | India | Encrust foliage |
| *Saissetia coffeae* (Wlk.) | Helmet Scale | Coccidae | India | Encrust foliage |
| *Laccifer communis* Mahd. | Lac Insect | Lacciferidae | India | Encrust twigs |
| *Retithrips syriacus* (Mayet) | Castor Thrips | Thripidae | India | Infest leaves |
| *Rhipiphorothrips cruentatus* Hood | Grapevine Thrips | Thripidae | India | Infest leaves |
| *Pyroderces falcetalla* St. | – | Cosmopterygidae | India | Larvae bore stems |
| *Dacus persicae* B. | Peach Fruit Fly | Tephritidae | India | Larvae inside fruit |
| *Dacus zonatus* (Saund.) | Peach Fruit Fly | Tephritidae | India | Larvae inside fruits |
| *Coccotrypes carpophagus* Horn | Seed Scolytid | Scolytidae | India | Adults bore seeds |

# DATE PALM (*Phoenix dactyliferae* – Palmae)

One of the earliest crop plants, having been cultivated for at least 5000 years; probably native to Arabia or India, but now naturalized throughout S.W. Asia and N. Africa. In habit a typical tall palm bearing clusters of fruit at the crown; the texture of the fruit varies with the variety. The Date Palm can grow in more arid areas than any other crop and hence is of great value in desert areas. The fruit has a high food value, with a sugar content of some 54% and 7% protein. Most of the commercial dates come from Iraq, the rest from N. Africa, Arabia, California and Arizona. Propagation is either by seed or by cuttings. (See also Butani, 1975*d*, and Carpenter & Elmer, 1978.)

## MAJOR PESTS

| | | | | |
|---|---|---|---|---|
| *Parlatoria blanchardii* (Targ.) | Date Palm Scale | Diaspididae | Africa, Asia Minor, India, Australia, S. America | Infest foliage & fruits |
| *Ephestia cautella* (Hb.) | Almond (Fig) Moth | Pyralidae | Pantropical | Larvae eat fruits |
| *Oryctes rhinoceros* (Oliv.) | Rhinoceros Beetle | Scarabaeidae | India | Adults damage crown |
| *Oligonychus* spp. | Date Mites | Tetranychidae | N. Africa, Iraq, Iran, S. USA | Scar leaves & fruits |

## MINOR PESTS

| | | | | |
|---|---|---|---|---|
| *Odontotermes obesus* (Rambur) | Scavenging Termite | Termitidae | India | Damage trunk |
| *Ommatissus binotatus* Fieb. | Dubas Bug | Tropiduchidae | N. Africa, Iran, Iraq, Egypt | Infest foliage |
| *Pseudococcus* spp. | Mealybugs | Pseudococcidae | Pantropical | Infest foliage |
| *Chrysomphalus aonidum* (L.) | Purple Scale | Diaspididae | Pantropical | Infest foliage |
| *Aonidiella orientalis* (Newst.) | Oriental Yellow Scale | Diaspididae | Pantropical | Infest foliage |
| *Aspidiotus destructor* Sign. | Coconut Scale | Diaspididae | India | Infest foliage & fruit |
| *Ischnaspis longirostris* (Sign.) | Black Line Scale | Diaspididae | Pantropical | Infest foliage |
| *Phoenicoccus marlatti* (Ckll.) | Red Date Scale | Diaspididae | Pantropical | Infest leaf bases |
| *Asterolecanium phoenicis* (Rao) | Green Date Scale | Asterolecaniidae | Egypt, Israel, Iran, Iraq | Infest foliage & fruit |
| *Arenipses sabella* (Hmpsn.) | Greater Date Moth | Pyralidae | N. Africa, Egypt, Iran, Iraq, India | Larvae damage fruit |
| *Paramyelois transitella* (Wlk.) | Navel Orangeworm | Pyralidae | USA (California) | Larvae damage fruit |
| *Batrachedra amydraula* (Meyr.) | Lesser Date Moth | Cosmopterygidae | Iran, Iraq, Yemen, Egypt, Arabia, India | Larvae damage fruit |
| *Nephantis serinopa* Meyr. | Black-headed Caterpillar | Xyloryctidae | India | Larvae defoliate |
| *Parasa* spp. | Stinging Caterpillars | Limacodidae | China | Larvae defoliate |
| *Oecophylla smaragdina* F. | Red Tree Ant | Formicidae | India | Nest in crown; attack workers |
| *Vespa* spp. | Common Wasps | Vespidae | Pantropical | Adults pierce fruit |
| *Carpophilus* spp. | Fig Beetle (etc.) | Nitidulidae | Pantropical | Damage ripening fruits |
| *Oryctes* spp. | Rhinoceros Beetles | Scarabaeidae | Africa, Iraq, India, Iran | Adults damage crown |

| | | | | |
|---|---|---|---|---|
| *Chalcophora japonica* Gory | Jewel Beetle | Buprestidae | China ⎫ | Larvae bore trunk |
| *Chrysobothris saccedanea* Saund. | Flat-headed Borer | Buprestidae | China ⎬ | |
| *Pseudophilus testaceus* Gah. | Palm Stem Borer | Cerambycidae | Egypt, Iran, Iraq, Arabia | Larvae bore trunk & petioles |
| *Rhynchophorus phoenicis* (F.) | African Palm Weevil | Curculionidae | Africa | Larvae bore crown & trunk |
| *Rhynchophorus ferrugineus* Oliv. | Asiatic Palm Weevil | Curculionidae | Iraq, India, Indonesia, Philippines | Larvae bore crown & trunk |
| *Diocalandra* spp. | Coconut Weevils | Curculionidae | Africa, Asia, Australasia | Larvae bore tissues |
| *Coccotrypes dactyliperda* (F.) | Date Stone Beetle | Scolytidae | N. Africa, Egypt, Israel, India, S. USA | Adults bore fruits |
| *Brevipalpus phoenicis* (Geijskes) | Red Crevice Tea Mite | Tenuipalpidae | Pantropical | Scarify foliage |
| *Raoiella indica* Hirst | Date Palm Scarlet Mite | Tenuipalpidae | Sudan, Egypt, India | Scarify foliage |

# DECCAN HEMP (*Hibiscus cannabinus* — Malvaceae)
## ( = Kenaf, etc.)

This tall annual herb yields a fibre known by many different common names, and has long been used in the Old World tropics as a substitute for jute and hemp in the manufacturing of coarse canvas, cordage, matting, fishing nets, etc. The fibres are 2–3 m in length and are usually extracted from the stems of the plants by retting. The seeds contain 20% oil which has certain commercial uses. It is a common wild plant in tropical Africa, which is probably its original home, but is now being grown widely throughout the tropics and sub-tropics as a fibre crop.

## MAJOR PESTS

| | | | | |
|---|---|---|---|---|
| *Maconellicoccus hirsutus* (Green) | Hibiscus Mealybug | Pseudococcidae | India | Infest foliage; stunt or kill shoots |

## MINOR PESTS

| | | | | |
|---|---|---|---|---|
| *Amrasca biguttula* Ish. | Leafhopper | Cicadellidae | India | Inhibit leaves |
| *Pinnaspis strachani* (Cooly) | Armoured Scale | Diaspididae | India | Encrust stems |
| *Dysdercus cingulatus* (Fb.) | Cotton Stainer | Pyrrhocoridae | India | Sap-sucker; toxic saliva |
| *Nezara viridula* L. | Green Stink Bug | Pentatomidae | India | Sap-sucker; toxic saliva |
| *Pectinophora gossypiella* (Saund.) | Pink Bollworm | Gelechiidae | India | Larvae bore fruits |
| *Earias insulana* Boisd. | Spiny Bollworm | Noctuidae | India | Larvae bore fruits |
| *Earias vittella* (F.) | Spotted Bollworm | Noctuidae | India | Larvae bore fruits |
| *Anomis flava* F. | Cotton Semi-looper | Noctuidae | India | Larvae defoliate |
| *Porthesia scintillans* Wlk. | Tussock Moth | Lymantriidae | India | Larvae defoliate |
| *Agrilus acutus* Thnb. | Jute Stem Borer | Buprestidae | India | Larvae bore stems |
| *Mylabris pustulata* Th. | Banded Blister Beetle | Meloidae | India | Adults deflower |
| *Podagrica* spp. | Cotton Flea Beetles | Chrysomelidae | India | Adults hole leaves |
| *Alcidodes affaber* F. | Stem Weevil | Curculionidae | India | Adult girdles stem |
| *Dereodus mastos* Hb. | Leaf Weevil | Curculionidae | India | Adults eat leaves |

# EGGPLANT (*Solanum melongena* – Solanaceae)
## ( = Brinjal; fruit called Aubergine)

Found wild and first cultivated in India; now cultivated throughout the tropics. It grows well up to 1000 m in altitude on light soils. It is a perennial, weakly erect herb, 0.5–1.5 m in height; with a fruit that is a large pendant berry, ovoid or oblong and 5–15 cm long, smooth in texture and usually black or purple when ripe. The fruit is eaten as a vegetable, boiled, fried, or stuffed. Propagation is by seed. It is grown throughout the tropics for local consumption but some countries (e.g. Kenya) are developing an export trade with Europe. (See also Butani & Varma, 1976*a*.)

## MAJOR PESTS

| | | | | |
|---|---|---|---|---|
| *Bemisia tabaci* (Genn.) | Tobacco Whitefly | Aleyrodidae | India | Infest foliage |
| *Phthorimaea operculella* (Zeller) | Potato Tuber Moth | Gelechiidae | Cosmopolitan | Larvae bore stem |
| *Euzophera perticella* (Rag.) | Brinjal Stem Borer | Pyralidae | India | Larvae bore stem |
| *Leucinodes orbonalis* Guen. | Eggplant Boring Caterpillar | Pyralidae | Thailand, Laos, Malaysia, India | Larvae bore stem |
| *Epicauta* spp. | Black Blister Beetles | Meloidae | Asia, USA | Adults eat flowers |
| *Epilachna* spp. | Epilachna Beetles | Coccinellidae | India, S.E. Asia | Adults & larvae eat leaves |

## MINOR PESTS

| | | | | |
|---|---|---|---|---|
| *Oxya japonica* (Thnb.) | Small Rice Grasshopper | Acrididae | India | Adults & nymphs eat leaves |
| *Aphis gossypii* Glov. | Cotton Aphid | Aphididae | Cosmopolitan | Infest leaves & stems; virus vectors |
| *Macrosiphum euphorbiae* (Thos.) | Potato Aphid | Aphididae | Cosmopolitan | |
| *Myzus persicae* (Sulz.) | Green Peach Aphid | Aphididae | S.E. Asia, India | |
| *Amrasca* spp. | Leafhoppers | Cicadellidae | India, S.E. Asia | Infest foliage |
| *Empoasca* spp. | Green Leafhoppers | Cicadellidae | S.E. Asia, India, Africa | Infest leaves & stems |
| *Quadraspidiotus destructor* Sign. | San José Scale | Diaspididae | India | Infest foliage |
| *Ferrisia virgata* (Ckll.) | Striped Mealybug | Pseudococcidae | S.E. Asia, India | Infest stems & leaves |
| *Orthezia insignis* Browne | Jacaranda Bug | Ortheziidae | India, Malaysia, Africa, N., C. & S. America | Infest stems & leaves |
| *Terentius nubifasciastus* | Treehopper | Membracidae | Papua NG | Sap-suckers |
| *Tricentrus bicolor* Dist. | Treehopper | Membracidae | India | Sap-suckers |
| *Nezara viridula* (L.) | Green Stink Bug | Pentatomidae | Africa, Asia | Toxic saliva; sap-suckers |
| *Creontiades pallidifer* Wlk. | Capsid Bug | Miridae | India | Toxic saliva; sap-suckers |
| *Cyrtopeltis tenuis* (Reut.) | Tobacco Capsid | Miridae | Pantropical | |
| *Dysdercus* spp. | Cotton Stainers | Pyrrhocoridae | S.E. Asia | Toxic saliva |
| *Urentius* spp. | Brinjal Lace Bugs | Tingidae | India | Toxic saliva |
| *Gargaphia solani* Heid. | Eggplant Lace Bug | Tingidae | USA | |
| *Thrips tabaci* Lind. | Onion Thrips | Thripidae | India | Infest foliage |
| *Thrips palmi* | Palm Thrips | Thripidae | S.E. Asia, China | |

| | | | | |
|---|---|---|---|---|
| *Scrobipalpa heliopa* (Lower) | Tobacco Stem Borer | Gelechiidae | India, S.E. Asia | Larvae bore stems |
| *Eublemma olivacea* (Wlk.) | Brinjal Leaf Roller | Tortricidae | India | Larvae roll leaves |
| *Homona coffearia* (Guen.) | Tea Tortrix | Tortricidae | Papua NG | Larvae roll leaves |
| *Spodoptera litura* (F.) | Rice Cutworm | Noctuidae | S.E. Asia, India | Larvae defoliate |
| *Spodoptera littoralis* (Boisd.) | Cotton Leafworm | Noctuidae | Africa | Larvae defoliate |
| *Heliothis assulta* Gn. | Cape Gooseberry Budworm | Noctuidae | India, S.E. Asia, Africa, Australasia | Larvae bore buds |
| *Agrotis ipsilon* (Hufn.) | Black Cutworm | Noctuidae | India | Larvae are cutworms |
| *Plusia orichalcea* F. | – | Noctuidae | India | Larvae eat leaves |
| *Heliothis armigera* (Hub.) | American Bollworm | Noctuidae | S.E. Asia | Larvae bore buds & fruit |
| *Antoba olivacea* (Wlk.) | – | Noctuidae | India | Larvae defoliate |
| *Phytometra chalcites* (Esp.) | Semi-looper | Noctuidae | Malaysia | Larvae defoliate |
| *Mythimna separata* (Wlk.) | Rice Ear-cutting Caterpillar | Noctuidae | S.E. Asia | Larvae defoliate |
| *Acherontia* spp. | Deaths Head Hawk Moths | Sphingidae | Laos, India | Larvae defoliate |
| *Oulema* spp. | Leaf Beetles | Chrysomelidae | India | Adults & larvae eat leaves |
| *Rhyparida coriacea* | Leaf Beetle | Chrysomelidae | Papua NG | Adults eat leaves |
| *Leptinotarsa decemlineata* (Say) | Colorado Beetle | Chrysomelidae | Europe, N. & C. America | Defoliate |
| *Psylliodes* spp. | Flea Beetles | Chrysomelidae | Philippines | Adults eat leaves |
| *Leucopholis irrorata* (Chevr.) | White Grub | Scarabaeidae | Philippines | Larvae eat roots |
| *Anomala* spp | White Grubs | Scarabaeidae | Philippines, India | Larvae eat roots |
| *Holotrichia consanguinea* Blanch. | Cockchafer | Scarabaeidae | India | Larvae eat roots |
| *Myllocerus* spp. | Grey Weevils | Curculionidae | India | Adults eat leaves |
| *Trichobaris trinotata* (Say) | Potato Stalk Borer | Curculionidae | USA | Larvae bore stalks |
| *Polyphagotarsonemus latus* (Banks) | Yellow Tea Mite | Tarsonemidae | S.E. Asia | Scrify foliage |
| *Tetranychus cinnabarinus* (Boisd.) | Carmine Spider Mite | Tetranychidae | S.E. Asia, India | Scarify foliage |
| *Tetranychus* spp. | Red Spider Mites | Tetranychidae | Cosmopolitan | Scarify leaves & web foliage |

# FIG (*Ficus carica* — Moraceae)

A native of Asia Minor and spread early into the Mediterranean Region; of great antiquity as a crop, being grown in Egypt before 4000 BC. The fruit may be eaten fresh, stewed, dried or now canned. The tree is of moderate size, up to 10 m in height with large palmate leaves. The main producing areas are around the Mediterranean (Turkey, Greece, Italy) and California, with some production in Australia. The plant does not flourish in the low wet tropics, but can be grown at higher elevations and drier parts of the tropics. The widely cultivated Smyrna fig variety requires pollination by the symbiotic wasp *Blastophaga psenes* for fruit development, but the common or Adriatic variety develops parthenocarpically. Wild species of *Ficus* are widespread and common throughout the tropics and several banyan species are grown as shade and avenue trees. (See also Butani, 1975*b*.)

## MAJOR PESTS

| | | | | |
|---|---|---|---|---|
| *Ceratitis capitata* (Wied.) | Medfly | Tephritidae | Europe, Africa, Australia, C. & S. America | Larvae inside fruits |
| *Chrysomphalus aonidum* (L.) | Purple Scale | Diaspididae | Cosmopolitan | Infest leaves |
| *Ephestia cautella* (Hb.) | Almond Moth | Pyralidae | S.E. Asia | Attack ripe & stored fruits |
| *Batocera* spp. | Longhorn Beetles | Cerambycidae | Med., E. Africa, India, Malaysia, China, Japan | Larvae bore trunk & branches |

## MINOR PESTS

| | | | | |
|---|---|---|---|---|
| *Velu caricae* Ghauri | Fig Leafhopper | Cicadellidae | India | Infest leaves |
| *Cosmocarta niteara* Dist. | Spittlebug | Cercopidae | India | Sap-suckers |
| *Planococcus citri* (Risso) | Root Mealybug | Pseudococcidae | Cosmopolitan | On roots & foliage |
| *Planococcus* spp. | Mealybugs | Pseudococcidae | India | Infest foliage |
| *Pseudococcus adonidum* (L.) | Long-tailed Mealybug | Pseudococcidae | Cosmopolitan | Infest foliage |
| *Pauropsylla depressa* Crawf. | Fig Leaf Psyllid | Psyllidae | India | Nymphs gall leaves |
| *Lepidosaphes conchiformis* (L. | Mussel Scale | Diaspididae | Europe, Africa, Asia, Japan, USA, S. America | Infest foliage |
| *Kerria fici* (Green) | Fig Lac Insect | Lacciferidae | India | Infest twigs |
| *Coccus hesperidium* L. | Soft Brown Scale | Coccidae | Cosmopolitan | Infest foliage |
| *Ceroplastes sinensis* Del G. | Chinese Waxy Scale | Coccidae | Almost cosmopolitan | Infest foliage |
| *Ceroplastes rubens* Mask. | Pink Waxy Scale | Coccidae | E. Africa, Asia | Infest foliage |
| *Ceroplastes rusci* (L.) | — | Coccidae | Med. | Infest foliage |
| *Saissetia oleae* Bern. | Black Scale | Coccidae | Cosmopolitan | Infest foliage |

| | | | | |
|---|---|---|---|---|
| *Drosicha mangiferae* (Green) | Mango Giant Mealybug | Margarodidae | India | Infest foliage |
| *Icerya purchasi* Mask. | Cottony Cushion Scale | Margarodidae | Cosmopolitan | Infest foliage |
| *Aspidiotus destructor* (Sign.) | Coconut Scale | Diaspididae | India | Infest foliage |
| *Thrips tabaci* Lind. | Onion Thrips | Thripidae | Cosmopolitan | Infest foliage |
| *Frankliniella* spp. | Flower Thrips | Thripidae | Cosmopolitan | Infest foliage |
| *Gigantothrips elegans* Zimm. | Giant Fig Thrips | Phlaeothripidae | India | Infest leaves |
| *Udumbaria nainiensis* Grover | Fig Midge | Cecidomyiidae | India | Larvae inside fruits |
| *Anjeerodiplosis peshawarensis* Mani | Fig Midge | Cecidomyiidae | India | Larvae inside fruits |
| *Dacus zonatus* Saund. | Peach Fruit Fly | Tephritidae | India | Larvae inside fruits |
| *Dacus orientalis* (Hend.) | Oriental Fruit Fly | Tephritidae | India | Larvae inside fruits |
| *Lonchaea aristella* Beck. | – | Lonchaeidae | Med. | Larvae inside fruits |
| *Zeuzera coffeae* Nietn. | Red Coffee Borer | Cossidae | S.E. Asia | Larvae bore branches |
| *Zeuzera pyrina* L. | Leopard Moth | Cossidae | Med. | Larvae bore branches |
| *Azochis gripusalis* Wlk. | – | Pyralidae | Brazil, Mexico | Larvae eat foliage |
| *Diaphania* spp. | – | Pyralidae | India | Larvae defoliate |
| *Paramyelois transitella* (Wlk.) | Navel Orange-worm | Pyralidae | USA (California) | Larvae bore fruits |
| *Latoia lepida* (Cram.) | Blue-striped Nettlegrub | Limacodidae | India | Larvae eat leaves |
| *Simaethis nemorona* | – | Glyphipterygidae | Med. | Larvae eat foliage |
| *Phycodes* spp. | – | Glyphipterygidae | India | Larvae web leaves |
| *Spodoptera litura* F. | Rice Cutworm | Noctuidae | India | Larvae defoliate |
| *Heliothis* spp. | – | Noctuidae | India | Larvae bore fruits |
| *Ocinara variana* Wlk. | Tussock Moth | Lymantriidae | India | Larvae defoliate |
| *Perina nuda* F. | Banyan Tussock Moth | Lymantriidae | India | Larvae defoliate |
| *Carpophilus hemipterus* (L.) | Dried Fruit Beetle | Nitidulidae | Cosmopolitan | Feed on dried figs |
| *Adoretus* spp. | Flower Beetles | Scarabaeidae | India | Adults eat leaves |
| *Cotinis* spp. | – | Scarabaeidae | S. USA | Adults defoliate |
| *Apriona* spp. | Longhorn Beetles | Cerambycidae | India, S. China | Larvae bore branches |
| *Hesperophanes* spp. | Longhorn Beetles | Cerambycidae | Med., S.E. Asia | Larvae bore branches |
| *Olenecamptus* spp. | Longhorn Beetles | Cerambycidae | India, China | Larvae bore branches |
| *Hypoborus ficus* Erichs. | Fig Twig Borer | Scolytidae | Med. | Adults bore twigs |
| *Eutetranychus orientalis* (Klein) | Oriental Mite | Tetranychidae | Africa, India | Scarify foliage |
| *Aceria ficus* (Cottee) | Fig Leaf Mite | Eriophyidae | India | Scarify foliage |
| *Eriophyes ficivorus* Ch. Bas. | Fig Gall Mite | Eriophyidae | India | Distort leaves |

# GINGER (*Zingiber officinale* – Zingiberaceae)

Native to S.E. Asia this plant is an erect perennial herb (1 m tall) with a thick scaly rhizome that branches digitately. The yellowish flowers are borne in a spike. Most cultivation is in small home gardens, in a moist tropical climate; the plant is propagated by the rhizomes. The rhizomes are pale yellow externally and yellowish green inside, and contain starch, oleoresin, gums and essential oils. Ginger is prepared either as preserved or green ginger, or dried (cured) ginger. It is widely used as a condiment (rather than a spice), in culinary preparation, beverages, and medicine. Grown chiefly in China, Japan, Sierra Leone, Indonesia, Queensland, Australia, and the W. Indies.

## MAJOR PESTS

## MINOR PESTS

| | | | | |
|---|---|---|---|---|
| *Pentalonia nigronervosa* Coq. | Banana Aphid | Aphididae | Pantropical | Infest leaf-bases |
| *Aspidiella hartii* (Ckll.) | Yam Scale | Diaspididae | India, W. Africa, W. Indies | Encrust rhizome |
| *Acrocercops irridians* Meyr. | Leaf Miner | Gracillariidae | India | Larvae mine leaves |
| *Dichocrocis punctiferalis* Guen. | Shoot Borer (Castor Capsule Borer) | Pyralidae | Malaysia, India, China | Larvae bore shoots |
| *Udaspes folus* (Cr.) | Grass Demon (Turmeric Skipper) | Hesperiidae | S. China, India, Malaysia | Larvae roll leaves |
| *Spodoptera litura* (F.) | Rice Cutworm | Noctuidae | Malaysia | Larvae defoliate |
| *Chalcidomyia atricornis* Mall. | Shoot Fly | Chloropidae | India, Bangladesh | Larvae bore shoots & rhizome |
| *Formosina flavipes* Mall. | Shoot Fly | Chloropidae | India, Bangladesh | Larvae bore shoots & rhizome |
| *Calobata* sp. | Rhizome Fly | Micropezidae | India | Larvae bore rhizome |
| *Mimegralla coerubifrons* Mall. | Rhizome Fly | Micropezidae | Bangladesh | Larvae bore rhizome |
| *Celyphus* sp. | Rhizome Fly | Celyphidae | India | Larvae bore rhizome |
| *Hedychorus rufofasciatus* M. | Leaf Weevil | Curculionidae | India | Adults nibble leaves |

# GRAPEVINE (*Vitus vinifera* – Vitaceae)

*V. vinifera* is the European or Wine Grape; there are several native American species with larger and more hardy fruit, and these are more resistant to diseases and pests. The European grapevine is one of the oldest cultivated plants, and probably originated in the Caspian Sea area of western Asia. They were spread all over Europe with the Roman civilization, and are now found in all temperate regions and in higher areas in many tropical countries. It is a woody, climbing, tendril-bearing vine, with large palmate leaves, small and insignificant flowers leading to large clusters of fruit. The fruit is technically a berry, and is either eaten fresh, dried as raisins, or currants, or the juice is pressed out to make wine. Raisins are dried wine grapes of high quality, and may be seedless like the Sultana variety. Currants are small dried grapes from a variety that grows in Greece. The chief grape-growing areas are Europe, USA, Argentina, Chile, Australia and S. Africa. Propagation is by stem cuttings. (See also Butani, 1974*b*.)

## MAJOR PESTS

| | | | | |
|---|---|---|---|---|
| *Erythroneura* spp. | Grape Leaf-hoppers | Cicadellidae | Canada, USA | Infest foliage; sap-suckers |
| *Viteus vitifolii* (Fitch) | Grape Phylloxera | Phylloxeridae | Europe, USA | Leaves galled |
| *Planococcus* spp. | Root Mealybugs | Pseudococcidae | Cosmopolitan | Infest rootstocks |
| *Pseudococcus maritimus* (Ehrh.) | Grape Mealybug | Pseudococcidae | Widespread | Infest stems |

## MINOR PESTS

| | | | | |
|---|---|---|---|---|
| *Odontotermes obesus* (Ramb.) | Scavenging Termite | Termitidae | India | Damage roots |
| *Aleurocanthus spiniferus* Quaint | Orange Spiny Whitefly | Aleyrodidae | S. & E. Asia | Infest foliage |
| *Aleurocanthus woglumi* (Ashby) | Citrus Blackfly | Aleyrodidae | India | Infest foliage |
| *Maconellicoccus hirsutus* (Green) | Hibiscus Mealybugs | Pseudococcidae | India | Kill shoots |
| *Ferrisia virgata* (Ckll.) | Striped Mealybug | Pseudococcidae | India | Infest foliage |
| *Lecanium* spp. | Soft Scales | Coccidae | India | Encrust foliage |
| *Saissetia oleae* (Colv.) | Black Scale | Coccidae | India | Encrust foliage |
| *Parasaissetia nigra* (Niet.) | Nigra Scale | Coccidae | India | Encrust foliage |
| *Parthenolecanium corni* (Bouche) | Plum Scale | Coccidae | Europe | Encrust foliage |
| *Kerria lacca* (Kerr.) | Lac Insect | Lacciferidae | India | Encrust stems |
| *Aspidiotus* spp. | Hard Scales | Diaspididae | India | Encrust foliage |
| *Aphis gossypii* (Glov.) | Cotton Aphid | Aphididae | India | Infest foliage |
| *Retithrips syriacus* (Mayet) | Black Vine Thrips | Thripidae | Med., India | Infest leaves |
| *Rhipiphorothrips cruentatus* Hood | Thrips | Thripidae | India | Infest leaves |
| *Scirtothrips dorsalis* Hood | Thrips | Thripidae | India | Infest foliage |
| *Lobesia botrana* (Schiff.) | Grape Berry Moth | Tortricidae | Europe, Japan, E. Africa | Larvae destroy fruit |

| *Paralobesia viteana* (Clem.) | Grape Berry Moth | Tortricidae | N. America | Larvae destroy fruit |
|---|---|---|---|---|
| *Cnephasia longana* (Haw.) | Omnivorous Leaf Roller | Tortricidae | Switzerland, S. USA | Larvae roll leaves |
| *Epiphyas postvittana* (Wlk.) | Light Brown Apple Moth | Tortricidae | Australia, NZ | Larvae eat leaves |
| *Hyles lineata* (F.) | Silver-striped Hawk Moth | Sphingidae | Pantropical | Larvae defoliate |
| *Sylepta lunalis* (Guen.) | Grape Leaf Roller | Pyralidae | India | Larvae roll leaves |
| *Paramyelois transitella* (Wlk.) | Navel Orange-worm | Pyralidae | USA | Larvae damage fruits |
| *Clysia ambiguella* (Hb.) | Vine Moth | Phalaonidae | Europe, China, Japan | Larvae bore stems |
| *Othreis fullonia* L. | Fruit-piercing Moth | Noctuidae | India | Adults pierce fruits |
| *Achaea janata* L. | Fruit-piercing Moth | Noctuidae | India | Adults pierce fruits |
| *Spodoptera litura* (F.) | Rice Cutworm | Noctuidae | India | Larvae eat leaves |
| *Xestia c-nigrum* (L.) | Spotted Cutworm | Noctuidae | Europe, Asia, N. America | Larvae eat leaves |
| *Euproctis* spp. | Tussock Moths | Lymantriidae | India | Larvae eat leaves |
| *Dacus cucurbitae* (Coq.) | Melon Fly | Tephritidae | India | Maggots bore fruits |
| *Vespa* spp. | Common Wasps | Vespidae | Cosmopolitan | Adults puncture berries |
| *Polistes olivaceous* (F.) | Paper Wasp | Vespidae | India | Adults puncture berries |
| *Colaspis brunnea* (F.) | Grape Colaspis | Chrysomelidae | S. USA | Adults eat leaves |
| *Scelodonta strigicollis* (Mot.) | Flea Beetle | Chrysomelidae | India | Adults eat shoots; larvae eat roots |
| *Nodostoma* spp. | Leaf Beetles | Chrysomelidae | India | Defoliate |
| *Oides decempunctata* (Billb.) | Ten-spotted Leaf Beetle | Chrysomelidae | S. China, India | Adults & larvae eat leaves |
| *Monolepta* spp. | Spotted Leaf Beetles | Chrysomelidae | India | Adults hole leaves |
| *Adoretus* spp. | Cockchafers | Scarabaeidae | India | Adults defoliate |
| *Anomala* spp. | Flower Beetles | Scarabaeidae | India | Adults eat leaves |
| *Brahmina coriacea* (Hope) | Cockchafer | Scarabaeidae | India | Adults defoliate |
| *Holotrichia longipennis* (Blanch.) | Cockchafer | Scarabaeidae | India | Adults eat leaves; larvae eat roots |
| *Sthenias grisator* F. | Girdler Beetle | Cerambycidae | India | Larvae girdle stems |
| *Myllocerus* spp. | Grey Weevils | Curculionidae | India | Adults eat leaves |
| *Xyleborus semiopacus* Eich. | Black Twig Borer | Scolytidae | India | Adults bore stems |
| *Tetranychus urticae* (Koch) | Temperate Red Spider Mite | Tetranychidae | Europe | Scarify foliage |
| *Eutetranychus orientalis* (Klein) | Oriental Mite | Tetranychidae | India | Scarify foliage |
| *Oligonychus* spp. | Spider Mites | Tetranychidae | India | Scarify foliage |

| *Brevipalpus californicus* (Banks) | – | Tenuipalpidae | Cosmopolitan | Scarify foliage |
| *Eriophyes vitis* (Pgst.) | Grape Gall Mite | Eriophyidae | USA | Erinea on leaves |
| *Calepitrimerus vitis* (Can.) | Grape Rust Mite | Eriophyidae | USA | Scarify foliage |

# GRASS (Many species – Gramineae)

It is not feasible to consider the different species and genera of grasses separately because of the number involved. Certain pest species are restricted to certain species of grass, but in general most of the pests listed are polyphagous and attack a wide range of host species, this being especially the case for the soil-dwelling larvae. Many of the pests listed under the various graminaceous crops (e.g. cereals and sugarcane) are to be found infesting species of grasses, and frequently grasses are the natural wild hosts from which the pests spread on to the cereal crops.

## MAJOR PESTS

| | | | | |
|---|---|---|---|---|
| *Locusta migratoria* sspp. | Migratory Locusts | Acrididae | Africa, Asia | Defoliate |
| Many species | Short-horned Grasshoppers | Acrididae | Cosmopolitan | Defoliate |
| *Gryllotalpa* spp. | Mole Crickets | Gryllotalpidae | Cosmopolitan | Destroy roots |
| *Gryllus* spp. | Field Crickets | Gryllidae | Cosmopolitan | Defoliate & destroy roots |
| *Hodotermes mossambicus* (Hagen) | Harvester Termite | Hodotermitidae | E. & S. Africa | Defoliate |
| *Odontotermes badius* (Hartland) | Crater Termite | Termitidae | Tropical Africa | Defoliate |
| *Nephotettix* spp. | Green Leafhoppers | Cicadellidae | Africa, Asia | Sap-suckers; virus vectors |
| Many species | Leafhoppers | Cicadellidae | Cosmopolitan | |
| Many species | Planthoppers | Delphacidae | Cosmopolitan | Sap-suckers; virus vectors |
| *Hysteroneura setariae* (Th.) | Rusty Plum Aphid | Aphididae | Pantropical & USA | |
| *Rhopalosiphum* spp. | Grass Aphids | Aphididae | Holarctic | Sap-suckers |
| *Schizaphis graminum* (Rond.) | Wheat Aphid | Aphididae | Cosmopolitan | Sap-sucker |
| *Macrosiphum* spp. | Grass Aphids | Aphididae | Cosmopolitan | Sap-suckers |
| *Metopolophium* spp. | Grass Aphids | Aphididae | Cosmopolitan | Sap-suckers |
| *Saissetia oleae* (Ol.) | Black Scale | Coccidae | Sub-tropical | Infest foliage |
| *Antonia graminis* (Mask.) | Rhodes Grass Scale | Diaspididae | Pantropical | Encrust foliage |
| *Brevennia rehi* (Ldgr.) | Rice Mealybug | Pseudococcidae | India, S.E. Asia, S. USA | Infest foliage |
| *Aptinothrips* spp. | Grass Thrips | Thripidae | Europe | Infest flowers |
| *Ostrinia* spp. | Corn Borers | Pyralidae | Cosmopolitan | Larvae bore stems |
| Several species | Grass Stem Borers | Pyralidae | Cosmopolitan | Larvae bore stems |
| *Diatraea saccharalis* (F.) | Sugarcane Borer | Pyralidae | S. USA, C. & S. America | Larvae bore stems |
| *Crambus* spp. | Grass Moths | (Crambidae) | Cosmopolitan | Larvae defoliate |
| *Hepialus* spp. | Swift Moths | Hepialidae | Palaearctic | Larvae in soil eat roots |

575

| | | | | |
|---|---|---|---|---|
| Several species | Skippers | Hesperiidae | Cosmopolitan | Larvae roll leaves |
| *Spodoptera frugiperda* (J.E.S.) | Fall Armyworm | Noctuidae | N., C. & S. America | Larvae defoliate |
| *Spodoptera exempta* (Wlk.) | African Armyworm | Noctuidae | Africa, India, S.E. Asia, Australia | Larvae defoliate |
| *Spodoptera spp.* | Armyworms, etc. | Noctuidae | Pantropical | Larvae defoliate |
| *Remigia repanda* (F.) | Guinea Grass Moth | Noctuidae | C. & S. America | Larvae defoliate |
| *Opomyza* spp. | Grass Flies | Opomyzidae | Palaearctic | Larvae bore shoots |
| *Geomyza* spp. | Grass Flies | Opomyzidae | Palaearctic | Larvae bore shoots |
| *Oscinella* spp. | 'Frit' Flies | Chloropidae | Palaearctic | Larvae gall stems |
| *Chlorops* spp. | Gout Flies | Chloropidae | Palaearctic | Larvae gall stems |
| *Tipula* spp. etc. | Leatherjackets | Tipulidae | Palaearctic | Larvae in soil eat roots |
| *Mayetiola* spp. | 'Flax' Midges | Cecidomyiidae | Europe | Larvae distort stems |
| *Contarinia* spp. | Grass Flower Midges | Cecidomyiidae | Europe | Larvae in flower head |
| *Amaurosoma* spp. | Grass Midges | Cordyluridae | Europe | Larvae gall spike |
| *Atherigona* spp. | Shoot Flies | Muscidae | Cosmopolitan | Larvae destroy shoot |
| Several species | Shoot Flies | Anthomyiidae | Cosmopolitan | Larvae destroy shoot |
| *Messor barbarus* L. | Harvester Ant | Formicidae | E. Africa | Defoliate |
| *Atta* spp. | Leaf-cutting Ants | Formicidae | W. Indies, C. & S. America | Defoliate |
| *Acromyrmex* spp. | | | | |
| *Agriotes* spp. etc. | Wireworms | Elateridae | Palaearctic | Larvae in soil eat roots |
| *Lacon* spp. etc. | Tropical Wireworms | Elateridae | Pantropical | |
| *Oulema* spp. | Cereal Leaf Beetles | Chrysomelidae | Cosmopolitan | Larvae mine leaves; adults eat strips |
| *Anomala* spp. | White Grubs | Scarabaeidae | Cosmopolitan | Larvae live in soil and feed on plant roots; especially serious to Gramineae & pastures. Adults may damage plant foliage |
| *Adoretus* spp. | White Grubs | Scarabaeidae | Pantropical | |
| *Cetonia* spp. | White Grubs | Scarabaeidae | Cosmopolitan | |
| *Heteronychus* spp. | Cereal Beetles | Scarabaeidae | Pantropical | |
| *Holotrichia* spp. | White Grubs | Scarabaeidae | Pantropical | |
| *Melolontha* spp. | Chafer Grubs | Scarabaeidae | Cosmopolitan | |
| *Leucopholis* spp. | White Grubs | Scarabaeidae | S.E. Asia | |
| *Schizonycha* spp. | White Grubs | Scarabaeidae | Africa | |
| *Protaetia* spp. | White Grubs | Scarabaeidae | Pantropical | |
| *Phyllobius* spp. | Common Leaf Weevils | Curculionidae | Palaearctic | Larvae in soil eat grass roots |
| Many species | Weevils | Curculionidae | Cosmopolitan | Larvae in soil eat grass roots |

# GROUNDNUT (*Arachis hypogaea* — Leguminosae)

This crop originated in the Grand Chaco area of S. America, and has been cultivated in Mexico and the W. Indies since pre-Columbian times. The 16th century Spaniards introduced it to W. Africa, Philippines, China, Japan, Malaya, India and Madagascar. Now it is grown in all tropical and sub-tropical countries, up to 40°N and S of the equator. It is a warm season crop and is killed by frost; mostly grown in areas of 100 cm or more rainfall: it needs 50 cm rain during the growing season, and dry weather for ripening. It is a small erect or trailing herb, 15–60 cm high. Seeds are produced underground in pods; the seeds are rich in oil (38–50%), protein, and vitamins B and C. The main production areas are India, China, Nigeria, Sudan, Senegal, Niger, Gambia, USA, Brazil and Argentina. (See also Feakin, 1973.)

## MAJOR PESTS

| | | | | |
|---|---|---|---|---|
| *Aphis craccivora* Koch | Groundnut Aphid | Aphididae | Pantropical | Infest foliage; virus vector |
| *Hilda patruelis* Stål | Groundnut Hopper | Tettigometridae | Africa | Subterranean sap-suckers |
| *Taeniothrips sjostedti* (Trybom) | Bean Flower Thrips | Thripidae | Africa | Infest flowers |
| *Frankliniella schulzei* (Trybom) | Cotton Flower Thrips | Thripidae | E. Africa, Sudan | Infest flowers |
| *Etiella zinckenella* (Triet.) | Pea Pod Borer | Pyralidae | Malaysia | Larvae bore pods |
| *Spodoptera littoralis* (Boisd.) | Cotton Leafworm | Noctuidae | Africa | Larvae defoliate |
| *Spodoptera exigua* Hub. | Lesser Armyworm | Noctuidae | Europe, Africa, India, Japan, USA | Larvae defoliate |
| *Heliothis armigera* (Hub.) | American Bollworm | Noctuidae | Old World tropics | Larvae damage pods & foliage |
| *Schizonycha* spp. | Chafer Grubs | Scarabaeidae | Africa | Larvae eat roots |
| *Diabrotica undecimpunctata* Mann. | Spotted Cucumber Beetle | Chrysomelidae | USA | Defoliate |
| *Ootheca mutabilis* (Salhb.) | Brown Leaf Beetle | Chrysomelidae | E. Africa, Nigeria | Defoliate |
| *Epicauta albovittata* (Gestro) | Striped Blister Beetle | Meloidae | E. Africa, Somalia | Adults eat flowers |
| *Epicauta* spp. | Black Blister Beetles | Meloidae | Asia, USA | Adults eat flowers |
| *Alcidodes dentipes* (Ol.) | Striped Sweet Potato Weevil | Curculionidae | Africa | Adult girdles stem; larvae gall stem |
| *Systates* spp. | Systates Weevils | Curculionidae | Africa | Adults eat leaves |
| *Graphognathus* spp. | White-fringed Weevils | Curculionidae | S. Africa, Australia, NZ, USA, S. America | Adults eat leaves; larvae eat roots |

## MINOR PESTS

| | | | | |
|---|---|---|---|---|
| *Locusta migratoria* sspp. | Migratory Locusts | Acrididae | Africa, Asia | Defoliate |
| *Odontotermes* spp. | Scavenging Termites | Termitidae | Africa, India | Damage roots & foliage |
| *Hodotermes mossambicus* (Hag.) | Harvester Termite | Hodotermitidae | S. & E. Africa | Defoliate |
| *Ferrisia virgata* (Ckll.) | Striped Mealybug | Pseudococcidae | Africa, India | Infest foliage |
| *Dysmicoccus brevipes* (Ckll.) | Pineapple Mealybug | Pseudococcidae | Pantropical | Infest foliage |
| *Pseudococcus* spp. | Mealybugs | Pseudococcidae | Africa, Australia, C. & S. America | Infest foliage |
| *Empoasca* spp. | Green Leafhoppers | Cicadellidae | Africa, India, USA, S. America | Infest foliage |
| *Cicadulina* spp. | Maize Leafhoppers | Cicadellidae | Africa | Infest foliage |
| *Nezara viridula* (L.) | Green Stink Bug | Pentatomidae | Cosmopolitan | Sap-sucker; toxic saliva |
| *Bagrada* spp. | Harlequin Bugs | Pentatomidae | Africa, Asia | Sap-suckers; toxic saliva |
| *Anoplocnemis phasiana* (F.) | Coreid Bug | Coreidae | Indo-China | Sap-sucker; toxic saliva |
| *Leptoglossus australis* (F.) | Leaf-footed Plant Bug | Coreidae | Africa, Asia, Australasia | Sap-sucker; toxic saliva |
| *Euborellia stali* Dohrn. | Earwig | Forficulidae | S. India | Damage pods |
| *Frankliniella fusca* Hinds. | Tobacco Thrips | Thripidae | USA | Infest flowers |
| *Caliothrips indicus* (Bagn.) | – | Thripidae | Africa, India | Infest foliage |
| *Parasa vivida* (Wlk.) | Stinging Caterpillar | Limacodidae | E. & W. Africa | Larvae defoliate |
| *Latoia lepida* (Cram.) | Blue-striped Nettlegrub | Limacodidae | India, S.E. Asia | Larvae defoliate |
| *Stegasta basqueella* (Chambers) | Red-necked Peanutworm | Gelechiidae | USA, Brazil | Larvae bore buds |
| *Stegasta variana* (Meyr.) | Peanutworm | Gelechiidae | Malaysia | Larvae bore buds |
| *Stomopteryx subsecivella* Zell. | Groundnut Leaf Miner | Gracillariidae | India, S.E. Asia | Larvae mine leaves |
| *Homona coffearia* (Nietn.) | Tea Tortrix | Tortricidae | Papua NG | Larvae roll leaves |
| *Archips micaceana* (Wlk.) | Tortrix | Tortricidae | Indo-China | Larvae roll leaves |
| *Maruca testulalis* (Geyer) | Mung Moth | Pyralidae | Cosmopolitan | Larvae bore pods |
| *Elasmopalpus lignosellus* (Zell.) | Lesser Cornstalk Borer | Pyralidae | S. USA, S. America | Larvae bore stems |
| *Lampides boeticus* (L.) | Pea Blue | Lycaenidae | Indo-China | Larvae bore pods |
| *Diacrisia obliqua* Wlk. | Tiger Moth | Arctiidae | India | Larvae defoliate |
| *Amsacta moorei* (Wlk.) | Tiger Moth | Arctiidae | India, Australasia | Larvae defoliate |

| | | | | |
|---|---|---|---|---|
| *Agrotis ipsilon* (Hfn.) | Black Cutworm | Noctuidae | Cosmopolitan | Larvae are cutworms |
| *Spodoptera frugiperda* (J.E. Smith) | Black Armyworm | Noctuidae | USA, C. & S. America | Larvae are cutworms |
| *Spodoptera litura* (F.) | Rice Armyworm | Noctuidae | India, S.E. Asia | Larvae defoliate |
| *Achaea finita* Gn. | Semi-looper | Noctuidae | Africa | Larvae defoliate |
| *Dorylus orientalis* Westw. | Oriental Army Ant | Formicidae | Indo-China | Defoliation |
| *Gonocephalum* spp. | Dusty Brown Beetles | Tenebrionidae | Africa | Adults damage plant |
| *Diabrotica* spp. | Leaf Beetles | Chrysomelidae | USA, S. America | Defoliate |
| *Schizonycha* spp. | Chafer Grubs | Scarabaeidae | Africa | Larvae eat roots |
| *Anomala* spp. | White Grubs | Scarabaeidae | S.E. Asia | Larvae eat roots |
| *Leucopholis* spp. | White Grubs | Scarabaeidae | Philippines | Larvae eat roots |
| *Eulepida mashona* Arr. | White Grub | Scarabaeidae | Africa | Larvae eat roots |
| *Strigoderma arboricola* F. | White Grub | Scarabaeidae | S. USA | Larvae eat roots |
| *Rhopaea magmicornis* Blkb. | Pasture White Grub | Scarabaeidae | Australia | Larvae eat roots |
| *Mylabris* spp. | Banded Blister Beetles | Meloidae | Pantropical | Adults eat flowers |
| *Caryedon serratus* (Oliv.) | Groundnut Bruchid | Bruchidae | W. Africa | Infest pods in field & stores |
| *Zygrita diva* Thoms. | Lucerne Crown Borer | Cerambycidae | Australia | Larvae bore stems |
| *Tetranychus* spp. | Red Spider Mites | Tetranychidae | Cosmopolitan | Scarify foliage |

# GUAVA (*Psidium guajava* – Myrtaceae)

Guava is indigenous to tropical America, but is now pantropical in distribution, mostly grown for local consumption, but areas of large production and export are India, Florida, Brazil and Guyana. It is grown throughout the tropics from sea-level to 2000 m in a wide range of soils and climate. A hardy, shallow-rooted shrub, or small tree, 3–10 m in height. The fruit is a large berry with seeds embedded in the edible pulp which is white or red in colour. The fruit is eaten raw or cooked, and is rich in vitamins C and A; can be used for jam, jelly, paste or juice, and also dried as a vitamin source. Fruits are produced when the tree is eight years old. (See also Butani, 1974c.)

## MAJOR PESTS

| | | | | |
|---|---|---|---|---|
| *Coccus alpinus* De Lotto | Soft Green Scale | Coccidae | E. Africa | Infest foliage |
| *Chloropulvinaria psidii* Mask. | Guava Scale | Coccidae | Pantropical | Infest foliage |
| *Saissetia coffeae* (Wlk.) | Helmet Scale | Coccidae | Philippines | Infest foliage |
| *Ferrisia virgata* (Ckll.) | Striped Mealybug | Pseudococcidae | Pantropical | Infest foliage |
| *Icerya purchasi* Mask. | Cottony Cushion Scale | Margarodidae | Cosmopolitan | Infest foliage |
| *Selenothrips rubrocinctus* (Giard) | Red-banded Thrips | Thripidae | Pantropical | Infest leaves |
| *Dacus dorsalis* (Hend.) | Oriental Fruit Fly | Tephritidae | Malaysia, Laos, Philippines, India | Larvae in fruit |

## MINOR PESTS

| | | | | |
|---|---|---|---|---|
| *Aleurocanthus woglumi* Ashby | Citrus Blackfly | Aleyrodidae | China | Infest foliage |
| *Aphis gossypii* (Glover) | Cotton Aphid | Aphididae | Malaysia, Laos, Philippines, India | Infest foliage |
| *Planococcus lilacinus* (Ckll.) | — | Pseudococcidae | S.E. Asia, India, Madagascar | Infest foliage |
| *Nipaecoccus nipae* (Mask.) | Nipa Mealybug | Pseudococcidae | India, Africa, C. America | Infest foliage |
| *Parthenolecanium persicae* (F.) | Peach Scale | Coccidae | Cosmopolitan | Infest foliage |
| *Coccus viridis* (Green) | Soft Green Scale | Coccidae | Pantropical | Infest foliage |
| *Coccus hesperidum* L. | Soft Brown Scale | Coccidae | Cosmopolitan | Infest foliage |
| *Icerya aegyptica* (Dgl.) | Egyptian Fluted Scale | Margarodidae | India, Africa, S.E. Asia, Australia | Infest foliage |
| *Drosicha mangiferae* Green | Mango (Giant) Mealybug | Margarodidae | India | Infest foliage |
| *Aspidiotus destructor* Sign. | Coconut Scale | Diaspididae | Pantropical | Infest foliage |
| *Chrysomphalum aonidum* (L.) | Florida Red Scale | Diaspididae | Philippines | Infest foliage |

| | | | | |
|---|---|---|---|---|
| *Aonidiella orientalis* Newst. | Oriental Yellow Scale | Diaspididae | India (Pantropical) | Infest foliage |
| *Hemiberlesia lataniae* (Sign.) | Latania Scale | Diaspididae | Cosmopolitan in warm areas | Infest foliage |
| *Helopeltis schoutedeni* Reuter | Cotton Helopeltis | Miridae | Africa | Sap-sucker; toxic saliva |
| *Helopeltis theobromae* (Miller) | Cocoa Helopeltis | Miridae | Malaysia | Toxic saliva |
| *Pseudotheraptus wayi* Brown | Coconut Bug | Coreidae | E. Africa | Toxic saliva |
| *Attacus atlas* (L.) | Atlas Moth | Saturniidae | Philippines | Larvae defoliate |
| *Microclona leucosticta* (Meyr.) | Guava Stem Borer | Gelechiidae | India | Larvae bore shoots |
| *Virachola isocrates* F. | Blue Butterfly | Lycaenidae | India | Larvae feed inside fruits |
| *Indarbela tetraonis* Moore | Wood-borer Moth | Metarbelidae | India, China | Larvae eat bark |
| *Othreis fullonia* (Clerke) | Fruit-piercing Moth | Noctuidae | Philippines, India | Larvae defoliate; adults pierce fruit |
| *Achaea* spp. | Fruit-piercing Moths | Noctuidae | Africa, India | |
| *Thosea sinensis* (Wlk.) | Slug Caterpillar | Limacodidae | Philippines | Larvae defoliate |
| *Cryptophlebia leucotreta* (Meyr.) | False Codling Moth | Tortricidae | Africa | Larvae bore fruits |
| *Metanestria hyrtaca* (Cram.) | – | Lasiocampidae | Philippines | Larvae defoliate |
| *Trabala irrorata* (Moore) | – | Lasiocampidae | Malaysia | Larvae defoliate |
| *Dichocrocis punctiferalis* (Guen.) | Castor Capsule Borer | Pyralidae | India | Larvae bore shoots |
| *Dacus zonatus* Saund. | Peach Fruit Fly | Tephritidae | India, China | Larvae bore fruit |
| *Dacus diversus* (Coq.) | Guava Fruit Fly | Tephritidae | India | Larvae inside fruit |
| *Dacus cucurbitae* Coq. | Melon Fly | Tephritidae | India, China | Larvae inside fruit |
| *Anastrepha fraterculus* (Wied.) | Fruit Fly | Tephritidae | C. & S. America | Larvae inside fruit |
| *Anastrepha mombinpraeoptans* Sein | West Indian Fruit Fly | Tephritidae | C. & S. America | Larvae inside fruit |
| *Ceratitis capitata* (Wied.) | Medfly | Tephritidae | Africa, Australia, C. & S. America | Larvae inside fruit |
| *Adoretus* spp. | Flower Beetles | Scarabaeidae | India | Adults eat leaves |
| *Aeolesthes holosericea* F. | Cherry Stem Borer | Curculionidae | India, S.E. Asia | Larvae bore stem |
| *Myllocerus* spp. | Grey Weevils | Curculionidae | India | Adults eat leaves |
| *Hypomeces squamosus* (F.) | Gold-dust Weevil | Curculionidae | Laos | Adults eat leaves |
| *Stephanoderes psidii* Hoph. | Guava Bark Borer | Scolytidae | Philippines | Adults bore bark |
| *Brevipalpus phoenicis* (Geijskes) | Red Crevice Tea Mite | Tenuipalpidae | India | Damage fruits |
| *Tenuipalpus puniicae* | – | Tenuipalpidae | India | Damage fruits |

# HEMP (*Cannabis sativa* — Cannabinaceae)
## (= Indian Hemp)

The term 'hemp' is confusing as it is applied loosely to include a number of quite unrelated plants and commercial fibres. But this is the true hemp; native to C. and W. Asia, but now widely cultivated in both tropical and temperature regions. The plant is a stout, bushy annual 2–4 m tall, dioecious, with hollow stem and palmate leaves. The fibre is a white bast which develops in the pericycle of the stem; it is long, strong, and durable, but being lignified is not very flexible. The fibre is used mainly for ropes, sacks, sailcloth and twine. The seeds contain an oil that can be used as a substitute for linseed oil. The dried flowering tops of the female plants are pressed into a solid mass and used as narcotic stimulant. The active principle is a resin containing three or four very powerful alkaloids; the extracted resin is usually called hashish, whereas the dried leaves are known as marijuana or bhang. The cultivation of hemp is usually subject to close supervision, if permitted at all, because of it being a source of a popular narcotic.

## MAJOR PESTS

| | | | | |
|---|---|---|---|---|
| *Empoasca* spp. | Green Leaf-hoppers | Cicadellidae | Thailand | Infest leaves |
| *Grapholitha delineata* Wlk. | Hemp Leaf-roller | Tortricidae | Yugoslavia | Larvae bore leaves, stems & fruits |

## MINOR PESTS

| | | | | |
|---|---|---|---|---|
| *Ostrinia nubilalis* (Hb.) | European Corn Borer | Pyralidae | Europe, N. America | Larvae bore stems |
| *Podagrica* sp. | Cotton Flea Beetle | Chrysomelidae | Thailand | Adults hole leaves |
| *Hypomeces squamosus* F. | Gold-dust Weevil | Curculionidae | Thailand | Adults eat leaves |

# HYACINTH BEAN (*Lablab niger* – Leguminosae)
## (= *Dolichos lablab*) (= Indian Bean; Bovanist Bean)

This is an herbaceous perennial herb, often grown as an annual, 1–6 m in height, either twining or as a bush; the young pods and tender beans are popular vegetables in India and elsewhere in the tropics. The dried seeds are used as a split pulse, for human food, and also for livestock feed. The haulms are used as livestock fodder. Sometimes the crop is grown as green manure or as a cover crop. It is grown as a dryland crop, being hardy and drought-resistant, and can be grown in areas with as little as 60–90 cm of rainfall. The species is widely grown throughout Asia.

## MAJOR PESTS

| | | | | |
|---|---|---|---|---|
| *Coptosoma cribraria* F. | Stink Bug | Pentatomidae | India | Sap-sucker; toxic saliva |
| *Adisura atkinsoni* Moore | Pod Borer | Noctuidae | India | Larvae bore pods |

## MINOR PESTS

| | | | | |
|---|---|---|---|---|
| *Bemisia tabaci* (Genn.) | Cotton Whitefly | Aleyrodidae | India | Infest foliage |
| *Trialeurodes rara* Singh | Castor Whitefly | Aleyrodidae | India | Infest foliage |
| *Aphis craccivora* Koch | Groundnut Aphid | Aphididae | India | Infest foliage |
| *Ceroplastes cajani* Mask. | Soft Scale | Coccidae | India | Infest foliage |
| *Ferrisia virgata* (Ckll.) | Striped Mealybug | Pseudococcidae | India | Infest foliage |
| *Haplothrips vernoniae* Pr. | Flower Thrips | Phlaeothripidae | India | Infest flowers |
| *Frankiniella sulphurea* Schm. | Flower Thrips | Thripidae | India | Infest flowers |
| *Ayyaria chaetophora* Ky. | Leaf Thrips | Thripidae | India | Infest leaves |
| *Cosmopteryx phaeogastra* Meyr. | Leaf Miner | Cosmopterygidae | India | Larvae mine leaves |
| *Thosea aperiens* Wlk. | Stinging Caterpillar | Limacodidae | India | Larvae defoliate |
| *Laspeyresia torodelta* Meyr. | Pod Borer | Tortricidae | India | Larvae bore pods |
| *Etiella zinckenella* Treit. | Pea Pod Borer | Pyralidae | India | Larvae bore pods |
| *Lamprosema indicata* F. | Webworm | Pyralidae | India | Larvae web leaves |
| *Diacrisia obliqua* Wlk. | Tiger Moth | Arctiidae | India | Larvae defoliate |
| *Acherontia* spp. | Death's Head Hawk Moths | Sphingidae | India | Larvae defoliate |
| *Anticarsa irrorata* F. | Green Leaf Caterpillar | Noctuidae | India | Larvae defoliate |
| *Heliothis armigera* Hb. | American Bollworm | Noctuidae | India | Larvae bore pods |
| *Ophiomyia phaseoli* (Coq.) | Bean Fly | Agromyzidae | India | Larvae bore stem |
| *Sagra nigrita* Oliv. | Stem Borer | Chrysomelidae | India | Larvae bore stems |
| *Alcidodes* spp. | Stem Weevils | Curculionidae | India | Adults girdle stems |

# JACKFRUIT (*Artocarpus heterophyllus* – Moraceae)

An Indo-Malaysian species, now widely dispersed in the tropics. It is a handsome evergreen tree 10–20 m tall, with entire leaves, and huge fruits up to 0.5 m long and weighing 10–20 kg, which are borne on the trunk and main branches (cauliflory). Although widely grown throughout the tropics, it is not used much except in Asia, especially southern India. (See also Butani, 1978*a*.)

## MAJOR PESTS

| | | | | |
|---|---|---|---|---|
| *Diaphania caesalis* (Wlk.) | Jackfruit Shoot Borer | Pyralidae | India, Malaysia, Philippines | Larvae bore shoots |
| *Dacus umbrosus* (F.) | Fruit Fly | Tephritidae | S.E. Asia | Larvae in fruit |
| *Ochyromera artocarpi* Mshl. | Jackfruit Bud Weevil | Curculionidae | India | Larvae bore buds |

## MINOR PESTS

| | | | | |
|---|---|---|---|---|
| *Greenidea artocarpi* (Wests.) | Jackfruit Aphid | Aphididae | India | Infest twigs |
| *Cosmocarta relata* Dist. | Spittlebug | Cercopidae | India | Feed on twigs |
| *Pealius schimae* Tak. | Blackfly | Aleyrodidae | India | Infest leaves |
| *Icerya seychellarum* (Westw.) | Seychelles Fluted Scale | Margarodidae | Philippines, India | Infest twigs |
| *Ferrisia virgata* (Ckll.) | Striped Mealybug | Pseudococcidae | India | Infest twigs |
| *Pseudococcus* spp. | Mealybugs | Pseudococcidae | Philippines | Infest foliage |
| *Ceroplastes rubens* Mask. | Pink Waxy Scale | Coccidae | India | Infest foliage |
| *Chloropulvinaria psidii* (Mask.) | Guava Scale | Coccidae | India | Infest foliage |
| *Aonidiella aurantii* (Mask.) | Red Scale | Diaspididae | Philippines | Infest leaves |
| *Hemiberlesia lantaniae* (Sign.) | Armoured Scale | Diaspididae | India | Infest foliage |
| *Oecophylla smaragdina* (F.) | Red Tree Ant | Formicidae | India | Attack workers |
| *Dacus dorsalis* (Hend.) | Oriental Fruit Fly | Tephritidae | India | Larvae in fruits |
| *Indarbela tetraonis* (Moore) | Wood-borer Moth | Metarbelidae | India | Larvae bore bark & wood |
| *Diaphania bivitralis* (Gn.) | Jackfruit Leaf-webber | Pyralidae | India | Larvae web leaves |
| *Thosea sinensis* Wlk. | Slug Caterpillar | Limacodidae | Laos | Larvae defoliate |
| *Perina nuda* F. | Banyan Tussock Moth | Lymantriidae | India | Larvae defoliate |
| *Apriona germari* | Jackfruit Longhorn | Cerambycidae | | ⎫ |
| *Apriona* spp. | Jackfruit Longhorns | Cerambycidae | S.E. Asia | ⎬ Larvae bore in trunk & branches |
| *Batocera* spp. | Longhorns | Cerambycidae | S.E. Asia | ⎭ |
| *Sthenias grisator* F. | Longhorn | Cerambycidae | India | |
| *Hypomeces squamosus* (F.) | Gold-dust Weevil | Curculionidae | Malaysia | Adults eat leaves |

# JUJUBE (*Zizyphus mauritiana* — Rhamnaceae)
## (= *Z. jujuba* (L.) Lam. *non* Mill.) (= Indian Jujube; Ber)

A small thorny evergreen tree, widespread throughout Africa and Asia in dried regions, and widely cultivated in India for the small edible fruit. The fruit is brown in colour, oval and about 2–3 cm in length, with edible acid pulp and a hard central stone. It is eaten fresh or dried and used as dessert, but is also candied and makes a very refreshing drink. The fruit is also a rich source of vitamin C.

*Zizyphus jujuba* Mill. is the Chinese Jujube, a tree of temperate climates and has been cultivated in China for at least 4000 years. (See also Butani, 1973.)

## MAJOR PESTS

| | | | | |
|---|---|---|---|---|
| *Drosichiella tamarandus* Green | Ber Mealybug | Margarodidae | India | Infest foliage |
| *Carpomyia vasuviana* Costa | Ber Fruit Fly | Tephritidae | India | Larvae inside fruits |
| *Adoretus* spp. | Flower Beetles | Scarabaeidae | India | Adults eat leaves; larvae eat roots |
| *Xanthochelus superciliosis* Gyll. | Ber Weevil | Curculionidae | India | Adults defoliate; larvae in soil eat roots |
| *Myllocerus* spp. | Leaf Weevils | Curculionidae | India | Adults eat leaves |

## MINOR PESTS

| | | | | |
|---|---|---|---|---|
| *Drosicha mangiferae* Green | Mango Giant Mealybug | Margarodidae | India | Infest foliage |
| *Nipaecoccus* spp. | Mealybugs | Pseudococcidae | India | Infest foliage |
| *Aonidiella* spp. | Armoured Scales | Diaspididae | India | Encrust twigs |
| *Aonidia ziziphi* Rah. | Ber Scale | Diaspididae | India | Encrust twigs |
| *Pulvinaria* spp. | Soft Scales | Coccidae | India | Infest foliage |
| *Urentius ziziphifolius* M. & H. | Ber Lacebug | Tingidae | India | Sap-sucker; toxic saliva |
| *Indarbela* spp. | Wood-boring Moths | Metarbelidae | India | Larvae eat bark & bore branches |
| *Virachola isocrates* F. | Anar Butterfly | Lycaenidae | India | Larvae bore fruits |
| *Porthmologa parclina* Meyr. | Leaf-eating Caterpillar | Pyralidae | India | Larvae eat leaves |
| *Thiacides postica* Wlk. | Hairy Caterpillar | Noctuidae | India | Larvae defoliate |
| *Dasychira* spp. | Tussock Moths | Lymantriidae | India | Larvae defoliate |
| *Euproctis* spp. | Tussock Moths | Lymantriidae | India | Larvae defoliate |
| *Dacus* spp. | Fruit Flies | Tephritidae | India | Larvae inside fruits |
| *Platypria andrewesi* Weise | Hispid Beetle | Chrysomelidae | India | Larvae mine leaves; adults eat strips |
| *Holotrichia insularis* Br. | Cockchafer | Scarabaeidae | India | Adults eat leaves |
| *Tanymecus* spp. | Leaf-eating Weevils | Curculionidae | India | Adults eat leaves |

# JUTE (*Corchorus* spp. — Tiliaceae)

Probably the most widely used fibre in the world (other than cotton); it is a bast fibre from the secondary phloem of two species of *Corchorus*. The plant is a tall, slender, half-shrubby annual, some 2–3 m in height, of Asiatic origin, and is now almost entirely an Indian crop, although some is grown in Brazil. (See also Dean, 1979.)

## MAJOR PESTS

| | | | | |
|---|---|---|---|---|
| *Dasychira mendosa* Hon. | Jute Hairy Caterpillar | Lymantriidae | India | Larvae defoliate |
| *Spilosoma obliqua* (Wlk.) | Jute Hairy Caterpillar | Lymantriidae | India | Larvae defoliate |

## MINOR PESTS

| | | | | |
|---|---|---|---|---|
| *Brachytrupes portentosus* (Licht.) | Large Brown Cricket | Gryllidae | Bangladesh | Cut seedlings |
| *Ferrisia virgata* (Ckll.) | Striped Mealybug | Pseudococcidae | India | Encrust foliage |
| *Pectinophora gossypiella* (Saund.) | Pink Bollworm | Gelechiidae | India | Larvae bore flowers |
| *Anomis sabulifera* (Gn.) | Jute Semi-looper | Noctuidae | India, Africa, S.E. Asia, Australasia | Larvae defoliate |
| *Achaea* spp. | Fruit-piercing Moths | Noctuidae | India | Larvae defoliate |
| *Spodoptera exigua* (Hb.) | Beet Armyworm | Noctuidae | India | Larvae defoliate |
| *Agrilus acutus* Thnb. | Jute Stem Borer | Buprestidae | India | Larvae bore stem |
| *Apion corchori* Marsh | Jute Stem Weevil | Apionidae | Bangladesh | Larvae bore stem |
| *Polyphagotarsonemus latus* (Banks) | Yellow Tea Mite | Tarsonemidae | India | Infest foliage |

# KAPOK (*Ciba pentandra* – Bombacaceae)

A lowland tropical tree, up to 30 m tall, probably originating in tropical American forests, now well established as different varieties in W. Africa and Asia. Kapok is the floss from the inner capsule wall of the fruit in which the seeds lie loosely when ripe. Each hair is a single cell, 1–3 cm long, waxy, elastic, light and water-repellant, and with a buoyancy five times that of cork. Because of these properties it is used for stuffing and insulating purposes. It is light (eight times lighter than cotton), has low thermal conductivity, and is a very effective sound absorber. Exporting countries include Thailand, Indonesia, E. Africa, India and Pakistan.

## MAJOR PESTS

| | | | | |
|---|---|---|---|---|
| *Crypticerya jacobsoni* (Green) | – | Margarodidae | Laos | Infest shoots & pods |
| *Mudaria variabilis* Roepke | Pod Moth | Pyralidae | Laos | Larvae bore pods |
| *Leucopholis irrorata* (Chevr.) | White Grub | Scarabaeidae | Philippines | Larvae eat roots |
| *Plocaederus obesus* Gahan | Longhorn | Cerambycidae | Thailand, Laos | Larvae bore trunk |

## MINOR PESTS

| | | | | |
|---|---|---|---|---|
| *Odontotermes* sp. | Scavenging Termite | Termitidae | Thailand | Workers damage trunk and roots |
| *Globitermes suphureus* Haviland | Termite | Termitidae | Thailand | |
| *Aphis gossypii* (Glov.) | Cotton Aphid | Aphididae | Philippines | Infest leaves |
| *Ferrisia virgata* (Ckll.) | Striped Mealybug | Pseudococcidae | Philippines | Encrust foliage |
| *Planococcoides njalensis* (Laing) | Cocoa Mealybug | Pseudococcidae | W. & C. Africa | Infest foliage |
| *Saissetia nigra* (Nietm.) | Nigra Scale | Coccidae | Philippines | Infest leaves |
| *Distaniella theobroma* (Dist.) | Cocoa Capsid | Miridae | W. Africa | Sap-sucker; toxic saliva |
| *Tectocoris diopthalmus* Thunb. | Red Shield Bug | Pentatomidae | Philippines | |
| *Odontophus nigricornis* Stal. | Leaf-sucking Bug | Pyrrhocoridae | Thailand | Sap-sucker; toxic saliva |
| *Dysdercus* spp. | Cotton Stainer | Pyrrhocoridae | S.E. Asia | Sap-sucker; toxic saliva |
| *Lygaeus hospes* F. | – | Lygaeidae | Solomon Isl. | |
| *Zeuzera coffeae* (Nietm.) | Red Coffee Borer | Cossidae | Philippines | Larvae bore branches |
| *Cryptothelea* spp. | Bagworms | Psychidae | Philippines | Larvae defoliate |
| *Dasychira mendosa* (Hub.) | Tussock Moth | Lymantriidae | Philippines | Larvae defoliate |
| *Anomala* spp. | White Grubs | Scarabaeidae | Philippines | Larvae eat roots |
| *Heterobostrychus acqualis* (Waterh.) | Kapok Borer | Bostrychidae | Philippines | Adults bore branches |
| *Batocera* sp. | Longhorn Beetle | Cerambycidae | Philippines, Mauritius | Larvae bore trunk |
| *Alcidodes obesus* Faust | Stem-boring Weevil | Curculionidae | Thailand, Laos, Malaysia | Larvae bore twigs |
| *Desmidophorus* spp. | Weevils | Curculionidae | Thailand, Laos | Adults eat leaves |
| *Hypomeces squamosus* (F.) | Gold-dust Weevil | Curculionidae | Laos, Malaysia | Adults eat leaves |

# KOLA (*Cola* spp. — Sterculiaceae)
## (= Kola-nut)

About 60 species grow naturally in tropical Africa, mostly W. Africa; a small to medium evergreen tree, leaves simple and entire. The seeds, in slightly fleshy pods borne in leaf axils, are chewed as a stimulating narcotic, and in addition to being used as a masticatory, the powdered seeds may be boiled in water and used as a beverage. Mostly planted in W. Africa where it is an important crop; four species are generally cultivated.

## MAJOR PESTS

| | | | | |
|---|---|---|---|---|
| *Toxoptera aurantii* B. de F. | Black Citrus Aphid | Aphididae | Pantropical | Infest foliage |
| *Parasaissetia nigra* (Nietn.) | Nigra Scale | Coccidae | Old World | Infest foliage |
| *Selenothrips rubrocinctus* Giard. | Red-banded Thrips | Thripidae | Pantropical | Infest foliage |
| *Helopeltis bergrothis* Reut. | Cocoa Mosquito Bug | Miridae | Tropical Africa | Sap-sucker; toxic saliva |
| *Sahlbergella singularis* Hagl. | Cocoa Capsid | Miridae | Africa | Sap-sucker; toxic saliva |
| *Balanogastris kolae* (Desbr.) | Kola Weevil | Curculionidae | W. Africa | Larvae bore nuts |
| *Paremydica inseperata* Faust | Weevil | Curculionidae | W. Africa | Larvae bore nuts |

## MINOR PESTS

| | | | | |
|---|---|---|---|---|
| *Brachytrupes membranaceus* Drury | Tobacco Cricket | Gryllidae | Africa | Seedling attacked |
| *Zonocerus variegatus* L. | Stink Grasshopper | Acrididae | Africa | Defoliate |
| *Planococcoides njalensis* (Laing) | Cocoa Mealybug | Pseudococcidae | Africa | Infest foliage |
| *Sylepta retractalis* Hmps. | Leaf-roller | Pyralidae | W. Africa | Larvae roll leaves |
| *Anaphe venata* Btlr. | Tent Caterpillar | Notodontidae | W. Africa, Zaïre | Gregarious larvae defoliate |
| *Characoma stictigrapha* Hmps. | Pod Husk Borer | Noctuidae | W. Africa | Larvae eat leaves & pods |
| *Ceratitis colae* Silv. | Kola Fruit Fly | Tephritidae | W. Africa | Larvae bore pods |
| *Apate monachus* F. | Black Borer | Bostrychidae | Africa | Adults bore branches |
| *Phosphorus gabonator* Thoms. | Longhorn Beetle | Cerambycidae | W. Africa, Zaïre | Larvae bore trunk |
| *Zyrcosa brunnea* Hust. | Leaf Weevil | Curculionidae | W. Africa | Adults eat leaves |

# LENTIL (*Lens esculenta* – Leguminosae)

One of the most ancient of food plants and also one of the most nutritious. Its origin was S.W. Asia. The plant is a slender, much-branched annual with short broad pods. The seeds are used mostly in soups, and are more easily digested than meat. The plants are sometimes used as fodder for animals. The main production areas are India, Pakistan, Ethiopia, Syria, Turkey and Spain.

## MAJOR PESTS

| | | | | |
|---|---|---|---|---|
| *Etiella zinckenella* (Treit.) | Pea Pod Borer | Pyralidae | India | Larvae bore pods |
| *Agrotis* spp. | Cutworms | Noctuidae | India | Larvae are cutworms |
| *Heliothis armigera* (Hb.) | American Bollworm | Noctuidae | India | Larvae eat young shoots & bore pods |

## MINOR PESTS

| | | | | |
|---|---|---|---|---|
| *Aphis craccivora* Koch. | Groundnut Aphid | Aphididae | India | Infest foliage |
| *Pseudococcus maritimus* (Ehrh.) | Grape Mealybug | Pseudococcidae | Widespread | Infest foliage |
| *Diachrysia orichalcea* (F.) | – | Noctuidae | India | Larvae eat leaves |
| *Spodoptera exigua* (Wlk.) | Lesser Armyworm | Noctuidae | India | Larvae eat leaves |
| *Callosobruchus chinensis* (L.) | Oriental Cowpea Bruchid | Bruchidae | India | Adults attack open pods in field |

589

# LETTUCE (*Lactuca sativa* – Compositae)

This is a native of S. Europe and W. Asia, and is descended from the wild lettuce (*L. scariola*), a common wasteland and roadside weed in both Old and New Worlds. It is another herbage vegetable of great antiquity; at the present time there are many varieties showing different horticultural characters. It has a milky sap and grows as a basal rosette of leaves, producing later in the season a stalk bearing the flowers. It has little food value in itself but does contain vitamins and iron salts. It grows best in a light sandy or loamy soil with a rather cool climate, and not too much sunshine; among the principal types grown are Cos, head, romains, and cut-leaf forms.

## MAJOR PESTS

| | | | | |
|---|---|---|---|---|
| *Aphis gossypii* Glov. | Melon/Cotton Aphid | Aphididae | Cosmopolitan | Infest foliage |
| *Macrosiphum euphorbiae* (Thos.) | Potato Aphid | Aphididae | Europe | Infest foliage |
| *Myzus ascalonicus* Don. | Shallot Aphid | Aphididae | Europe | Infest foliage |
| *Myzus persicae* (Sulz.) | Peach–Potato Aphid | Aphididae | Cosmopolitan | Infest foliage |
| *Nasonovia ribis-nigri* (Mosley) | Lettuce Aphid | Aphididae | Europe | Infest foliage |
| *Pemphigus bursarius* (L.) | Lettuce Root Aphid | Pemphigidae | Europe | Infest roots |

## MINOR PESTS

| | | | | |
|---|---|---|---|---|
| *Cnephasia* spp. | Tortrix Moths | Tortricidae | Europe | Larvae defoliate |
| *Agrotis segetum* (D. & S.) | Common Cutworm | Noctuidae | Cosmopolitan in Old World | Larvae cutworms |
| *Agrotis dahli* (Hub.) | Cutworm | Noctuidae | Malaysia | Larvae are cutworms |
| *Noctua pronuba* (L.) | Large Yellow Underwing | Noctuidae | Europe | Larvae are cutworms |
| *Plusia orichalcea* | – | Noctuidae | S.E. Asia | Larvae defoliate |

# LITCHI (*Litchi chinensis* – Sapindaceae)
## (=Lychee)

A native of southern China, the Litchi is a dense polygamous tree, evergreen and about 10 m high. It has been widely introduced throughout the tropics, but only flourishes at the higher altitudes. The tree is also used as an ornamental. The leaves are compound, elongate, leathery, and shiny with an indistinct venation. The fruit is distinctive; round, 2–5 cm in diameter, and is borne in loose clusters. The pericarp is bright red and leathery, becoming brown and brittle on drying; the translucent white flesh surrounds a single large seed. The fresh fruit is regarded as a great delicacy, especially in China, and some are now canned for the export trade. The main areas of production are China, Taiwan and India. (See also Butani, 1977.)

## MAJOR PESTS

| | | | | |
|---|---|---|---|---|
| *Tessaratoma papillosa* (Dru.) | Litchi Stink Bug | Pentatomidae | S. China | Sap-sucker; toxic saliva |

## MINOR PESTS

| | | | | |
|---|---|---|---|---|
| *Odontotermes formosanus* (Shir.) | Scavenging Termite | Termitidae | China, Taiwan | Workers damage tree |
| *Aleurocanthus husaini* Corbett | Blackfly | Aleyrodidae | India | Infest leaves |
| *Aphis gossypii* Glov. | Cotton Aphid | Aphididae | S. China | Infest shoots |
| *Toxoptera aurantii* (B. de F.) | Black Citrus Aphid | Aphididae | India | Infest shoots |
| *Pseudococcus comstocki* (Kuw.) | Comstock's Mealybug | Pseudococcidae | China, Japan, USA | Infest foliage |
| *Saissetia coffeae* (Wlk.) | Helmet Scale | Coccidae | India | Infest foliage |
| *Aulacaspis* spp. | Armoured Scales | Diaspididae | India | Infest foliage |
| *Pyrops candelaria* (L.) | Lantern Bug | Fulgoridae | S. China | Infest branches |
| *Halys dentatus* F. | Mulberry Bug | Pentatomidae | India | Sap-sucker; toxic saliva |
| *Tessaratoma javanica* Thnb. | Stink Bug | Pentatomidae | India | Sap-sucker; toxic saliva |
| *Chrysocoris stolii* (Wolff.) | Lychee Shield Bug | Pentatomidae | India | Cause fruit-drop |
| *Taeniothrips distalis* Karny | Thrips | Thripidae | India | Infest leaves & flowers |
| *Indarbela* spp. | Wood-borer Moths | Metarbelidae | China, India | Larvae eat bark & bore wood |
| *Acrocercops* spp. | Leaf-miners | Gracillariidae | India | Larvae mine leaves & damage fruits |
| *Adoxophyes orana* (F.R.) | Summer Fruit Tortrix | Tortricidae | S. China | Larvae eat leaves & fruit surface |
| *Archips micaceana* (Wlk.) | – | Tortricidae | S. China | Larvae roll leaves |

| | | | | |
|---|---|---|---|---|
| *Argyroploce illepida* Butler | Seed (Fruit) Borer | Tortricidae | India | Larvae bore seeds |
| *Cryptophlebia ombrodelta* (Lower) | Macadamia Nut Borer | Tortricidae | China, Japan, S.E. Asia, India | Larvae bore fruits |
| *Homona coffearia* (Niet.) | Tea Tortrix | Tortricidae | S. China | Larvae roll leaves |
| *Olethreutes leucaspis* (Meyr.) | Litchi Leaf-roller | Tortricidae | China, India | Larvae roll leaves |
| *Chlumetia transversa* Wlk. | Mango Shoot Borer | Noctuidae | India | Larvae bore shoots |
| *Lymantria mathura* Moore | Tussock Moth | Lymantriidae | India | Larvae damage flowers & bark |
| *Oecophylla smaragdina* (F.) | Red Tree Ant | Formicidae | China, India | Nest in foliage; attack workers |
| *Myllocerus* spp. | Leaf Weevils | Curculionidae | India | Adults eat leaves |
| *Amblyrrhinus poricollis* Boh. | Plum Leaf Weevil | Curculionidae | India | Adults eat leaves |
| *Aceria litchi* (Keif.) | Litchi Gall Mite | Eriophyidae | India, China, Taiwan | Young leaves galled & rolled |

# LONGAN (*Euphoria longana* — Sapindaceae)

A native of S. China, little cultivated away from this region. A dense evergreen tree up to 10 m tall, similar in appearance to Litchi but the leaf veins are prominent, the leaf less elongate, and the leaf surface rather rough. The fruit is smaller and less succulent than that of Litchi and it is grown for local consumption in China and Taiwan.

## *MAJOR PESTS*

| | | | | |
|---|---|---|---|---|
| *Tessaratoma papillosa* (Dru.) | Litchi Stink Bug | Pentatomidae | S. China, Laos, Malaysia | Sap-sucker; toxic saliva |

## *MINOR PESTS*

| | | | | |
|---|---|---|---|---|
| *Pyrops candelaria* (L.) | Lantern Bug | Fulgoridae | S. China, Laos | Sap-sucker |
| *Leptocentrus terminalis* Wlk. | Treehopper | Membracidae | S. China | Infests twigs |
| *Chelaria* sp. | — | Gelechiidae | S. China | Larvae bore leaves |
| *Clania* spp. | Bagworms | Psychidae | Laos | Larvae defoliate |
| *Olethreutes discana* Feld. | Tortrix | Tortricidae | Laos | Larvae roll young leaves |
| *Archips micaceana* (Wlk.) | Tortrix | Tortricidae | Laos | Larvae roll leaves |
| *Setora nitens* (Wlk.) | Stinging Caterpillar | Limacodidae | Laos | Larvae eat leaves |
| Gen. & sp. indet. | Longan Gall Midge | Cecidomyiidae | S. China | Larvae gall leaves |
| *Aspidomorpha sanctaecrusis* F. | Tortoise Beetle | Chrysomelidae | Laos | Adults & larvae eat leaves |
| *Adoretus tenuimaculatus* Waterh. | Rose Beetle | Scarabaeidae | S. China | Adults eat leaves |

# LOQUAT (*Eriobotrya japonica* — Rosaceae)

One of the few tropical fruits belonging to the Rosaceae; it is a native of China, now grown in most tropical and sub-tropical countries. A small evergreen tree with broad leaves and fragrant white flowers. The fruits are round, small, downy and yellowish orange in colour. The flesh is rather tart and highly esteemed in the Orient where it has been grown since antiquity. The fruit is used fresh, and is made into jellies, sauces and pies. (See also Butani, 1974*a*.)

## MAJOR PESTS

| | | | | |
|---|---|---|---|---|
| *Dacus dorsalis* (Hend.) | Oriental Fruit Fly | Tephritidae | India | Larvae inside fruits |
| *Indarbela quadrinotata* Wlk. | Wood Boring Moth | Metarbelidae | India | Larvae eat bark & bore stems |

## MINOR PESTS

| | | | | |
|---|---|---|---|---|
| *Platypleura kaempferi* (F.) | Speckled Brown Cicada | Cicadidae | S. Japan | Adults pierce trunk; sap exudation & fungus attack |
| *Toxoptera aurantii* (B. de F.) | Black Citrus Aphid | Aphididae | India | Infest foliage |
| *Aphis malvae* (Koch) | Mallow Aphid | Aphididae | India | Infest foliage |
| *Chloropulvinaria psidii* (Mask.) | Guava Mealy Scale | Coccidae | India | Infest foliage |
| *Saissetia coffeae* (Wlk.) | Helmet Scale | Coccidae | India | Infest foliage |
| *Coccus viridis* (Green) | Soft Green Scale | Coccidae | India | Infest foliage |
| *Parlatoria* spp. | Armoured Scales | Diaspididae | India | Infest twigs |
| *Haplothrips* sp. | Leaf-curling Thrips | Phlaeothripidae | India | Distort young leaves |
| *Heliothrips* sp. | Flower Thrips | Thripidae | India | Infest flowers |
| *Zeuzera coffeae* Nietn. | Red Coffee Borer | Cossidae | India | Larvae bore branches |
| *Virachola isocrates* F. | Anar Butterfly | Lycaenidae | India | Larvae bore fruits |
| *Megachile anthracina* Smith | Leaf-cutter Bee | Megachilidae | India | Adults defoliate |
| *Adoretus* spp. | Flower Beetles | Scarabaeidae | India | Adults eat leaves |
| *Myllocerus* spp. | Grey Weevils | Curculionidae | India | Adults eat leaves |

# MACADAMIA (*Macadamia terhifolia* – Proteaceae)
## (= Queensland Nut)

Macadamia nuts, also known as Queensland nuts being native to that part of Australia, are commercially the most expensive nuts known. The tree is small, some 5–10 m in height, and is now introduced into many parts of the tropical world, such as Kenya and southern USA, and is of particular importance in Hawaii. Both thick-shelled and thin-shelled varieties are grown; the kernels are rich in oil and have a sweet flavour, and are generally regarded as the most delicious of nuts. Species currently being grown in Queensland are described as *M. tetraphylla* and *M. integrifolia*. (See also Ironside, 1973.)

## MAJOR PESTS

| | | | | |
|---|---|---|---|---|
| *Eriococcus ironsidei* Williams | Macadamia Felted Coccid | Eriococcidae | E. Australia | Distort young shoots |
| *Amblypelta nitida* Stal | Fruit-spotting Bug | Coreidae | E. Australia | Premature nut-fall & kernel damage |
| *Amblypelta lutescens* (Dist.) | Banana-spotting Bug | Coreidae | N. Australia | |
| *Neodrepta luteotactella* (Wlk.) | Macadamia Twig Girdler | Xyloryctidae | E. Australia | Larvae ring-bark twigs |
| *Acrocercops chionosema* Turner | Macadamia Leaf Miner | Gracillariidae | E. Australia | Larvae mine young leaves |
| *Cryptophlebia ombrodelta* (Lower) | Macadamia Nut Borer | Tortricidae | E. & N. Australia (India, Japan, S.E. Asia) | Larvae bore nuts & kernels |
| *Homoeosoma vagella* Zell. | Macadamia Flower Caterpillar | Pyralidae | E. Australia | Larvae destroy flowers |

## MINOR PESTS

| | | | | |
|---|---|---|---|---|
| *Toxoptera aurantii* (B. de F.) | Black Citrus Aphid | Aphididae | E. Australia | Damage developing racemes |
| *Ulonemia concava* Drake | Macadamia Lace Bug | Tingidae | S.E. Queensland | Adults & nymphs kill buds & flowers |
| *Erysichton lineata* (Murray) | Hairy Line Blue Butterfly | Lycaenidae | E. Australia | Larvae bore buds |
| *Deudorix epijarbas diovis* Hew. | Cornelian Butterfly | Lycaenidae | E. Australia | Larvae bore nuts |
| *Isotenes miserana* (Wlk.) | Orange Fruit Borer | Tortricidae | E. Australia | Larvae roll leaves; bore nuts |
| *Comana fasciata* (Wlk.) | Macadamia Cup Moth | Limacodidae | N. & E. Australia | Larvae defoliate |
| *Anthela varia* (Wlk.) | Variegated Hairy Caterpillar | Anthelidae | S.E. Queensland | Larvae defoliate |
| *Dichocrocis punctiferalis* (Guen.) | Yellow Peach Moth | Pyralidae | E. & N. Australia | Larvae tunnel nuts |

| *Cateremna* sp. | Macadamia Kernel Grub | Pyralidae | S.E. Queensland | Larvae damage kernels |
| *Ephestia cautella* (Wlk.) | Tropical Warehouse Moth | Pyralidae | S.E. Queensland | Larvae infest damaged nuts in storage |
| *Lophodes sinistraria* Guen. | Brown Looper | Geometridae | S.E. Queensland | Larvae defoliate |
| *Olene mendosa* (Hbn.) | Brown Tufted Caterpillar | Lymantriidae | S.E. Queensland | Larvae defoliate |
| *Orgyia australis* Wlk. | Macadamia Tufted Caterpillar | Lymantriidae | S.E. Queensland | Larvae defoliate |
| *Monolepta australis* (Jac.) | Red-shouldered Leaf Beetle | Chrysomelidae | E. Australia | Adults damage foliage, flowers, young nuts |

# MAIZE (*Zea mays* — Gramineae)
## (= Sweet Corn, when unripe; Corn (in USA))

Maize originated in America and is now the principal cereal in the tropics and sub-tropics. It is also being grown for fodder and as a vegetable in Europe and northern N. America. It needs a good summer temperature for the grain to ripen, and grows best in lowlands with a good soil cover; it can withstand some drought once established. It is a tall, broad-leaved cereal; a single stem usually (4–5 m high in some varieties), with the male flower terminal and one or two cobs per stalk. Some varieties tiller more than others. The main production areas are S. America, parts of the USA, E. & S. Africa.

## MAJOR PESTS

| | | | | |
|---|---|---|---|---|
| *Homorocoryphus nitidulus* Wlk. | Grasshopper | Tettigoniidae | E. Africa | Defoliate |
| *Patanga succincta* (L.) | Bombay Locust | Acrididae | India, S.E. Asia | Defoliate |
| *Cicadulina mbila* Naude | Maize Leaf-hopper | Cicadellidae | E. & S. Africa | Sap-sucker; virus vector |
| *Rhopalosiphum maidis* (Fitch) | Corn Leaf Aphid | Aphididae | Cosmopolitan | Infest foliage |
| *Cryptophlebia leucotreta* (Meyr.) | False Codling Moth | Tortricidae | Africa | Larvae damage cobs |
| *Chilo partellus* (Swinhoe) | Spotted Stalk Borer | Pyralidae | Africa, India, S.E. Africa | Larvae bore stalk |
| *Chilo orichalcociliella* (Strand) | Coastal Stalk Borer | Pyralidae | Africa | Larvae bore stalk |
| *Chilo suppressalis* (Wlk.) | Striped Rice Borer | Pyralidae | Spain, S.E. Asia, China, Japan, Australia | Larvae bore stalk |
| *Eldana saccharina* Wlk. | Sugarcane Stalk Borer | Pyralidae | Africa | Larvae bore stalk |
| *Ostrinia nubilalis* (Hb.) | European Corn Borer | Pyralidae | Europe, N. Africa, USA, S. Canada | Larvae bore stalk |
| *Ostrinia furnacalis* (Gn.) | Asiatic Corn Borer | Pyralidae | India, S.E. Asia, Japan, Australasia | Larvae bore stalk |
| *Marasmia trapezalis* (Gn.) | Maize Webworm | Pyralidae | Pantropical | Larvae web leaves |
| *Heliothis armigera* (Hb.) | American Bollworm | Noctuidae | Cosmopolitan in Old World | Larvae feed on cobs |
| *Heliothis zea* (Boddie) | Cotton Bollworm (Corn Earworm) | Noctuidae | N., C. & S. America | Larvae feed on cobs |
| *Agrotis* spp. | Cutworms | Noctuidae | Cosmopolitan | Larvae cutworms |
| *Busseola fusca* (Fuller) | Maize Stalk Borer | Noctuidae | Africa | Larvae bore stalk |
| *Euxoa* spp. | Cutworms | Noctuidae | N. America | Larvae cutworms |
| *Sesamia calamistis* Hmps. | Pink Stalk Borer | Noctuidae | Africa | Larvae bore stalk |
| *Spodoptera exempta* (Wlk.) | African Armyworm | Noctuidae | Africa, India, Australasia | Larvae defoliate |

| *Mythimna unipuncta* (Haw.) | Rice Armyworm | Noctuidae | Europe, E. & W. Africa, USA, C. & S. America | Larvae defoliate |
|---|---|---|---|---|
| *Mythimna separata* (Wlk.) | Rice Ear-cutting Caterpillar | Noctuidae | S.E. Asia | Larvae defoliate |
| *Atherigona soccata* Rond. | Sorghum Shoot Fly | Muscidae | Africa, India | Larvae bore shoots |
| *Atherigona oryzae* Man. | Rice Shoot Fly | Muscidae | Malaysia | Larvae bore shoots |
| *Delia platura* (Meign.) | Bean Seed Fly | Anthomyiidae | USA, Europe | Larvae bore sown seeds |
| *Schizonycha* spp. | Chafer Grubs | Scarabaeidae | Africa | Larvae eat roots |
| *Heteronychus licas* (Klug) | Black Maize Beetle | Scarabaeidae | Africa | Adults bite stems underground |
| *Epilachna similis* (Thn.) | Epilachna Beetle | Coccinellidae | Africa | Adults & larvae eat leaves |
| *Monolepta bifasciata* (Hornst.) | Maize Silk Beetle | Chrysomelidae | Philippines | Adults eat silks |
| *Megalognatha rufiventris* Baly | Maize Tassel Beetle | Chrysomelidae | E. Africa | Adults eat tassel |
| *Nematocerus* spp. | Weevils | Curculionidae | Africa | Adults eat leaves |
| *Sitophilus zeamais* Mot. | Maize Weevil | Curculionidae | Cosmopolitan | Attack ripe seeds |
| *Sitophilus oryzae* L. | Rice Weevil | Curculionidae | S.E. Asia | Attack ripe seeds |

## MINOR PESTS

| *Acheta testaceus* Wlk. | Field Cricket | Gryllidae | S.E. Asia | Seedling pest |
|---|---|---|---|---|
| *Gastrimargus marmoratus* (Thnb.) | – | Acrididae | S.E. Asia | Defoliate |
| *Hieroglyphus banian* (F.) | Large Rice Grasshopper | Acrididae | S.E. Asia | Defoliate |
| *Phymateus aegrotus* Gerst. | – | Acrididae | Africa | Defoliate |
| *Oxya* spp. | Small Rice Grasshoppers | Acrididae | S.E. Asia | Defoliate |
| *Microtermes* spp. | – | Termitidae | Africa | Collect plant material |
| *Hodotermes mossambicus* (Hagen) | Harvester Termite | Hodotermitidae | E. Africa | Defoliate |
| *Cicadulina zeae* China | Maize Leaf-hopper | Cicadellidae | Africa | Sap-sucker; virus vector |
| *Graminella nigrifrons* | Black-faced Leafhopper | Cicadellidae | S. USA | Sap-sucker; virus vector |
| *Baldulus maidis* D. & W. | Corn Leafhopper | Cicadellidae | S. USA, C. & S. America | Sap-sucker virus vector |
| *Pyrilla perpusilla* Wlk. | Indian Sugarcane Leafhopper | Lophopidae | India, Sri Lanka | Sap-sucker |
| *Peregrinus maidis* Ashm. | Maize Plant-hopper | Dephacidae | Pantropical | Sap-sucker |
| *Laodelphax striatella* (Fall.) | Small Brown Planthopper | Delphacidae | Europe, Asia | Sap-sucker; virus vector |

| | | | | |
|---|---|---|---|---|
| *Schizaphis graminum* (Rond.) | Wheat Aphid | Aphididae | Old World | Sap-sucker |
| *Nezara viridula* (L.) | Stink Bug | Pentatomidae | S.E. Asia | Sap-sucker; toxic saliva |
| *Blissus leucopterus* (Say) | Chinch Bug | Pentatomidae | Canada, USA | |
| *Frankliniella williamsi* Hood | Flower Thrips | Thripidae | Thailand | Infest flowers |
| *Chilo polychrysus* (Meyr.) | Dark-headed Rice Borer | Pyralidae | India, S.E. Asia | Larvae bore stalks |
| *Diatraea saccharalis* (F.) | Sugarcane Borer | Pyralidae | N. & S. America | Larvae bore stalks |
| *Nacoleia octasema* (Meyr.) | Banana Scab Moth | Pyralidae | Indonesia, Australasia | Larvae damage cobs |
| *Mythimna loreyi* Dup. | Rice Armyworm | Noctuidae | E. Africa | Larvae damage cobs |
| *Remigia repanda* (F.) | Guinea Grass Moth | Noctuidae | C. & S. America | Larvae defoliate |
| *Spodoptera exigua* (Hub.) | Lesser Armyworm | Noctuidae | Thailand | Larvae defoliate |
| *Spodoptera littoralis* (Boisd.) | Cotton Leafworm | Noctuidae | Africa | Larvae defoliate |
| *Spodoptera litura* (F.) | Rice Cutworm | Noctuidae | Asia, Australasia | Larvae defoliate |
| *Spodoptera frugiperda* (J.E. Smith) | Black Armyworm | Noctuidae | N., C. & S. America | Larvae defoliate |
| *Sesamia cretica* Led. | Sorghum Stalk Borer | Noctuidae | S. Europe, Africa, India | Larvae bore stalks |
| *Sesamia inferens* (Wlk.) | Purple Stem Borer | Noctuidae | Asia, Australasia | Larvae bore stalks |
| *Sesamia nonagrioides* (Lef.) | – | Noctuidae | Africa, Med. | Larvae bore stalks |
| *Borbo cinnara* (Wlk.) | Formosan Swift | Hesperiidae | S.E. Asia | Larvae roll leaves |
| *Sitotroga cerealella* (Ol.) | Angoumois Grain Moth | Gelechiidae | Cosmopolitan | Larvae attack ripe grains |
| *Oscinella frit* (L.) | Frit Fly | Chloropidae | Europe | Larvae gall stem |
| *Delia arambourgi* Seguy | Barley Fly | Anthomyiidae | Africa | Larvae destroy shoot |
| *Mylabris* spp. | Banded Blister Beetles | Meloidae | Europe, Africa, India, S.E. Asia | Adults eat flowers |
| *Monolepta* spp. | Corn Silk Beetles | Chrysomelidae | S.E. Asia | Adults eat silks |
| *Diabrotica undecimpunctata* Mann. | Spotted Cucumber Beetle | Chrysomelidae | N. America | Larvae bore stems underground; adults eat silks |
| *Diabrotica* spp. | Corn Rootworms | Chrysomelidae | N. America | |
| *Adoretus* spp. | White Grubs (Rose Chafers) | Scarabaeidae | S.E. Asia | Larvae eat roots in soil; adults may damage flowers or leaves |
| *Anomala* spp. | White Grubs | Scarabaeidae | S.E. Asia | |
| *Leucophilis irrorata* (Chevr.) | White Grub | Scarabaeidae | Philippines | |
| *Calomycterus* sp. | Corn Seedling Weevil | Curculionidae | Thailand | Larvae kill seedlings |
| *Graphognathus* spp. | White-fringed Weevils | Curculionidae | Australia, USA, S. America | Larvae eat roots; adults leaves |
| *Protostrophus* spp. | Ground Weevils | Curculionidae | Africa | Adults eat foliage; larvae eat roots |

# MANGO (*Mangifera indica* – Anacardiaceae)

The centre of origin is the Indo-Burma region, and it grows wild in the forests of N.E. India; now it is grown widely throughout the tropics for fruit, and in the sub-tropics as an ornamental or shade tree. The main production areas are India, Florida, Egypt, Natal, the E. Africa coast, W. Indies and the Philippines. It is grown from sea-level to 1500 m, but grows best below 1000 m in climates with strongly marked seasons; dry weather is required for flowering. The tree is susceptible to frost, and the preferred temperature is 25–30°C. In habit it is a tree, large and evergreen, from 10–40 m in height, and can live for 100 years or more; fruit-bearing is often biennial. The fruit is a large, fleshy, delicious drupe, in size up to 20 cm long, yellow or red when ripe. The fruit is eaten fresh, or canned, and is also used in chutney and pickles to be eaten with curries.

## MAJOR PESTS

| | | | | |
|---|---|---|---|---|
| *Idiocerus* spp. | Mangohoppers | Cicadellidae | India, S.E. Asia | Cause fruit-fall |
| *Aspidiotus destructor* Sign. | Coconut Scale | Diaspididae | Pantropical | Infest foliage |
| *Ceratitis cosyra* (Wlk.) | Natal Fruit Fly | Tephritidae | Africa | Larvae in fruits |
| *Dacus dorsalis* Hend. | Oriental Fruit Fly | Tephritidae | S.E. Asia | Larvae in fruits |
| *Selenothrips rubrocinctus* (Giard.) | Red-banded Thrips | Thripidae | Pantropical | Leaves scarified |
| *Plocaederus* spp. | Longhorn Beetles | Cerambycidae | Indo-China | Larvae bore trunk |
| *Niphonoclea* spp. | Mango Twig Borers | Cerambycidae | Philippines | Larvae bore twigs |
| *Sternochetus mangiferae* (F.) | Mango Seed Weevil | Curculionidae | S.E. Asia, India | Larvae bore seed in fruit |
| *Deporaus marginatus* (Pasc.) | Leaf Weevil | Curculionidae | Malaysia, Sri Lanka | Adults eat leaves |

## MINOR PESTS

| | | | | |
|---|---|---|---|---|
| *Microcerotermes edentatus* Wasm. | – | Termitidae | E. Africa | Remove foliage |
| *Idiocerus atkinsoni* Leth. | Mangohopper | Cicadellidae | India | Infest shoots & flowers; cause fruit loss |
| *Idiocerus clypealis* Leth. | Mangohopper | Cicadellidae | India, Philippines | |
| *Idiocerus nitidulus* Wlk. | Mangohopper | Cicadellidae | Malaysia, Laos | |
| *Apsylla cistellata* Buckt. | Mango Shoot Psyllid | Psyllidae | India | Gall buds |
| *Aleurocanthus woglumi* Ashby | Citrus Blackfly | Aleyrodidae | Asia, S. America | Infest leaves |
| *Pseudococcus adonidum* (L.) | Long-tailed Mealybug | Pseudococcidae | S.E. Asia, India, Africa, S. America | Infest foliage |
| *Aspidiotus nerii* Bche. | Oleander Scale | Diaspididae | Australasia, Africa, Asia | Infest foliage |
| *Chrysomphalus aonidum* (L.) | Purple Scale (Florida Red Scale) | Diaspididae | S.E. Asia | On leaves & twigs |

600

| | | | | |
|---|---|---|---|---|
| *Chrysomphalus dictyospermi* (Morg.) | Spanish Red Scale | Diaspididae | Cosmopolitan | On leaves & twigs |
| *Ischnaspis longirostris* (Sign.) | Black Line Scale | Diaspididae | Pantropical | On leaves, shoots & fruits |
| *Parlatoria crypta* McKenz. | Mango White Scale | Diaspididae | India, Africa, Iran, Iraq | On leaves & shoots |
| *Pseudaonidia trilobitiformis* (Green) | Trilobite Scale | Diaspididae | Pantropical | Encrust foliage |
| *Coccus mangiferae* Green | Mango Soft Scale | Coccidae | India, Indonesia, Africa, S. America | On leaves & buds |
| *Ceroplastes rubens* Mask. | Pink Waxy Scale | Coccidae | E. Africa, Asia | Infest foliage |
| *Chloropulvinaria psidii* Mask. | Guava Scale | Coccidae | Pantropical | Infest foliage |
| *Saissetia coffeae* (Wlk.) | Helmet Scale | Coccidae | S.E. Asia | Infest foliage |
| *Icerya purchasi* Mask. | Cottony Cushion Scale | Margarodidae | S.E. Asia | Infest twigs & foliage |
| *Icerya seychellarum* Westw. | Seychelles Fluted Scale | Margarodidae | E. Africa, India, S.E. Asia, China, Japan | Infest twigs & foliage |
| *Drosicha stebbingii* Stebb. | Giant Mealybug | Margarodidae | S.E. Asia | Infest foliage |
| *Drosicha mangiferae* Green | Mango Giant Mealybug | Margarodidae | S.E. Asia, India, China | Infest foliage |
| *Pseudotheraptus wayi* Brown | Coconut Bug | Coreidae | E. Africa | Sap-sucker |
| *Mictis longicornis* Westw. | Coreid Bug | Coreidae | Malaysia | Sap-sucker; toxic saliva |
| *Indarbela* spp. | Wood-boring Moths | Metarbelidae | India | Larvae eat bark |
| *Orthaga incarusalis* | Mango Webworm | Pyralidae | Malaysia, Laos ⎫ | Larvae web flowers |
| *Orthaga exvinacea* Mi. | Mango Webworm | Pyralidae | India, Laos ⎬ | & fruits |
| *Clania* spp. | Bagworms | Psychidae | Laos | Larvae defoliate |
| *Autoba* spp. | – | Noctuidae | Laos | Larvae attack flowers & leaves |
| *Bombotelia jocosatrix* Guen. | – | Noctuidae | Laos | Larvae defoliate |
| *Heliothis armigera* (Hb.) | American Bollworm | Noctuidae | Australasia | Larvae deflower |
| *Othreis fullonia* (Cl.) | Fruit-piercing Moth | Noctuidae | S.E. Asia | Adults pierce fruit |
| *Chlumetia transversa* (Wlk.) | Mango Shoot Borer | Noctuidae | Philippines | Larvae bore shoots |
| *Parasa lepida* Cram. | Blue-striped Nettlegrub | Limacodidae | S.E. Asia | Larvae defoliate |
| *Attacus atlas* (L.) | Atlas Moth | Saturniidae | Philippines | Larvae defoliate |
| *Orgyia postica* (Wlk.) | Tussock Moth | Lymantriidae | Laos, Philippines | Larvae attack flower stalks |
| *Dacus zonatus* Saund. | Peach Fruit Fly | Tephritidae | India | Larvae in fruit |
| *Dacus cucurbitae* Coq. | Melon Fly | Tephritidae | Philippines, India | Larvae in fruit |
| *Anastrepha fraterculus* (Wied.) | S. American Fruit Fly | Tephritidae | S. America | Larvae in fruit |
| *Anastrepha mombinpraeoptans* Sein. | West Indian Fruit Fly | Tephritidae | C. America | Larvae in fruit |

| | | | | |
|---|---|---|---|---|
| *Ceratitis capitata* (Wied.) | Medfly | Tephritidae | Africa, Australia, C. & S. America | Larvae in fruit |
| *Anastrepha ludens* (Loew.) | Mexican Fruit Fly | Tephritidae | Mexico, C. America | Larvae in fruit |
| *Erosomiyia indica* Grov. | Flower Gall Midge | Cecidomyiidae | India | Larvae damage flowers |
| *Raodiplosis orientalis* | Gall Midge | Cecidomyiidae | Burma, Laos | Larvae gall leaves |
| *Oecophylla smaragdina* (F.) | Red Tree Ant | Formicidae | India | Nest in foliage; attack workers |
| *Monolepta bifasciata* (Hornst.) | Corn Silk Beetle | Chrysomelidae | Philippines | Adults eat flowers |
| *Leucopholis irrorata* (Chevr.) | White Grub | Scarabaeidae | Philippines | Larvae eat roots of seedlings |
| *Pachnoda sinuata* (F.) | Rose Beetle | Scarabaeidae | Africa | Adults eat flowers |
| *Protaetia* spp. | Rose Chafers | Scarabaeidae | Philippines | Adults attack fruits |
| *Batocera rubus* (L.) | White-spotted Longhorn | Cerambycidae | S.E. Asia, India, Mauritius | Larvae bore trunk |
| *Batocera rufomaculatus* (De Geer) | Red-spotted Longhorn | Cerambycidae | S.E. Asia, India | Larvae bore trunk |
| *Plocaederus fulvicornis* Guer. | Mango Bark Borer | Cerambycidae | Philippines | Larvae bore bark |
| *Olenecamptus* spp. | Longhorn Beetles | Cerambycidae | Laos | Larvae bore trunk |
| *Deporaus marginatus* (Pasc.) | Mango Leaf Weevil | Curculionidae | India, Sri Lanka | Adults eat leaves |
| *Hypomeces squamosus* (F.) | Gold-dust Weevil | Curculionidae | S.E. Asia | Adults eat leaves |
| *Myllocerus* spp. | Grey Weevils | Curculionidae | India | Adults eat leaves |
| *Sternochetus frigidus* F. | Mango Weevil | Curculionidae | S.E. Asia | Larvae in fruit pulp |
| *Polyphagotarsonemus latus* (Banks) | Yellow Tea Mite | Tarsonemidae | Pantropical | Distort leaves |
| *Oligonychus mangiferus* (R. & S.) | Mango Red Spider Mite | Tetranychidae | India | Damage leaves |
| *Aceria mangifera* Sayed | Mango Bud Mite | Eriophyidae | India | Distort buds |

# MANILA HEMP (*Musa textilis* — Musaceae)
## (= Abaca)

This is the world's premier cordage material, obtained from several species of wild banana. *Musa textilis* is the main species concerned and looks like the true banana, but with narrower leaves and small, inedible fruit. The fibre is made from the outer part of the leaf stalks. The crop is commercially important in the Philippines, Borneo and Sumatra, and now in parts of C. America.

### MAJOR PESTS

| | | | | |
|---|---|---|---|---|
| *Cosmopolites sordidus* Germ. | Banana Weevil | Curculionidae | S.E. Asia | Larvae bore rhizome |

### MINOR PESTS

| | | | | |
|---|---|---|---|---|
| *Aphis gossypii* Glov. | Cotton Aphid | Aphididae | Cosmopolitan | Infest foliage; virus vector |
| *Pentalonia nigronervosa* Coq. | Banana Aphid | Aphididae | S.E. Asia (Pantropical) | Infest leaf-bases |
| *Nacoleia octasema* (Meyr.) | Banana Scab Moth | Pyralidae | S.E. Asia | Larvae feed on flowers |

603

# MILLETS (Gramineae)

## (*Pennisetum typhoides* – Bulrush (Pearl) Millet)
## (*Elusine coracana* – Finger Millet)
## (*Panicum miliaceum* – Common Millet)
## (*Setaria italica* – Foxtail Millet) etc.

The millets are a somewhat heterogeneous assemblage of cereals with certain common characteristics, lumped together here for convenience. The four main species are listed above, but there are a few others not mentioned. Bulrush Millet is African in origin and is an important crop in Sudan, Nigeria, and around the southern edge of the Sahara. Finger Millet is native to India, and mostly grown there, but it is also grown widely in Africa south of the Sahara. All four species are grown extensively in India, and Foxtail Millet is in addition an important crop in China. Generally they are dry area crops, resistant to desiccation, and the grains store well. In the different regions of cultivation the different millets generally have quite similar pest spectra.

## MAJOR PESTS

| | | | | |
|---|---|---|---|---|
| *Zonocerus* spp. | Variegated Grasshoppers | Acrididae | Africa | Defoliate & eat panicle |
| *Pyrilla perpusilla* Wlk. | Indian Sugarcane Leafhopper | Lophopidae | India | Suck sap |
| *Laodelphax striatella* (Fall.) | Small Brown Planthopper | Delphacidae | Europe, Asia | Suck sap |
| *Taylorilygus vosseleri* (Popp.) | Cotton Lygus | Miridae | Africa | Sap-sucker; toxic saliva |
| *Chilo partellus* (Swinh.) | Spotted Stalk Borer | Pyralidae | Africa, India, S.E. Asia | Larvae bore stalks |
| *Sesamia calamistis* Hamp. | Pink Stalk Borer | Noctuidae | Africa | Larvae bore stalks |
| *Spodoptera litura* (F.) | Rice Cutworm | Noctuidae | India | Larvae defoliate; destroy seedlings |
| *Spodoptera littoralis* (Boisd.) | Cotton Leafworm | Noctuidae | Africa | Larvae defoliate; destroy seedlings |
| *Atherigona soccata* Rond. | Sorghum Shoot Fly | Muscidae | Africa, India | Larvae bore seedlings |
| *Epilachna similis* (Thnb.) | Epilachna Beetle | Coccinellidae | Africa | Adults & larvae eat foliage |
| *Oulema* spp. | Leaf Beetles | Chrysomelidae | Manchuria, China, India | Larvae mine leaves; adults strip leaves |

## MINOR PESTS

| | | | | |
|---|---|---|---|---|
| *Homorocoryphus nitidulus* Wlk. | Edible Grasshopper | Tettigoniidae | E. Africa | Defoliate & eat panicle |
| *Patanga succincta* (L.) | Bombay Locust | Acrididae | India | Defoliate & eat panicle |
| *Colemania sphenarioides* Bol. | Deccan Wingless Grasshopper | Acrididae | India | Defoliate & eat panicle |

| | | | | |
|---|---|---|---|---|
| *Cicadulina* spp. | Leafhoppers | Cicadellidae | India | Suck sap |
| *Rhopalosiphum* spp. | Cereal Aphids | Aphididae | India | Infest foliage |
| *Schizaphis graminum* Rond. | Wheat Aphid | Aphididae | India | Infest foliage |
| *Peregrinus maidis* Ashm. | Maize Plant-hopper | Delphacidae | India | Infest foliage |
| *Leptocorisa acuta* (Thnb.) | Rice Bug | Coreidae | India | Sap-sucker; toxic saliva |
| *Nezara viridula* L. | Green Stink Bug | Pentatomidae | India | Sap-sucker; toxic saliva |
| *Bagrada hilaris* (Burm.) | Harlequin Bug | Pentatomidae | Africa, India | |
| *Anaphothrips sudanensis* Trybom | Thrips | Thripidae | India, Africa | Infest foliage |
| *Thrips hawaiiensis* (Morg.) | Thrips | Thripidae | India | Infest foliage |
| *Caliothrips indicus* (Bagn.) | Thrips | Thripidae | India | Infest foliage |
| *Amsacta* spp. | Tiger Moths | Arctiidae | India | Larvae defoliate |
| *Diacrisia obliqua* Wlk. | Tiger Moth | Arctiidae | India | Larvae defoliate |
| *Chilo infuscatellus* Sn. | Yellow Top-borer | Pyralidae | Southern Asia | Larvae bore stem tops |
| *Chilo orichalcociliella* (Strand) | Coastal Stalk Borer | Pyralidae | Africa | Larvae bore stalks |
| *Marasmia trapezalis* Guen. | Maize Webworm | Pyralidae | India | Larvae web panicle |
| *Heliothis armigera* (Hb.) | American Bollworm | Noctuidae | India, Africa | Larvae feed on panicle |
| *Sesamia inferens* Wlk. | Purple Stem Borer | Noctuidae | India | Larvae bore stalks |
| *Sesamia nonagriodes* (Lef.) | Rice Ear-cutting Caterpillar | Noctuidae | India | Larvae feed on panicle |
| *Mythimna separata* (Wlk.) | Rice Ear-cutting Caterpillar | Noctuidae | India | Larvae feed on panicle |
| *Remigia repanda* (F.) | Guinea Grass Moth | Noctuidae | C. & S. America | Larvae defoliate |
| *Atherigona* spp. | Shoot Flies | Muscidae | India | Larvae bore shoots |
| *Itonida* spp. | Gall Midges | Cecidomyiidae | India | Larvae in inflorescence |
| *Holotrichia* spp. | Chafer Grubs | Scarabaeidae | India | Larvae eat roots |
| *Mylabris* spp. | Banded Blister Beetles | Meloidae | Asia, Africa | Adults eat inflorescence |
| *Epicauta* spp. | Black Blister Beetles | Meloidae | Africa, India, USA | |
| *Chaetocnema* spp. | Flea Beetles | Chrysomelidae | India | Adults hole leaves |
| *Nematocerus* spp. | Nematocerus Weevils | Curculionidae | Africa | Adults eat leaves |
| *Myllocerus* spp. | Grey Weevils | Curculionidae | India | Adults eat leaves |

# MULBERRY (*Morus* spp. — Moraceae)

*Morus alba* (White Mulberry) is native to China and grown to a limited extent in parts of the tropics for its edible fruits, for its leaves as food for silkworms, and for its wood used in making certain sports goods such as hockey sticks and tennis racquets. It is a small tree, up to 5 m in height, and the fruit is a syncarp.

*Morus nigra* (Black Mulberry) grows well only at higher elevations in the tropics; native to Iran. (See also Butani, 1978*c*.)

## MAJOR PESTS

| | | | | |
|---|---|---|---|---|
| *Pealius mori* (Tak.) | Mulberry Whitefly | Aleyrodidae | Thailand | Infest leaves |
| *Batocera rufomaculata* (De Geer) | Red-spotted Longhorn | Cerambycidae | India, E. Africa, Malaysia | Larvae bore trunk |
| *Apriona germari* (Hope) | Jackfruit Longhorn | Cerambycidae | Thailand | Larvae bore branches |

## MINOR PESTS

| | | | | |
|---|---|---|---|---|
| *Aleurolobus marlatti* (Quaint.) | Marlatt Blackfly | Aleyrodidae | India | Infest leaves |
| *Icerya aegyptica* (Dgl.) | Fluted Scale | Margarodidae | Africa, Asia | Infest foliage |
| *Drosicha mangiferae* (Green) | Giant Mealybug | Margarodidae | India | Infest foliage |
| *Perissopneumon tamarinda* (Green) | Mealybug | Pseudococcidae | India | Infest foliage |
| *Pseudococcus comstocki* (Kum.) | Comstock's Mealybug | Pseudococcidae | Asia, China, Japan, USA | Infest foliage |
| *Maconellicoccus hirsutus* (Green) | Mealybug | Pseudococcidae | Thailand, Laos | Infest shoots |
| *Quadraspidiotus perniciousus* (Comst.) | San José Scale | Diaspididae | India | Infest foliage |
| *Aonidiella aurantii* (Mask.) | Red Scale | Diaspididae | India | Infest foliage |
| *Pseudaulacaspis pentagona* (Targ.) | White Scale | Diaspididae | Cosmopolitan | Infest foliage |
| *Chrysomphalus aonidum* (L.) | Purple Scale | Diaspididae | India | Infest foliage |
| *Halys dentatus* F. | Mulberry Bug | Pentatomidae | India | Sap-sucker |
| *Tryphactothrips rutherfordi* Bagnall | Thrips | Thripidae | India | Infest leaves |
| *Indarbela* spp. | Wood-borer Moths | Metarbelidae | India, China | Larvae eat bark |
| *Archips micaceana* (Wlk.) | Leaf-roller | Tortricidae | Thailand | Larvae roll leaves |
| *Latoia lepida* (Cramer) | Nettlegrub | Limacodidae | India | Larvae defoliate |
| *Dichocrocis punctiferalis* (Guen.) | Shoot Borer | Pyralidae | India | Larvae bore shoots |
| *Spodoptera litura* F. | Rice Cutworm | Noctuidae | India | Larvae eat leaves |
| *Bombyx mori* (L.) | Silkworm | Bombycidae | Laos | Larvae defoliate |
| *Dacus tau* (Wlk.) | Fruit Fly | Tephritidae | India, S.E. Asia | Larvae in fruits |
| *Vespa* spp. | Oriental Wasps | Vespidae | India, S.E. Asia | Adults pierce ripe fruits |
| *Mimastra cyanema* Hope | Almond Beetle | Chrysomelidae | India | Adults defoliate |
| *Apriona cinerea* Chevr. | Longhorn Beetle | Cerambycidae | India | Larvae bore trunk |
| *Batocera* spp. | Longhorns | Cerambycidae | India | Larvae bore trunk |
| *Sthenias grisator* F. | Stem-Girdler | Cerambycidae | S. India | Larvae girdle stems |
| *Hypomeces squamosus* (F.) | Gold-dust Weevil | Curculionidae | Thailand, Laos | Adults eat leaves |

# NUTMEG (*Myristica fragrans* — Myristicaceae)

This large evergreen tree, 10–20 m tall, is native to the Moluccas (or Spice Islands) but is now widely grown throughout the hot wet tropics. The ripe fruits are yellow and plum-like, and when ripe the husk splits open revealing the brown seed covered by the red, branching aril. The kernel of the seed is the nutmeg of commerce, and the aril is the source of the spice mace. The main production areas are the W. Indies (Granada), Indonesia and Malaysia.

## MAJOR PESTS

| | | | | |
|---|---|---|---|---|
| *Xyleborus fornicatus* Eichh. | Tea Shot-hole Borer | Scolytidae | Malaysia | Adults bore branches |
| *Phloeosomus cribratus* Bland. | Shot-hole Borer | Scolytidae | Malaysia (Penang) | Adults bore branches |

## MINOR PESTS

| | | | | |
|---|---|---|---|---|
| *Coccus mangiferae* Green | Mango Soft Scale | Coccidae | Malaysia | Encrust leaves |
| *Saissetia nigra* (Nietn.) | Nigra Scale | Coccidae | India | Encrust leaves |
| *Oryzaephilus mercator* Fauv. | Merchant Grain Beetle | Cucujidae | Malaysia | Attack stored seeds |
| *Araeocerus fasciculatus* (De Geer) | Nutmeg Weevil | Brenthidae | Cosmopolitan | Larvae bore inside the kernel; more important as a storage pest |

# OIL PALM (*Elaeis guineensis* – Palmae)

The centre of origin is western tropical Africa, where it is found wild. It is now established as a plantation crop in W. Africa, Malaysia, and Indonesia. It thrives only where rainfall is high, but will grow on poor soils. A typical palm tree in appearance, up to 10–15 m high at maturity. Fruit-bearing starts at five years, but full potential is not realized until the palm is ten years old. Oil is extracted from the mesocarp of the fruit; palm oil contains vitamin A; the oil is used in industry and to make soap. Kernel oil is of a higher quality and is used for margarine and other foodstuffs. The oil cake residue is used for livestock food. (See also Wood, 1968.)

## MAJOR PESTS

| | | | | |
|---|---|---|---|---|
| *Pseudococcus adonidum* (L.) | Long-tailed Mealybug | Pseudococcidae | Pantropical | Infest foliage |
| *Mahasena corbetti* Tams | Coconut Case Caterpillar | Psychidae | S.E. Asia | Larvae defoliate |
| *Rhynchophorus phoenicis* (F.) | African Palm Weevil | Curculionidae | Africa | Larvae bore crown |
| *Rhynchophorus ferrugineus* (Oliv.) | Asiatic Palm Weevil | Curculionidae | India, S.E. Asia, Papua NG | Larvae bore crown |
| *Rhynchophorus palmarum* (L.) | South American Palm Weevil | Curculionidae | C. & S. America | Larvae bore crown |
| *Diocalandra frumenti* (F.) | Four-spotted Coconut Weevil | Curculionidae | E. Africa, India, S.E. Asia, Papua NG | Larvae bore plant body |
| *Oryctes monoceros* (Ol.) | Rhinoceros Beetle | Scarabaeidae | Africa | Adults damage crown |
| *Oryctes rhinoceros* (L.) | Rhinoceros Beetle | Scarabaeidae | Asia, Papua NG | Adults damage crown |

## MINOR PESTS

| | | | | |
|---|---|---|---|---|
| *Brachytrupes* spp. | Brown Crickets | Gryllidae | S.E. Asia | Important pests in nurseries |
| *Gryllotalpa africana* Beau. | Mole Cricket | Gryllotalpidae | S.E. Asia | Important pests in nurseries |
| *Aularches miliaris* L. | Spotted Grass-hopper | Acrididae | Thailand | Adults and nymphs eat leaves |
| *Valanga nigricornis* (Burm.) | – | Acrididae | Malaysia | Adults and nymphs eat leaves |
| *Gastrimargus marmoratus* (Thnb.) | – | Acrididae | Malaysia | Adults and nymphs eat leaves |
| *Coptotermes curvignathus* Holmg. | – | Rhinotermitidae | Malaysia | Workers tunnel live trunk |
| *Ricania speculum* Wlk. | Black Plant-hopper | Ricaniidae | Malaysia | Suck sap |
| *Cerataphis lataniae* Boisd. | Latania Aphid | Aphididae | Malaysia | Suck sap; infest foliage |
| *Dysmicoccus brevipes* (Ckll.) | Pineapple Mealybug | Pseudococcidae | Malaysia | Infest foliage |
| *Pseudococcus* spp. | Mealybugs | Pseudococcidae | Malaysia | Infest fruits & foliage |

608

| Species | Common name | Family | Distribution | Notes |
|---|---|---|---|---|
| *Pinnaspis buxi* (Bch.) | — | Diaspididae | Pantropical | Infest foliage |
| *Ischnaspis longirostris* (Sign.) | Black Line Scale | Diaspididae | Pantropical | Infest foliage |
| *Chrysomphalus dictyospermi* (Morg.) | Spanish Red Scale | Diaspididae | Pantropical | Infest foliage |
| *Leptoglossus australis* (F.) | Leaf-footed Plant Bug | Coreidae | Malaysia | Sap-sucker; toxic saliva |
| *Artona catoxantha* (Hmps.) | Coconut Leaf Skeletonizer | Zygaenidae | Thailand | Larvae skeletonize leaves |
| *Tirathaba mundella* Wlk. | Oil Palm Bunch Moth | Pyralidae | Malaysia, Sumatra | Larvae feed on fruit bunches |
| *Pimelephila ghesquierei* Tams | Palm Moth | Pyralidae | Africa (Zaïre) | Larvae bore leaves |
| *Spodoptera litura* (F.) | Rice Cutworm | Noctuidae | Malaysia | Larvae eat leaves |
| *Parasa vivida* (Wlk.) | Nettlegrub | Limacodidae | Africa | Larvae with protruding setae with urticating properties; larvae defoliate |
| *Parasa lepida* Cramer | Nettlegrub | Limacodidae | Thailand | |
| *Setora nitens* Wlk. | Nettle Caterpillar | Limacodidae | S.E. Asia | |
| *Thosea sinensis* Wlk. | Nettle Caterpillar | Limacodidae | Thailand, Sumatra | |
| *Cremastopsyche pendula* Joannis | Bagworm | Psychidae | S.E. Asia | Larvae are bagworms & construct cases from plant leaf material; defoliators |
| *Clania* spp. | Bagworms | Psychidae | S.E. Asia | |
| *Metisa plana* Wlk. | Bagworm | Psychidae | S.E. Asia | |
| *Erionota thrax* L. | Banana Skipper | Hesperiidae | Thailand, Malaysia | Larvae cut leaf rolls |
| *Hidari irava* Moore | Leaf-binder | Hesperiidae | Thailand | Larvae fold leaves |
| *Lotongus calathus* How. | Leaf-binder | Hesperiidae | Thailand | Larvae fold leaves |
| *Cephrenes chrysozona* Plotz. | Skipper | Hesperiidae | Malaysia | Larvae fold leaves |
| *Dasychira* spp. | Tussock Moths | Lymantriidae | S.E. Asia | Larvae defoliate |
| *Orgyia turbata* Butler | Tussock Moth | Lymantriidae | Malaysia | Larvae defoliate |
| *Pachnoda* spp. | Rose Beetles | Scarabaeidae | Africa | Adults eat leaves of young palms; serious in nurseries |
| *Adoretus* spp. | Flower Beetles | Scarabaeidae | Malaysia, Sumatra | |
| *Apogonia* spp. | Brown Flower Beetles | Scarabaeidae | Malaysia, Sumatra | |
| *Xylotrupes gideon* L. | Unicorn Beetle | Scarabaeidae | Malaysia | Adults gnaw foliage |
| *Leucopholis rorida* (F.) | White Grub | Scarabaeidae | Malaysia, Sumatra | Larvae eat roots |
| *Straegus aloeus* L. | — | Curculionidae | C. & S. America | Larvae bore plant |
| *Temnoschoita quadripustulata* Gyll. | — | Curculionidae | Africa | Larvae bore plant |
| *Coelaenomenodera elaeidis* Wlk. | Leaf Miner | Chrysomelidae | W. & C. Africa | Larvae mine leaves, adults eat long strips of leaf material |
| *Hispoleptis elaeidis* | Leaf Miner | Chrysomelidae | Ecuador | |
| *Promecotheca cumingi* Baly | Leaf Miner | Chrysomelidae | Malaysia | |
| *Plesispa reichei* Chapuis | Coconut Hispid | Chrysomelidae | Thailand | |

609

| *Xyleborus similis* Ferrari | Shot-hole Borer | Scolytidae | Thailand | Adults bore plant |
| *Oligonychus* spp. | Red Spider Mites | Tetranychidae | Malaysia ⎫ | Adults & nymphs |
| *Tetranychus* spp. | Red Spider Mites | Tetranychidae | Malaysia ⎭ | scarify leaves |

# OKRA (*Hibiscus esculentus* – Malvaceae)
### (= Ladies' Fingers)

Okra is native to tropical Africa but is now widespread throughout the tropics. It grows well in the lowland tropics on most types of soil, but the best crops are produced on well-manured loams. The plant is a robust erect herb 1–2 m tall, and the fruit is a beaked pyramidal capsule 10–30 cm long by 2–3 cm broad, with a high mucilage content, and is used as a vegetable either boiled or fried. The ripe seeds contain 20% edible oil. Okra is grown on a pantropical basis, but only for local fresh consumption; a little canning is done.

## MAJOR PESTS

| | | | | |
|---|---|---|---|---|
| *Aphis gossypii* Glov. | Cotton Aphid | Aphididae | Cosmopolitan | Infest foliage |
| *Empoasca* spp. | Green Leaf-hoppers | Cicadellidae | S.E. Asia, Africa, India | Sap-suckers on leaves |
| *Ferrisia virgata* (Ckll.) | Striped Mealybug | Pseudococcidae | S.E. Asia, India | Infest foliage |
| *Dysdercus* spp. | Cotton Stainers | Pyrrhocoridae | S.E. Asia, Africa | Suck sap from seeds |
| *Oxycarenus hyalipennis* Costa | Cotton Seed Bug | Lygaeidae | India, S.E. Asia, Africa | Suck sap from seeds |
| *Earias vittella* Stoll. | Spiny Bollworm | Noctuidae | Malaysia | Larvae attack & bore fruit capsules |
| *Earias biplaga* Wlk. | Spiny Bollworm | Noctuidae | Africa | |
| *Earias insulana* (Boisd.) | Spiny Bollworm | Noctuidae | India, S.E. Asia, Africa | |
| *Heliothis armigera* (Hub.) | American Bollworm | Noctuidae | Cosmopolitan in Old World | |

## MINOR PESTS

| | | | | |
|---|---|---|---|---|
| *Atractomorpha crenulata* F. | Grasshopper | Acrididae | India | Defoliate |
| *Heteroplernis obscurella* | Grasshopper | Acrididae | Papua NG | Defoliate |
| *Amrasca biguttula* (Ishida) | Leafhopper | Cicadellidae | India | Sap-sucker |
| *Saissetia coffeae* (Wlk.) | Helmet Scale | Coccidae | Philippines, India | On leaves & stems |
| *Bemisia tabaci* (Genn.) | Tobacco Whitefly | Aleyrodidae | India | Infest foliage |
| *Colgar* sp. | Moth Bug | Flattidae | Papua NG | Sap-sucker |
| *Calidea dregii* Germ. | Blue Bug | Pentatomidae | Africa | Sap-sucker; toxic saliva |
| *Nezara viridula* (L.) | Green Stink Bug | Pentatomidae | S.E. Asia | Sap-sucker; toxic saliva |
| *Tarundia glaucesenus* | Planthopper | Ricaniidae | Papua NG | Sap-sucker |
| *Frankliniella sulphurea* Schmutz | Thrips | Thripidae | India | Infest flowers |
| *Thrips tabaci* Lind. | Onion Thrips | Thripidae | India | Infest flowers |
| *Pectinophora gossypiella* (Saund.) | Pink Bollworm | Gelechiidae | Cosmopolitan | Larvae bore fruits |

| | | | | |
|---|---|---|---|---|
| *Homona coffearia* (Nietn.) | Tea Tortrix | Tortricidae | Papua NG | Larvae roll leaves |
| *Adoxophyes* sp. | Leaf Roller | Tortricidae | Papua NG | Larvae roll leaves |
| *Sylepta derogata* (F.) | Cotton Leaf Roller | Pyralidae | S.E. Asia, India, Africa | Larvae roll leaves |
| *Zeuzera coffeae* Nietn. | Red Coffee Borer | Cossidae | Philippines | Larvae bore stems |
| *Agrotis ipsilon* (Hfn.) | Black Cutworm | Noctuidae | India | Larvae are cutworms |
| *Acontia* spp. | – | Noctuidae | India | Larvae eat foliage |
| *Spodoptera litura* (F.) | Rice Cutworm | Noctuidae | S.E. Asia, India, Australasia | Larvae eat foliage |
| *Spodoptera littoralis* (Boisd.) | Cotton Leaf-worm | Noctuidae | Africa | Larvae defoliate |
| *Xanthodes transversus* | – | Noctuidae | Papua NG | Larvae defoliate |
| *Chrysodeixis chalcites* (Esp.) | Cabbage Semi-looper | Noctuidae | Malaysia | Larvae defoliate |
| *Anomis flava* (F.) | Cotton Semi-looper | Noctuidae | Old World tropics | Larvae defoliate |
| *Latoia lepida* (Cram.) | Blue-striped Nettlegrub | Limacodidae | Philippines | Larvae defoliate |
| *Solenopsis geminata* (F.) | Fire Ant | Formicidae | Philippines | Attack workers |
| *Mylabris* spp. | Banded Blister Beetles | Meloidae | India, S.E. Asia, Europe, Africa | Adults deflower |
| *Anomala* spp. | White Grubs | Scarabaeidae | Philippines | Larvae eat roots |
| *Oxycetonia* spp. | Rose Chafers | Scarabaeidae | India | Adults eat flowers |
| *Leucopholis irrorata* (Chevr.) | White Grub | Scarabaeidae | Philippines | Larvae eat roots |
| *Holotrichia insularis* Brenske | White Grub | Scarabaeidae | India | Larvae eat roots |
| *Epilachna* spp. | Epilachna Beetles | Coccinellidae | S.E. Asia | Adults & larvae eat leaves |
| *Podagrica bowringi* B. | Flea Beetle | Chrysomelidae | India | Adults hole leaves |
| *Nisotra gemella* Erichs. | Flea Beetle | Chrysomelidae | Philippines | Adults hole leaves |
| *Monolepta bifasciata* Hornst. | Corn Silk Beetle | Chrysomelidae | Philippines | Adults eat leaves |
| *Paratrachys* sp. | Jewel Beetle | Buprestidae | India | Larvae bore stems |
| *Trachys herilla* Obenb. | Jewel Beetle | Buprestidae | India | Larvae bore stems |
| *Sphenoptera gossypii* Banks | Cotton Stem Borer | Buprestidae | India | Larvae bore stems |
| *Alcidodes affaber* F. | Stem-girdling Weevil | Curculionidae | India | Adults eat shoots; larvae bore stems |
| *Myllocerus* spp. | Grey Weevils | Curculionidae | India | Adults eat leaves; larvae eat roots |

# OLIVE (*Olea europaea* — Oleaceae)

The Olive is one of the oldest of fruits and has been grown since prehistoric times. It was known in Egypt in the 17th century BC. It is widely cultivated throughout the Mediterranean region and has now been introduced extensively throughout the tropics and sub-tropics. The main areas of commercial production are California, Spain, Italy, Portugal, Turkey, Tunisia and Greece. The fruits have a high oil content, and are eaten both green and when ripe (black), but are most important as the source of olive oil. The tree is a small evergreen 8–12 m in height, with small leathery leaves. A deep fertile soil is required, together with temperatures of about 14 °C (average) and not falling below − 10 °C. Irrigation is often required for successful groves.

## MAJOR PESTS

| | | | | |
|---|---|---|---|---|
| *Parlatoria oleae* (Colv.) | Olive Scale | Diaspididae | Cosmopolitan | Encrust twigs |
| *Saissetia oleae* (Bern.) | Black Scale | Coccidae | Cosmopolitan | Encrust foliage |
| *Dacus oleae* (Gmel.) | Olive Fruit Fly | Tephritidae | Med., S. Afric | Larvae in fruits |

## MINOR PESTS

| | | | | |
|---|---|---|---|---|
| *Aphis spiraecola* Patch | Apple Aphid | Aphididae | Cosmopolitan | Infest foliage |
| *Euphyllura olivina* Costa | Olive Psyllid | Psyllidae | Med. | Sap-suckers; infest leaf axils |
| *Aleurolobus olivinus* Silv. | Olive Whitefly | Aleyrodidae | Italy, Africa | Infest foliage |
| *Ceroplastes rusci* (L.) | Fix Wax Scale | Coccidae | Med. | Infest foliage |
| *Aspidiotus nerii* Bch. | Oleander Scale | Diaspididae | Cosmopolitan | Encrust twigs |
| *Teleonemia australis* Dist. | Lace Bug | Tingidae | Southern Africa | Sap-sucker |
| *Dasyneura oleae* (Lw.) | Olive Midge | Cecidomyiidae | Med. | Larvae make leaf galls |
| *Ceratitis capitata* (Wied.) | Mediterranean Fruit Fly | Tephritidae | Cosmopolitan | Larvae in fruits |
| *Liothrips oleae* Costa | Olive Thrips | Thripidae | Med., Africa | Infest foliage |
| *Prays oleae* F. | Olive Moth | Yponomeutidae | Med., S. Africa | Larvae bore fruits |
| *Cacoecimorpha pronubana* Hb. | Carnation Leaf-roller | Tortricidae | Med., Europe | Larvae roll leaves |
| *Metriochroa latifoliella* (Milliere) | – | Gracillariidae | Med. | Larvae mine leaves |
| *Acherontia atropos* L. | Death's Head Hawk Moth | Sphingidae | Europe, Africa, S.E. Asia | Larvae defoliate |
| *Hyles lineata* (F.) | Silver-striped Hawk Moth | Sphingidae | Med. | Larvae defoliate |
| *Zeuzera pyrina* L. | Leopard Moth | Cossidae | Med. | Larvae bore branches |
| *Otiorhynchus cribricollis* Gylh. | Apple Weevil | Curculionidae | Med., W. USA, Australia | Larvae eat roots; adults eat leaves |
| *Rhynchites cribripennis* (Desbr.) | Twig Cutter | Curculionidae | Med. | Adults cut twigs |
| *Mylabris oleae* Chevr. | Olive Blister Beetle | Meloidae | N. Africa | Adults eat flowers |

# ONIONS (*Allium* spp. — Amaryllidaceae)
## (Onions, Shallot, Garlic, Chives, Leek)

Onions are a crop of great antiquity, unknown in the wild state, but probably originating in S. Asia, or the Mediterranean region. They are mainly temperate crops but are quite widely grown in sub-tropical areas, and also in some parts of the tropics, although they generally prefer light sandy soils in cool moist regions. Onions are grown either as bulbs for drying (ware onions), or for pickling, or as salad onions. Leeks are more temperate and grown as a vegetable for cooking. Garlic and chives are used in cooking for flavouring and garnishing purposes; the local equivalents in the Far East are different species from the European ones. (See also Butani & Varma, 1976*b*.)

## MAJOR PESTS

| | | | | |
|---|---|---|---|---|
| *Thrips tabaci* Lind. | Onion Thrips | Thripidae | Cosmopolitan | Infest leaves |
| *Delia antiqua* (Meign.) | Onion Fly | Anthomyiidae | Cosmopolitan | Larvae inside bulb |
| *Delia platura* (Meign.) | Bean Seed Fly | Anthomyiidae | Cosmopolitan | Larvae destroy seed or invade bulb |
| *Spodoptera exigua* (Hub.) | Beet Armyworm | Noctuidae | S. Europe, Asia | Larvae eat leaves |

## MINOR PESTS

| | | | | |
|---|---|---|---|---|
| *Euborellia annulipes* Lucas | Earwig | Forficulidae | India | Damage seedlings |
| *Myzus ascalonicus* Don. | Shallot Aphid | Aphididae | Europe | Infest foliage |
| *Myzus persicae* (Sulz.) | Peach—Potato Aphid | Aphididae | Cosmopolitan | Infest foliage |
| *Lipaphis erysimi* (Kalt.) | Turnip Aphid | Aphididae | Cosmopolitan | Infest foliage |
| *Aeolothrips* spp. | Thrips | Aeolothripidae | India | Infest foliage |
| *Caliothrips indicus* (Bag.) | Groundnut Thrips | Thripidae | India | Infest foliage |
| *Acrolepia assectella* (Zell.) | Leek Moth | Yponomeutidae | Europe, Asia | Larvae defoliate |
| *Cnephasia* spp. | — | Tortricidae | Europe | Larvae defoliate |
| *Noctua pronuba* (L.) | Large Yellow Underwing | Noctuidae | Europe | Larvae are cutworms |
| *Agrotis ipsilon* (Roth.) | Black Cutworm | Noctuidae | Cosmopolitan | |
| *Heliothis assulta* | — | Noctuidae | Papua NG | Larvae defoliate |
| *Heliothis armigera* (Hub.) | American Bollworm | Noctuidae | Old World | Larvae inside leaves |
| *Spodoptera litura* F. | Rice Cutworm | Noctuidae | India, S.E. Asia | Larvae defoliate |
| *Spodoptera littoralis* (Boisd.) | Cotton Leafworm | Noctuidae | Africa, S. Europe | Larvae defoliate |
| *Phytobia cepae* (Her.) | Onion Leaf Miner | Agromyzidae | C. Europe, Malaya, China, Japan | Larvae mine leaves |
| *Anthrenus* spp. | — | Dermestidae | India | Attack stored bulbs |
| *Aceria tulipae* (K.) | Onion Mite | Eriophyidae | USA | Infest bulb |

# OPIUM POPPY (*Papaver somniferum* — Papaveraceae)

Opium is one of the oldest narcotics known to man, originating probably in Asia Minor, and mostly cultivated in this region, and parts of India, S.E. Asia and China. Opium is actually the dried exudate (juice) from injured capsules of the opium poppy. Properly utilized, optium and its alkaloid derivatives (morphine, and heroin) are invaluable to mankind medicinally for the relief of pain. However, this drug has been used for centuries as a narcotic and its inevitable abuse leads to habituation and both physical and mental deterioration and degradation. Thus its cultivation is subject to rigorous international legislation in most parts of the world.

## MAJOR PESTS

| | | | | |
|---|---|---|---|---|
| *Stenocarus fuliginosus* | Poppy Root Weevil | Curculionidae | Romania | Larvae eat roots |

## MINOR PESTS

| | | | | |
|---|---|---|---|---|
| *Chrotogonus* sp. | Surface Grasshopper | Acrididae | India | Defoliate |
| *Frankliniella sulphurea* Schm. | Flower Thrips | Thripidae | India | Infest flowers |
| *Agrotis ipsilon* (Hfn.) | Black Cutworm | Noctuidae | India | Larvae are cutworms |
| *Agrotis segetum* (Schff.) | Common Cutworm | Noctuidae | India | Larvae are cutworms |
| *Heliothis armigera* Hb. | American Bollworm | Noctuidae | India | Larvae bore capsule |
| *Spodoptera litura* (F.) | Rice Cutworm | Noctuidae | India | Larvae defoliate |
| *Dasyneura papaveris* (Winn.) | Poppy Capsule Midge | Cecidomyiidae | Europe (not UK) | Larvae infest capsules |
| *Carpodiplosis papaveris* Kjell. | – | Cecidomyiidae | Europe (not UK) | Larvae in capsule wall |

# PAPAYA (*Carica papaya* – Caricaceae)
## (= Pawpaw; Papita)

Papaya has never been found wild, but probably originated in S. Mexico and Costa Rica. It was spread to the W. Indies and Philippines in the 16th century, and to E. Africa by the 19th century. It is essentially a tropical plant, grown mainly between 32°N and 32°S, and is killed by frost. It can be grown from sea-level to 2000 m near the Equator, and it needs sunshine and high temperatures, with well-drained soil. It can be grown under irrigation. The plant is a short-lived, quick-growing, fleshy tree with few branches, 2–10 m tall. Latex vessels run in all parts of the plant body. The fruit is a large fleshy berry 7–30 cm in length, weighing up to 9 kg, oblong to spherical in shape, yellow when ripe. The edible reddish flesh is eaten for breakfast and dessert, also for jams, flavouring and canning; papain, a proteolytic enzyme, is extracted and used for tenderizing meat. The fruit has a high vitamin A and B content.

## MAJOR PESTS

| | | | | |
|---|---|---|---|---|
| *Ceratitis capitata* (Wied.) | Medfly | Tephritidae | Africa, C. & S. America, Med. | Larvae inside fruits |
| *Planococcus citri* Risso | Root Mealybug | Pseudococcidae | Pantropical | Encrust leaves & fruits |
| *Aspidiotus destructor* Sign. | Coconut Scale | Diaspididae | Pantropical | Encrust leaves & fruits |
| *Tetranychus cinnabarinus* (Boisd.) | Tropical Red Spider Mite | Tetranychidae | India, China, Hawaii | Scarify foliage |

## MINOR PESTS

| | | | | |
|---|---|---|---|---|
| *Poecilocerus pictus* F. | Ak Grasshopper | Acrididae | India | Defoliate |
| *Gryllotalpa africana* Pal. | African Mole Cricket | Gryllotalpidae | S.E. Asia | Attack roots |
| *Aphis gossypii* Glov. | Cotton Aphid | Aphididae | India | Sap-suckers; vector of Papaya Mosaic Virus disease |
| *Pergandeida robiniae* Macc. | – | Aphididae | Pantropical | |
| *Myzus persicae* (Sulz.) | Green Peach Aphid | Aphididae | Cosmopolitan | |
| *Ferrisia virgata* (Ckll.) | Striped Mealybug | Pseudococcidae | Philippines | Infest foliage |
| *Drosicha mangiferea* Green | Mango Giant Mealybug | Margarodidae | India | Encrust leaves |
| *Saissetia nigra* (Nietn.) | Nigra Scale | Coccidae | Philippines | Infest leaves |
| *Aspidiotus orientalis* New. | Oriental Scale | Diaspididae | Malaysia, China | Encrust foliage |
| *Aonidiella orientalis* New. | Oriental Yellow Scale | Diaspididae | Pantropical | Encrust foliage |
| *Morganella longispina* Morg. | – | Diaspididae | Pantropical | Infest foliage |
| *Dacus dorsalis* Hend. | Oriental Fruit Fly | Tephritidae | Philippines | Larvae in fruit |
| *Dacus ferrugineus* F. | Fruit Fly | Tephritidae | Malaysia | Larvae in fruit |
| *Dacus pedestris* Bezzi | Fruit Fly | Tephritidae | India, Sri Lanka | Larvae in fruit |

616

| | | | | |
|---|---|---|---|---|
| *Toxotrypana curvicauda* Gerst. | Fruit Fly | Tephritidae | India, S. America | Larvae in fruit |
| *Ptecticus elongatus* F. | – | Therevidae | E. Africa | Larvae in fruit |
| *Dichocrosis punctiferalis* (Guen.) | Leaf-folder | Pyralidae | Philippines | Larvae fold leaf |
| *Diacrisia investigatorum* Karsch. | Tiger Moth | Arctiidae | Africa | Larvae defoliate |
| *Othreis fullonia* (Cl.) | Fruit-piercing Moth | Noctuidae | Old World tropics | Adults pierce fruits |
| *Rhabdoscelis obscurus* Boisd. | Cane Weevil Borer | Curculionidae | Australasia, Fiji, Hawaii, W. Indies | Larvae bore stem |
| *Protaetia* spp. | Rose Chafers | Scarabaeidae | S.E. Asia | Adults attack ripe fruits |
| *Dihammus vastator* Newm. | Longhorn Beetle | Cerambycidae | Philippines | Larvae bores stem |
| *Brevipalpus phoenicis* (Geij.) | Red Crevice Mite | Tenuipalpidae | Hawaii | Scarify foliage |
| *Tenuipalpus bioculatus* McG. | – | Tenuipalpidae | Hawaii | Scarify foliage |
| *Tetranychus cinnabarinus* (Boisd.) | Tropical Red Spider Mite | Tetranychidae | S.E. Asia | Scarify foliage |
| *Polyphagotarsonemus latus* (Banks) | Yellow Tea Mite | Tarsonemidae | Hawaii | Scarify foliage |

# PASSION FRUIT (*Passiflora edulis* – Passifloraceae)
## (*P. quadrangularis*)
## (= Grenadilla & Giant Granadilla)

These are both native to S. America, the latter species being grown mostly for local consumption. Both are now pan-tropical in distribution. They are grown commercially in S. Africa, Kenya, Australia, New Zealand, and Hawaii. The former species occurs as two varieties, one purple and the other yellow in colour. The plant body is a vigorous woody perennial climber, up to 15 m long. The purple passion fruit does best in the highlands of the tropics, whereas the yellow variety tolerates lower altitudes. In regions of heavy rainfall, however, pollination is often poor; a rainfall of less than 80 cm is preferred, and it will grow on most soils so long as they are not waterlogged. The globular fleshy berry is eaten fresh, but there is now a great demand for passion fruit juice which is delicious in taste and very rich in vitamin C.

## MAJOR PESTS

| | | | | |
|---|---|---|---|---|
| *Planococcus kenyae* (Le Pell.) | Kenya Mealybug | Pseudococcidae | E. & W. Africa | Encrust foliage |
| *Ceratitis capitata* (Wied.) | Medfly | Tephritidae | Africa, Hawaii | Larvae inside fruit |
| *Dacus umbrosus* F. | Fruit Fly | Tephritidae | Malaysia | Larvae in fruit |
| *Brevipalpus phoenicis* (Geijskes) | Red Crevice Tea Mite | Tenuipalpidae | Pantropical | Invest foliage |

## MINOR PESTS

| | | | | |
|---|---|---|---|---|
| *Scolypopa australis* Wlk. | Planthopper | Ricaniidae | Australia | Infest foliage |
| *Aphis gossypii* Glov. | Cotton Aphid | Aphididae | S.E. Asia, Australia | Infest foliage |
| *Macrosiphum euphorbiae* (Ths.) | Potato Aphid | Aphididae | S.E. Asia | Infest foliage |
| *Pseudaonidia trilobitiformis* (Green) | Trilobite Scale | Diaspididae | Pantropical | Encrust foliage |
| *Leptoglossus australis* (F.) | Leaf-footed Plant Bug | Coreidae | S.E. Asia, Kenya | Sap-sucker; toxic saliva |
| *Megymenum brevicorne* F. | Shield Bug | Pentatomidae | Malaysia | Sap-sucker; toxic saliva |
| *Ceratitis catoirii* G.-M. | Fruit Fly | Tephritidae | Mauritius | Larvae in fruits |
| *Dacus dorsalis* Hend. | Oriental Fruit Fly | Tephritidae | S.E. Asia, Hawaii | Larvae in fruits |
| *Dacus cucurbitae* Coq. | Melon Fly | Tephritidae | S.E. Asia, Hawaii | Larvae in fruits |
| *Agraulis vanillae* (L.) | Vanilla Butterfly | Nymphalidae | Hawaii, Colombia | Larvae eat leaves |
| *Porthesia scintillans* Wlk. | Tussock Moth | Lymantriidae | Malaysia | Larvae defoliate |
| *Heliothrips haemorrhoidalis* (Bouché) | Black Tea Thrips | Thripidae | S.E. Asia | Infest foliage |
| *Dihammus vastator* Newm. | Longhorn Beetle | Cerambycidae | Australia | Larvae bore vines |
| *Brevipalpus phoenicis* (Banks) | Red Crevice Mite | Tenuipalpidae | Cosmopolitan | Infest foliage |

# PEA (*Pisum sativum* – Leguminosae)
## (Garden Pea; Field Pea)

The common pea is native to S. Europe and has been cultivated since prehistoric times, and was taken to America by the earliest colonists. Although indigenous to warm climates, it grows well under cool, moist, summer conditions, and thrives in N. Europe, N. America and Canada. The two main groups of varieties are Field Peas, grown for their dried seeds, the plants used for silage, forage and green manuring, and Garden Peas grown as a vegetable, the seeds eaten green, either fresh or frozen, or used in canning, and in some varieties the young pod is cooked whole. The plant body is used as livestock feed. Most peas are now grown as field crops, with a bushy habit, and harvested mechanically. Peas are important field and garden crops in all temperate countries, and in some tropical countries at high altitudes or in the cool season.

## MAJOR PESTS

| | | | | |
|---|---|---|---|---|
| *Acyrthosiphon pisum* (Harris) | Pea Aphid | Aphididae | Cosmopolitan | Infest foliage |
| *Cydia nigricana* (F.) | Pea Moth | Tortricidae | Europe, USA, Canada | Larvae in pods |
| *Maruca testulalis* (Geyer) | Mung Moth | Pyralidae | Pantropical | Larvae bore pods |
| *Etiella zinckenella* (Triet.) | Pea Pod Borer | Pyralidae | Pantropical | Larvae bore pods |
| *Contarinia pisi* (Winn.) | Pea Midge | Cecidomyiidae | Europe | Larvae infest shoots |
| *Delia platura* (Meig.) | Bean Seed Fly | Anthomyiidae | Cosmopolitan | Larvae destroy seed in soil |
| *Apion* spp. | Apion Weevils | Apionidae | Cosmopolitan | Adults infest flowers |

## MINOR PESTS

| | | | | |
|---|---|---|---|---|
| *Aphis craccivora* Koch. | Groundnut Aphid | Aphididae | Cosmopolitan | Infest foliage |
| *Bemisia tabaci* (Genn.) | Tobacco Whitefly | Aleyrodidae | India | Infest foliage |
| *Empoasca* spp. | Green Leaf-hoppers | Cicadellidae | Cosmopolitan | Infest foliage |
| *Creontiades* spp. | Capsid Bugs | Miridae | India | Sap-suckers |
| *Thrips angusticeps* Uzel | Cabbage Thrips | Thripidae | Europe | Infest flowers & shoots |
| *Kakothrips robustus* (Uzel) | Pea Thrips | Thripidae | Europe | Scarifies young pods |
| *Cnephasia* spp. | Leaf-rollers | Tortricidae | Europe | Larvae roll leaves |
| *Adoxophyes* sp. | Leaf-roller | Tortricidae | Papua NG | Larvae roll leaves |
| *Leucinodes orbonalis* Gn. | Eggplant Boring Caterpillar | Pyralidae | India | Larvae bore shoots |
| *Lampides boeticus* (L.) | Pea Blue Butterfly | Lycaenidae | Asia | Larvae bore pods to eat seeds |
| *Euchrysops cnejus* (F.) | Blue Butterfly | Lycaenidae | India | |

| | | | | |
|---|---|---|---|---|
| *Heliothis armigera* (Hub.) | American Bollworm | Noctuidae | Old World | Larvae eat leaves & pods |
| *Spodoptera litura* (F.) | Rice Cutworm | Noctuidae | India | Larvae eat leaves |
| *Spodoptera exigua* (Hbn.) | Lesser Armyworm | Noctuidae | India | Larvae eat leaves |
| *Mythimna separata* (Wlk.) | Rice Ear-cutting Caterpillar | Noctuidae | India | Larvae defoliate |
| *Plusia orichalcea* F. | Pea Semi-looper | Noctuidae | India | Larvae eat leaves |
| *Plusia signata* F. | Semi-looper | Noctuidae | Malaysia | Larvae eat leaves |
| *Diacrisia obliqua* Wlk. | Tiger Moth | Arctiidae | India | Larvae defoliate |
| *Phytomyza horticola* Goureau | Pea Leaf Miner | Agromyzidae | Old World | Larvae mine leaves |
| *Ophiomyia phaseoli* (Tryon) | Bean Fly | Agromyzidae | India | Larvae bore stem |
| *Tipula* spp. | Leatherjackets | Tipulidae | Europe | Larvae eat roots |
| *Anomala* spp. | White Grubs | Scarabaeidae | Philippines | Larvae eat roots |
| *Epicauta* spp. | Black Blister Beetle | Meloidae | Asia, USA, Canada | Adults eat flowers |
| *Bruchus pisorum* L. | Pea Pod Beetle | Bruchidae | Bangladesh, Europe, Canada | Attack ripe pods |
| *Callosobruchus chinensis* L. | Oriental Cowpea Bruchid | Bruchidae | Bangladesh | Attack ripe pods |
| *Chaetocnema concinnipennis* Baly | Flea Beetle | Chrysomelidae | Bangladesh | Adults hole leaves |
| *Alcidodes* spp. | Striped Weevils | Curculionidae | Bangladesh | Adults eat leaves |
| *Sitona* spp. | Pea & Bean Weevils | Curculionidae | Europe | Adults notch leaf margins |
| *Tanymecus indicus* Fst. | Surface Weevil | Curculionidae | India | Adults cut seedling stems |
| *Hylastinus obscurus* (Mar.) | Clover Rot Borer | Scolytidae | USA | Larvae bore roots |
| *Tetranychus* spp. | Red Spider Mites | Tetranychidae | Cosmopolitan | Scarify foliage |

# PEACH (*Prunus persicae* – Rosaceae)

This is a tree native to China, which is grown commercially in temperate and sub-tropical areas. The more important centres of cultivation are S. USA, southern Europe, S. Africa, Australia, and Japan. The tree is small, rather short-lived, and susceptible to frost injury and low temperatures. The fruit has a soft velvety skin, a pitted compressed stone, is used mainly as a table fruit. Because of their delicate perishable nature, the fruits are difficult to transport and store, but they are the most popular fruit for canning, and large quantities are also dried. Nectarine is var. *nectarina* and clearly very closed allied to peach, but with a smaller, smooth fruit.

## MAJOR PESTS

| | | | | |
|---|---|---|---|---|
| *Myzus persicae* (Sulz.) | Peach–Potato Aphid (Green Peach Aphid) | Aphididae | Cosmopolitan | Infest foliage |
| *Quadraspidiotus perniciosus* (Comst.) | San José Scale | Diaspididae | Pantropical | Encrust foliage |
| *Cydia molesta* (Busck.) | Oriental Fruit Moth | Tortricidae | Europe, E. Asia, Australia, N. & S. America | Larvae bore fruit |
| *Cydia pomonella* (L.) | Codling Moth | Tortricidae | Europe, S. Africa, China, Australia, NZ, N. & S. America | Larvae bore fruits |
| *Ceratitis capitata* (Wied.) | Mediterranean Fruit Fly | Tephritidae | Europe, Africa, Asia, C. & S. America | Larvae inside fruit |
| *Ceratitis rosa* Karsch | Natal Fruit Fly | Tephritidae | S. & E. Africa | Larvae inside fruit |
| *Panonychus citri* (McG.) | Citrus Red Spider Mite | Tetranychidae | S. Africa | Scarify foliage |

## MINOR PESTS

| | | | | |
|---|---|---|---|---|
| *Aphis spiraecola* Patch | Apple Aphid | Aphididae | Cosmopolitan | Infest foliage |
| *Appelia schwartzi* Borner | Peach Aphid | Aphididae | Europe | Infest foliage |
| *Brachycaudus persicae* (Pass.) | Black Peach Aphid | Aphididae | Europe | Infest foliage |
| *Hyalopterus pruni* Geoff. | Mealy Plum Aphid | Aphididae | Europe | Infest foliage |
| *Hysteroneura setariae* (Thomas) | Rusty Plum Aphid | Aphididae | Pantropical (esp. USA) | Infest foliage |
| *Pterochloroides persicae* (Chol.) | Peach Aphid | Aphididae | Med. C. Asia | Infest foliage |
| *Parthenolecanium corni* (Bch.) | Plum Scale | Coccidae | Europe, W. Asia, Med. | Infest foliage |
| *Parthenolecanium persicae* (F.) | Peach Scale | Coccidae | Cosmopolitan | Infest foliage |
| *Pseudococcus maritimus* (Ehrh.) | Grape Mealybug | Pseudococcidae | Almost cosmopolitan | Infest foliage |
| *Drosicha mangiferae* Green | Mango Giant Mealybug | Margarodidae | India | Encrust foliage |

| | | | | |
|---|---|---|---|---|
| *Pseudaulacaspis pentagona* (Targ.) | White Peach Scale | Diaspididae | Cosmopolitan | Encrust twigs |
| *Lygocoris pabulinus* (L.) | Common Green Capsid | Miridae | Europe | Sap-sucker; toxic saliva |
| *Lygus oblineatus* (Say) | Tarnished Plant Bug | Miridae | N. America | |
| *Dacus dorsalis* Hend. | Oriental Fruit Fly | Tephritidae | India, S.E. Asia | Larvae inside fruits |
| *Dacus zonatus* Saund. | Peach Fruit Fly | Tephritidae | India | Larvae inside fruits |
| *Ceratitis cosyra* (Wlk.) | Mango Fruit Fly | Tephritidae | Africa, Australia, C. & S. America | Larvae inside fruits |
| *Pardalaspis quinaria* Bez. | Rhodesian Fruit Fly | Tephritidae | S. & N.W. Africa | Larvae inside fruits |
| *Anastrepha fraterculus* (Wied.) | – | Tephritidae | C. & S. America | Larvae inside fruits |
| *Rhagoletis completea* Cress. | Walnut Husk Fly | Tephritidae | USA | Larvae inside fruits |
| *Dichocrocis punctiferalis* (Guen.) | Peach Moth | Pyralidae | Thailand, India | Larvae bore buds & fruits |
| *Paramyelois transitella* (Wlk.) | Navel Orange-worm | Pyralidae | USA (California) | Larvae bore fruits |
| *Spilonota ocellana* (D. & S.) | Eye-spotted Bud Moth | Tortricidae | Europe, Asia, N. America | Larvae bore buds |
| *Anarsia lineatella* Zell. | Peach Twig Borer | Gelechiidae | Europe, Asia, S. China, N. America | Larvae bore twigs |
| *Conotrachelus nenuphar* (Hbst.) | Plum Curculio | Curculionidae | USA | Larvae bore fruits |
| *Tetranychus urticae* (Koch) | Temperate Red Spider Mite | Tetranychidae | Europe | Scarify foliage |
| *Panonychus ulmi* (Koch) | Fruit Tree Red Spider Mite | Tetranychidae | Europe | Scarify foliage |
| *Vasates cornutus* Banks | Silver Mite | Eriophyidae | USA | Scarify foliage |
| *Tarsonemus waitei* Banks | – | Tarsonemidae | USA | Scarify foliage |

# PECAN (*Carya illinoensis* – Juglandaceae)
## (Hickory Nut)

Pecan is a native of the southeastern USA and Mexico. The nuts were originally harvested from wild trees but they have so increased in popularity that the trees are now being extensively cultivated in the southern States, particularly in Texas and Oklahoma, and with the development of new varieties the area of cultivation is spreading farther north-wards. The trees start to bear nuts within four years of setting out, and paper-shelled varieties have now been developed. About half the annual crop is now marketed in the shell. Pecans have a higher fat content than any other vegetable product (over 70%). The nuts are used for dessert, and in icecream, cakes, candy, etc.

*MAJOR PESTS*

*MINOR PESTS*

| | | | | |
|---|---|---|---|---|
| *Melanocallis caryaefoliae* (Davis) | Black Pecan Aphid | Aphididae | USA | Infest foliage |
| *Phylloxera notabilis* Pergande | Pecan Leaf Phylloxera | Phylloxeridae | USA | Infest leaves |
| *Clastoptera achatina* Germ. | Pecan Spittlebug | Cercopidae | USA | Sap-sucker |
| *Gretchena bolliana* (Sling.) | Pecan Bud Borer | Tortricidae | USA | Larvae bore buds |
| *Paramyelois transitella* (Wlk.) | Navel Orangeworm | Pyralidae | USA (California) | Larvae bore young fruits |
| *Cossula magnifica* (Streck.) | Pecan Carpenter-worm | Cossidae | USA | Larvae bore branches |
| *Acrobasis juglandis* (Le Baron) | Pecan Leaf Casebearer | Coleophoridae | USA | Larvae defoliate |
| *Acrobasis caryae* Grote | Pecan Nut Casebearer | Coleophoridae | USA | Larvae defoliate |
| *Coleophora caryaefoliella* Clemens | Pecan Cigar Casebearer | Coleophoridae | USA | Larvae defoliate |
| *Curculio caryae* (Horn) | Pecan Weevil | Curculionidae | USA | Larvae bore kernel |
| *Tetranychus* spp. | Red Spider Mites | Tetranychidae | USA | Scarify foliage |
| *Aceria caryae* Pergande | Pecan Leafroll Mite | Eriophyidae | USA | Distort young leaves |

# PEPPER (*Piper nigrum* – Piperaceae)

Black pepper is the dried unripe fruits (berries) of a weak vine, native to the Indo-Malaysian region, but now widely cultivated in the hot, humid eastern tropics. The plants are supported on posts or on living trees, and yield a crop after 2–3 years, reaching full bearing in seven years. White pepper comes from berries that are almost ripe, and is less pungent in nature.

## MAJOR PESTS

| | | | | |
|---|---|---|---|---|
| *Lepidosaphes piperis* G. | Pepper Scale | Diaspididae | India | Infest foliage |
| *Dasynus piperis* Chin. | Pepper Bug | Coreidae | Indonesia, Malaysia | Sap-sucker; toxic saliva |
| *Longitarsus nigripennis* Motsch. | Pepper Flea Beetle | Chrysomelidae | India | Larvae bore berries |
| *Lophobaris piperis* M. | Pepper Bark Weevil | Curculionidae | Indonesia, Malaysia | Larvae bore bark |

## MINOR PESTS

| | | | | |
|---|---|---|---|---|
| *Aleurocanthus piperis* Mask. | Pepper Whitefly | Aleyrodidae | India | Infest foliage; sap-suckers |
| *Amrasca devastans* (Dist.) | Cotton Leaf-hopper | Cicadellidae | India, S.E. Asia | |
| *Ferrisia virgata* Ckll. | Striped Mealybug | Pseudococcidae | S.E. Asia, India | Encrust foliage |
| *Saissetia coffeae* Wlk. | Helmet Scale | Coccidae | S.E. Asia | Encrust foliage |
| *Aspidiotus destructor* Sign. | Coconut Scale | Diaspididae | India | Encrust foliage |
| *Pinnaspis* spp. | Armoured Scales | Diaspididae | India | Encrust foliage |
| *Disphinctus maesarum* Kirk. | Capsid Bug | Miridae | India, Sri Lanka | Sap-sucker; toxic saliva |
| *Elasmognathus greeni* Kby. | Lace Bug | Tingidae | Indonesia, Sri Lanka | Sap-sucker; toxic saliva |
| *Diplogomphus hewetti* Dist. | Lace Bug | Tingidae | Indonesia | |
| *Gnorimoschema gudmannella* Wals. | Pepper Flower-bud Moth | Gelechiidae | C. America | Larvae bore flower buds |
| *Gynaikothrips karny* Bagn. | Leaf-roller Thrips | Phlaeothripidae | India | Cause leaf galls |
| *Gynaikothrips* spp. | Thrips | Phlaeothripidae | India | Cause leaf galls |
| *Thosea sinensis* Wlk. | Slug Caterpillar | Limacodidae | India, Indonesia, China | Larvae defoliate |
| *Cricula trifenestrata* Helf. | Hairy Mango Caterpillar | Saturniidae | India | Larvae defoliate |
| *Laspeyresia hemidoxa* Meyr. | Pepper Top Shoot Borer | Tortricidae | India | Larvae bore shoots |
| *Cecidomyia malabarensis* Felt | Pepper Gall Midge | Cecidomyiidae | India | Larvae bore berries |
| *Pagria costatipennis* Jacoby | Flea Beetle | Chrysomelidae | India | Damage foliage |
| *Neculla pollinaria* Baly. | Flea Beetle | Chrysomelidae | India | Damage foliage |
| *Eugnathus curvus* Faust. | Pepper Weevil | Curculionidae | India | Damage foliage |

# PIGEON PEA (*Cajanus cajan* – Leguminosae)
## (= Cajan Pea; Red Gram; Dhal; Tur)

This crop originated either in Africa or India, but is now widely cultivated in warmer regions. The plant is an erect shrub, and both immature and ripe seeds are used for human and animal food. It is now being used as a forage crop and rivals alfalfa in importance, being drought-resistant and capable of growing in almost any type of soil. It is one of the most promising legumes at the present time. The main areas of production are India, W. Indies, E. Indies, Africa and S. America.

## MAJOR PESTS

| | | | | |
|---|---|---|---|---|
| *Heliothis armigera* Hbn. | American Bollworm | Noctuidae | India, W. Indies | Larvae hole pods |
| *Melanagromyza obtusa* Mall. | Bean Pod Fly | Agromyzidae | India, S.E. Asia | Larvae mine leaves, bore pods & eat seeds |

## MINOR PESTS

| | | | | |
|---|---|---|---|---|
| *Aphis craccivora* Koch | Groundnut Aphid | Aphididae | India | Infest foliage |
| *Coccus* spp. | Soft Scales | Coccidae | India | Infest foliage |
| *Margarodes* spp. | 'Mealybugs' | Margarodidae | India | Infest foliage |
| *Coptosoma* spp. | Stink Bugs | Pentatomidae | Malaysia, India | Sap-sucker; toxic saliva |
| *Clavigralla* spp. | Gram Pod Bugs | Coreidae | India } | Sap-suckers on pods & seeds |
| *Riptortus* spp. | Coreid Bugs | Coreidae | India } | |
| *Stomopteryx subsecivella* Zell. | Leaf Miner | Gracillariidae | India | Larvae mine leaves |
| *Eucosma* spp. | Leaf Rollers | Tortricidae | India | Larvae roll leaves |
| *Elasmopalpus rubedinellus* Zell. | Pod Borer | Pyralidae | W. Indies | Larvae bore pods |
| *Etiella zinckenella* Treit. | Pea Pod Borer | Pyralidae | India | Larvae bore pods |
| *Ancylostomia stercorea* (Zell.) | – | Pyralidae | W. Indies | Larvae eat pods |
| *Catochrysops* spp. | Blue Butterflies | Lycaenidae | India | Larvae bore pods |
| *Amsacta* spp. | Tiger Moths | Arctiidae | India, Malaysia | Larvae defoliate |
| *Heliothis virescens* (F.) | Tobacco Budworm | Noctuidae | W. Indies | Larvae bore pods |
| *Dasychira mendosa* Hbn. | Tussock Moth | Lymantriidae | Malaysia, India | Larvae defoliate |
| *Megachile* spp. | Leaf Cutter Bees | Megachilidae | India | Adults defoliate |
| *Solenopsis geminata* F. | Fire Ant | Formicidae | India | Attack workers |
| *Mylabris pustulata* Th. | Blister Beetle | Meloidae | India | Adults eat flowers |
| *Callosobruchus chinensis* (L.) | Cowpea Bruchid | Bruchidae | India | Attack ripe seeds |
| *Sphenoptera perotetti* G. | Stem Borer | Buprestidae | India | Larvae bore stems |

| | | | | |
|---|---|---|---|---|
| *Ceutorhynchus asperulus* Fst. | Bud Weevil | Curculionidae | India | Larvae destroy bud |
| *Eucolobes* sp. | Weevil | Curculionidae | Malaysia | Damage foliage |
| *Myllocerus* spp. | Grey Weevils | Curculionidae | India | Adults eat leaves |

# PINEAPPLE (*Ananas cosmosus* – Bromeliaceae)

The country of origin was S. America, but this crop is now grown widely throughout the tropics, and can be grown in heated greenhouses in temperate countries. It is grown most successfully in tropical lowlands, but requires a fertile soil, although it can survive a low rainfall; it can be grown successfully under irrigation. In habit it is a rosette plant with strong spiky leaves, about a metre in height, in appearance quite similar to small *Agave* plants. The fruit is a multiple organ formed from the coalescence of 100 or more individual flowers, with a very high sugar content, and is rich in vitamins A and C. The fruit is orange when ripe, with yellow flesh, and is eaten fresh as a table fruit, canned, or crushed for juice. It can be shipped unripe easily. Propagation is by slips, suckers and fruit crowns. The main production areas are Hawaii, Malaysia, Cuba, Brazil, Australia, S. Africa, and Kenya.

## MAJOR PESTS

| | | | | |
|---|---|---|---|---|
| *Dysmicoccus brevipes* (Ckll.) | Pineapple Mealybug | Pseudococcidae | India, S.E. Asia, Africa, C. & S. America | Infest leaf-bases & roots |
| *Aspidiotus destructor* Sign. | Coconut Scale | Diaspididae | Pantropical | Infest leaves |
| *Thrips tabaci* Lind. | Onion Thrips | Thripidae | India | Scarify leaves; virus vector |
| *Leucopholois irrorata* (Chevr.) | White Grub | Scarabaeidae | Philippines | Larvae eat roots |

## MINOR PESTS

| | | | | |
|---|---|---|---|---|
| *Rhinotermes intermedius* Br. | Termite | Rhinotermitidae | Australia | Damage plants |
| *Acheta bimaculata* Deggar | Two-spotted Cricket | Gryllidae | China | Damage plants |
| *Planococcus citri* Risso | Root Mealybug | Pseudococcidae | Pantropical | Infest roots |
| *Aonidiella aurantii* (Mask.) | California Red Scale | Diaspididae | Philippines | Infest leaves & fruit |
| *Diaspis bromeliae* (Kern.) | Pineapple Scale | Diaspididae | Malaysia | Infest leaves & fruit |
| *Tmoleus echion* L. | Pineapple Caterpillar | Lycaenidae | Hawaii, C. & S. America | Larvae bore fruit |
| *Thecla basilides* (Geyer) | Fruit-boring Caterpillar | Lycaenidae | S. America | Larvae bore fruit |
| *Castnia licas* (Drury) | Giant Moth Borer | Castniidae | Brazil | Larvae bore stem |
| *Hoplothrips anansi* Da C.L. | Pineapple Thrips | Thripidae | S. America | Infest shoots |
| *Hercinothrips femoralis* (Reut.) | Banded Greenhouse Thrips | Thripidae | Cosmopolitan | Infest foliage |
| *Atherigona* spp. | Fruit Maggots | Muscidae | India, Malaysia | Larvae bore fruit |
| *Ahasverus advena* (Waltl) | Foreign Grain Beetle | Cucujidae | Malaysia | Larvae damage fruit |
| *Baris* spp. | – | Curculionidae | S. America | Larvae mine stem |

627

| *Metamasius ritchei* Bs. | Weevil Borer | Curculionidae | W. Indies | Larvae bore stem or fruit |
| *Tarsonemus ananas* Tryon | Pineapple Mite | Tarsonemidae | Australia | Damage shoot & young leaves |

# PISTACHIO (*Pistacia vera* – Anacardiaceae)
## (= Green Almond)

This is a small tree, native to western Asia and cultivated in the Mediterranean region for more than 4000 years. It is now grown in Iran, Afghanistan, and in California and other southern states of the USA. The fruit is a drupe; the seeds contain two large green cotyledons with a reddish covering. The 'nuts' are salted in brine while still in the slightly opened shell; they are highly prized for their flavour and colour, and are sold widely in cans or in mixed nuts, and as a flavouring in icecream and candy.

Other species of *Pistacia* yield a high-grade, and very expensive, resin called mastic. (See also Hammad & Mohamed, 1965.)

## MAJOR PESTS

| | | | | |
|---|---|---|---|---|
| *Recurvaria pistaciicola* Danil | Pistachio Moth | Olethreutidae | Syria | Larvae web leaves & defoliate |
| *Eurytoma* sp. | Gall Wasp | Eurytomidae | Syria | Larvae gall kernel in young fruits |
| *Hylesinus vestitus* Muls-Rey | Shot-hole Borer | Scolytidae | Syria | Adults bore twigs & branches |

## MINOR PESTS

| | | | | |
|---|---|---|---|---|
| *Agonoscena targionii* (Litch.) | Psyllid | Psyllidae | Syria | Nymphs & adults suck sap |
| *Idiocerus stali* Fieb. | Leafhopper | Cicadellidae | Syria | Sap-sucker; infests foliage |
| *Anapulvinaria pistaciae* (Bod.) | Pistachio Brown Scale | Coccidae | Syria | Encrust foliage |
| *Ceroplastes rusci* (L.) | Fig Wax Scale | Coccidae | Med. | Encrust foliage |
| *Salicicola pistaciae* Lind. | Pistachio Scale | Coccidae | Syria | Encrust foliage |
| *Melanaspis inopinatus* Leon. | Armoured Scale | Diaspididae | Syria | Encrust foliage |
| *Lygaeus pandurus* Scop. | Lygaeid Bug | Lygaeidae | India | Sap-sucker; toxic saliva |
| *Retithrips syriacus* Mayet | Black Vine Thrips | Thripidae | Syria, Israel | Infest foliage |
| *Paramyelois transitella* (Wlk.) | Navel Orange-worm | Pyralidae | USA (California) | Larvae bore young fruits |
| *Pachypasa otus* (Dru.) | – | Lasiocampidae | Syria | Larvae defoliate |
| *Capnodis* spp. (6) | Jewel Beetles | Buprestidae | Syria | Larvae bore branches & trunk, sometimes roots |

# PLUM (*Prunus domestica* – Rosaceae)

The plums of commerce come from three main sources; the European plums, native American species, and Japanese species. However, the bulk of fruit comes from the European Plum which is now very widely cultivated throughout the temperate parts of the world. It has been cultivated for over 2000 years, and was taken to America by the colonists. It is a large tree, 10–16 m in height, with variously coloured fruits; over 900 varieties are cultivated, although certain varieties are only used for cooking or for drying as prunes. The fruit is soft and fleshy, the stone smooth and flattened.

## MAJOR PESTS

| | | | | |
|---|---|---|---|---|
| *Brachycaudus helichrysi* (Kalt.) | Plum Aphid | Aphididae | Europe, Australia | Infest foliage |
| *Phorodon humuli* (Schrank) | Damson Aphid | Aphididae | Europe, USA, Canada | Infest foliage |
| *Hyalopterus pruni* (Geoff.) | Mealy Aphid | Aphididae | Europe | Infest foliage |
| *Parthenolecanium corni* (Bch.) | Plum Scale | Coccidae | Europe | Infest foliage |
| *Cydia funebrana* (Treits.) | Red Plum Maggot | Tortricidae | Europe, Asia | Larvae bore fruits |
| *Operophtera brumata* (L.) | Winter Moth | Geometridae | Europe | Larvae defoliate |
| *Caliroa cerasi* (L.) | Pear Slug Sawfly | Tenthredinidae | Cosmopolitan | Larvae skeletonize leaves |
| *Hoplocampa flava* (L.) | Plum Sawfly | Tenthredinidae | Europe | Larvae defoliate |
| *Panonychus ulmi* (Koch) | Fruit Tree Red Spider Mite | Tetranychidae | Cosmopolitan | Scarify leaves |

## MINOR PESTS

| | | | | |
|---|---|---|---|---|
| *Macropsis trimaculata* (Fitch) | Plum Leafhopper | Cicadellidae | USA | Infest foliage |
| *Hysteroneura setariae* (Th.) | Rusty Plum Aphid | Aphididae | Cosmopolitan | Infest foliage |
| *Lygocoris pabulinus* (L.) | Green Capsid | Miridae | Europe | Sap-sucker |
| *Taeniothrips inconsequens* (Uzel) | Pear Thrips | Thripidae | Europe | Infest foliage |
| *Hoplocampa minuta* (Christ) | Plum Sawfly | Tenthredinidae | Europe | Larvae defoliate |
| *Neurotoma inconspicua* (Nort.) | Plum Sawfly | Tenthredinidae | USA | Larvae defoliate |
| *Vespula/Vespa* spp. | Common Wasps | Vespidae | Cosmopolitan | Adults puncture ripe fruits |
| *Spilonota ocellana* (D. & S.) | Bud Moth | Tortricidae | Cosmopolitan | Larvae bore buds |
| *Hedya pruniana* (Hubn.) | Plum Tortrix | Tortricidae | Europe | Larvae bore shoots |
| *Paramyelois transitella* (Wlk.) | Orangeworm | Pyralidae | USA | Larvae bore fruits |
| *Antheraea polyphemus* | Emperor Moth | Saturniidae | USA | Larvae defoliate |
| *Euproctis* spp. | Browntail Moths | Lymantriidae | Cosmopolitan | Larvae defoliate |
| *Anthonomus scutellaris* Le Conte | Plum Gouger | Curculionidae | USA | Larvae bore fruit |
| *Conotrachelus nenuphar* Herbst | Plum Weevil | Curculionidae | USA | Larvae bore fruit |
| *Otiorhynchus cribricollis* Gylh. | Apple Weevil | Curculionidae | Med., W. USA, Australia | Larvae eat roots; adults eat leaves |
| *Scolytes mali* Bht. | Large Fruit Bark Beetle | Scolytidae | Europe | Adults bore under bark |

| *Scolytes rugulosus* (Muller) | Fruit Bark Beetle | Scolytidae | Eurasia, N. & S. America | Adults bore under bark |
| *Eriophyes pyri* (Pgst.) | Blister Mite | Eriophyidae | Cosmopolitan | Blister leaves |
| *Vasates fockeui* (N. & T.) | Plum Mite | Eriophyidae | Canada, USA | Distort leaves |

# POMEGRANATE (*Punica granatum* – Punicaceae)

A native of Iran, this plant has been cultivated in the Mediterranean region for centuries, and was early taken to India, S.E. Asia and China. It is now grown in most parts of the tropics and sub-tropics where the climate is not too humid. The best quality fruits are produced in areas with hot dry summers and cool winters. The plant is a bush or small tree, 2–4 m in height, deciduous in cooler regions, and the fruit is a round brown berry, 5–12 cm in diameter, containing many seeds embedded in a pink juicy pulp. The acid pulp is the edible part of the fruit, and may be eaten as a dessert fruit, as a salad, or in beverages. The roots, rinds, and seeds are used medicinally. Propagation is usually by cuttings. Areas of some commercial production are California, Arizona, and New Mexico. (See also Butani, 1976*a*.)

## MAJOR PESTS

| | | | | |
|---|---|---|---|---|
| *Indarbela* spp. | Bark Moths | Metarbelidae | India | Larvae eat bark |
| *Virachola isocrates* (F.) | Pomegranate Butterfly | Lycaenidae | India | Larvae bore fruits |

## MINOR PESTS

| | | | | |
|---|---|---|---|---|
| *Aleurocanthus woglumi* Ashby | Citrus Blackfly | Aleyrodidae | Pantropical | Infest leaves |
| *Siphoninus finitimus* Silv. | Whitefly | Aleyrodidae | India | Infest leaves |
| *Aphis punicae* | Aphid | Aphididae | India | Infest foliage |
| *Drosicha mangiferae* (Green) | Giant Mealybug | Margarodidae | India | Infest foliage |
| *Ferrisia virgata* (Ckll.) | Striped Mealybug | Pseudococcidae | India | Infest foliage |
| *Planococcus lilacinus* (Ckll.) | Cocoa Mealybug | Pseudococcidae | India | Infest foliage |
| *Parlatoria oleae* (Col.) | Olive Scale | Diaspididae | India | Infest twigs |
| *Aspidiotus rossi* (Mask.) | Pomegranate Scale | Diaspididae | India | Infest twigs |
| *Retithrips syriacus* (Mayet) | Thrips | Thripidae | India | Infest foliage |
| *Zeuzera coffeae* Neitn. | Red Coffee Borer | Cossidae | India | Larvae bore stems |
| *Clania crameri* Westw. | Bagworm | Psychidae | India | Larvae defoliate |
| *Deudorix epijarbas* (Moore) | Pomegranate Borer | Pyralidae | India | Larvae bore fruits |
| *Dichocrocis punctiferalis* (Guen.) | Castor Capsule Borer | Pyralidae | India | Larvae bore fruits |
| *Latoia lepida* (Cram.) | Blue-striped Nettlegrub | Limacodidae | India | Larvae defoliate |
| *Euproctis* spp. | Tussock Moths | Lymantriidae | India | Larvae defoliate |
| *Othreis fullonia* (L.) | Fruit-piercing Moth | Noctuidae | India | Adults pierce ripe fruits |
| *Dacus* spp. | Fruit Flies | Tephritidae | India | Larvae inside fruits |
| *Anomala* spp. | Flower Beetles | Scarabaeidae | India | Adults eat leaves |
| *Olenecamptus bilobus* F. | Stem Borer | Cerambycidae | India | Larvae bore stems |
| *Myllocerus* spp. | Grey Weevils | Curculionidae | India | Adults eat leaves |

# POTATO (*Solanum tuberosum* — Solanaceae)
## (= Irish Potato)

Wild species are found from S. USA to S. Chile, with the centre of diversity in the Andes between 10°N and 20°N at altitudes above 2000 m. They spread slowly, to the Philippines and India in the 17th century, later to Europe, Japan, Java, and E. Africa. It is essentially a temperate crop, and is grown in the tropics only at high altitudes, using shorter-day cultivars. The usual cultivars are long-day forms. Optimum temperatures for tuber development are about 16 °C (not above 27 °C), and the crop can be grown successfully under irrigation. It is a herbaceous branched annual 0.3–1 m in height; the swollen stem tubers contain about 2% protein, 17% starch. It is grown more universally than any other crop.

## MAJOR PESTS

| | | | | |
|---|---|---|---|---|
| *Aulacorthum solani* (Kalt.) | Potato Aphid | Aphididae | Cosmopolitan ⎫ | Infest foliage; suck |
| *Myzus persicae* (Sulz.) | Peach–Potato Aphid | Aphididae | Cosmopolitan ⎭ | sap; virus vectors |
| *Phthorimaea operculella* (Zeller) | Potato Tuber Moth | Gelechiidae | Pantropical | Larvae bore stems & tubers |
| *Epilachna* spp. | Epilachna Beetles | Coccinellidae | Africa, China, India | Adults & larvae eat leaves |
| *Leptinotarsa decemlineata* (Say) | Colorado Beetle | Chrysomelidae | Europe (not UK), N. & C. America | Adults & larvae eat leaves |
| *Agriotes* spp. | Wireworms | Elateridae | Europe, Asia, USA | Larvae in soil eat roots & bore tubers |

## MINOR PESTS

| | | | | |
|---|---|---|---|---|
| *Amrasca devastans* (Dist.) | Leafhopper | Cicadellidae | India, S.E. Asia ⎫ | |
| *Empoasca* spp. | Green (Potato) Leafhoppers | Cicadellidae | Europe, Africa, Asia, Americas | Infest foliage; stunt plants |
| *Cicadella aurata* (L.) | Leafhopper | Cicadellidae | Europe | |
| *Typhlocyba jucunda* Herr.-Schafff. | Leafhopper | Cicadellidae | Europe ⎭ | |
| *Macrosiphum euphorbiae* (Thos.) | Potato Aphid | Aphididae | Cosmopolitan ⎫ | |
| *Rhopalosiphoninus latysiphon* (Davids.) | Bulb & Potato Aphid | Aphididae | Europe, India | Infest foliage; suck sap; virus vectors |
| *Aphis nasturtii* (Borner) | Buckthorn–Potato Aphid | Aphididae | Europe, India ⎭ | |
| *Planococcus citri* (Risso) | Root Mealybug | Pseudococcidae | Pantropical | Infest roots or foliage |
| *Ferrisia virgata* (Ckll.) | Striped Mealybug | Pseudococcidae | Pantropical | Infest foliage |
| *Pseudococcus maritimus* (Ehrh.) | Grape Mealybug | Pseudococcidae | Cosmopolitan | Infest foliage |
| *Lygocoris pabulinus* (L.) | Green Capsid | Miridae | Europe ⎫ | Sap-suckers; |
| *Calocoris norvegicus* (Gmel.) | Potato Capsid | Miridae | Europe, Australia ⎭ | toxic saliva |

| | | | | |
|---|---|---|---|---|
| *Lygus rugulipennis* Popp. | Tarnished Plant Bug | Miridae | Europe | Sap-sucker; toxic saliva |
| *Nezara viridula* (L.) | Green Stink Bug | Pentatomidae | Cosmopolitan | Sap-suckers; toxic saliva |
| *Bagrada* spp. | Harlequin Bugs | Pentatomidae | Africa, Asia | |
| *Hepialus* spp. | Swift Moths | Hepialidae | Europe | Larvae in soil eat roots and bore tubers |
| *Leucinodes orbonalis* Guen. | Eggplant Fruit Borer | Pyralidae | Africa, S. Asia | Larvae eat foliage |
| *Noctua pronuba* (L.) | Large Yellow Underwing | Noctuidae | Europe | Larvae are cutworms; live in soil and eat into tubers |
| *Agrotis ipsilon* (Hfn.) | Black Cutworm | Noctuidae | Cosmopolitan | |
| *Agrotis segetum* (D. & S.) | Common Cutworm | Noctuidae | Cosmopolitan in Old World | |
| *Agrotis exclamationis* (L.) | Heart & Dart Moth | Noctuidae | Europe | |
| *Hydraecia micacea* (Esp.) | Rosy Rustic Moth | Noctuidae | Europe, USA | Larvae bore stems |
| *Heliothis armigera* (Hb.) | American Bollworm | Noctuidae | India | Larvae defoliate |
| *Acherontia atropos* (L.) | Death's Head Hawk Moth | Sphingidae | Europe, Africa, Asia | Larvae defoliate |
| *Tipula* spp. | Leatherjackets | Tipulidae | Europe | Larvae in soil eat roots |
| *Nephrotoma maculata* Meig. | Spotted Cranefly | Tipulidae | Europe, Asia Canada | |
| *Melolontha* spp. | Cockchafers | Scarabaeidae | Europe | Larvae are white grubs or chafers; live in soil and eat into tubers |
| *Phyllopertha horticola* (L.) | Garden Chafer | Scarabaeidae | Europe | |
| *Amphimallon solstitialis* (L.) | Summer Chafer | Scarabaeidae | Europe | |
| *Cetonia aurata* (L.) | Rose Chafer | Scarabaeidae | Europe | |
| *Serica* spp. | Brown Chafers | Scarabaeidae | Europe, Asia | |
| *Aspidomorpha* spp. | Tortoise Beetles | Chrysomelidae | Africa | Adults & larvae eat leaves |
| *Psylliodes affinis* (Payk.) | Potato Flea Beetle | Chrysomelidae | Europe | Adults hole leaves |
| *Epitrix* spp. | Flea Beetles | Chrysomelidae | Canada, USA | |
| *Drasterius* spp. | Wireworms | Elateridae | Asia | Larvae in soil eat roots and bore tubers |
| *Limonius* spp. | Wireworms | Elateridae | USA, Canada | |
| *Hypolithus* spp. | Wireworms | Elateridae | USA, Canada | |
| *Epicauta* spp. | Striped Blister Beetles | Meloidae | Africa, USA, Canada | Adults destroy flowers |
| *Polyphagotarsonemus latus* (Banks) | Yellow Tea Mite | Tarsonemidae | Cosmopolitan | Scarify foliage |
| *Tetranychus* spp. | Red Spider Mites | Tetranychidae | Cosmopolitan | Scarify & web foliage |

# PYRETHRUM (*Chrysanthemum cinerariifolium* – Compositae)

The origin of this plant is the Dalmatian coast of Yugoslavia; it was introduced into Japan in 1881, and to Kenya in 1929. The main production areas are Kenya, Uganda, Tanzania, New Guinea, Ecuador, Brazil, Japan, China, and India. It is a small tufted perennial herb, about 0.5 m in height, which thrives in areas of moderate rainfall (100–160 cm); in the tropics successful cultivation only occurs at high altitudes (e.g. 2000–3000 m in Kenya); chilling is required to initiate flower buds. The flowers are typical of composites, white with yellow centres. The dried flower heads contain pyrethrins (1.0–1.3%) which are very useful insecticides because of their effective 'knockdown' properties; 90% of the pyrethrins are in the ovary and the developing achenes. This crop is to some extent losing its importance with the recent successful development of several synthetic pyrethrins, and pyrethroids.

## MAJOR PESTS

| | | | | |
|---|---|---|---|---|
| *Thrips nigropilosus* Uzel | Pyrethrum Thrips | Thripidae | Europe, Africa, Canada, USA | Infest foliage & scarify |

## MINOR PESTS

| | | | | |
|---|---|---|---|---|
| *Myzus persicae* (Sulz.) | Green Peach Aphid | Aphididae | Cosmopolitan | Infest foliage |
| *Brachycaudus helichrysi* (Kalt.) | Leaf-curling Plum Aphid | Aphididae | Cosmopolitan | Infest foliage |
| *Nysius* sp. | Lygus Bug | Lygaeidae | E. Africa | Sap-sucker; toxic saliva |
| *Thrips tabaci* Lind. | Onion Thrips | Thripidae | Cosmopolitan | Infest and damage flowers |
| *Haplothrips gowdeyi* (Frankl.) | – | Thripidae | Africa, S. America | Scarify foliage |
| *Tetranychus ludeni* (Zacher) | Red Spider Mite | Tetranychidae | Widespread | Scarify foliage |

# QUINCE (*Cydonia oblonga* — Rosaceae)

A small tree from 2–5 m in height, cultivated since ancient times but apparently little changed during cultivation; wild trees may be found in western Asia. It has a bushy habit with many branches. The leaves are woolly as are the young fruit. The flower is an attractive pink colour, and the large oblong fruit has golden-yellow flesh that is hard and rather unpalatable; it is used mainly to make jelly and marmalade, and is sometimes canned. The main areas of commercial production are the southern states of the USA. In Europe and Asia it is grown either for local consumption or as an ornamental shrub. It is not really a tropical species as it is usually grown in warm temperate or sub-tropical areas.

## MAJOR PESTS

| | | | | |
|---|---|---|---|---|
| *Cydia molesta* (Busck) | Oriental Fruit Moth | Tortricidae | China, Japan, N. America | Larvae bore in fruits |
| *Cydia pomonella* (L.) | Codling Moth | Tortricidae | Europe, N. America | Larvae bore in fruits |

## MINOR PESTS

| | | | | |
|---|---|---|---|---|
| *Eriosoma lanigerum* (Hsm.) | Woolly Apple Aphid | Pemphigidae | Cosmopolitan | Encrust twigs |
| *Aleurocanthus woglumi* Ashby | Citrus Butterfly | Aleyrodidae | India | Encrust foliage |
| *Pseudococcus maritimus* (Ehrh.) | Grape Mealybug | Pseudococcidae | Widespread | Infest foliage |
| *Eulecanium corylii* L. | Soft Scale | Coccidae | India | Encrust foliage |
| *Parthenolecanium persicae* (F.) | Peach Scale | Coccidae | Cosmopolitan | Infest foliage |
| *Cacaecia sarcostega* Meyr. | Fruit Tree Tortrix | Tortricidae | India | Larvae damage fruits |
| *Spilonota ocellana* (D. & S.) | Eye-spotted Bud Moth | Tortricidae | India, Europe, Asia, N. America | Larvae bore buds |
| *Zeuzera* spp. | Leopard Moths | Cossidae | India, Europe | Larvae bore branches |
| *Euproctis fraterna* (Moore) | Tussock Moth | Lymantriidae | India | Larvae defoliate |
| *Dacus dorsalis* Hend. | Oriental Fruit Fly | Tephritidae | India | Larvae inside fruits |
| *Aeolesthes sarta* Solsky | Longhorn Beetle | Cerambycidae | India | Larvae bore trunk & branches |
| *Conotrachelus nenuphar* (Herbst.) | Plum Weevil (Curculio) | Curculionidae | USA | Larvae bore in fruits |

# RAMBUTAN (*Nephelium lappaceum* — Sapindaceae)

An evergreen bushy tree, up to 20 m tall, dioecious, native to the lowlands of Malaysia where it is widely cultivated. For reasons not known it is seldom successfully grown away from the indigenous area. The fruits are red-coloured, softly spiky, and borne in clusters; in good cultivars the edible aril is sweet, juicy and delicious, and is highly esteemed locally.

## MAJOR PESTS

| | | | | |
|---|---|---|---|---|
| *Tessaratoma papillosa* (Dru.) | Litchi Stink Bug | Pentatomidae | Malaysia, Philippines | Adults & nymphs suck sap; toxic saliva |
| *Porthesia scintillans* Wlk. | Tussock Moth | Lymantriidae | Malaysia | Larvae defoliate |

## MINOR PESTS

| | | | | |
|---|---|---|---|---|
| *Planococcus lilacinus* (Ckll.) | Mealybug | Pseudococcidae | Philippines | Encrust leaves |
| *Thalassodes depulsata* Wlk. | Looper Cater-pillar | Geometridae | Malaysia | Larvae defoliate |
| *Acrocercops cramerella* Sn. | Cocoa Pod Borer | Gracillariidae | Philippines | Larvae bore fruit |
| *Pteroma plagiophleps* Hmps. | Bagworm | Psychidae | Malaysia | Larvae defoliate |
| *Stauropus alternus* Wlk. | — | Notodontidae | Philippines | Larvae defoliate |
| *Helina propinqua* Stein. | — | Anthomyiidae | Malaysia | Larvae bore fruit |
| *Adoretus compressus* Weber. | Flower Beetle | Scarabaeidae | Malaysia | Adults damage fruit |
| *Niphonoclea* spp. | Twig Borers | Cerambycidae | Philippines | Larvae bore twigs |
| *Hypomeces squamosus* (F.) | Gold-dust Weevil | Curculionidae | Thailand | Adults eat leaves |

# RICE (*Oryza sativa* – Gramineae)

Rice probably originated in China, but spread very early to India, and is now grown extensively throughout Asia, and is rapidly increasing in Africa, S. America, USA, Australia and southern Europe. Approximately 10% of the world rice-growing area is 'upland' rice which is grown dry like an ordinary cereal. 'Lowland' or 'padi' rice is grown in shallow standing water, either impounded rain water or irrigation water. Many varieties are cultivated for different culinary purposes or to meet different requirements of cultivation; long grain and short grain rice have quite different cooking qualities. The most important production areas for export purposes are Thailand, Burma, China, and the USA; in most other countries the crop is largely for home consumption. Several plant breeding projects are in progress for selection of pest and disease resistance, the largest of which is at IRRI, Philippines. (See also Grist & Lever, 1969.)

## MAJOR PESTS

| | | | | |
|---|---|---|---|---|
| *Homorocoryphus nitidulus* Wlk. | Edible Grass-hopper | Tettigoniidae | E. Africa | Defoliate |
| *Oxya chinensis* (Thunb.) | Small Rice Grasshopper | Acrididae | S.E. Asia, India, China | Defoliate |
| *Nephotettix nigropictus* (Stal) | Green Rice Leafhopper | Cicadellidae | India, S.E. Asia, China, Japan | Sap-sucker; virus vector |
| *Nephotettix virescens* (Dist.) | Green Rice Leafhopper | Cicadellidae | India, S.E. Asia, China | Sap-sucker; virus vector |
| *Recilia dorsalis* (Mot.) | Zig-zag Rice Leafhopper | Cicadellidae | India, S.E. Asia, China, Japan | Sap-sucker; virus vector |
| *Laodelphax striatella* (Fall.) | Small Brown Planthopper | Delphacidae | Philippines, Japan, Korea, Europe | Sap-sucker; virus vector |
| *Sogatella furcifera* (Horv.) | White-backed Planthopper | Delphacidae | India, S.E. Asia, Korea, Japan | Sap-sucker; virus vector |
| *Nilaparvata lugens* (Stal) | Brown Rice Planthopper | Delphacidae | S.E. Asia | Sap-sucker; virus vector |
| *Leptoglossus australis* (F.) | Leaf-footed Plant Bug | Coreidae | S.E. Asia, Africa | Sap-sucker; toxic saliva |
| *Leptocorisa acuta* (Thnb.) | Rice Bug | Coreidae | India, S.E. Asia, China, Japan, Australia | Sap-sucker; toxic saliva |
| *Stenocoris southwoodi* Ahmad | African Rice Bug | Coreidae | Africa | Sap-sucker |
| *Scotinophara coarctata* (F.) | Black Rice Bug | Pentatomidae | India, S.E. Asia | Sap-sucker; toxic saliva |
| *Diploxys fallax* Stal | Rice Shield Bug | Pentatomidae | Africa | |
| *Nezara viridula* (L.) | Green Stink Bug | Pentatomidae | Cosmopolitan | Sap-sucker; toxic saliva |
| *Oebalus pugnax* (F.) | Rice Stink Bug | Pentatomidae | USA, Dominican Rep. | |
| *Baliothrips biformis* (Bag.) | Rice Thrips | Thripidae | India, S.E. Asia, Japan | Withered leaves |

| | | | | |
|---|---|---|---|---|
| *Chilo partellus* (Swinhoe) | Spotted Stalk Borer | Pyralidae | India, Thailand, E. Africa | Larvae bore stalks |
| *Chilo polychrysus* (Meyr.) | Dark-headed Rice Borer | Pyralidae | E. Pakistan, India, S.E. Asia | Larvae bore stalks |
| *Chilo suppressalis* (Wlk.) | Striped Rice Borer | Pyralidae | India, S.E. Asia, China, Japan, Australia, Spain | Larvae bore stalks |
| *Maliarpha separatella* Rag. | White Rice Borer | Pyralidae | Burma, China, Africa | Larvae bore stalks |
| *Nymphula depunctalis* (Gn.) | Rice Caseworm | Pyralidae | India, S.E. Asia, Australia | Larvae eat leaves; make leaf cases |
| *Cnaphalocrocis medinalis* (Guen.) | Rice Leaf Folder | Pyralidae | India, S.E. Asia, Korea | Larvae fold leaves |
| *Tryporyza incertulus* (Wlk.) | Yellow Rice Borer | Pyralidae | India, S.E. Asia | Larvae bore stalks |
| *Tryporyza innotata* (Wlk.) | White Rice Borer | Pyralidae | India, S.E. Asia | Larvae bore stalks |
| *Scirpophaga* spp. | Small Rice Borers | Pyralidae | S.E. Asia | Larvae bore stalks |
| *Telicota augias* (L.) | Rice Skipper | Hesperiidae | Java, Philippines, Malaysia | Larvae fold leaves |
| *Mythimna unipuncta* (Haw.) | Rice Armyworm | Noctuidae | Europe, W. & E. Africa, USA, C. & S. America | Larvae defoliate |
| *Mythimna separata* (Wlk.) | Rice Ear-cutting Caterpillar | Noctuidae | India, S.E. Asia, Australasia | Larvae cut stems |
| *Mythimna loreyi* (Dup.) | Rice Armyworm | Noctuidae | S.E. Asia, Australasia | Larvae defoliate |
| *Sesamia calamistis* Hmps. | Pink Stalk Borer | Noctuidae | Africa | Larvae bore stems |
| *Sesamia inferens* (Wlk.) | Pink Borer | Noctuidae | India, S.E. Asia, China, Japan | Larvae bore stalks |
| *Spodoptera exempta* (Wlk.) | African Armyworm | Noctuidae | S.E. Asia, Africa | Larvae defoliate |
| *Spodoptera exigua* (Hb.) | Lesser Armyworm | Noctuidae | S.E. Asia, Africa, Europe, Australia, USA | Larvae defoliate |
| *Spodoptera mauritia* (Boisd.) | Rice Armyworm | Noctuidae | India, S.E. Asia, Africa, USA, Australia | Larvae defoliate |
| *Spodoptera litura* (F.) | Rice Cutworm | Noctuidae | India, S.E. Asia, Australasia | Larvae defoliate |
| *Spodoptera littoralis* (Boisd.) | Cotton Leafworm | Noctuidae | Africa, Near East | Larvae defoliate |
| *Spodoptera pecten* (Guen.) | – | Noctuidae | Malaysia | Larvae defoliate |
| *Orseolia oryzae* (W-M) | Rice Gall Midge | Cecidomyiidae | India, S.E. Asia, Africa | Larvae gall stems |
| *Atherigona oryzae* Mall. | Rice Shoot Fly | Muscidae | India, S.E. Asia, Japan | Larvae destroy shoots |
| *Diopsis* spp. | Stalk-eyed Flies | Diopsidae | Africa | Larvae bore shoots |

639

| *Hydrellia griseola* Fall. | Rice Whorl Maggot | Ephydridae | Europe, N. Africa, S.E. Asia, Japan, USA, S. America | Larvae mine leaves |
|---|---|---|---|---|
| *Dicladispa armigera* (Ol.) | Paddy Hispid | Chrysomelidae | India, S.E. Asia, China | Larvae mine leaves |
| *Trichispa serica* (Gn.) | Rice Hispid | Chrysomelidae | Africa | Larvae mine leaves |
| *Oulema oryzae* Kuw. | Brown Leaf Beetle | Chrysomelidae | China, Japan | Adults strip, larvae mine leaves |
| *Colaspis brunnea* (F.) | Grape Colaspis | Chrysomelidae | S. USA | Larvae eat roots |
| *Sitophilus oryzae* (L.) | Rice Weevil | Curculionidae | S.E. Asia, China | Attack ripe seeds |
| *Lissorhoptrus oryzophilus* Kusch. | Rice Water Weevil | Curculionidae | USA, Japan | Larvae eat roots, adults leaves |
| *Graphognathus* spp. | White-fringed Weevils | Curculionidae | Australia, S. USA, S. America | Larvae eat roots |

## MINOR PESTS

| *Patanga succincta* (L.) | Bombay Locust | Acrididae | India, S.E. Asia | Defoliate |
|---|---|---|---|---|
| *Gastrimargus* spp. | Grasshoppers | Acrididae | S.E. Asia | Defoliate |
| *Oxya* spp. | Small Rice Grasshoppers | Acrididae | India, S.E. Asia, China | Defoliate |
| *Hieroglyphus banian* (F.) | Large Rice Grasshopper | Acrididae | India, Indonesia, Thailand | Defoliate |
| *Zonocerus* spp. | Variegated Grasshoppers | Acrididae | E. & W. Africa | Defoliate |
| *Locusta migratoria* sspp. | Locusts | Acrididae | Africa, Asia | Defoliate |
| *Gryllotalpa africana* Pal. | Mole Cricket | Gryllotalpidae | Cosmopolitan | Damage roots |
| *Balclutha viridis* (Mat.) | Green-splashed Leafhopper | Cicadellidae | Thailand, Laos | Sap-sucker |
| *Nephotettix* spp. | Rice Green Leafhoppers | Cicadellidae | India, S.E. Asia, China | Sap-suckers; virus vectors |
| *Cicadella spectra* (Dist.) | White Jassid | Cicadellidae | Africa, India, S.E. Asia | |
| *Sogatodes* spp. | Rice Plant-hoppers | Delphacidae | S. USA, C. America | |
| *Rhopalosiphum padi* (L.) | Rice Aphid | Aphididae | S.E. Asia | Infest foliage |
| *Rhopalosiphum rufiabdominalis* (Sasaki) | Rice Root Aphid | Aphididae | S.E. Asia | Infest roots |
| *Rhopalosiphum maidis* (Fitch) | Corn Aphid | Aphididae | Cosmopolitan | Infest foliage |
| *Nisia atrovenosa* Leth. | Rice Leaf-hopper | Flattidae | India, China, Japan, Africa, Australia | Sap-sucker |
| *Brevennia rehi* (Ldgr.) | Rice Mealybug | Pseudococcidae | India, Thailand, Java | Infest foliage |
| *Saccharicoccus sacchari* (Ckll.) | Sugarcane Mealybug | Pseudococcidae | Pantropical | Infest leaf bases |

| | | | | |
|---|---|---|---|---|
| *Leptocorisa* spp. | Rice Bugs | Coreidae | S.E. Asia | Sap-suckers; toxic saliva |
| *Blissus gibbus* F. | Lygaeid Bug | Lygaeidae | Laos ⎫ | Sap-sucker; toxic saliva |
| *Oebalus poecilus* | Rice Stink Bug | Pentatomidae | Indonesia ⎭ | |
| *Haplothrips aculentus* (F.) | Thrips | Thripidae | Palaearctic | Infest foliage |
| *Diatraea saccharalis* (F.) | Sugarcane Borer | Pyralidae | N. & S. America | Larvae bore stems |
| *Chilo auricilius* Dudgeon | Gold-fringed Borer | Pyralidae | Thailand | Larvae bore stalks |
| *Chilo infuscatellus* Sn. | — | Pyralidae | India, S.E. Asia | Larvae bore stalks |
| *Chilo* spp. | Stem Borers | Pyralidae | Africa, Asia | Larvae bore stems |
| *Maruca testulalis* (Geyer) | Mung Moth | Pyralidae | Pantropical | Larvae eat panicle |
| *Tryporyza nivella* (F.) | White Tip Borer | Pyralidae | Thailand | Larvae bore top of stems |
| *Marasmia trapezalis* (Gn.) | Maize Webworm | Pyralidae | Pantropical | Larvae web panicle |
| *Eldana saccharina* Wlk. | Sugarcane Stalk Borer | Pyralidae | Africa | Larvae bore stems |
| *Rupela albinella* (Cram.) | South American White Borer | Pyralidae | N. & S. America | Larvae bore stems |
| *Agrotis ipsilon* (Hfn.) | Black Cutworm | Noctuidae | Cosmopolitan | Larvae cutworms |
| *Remigia repanda* (F.) | Guinea Grass Moth | Noctuidae | C. & S. America | Larvae defoliate |
| *Sesamia* spp. | Stem Borers | Noctuidae | Africa, Asia | Larvae bore stems |
| *Spodoptera* spp. | Armyworms | Noctuidae | Pantropical | Larvae defoliate |
| *Sitotroga cerealella* (Ol.) | Angoumois Grain Moth | Gelechiidae | Indo-China | Larvae infest panicle |
| *Parnara guttata* B. & C. | Rice Skipper | Hesperiidae | India, Indonesia, China, Japan | Larvae fold leaves |
| *Pelopidas mathias* (F.) | Rice Skipper | Hesperiidae | India, S.E. Asia, China, Africa | Larvae fold leaves |
| *Ampittia dioscorides* (F.) | Small Rice Skipper | Hesperiidae | India, Malaysia, S. China | Larvae fold leaves |
| *Melanitis leda* (L.) | Brown Butterfly | Nymphalidae | India, Philippines, Thailand | Larvae eat leaves |
| *Hydrellia* spp. | Leaf Maggots | Ephydridae | Europe, Japan, Asia, USA | Larvae mine leaves |
| *Atherigona* spp. | Shoot Flies | Muscidae | India, S.E. Asia, Japan | Larvae destroy apical shoot |
| *Chlorops oryzae* (Mats.) | Stem Maggot | Chloropidae | Japan | Larvae gall shoot |
| *Heteronychus* spp. | Cereal Beetles | Scarabaeidae | Africa, Australia | Adults bite stems; larvae eat roots |
| *Lepidiota* spp. | White Grubs | Scarabaeidae | S.E. Asia | Larvae eat roots |
| *Anomala* spp. | White Grubs | Scarabaeidae | S.E. Asia | Larvae eat roots |
| *Monolepta* sp. | Corn Silk Beetle | Chrysomelidae | Philippines | Adults eat flowers |
| *Echinocnemus oryzae* (Marshall) | Rice Root Weevil | Curculionidae | India | Larvae eat roots under water |

# ROSE APPLE (*Eugenia jambos* – Myrtaceae)
## (= *Syzygium jambos*)

This has long been cultivated in the Indo-Malaysian region for its reddish, rose-scented fruits which may be eaten fresh or made into preserves, and is now grown widely throughout the tropics, including Florida in the USA. The tree is evergreen, with lanceolate leaves characteristic multi-staminate flowers and grows to a height of 10 m.

## MAJOR PESTS

## MINOR PESTS

| | | | | |
|---|---|---|---|---|
| *Megatrioza vitiensis* (Kirk.) | Leaf Gall Psyllid | Psyllidae | India, Malaysia, S. China | Nymphs make leaf-pits |
| *Trioza jambolanae* C. | Leaf Gall Psyllid | Psyllidae | India | Nymphs make leaf-pits |
| *Dialeurodes eugeniae* Mask. | Rose Apple Whitefly | Aleyrodidae | India | Infest foliage |
| *Dialeurodes vulgaris* Singh | Whitefly | Aleyrodidae | India | Infest foliage |
| *Aleurocanthus rugosa* Singh | Blackfly | Aleyrodidae | India | Infest foliage |
| *Chloropulvinaria psidii* Mask. | Guava Scale | Coccidae | India | Infest foliage |
| *Aonidiella orientalis* (Newst.) | Oriental Scale | Diaspididae | India | Encrust twigs |
| *Aspidiotus destructor* Sign. | Coconut Scale | Diaspididae | India | Encrust twigs |
| *Thrips florum* Sch. | Flower Thrips | Thripidae | India | Infest flowers |
| *Teuchothrips eugeniae* S. & A. | Rose Apple Thrips | Thripidae | India | Infest leaves |
| *Mallothrips indicus* R. | Leaf Thrips | Thripidae | India | Infest leaves |
| *Acrocercops* spp. | Leaf Miners | Gracillariidae | India | Larvae mine leaves |
| *Indarbela* spp. | Bark-eating Caterpillars | Metarbelidae | India, S. China | Larvae eat bark & bore branches |
| *Argyroploce aprobola* Meyr. | Leaf Roller | Tortricidae | India | Larvae roll leaves |
| *Argyroploce mormopa* Meyr. | Leaf Miner | Tortricidae | India | Larvae mine leaves |
| *Metanastria hyrtace* C. | Tent Caterpillar | Lasiocampidae | India | Gregarious larvae defoliate |
| *Oenospila flavifusata* Wlk. | Looper Caterpillar | Geometridae | India | Larvae defoliate |
| *Carea subtilis* W. | Leaf Caterpillar | Noctuidae | India | Larvae defoliate |
| *Euproctis fraterna* M. | Tussock Moth | Lymantriidae | India | Larvae defoliate |
| *Dacus* spp. | Fruit Flies | Tephritidae | India | Larvae inside fruits |
| *Holotrichia insularis* Bren. | Cockchafer | Scarabaeidae | India | Adults eat leaves |
| *Balaninus c-album* | Fruit Weevil | Curculionidae | India | Larvae inside fruits |

642

# ROSELLE (*Hibiscus sabdariffa* — Malvaceae)
## (= Jamaican Sorrel; Rama)

This occurs as two distinct botanical varieties. Var. *sabdariffa* is a bushy branched shrub with elongate, pale yellow, edible calyces which are boiled to make a drink, or used in sauces, curries, chutneys, preserves; the leaves may also be used as salad or as a pot herb. Var. *altissima* is a tall, vigorous plant practically without branching 3–5 m tall, grown for fibre in India, Java and the Philippines. The species probably originated in W. Africa, but is now cultivated throughout the tropics, being taken to the New World by the slave trade. The fibres are soft, silky and lustrous, pale brown in colour, and are obtained from the bark of the plant. Roselle grows in any well-drained fertile soil which receives at least 50 cm of rain. The fibre is used as a substitute for jute.

## *MAJOR PESTS*

## *MINOR PESTS*

| | | | | |
|---|---|---|---|---|
| *Maconellicoccus hirsutus* (Green) | Hibiscus Mealybug | Pseudococcidae | India | Infest shoots; suck sap; stunt or kill shoots |
| *Cerococcus hibisci* Green | Hibiscus Scale | Coccidae | India | Infest foliage |
| *Earias insulana* Boisd. | Spiny Bollworm | Noctuidae | India | Larvae bore fruits |
| *Earias vittella* (F.) | Spotted Bollworm | Noctuidae | India | Larvae bore fruits |
| *Porthesia scintillans* Wlk. | Tussock Moth | Lymantriidae | India | Larvae defoliate |
| *Agrilus acutus* Thnb. | Jute Stem Borer | Buprestidae | India | Larvae bore stems |

# RUBBER (*Hevea brasiliensis* — Euphorbiaceae)

As the name suggests trees are found wild in the tropical rain forests of Brazil in the Amazon basin, adjoining Bolivia and Peru. It was introduced into India, Sri Lanka, Java, Singapore, and Malaysia in the late 19th century, and Africa early in the 20th century. The main areas of production now are S. and C. America, Philippines, Malaysia, C. & W. Africa. The most suitable areas are lowland, hot, wet forests in the tropics, between 15°N and 10°S. It is a quick-growing tree of some 25 m in height in plantations, but growing up to 40 m in the wild. The tree has copious latex in all parts, for which it is cultivated. The latex vessels (modified sieve tubes) under the bark of the trunk are cut diagonally and the exudation collected in cups fastened to the trunk; from this 'tapped' latex is produced natural rubber. (See also Rao, 1965.)

## MAJOR PESTS

| | | | | |
|---|---|---|---|---|
| *Melolontha verex* Shp. | White Grub | Scarabaeidae | Malaysia | Larvae eat roots; nursery pest |
| *Tetranychus cinnabarinus* (Boisd.) | Tropical Red Spider Mite | Tetranychidae | Pantropical | Scarify foliage |

## MINOR PESTS

| | | | | |
|---|---|---|---|---|
| *Brachytrupes portentosus* (Licht.) | Large Brown Cricket | Gryllidae | Malaysia | Seedlings attacked |
| *Valanga nigricornis* (Burm.) | Grasshopper | Acrididae | Malaysia | Defoliate |
| *Macrotermes* spp. | Bark-eating Termites | Termitidae | Malaysia | } Bark feeders |
| *Pseudacanthotermes militaris* (Hagen) | Sugarcane Termite | Termitidae | Africa | |
| *Coptotermes curvignathus* Holmgr. | Rubber Tree Termite | Rhinotermitidae | Malaysia, S.E. Asia | } Bore inside live tree trunk |
| *Coptotermes testaceus* (L.) | Rubber Tree Termite | Rhinotermitidae | W. Indies, S. America | |
| *Aleuroplatus malayanus* | Whitefly | Aleyrodidae | Malaysia | Infest foliage |
| *Planococcus citri* Risso | Root Mealybug | Pseudococcidae | Cosmopolitan | Infest foliage |
| *Ferrisia virgata* (Ckll.) | Striped Mealybug | Pseudococcidae | Pantropical | Infest foliage |
| *Saissetia* spp. | Soft Brown Scales | Coccidae | Pantropical | Infest foliage |
| *Coccus viridis* (Green) | Soft Green Scale | Coccidae | Pantropical | Infest foliage |
| *Pulvinaria maxima* | — | Coccidae | Malaysia | Infest foliage |
| *Aspidiotus destructor* Sign. | Coconut Scale | Diaspididae | Pantropical | Encrusts foliage |
| *Lepidosaphes cocculi* | — | Diaspididae | Malaysia | Encrusts foliage |
| *Laccifer greeni* | Lac Insect | Lacciferidae | Malaysia, Sumatra | Infest twigs; protected by ants |
| *Lawana candida* F. | Moth Bug | Flattidae | Malaysia | Infest twigs |
| *Scirtothrips dorsalis* (Hood) | Chilli Thrips | Thripidae | India, Malaysia | Infest leaves |

| *Heliothrips haemorrhoidalis* (Bché.) | Black Tea Thrips | Thripidae | Pantropical | Infest leaves |
|---|---|---|---|---|
| *Acrocercops* spp. | Leaf Miners | Gracillariidae | Malaysia | Larvae mine leaves |
| *Thosea sinensis* (L.) | Slug Caterpillar | Limacodidae | Malaysia | Larvae defoliate |
| *Spodoptera litura* (F.) | Rice Cutworm | Noctuidae | Malaysia | Larvae defoliate |
| *Achaea janata* (L.) | Castor Semi-looper | Noctuidae | S.E. Asia | Larvae defoliate |
| *Tiracola plagiata* (Wlk.) | Banana Fruit Caterpillar | Noctuidae | Malaysia | Larvae defoliate |
| *Adoxophyes privatana* | Shoot Tortrix | Tortricidae | Malaysia | Larvae web shoots |
| *Leucopholis* spp. | White Grubs | Scarabaeidae | Malaysia | } Larvae eat roots; nursery pests |
| *Lepidiota* spp. | White Grubs | Scarabaeidae | Malaysia | |
| *Batocera rufomaculata* Deg. | Red-spotted Longhorn | Cerambycidae | India, Sri Lanka | Larvae bore trunk |
| *Hypomeces squamosus* (F.) | Gold-dust Weevil | Curculionidae | Malaysia | Adults eat leaves |
| *Xyleborus* spp. | Black Twig Borers | Solytidae | Pantropical | Adults bore twigs |
| *Polyphagotarsonemus latus* (Bank) | Yellow Tea Mite | Tarsonemidae | Pantropical | } Scarify foliage |
| *Brevipalpus phoenicis* (Geijskes) | Red Crevice Tea Mite | Tenuipalpidae | Pantropical | |
| *Eutetranychus orientalis* (Klein.) | Oriental Mite | Tetranychidae | Pantropical | |

# SAFFLOWER (*Carthamus tinctorius* — Compositae)

This glabrous, somewhat spiny, herbaceous annual looks rather like a thistle with a yellow or orange-coloured flowerhead, and is well known as one of the great tropical crops. A native of India, it is now widely distributed in most warm countries where the climate is not too wet and humid. It is usually grown as a rain-fed crop, and it shows considerable resistance to drought and wind. The flowers are a source of dye (used chiefly in colouring food), the seeds provide an edible oil, and the leaves are used as a salad vegetable. The dye is called safflower carmin, but is fugitive, and now is mostly replaced by aniline dyes. (The dye should not be confused with saffron, which comes from the anthers of a crocus (*Crocus sativus*) in Asia Minor and India.) Most production of safflower is now just for the seed oil (USA), but in Bengal and S. France it is still grown for the dye.

## MAJOR PESTS

| | | | | |
|---|---|---|---|---|
| *Acanthiophilus helianthi* Rossi | Safflower Fly | Tephritidae | India, S. Europe | Larvae bore flower head |

## MINOR PESTS

| | | | | |
|---|---|---|---|---|
| *Typhlocyba* sp. | Leafhopper | Cicadellidae | India | Infest foliage |
| *Empoasca* sp. | Leafhopper | Cicadellidae | India | Infest foliage |
| *Aphis gossypii* Glov. | Cotton Aphid | Aphididae | India | Infest foliage |
| *Macrosiphum sonchi* L. | Thistle Aphid | Aphididae | India | Infest foliage |
| *Dactynotus carthami* HRL | Safflower Aphid | Aphididae | India | Infest foliage |
| *Bemisia tabaci* (Genn.) | Cotton Whitefly | Aleyrodidae | India | Infest foliage |
| *Ferrisia virgata* (Ckll.) | Striped Mealybug | Pseudococcidae | India | Infest foliage |
| *Monanthia globulifera* W. | Lace Bug | Tingidae | India | Sap-suckers; with toxic saliva |
| *Lygaeus* spp. | Lygaeid Bugs | Lygaeidae | India | |
| *Nezara viridula* L. | Green Stink Bug | Pentatomidae | India | |
| *Thrips* spp. | Flower Thrips | Thripidae | India | Infest flower heads |
| *Microcephalothrips abdominalis* (Crawf.) | Thrips | Thripidae | India | Infest flower heads |
| *Frankliniella sulphurea* Schm. | Flower Thrips | Thripidae | India | Infest flower heads |
| *Eublemma rivula* Moore | Safflower Semi-looper | Noctuidae | India | Larvae defoliate |
| *Heliothis armigera* Hb. | American Bollworm | Noctuidae | India | Larvae bore fruits |
| *Plusia orichalcea* Hb. | Cabbage Semi-looper | Noctuidae | India | Larvae defoliate |
| *Spodoptera exigua* (Hb.) | Lesser Armyworm | Noctuidae | India | Larvae defoliate |
| *Trichoplusia ni* Hb. | Cabbage Semi-looper | Noctuidae | India | Larvae defoliate |
| *Phytomyza atricornis* (Meign.) | Leaf Miner | Agromyzidae | India | Larvae mine leaves |
| *Melanagromyza obtusa* (Mall.) | Bean Pod Fly | Agromyzidae | India | Larvae in flower head |
| *Tanymecus indicus* Fst. | Surface Weevil | Curculionidae | India | Adults eat leaves |

# SANN HEMP (*Crotalaria juncea* – **Leguminosae**)
## (= **Sunn Hemp**)

An important Asiatic fibre plant, second only in importance to jute as a bast fibre in India where it has been cultivated since ancient times. The bast yields a fibre that is stronger than jute, and more durable. It is essentially a cordage fibre and used in the manufacture of twine, cord, sacking, fishing nets etc. The plant is a shrubby annual legume growing up to 3–4 m in height, with bright yellow flowers. It is very extensively grown in India, about 30% of the crop being exported to the UK and USA. This crop is now being grown extensively throughout the tropics as a green manure and cover crop in orchards, and the dried foliage may be used as livestock fodder, although there are reports that the seeds may be toxic.

## MAJOR PESTS

| | | | | |
|---|---|---|---|---|
| *Laspeyresia pseudonectis* M. | Stem Borer | Tortricidae | India | Larvae bore stem & shoots |
| *Uthetheisa pulchella* L. | Sann Hemp Moth | Arctiidae | India | Larvae defoliate |

## MINOR PESTS

| | | | | |
|---|---|---|---|---|
| *Bemisia tabaci* (Genn.) | Cotton Whitefly | Aleyrodidae | India | Infest foliage |
| *Pinnaspis temporaria* Ferris | Armoured Scale | Diaspididae | India | Encrust foliage |
| *Ragmus importunitas* Dist. | Sann Hemp Mirid Bug | Miridae | India | Sap-sucker; toxic saliva |
| *Etiella zinckenella* Treit. | Pea Pod Borer | Pyralidae | India | Larvae bore pods |
| *Lamprosema indicata* F. | – | Pyralidae | India | Larvae eat leaves |
| *Lampides boeticus* L. | Pea Blue Butterfly | Lycaenidae | India | Larvae bore pods |
| *Argina* spp. | Tiger Moths | Arctiidae | India | Larvae defoliate |
| *Diacrisia obliqua* Wlk. | Tiger Moth | Arctiidae | India | Larvae defoliate |
| *Spodoptera litura* (F.) | Rice Cutworm | Noctuidae | India | Larvae defoliate |
| *Plusia eriosoma* D. | Semi-looper | Noctuidae | India | Larvae defoliate |
| *Dasychira mendosa* Hb. | Tussock Moth | Lymantriidae | India | Larvae defoliate |
| *Bruchus pisorum* L. | Pod Bruchid | Bruchidae | India | Attack ripe pods |
| *Longitarsus belagaumensis* F. | Flea Beetle | Chrysomelidae | India | Adults hole leaves |
| *Exora* spp. | Leaf Beetles | Chrysomelidae | C. Africa | Adults eat leaves; larvae eat roots |

# SAPODILLA (*Achras zapota* — Sapotaceae)
## (= Chiku; Chikoo; Sapota)

A large evergreen tropical forest tree, up to 20 m high, native to Mexico and C. America, known for its delicious fruit and for the milky latex which is the source of chicle, used to make chewing gum. It was taken to the Philippines by the Spaniards, and from there spread to Malaysia and India. Now it is widely cultivated throughout the tropics and sub-tropics of the Old World, as well as Florida and C. America. The fruit is large, rough, and brown in colour, measuring some 8–10 cm in diameter.

A close relative is *Palaquium gutta* in Malaysia, the source of gutta-percha. (See also Butani, 1975c.)

## MAJOR PESTS

| | | | | |
|---|---|---|---|---|
| *Chloropulvinaria psidii* (Mask.) | Guava Mealy Scale | Coccidae | India | Infest foliage |
| *Phenacoccus iceryoides* Green | Mealybug | Pseudococcidae | India | Infest foliage |
| *Planococcus lilacinus* (Ckll.) | Mealybug | Pseudococcidae | India | Infest foliage |
| *Dacus* spp. | Fruit Flies | Tephritidae | India | Larvae inside fruits |
| *Nephopteryx eugraphella* Rag. | Chikoo Moth | Pyralidae | India | Larvae web shoots; eat buds & leaves |

## MINOR PESTS

| | | | | |
|---|---|---|---|---|
| *Idioscopus* spp. | Mangohoppers | Cicadellidae | India | Infest foliage |
| *Trialeurodes ricini* Misra | Castor Whitefly | Aleyrodidae | India | Infest foliage |
| *Icerya* spp. | Fluted Scales | Margarodidae | India | Infest foliage |
| *Planococcus citri* (Risso) | Citrus Mealybug | Pseudococcidae | India | Infest foliage |
| *Coccus longulum* (Dougl.) | Soft Green Scale | Coccidae | India | Infest leaves |
| *Saissetia oleae* Ber. | Black Scale | Coccidae | India | Infest foliage |
| *Frankliniella sulphurea* Sch. | Flower Thrips | Thripidae | India | Infest flowers |
| *Indarbela* spp. | Bark-eating Caterpillars | Metarbelidae | India | Larvae eat bark & bore wood |
| *Acrocercops gemoniella* (Stnt.) | Leaf Miner | Gracillariidae | India | Larvae mine leaves |
| *Virachola isocrates* (F.) | Anar Butterfly | Lycaenidae | India | Larvae bore fruits |
| *Rhodoneura* spp. | Leaf Webworms | Thyrididae | India | Larvae web leaves |
| *Acrobasis romonella* | Bud Borer | Pyralidae | India | Larvae bore buds |
| *Metanastria hyrtaca* Cramer | Hairy Caterpillar | Lasiocampidae | India | Larvae defoliate |
| *Phyllophaga consanguinea* (Blanch.) | Cockchafer | Scarabaeidae | India | Adults defoliate |
| *Myllocerus undecimpustulatus* Faust | Grey Weevil | Curculionidae | India | Adults eat leaves |

# SESAME (*Sesamum indicum* – Pedaliaceae)
## (= Simsim; Til; Beniseed; Gingelly)

Sesame is native to Africa, but was taken very early to India. It grows essentially in hot, dry tropical areas of annual rainfall 50–100 cm; it is drought-tolerant. It grows well on poor soils, but prefers sandy-loams. It is sensitive to day-length, and both short- and long-day varieties occur. The plant is a variable erect annual herb, 1–2 m tall, producing capsules containing the small white, red, or black seeds. The seeds contain 45–55% protein; the oil is used for salads and cooking, in soaps, paint, medicines, perfumes, and as a synergist for pyrethrum. The main production areas are India, China, Burma, Sudan, Mexico, Pakistan, Turkey, Venezuela, Uganda, and Nigeria.

## MAJOR PESTS

| | | | | |
|---|---|---|---|---|
| *Antigastra catalaunalis* (Dup.) | Sesame Webworm | Pyralidae | India, S.E. Asia, S. Europe, Africa | Larvae web & bore pods |
| *Asphondylia sesami* Felt | Sesame Gall Midge | Cecidomyiidae | India, E. Africa | Larvae gall pods |

## MINOR PESTS

| | | | | |
|---|---|---|---|---|
| *Myzus persicae* (Sulz.) | Peach Aphid | Aphididae | Cosmopolitan | Infest foliage |
| *Bemisia tabaci* Genn. | Whitefly | Aleyrodidae | India | Infest foliage |
| *Anoplocnemis curvipes* (F.) | Coreid Bug | Coreidae | Africa | Sap-sucker; toxic saliva |
| *Cyrtopeltis tenuis* Reut. | Tomato Mirid | Miridae | India, S.E. Asia | Sap-sucker; toxic saliva |
| *Taylorilygus vosseleri* (Popp.) | Cotton Lygus | Miridae | Africa | |
| *Nezara viridula* (L.) | Green Stink Bug | Pentatomidae | Cosmopolitan | Sap-sucker; toxic saliva |
| *Agonoscelis pubescens* (Thunb.) | Cluster Bugs | Pentatomidae | Africa | |
| *Teleonemia scrupulosa* Stal | Lantana Bug | Tingidae | E. Africa | |
| *Stomopteryx subsecivella* Zell. | Groundnut Leaf Miner | Gracillariidae | Malaysia | Larvae mine leaves |
| *Maruca testulalis* Geyer | Mung Moth | Pyralidae | S.E. Asia, India | Larvae bore pods |
| *Acherontia styx* Wst. | Death's Head Hawk Moth | Sphingidae | Malaysia | Larvae defoliate |
| *Heliothis armigera* (Hb.) | American Bollworm | Noctuidae | Cosmopolitan in Old World | Larvae bore pods |
| *Spodoptera litura* (F.) | Rice Cutworm | Noctuidae | India | Larvae defoliate |
| *Anomis flava* (F.) | Semi-looper | Noctuidae | Asia, Africa | Larvae defoliate |
| *Thrips* spp. | Thrips | Thripidae | E. Africa | Infest flowers |
| *Henosephilachna elateria* (Rossi) | Epilachna Beetle | Coccinellidae | Med. | Defoliate |
| *Ootheca mutabilis* (Sahlb.) | Leaf Beetle | Chrysomelidae | Africa | Defoliate |
| *Tetranychus cinnabarinus* (Boisd.) | Carmine Mite | Tetranychidae | S.E. Asia | Scarify foliage |
| *Polyphagotarsonemus latus* Banks | Yellow Tea Mite | Tarsonemidae | India | Scarify foliage |

# SISAL (*Agava sisalana* – Agavaceae)

A native of Mexico and C. America, Sisal is now cultivated in Hawaii, W. Indies, S.E. Asia and many parts of Africa. It is a short-stemmed plant with rows of stiff, sword-like, sharply pointed leaves arranged in a rosette. The fibres for which the plant is cultivated are obtained from the leaves. The plant is very drought-resistant and requires little cultivation, but the first crop is not harvested for seven years. Other species of *Agava* yield inferior fibres and are not grown much, except as ornamentals. (See also Anon, 1965c.)

## MAJOR PESTS

| | | | | |
|---|---|---|---|---|
| *Dysmicoccus brevipes* (Ckll.) | Pineapple Mealybug | Pseudococcidae | Malaysia, S. America | Infest leaf bases |
| *Scyphophorus interstitialis* Gyll. | Sisal Weevil | Curculionidae | Java, Sumatra, E. Africa, C. America | Larvae bore stem |

## MINOR PESTS

| | | | | |
|---|---|---|---|---|
| *Nastonotus reductus* (Brunn.) | Grasshopper | Tettigoniidae | Venezuela | Damage leaves |
| *Coccus discrepans* Gr. | Soft Scale | Coccidae | Malaysia | Infest leaves |
| *Aonidiella* sp. | Red Scale | Diaspididae | E. Africa | Infest leaves |
| *Aspidiotus* sp. | – | Diaspididae | E. Africa | Infest leaves |
| *Lepidosaphes* sp. | Mussel Scale | Diaspididae | E. Africa | Infest leaves |
| *Oryctes rhinoceros* (L.) | Rhinoceros Beetle | Scarabaeidae | Malaysia | Adults attack shoots |

650

# SORGHUM (*Sorghum bicolor* — Gramineae)
## (= Great Millet; Guinea Corn; Kaffir Corn; Durra; Milo; Jola)

Sorghum originated in Africa but has long been cultivated in Asia, and is now grown widely in Africa, India, China, Australia and the USA. It is a plant that will grow in semi-desert conditions and so can be grown in areas where maize would fail to establish. It is now the fourth most important cereal crop in the world, following wheat, rice and maize.

The plant habit is 1–5 m tall, and like the millets it bears the seed-carrying panicle at the apex of the stem. Red-grained varieties are used to make beer, white-grained ones for flour. It is also used extensively for livestock food. Most is grown for local consumption.

## MAJOR PESTS

| | | | | |
|---|---|---|---|---|
| *Homorocoryphus nitidulus* Wlk. | Edible Grass-hopper | Tettigoniidae | E. Africa | Defoliate & attack panicle |
| *Rhopalosiphum maidis* (Fitch) | Corn Leaf Aphid | Aphididae | Cosmopolitan | Infest foliage |
| *Rhopalosiphum sacchari* (Zehn.) | Sorghum Aphid | Aphididae | Thailand | Infest foliage |
| *Taylorilygus vosseleri* (Popp.) | Cotton Lygus | Miridae | Africa ⎱ | Suck sap from seeds in panicle |
| *Calidea* spp. | Blue Bugs | Pentatomidae | Africa ⎰ | |
| *Chilo orichalcociliella* (Strand) | Coastal Stalk Borer | Pyralidae | Africa | Larvae bore stalks |
| *Chilo partellus* (Swinhoe) | Spotted Stalk Borer | Pyralidae | India, S.E. Asia | Larvae bore stalks |
| *Eldana saccharina* Wlk. | Sugarcane Stalk Borer | Pyralidae | Africa | Larvae bore stalks |
| *Heliothis zea* (Boddie) | Cotton Bollworm | Noctuidae | N., C. & S. America ⎱ | Larvae feed on panicle |
| *Heliothis armigera* (Hub.) | American Bollworm | Noctuidae | Old World ⎰ | |
| *Chloridea obsoleta* F. | – | Noctuidae | Malaysia | Larvae defoliate |
| *Sesamia calamistis* Hmps. | Pink Stalk Borer | Noctuidae | Africa | Larvae bore stalks |
| *Spodoptera exempta* (Wlk.) | African Armyworm | Noctuidae | S.E. Asia | Larvae defoliate |
| *Busseola fusca* (Fuller) | Maize Stalk Borer | Noctuidae | Africa | Larvae bore stalks |
| *Mythimna separata* (Wlk.) | Rice Ear-cutting Caterpillar | Noctuidae | Thailand, Laos, Philippines | Larvae feed on panicle |
| *Contarinia sorghicola* (Coq.) | Sorghum Midge | Cecidomyiidae | S.E. Asia | Larvae destroy seeds |
| *Diopsis* spp. | Stalk-eyed Flies | Diopsidae | Africa | Larvae bore shoots |
| *Atherigona soccata* Rond. | Sorghum Shoot Fly | Muscidae | India, Thailand | Larvae bore shoots |
| *Atherigona excisa* Thomas | Sorghum Shoot Fly | Muscidae | Thailand | Larvae bore shoots |

| | | | | |
|---|---|---|---|---|
| *Epilachna similis* (Thunb.) | Epilachna Beetle | Coccinellidae | Africa | Adults & larvae eat leaves |
| *Schizonycha* spp. | Chafer Grubs | Scarabaeidae | Africa | Larvae eat roots |

## MINOR PESTS

| | | | | |
|---|---|---|---|---|
| *Locusta migratoria* sspp. | Migratory Locusts | Acrididae | Africa, India, Asia | Adults & nymphs eat leaves & attack panicle |
| *Patanga succincta* (L.) | Bombay Locust | Acrididae | India, S.E. Asia | |
| *Colemania sphenarioides* Bol. | Wingless Grass-hopper | Acrididae | India | |
| *Acheta testaceus* Wlk. | Field Cricket | Gryllidae | Philippines | Destroy roots & seedlings |
| *Melanaphis sacchari* (Zehnt.) | Sugarcane Aphid | Aphididae | Pantropical | Infest foliage |
| *Schizaphis graminum* (Rond.) | Wheat Aphid | Aphididae | S.E. Asia, Africa, USA, S. America | Infest foliage |
| *Saccharicoccus sacchari* (Ckll.) | Sugarcane Mealybug | Pseudococcidae | Pantropical | Infest leaf bases |
| *Peregrinus maidis* Ashm. | Maize Plant-hopper | Delphacidae | Pantropical | Infest foliage |
| *Ricania speculum* (Wlk.) | Black Plant-hopper | Ricaniidae | Philippines | Infest stems |
| *Leptocorisa* spp. | Rice Bugs | Miridae | Philippines | Attack grains |
| *Blissus leucopterus* (Say) | Chinch Bug | Miridae | USA | Attack grains |
| *Dysdercus* spp. | Cotton Stainers | Pyrrhocoridae | Africa | Attack grains |
| *Frankliniella williamsi* Hood | Flower Thrips | Thripidae | Thailand | Infest flowers |
| *Cryptophlebia leucotreta* (Meyr.) | False Codling Moth | Tortricidae | Africa | Larvae in panicle |
| *Ephestia* sp. | Webworm | Pyralidae | Thailand | Larvae web panicle |
| *Ostrinia furnacalis* (Gn.) | Asian Corn Borer | Pyralidae | Malaysia | Larvae bore stalks |
| *Ostrinia nubilalis* (Hb.) | European Corn Borer | Pyralidae | Med. | Larvae bore stalks |
| *Diatraea saccharalis* (F.) | Sugarcane Borer | Pyralidae | N. & S. America | Larvae bore stalks |
| *Chilo infuscatellus* Sn. | – | Pyralidae | India, S.E. Asia | Larvae bore stalks |
| *Chilo sacchariphagus* (Boyer) | Sugarcane Stalk Borer | Pyralidae | Malaysia | Larvae bore stalks |
| *Marasmia trapezalis* (Gn.) | Maize Webworm | Pyralidae | Pantropical | Larvae web panicle |
| *Stenachroida elongella* Wlk. | Sorghum Webworm | Pyralidae | Thailand | Larvae web panicle |
| *Chloridea obsoleta* F. | – | Noctuidae | Malaysia | Larvae eat foliage |
| *Celama sorghiella* Riley | Sorghum Webworm | Noctuidae | USA (Texas) | Larvae web panicle |
| *Sesamia nonagrioides* (Lef.) | – | Noctuidae | Africa | Larvae bore stalk |
| *Sesamia inferens* (Wlk.) | Purple Stem Borer | Noctuidae | S.E. Asia | Larvae bore stalks |

| | | | | |
|---|---|---|---|---|
| *Sesamia cretica* Led. | Sorghum Stem Borer | Noctuidae | Europe, Africa | Larvae bore stalks |
| *Spodoptera litura* (F.) | Rice Cutworm | Noctuidae | S.E. Asia ⎞ | Larvae defoliate & attack panicle |
| *Spodoptera littoralis* (Boisd.) | Cotton Leafworm | Noctuidae | Africa ⎠ | |
| *Spodoptera mauritia* (Boisd.) | Paddy Armyworm | Noctuidae | Philippines | Larvae defoliate |
| *Elachiptera* spp. | Shoot Flies | Chloropidae | E. Africa ⎞ | Larvae attack young shoots; dead-hearts |
| *Scoliophthalmus* spp. | Shoot Flies | Chloropidae | Africa ⎠ | |
| *Oscinella* spp. | Frit Flies | Oscinellidae | Africa | Larvae gall stems |
| *Mylabris* spp. | Banded Blister Beetles | Meloidae | India, S.E. Asia, Europe, Africa | Adults eat flowers |
| *Sitophilus oryzae* (L.) | Rice Weevil | Curculionidae | Pantropical | Infest panicle |
| *Tanymecus dilaticollis* | Southern Grey Weevil | Curculionidae | E. Europe | Damage foliage |
| *Oligonychus indicus* Hirst | — | Tetraynchidae | India | Scarify leaves |

# SOYBEAN (*Glycine max* — Leguminoseae)
## (= Soya Bean)

Soybean cultivation originated in the Far East, where it is now the most important legume crop. Much of the world production is for stock feed, but the bean is being increasingly used as a high protein source in human diet. Cultivation in Africa, India, and the Americas is now extensive and increasing; it will grow, as different cultivars, under a wide range of climatic conditions. It is a small bushy, erect, annual which does not produce a tangled growth, with long pendant pods. The seed is the richest natural vegetable food known, and has manifold culinary and agricultural uses. (See also Turnipseed & Kogan, 1976.)

## MAJOR PESTS

| | | | | |
|---|---|---|---|---|
| *Aphis fabae* Scop. | Black Bean Aphid | Aphididae | Widespread | Infest foliage |
| *Empoasca* spp. | Green Leaf-hoppers | Cicadellidae | S.E. Asia, S.E. USA | Infest foliage |
| *Etiella zinckenella* (Treit.) | Pea Pod Borer | Pyralidae | Indonesia, China, Laos, Malaysia | Larvae bore pods |
| *Maruca testulalis* (Geyer) | Mung Moth | Pyralidae | Pantropical | Larvae bore pods |
| *Epicauta* spp. | Blister Beetles | Meloidae | S.E. Asia, USA, Africa, China | Adults eat flowers |
| *Epilachna* spp. | Epilachna Beetles | Coccinellidae | S.E. Asia, Africa, USA, S. America | Adults & larvae defoliate |
| *Callosobruchus* spp. | Cowpea Bruchids | Bruchidae | Pantropical | Infest ripe pods |

## MINOR PESTS

| | | | | |
|---|---|---|---|---|
| *Aphis glycines* Mats. | Soybean Aphid | Aphididae | S.E. Asia | Infest foliage |
| *Pseudococcus* spp. | Mealybugs | Pseudococcidae | Pantropical | Infest foliage |
| *Dysmicocccus brevipes* (Ckll.) | Pineapple Mealybug | Pseudococcidae | Pantropical | Infest foliage |
| *Icerya purchasi* Mask. | Cottony Cushion Scale | Margarodidae | Pantropical | Infest foliage |
| *Riptortus linearis* (L.) | Coreid Bug | Coreidae | Laos | Sap-sucker; toxic saliva |
| *Acanthomia* spp. | Spiny Brown Bugs | Coreidae | Africa | Sap-suckers, toxic saliva |
| *Nezara viridula* (L.) | Green Stink Bug | Pentatomidae | Cosmopolitan | Sap-sucker; toxic saliva |
| *Thrips palmae* | Palm Thrips | Thripidae | China, S.E. Asia | Infest foliage |
| *Sylepta derogata* (F.) | Cotton Leaf-roller | Pyralidae | S.E. Asia | Larvae roll leaves |
| *Elasmopalpus lignosellus* (Zell.) | Lesser Cornstalk Borer | Pyralidae | S. USA, C. & S. America | Larvae bore pods |

| | | | | |
|---|---|---|---|---|
| *Lamprosema diemenalis* (Gn.) | — | Pyralidae | Laos | Larvae roll leaves |
| *Heliothis armigera* (Hb.) | American Bollworm | Noctuidae | Old World tropics | Larvae bore pods |
| *Heliothis zea* (Boddie) | Cotton Bollworm | Noctuidae | USA, C. & S. America | Larvae bore pods |
| *Anticarsia gemmatalis* (Hub.) | Velvetbean Caterpillar | Noctuidae | S. USA | Larvae bore pods |
| *Agrotis segetum* (D. & S.) | Common Cutworm | Noctuidae | Cosmopolitan in Old World | Larvae are cutworms |
| *Agrotis ipsilon* (Hfn.) | Black Cutworm | Noctuidae | Cosmopolitan | Larvae are cutworms |
| *Spodoptera litura* (F.) | Rice Cutworm | Noctuidae | Malaysia, Laos | Larvae defoliate |
| *Spodoptera littoralis* (Boisd.) | Cotton Leafworm | Noctuidae | Africa | Larvae defoliate |
| *Plusia orichalcea* (F.) | — | Noctuidae | India, Ethiopia, Israel | Larvae defoliate |
| *Trichoplusia ni* Hb. | Cabbage Semi-looper | Noctuidae | China | Larvae defoliate |
| *Homona coffearia* (Neitn.) | Tea Tortrix | Tortricidae | Papua NG | Larvae roll leaves |
| *Laspeyresia glycinivorella* Mata. | Soybean Pod Borer | Tortricidae | S.E. Asia | Larvae bore pods |
| *Adoxophyes* sp. | Leaf-roller | Tortricidae | Papua NG | Larvae roll leaves |
| *Cydia ptychora* Meyr. | African Pea Moth | Tortricidae | Africa | Larvae bore pods |
| *Agrius convolvuli* (L.) | Convolvulus Hawk Moth | Sphingidae | China | Larvae defoliate |
| *Porthesia scintillans* Wlk. | Tussock Moth | Lymantriidae | Malaysia | Larvae defoliate |
| *Melanagromyza sojae* (Zehn.) | Bean Fly | Agromyzidae | S.E. Asia | Larvae bore seedling stems & leaf petioles |
| *Ophiomyia phaseoli* (Tryon) | Bean Fly | Agromyzidae | S.E. Asia, Africa, Australasia | Larvae stem-bore seedlings & leaf petioles |
| *Epilachna varivestis* Mulsant | Mexican Bean Beetle | Coccinellidae | USA, Mexico | Defoliate |
| *Popillia japonica* Newm. | Japanese Beetle | Scarabaeidae | China, USA, Canada | Adults defoliate |
| *Mylabris* spp. | Banded Blister Beetles | Meloidae | S.E. Asia | Adults deflower |
| *Plagiodera inclusa* Stal | Leaf Beetle | Chrysomelidae | S.E. Asia | Adults defoliate |
| *Monolepta nigroapicata* | Leaf Beetle | Chrysomelidae | Papua NG | Adults attack foliage |
| *Tetranychus* spp. | Red Spider Mites | Tetranychidae | India, China | Scarify leaves |

# SUGARCANE (*Saccharum officinarum* – Gramineae)

The country of origin is not certain but it is probably somewhere in S.E. Asia; by 327 BC it had become an important crop in India, and was later spread to Egypt and then Spain. It was taken to the New World by Columbus, and is historically important as the original basis of the plantation industry in the tropics and the associated slave trade from Africa to the New World. Now it is grown throughout the tropics and sub-tropics. It needs a high rainfall (or irrigation) with very fertile soil for the best yields. It is a tall grass, up to 5 m height, with a thick bluish stem and short solid internodes. The cane (stem) has a high sucrose content and is cut annually, the leaves being removed before harvest, by cutting, burning or chemical defoliation. Two or three ratoon crops can be taken before replanting from setts (stem cuttings) is required. It produces more food per hectare than any other crop. The main areas of production are Brazil, India, Cuba, Hawaii, Puerto Rico, Barbados, Guyana, Mauritius and E. Africa. (See also Long & Hensley (1972) and Box (1953).)

## MAJOR PESTS

| | | | | |
|---|---|---|---|---|
| *Hieroglyphus banian* (F.) | Large Rice Grasshopper | Acrididae | Thailand, Laos | Defoliate |
| *Pseudacanthotermes militaris* (Hagen) | Sugarcane Termite | Termitidae | E. Africa | Damage stems |
| *Microtermes obesi* Holmgren | Stem-eating Termite | Termitidae | Thailand | Workers damage stems |
| *Ceratovacuna lanigera* Zhnt. | Sugarcane Woolly Aphid | Pemphigidae | China, Taiwan | Sap-suckers |
| *Aulacaspis tegalensis* (Zehn.) | Sugarcane Scale | Diaspididae | S.E. Asia, E. Africa | Encrust stems |
| *Saccharicoccus sacchari* (Ckll.) | Sugarcane Mealybug | Pseudococcidae | Pantropical | Infest stems |
| *Perkinsiella saccharicida* (Ckll.) | Sugarcane Planthopper | Delphacidae | S.E. Asia, S. Africa, Madagascar, Australia | Sap-sucker; virus vector |
| *Pyrilla perpusilla* Wlk. | Indian Sugarcane Planthopper | Lophopidae | India, Thailand | Sap-sucker |
| *Tomaspis* spp. | Sugarcane Spittlebugs | Cercopidae | C. & S. America | Sap-suckers |
| *Aeneolamia* spp. | Sugarcane Spittlebugs | Cercopidae | C. & S. America | Sap-suckers |
| *Chilo orichalcociliella* (Strand) | Coastal Stalk Borer | Pyralidae | Africa | Larvae bore stems |
| *Chilo partellus* (Swinhoe) | Spotted Stalk Borer | Pyralidae | India, S.E. Asia, Africa, Pakistan | Larvae bore stems |
| *Chilo sacchariphagus* (Boyer) | Sugarcane Stalk Borer | Pyralidae | S.E. Asia, China | Larvae bore stems |
| *Eldana saccharina* Wlk. | Sugarcane Stalk Borer | Pyralidae | Africa | Larvae bore stems |
| *Diatraea saccharalis* (F.) | Sugarcane Borer | Pyralidae | N. & S. America | Larvae bore stems |
| *Sesamia inferens* (Wlk.) | Purple Stalk Borer | Noctuidae | India, S.E. Asia, China | Larvae bore stems |

| | | | | |
|---|---|---|---|---|
| *Sesamia calamistis* (F.) | Pink Stalk Borer | Noctuidae | Africa | Larvae bore stems |
| *Leucopholis* spp. | White Grubs | Scarabaeidae | Philippines | Larvae eat roots |
| *Dermolepida albohirtum* Waterh. | Grey-back Cane Beetle | Scarabaeidae | Australia (Queensland) | Larvae eat roots |
| *Cochliotis melolonthoides* (Ger.) | Sugarcane White Grub | Scarabaeidae | E. Africa | Larvae eat roots |
| *Schizonycha* spp. | Chafer Grubs | Scarabeidae | Africa | Larvae eat roots |
| *Dorystenes buqueti* Gn. | Longhorn Beetle | Cerambycidae | Thailand | Larvae bore stems |

## MINOR PESTS

| | | | | |
|---|---|---|---|---|
| *Acheta testaceus* (Wlk.) | Field Cricket | Gryllidae | Philippines | Attack roots |
| *Gryllotalpa africana* Pal. | African Mole Cricket | Gryllotalpidae | S.E. Asia | Attack roots |
| *Macrotermes* spp. | Bark Termites | Termitidae | Africa, India, S.E. Asia | Damage foliage |
| *Heterotermes* spp. | Moist Wood Termite | Rhinotermitidae | Philippines, C. & S. America | Damage roots & stems |
| *Microcerotermes annandalei* Silvestri | — | Termitidae | Thailand | Damage roots & stems |
| *Odontotermes obesus* Ramb. | Scavenging Termite | Termitidae | India, Pakistan | Damage foliage |
| *Odontotermes* spp. | Scavenging Termites | Termitidae | India, S.E. Asia, China, Africa | Damage foliage |
| *Mogannia hebes* Wlk. | Grass Cicada | Cicadidae | Taiwan, Okinawa | Nymphs suck sap from roots |
| *Rhopalosiphum maidis* (Fitch) | Corn Leaf Aphid | Aphididae | Pantropical | Infest foliage |
| *Melanaphis sacchari* Zehnt. | Sugarcane Aphid | Aphididae | Pantropical | Infest foliage |
| *Oregma* sp. | Woolly Aphid | Pemphigidae | Thailand | Infest foliage |
| *Cicadella spectra* (Dist.) | White Jassid | Cicadellidae | Africa, India, S.E. Asia | Infest foliage |
| *Cicadulina mbila* (Naude) | Maize Leaf-hopper | Cicadellidae | Africa | Infest foliage |
| *Matsumuratettrix hiroglyphicus* (Mats.) | Sugarcane Leafhopper | Cicadellidae | Taiwan | Infest foliage |
| *Laodelphax striatella* Fall. | Small Brown Planthopper | Delphacidae | S.E. Asia | Sap-suckers; virus vectors |
| *Saccharosydne* spp. | Sugarcane Planthoppers | Delphacidae | China, W. Indies | |
| *Numicia viridis* Muir | — | Trophiduchidae | S. Africa | Infest foliage |
| *Phenice moesta* Westw. | — | Fulgoridae | Malaysia | Sap-sucker |
| *Aleurolobus barodensis* Mask. | Sugarcane Whitefly | Aleyrodidae | India | Sap-sucker |
| *Brevennia rehi* (Ldgr.) | Rice Mealybug | Pseudococcidae | India, S.E. Asia, S. USA | Infests roots |

| | | | | |
|---|---|---|---|---|
| *Dysmicoccus brevipes* (Ckll.) | Pineapple Mealybug | Pseudococcidae | Egypt, Pacific | |
| *Dysmicocccus boninsis* (Kuw.) | Grey Sugarcane Mealybug | Pseudococcidae | Pantropical | Sap-suckers; in colonies under leaf sheaths |
| *Ferrisia virgata* (Ckll.) | Striped Mealybug | Pseudococcidae | Pantropical | |
| *Pseudococcus adonidum* (L.) | Long-tailed Mealybug | Pseudococcidae | Pantropical | |
| *Planococcus kenyae* (Le Pell.) | Kenya Mealybug | Pseudococcidae | E. Africa | |
| *Aspidiotus destructor* Sign. | Coconut Scale | Diaspididae | Pantropical | Encrust stems |
| *Nisia atrovenosa* (Leth.) | – | Menopliidae | Laos | Sap-sucker |
| *Diostrombus dilatatus* West. | – | Derbidae | E. Africa | Sap-sucker |
| *Hercinothrips femoralis* (Reut.) | Banded Greenhouse Thrips | Thripidae | Cosmopolitan | Infest foliage |
| *Phragmataecia castaneae* Hb. | Moth Borer | Cossidae | Malaysia | Larvae bore stems |
| *Castnia licas* (Drury) | Giant Moth Borer | Castniidae | S. America | Larvae bore stems |
| *Eucosma isogramma* Meyr. | Bud Moth | Eucosmidae | Malaysia | Larvae bore shoots |
| *Marasmia trapezalis* (Gn.) | Maize Webworm | Pyralidae | Pantropical | Larvae defoliate |
| *Tryporyza nivella* (F.) | White Tip Borer | Pyralidae | India, S.E. Asia, China | Larvae bore shoots |
| *Tryporyza* spp. | Stem Borers | Pyralidae | S.E. Asia | Larvae bore stems |
| *Chilo polychrysa* (Meyr.) | Spotted Stalk Borer | Pyralidae | India, S.E. Asia | Larvae bore stems |
| *Chilo infuscatellus* Sn. | Yellow Tip Borer | Pyralidae | India, S.E. Asia | Larvae bore shoots |
| *Sesamia nonagrioides* (Lef.) | – | Noctuidae | Africa | Larvae bore stalks |
| *Spodoptera litura* (F.) | Rice Cutworm | Noctuidae | Malaysia | Larvae defoliate |
| *Spodoptera mauritia* (Boisd.) | Paddy Armyworm | Noctuidae | S.E. Asia, Australasia | Larvae defoliate |
| *Remigia repanda* (F.) | Guinea Grass Moth | Noctuidae | C. & S. America | Larvae defoliate |
| *Mythimna* spp. | Rice Armyworms | Noctuidae | Cosmopolitan | Larvae defoliate |
| *Panara* spp. | Skipper Butterflies | Hesperiidae | Philippines | Larvae roll leaves |
| *Telicota augias* (L.) | Rice Skipper | Hesperiidae | India, S.E. Asia, Australasia | Larvae fold leaves |
| *Heteronychus* spp. | Black Cereal Beetles | Scarabaeidae | Africa | Adults eat young stems, larvae eat roots |
| *Adoretus* spp. | White Grubs | Scarabaeidae | S.E. Asia | |
| *Anomala* spp. | White Grubs | Scarabaeidae | S.E. Asia | Larvae live in soil and eat roots; especially damaging to young plants |
| *Lepidiota stigma* F. | Cane Grub | Scarabaeidae | Thailand, Laos | |
| *Lepidiota discendens* Sharp | Cane Grub | Scarabaeidae | Laos | |
| *Lepidiota* spp. | White Grubs | Scarabaeidae | S.E. Asia, Australasia | |
| *Protaetia fusca* Hbst. | Flower Beetle | Scarabaeidae | Malaysia, Indonesia | Adults feed on flowers |

| | | | | |
|---|---|---|---|---|
| *Oryctes* spp. | Rhinoceros Beetles | Scarabaeidae | Africa, India | Adults damage shoots |
| *Opatrum* spp. | False Wireworms | Tenebrionidae | Philippines | Adults eat leaves; larvae soil pests |
| *Lacon* spp. | Sugarcane Wireworms | Elateridae | Australia | Larvae eat roots |
| *Melanotus tamsuyensis* Bates | Sugarcane Wireworm | Elateridae | Taiwan | Larvae eat roots |
| *Chlorophorus annularis* (F.) | Bamboo Longhorn Beetle | Cerambycidae | S. China | Larvae bore stem |
| *Rhabdoscelis obscurus* Boisd. | Cane Weevil Borer | Curculionidae | Australasia | Larve bore stems |
| *Metamasius hemipterus* L. | West Indian Cane Weevil | Curculionidae | Africa, W. Indies | Larvae bore stems |
| *Sepiomus* sp. | Leaf-eating Weevil | Curculionidae | Thailand | Adults eat leaves |
| *Paratetranychus exsiccator* Zehnt. | Spider Mite | Tetranychidae | Indonesia, Philippines | Scarify foliage |
| *Tarsonemus bancrofti* Mich. | Cane Blister Mite | Tarsonemidae | Indonesia, Philippines, Australia, Hawaii, Florida, S. America | Shoots damaged |

659

# SUNFLOWER (*Helianthus annuus* – Compositae)

Sunflower is not known in the truly wild state, but possibly originated in Utah, Arizona or S. California. It was taken to Europe in 1510 and to Russia in the 18th century, and is now grown in most tropical and temperate countries. It is grown from the Equator to 55°N; and can withstand a slight frost. In the tropics it grows best at medium to high elevations; it is not suited to the wet tropics. It can grow in very dry areas in a variety of soils. It is a variable annual herb 0.7–3.5 m tall. Giant, semi-dwarf, and dwarf varieties are grown. The large flower head, with yellow petals, produces a disc of ovoid achenes, white, black, or striped in colour. The seeds are rich in oil (25–35%) and linoleic acid, and 13–20% protein. The decorticated cake is also rich in protein. Seeds can be eaten raw, roasted, or fed to stock; the oil is used in cooking and for margarine. The main areas of production are USSR, Argentina, Romania, Bulgaria, Hungary, Yugoslavia, Turkey, S. Africa, Uruguay, Tanzania, Kenya, Zimbabwe, and Australia. (See also Rajamohan, 1976.)

## MAJOR PESTS

| | | | | |
|---|---|---|---|---|
| *Calidea* spp. | Blue Bugs | Pentatomidae | Africa | Sap-suckers; toxic saliva |
| *Heliothis armigera* (Hub.) | American Bollworm | Noctuidae | Cosmopolitan in Old World | Larvae feed on seeds |
| *Nezara viridula* (L.) | Green Stink Bug | Pentatomidae | Australia | Suck sap from seeds |
| *Schizonycha* spp. | Chafer Grubs | Scarabaeidae | Africa | Larvae eat roots |

## MINOR PESTS

| | | | | |
|---|---|---|---|---|
| *Aphis gossypii* Glov. | Cotton Aphid | Aphididae | Australia | Infest foliage |
| *Clastoptera xanthocephala* Germ. | Spittlebug | Cercopidae | USA | Sap-suckers |
| *Amrasca terraereginae* (Paoli) | Leafhopper | Cicadellidae | Australia | Infest foliage |
| *Empoasca* spp. | Leafhoppers | Cicadellidae | Australia | Infest foliage |
| *Nysius* spp. | – | Lygaeidae | Australia ⎫ | Sap-suckers; toxic saliva |
| *Agonoscelis pubescens* Thnb. | Cluster Bug | Pentatomidae | Africa ⎭ | |
| *Homoeosoma electellum* (Hulst) | Sunflower Moth | Phycitidae | USA | Larvae eat seeds |
| *Agrius convolvuli* (L.) | Convolvulus Hawk | Sphingidae | Old World | Larvae defoliate |
| *Spodoptera littoralis* (Boisd.) | Cotton Leafworm | Noctuidae | Africa | Larvae defoliate |
| *Spodoptera litura* (F.) | Rice Cutworm | Noctuidae | Asia, Australasia | Larvae defoliate |
| *Heliothis punctigera* Wllgr. | Native Budworm | Noctuidae | Australia | Larvae bore buds |
| *Lasioptera murtfeldtiana* Felt | Sunflower Seed Midge | Cecidomyiidae | USA | Larvae gall seeds |
| *Dacus cucurbitae* Coq. | Melon Fly | Tephritidae | Africa, Asia, Australasia | Larvae in flower head |
| *Strauzia longipennis* (Wied.) | Sunflower Maggot | Tephritidae | USA | Larvae in flower head |

| *Epilachna* spp. | Epilachna Beetles | Coccinellidae | Africa | Defoliate |
| *Zygospila exclamationis* (F) | Sunflower Beetle | Chrysomelidae | USA | Defoliate |
| *Gonocephalum macleayi* Blkb. | Southern False Wireworm | Tenebrionidae | Australia | Larvae eat roots |

# SWEET POTATO (*Ipomoea batatas* — Convolvulaceae)

This is not known in the wild state but is thought to have come from C. or S. America. It is now widely cultivated throughout the tropics from about 40°N to 32°S, from sea-level to about 500 m mostly. Best growth is where the average temperature is 24 °C or over, with a well-distributed rainfall of 75–125 cm per annum. It is a short-day plant, with a photoperiod of 11 hours or less promoting flower formation. The plant is a perennial herb cultivated as an annual vine with trailing stems 1–5 m long. It produces about 10 tubers per plant in the top 20 cm of soil by secondary thickening of the roots. The tubers do not store well so are usually harvested gradually as required. The tubers are an important staple food, and may also be processed for starch, glucose or alcohol. Leaves and vines are used for cattle food. Most cultivation is in Africa, but it is also extensive in China, Japan, USA and New Zealand. (See also Butani & Varma, 1976c.)

## MAJOR PESTS

| | | | | |
|---|---|---|---|---|
| *Bemisia tabaci* (Genn.) | Tobacco Whitefly | Aleyrodidae | Cosmopolitan | Infest foliage |
| *Synanthedon dasysceles* Bradley | Sweet Potato Clearwing | Sesiidae | Africa | Larvae bore vines |
| *Omphisa anastomosalis* Guen. | Sweet Potato Stem Borer | Pyralidae | Malaysia, Laos, India, Indonesia, China | Larvae bore vines |
| *Agrius convolvuli* (L.) | Convolvulus Hawk | Sphingidae | Old World | Larvae defoliate |
| *Acraea acerata* Hew. | Sweet Potato Butterfly | Nymphalidae | E. Africa, Zaïre | Larvae defoliate |
| *Aspidomorpha* spp. | Tortoise Beetles | Chrysomelidae | Africa, Asia | Adults & larvae defoliate |
| *Metriona circumdata* (Herbst.) | Green Tortoise Beetle | Chrysomelidae | S.E. Asia, China, India | Adults & larvae defoliate |
| *Alcidodes dentipes* (Ol.) | Sweet Potato Weevil | Curculionidae | Africa | Adults eat stems; larvae gall stems |
| *Cylas formicarius* (F.) | Sweet Potato Weevil | Apionidae | Pantropical | Adults & larvae bore tubers |
| *Cylas puncticollis* Boh. | Sweet Potato Weevil | Apionidae | Africa | Adults & larvae bore tubers |

## MINOR PESTS

| | | | | |
|---|---|---|---|---|
| *Locusta migratoria* sspp. | Migratory Locusts | Acrididae | Africa, India, S.E. Asia | Adults & nymphs defoliate |
| *Oxya* spp. | Small Rice Grasshoppers | Acrididae | S.E. Asia, China | |
| *Catantops humilis* Serv. | Grasshopper | Acrididae | Malaysia | |
| *Zonocerus variegatus* L. | Variegated Grasshopper | Acrididae | Africa | |

| | | | | |
|---|---|---|---|---|
| *Empoasca* spp. | Green Leaf-hoppers | Cicadellidae | S.E. Asia, China, Africa | Infest foliage |
| *Aphis gossypii* Glov. | Cotton Aphid | Aphididae | Cosmopolitan | Infest foliage |
| *Ferrisia virgata* (Ckll.) | Striped Mealybug | Pseudococcidae | Pantropical | Infest foliage |
| *Planococcus kenyae* (Le Pell.) | Kenya Mealybug | Pseudococcidae | E. & W. Africa | Infest foliage |
| *Geococcus coffeae* Green | Root Mealybug | Pseudococcidae | India | Infest roots |
| *Orthezia insignis* Browne | Jacaranda Bug | Orthezidae | Pantropical | Infest foliage |
| *Helopeltis* spp. | Helopeltis Bugs | Miridae | Africa | ⎫ |
| *Leptoglossus australis* (F.) | Leaf-footed Plant Bug | Coreidae | S.E. Asia | Sap-suckers; toxic saliva |
| *Anoplocnemis* sp. | Coreid Bug | Coreidae | S.E. Asia | |
| *Nezara viridula* (L.) | Green Stink Bug | Pentatomidae | Cosmopolitan | ⎭ |
| *Frankliniella schulzei* (Trybom) | Flower Thrips | Thripidae | Africa | Infest flowers |
| *Synanthedon* spp. | Sweet Potato Clearwings | Sesiidae | Africa | Larvae bore vines |
| *Spodoptera litura* (F.) | Rice Cutworm | Noctuidae | S.E. Asia, China, Australasia | Larvae defoliate |
| *Spodoptera littoralis* (Boisd.) | Cotton Leafworm | Noctuidae | Africa | Larvae defoliate |
| *Mythimna* spp. | Rice Armyworms | Noctuidae | S.E. Asia | Larvae defoliate |
| *Diacrisia* spp. | Tiger Moths | Arctiidae | India, S.E. Asia, Africa | Larvae defoliate |
| *Hyles lineata* (Esp.) | Striped Hawk | Sphingidae | Africa | Larvae defoliate |
| *Hippotion celerio* (L.) | Silver-striped Hawk Moth | Sphingidae | Africa | Larvae defoliate |
| *Adoxophyes* sp. | Leaf Roller | Tortricidae | Papua NG | Larvae roll leaves |
| *Ascotis reciprocaria* (Wlk.) | Coffee Looper | Geometridae | E. & S. Africa | Larvae defoliate |
| *Bedellia* spp. | Leaf Miners | Lyonetidae | Africa, Hawaii | Larvae mine leaves |
| *Aciptilia* sp. | Vine Borer | Pterophoridae | S. China | Larvae bore vines |
| *Ochyrotica* sp. | Leaf Roller | Pterophoridae | S. China | Larvae roll leaves |
| *Parasa vivida* (Wlk.) | Nettlegrub | Limacodidae | Africa | Larvae defoliate |
| *Euproctis* spp. | Tussock Moths | Lymantriidae | Africa | Larvae defoliate |
| *Epilachna* spp. | Epilachna Beetles | Coccinellidae | S.E. Africa | Adults & larvae defoliate |
| *Anomala* spp. | White Grubs | Scarabaeidae | S.E. Asia | ⎫ |
| *Apogonia cribricollis* Burm. | Chafer Beetle | Scarabaeidae | Malaysia | Larvae attack roots & tubers; adults may eat leaves |
| *Leucopholis* spp. | White Grubs | Scarabaeidae | Philippines | |
| *Lachnosterna* sp. | June Beetle | Scarabaeidae | USA, C. & S. America | ⎭ |
| *Mylabris* spp. | Blister Beetles | Meloidae | S.E. Asia | Adults deflower |
| *Lacoptera chinensis* (F.) | Tortoise Beetle | Chrysomelidae | S. China | Adults & larvae defoliate |
| *Diabrotica balteata* Lec. | Cucumber Beetle | Chrysomelidae | Pacific, C. & S. America | Larvae eat roots |
| *Heteroderes laurenti* Guer. | Gulf Wireworm | Elateridae | S. USA, C. & S. America | Larvae damage roots & defoliate |

663

| | | | | |
|---|---|---|---|---|
| *Typophorus viridicyanus* (Crotch) | Leaf Beetle | Chrysomelidae | S. USA | Adults & larvae eat leaves |
| *Euscepes postfasciatus* Fairm. | West Indian Sweet Potato Weevil | Curculionidae | Pacific, W. Indies, S. America | Larvae bore tubers |
| *Blosyrus ipomocae* Mahl. | Sweet Potato Leaf Weevil | Curculionidae | China, India, Africa | Adults eat leaves |
| *Tetranychus cinnabarinus* (Boisd.) | Carmine Mite | Tetranychidae | Pantropical | Scarify leaves |

# TAMARIND (*Tamarindus indica* — Caesalpiniaceae)

This large leguminous tree, growing to a height of 25 m, probably originated in tropical Africa or southern Asia in semi-arid regions; it is now widely grown throughout the drier tropics, both for the pods and as a shade or ornamental tree. The curved oblong pod, about 8 × 2 cm, is characteristically constricted and contains 2–5 brown obovate seeds (sometimes more). The pulp around the seeds is tart and brown in colour, and may be eaten fresh, mixed with sugar to make a sweetmeat, used in seasoning as well as curries, sauces, preserves, and chutneys, and also makes a refreshing acid drink. The seeds may be eaten after the removal of the test and boiling or roasting. In India the seeds may be used to make flour, and as a source of carbohydrate for sizing cloth and for vegetable gum. Most cultivation takes place in India, W. Indies and Florida. (See also Butani, 1978b.)

## MAJOR PESTS

| | | | | |
|---|---|---|---|---|
| *Drosichiella tamarindus* (Gr.) | Tamarind Mealybug | Margarodidae | India | Encrust foliage |
| *Planococcus lilacinus* (Ckll.) | Mealybug | Pseudococcidae | India | Encrust foliage |
| *Saissetia oleae* Ber. | Black Scale | Coccidae | India | Encrust foliage |
| *Aonidiella orientalis* (New.) | Oriental Yellow Scale | Diaspididae | India | Encrust foliage |
| *Aspidiotus destructor* Sign. | Coconut Scale | Diaspididae | India | Encrust foliage |

## MINOR PESTS

| | | | | |
|---|---|---|---|---|
| *Toxoptera aurantii* (Fon.) | Black Citrus Aphid | Aphididae | India | Infest foliage |
| *Nipaecoccus* spp. | Mealybugs | Pseudococcidae | India | Encrust foliage |
| *Aspidiotus tamarindi* Gr. | Tamarind Scale | Diaspididae | India | Encrust foliage |
| *Pinnaspis* spp. | Armoured Scales | Diaspididae | India | Encrust foliage |
| *Kerria lacca* (Kerr) | Lac Insect | Lacciferidae | India | Encrust twigs |
| *Drosicha stebbingi* (Gr.) | Giant Mealybug | Margarodidae | India | Encrust foliage |
| *Scirtothrips dorsalis* Hood | Chilli Thrips | Thripidae | India | Infest buds & flowers |
| *Laspeyresia palamedes* Meyr. | Flower Webber | Tortricidae | India | Larvae bore buds; web flowers & buds |
| *Argyroploce illepida* Bult. | Fruit Borer | Tortricidae | India | Larvae bore pods |
| *Virachola isocrates* (F.) | Anar Butterfly | Lycaenidae | India | Larvae bore pods |
| *Dichocrocis punctiferalis* (Guen.) | Castor Capsule Borer | Pyralidae | India | Larvae bore pods |
| *Assara albicostalis* Wlk. | — | Pyralidae | India | Larvae bore pods |
| *Etiella zinckenella* Treit. | Pea Pod Borer | Pyralidae | India | Larvae bore pods |
| *Eublemma angulifera* Moore | Flower Webber | Noctuidae | India | Larvae damage flowers |
| *Alphitobius laevigiatus* (F.) | — | Tenebrionidae | India | Larvae bore dry pods |
| *Ulomo* sp. | — | Tenebrionidae | India | Larvae inside pods |

# TARO (*Colocasia esculenta* – Araceae)
## (= Cocoyam; Dasheen)

*C. esculenta* occurs wild in S.E. Asia, and in early times was taken to China and Japan; it was taken to the Mediterranean region in biblical times and then spread to W. Africa. It is an important staple crop throughout the Pacific region, and in the W. Indies and W. Africa. The corms are roasted, baked, or boiled, and have a high content of tiny, easily digestible, starch grains; the young leaves are sometimes eaten as a vegetable. It occurs as two fairly distinct varieties and a large number of clones, and is probably best regarded as a polymorphic single species. The plant is a herb, 1–2 m tall, with an underground starchy corm; the leaves are large and spade-shaped, with the margin entire, and a long fleshy petiole. Many cultivars do not flower, and propagation is generally vegetative; in the wild pollination is probably effected by flies.

## MAJOR PESTS

| | | | | |
|---|---|---|---|---|
| *Tarophagus proserpina* (Kir.) | Taro Leafhopper | Cicadellidae | S.E. Asia, Pacific Isl. | Infest leaves |
| *Spodoptera litura* (F.) | Rice Cutworm | Noctuidae | India, S.E. Asia | Larvae eat leaves |
| *Papuana laevipennis* Arrow | Taro Beetle | Scarabaeidae | Papua NG, Solomon Isl. | Larvae bore tubers |
| *Monolepta signata* Oliv. | White-spotted Flea Beetle | Chrysomelidae | India | Adults hole leaves |

## MINOR PESTS

| | | | | |
|---|---|---|---|---|
| *Gesonia* spp. | Aquatic Grass-hoppers | Acrididae | Papua NG, India | Defoliate |
| *Cicadella* sp. | Leafhopper | Cicadellidae | Papua NG | Infest foliage |
| *Aphis gossypii* Glov. | Cotton Aphid | Aphididae | S.E. Asia, India, | Infest foliage |
| *Pentalonia nigronervosa* Coq. | Banana Aphid | Aphididae | S.E. Asia, India | Infest foliage |
| *Dysmicoccus brevipes* (Ckll.) | Pineapple Mealybug | Pseudococcidae | Philippines | Infest foliage |
| *Aspidiotus destructor* Sign. | Coconut Scale | Diaspididae | Philippines | Infest foliage |
| *Stephanitis typicus* Dist. | Banana Lace Bug | Tingidae | India | Sap-sucker; toxic saliva |
| *Heliothrips haemorrhoidalis* (Bché.) | Black Tea Thrips | Thripidae | India | Infest foliage |
| *Caliothrips indicus* (Bagn.) | Leaf Thrips | Thripidae | India | Infest foliage |
| *Agrius convolvuli* (L.) | Sweet Potato Hawk Moth | Sphingidae | Philippines | Larvae defoliate |
| *Hyloicus pinastri* (L.) | Pine Hawk Moth | Sphingidae | Papua NG | Larvae defoliate |
| *Hippotion celerio* (L.) | Silver-striped Hawk | Sphingidae | S.E. Asia, India, Africa | Larvae defoliate |
| *Papuana huebneri* | Taro Beetle | Scarabaeidae | Papua NG | Larvae bore tubers |

666

# TEA (*Thea sinensis* — Theaceae)
## (= *Camellia sinensis*)

Tea originated near the source of the river Irriwaddy and spread to S.E. China, Indo-China and Assam where wild teas can still be found. The main centres of early cultivation were in S.E. Asia and China, but now the crop is grown in many parts, mainly in the sub-tropics and in the mountain regions of the tropics (e.g. at 1–2000 m at the Equator). It needs equable temperatures, moderate to high rainfall and high humidity all the year round, and cannot tolerate frost. It is a small evergreen tree which can grow to 15 m high, but is pruned to a bush of 0.5–1.5 m. The leaves and buds are picked and dried and treated in various ways according to which type of tea is being produced. The leaves contain caffeine, polyphenols and essential oils. The main production areas are India, Sri Lanka, China, Indonesia, Taiwan, Japan, Kenya, Malawi, Uganda, Tanzania, Mozambique, USSR and Argentina. (See also Cranham, 1966*b*.)

## MAJOR PESTS

| | | | | |
|---|---|---|---|---|
| *Brachytrupes membranaceus* (Drury) | Tobacco Cricket | Gryllidae | Africa | Adults & nymphs destroy seedlings; nursery pests mostly |
| *Brachytrupes portentosus* Licht. | Large Brown Cricket | Gryllidae | India, China | |
| *Gryllotalpa africana* Pal. | African Mole Cricket | Gryllotalpidae | S.E. Asia, Africa, Australasia | |
| *Helopeltis schoutedeni* Reuter | Cotton Helopeltis | Miridae | Africa | Sap-sucker; toxic saliva |
| *Heliothrips haemorrhoidalis* (Bouché) | Black Tea Thrips | Thripidae | Cosmopolitan | Scarify leaves |
| *Homona coffearia* Nietn. | Tea Tortrix | Tortricidae | India, Sri Lanka, Indonesia | Larvae roll leaves |
| *Aperitmetus brunneus* (Hust.) | Tea Root Weevil | Curculionidae | Kenya, Somalia | Larvae bore roots |
| *Xyleborus fornicatus* Eichh. | Tea Shot-hole Borer | Scolytidae | India, Sri Lanka, S.E. Asia | Adults bore stems |
| *Brevipalpus phoenicis* (Geijskes) | Red Crevice Tea Mite | Tenuipalpidae | Pantropical | Distort leaves |
| *Oligonychus coffeae* (Neitn.) | Red Coffee Mite | Tetranychidae | India, S.E. Asia, Australasia, C. & S. America | Scarify leaves |
| *Polyphagotarsonemus latus* (Banks) | Yellow Tea Mite | Tarsonemidae | India, S.E. Asia, Europe, USA, C. America | Distort leaves |

## MINOR PESTS

| | | | | |
|---|---|---|---|---|
| *Odontotermes* spp. | Scavenging Termites | Termitidae | India, Malaysia, China, Africa | Eat bark and leaves |
| *Microcerotermes* spp. | Live Wood-eating Termites | Termitidae | India, Malaysia, Java | Bore stems & eat live wood |

667

| | | | | |
|---|---|---|---|---|
| *Kalotermes* spp. | Dry-wood Termites | Kalotermitidae | India, Sri Lanka | Live in dead wood; sometimes pests |
| *Chlorita onukii* | Tea Green Leafhopper | Cicadellidae | Japan | Infest foliage |
| *Toxoptera aurantii* (B. de F.) | Black Citrus Aphid | Aphididae | Cosmopolitan | Infest foliage |
| *Ceroplastes rubens* Mask. | Pink Waxy Scale | Coccidae | India, S.E. Asia, China | Infest twigs and leaves |
| *Coccus viridis* Green | Soft Green Scale | Coccidae | Pantropical | Infest twigs & leaves |
| *Coccus hesperidum* L. | Soft Brown Scale | Coccidae | Cosmopolitan | Infest twigs & leaves |
| *Saissetia coffeae* (Wlk.) | Helmet Scale | Coccidae | Cosmopolitan ⎫ | Infest twigs & leaves |
| *Pseudococcus maritimus* (Ehrh.) | Grape Mealybug | Pseudococcidae | Widespread ⎬ | |
| *Aspidiotus destructor* Sign. | Coconut Scale | Diaspididae | Pantropical | Encrust foliage |
| *Selanaspidus* spp. | Armoured Scales | Diaspididae | E. Africa | Encrust foliage |
| *Fiorinia theae* Green | Tea Scale | Diaspididae | India | Encrust foliage |
| *Helopeltis antonii* Sign. | Tea Mosquito Bug | Miridae | India, S.E. Asia | |
| *Helopeltis theivora* Waterh. | Tea Capsid Bug | Miridae | India, S.E. Asia | Sap-suckers; toxic saliva |
| *Helopeltis bergrothi* Reut. | Cocoa Mosquito Bug | Miridae | Africa | |
| *Poecilocoris latus* Dall. | Tea Seed Bug | Pentatomidae | India, Indo-China | |
| *Zeuzera coffeae* Neitn. | Red Coffee Borer | Cossidae | Sri Lanka, Malaysia | Larvae bore stems |
| *Gracillaria theivora* (Wlsm.) | Tea Leaf Roller | Gracillariidae | India | Larvae roll leaves |
| *Adoxophyes orana* (F.v.S.) | Smaller Tea Tortrix | Tortricidae | Japan | Larvae defoliate |
| *Cryptophlebia leucotreta* (Meyr.) | False Codling Moth | Tortricidae | Africa | Larvae roll leaves |
| *Cydia leucostoma* (Meyr.) | Tea Flushworm | Tortricidae | India, Sumatra, Java, Taiwan | Larvae defoliate |
| *Tortrix dinota* Meyr. | Brown Tortrix | Tortricidae | Malawi | Larvae defoliate |
| *Clania cramerii* Westw. | Bagworm | Psychidae | India, China | Larvae defoliate |
| *Caloptilia theivora* | Tea Leaf Roller | Pyralidae | Japan | Larvae roll leaves |
| *Andraca bipunctata* Wlk. | Bunch Cater-pillar | Bombycidae | India | Larvae defoliate |
| *Eterusia magnifica* Butl. | Red Slug Cater-pillar | Zygaenidae | India, S.E. Asia | Larvae defoliate |
| *Euproctis* spp. | Tussock Moths | Lymantriidae | India, Japan, China | Larvae defoliate |
| *Agrotis segetum* (D. & S.) | Common Cutworm | Noctuidae | Old World | Nursery pest |
| *Tiracola plagiata* (Wlk.) | Banana Fruit Caterpillar | Noctuidae | S.E. Asia | Larvae defoliate |
| *Biston suppressaria* Guen. | Common Looper | Geometridae | India | Larvae defoliate |

| | | | | |
|---|---|---|---|---|
| *Niphadolepis alianta* Karsch. | Jelly Grub | Limacodidae | Malawi | Larvae defoliate |
| *Setora nitens* (Wlk.) | Stinging Caterpillar | Limacodidae | S.E. Asia | Larvae defoliate |
| *Parasa vivida* (Wlk.) | Stinging Caterpillar | Limacodidae | E. & W. Africa | Larvae defoliate |
| *Parasa lepida* Cram. | Blue-striped Nettlegrub | Limacodidae | India, S.E. Asia | Larvae defoliate |
| *Thosea* spp. | Stinging Caterpillars | Limacodidae | India | Larvae defoliate |
| *Attacus atlas* L. | Atlas Moth | Saturniidae | India, Indonesia, S. China | Larvae defoliate |
| *Tropicomyia theae* (Cotes) | Tea Leaf Miner | Agromyzidae | India, Java, China | Larvae mine leaves |
| *Gonocephalum simplex* (F.) | Dusty Brown Beetle | Tenebrionidae | Africa | Adults damage stems |
| *Holotrichia seticollis* Moser | Chafer Grub | Scarabaeidae | India | Larvae eat roots; nursery pest |
| *Brevipalpus californicus* (Banks) | – | Tenuipalpidae | Cosmopolitan | Scarify leaves |
| *Eriophyes theae* Watt. | Tea Blister Mite | Eriophyidae | India, Indonesia | Form blisters (erinia) on leaves |
| *Calacarus carinatus* (Green) | Purple Mite | Eriophyidae | India, Japan, S.E. Asia | Distort leaves |

# TOBACCO (*Nicotiana tabacum* – Solanaceae)

Probably Tobacco originated in N.W. Argentina, but it was cultivated in pre-Columbian times in the W. Indies, Mexico, C. and S. America. By the 17th century it had been spread to India, Africa, Japan, Philippines and the Middle East. It is now very widely cultivated throughout the warmer parts of the world, from central Sweden in the north down to southern Australia. The crop needs 90–120 frost-free days from transplanting to harvest. The optimum mean temperature for growth is 20–26 °C; strong illumination is needed; it can grow in as little as 25 cm of rain, but prefers 50 cm. Dry weather is essential for ripening and harvest. The plant is a perennial herb 1–3 m tall, usually grown as an annual for its leaves which are cured to make tobacco and snuff. The main production areas are USA, Brazil, Japan, Canada, Pakistan, India, Greece, Turkey and Zimbabwe.

## MAJOR PESTS

| | | | | |
|---|---|---|---|---|
| *Brachytrupes membranaceus* Drury | Tobacco Cricket | Gryllidae | Africa | Destroy seedlings |
| *Gryllotalpa* spp. | Mole Crickets | Gryllotalpidae | Cosmopolitan | Eat roots |
| *Bemisia tabaci* (Genn.) | Whitefly | Aleyrodidae | Cosmopolitan | Infest leaves |
| *Thrips tabaci* Lind. | Onion Thrips | Thripidae | Cosmopolitan | Scarify leaves |
| *Heliothis armigera* (Hbn.) | American Bollworm | Noctuidae | Cosmopolitan in Old World | Larvae bore buds, & eat leaves |
| *Heliothis zea* (Boddie) | Cotton Bollworm | Noctuidae | N., C. & S. America | Larvae eat leaves |
| *Agrotis ipsilon* (Hfn.) | Black Cutworm | Noctuidae | Cosmopolitan | Larvae cutworms |
| *Manduca sexta* (L.) | Tobacco Hornworm | Sphingidae | USA, C. & S. America | Larvae defoliate |
| *Lasioderma serricorne* (F.) | Tobacco Beetle | Anobiidae | Pantropical | Attack dried leaves |

## MINOR PESTS

| | | | | |
|---|---|---|---|---|
| *Atractomorpha crenula* F. | Grasshopper | Acrididae | Malaysia | Defoliate |
| *Myzus persicae* (Sulz.) | Peach Aphid | Aphididae | Cosmopolitan | |
| *Myzus ascalonicus* Don. | Shallot Aphid | Aphididae | Europe, N. America | Infest foliage; suck sap; virus vectors |
| *Aphis gossypii* Glov. | Cotton Aphid | Aphididae | S.E. Asia | |
| *Aulacorthum solani* (Malt.) | Potato Aphid | Aphididae | Cosmopolitan | |
| *Rhopalosiphum maidis* (Fitch) | Corn Leaf Aphid | Aphididae | Cosmopolitan | |
| *Empoasca* spp. | Leafhoppers | Cicadellidae | S.E. Asia | Infest foliage |
| *Ferrisia virgata* (Ckll.) | Striped Mealybug | Pseudococcidae | S.E. Asia | Infest foliage |
| *Planococcus citri* Risso | Citrus (Root) Mealybug | Pseudococcidae | Widespread | Infest foliage |
| *Saissetia coffeae* (Wlk.) | Helmet Scale | Coccidae | Philippines | Infest stem |
| *Cyrtopeltis* spp. | Tomato Mirids | Miridae | India, S.E. Asia | Sap-suckers; toxic saliva |
| *Acanthocoris* spp. | Coreid Bugs | Coreidae | Philippines | |
| *Nezara viridula* (L.) | Green Stink Bug | Pentatomidae | Cosmopolitan | |

| | | | | |
|---|---|---|---|---|
| *Frankliniella* spp. | Flower Thrips | Thripidae | Cosmopolitan | Infest flowers & foliage |
| *Lamprosema diamenalis* Guen. | – | Pyralidae | Malaysia | Larvae eat leaves |
| *Maruca testulalis* (Geyer) | Mung Moth | Pyralidae | Widespread | Larvae eat leaves |
| *Sylepta derogata* (F.) | Cotton Leaf-roller | Pyralidae | S.E. Asia | Larvae roll leaves |
| *Agrotis segetum* (D. & S.) | Common Cutworm | Noctuidae | India, S.E. Asia | Larvae destroy seedlings |
| *Tiracola plagiata* (Wlk.) | Banana Fruit Caterpillar | Noctuidae | S.E. Asia | Larvae defoliate |
| *Plusia* spp. | Semi-loopers | Noctuidae | S.E. Asia | Larvae defoliate |
| *Chrysodeixis chalcites* (Esp.) | Cabbage Semi-looper | Noctuidae | Old World | Larvae defoliate |
| *Heliothis assulta* Gn. | Cape Gooseberry Budworm | Noctuidae | India, S.E. Asia | Larvae bore buds |
| *Heliothis virescens* (F.) | Tobacco Budworm | Noctuidae | N., C. & S. America | Larvae bore buds |
| *Heliothis punctigera* Wilgr. | Native Budworm | Noctuidae | Australia | Larvae bore buds |
| *Mythimna* spp. | Rice Armyworms | Noctuidae | Pantropical | Larvae defoliate |
| *Xestia c-nigrum* (L.) | Spotted Cutworm | Noctuidae | Europe. Asia, N. America | Larvae are cutworms |
| *Spodoptera litura* (F.) | Rice Cutworm | Noctuidae | S.E. Asia | Larvae defoliate |
| *Spodoptera littoralis* (Boisd.) | Cotton Leafworm | Noctuidae | Africa, S. Europe | Larvae defoliate |
| *Spodoptera exigua* (Hb.) | Lesser Armyworm | Noctuidae | Widespread in warmer regions | Larvae defoliate |
| *Manduca quinquemaculata* (Haw.) | Tomato Hornworm | Sphingidae | N., C. & S. America | Larvae defoliate |
| *Agrius convolvuli* (L.) | Convolvulus Hawk | Sphingidae | Old World | Larvae defoliate |
| *Phthorimaea operculella* (Zell.) | Potato Tuber Moth | Gelechiidae | Cosmopolitan | Larvae bore stem |
| *Scrobipalpa heliopa* (Lower) | Tobacco Stem Borer | Gelechiidae | India, S.E. Asia | Larvae bore stem |
| *Delia platura* (Meign.) | Bean Seed Fly | Anthomyiidae | Cosmopolitan | Larvae bore sown seeds & seedlings |
| *Solenopsis geminata* (F.) | Fire Ant | Formicidae | India, S.E. Asia | Attack workers |
| *Gonocephalum* spp. | Dusty Brown Beetles | Tenebrionidae | Philippines, Africa | Adults damage stem |
| *Epilachna* spp. | Epilachna Beetles | Coccinellidae | Philippines, Asia, USA | Adults & larvae defoliate |
| *Psylliodes* spp. | Tobacco Flea Beetles | Chrysomelidae | Philippines | Adults hole leaves |
| *Leptinotarsa decemlineata* Say | Colorado Beetle | Chrysomelidae | Europe, USA, C. America | Adults & larvae defoliate |
| *Epitrix hirtipennis* (Melsh.) | Tobacco Flea Beetle | Chrysomelidae | USA (Florida) | Adults hole leaves |
| *Oulema bilineata* (Germ.) | Tobacco Leaf Beetle | Chrysomelidae | Africa, S. America | Adults eat leaves |
| *Orthaulaca similis* Ol. | Tobacco Leaf Beetle | Chrysomelidae | Philippines | Adults eat leaves |

| | | | | |
|---|---|---|---|---|
| *Agriotes* spp. | Wireworms | Elateridae | Europe, Asia, USA ⎫ | |
| *Conoderus* spp. | Tobacco Wireworms | Elateridae | USA ⎬ Larvae eat roots | |
| *Anomala* spp. | White Grubs | Scarabaeidae | Philippines ⎭ | Larvae eat roots |
| *Leucopholis irrorata* (Chevr.) | White Grub | Scarabaeidae | Philippines | Larvae eat roots |
| *Polyphagotarsonemus latus* (Banks) | Yellow Mite | Tarsonemidae | S.E. Asia | Scarify leaves |
| *Tetranychus cinnabarinus* (Boisd.) | Carmine Mite | Tetranychidae | S.E. Asia | Scarify leaves |

# TOMATO (*Lycopersicum esculentum* – Solanaceae)

Tomato originated in S. America in the Peru/Ecuador region, and was taken to the Philippines and Malaya by 1650. For some time it has been cultivated in the temperate regions of America and Europe, but it was not cultivated in the tropics until the 20th century, but now is grown very widely throughout the world. It can be grown in the open wherever there is more than three months of frost-free weather, but needs even rainfall and long sunny periods for best results. It can be grown at sea-level but usually does better at higher altitudes. It is a variable annual herb, 0.7–2 m high, and the fruit for which it is grown is a fleshy berry, red or yellow when ripe, containing vitamins A and C. The fruit is used as a vegetable, raw or cooked, made into soup, sauce, juice, ketchup, paste, puree, powder or may be canned; also used unripe in chutneys. The main production areas are in the USA, Mexico and Italy, but most tropical countries have a large local production and consumption, and most temperate countries produce quantities under glass or polythene covers for local fresh consumption.

## MAJOR PESTS

| | | | | |
|---|---|---|---|---|
| *Brachytrupes membranaceus* (Drury) | Tobacco Cricket | Gryllidae | Africa | Destroy seedlings |
| *Brachytrupes portentosus* Litch. | Large Brown Cricket | Gryllidae | S.E. Asia | Destroy seedlings |
| *Bemisia tabaci* (Genn.) | Tobacco Whitefly | Aleyrodidae | Cosmopolitan | Infest foliage |
| *Cyrtopeltis tenuis* Reut. | Tomato Mirid | Miridae | Philippines | Sap-sucker; toxic saliva |
| *Nezara viridula* (L.) | Green Stink Bug | Pentatomidae | Cosmopolitan | Fruit spotted by feeding scars |
| *Heliothis armigera* (Hub.) | American Bollworm (Tomato Fruitworm) | Noctuidae | Cosmopolitan in Old World | Larvae bore fruits |
| *Heliothis zea* Boddie | Cotton Bollworm | Noctuidae | N., C. & S. America | Larvae bore fruits |
| *Thrips tabaci* Lind. | Onion Thrips | Thripidae | Cosmopolitan | Infest foliage |
| *Tetranychus urticae* (Koch) | Temperate Red Spider Mite | Tetranychidae | Europe, N. America | Scarify foliage |

## MINOR PESTS

| | | | | |
|---|---|---|---|---|
| *Phymateus aegrotus* Gerst. | – | Acrididae | Africa | Defoliate |
| *Grylloptalpa africana* (Pal.) | African Mole Cricket | Gryllotalpidae | S.E. Asia | Eat roots |
| *Aulacorthum solani* (Kalt.) | Glasshouse/ Potato Aphid | Aphididae | Europe, USA | Infest foliage; virus vector |
| *Aphis gossypii* Glov. | Cotton Aphid | Aphididae | Cosmopolitan | Infest foliage; virus vector |
| *Trialeurodes vaporariorum* (Westw.) | Glasshouse Whitefly | Aleyrodidae | Europe | Infest foliage; virus vector |

| *Myzus persicae* (Sulz.) | Green Peach Aphid (Peach– Potato Aphid) | Aphididae | Cosmopolitan | Sap-sucker; virus vector |
|---|---|---|---|---|
| *Empoasca* spp. | Green Leaf-hoppers | Cicadellidae | S.E. Asia | Sap-sucker; infest foliage |
| *Zygina pallidifrons* | Glasshouse Leafhopper | Cicadellidae | Europe | Infest foliage |
| *Ferrisia virgata* (Ckll.) | Striped Mealybug | Pseudococcidae | Pantropical | Infest foliage |
| *Pinnaspis minor* Mask. | Armoured Scale | Diaspididae | Malaysia | Infest foliage |
| *Anthocoris* spp. | Coreid Bugs | Coreidae | Philippines | Sap-suckers; toxic saliva |
| *Frankliniella schulzei* (Trybom) | Cotton Bud Thrips | Thripidae | Africa | Infest flowers |
| *Dacus* spp. | Fruit Flies | Tephritidae | Malaysia, Philippines | Larvae in fruit |
| *Liriomyza bryoniae* (Kalt.) | Tomato Leaf Miner | Agromyzidae | Europe, N. Africa, USSR | Larvae mine leaves |
| *Keiferia lycopersicella* | Tomato Pinworm | Gelechiidae | USA, W. Indies | Larvae mine leaves, bore stems & fruits |
| *Phthorimaea perculella* (Zeller) | Potato Tuber Moth | Gelechiidae | Cosmopolitan | |
| *Leucinodes orbonalis* Guen. | Eggplant Boring Caterpillar | Pyralidae | Africa, India, S.E. Asia | Larvae bore fruit |
| *Lacanobia oleracea* (L.) | Tomato Moth | Noctuidae | Europe | Larvae bore fruits |
| *Agrotis ipsilon* Roth. | Black Cutworm | Noctuidae | S.E. Asia | Larvae are cutworms |
| *Heliothis virescens* (F.) | Tobacco Budworm | Noctuidae | N., C. & S. America | Larvae bore fruits |
| *Heliothis assulta* Gn. | Cape Gooseberry Budworm | Noctuidae | India, S.E. Asia Australasia, Africa | Larvae bore fruits |
| *Othreis fullonia* (Cl.) | Fruit-piercing Moth | Noctuidae | Old World tropics | Adults pierce fruits |
| *Mythimna separata* (Wlk.) | Rice Ear-cutting Caterpillar | Noctuidae | S.E. Asia | Larvae defoliate |
| *Spodoptera litura* (F.) | Rice Cutworm | Noctuidae | S.E. Asia | Larvae defoliate |
| *Xestia c-nigrum* (L.) | Spotted Cutworm | Noctuidae | Europe, Asia, N. America | Larvae are cutworms |
| *Spodoptera littoralis* (Boisd.) | Cotton Leafworm | Noctuidae | Africa, S. Europe | Larvae defoliate |
| *Spodoptera exigua* (Hb.) | Lesser Army-worm | Noctuidae | Widespread in warmer regions | Larvae defoliate |
| *Chrysodeixis chalcites* (Esp.) | Cabbage Semi-looper | Noctuidae | Old World | Larvae defoliate |
| *Plusia* spp. | Semi-loopers | Noctuidae | Cosmopolitan | Larvae defoliate |
| *Anomis flava* (F.) | Cotton Semi-looper | Noctuidae | Old World tropics | |
| *Manduca quinquemaculata* (Haw.) | Tomato Hornworm | Sphingidae | USA, C. & S. America | Larvae defoliate |

674

| | | | | |
|---|---|---|---|---|
| *Manduca sexta* (L.) | Tobacco Hornworm | Sphingidae | USA, C. & S. America | Larvae defoliate |
| *Acherontia atropos* (L.) | Death's Head Hawk Moth | Sphingidae | Africa | |
| *Agrius convolvuli* (L.) | Sweet Potato Hawk Moth | Sphingidae | Cosmopolitan in Old World | Larvae defoliate |
| *Leucopholis irrorata* (Chevr.) | White Grub | Scarabaeidae | Philippines | Larvae eat roots |
| *Anomala* spp. | White Grubs | Scarabaeidae | Philippines | Larvae eat roots |
| *Epilachna* spp. | Epilachna Beetles | Coccinellidae | S.E. Asia | Adults & larvae defoliate |
| *Leptinotarsa decemlineata* (Say) | Colorado Beetle | Chrysomelidae | Europe (not UK), N. & C. America | Adults & larvae defoliate |
| *Psylliodes* spp. | Tobacco Flea Beetles | Chrysomelidae | Philippines | Adults hole leaves |
| *Agriotes* spp. | Wireworms | Elateridae | Europe | Larvae eat roots |
| *Epicauta albovittata* (Gestro) | Striped Blister Beetle | Meloidae | Africa | Adults eat flowers |
| *Polyphagotarsonemus latus* (Banks) | Yellow Tea Mite | Tarsonemidae | Cosmopolitan | Scarify leaves & fruits |
| *Tetranychus cinnabarinus* (Boisd.) | Tropical Red Spider Mite | Tetranychidae | Pantropical | Scarify leaves & fruits; web foliage |
| *Tetranychus* spp. | Red Spider Mites | Tetranychidae | Cosmopolitan | |
| *Aculus lycopersici* (Massee) | Tomato Russet Mite | Eriophyidae | Cosmopolitan | Scarify foliage |

# TURMERIC (*Curcuma domestica* – Zingiberaceae)
### (= *C. longa*)

A rhizomatous herb, native to the tropical rain forests of India, cultivated for its tubers which yield a natural reddish dye, sometimes yellowish. It is an important spice amongst the rice-eating peoples of India, S.E. Asia, and Indonesia and it is indispensible in the preparation of curry powder; it gives the musky flavour and yellow colour to curries. Most curry powders contain about 24% turmeric powder, and this is the main present-day use for this plant. As a dye for cloth it was important until the discovery of the aniline dyes, but it is still used in India for this purpose, and in addition it has quite strong religious associations in that country.

## MAJOR PESTS

| | | | | |
|---|---|---|---|---|
| *Dichocrocis punctiferalis* Guen. | Castor Capsule Borer | Pyralidae | India | Larvae bore shoots & fruits |

## MINOR PESTS

| | | | | |
|---|---|---|---|---|
| *Aspidiella hartii* Ckll. | Yam (Turmeric) Scale | Diaspididae | India, W. Africa, W. Indies | Encrust foliage & rhizomes |
| *Stephanitis typicus* Dist. | Banana Lace Bug | Tingidae | India | Sap-sucker; toxic saliva |
| *Anaphothrips sudanensis* Trybom | Leaf Thrips | Thripidae | India | Infest foliage |
| *Panchaetothrips indicus* Bagn. | Thrips | Thripidae | India | Infest flowers |
| *Udaspes folus* Cr. | Skipper Butterfly | Hesperiidae | India | Larvae roll leaves |
| *Diacrisia obliqua* Wlk. | Tiger Moth | Arctiidae | India | Larvae defoliate |

# VANILLA (*Vanilla fragrans* — Orchidaceae)
## (= *V. planifolia*)

This important and popular flavouring material and spice comes from the fruits of a climbing orchid, native to the hot humid rain forests of tropical America. The fruits are harvested when fully grown but still unripe, and then they are fermented and cured; they are called vanilla beans. Vanilla extract is made by macerating the cured beans in alcohol; it is used mostly to flavour icecream, chocolate, beverages, cakes, puddings, and other confectionery. When cultivated, in hot wet tropical regions, the vine can be trained to grow up poles or else tree trunks; island climates are often the most suitable for vanilla production, but in general it is grown throughout the hot wet tropics.

## MAJOR PESTS

## MINOR PESTS

| | | | | |
|---|---|---|---|---|
| *Mertilanidea fasciata* Ghauri | Capsid Bug | Miridae | Papua NG | Sap-sucker; toxic saliva |
| *Agraulis vanillae* (L.) | Vanilla Butterfly | Nymphalidae | Colombia, Hawaii | Larvae eat leaves & fruits |
| *Saula ferruginea* Gerst. | — | ? | India | Adults eat ventral leaf lamina |

# WALNUT (*Juglans regia* – Juglandaceae)
## (= English Walnut)

Despite its name this is a native of Iran and is now extensively cultivated in Europe (particularly France), China, N. India, other parts of Asia and also the USA. The trees are often used in an additional capacity as ornamentals for they are large (up to 20 m in height), and with a pleasing regular shape. The kernels are characteristically furrowed and are easily freed from the pericarp; they represent the cotyledons of the seed, no endosperm being present. The kernels are eaten raw, used in cakes and confectionery, and also an oil can be extracted which is excellent for table use. *J. nigra* is the Black Walnut of the eastern deciduous forest region of the USA, but the shell is so hard that its use is more or less limited to the confectionery industry. *J. cinerea* is the Butternut of eastern USA and Canada, with a higher fat content than walnuts, and is used mainly for confectionery.

## MAJOR PESTS

| | | | | |
|---|---|---|---|---|
| *Paramyelois transitella* Wlk. | Navel Orange-worm | Pyralidae | USA (California) | Larvae bore fruits |
| *Batocera horsfieldi* Hope | Longhorn Beetle | Cerambycidae | India | Larvae bore trunk |

## MINOR PESTS

| | | | | |
|---|---|---|---|---|
| *Arytania fasciata* Laing | Walnut Psylla | Psyllidae | India | Nymphs gall leaves |
| *Aphis pomi* de Geer | Apple Aphid | Aphididae | India | Adults & nymphs infest foliage, suck sap & make leaves curl |
| *Callipterus juglandis* Goeze | Walnut Aphid | Aphididae | Europe, India | |
| *Chromaphis juglandicola* (Kltb.) | Walnut Aphid | Aphididae | Europe, India, USA | |
| *Parthenolecanium corni* (Bch.) | Plum Scale | Coccidae | Europe, W. Asia | Infest foliage |
| *Quadraspidiotus perniciosus* (Comst.) | San José Scale | Diaspididae | India | Infest foliage |
| *Aspidiotus juglansregiae* Comst. | Walnut Scale | Diaspididae | USA | Infest foliage |
| *Icerya purchasi* Mask. | Cottony Cushion Scale | Margarodidae | India | Infest foliage |
| *Pseudococcus maritimus* (Ehrh.) | Grape Mealybug | Pseudococcidae | Widespread | Infest foliage |
| *Cydia pomonella* (L.) | Codling Moth | Tortricidae | Europe | Larvae bore fruits |
| *Cydia funebrana* (Treit.) | Red Plum Maggot | Tortricidae | Europe, Asia | Larvae bore fruits |
| *Zeuzera* spp. | Leopard Moths | Cossidae | Europe, India | Larvae bore branch |
| *Datana integerrima* G. & R. | Walnut Cater-pillar | Notodontidae | USA | Larvae defoliate |
| *Cressonia juglandis* (J.E. Smith) | Walnut Sphinx | Sphingidae | USA | Larvae defoliate |
| *Arctias selene* (Hb.) | Moon Moth | Saturniidae | India | Larvae eat leaves |
| *Phalera bucephala* L. | Buff-tip Moth | Noctuidae | Europe | Larvae defoliate |
| *Malacosoma indica* Wlk. | Tent Cater-pillar | Lasiocampidae | India | Larvae defoliate |
| *Rhagoletis completa* Cresson | Walnut Husk Fly | Tephritidae | USA | Larvae bore fruit husk |

| | | | | |
|---|---|---|---|---|
| *Holotrichia longipennis* (Blanch.) | Cockchafer | Scarabaeidae | India | Larvae eat roots; adults eat leaves |
| *Anomala* spp. | Flower Beetles | Scarabaeidae | India | Adults eat leaves |
| *Mimela pusilla* Hope | Flower Beetle | Scarabaeidae | India | Adults eat leaves |
| *Dorysthenus hugelii* Redt. | Root Borer | Scarabaeidae | India | Larvae eat roots of young trees; adults eat leaves |
| *Hispa dama* Chap. | Hispid Beetle | Chrysomelidae | India | Larvae mine leaves; adults eat leaves |
| *Altica cerulescens* (Baly) | Flea Beetle | Chrysomelidae | India | Adults & larvae eat leaves |
| *Monolepta erythrocephala* Baly | Leaf Beetle | Chrysomelidae | India | Defoliate |
| *Aeolesthes holoserica* F. | Cherry Stem Borer | Cerambycidae | Pakistan, India | Larvae bore trunk & branches |
| *Aeolesthes sarta* Solsky | Quetta Borer | Cerambycidae | Pakistan, India | Larvae bore trunk & branches |
| *Batocera rufomaculata* (de Geer) | Red-spotted Longhorn | Cerambycidae | India | Larvae bore trunk |
| *Alcidodes porrectirostris* Mshll. | Walnut Weevil | Curculionidae | India | Larvae bore fruits |
| *Myllocerus viridianus* F. | Grey Weevil | Curculionidae | India | Adults eat leaves |
| *Diapus pusillimus* Chapuis | Walnut Pinhole Borer | Scolytidae | Australia | Adults bore trunk & branches |
| *Scolytus juglandis* | Walnut Pinhole Borer | Scolytidae | India | Adults bore branches |
| *Aceria erinea* (Nal.) | Walnut Blister Mite | Eriophyidae | Europe, USA, Australia | Erinea on leaves |

# WATERCRESS (*Nasturtium officinale* — Crucifereae)

This is regarded as one of the minor herbage vegetables, grown widely but only for immediate local consumption. It occurs in the wild state in the UK and S. Europe, and is now introduced into many parts of the world. The plant is an aquatic perennial herb, and the tips of the leafy stems are used as salad, or it may be cooked as a vegetable. It grows best in clear running shallow water; the distal parts of the plant usually protrude from the water, and flowers are aerial.

### MAJOR PESTS

| | | | | |
|---|---|---|---|---|
| *Plutella xylostella* L. | Diamond-back Moth | Yponomeutidae | Malaysia | Larvae eat leaves |

### MINOR PESTS

| | | | | |
|---|---|---|---|---|
| *Myzus persicae* (Sulz.) | Green Peach Aphid | Aphidiae | S. China | Adults & nymphs infest aerial foliage; suck sap |
| *Lipaphis erysimi* (Kalt.) | Turnip Aphid | Aphididae | S. China | |
| *Rhopalosiphum rufiabdominalis* (Sasaki) | Rice Root Aphid | Aphididae | S. China | |
| *Phaedon aeruginosus* Suff. | Watercress Leaf Beetle | Chrysomelidae | USA | Adults eat leaves |

# WHEAT (*Triticum sativum* – Gramineae)
## (including Barley and Oats)

Wheat is the chief cereal of temperate regions, and is the most widely grown cereal. As a crop it is of great antiquity, and its native home uncertain, although it is thought to be somewhere in C. or S.W. Asia. It was introduced into the New World in 1529 by the Spaniards who took it to Mexico. In the tropics this is not an important crop, but it is grown ever increasingly in the higher and cooler parts of some tropical countries (e.g. Kenya, India). In general, in the tropics, Wheat and Barley do not have a comparable pest spectrum to that which can cause considerable concern in most temperate countries. In temperate regions there are important differences between the pest spectra of Wheat, Barley, Oats and Rye.

## MAJOR PESTS

| | | | | |
|---|---|---|---|---|
| *Homorocoryphus nitidulus* Wlk. | Edible Grass-hopper | Tettigoniidae | E. Africa | Defoliate |
| *Schizaphis graminum* Rond. | Wheat Aphid | Aphididae | Africa, Asia, USA, S. America | Infest foliage |
| *Delia arambourgi* Seguy | Barley Fly | Anthomyiidae | Africa | Destroy seedlings |
| *Delia coarctata* (Fall.) | Wheat Bulb Fly | Anthomyiidae | Europe | Larvae bore stems of seedlings |
| *Oscinella frit* (L.) | Frit Fly | Chloropidae | Europe | Larvae gall stems |
| *Mayetiola destructor* (Say) | Hessian Fly | Cecidomyiidae | Europe, N. America | Larvae gall stems |
| *Sesamia inferens* (Wlk.) | Purple Stem Borer | Noctuidae | India, Pakistan, China, S.E. Asia, Japan, Indonesia, Philippines | Larvae bore stems |
| *Nematocerus* spp. | Nematocerus Weevils | Curculionidae | E. Africa | Adults eat leaves; larvae eat roots |
| *Heteronychus consimilis* Kolbe | Black Wheat Beetle | Scarabaeidae | E. Africa | Adults eat stem |
| *Epilachna similis* (Thun.) | Epilachna Beetle | Coccinellidae | Africa | Adults & larvae eat leaves |

## MINOR PESTS

| | | | | |
|---|---|---|---|---|
| *Rhopalosiphum maidis* (Fitch) | Corn Leaf Aphid | Aphididae | Cosmopolitan | |
| *Rhopalosiphum padi* (L.) | Bird-cherry Aphid | Aphididae | Europe | |
| *Rhopalosiphum insertum* (Wlk.) | Apple–Grass Aphid | Aphididae | Europe | |
| *Macrosiphum fragariae* Wlk. | Blackberry Aphid | Aphididae | Cosmopolitan | Adults & nymphs infest foliage; sap-suckers & virus vectors |
| *Macrosiphum avenae* (F.) | Grain Leaf Aphid | Aphididae | Europe | |
| *Metopolophium dirhodum* (Wlk.) | Rose–Grain Aphid | Aphididae | Europe | |
| *Metopolophium festucae* (Theob.) | Fescue Aphid | Aphididae | Europe | |
| *Melanaphis sacchari* (Zhnt.) | Sugarcane Aphid | Aphididae | Pantropical | |

| *Laodelphax striatella* (Fall.) | Small Brown Leafhopper | Delphacidae | Europe, Asia, China, Japan, S.E. Asia | Sap-sucker; virus vector |
|---|---|---|---|---|
| *Delphacodes pellucida* (F.) | Cereal Leaf-hopper | Delphacidae | Europe | Sap-sucker |
| *Eurygaster* spp. | Wheat Shield Bugs | Pentatomidae | Europe, W. Asia | Sap-suckers |
| *Blissus leucopterus* (Say) | Chinch Bug | Lygaeidae | N. & S. America | Sap-sucker |
| *Pyrilla perpusilla* Wlk. | Indian Sugar-cane Leafhopper | Lophopidae | India | Sap-sucker |
| *Crambus* spp. | Grass Moths | Crambidae | Europe, USA | Larvae bore stems |
| *Marasmia trapezalis* (Gn.) | Maize Webworm | Pyralidae | Africa, India, S.E. Asia, Australasia, C. & S. America | Larvae eat leaves |
| *Sitotroga cerealella* (Ol.) | Angoumois Grain Moth | Gelechiidae | Widespread | Larvae eat grains |
| *Hepialus* spp. | Swift Moths | Hepialidae | Europe | Larvae eat roots |
| *Spodoptera frugiperda* (J.E. Smith) | Fall Armyworm | Noctuidae | N., C. & S. America | Larvae defoliate |
| *Luperina testacea* (Schiff.) | Flounced Rustic Moth | Noctuidae | Europe ⎫ | |
| *Mesapamea secalis* (L.) | Common Rustic Moth | Noctuidae | Europe ⎬ | Larvae bore stems |
| *Mythimna* spp. | Cereal Army-worms | Noctuidae | Cosmopolitan | Larvae defoliate |
| *Hydrellia griseola* (Fall.) | Cereal Leaf Miner | Ephydridae | Europe, Asia, N. Africa, USA, S. America | Larvae mine leaves |
| *Atherigona oryzae* Mall. | Rice Shoot Fly | Muscidae | S. & E. Asia | Larvae bore young shoot |
| *Delia platura* (Meign.) | Bean Seed Fly | Anthomyiidae | Cosmopolitan | Larvae eat germinating seed |
| *Phorbia genitalis* Tiens. | Late Wheat Shoot Fly | Anthomyiidae | Europe | Larvae bore shoots |
| *Agromyza ambigua* Fall. | Cereal Leaf Miner | Agromyzidae | Europe, USA, Canada | Larvae mine leaves |
| *Chlorops pumilionis* (Bjerk.) | Gout Fly | Chloropidae | Europe | Larvae gall shoots |
| *Geomyza* spp. | Grass Flies | Geomyzidae | Europe | Larvae bore shoots |
| *Opomyza* spp. | Grass Flies | Opomyzidae | Europe | Larvae bore shoots |
| *Bibio marci* (L.) | St. Mark's Fly | Bibionidae | Europe | Larvae eat roots |
| *Haplodiplosis equestris* (Wagn.) | Saddle Gall Midge | Cecidomyiidae | Europe | Larvae gall stems |
| *Mayetiola* spp. | 'Flax' Flies | Cecidomyiidae | Cosmopolitan | Larvae gall stems |
| *Mayetiola avenae* (March.) | Oat Stem Midge | Cecidomyiidae | Europe | Larvae gall stems |
| *Contarinia tricici* (Kirby) | Yellow Wheat Blossom Midge | Cecidomyiidae | Europe | Larvae infest flower head |

682

| | | | | |
|---|---|---|---|---|
| *Sitodiplosis mossellana* (Gehin) | Orange Wheat Blossom Midge | Cecidomyiidae | Europe | Larvae infest flower head |
| *Limothrips cerealium* Hal. | Grain Thrips | Thripidae | Europe | Infest flowers |
| *Aptinothrips* spp. | Grass Thrips | Thripidae | Europe | Infest flowers |
| *Stenothrips graminum* Uzel | Oat Thrips | Thripidae | Europe | Infest flowers |
| *Thrips nigropilosus* Uzel | Chrysanthemum Thrips | Thripidae | Europe, E. Africa, N. America | Infest flowers |
| *Tipula* spp. | Leatherjackets | Tipulidae | Europe ⎞ | Larvae eat roots |
| *Nephrotoma* spp. | Leatherjackets | Tipulidae | Europe ⎠ | |
| *Cephus pygmaeus* (L.) | Wheat Stem Sawfly | Cephidae | Europe | Larvae gall stem |
| *Dolerus* spp. | Leaf Sawflies | Tenthredinidae | Europe | Larvae eat leaves |
| *Agriotes* spp. | Wireworms | Elateridae | Europe | Larvae eat roots |
| *Athous* spp. | Garden Wireworms | Elateridae | Europe | Larvae eat roots |
| *Corymbites* spp. | Upland Wireworms | Elateridae | Europe | Larvae eat roots |
| *Helophorus nubilus* F. | Wheat Shoot Beetle | Hydrophilidae | Europe | Larvae bite young shoots |
| *Oulema melanopa* (L.) | Cereal Leaf Beetle | Chrysomelidae | Europe, USA | Larvae mine leaves; adults eat leaves |
| *Phyllotreta vittula* Redt. | Barley Flea Beetle | Chrysomelidae | Europe | Adults eat leaves |
| *Chaetocnema hortensis* (Geoff.) | Cereal Flea Beetle | Chrysomelidae | Europe | Adults eat leaves |
| *Crepidodera ferruginea* (Scop.) | Wheat Flea Beetle | Chrysomelidae | Europe | Adults eat leaves |
| *Melolontha melolontha* (L.) | Cockchafer | Scarabaeidae | Europe | Larvae eat roots |
| *Amphimallon solstitialis* (L.) | Summer Chafer | Scarabaeidae | Europe | Larvae eat roots |
| *Serica brunnea* (L.) | Brown Chafer | Scarabaeidae | Europe | Larvae eat roots |
| *Schizonycha* spp. | Chafer Grubs | Scarabaeidae | Europe | Larvae eat roots |
| *Gonocephalum simplex* (F.) | Dusty Brown Beetle | Tenebrionidae | Africa | Adult damages stem; larvae eat roots |
| *Tanymecus dilaticollis* Gylh. | Southern Grey Weevil | Curculionidae | E. Europe | Damage foliage |
| *Stenotarsonemus spirifex* (March.) | Oat Spiral Mite | Tarsonemidae | Europe | Damage flowers |

# YAM (*Dioscorea esculenta* — Dioscoreaceae)

Yams are native to the Old World tropics, with wild species being found in parts of both Africa and Asia. Now the crop is really only of importance in W. Africa and parts of Vietnam, Cambodia and Laos. It needs a high tropical rainfall. The yam itself is a swollen tuber of a climbing vine, and contains little food except starch. The tubers can be stored either in the ground or on racks in farm stores. About a dozen different species of yams are known in addition to the common one (*D. esculenta*).

## MAJOR PESTS

| | | | | |
|---|---|---|---|---|
| *Prionoryctes caniculus* Arr. | Yam Beetle | Scarabaeidae | Africa | Larvae bore tubers |
| *Heteroligus meles* (Billb.) | Greater Yam Beetle | Scarabaeidae | W. Africa | Larvae bore tubers |

## MINOR PESTS

| | | | | |
|---|---|---|---|---|
| *Gymnogryllus lucens* (W.) | Cricket | Gryllidae | Nigeria | Damage roots & tubers |
| *Aphis gossypii* Glov. | Cotton Aphid | Aphididae | Nigeria | Infest foliage; virus vector |
| *Planococcus kenyae* (Le Pelley) | Kenya Mealybug | Pseudococcidae | E. & W. Africa | Infest foliage |
| *Planococcus citri* (Risso) | Citrus (Root) Mealybug | Pseudococcidae | S.E. Asia | Infest foliage & roots |
| *Ptyelus grossus* F. | Spittle Bug | Cercopidae | Africa | Larvae infest foliage |
| *Aspidiella hartii* (Ckll.) | Yam Scale | Diaspididae | India, W. Indies | Infest foliage |
| *Aspidiotus destructor* Sign. | Coconut Scale | Diaspididae | S.E. Asia | Infest foliage |
| *Quadraspidiotus perniciosus* (Comst.) | San José Scale | Diaspididae | S.E. Asia | Infest foliage |
| *Helopeltis* sp. | Mosquito Bug | Miridae | Malaysia | Sap-sucker; toxic saliva |
| *Leptoglossus australis* (F.) | Leaf-footed Plant Bug | Coreidae | S.E. Asia | Sap-sucker; toxic saliva |
| *Tagiades litigiosa* | Yam Skipper (Water Snow Flat) | Hesperiidae | S.E. Asia | Larvae roll leaf edges |
| *Schizonycha* sp. | Chafer Grub | Scarabaeidae | Africa | Larvae eat roots |
| *Heteronychus* spp. | Black Cereal Beetles | Scarabaeidae | Africa | Larvae eat roots |
| *Heteroligus appius* (Burm.) | Lesser Yam Beetle | Scarabaeidae | W. Africa | Larvae bore tubers |
| *Prionoryctes rufopiceus* Arr. | — | Scarabaeidae | W. Africa | Larvae bore tubers |
| *Lepidiota reichei* (J. Thom.) | — | Scarabaeidae | W. Africa | Larvae bore tubers |
| *Crioceris livida* Dalm. | Leaf Beetle | Chrysomelidae | W. Africa | Adults & larvae eat leaves |
| *Apomecyna parumpunctata* Chrvr. | Longhorn Beetle | Cerambycidae | Nigeria | Larvae bore stems |
| *Palaeopus dioscorae* Pierce | Yam Weevil | Curculionidae | Jamaica | Larvae bore tubers |

# PEST OF SEEDLINGS AND GENERAL PESTS

Many of the pests already referred to are particularly damaging to seedlings and young crop plants, and many are also polyphagous and recorded from many different crops, both as major and minor pests. In general soil-dwelling pests are not host specific but will attack (eat) the roots and underground stems of almost anything rather small, and not too woody, growing in that soil. Similarly, some general sap-suckers (Hemiptera; Tetranychidae) and leaf-eaters (Acrididae; Noctuidae; Curculionidae) do not appear to be particularly selective as to host. Some of these general pests, such as locusts, are more restricted geographically than by host preferences.

## MAJOR PESTS

| | | | | |
|---|---|---|---|---|
| *Sminthurus viridis* (L.) | Lucerne 'Flea' | Sminthuridae | Europe | Damage seedlings, especially ones with soft stems |
| *Bourletiella hortensis* (Fitch) | Garden Springtail | Sminthuridae | Europe, USA | |
| *Onychiurus* spp. | Springtails | Onychiuridae | Cosmopolitan | |
| *Patanga succincta* (L.) | Bombay Locust | Acrididae | India, S.E. Asia, China | General defoliator |
| *Zonocerus* spp. | Variegated Grasshoppers | Acrididae | Africa | General defoliators |
| *Chortoicetes terminifera* (Wlk.) | Australian Plague Locust | Acrididae | Australia | General defoliator |
| *Dociostaurus maroccanus* (Thnb.) | Mediterranean Locust | Acrididae | Med. | General defoliator |
| *Schistocerca gregaria* (Forsk.) | Desert Locust | Acrididae | Africa to India | General defoliator |
| *Nomadacris septemfasciata* (Serv.) | Red Locust | Acrididae | C. & Southern Africa | General defoliator |
| *Locusta migratoria migratoria* | Asiatic Migratory Locust | Acrididae | C. Asia | General defoliator |
| *Locusta m. migratorioides* (R. & F.) | African Migratory Locust | Acrididae | Tropical Africa | General defoliator |
| *Locusta m. maniliensis* (Meyr.) | Oriental Migratory Locust | Acrididae | S.E. Asia, Philippines, Borneo, China | General defoliator |
| *Gryllotalpa* spp. | Mole Crickets | Gryllotalpidae | Cosmopolitan | Soil feeder |
| *Acheta* spp. | Field Crickets | Gryllidae | Cosmopolitan | Nest in soil; seedling pests; eat roots of older plants |
| *Brachytrupes membranaceus* (Drury) | Tobacco Cricket | Gryllidae | Africa | |
| *Brachytrupes portentosus* Licht. | Large Brown Cricket | Gryllidae | India, S.E. Asia, China | |
| *Macrotermes* spp. | Mound-building Termites | Termitidae | Pantropical | Remove bark & foliage for fungus gardens |

| | | | | |
|---|---|---|---|---|
| *Odontotermes* spp. | Scavenging Termites | Termitidae | Pantropical | Collect vegetable material |
| *Aphis* spp. | Aphids | Aphididae | Cosmopolitan | Polyphagous sap-suckers |
| *Coccus* spp. | Scale Insects | Coccidae | Pantropical | |
| *Nezara viridula* (L.) | Green Stink Bug | Pentatomidae | Cosmopolitan | |
| *Lygocoris pabulinus* (L.) | Common Green Capsid | Miridae | Europe | Polyphagous sap-suckers; toxic saliva |
| *Lygus rugulipennis* Popp. | Tarnished Plant Bug | Miridae | Europe | |
| *Helopeltis* spp. | Mosquito Bugs | Miridae | Pantropical | |
| *Forficula auricularia* L. | Common Earwig | Forficulidae | Cosmopolitan | Pest of small seedlings only |
| *Hepialus* spp. | Swift Moths | Hepialidae | Cosmopolitan in temperate regions | Larvae in soil eat plant roots |
| *Agrotis* spp. *Euoxa* spp. *Spodoptera* spp. | Cutworms | Noctuidae | Cosmopolitan | Larvae are cutworms in soil |
| *Noctua pronuba* (L.) | Large Yellow Underwing | Noctuidae | Europe, Asia, | Larvae are cutworms |
| *Heliothis* spp. *Spodoptera* spp. | Leafworms, etc. | Noctuidae | Cosmopolitan | Larvae polyphagous leaf-eaters |
| *Spodoptera* spp. *Mythimna* spp. | Armyworms | Noctuidae | Pantropical | Larvae polyphagous leaf-eaters; gregarious habits |
| *Plusia* spp. | Semi-loopers | Noctuidae | Cosmopolitan | Larvae polyphagous defoliators |
| *Cnephasia* spp. | Polyphagous Leaf Tiers | Tortricidae | Europe, Asia N. America | Larvae polyphagous leaf eaters |
| *Delia platura* (Meign.) | Bean Seed Fly | Anthomyiidae | Cosmopolitan | Larvae eat germinating seeds & seedlings |
| *Tipula* spp. | Leatherjackets (Common Crane Flies) | Tipulidae | Europe, Asia | Larvae polyphagous soil pests; eat roots & seedlings |
| *Nephrotoma* spp. | Spotted Crane Flies | Tipulidae | Europe, USA, Canada | |
| *Solenopsis geminata* (F.) | Fire Ant | Formicidae | Pantropical | Biting ant; nests in plantations & attacks farm workers |
| *Atta* spp. | Leaf-cutting Ants | Formicidae | C. & S. America | Adults are polyphagous defoliators |
| *Acromyrmex* spp. | Leaf-cutting Ants | Formicidae | C. & S. America | |
| *Vespa* spp. | Common Wasps | Vespidae | Cosmopolitan | Adults pierce ripe fruits |
| *Agriotes* spp. | Wireworms | Elateridae | Europe, etc. | Larvae polyphagous soil pests |

| | | | | |
|---|---|---|---|---|
| *Popillia japonica* Newm. | Japanese Beetle | Scarabaeidae (Rutelinae) | E. Asia, N. Europe, USA | Adults defoliate |
| *Anomala* spp. | Flower Beetles (White Grubs) | Scarabaeidae | Cosmopolitan | Larvae in soil eat plant roots; adults eat leaves |
| *Adoretus* spp. | Flower Beetles | Scarabaeidae | Pantropical | |
| *Cetonia* spp. | Rose Chafers (White Grubs) | Scarabaeidae (Cetonidae) | Cosmopolitan | Larvae in soil eat roots; adults feed on nectar & ripe fruits only |
| *Protaetia* spp. | Rose Chafers | Scarabaeidae | Pantropical | |
| *Melolontha* spp. *Holotrichia* spp. *Serica* spp. *Leucopholis* spp. *Schizonycha* spp. | Cockchafers (Chafers) (Chafer Grubs) (White Grubs) | Scarabaeidae (Melolonthinae) | Cosmopolitan | Larvae in soil; polyphagous pests; adults eat leaves, & unripe fruits sometimes |
| *Otiorhynchus* spp. | Clay-coloured Weevils | Curculionidae | Europe | Larvae eat roots in soil; adults eat leaf edges |
| *Hypomeces squamosus* (F.) | Gold-dust Weevil | Curculionidae | India, S.E. Asia, China | Adults eat leaves |
| *Systates* spp. | Systates Weevils | Curculionidae | Africa | Adults eat leaves |
| *Myllocerus* spp. | Grey Weevils | Curculionidae | India | Adults eat leaves |
| *Tetranychus cinnabarinus* (Boisd.) | Tropical Red Spider Mite | Tetranychidae | Pantropical | Adults & nymphs scarify foliage |
| *Tetranychus urticae* (Koch) | Temperate Red Spider Mite | Tetranychidae | Cosmopolitan in temperate countries | |
| *Panonychus ulmi* (Koch) | Fruit Tree Red Spider Mite | Tetranychidae | Cosmopolitan in temperate countries | |

687

# INSECT PESTS OF STORED PRODUCTS

The number of types of foodstuffs, plant and animal material stored, and the vast range of on-farm stores, barns, warehouses and godowns, is so great and varied that it is not feasible to make generalizations about either the stored products or the stores. But the range of pests encountered in the different parts of the world is much the same irrespective of the precise locality, and the pests do show certain definite preferences in relation to their choice of food; some species will only feed on pulses, others only on dried animal material, or grains, or flours. Some species are primary pests in that they can attack intact seeds and grains; the secondary pests are unable to do this and only feed on damaged grains. A few species are definitely tropical and some others equally temperate, so their distribution is not so wide as other species. Often the recognition and control of stored products pests does not come within the responsibility of the agricultural entomologist, but sometimes it does.

## *MAJOR PESTS*

| | | | | |
|---|---|---|---|---|
| *Sitotroga cerealella* (Ol.) | Angoumois Grain Moth | Gelechiidae | Widespread | Grains & foodstuffs |
| *Ephestia cautella* (Hb.) | Dried Currant Moth | Pyralidae | Widespread | Dried fruits |
| *Ephestia elutella* (Hb.) | Warehouse Moth | Pyralidae | Widespread | Dried fruits, cocoa beans, tobacco etc. |
| *Lasioderma serricorne* (F.) | Tobacco Beetle (Cigarette Beetle) | Anobiidae | Tropical | Dried tobacco, foodstuffs etc. |
| *Tribolium castaneum* (Herbst) | Red Flour Beetle | Tenebrionidae | Widespread | Flours & grain products |
| *Oryzaephilus surinamensis* (L.) | Saw-toothed Grain Beetle | Silvaniidae | Tropical | Secondary pests of foodstuffs |
| *Oryzaephilus mercator* (Fauvel) | Merchant Grain Beetle | Silvaniidae | Tropical | |
| *Rhizopertha dominica* (F.) | Lesser Grain Borer | Bostrychidae | Tropical | Grains |
| *Acanthoscelides obtectus* (Say) | Bean Bruchid | Bruchidae | Tropical | Pulses only |
| *Callosobruchus maculatus* (F.) | Spotted Cowpea Bruchid | Bruchidae | Tropical | Pulses only |
| *Callosobruchus chinensis* (L.) | Oriental Cowpea Bruchid | Bruchidae | Tropical | Pulses only |
| *Caryedon serratus* (Oliv.) | Groundnut Borer | Bruchidae | Tropical | Pulses, groundnuts |
| *Dermestes lardarius* L. | Larder Beetle | Dermestidae | Widespread | Dried animal matter |
| *Dermestes maculatus* Deg. | Hide Beetle | Dermestidae | Widespread | Dried animal matter |
| *Trogoderma granarium* Everts. | Khapra Beetle | Dermestidae | Tropical | Grains, groundnut |
| *Sitophilus oryzae* (L.) | Rice Weevil | Curculionidae | Widespread | Rice, maize, foodstuffs etc. |
| *Sitophilus zeamais* Motsch. | Maize Weevil | Curculionidae | Widespread | Maize, rice, foodstuffs etc. |
| *Araeocerus fasciculatus* (Deg.) | Coffee Bean Weevil | Brenthidae | Tropical | Coffee beans, seeds etc. |

## MINOR PESTS

| | | | | |
|---|---|---|---|---|
| *Lepisma saccharina* L. | 'Silverfish' | Thysanura | Cosmopolitan | Scavenger; polyphagous |
| *Periplaneta americana* (L.) | American Cockroach | Blattidae | Cosmopolitan | Scavenger; polyphagous |
| *Blatta orientalis* L. | Oriental Cockroach | Blattidae | Temperate | Scavenger; polyphagous |
| *Blatella germanica* (L.) | German Cockroach | Blattidae | Cosmopolitan | Scavenger; polyphagous |
| *Acheta domesticus* (L.) | House Cricket | Gryllidae | Temperate | Scavenger; polyphagous |
| *Ephestia kuehniella* (Zell.) | Mediterranean Flour Moth | Pyralidae | Sub-tropical | Flours mostly |
| *Plodia interpunctella* (Hub.) | Indian Meal Moth | Pyralidae | Tropical | Dried fruits, meals, flours etc. |
| *Carpophilus hemipterus* (L.) | Dried Fruit Beetle | Nitidulidae | Widespread | Dried fruits |
| *Attagenus piceus* Oliv. | Black Carpet Beetle | Dermestidae | Widespread | Dried animal matter |
| *Ahasverus advena* (Waltl.) | Foreign Grain Beetle | Silvanidae | Widespread | Fungus feeder |
| *Necrobia rufipes* (Deg.) | Copra Beetle | Cleridae | Tropical | Copra, oil seeds, dried meats |
| *Typhaea stercorea* (L.) | Hairy Fungus Beetle | Mycetophagidae | Widespread | Fungus feeder |
| *Stegobium paniceum* (L.) | Drug Store Beetle | Anobiidae | Temperate | Foodstuffs |
| *Cryptolestes ferrugineas* (Steph.) | Rust-red Grain Beetle | Cucujidae | Widespread | Grains |
| *Ptinus* spp. | Spider Beetles | Ptinidae | Tropical | Miscellaneous foodstuffs |
| *Tenebrio molitor* L. | Yellow Mealworm Beetle | Tenebrionidae | Widespread | Flours & foodstuffs |
| *Tribolium confusum* J. du V. | Confused Flour Beetle | Tenebrionidae | Widespread | Flours mostly |
| *Tenebriodes mauritanicus* (L.) | Cadelle | Tenebrionidae | Widespread | Miscellaneous foodstuffs |
| *Sitophilus granarius* (L.) | Grain Weevil | Curculionidae | Temperate | Wheat, other grains, foodstuff etc. |
| *Acarus siro* | Flour Mite | Acaridae | Widespread | Flours, meals etc. |

# 9  General bibliography

Akehurst, B. C. (1968). *Tobacco*, 551 pp. Longmans: London.

Allee, W. C., A. E. Emerson, O. Park, T. Park & K. P. Schmidt (1955). *Principles of Animal Ecology*, 835 pp. W. B. Saunders: Philadelphia.

Amsden, R. C. & C. P. Lewins (1966). Assessment of wettability of leaves by dipping in crystal violet. *World Rev. Pest Control* 5, 187–94.

Andrewartha, H. G. & L. C. Birch (1954 & 1961). *The Distribution and Abundance of Animals*, 782 pp. University of Chicago Press: Chicago & London.

Angus, A. (1962). *Annotated list of plant pests, diseases and fungi in Northern Rhodesia, recorded at the Plant Pathology Laboratory, Mount Makulu Research Station*, parts 1 to 7, and supplement (cyclostyled), *c.* 600 pp.

Anon. (1952). *Agriculture (Poisonous Substances) Act, 1952*, 9 pp. HMSO: London.

Anon. (1961). Farm sprayers and their use. *MAFF Bulletin no. 182*, 99 pp. HMSO: London.

Anon. (1964). *Bibliography on Insect Pest Resistance in Plants*, 39 pp. Imp. Bur. Pl. Breed. Genetics: Cambridge.

Anon. (1964). *A Handbook on Arabica Coffee in Tanganyika*, 182 pp. Tanzania Coffee Board: Lyamungu, Tanzania.

Anon. (1965*a*). *An Atlas of Coffee Pests and Diseases*, 146 pp. Coffee Res. Found.: Ruiru, Kenya.

Anon. (1965*b*). Conversion tables for research workers in forestry and agriculture. *Forestry Commission Booklet no. 5*, 64 pp. HMSO: London.

Anon. (1965*c*). *A Handbook for Sisal Planters*, *c.* 100 pp. Tanganyika Sisal Growers Assoc.

Anon. (1967). *Coffee Pests and their Control*, 90 pp. Coffee Res. Found.: Ruiru, Kenya.

Anon. (1971). *VIIth Int. Congr. Plant Protection*, 866 pp. Sec. Gen.: Paris.

Anti-Locust Research Centre (1966). *The Locust Handbook*, 276 pp. Anti-Locust Research Centre: London.

Apple, J. L. & R. F. Smith (1976). *Integrated Pest Management*, 200 pp. Plenum Publishing Corporation: New York.

Ashworth, R. de B. & G. A. Lloyd (1961). Laboratory and field tests for evaluating the efficiency of wetting agents used in agriculture. *J. Sci. Food Agric.* 12, 234–40.

Avidoz, Z. & I. Harpaz (1969). *Plant Pests of Israel*, 549 pp. Israel University Press: Jerusalem.

Bailey, S. F. (1938). Thrips of economic importance in California. *Agric. Expt. Sta., Berkeley, California*, circ. 346, 77 pp.

Bailey, S. F. (1964). A revision of the genus *Scirtothrips* Shull (Thysanoptera: Thripidae). *Hilgardia* 35, 329–62.

Baker, E. W. & A. E. Pritchard (1960). The Tetranychoid mites of Africa. *Hilgardia* 29, 455–574.

Baker, E. W. & G. W. Wharton (1964). *An Introduction to Acarology*, 465 pp. Macmillan: New York.

Balachowsky, A. S. (ed.) (1962). *Entomologie Appliquée à l'Agriculture*, 8 vols. Masson: Paris.

Balachowsky, A. S. & L. Mesnil (1935). *Les Insectes Nuisibles aux Plantes Cultivées*, 2 vols., 1921 pp. Min. Agric.: Paris.

Bals, E. J. (1970). Ultra low volume and ultra low dosage spraying. *Cott. Gr. Rev.* 47, 217–21.

Banerjee, B. (1981). An analysis of the effect of latitude, age and area on the number of arthropod pest species of tea. *J. Appl. Ecol.* 18, 339–42.

Barnes, H. F. (1939). Gall midges (Cecidomyiidae) associated with coffee. *Rev. Zool. Bot. Afr.* 32, 324–36.

Barnes, H. F. (1946–59). *Gall Midges of Economic Importance*, Crosby Lockwood & Son Ltd.: London.
(1946) Vol. I, *Root and Vegetable Crops*, 104 pp.
(1946) Vol. II, *Fodder Crops*, 160 pp.
(1948) Vol. III, *Fruit*, 184 pp.
(1948) Vol. IV, *Ornamental Plants and Shrubs*, 165 pp.
(1951) Vol. V, *Trees*, 270 pp.
(1949) Vol. VI, *Miscellaneous Crops*, 229 pp.
(1956) Vol. VII, *Cereal Crops*, 261 pp.
(W. Nijveldt) (1969) Vol. VIII, *Miscellaneous*, 221 pp.

Barrass, R. (1964). *The Locust*, 59 pp. Butterworths: London.

Bateman, M. A. (1972). The ecology of fruit flies. *Ann. Rev. Entomol.* 17, 493–518.

Baum, H. (1968). The coffee root mealybug complex. *Kenya Coffee*, 1–4.

'Bayer' (1965). *Manual of Plant Protection in Coffee*, 43 pp. Bayer Co.: Leverkusen.

'Bayer' (1965). *Manual of Crop Protection in Cotton*, 61 pp. Bayer Co.: Leverkusen.

'Bayer' (1968). *Bayer Crop Protection Compendium*, 2 vols., 511 pp. Bayer Co.: Leverkusen.

Beardsley, J. W. & R. H. Gonzalez (1975). The biology and ecology of armoured scales. *Ann. Rev. Entomol.* **20**, 47–73.

Beirne, B. P. (1967). *Pest Management*, 123 pp. Leonard Hill: London.

Bell, T. R. D. & F. B. Scott (1937). *The Fauna of British India*, Vol. 5. *Moths*, 537 pp. Taylor & Francis: London.

Bellotti, A. & A. van Schoonhoven (1978). Mite and insect pests of cassava. *Ann. Rev. Entomol.* **23**, 39–67.

Bennett, F. D. (1971). Current status of biological control of the small moth borers of sugarcane *Diatraea* spp. (Lep. Pyralidae). *Entomophaga* **16**, 111–24.

Berger, R. S. (1968). Sex pheromone of the Cotton Leafworm. *J. econ. Ent.* **61**, 326–7.

Berger, R. S., J. M. McGough & D. F. Martin (1965). Sex attractants of *Heliothis zea* and *H. virescens. J. econ. Ent.* **58**, 1023–4.

Beroza, M. (1964). Insect sex attractants and their use. *Proc. 2nd Int. Congr. Endocrinol.* 203–8.

Beroza, M. (ed.) (1970). *Chemicals Controlling Insect Behaviour*, 182 pp. Academic Press: New York.

Bigger, M. (1966). The biology and control of termites damaging field crops in Tanganyika. *Bull. ent. Res.* **56**, 417–44.

Bleszynski, S. (1970). A revision of the world species of *Chilo* Zincken (Lep.: Pyralidae). *Bull. Br. Mus. Nat. His.* (B) **25**, 97 pp.

Blood, P. B. & A. Bishop (1975). Biological control of cotton looper *Anomis flava* (Fabr.) larval populations in untreated cotton in S.E. Queensland 1973–74 *I.P.M.U. Res. Paper 1975/4.*

Bodenheimer, F. S. (1951). *Citrus Entomology in the Middle East*, 664 pp. W. Junk: The Hague.

Bodenheimer, F. S. & E. Swirski (1957). *The Aphidoidea of the Middle East*, 378 pp. Weizman Sci. Press: Jerusalem.

Borror, D. J. & D. M. DeLong (1971). *An Introduction to the Study of Insects*, 3rd ed., 812 pp. Holt, Rinehart & Winston: New York.

Box, H. E. (1953). *List of Sugar Cane Insects*, 101 pp. CIE: London.

Bradley, J. D. (1967). Some Lepidoptera of economic importance in Commonwealth countries. *Acta Universitatis Agriculturae* **15**, 501–19.

Bradley, J. D. (1968). Two new species of clearwing moths (Lepidoptera: Sesiidae) associated with sweet potato (*Ipomoea batatas*) in East Africa. *Bull. ent. Res.* **58**, 47–53.

Brooks (1980). *See* Int. Congress of Entomology (1980).

Brooks, A. R. (1951). Identification of the root maggots (Diptera: Anthomyiidae) attacking cruciferous garden crops in Canada, with notes on biology and control. *Canad. Ent.* **83**, 109–20.

Brown, A. W. A. & R. Pal (1971). *Insecticide Resistance in Arthropods*, 2nd ed. WHO: Geneva.

Brown, E. S. (1954). The biology of the coconut pest *Melittomma insulare* (Col., Lymexylonidae), and its control in the Seychelles. *Bull. ent. Res.* **45**, 1–66.

Brown, E. S. (1955). *Pseudotherapterus wayi*, a new genus and species of coreid (Hemiptera) injurious to coconuts in East Africa. *Bull. ent. Res.* **46**, 221–40.

Brown, E. S. (1962). *The African Armyworm* Spodoptera exempta (*Walker*) (*Lepidoptera, Noctuidae*): *a review of the literature*, 69 pp. CIE: London.

Brown, E. S. (1972). Armyworm control. *PANS* **18**, 197–204.

Brown, E. S., E. Betts & R. C. Rainey (1969). Seasonal changes in distribution of the African Armyworm, *Spodoptera exempta* (Wlk.) (Lep., Noctuidae), with special reference to eastern Africa. *Bull. ent. Res.* **58**, 661–728.

Brown, E. S. & C. F. Dewhurst (1975). The genus *Spodoptera* (Lepidoptera, Noctuidae) in Africa and the Near East. *Bull. ent. Res.* **65**, 221–62.

Brown, F. G. (1968). *Pests and Diseases of Forest Plantation Trees: an annotated list of the principle species occurring in the British Commonwealth*, 1330 pp. Oxford: Clarendon Press.

Brown, K. W. (1967). *Forest Insects of Uganda (an annotated list)*, 98 pp. Govt. Printer: Entebbe.

BSI (1969). *Recommended Common Names for Pesticides*, 4th revision, 108 pp. Brit. Stand. Inst.: London.

Bucher, G. E. & H. H. Cheng (1970). Use of trap crops for attracting cutworm larvae. *Canad. Ent.* **102**, 797–8.

Bullock, J. A. (1965). The control of *Hylemya arambourgi* Seguy (Dipt., Anthomyiidae) on barley. *Bull. ent. Res.* **55**, 645–61.

Burges, H. D. (ed.) (1981). *Microbial Control of Pests and Plant Diseases 1970–1980*, 914 pp. Academic Press: London.

Burges, H. D. & N. W. Hussey (1971). *Microbial Control of Insects and Mites*, 861 pp. Academic Press: London.

Busvine, J. R. (1966). *Insects and Hygiene*, 2nd ed. 467 pp. Methuen: London.

Busvine, J. R. (1971*a*). *A Critical Review of the Techniques for Testing Insecticides*, 2nd ed., 345 pp. CIE: London.

Busvine, J. R. (1971*b*). The biochemical and genetic bases of insecticidal resistance. *PANS* **17**, 135–46.

Butani, D. K. (1970). Insect pests of cotton, Western Herbaceum Region of India. *PANS* **16**, 56–64.

Butani, D. K. (1973). Insect pests of fruit crops and their control – 1: Ber. *Pesticides* **7**, 33–5.

Butani, D. K. (1974*a*). Pests of fruit crops in India and their control – 12: Loquat. *Pesticides* **8**, 17–18.

Butani, D. K. (1974*b*). Insect pests of fruit crops and their control – 10: Grapes. *Pesticides* **8** (**10**), 25–9.

Butani, D. K. (1974*c*). Insect pests of fruit crops and their control – 11: Guava. *Pesticides* **8** (**11**), 26–30.

Butani, D. K. (1975*a*). Crop pests and their control – 1: Cotton. *Pesticides* **9**, 21–9.

Butani, D. K. (1975*b*). Insect pests of fruit crops and their control – 16: Fig. *Pesticides* **9**, 32–6.

Butani, D. K. (1975*c*). Insect pests of fruit crops and their control – 17: Sapota. *Pesticides* **9**, 37–9.

Butani, D. K. (1975*d*). Insect pests of fruit crops and their control – 15: Date Palm. *Pesticides* **9**, 40–2.

Butani, D. K. (1976*a*). Insect pests of fruit crops and their control – 21: Pomegranate. *Pesticides* **10**, 23–6.

Butani, D. K. (1976*b*). Insect pests of fruit crops and their control – 20: Custard Apple. *Pesticides* **10**, 27–9.

Butani, D. K. (1976*c*). Pests and diseases of chillies and their control. *Pesticides* **10**, 38–41.

Butani, D. K. (1977). Pests of fruit crops and their control: Litchi. *Pesticides* **11**, 43–8.

Butani, D. K. (1978*a*). Pests and diseases of Jackfruit in India and their control. *Fruits* **33**, 351–7.

Butani, D. K. (1978*b*). Insect pests of Tamarind and their control. *Pesticides* **12**, 34–41.

Butani, D. K. (1978*c*). Insect pests of fruit crops and their control – 25: Mulberry. *Pesticides* **12**, 53–9.

Butani, D. K. (1979*a*). *Insects and Fruits*, 415 pp. Periodical Expert Book Agency: New Delhi.

Butani, D. K. (1979*b*). Insect pests of *Citrus* and their control. *Pesticides* **13**, 27–33.

Butani, D. K. & S. Varma (1976*a*). Pests of vegetables and their control: – Brinjal. *Pesticides* **10** (**2**) 32–5.

Butani, D. K. & S. Varma (1976*b*). Insect pests of vegetables and their control: Onion and Garlic. *Pesticides* **10** (**11**) 33–5.

Butani, D. K. & S. Varma (1976*c*). Pests of vegetables and their control: Sweet Potato. *Pesticides* **10**, 36–8.

Butani, D. K. & S. Varma (1977). Pests of vegetables and their control: Cucurbits. *Pesticides* **11**, 37–41.

Buyckx, E. J. E. (1962). *Précis des Maladies et des Insectes nuisibles recontrées sur les Plants Cultivées au Congo, au Rwanda et au Burundi*, 708 pp. INEAC.

Byass, J. B. & J. Holroyd (eds.) (1970). Proceedings of a Symposium for Research Workers on Pesticide Application. *Br. Crop. Prot. Council, Mon. no. 2*, 139 pp. Boots Pure Drug Co.: Nottingham.

CAB (1951–81). *Distribution Maps of Insect Pests, Series A (Agricultural)*, nos. 1–423, with index (1–234). CIE: London.

CAB (1961–7). *CIBC Technical Bulletins, 1–8*. CAB: London.

CAB (1969). *List of Research Workers in the Agricultural Sciences in the Commonwealth and in the Republic of Ireland*, 3rd ed., 732 pp. CAB: London.

CAB (1980). *Perspectives in World Agriculture*, 532 pp. CAB: Slough.

Campion, D. G. (1969). Factors affecting the use of a chemosterilising bait-station for control of the red bollworm *Diaparopis castanea* (Hmps). *PANS* **15**, 535–41.

Campion, D. G. (1972). Insect chemosterilants: a review. *Bull. ent. Res.* **61**, 577–635.

Caresche, L., G. S. Cotterell, J. E. Peachey, R. W. Rayner & H. Jacques-Felix (1969). *Handbook for Phytosanitary Inspectors in Africa*, 444 pp. OAU/STRC: Lagos.

Carpenter, J. B. & H. S. Elmer (1978). *Pests and Diseases of* the *Date Palm*, 42 pp. U.S. Dept. Agric., Agric. Hdbk No. 527.

Caswell, G. H. (1962). *Agricultural Entomology in the Tropics*, 152 pp. Edward Arnold: London.

Chapman, R. F. (1970). *The Insects – structure and function*, 819 pp. English Universities Press: London.

Cherrett, J. M., J. B. Ford, I. V. Herbert & A. J. Probert (1971). *The Control of Injurious Animals*, 210 pp. English Universities Press: London.

Cherrett, J. M. & T. Lewis (1974). *Control of insects by exploiting their behaviour*, pp. 130–46. In: *Biology in Pest and Disease Control,* eds. Price Jones, D. & M. E. Solomon, Blackwell: Oxford.

Cherrett, J. M. & D. J. Peregrine (1976). A review of the status of leaf-cutting ants and their control. *Ann. appl. Biol.* **84**, 128–33.

Cherrett, J. M. & G. R. Sagar (1977). *Origins of Pest, Parasite, Disease and Weed Problems*, (18th Symp. Brit. Ecol. Soc.) 413 pp. Blackwell: Oxford.

Child, R. (1964). *Coconuts*, 216 pp. Longmans: London.

Chu (1980). *See* Int. Congress of Entomology (1980).

Clark, L. R., P. W. Geier, R. D. Hughes & R. F. Morris (1967). *The Ecology of Insect Populations in Theory and Practice*, 232 pp. Methuen: London.

Clausen, C. P. (1940). *Entomophagous Insects*, 1st ed., 688 pp. McGraw-Hill: New York.

Clayphon, J. E. (1971). Comparison trials of various motorised knapsack mist-blowers at the Cocoa Research Institute of Ghana, *PANS* **17**, 209–25.

Clearwater, J. R. (1981). Practical identification of the females of five species of *Atherigona* Rondani (Diptera, Muscidae) in Kenya. *Tropical Pest Management* (formerly PANS) **27**, 303–12.

Coaker, T. H. (1958). Experiments with a virus disease of the Cotton Bollworm *Heliothis armigera* (Hb.) *Ann. appl. Biol.* **46**, 537–41.

Coaker, T. H. (1959). Investigations on *Heliothis armigera* (Hb.) in Uganda. *Bull ent. Res.* **50**, 487–506.

Coaker, T. H. & S. Finch (1971). The cabbage root fly, *Erioischia brassicae* (Bouché). *Rep. natn. Veg. Res. Stn.* 1970, 23–42.

Coffee, R. (1973). Electrostatic crop spraying. *New Scientist* **84**, 194–6.

Conway, G. R. (1972*a*). *Pests of Cocoa in Sabah, Malaysia,* Bull. Dept. Agric.: Malaysia.

Conway, G. R. (1972*b*). *Ecological Aspects of Pest Control in Malaysia.* In: J. Milton (ed.), *The Careless Technology; Ecological Aspects of International Development*, Nat. Hist. Press.

Conway, G. R. & E. B. Tay (1969). *Crop Pests in Sabah, Malaysia and their Control*, 73 pp. St. Min. Agric. Fish., Sabah, Malaysia.

Cope, O. B. (1971). Interactions between pesticides and wild life. *Ann. Rev. Ent.* **16**, 325–64.

COPR (1978). *Pest Control in Tropical Root Crops*, 235 pp. (PANS manual No. 4) COPR: London.

COPR (1981). *Pest Control in Tropical Grain Legumes,* 206 pp. COPR: London.

Corpuz, L. R. (1969). The biology, host range, and natural enemies of *Nezara viridula* L. (Pentatomidae, Hemiptera). *Philipp. Ent.* **1**, 227–39.

Cotterell, G. S. (1963). The more important insect pests of limited distribution in Africa which attack economic plants, and their world distribution. *IAPSC Doc. no. 63* (3).

Coursey, D. G. (1972). *Yams*, 2nd ed., 230 pp. Longmans: London. (1st ed., 1969).

Cramer, H. H. (1967). *Plant Protection and World Crop Production*, 254 pp. Bayer Pflanzenschutz: Leverkusen.

Cranham, J. E. (1966*a*). *Insect and Mite Pests of Tea in Ceylon and their Control*, 122 pp. Tea Res. Inst. Ceylon: Talawakelle.

Cranham, J. E. (1966*b*). Tea pests and their control. *Ann. Rev. Entomol.* **11**, 491–510.

Crowe, T. J. (1960). The leaf skeletonizer. *Kenya Coffee*, 1–2.

Crowe, T. J. (1962*a*). The biology and control of *Diphya nigricornis* (Oliver), a pest of coffee in Kenya (Coleoptera: Cerambycidae). *J. ent. Soc. S. Afr.* **25**, 304–12.

Crowe, T. J. (1962*b*). The white waxy scale. *Kenya Coffee*, 1–3.

Crowe, T. J. (1962*c*). The star scale. *Kenya Coffee*, 1–4.

Crowe, T. J. (1964). Coffee Leafminers in Kenya. II – Causes of outbreaks. III – Control measures. *Kenya Coffee*, 1–4, 1–5.

Crowe, T. J. (1967*a*). Common names for agricultural and forestry insects and mites in East Africa. *E. Afr. Agric. for. J.* **33**, 55–63.

Crowe, T. J. (1967*b*). *Cotton Pests and their Control,* c. 20 pp. Dept. of Agric.: Nairobi, Kenya.

Crowe, T. J. & J. Leeuwangh (1965). The green looper. *Kenya Coffee*, 1–4.

CSCPRC (1977). *Insect Control in the People's Republic of China*, 218 pp. National Academy of Sciences: Washington.

CSIRO (1973). Scientific and common names of insects and allied forms occurring in Australia. *CSIRO Bull.* 287, 47 pp.

CSIRO (*see also* Mackerras).

Danilevskii, A. S. (1965). *Photoperiodism and Seasonal Development of Insects*, 283 pp. (Translation of 1961 edition in Russian.) Oliver & Boyd: London.

Darlington, A. (1968). *The Pocket Encyclopedia of Plant Galls in Colour*, 191 pp. Blandford Press: London.

Davidson, R. H. & L. M. Peairs (1966). *Insect Pests of Farm, Garden and Orchard*, 6th ed., 675 pp. John Wiley: New York.

Dean, G. J. (1979). The major pests of rice, sugarcane and jute in Bangladesh. *PANS* 25, 378–85.

DeBach, P. (1964). *Biological Control of Insect Pests and Weeds*, 844 pp. Chapman & Hall: London.

DeBach, P. (1969). Biological control of Diaspine scale insects on Citrus in California. *Proc. 1st Int. Citrus Symp.* 2, 801–16.

DeBach, P. (1971). *Fortuitous biological control from ecesis of natural enemies*. In: *Entomological Essays to Commemorate the Retirement of Professor K. Yasumatsu.* pp. 293–307, Hokuryukan Pub. Co. Ltd.: Tokyo.

DeBach, P. (1974). *Biological Control by Natural Enemies*, 323 pp. Cambridge University Press: Cambridge.

DeBach, P. & C. B. Huffaker (1971). Experimental techniques for evaluation of the effectiveness of natural enemies. In: C. B. Huffaker (ed.), *Biological Control*, pp. 113–40. Plenum: New York.

DeBach, P., D. Rosen & C. E. Kennett (1971). Biological control of coccids by introduced natural enemies. In: C. B. Huffaker (ed.), *Biological Control*, pp. 165–94. Plenum: New York.

Deeming, J. C. (1971). Some species of *Atherigona* Rondani (Diptera; Muscidae) from Northern Nigeria, with special reference to those injurious to cereal crops. *Bull. ent. Res.* 61, 133–90.

Dekle, G. W. (1970). Oleander Scale (*Phenacaspis cockerelli* (Cooley)) (Homoptera: Diaspididae). *Florida Dept. Agric., Ent. Circ. No. 95,* 2 pp.

Dekle, G. W. (1971). Red Wax Scale (*Ceroplastes rubens* Maskell) Coccidae – Homoptera. *Florida Dept. Agric., Ent. Circ. No. 115,* 2 pp.

De Long, D. (1971). The bionomics of leafhoppers. *Ann. Rev. Ent.* 16, 179–210.

De Lotto, G. (1967). The soft scales (Homoptera, Coccidae) of South Africa. I. *S. Afr. J. Agric. Sci.* 10, 781–810.

Den Otter (1980). *See* Int. Congress of Entomology (1980).

Doggett, H. (1970). *Sorghum*, 403 pp. Longmans: London.

Drew, R. A. I., G. H. S. Hooper & M. A. Bateman (1978). *Economic Fruit Flies of the South Pacific Region*, 137 pp. Dept. Primary Industries: Queensland.

Duffey, E. A. J. (1957). *African Timber Beetles*, 338 pp. British Museum (NH): London.

Dunbar, A. R. (1969). *The Annual Crops of Uganda*, 189 pp. E. Afr. Lit. Bur.: Kampala.

Duval, C. T. (1970). Some introductory aspects of the chemical relationships and nomenclature of synthetic organic insecticides. *PANS* 16, 11–35.

Eastop, V. F. (1958). A study of the Aphididae (Homoptera) of East Africa. *Col. Res. Pub. no. 20*, 77 pp. HMSO: London.

Eastop, V. F. (1961). *A study of the Aphididae (Homoptera) of West Africa*, 93 pp. British Museum (NH): London.

Eastop, V. F. (1966). A taxonomic study of Australian Aphidoidea (Homoptera). *Aust. J. Zool.* 14, 399–592.

Eastop, V. F. (1971). Keys for identification of *Acyrthosiphon* (Hemiptera: Aphididae). *Bull. Br. Mus. Nat. Hist., Ent.* 26, 1–115.

Ebbels, D. L. & J. E. King (1979). *Plant Health*, 322 pp. Blackwell: Oxford.

Ebeling, W. (1959). *Subtropical Fruit Pests*, 2nd ed., 436 pp. University of California Press: California.

Ebeling, W. (1971). Sorptive dusts for pest control. *Ann. Rev. Entomol.* 16, 123–58.

Ebeling, W. (1975). *Urban Entomology*, 695 pp. University of California, Division of Agricultural Sciences: California.

Eden, T. (1965). *Tea*, 2nd ed., 205 pp. Longmans: London.

Edwards, C. A. (1970a). Problem of insecticidal residues in agricultural soils. *PANS* 16, 271–6.

Edwards, C. A. (1970b). *Persistent Pesticides in the Environment*, 77 pp. Butterworths: London.

Edwards, C. A. & G. W. Heath (1964). *Principles of Agricultural Entomology*, 418 pp. Chapman & Hall: London.

Elton, C. S. (1958). *The Ecology of Invasions by Animals and Plants*, 181 pp. Methuen: London.

Emden, H. F. van, V. F. Eastop, R. D. Hughes & M. J. Way (1969). The ecology of *Myzus persicae*. *Ann. Rev. Entomol.* **14**, 197–270.

Entwhistle, P. F. (1972). *Pests of Cocoa*, 804 pp. Longmans: London.

EPA (1976). *List of Insects and other Organisms*, (3rd edition) Parts I, II, III, and IV. EPA: Washington.

EPPO (1970). Report of the International Conference on Methods for Forecasting, Warning, Pest Assessment and Detection of Infestation. *EPPO Pub. Ser.* no. 57, 206 pp. EPPO: Paris.

Evans, D. E. (1967). Insecticide field trials against coffee thrips (*Diarthrothrips coffeae* Williams) in Kenya. *Turrialba* **17**, 376–80.

Evans, D. E., V. Andrade & W. M. Mathenge (1968). The biology and control of *Archips occidentalis* (Wals) and *Tortrix dinota* Meyr. (Lepidoptera: Tortricidae) on coffee in Kenya. *J. ent. Soc. S. Afr.* **31**, 133–40.

Evans, J. W. (1952). *Injurious Insects of the British Commonwealth*, 242 pp. CIE: London.

FAO (1966). *Proceedings of the FAO Symposium on Integrated Pest Control* (11–15 October 1965). 1, 91 pp. 2, 186 pp. 3, 129 pp. FAO: Rome.

FAO/CAB (1971). *Crop Loss Assessments Methods*. FAO Manual on the evaluation of losses by pests, diseases and weeds, *c*. 130 pp. FAO: Rome. Supplement 1 (1973), 2 (1977).

Feakin, S. D. (ed.) (1971). *Pest Control in Bananas*. PANS Manual no. 1, 2nd ed., 128 pp. Centre for Overseas Pest Research: London.

Feakin, S. D. (ed.) (1973). *Pest Control in Groundnuts*. PANS Manual no. 2, 3rd ed., 197 pp. Centre for Overseas Pest Research: London.

Feakin, S. D. (1976). *Pest Control in Rice*. PANS manual no. 3, 2nd ed., 295 pp. COPR: London.

Fennah, R. G. (1947). *The Insect Pests of Food Crops in the Lesser Antilles*, 207 pp. Dept. Agric.: Antigua, BWI.

Fennah, R. G. (1963). The species of *Pyrilla* (Fulgoroidea: Lophopidae) in Ceylon and India. *Bull. ent. Res.* **53**, 715–35.

Ferro, D. N. (ed.) (1976). *New Zealand Insect Pests*, 311 pp. Lincoln Univ. Coll. Agric.: Canterbury.

Fichter, G. S. (1968). *Insect Pests*, 160 pp. Paul Hamlyn: London.

Firman, I. D. (1970). Crop protection problems of bananas in Fiji. *PANS* **16**, 625–31.

Fletcher, W. W. (1974). *The Pest War*, 218 pp. Blackwell: Oxford.

Florkin, M. & B. T. Scheer (1970). *Chemical Zoology*, Vol. V. *Arthropoda*, A, 460 pp. Academic Press: New York.

Florkin, M. & B. T. Scheer (1971). *Chemical Zoology*, Vol. VI. *Arthropoda*, B, 484 pp. Academic Press: New York.

Forsyth, J. (1966). *Agricultural Insects of Ghana*, 163 pp. Ghana University Press.

Fox Wilson, G. (1960). *Horticultural Pests – Detection and Control*, 2nd ed., 240 pp. Crosby Lockwood: London.

Free, J. B. (1970). *Insect Pollination of Crops*, 544 pp. Academic Press: London.

Free, J. B. & I. H. Williams (1977). *The Pollination of Crops by Bees*, 14 pp. Apimondia: Bucharest & Int. Bee Res. Assoc., UK.

Freeman, G. H. (1967). Problems in plant pathology and entomology field trials. *Exp. Agric.* **3**, 351–8.

Freeman, P. (1939). A contribution to the study of the genus *Calidea* Laporte (Hemiptera–Heteropt., Pentatomidae). *Trans. R. ent. Soc. Lond.* **88**, 139–60.

Freeman, P. (1940). A contribution to the study of the genus *Nezara* A. & S. (Hemiptera, Pentatomidae). *Trans. R. ent. Soc. Lond.* **90**, 351–74.

Freeman, P. (1947). A revision of the genus *Dysdercus* Boisduval (Hemiptera, Pyrrhocoridae) excluding the American species. *Trans. R. ent. Soc. London.* **98**, 373–424.

Frohlich, G. & W. Rodewald (1970). *General Pests and Diseases of Tropical Crops and their Control*, 366 pp. Pergamon Press: London.

Gair, R. (1968). The conduct of field variety trials involving fertilisers and pesticides or both. *PANS* (A) **14**, 216–30.

Geering, Q. A. (1953). The sorghum midge, *Contarinia sorghicola* (Coq.) in East Africa. *Bull. ent. Res.* **44**, 363–6.

Geier, P. W. (1966). Management of insect pests. *Ann. Rev. Entomol.* **11**, 471–90.

Geier, P. W., L. R. Clark, D. J. Anderson & H. A. Nix (eds.)

(1973). *Insects: studies in population management*, 294 pp. Ecol. Soc. Australia, Memoir 1: Canberra.

Gentry, J. W. (1965). *Crop Pests of Northeast Africa–Southwest Asia*. USDA.

Ghauri, M. S. K. (1971). Revision of the genus *Nephotettix* Matsumura (Homoptera: Cicadelloidea: Euscelidae) based upon the type material. *Bull. ent. Res.* 60, 481–512.

Glass, E. H. (co-ordinator) (1975). *Integrated Pest Management: rationale, potential, needs and implementation*, 141 pp. Ent. Soc. Amer., Special Pub. 75 – 2.

Gram, E., P. Bovien & C. Stapel (1969). *Recognition of Diseases and Pests of Farm Crops*, 2nd ed., 128 pp. Blandford Press: London.

Gray, B. (1972). Economic tropical forest entomology. *Ann. Rev. Entomol.* 17, 313–54.

Greathead, D. J. (1963). A review of the insect enemies of Acridoidea (Orthoptera). *Trans. R. ent. Soc. Lond.* 114, 437–517.

Greathead, D. J. (1966). A taxonomic study of the species of *Antestiopsis* (Hemiptera: Pentatomidae) associated with *Coffea arabica* in Africa. *Bull. ent. Res.* 56, 514–54.

Greathead, D. J. (1971). A review of biological control in the Ethiopian Region. *Tech. Commun. CIBC*, no. 5, 162 pp. CAB: London.

Greathead, D. J. (1972). Dispersal of the sugarcane scale *Aulacaspis tegalensis* (Zhnt.) (Hem., Diaspididae) by air currents. *Bull. ent. Res.* 61, 547–58.

Grist, D. H. (1970). *Rice*, 4th ed., 548 pp. Longmans: London.

Grist, D. H. & R. J. A. W. Lever (1969). *Pests of Rice*, 520 pp. Longmans: London.

Hagen, K. S. & R. van den Bosch (1968). Impact of pathogens, parasites and predators on aphids. *Ann. Rev. Entomol.* 13, 325–84.

Hainsworth, E. (1952). *Tea Pests and Diseases*, 130 pp. Heffers: Cambridge.

Halstead, D. G. H. (1964). The separation of *Sitophilus oryzae* (L.) and *S. zeamais* Motschulsky (Col., Curculionidae), with a summary of their distribution. *Ent. mon. Mag.* 99, 72–4.

Halstead, D. G. H. (1980). A revision of the genus *Oryzaephilus* Ganglbauer, including descriptions of related genera (Coleoptera: Silvanidae). *Zool. J. Linn. Soc.* 69, 271–374.

Hammad, S. M. & M. H. Mohamed (1965). Insect pests of *Pistacia* in the Aleppo District (Syria). *Bull. Soc. ent. Egypte* 49, 1–5.

Hanover, J. W. (1975). Physiology of tree resistance to insects. *Ann. Rev. Entomol.* 20, 75–95.

Harcourt, D. G. (1963). Major mortality factors in the population dynamics of the Diamondback moth, *Plutella maculipennis* (Curt.) (Lepidoptera: Plutellidae). *Mem. ent. Soc. Canada* 32, 55–66.

Harcourt, D. G. (1969). The development and use of life tables in the study of natural insect populations. *Ann. Rev. Ent.* 14, 175–96.

Hardwick, D. F. (1965). The Corn Earworm complex. *Mem. ent. Soc. Canada* 40, 1–247.

Harris, C. R. (1972). Factors influencing the effectiveness of soil insecticides. *Ann. Rev. Entomol.* 17, 177–98.

Harris, K. F. & K. Maramorosch (eds.) (1977). *Aphids as Virus Vectors*, 570 pp. Academic Press: New York.

Harris, K. F. & K. Maramorosch (eds.) (1980). *Vectors of Plant Pathogens*, 480 pp. Academic Press: New York.

Harris, K. M. (1961). The sorghum midge, *Contarinia sorghicola* Coq, in Nigeria. *Bull. ent. Res.* 52, 129–46.

Harris, K. M. (1962). Lepidopterous stem borers of cereals in Nigeria. *Bull. ent. Res.* 53, 139–71.

Harris, K. M. (1964*a*). Annual variations of dry-season populations of larvae of *Busseola fusca* (Fuller) in Northern Nigeria. *Bull. ent. Res.* 54, 643–7.

Harris, K. M. (1964*b*). The sorghum midge complex (Diptera, Cecidomyiidae). *Bull. ent. Res.* 55, 233–47.

Harris, K. M. (1966). Gall midge genera of economic importance (Diptera: Cecidomyiidae). Part I: Introduction and subfamily Cecidomyiidae: supertribe Cecidomyiidi. *Trans. R. ent. Soc. Lond.* 118, 313–58.

Harris, K. M. (1968). A systematic revision and biological review of the cecidomyiid predators (Diptera: Cecidomyiidae) on world Coccoidea (Hemiptera: Homoptera). *Trans. R. ent. Soc. Lond.* 119, 401–94.

Harris, K. M. (1970). The Sorghum Midge. *PANS* 46, 36–42.

Harris, K. M. & E. Harris (1968). Losses of African grain sorghums to pests and diseases. *PANS* 14, 48–54.

Harris, W. V. (1969). *Termites as Pests of Crops and Trees*, 41 pp. CIE: London.

Harris, W. V. (1971). *Termites, their Recognition and Control*, 2nd ed., 186 pp. Longmans: London.

Hartley, C. W. S. (1969). *The Oil Palm*, 706 pp. Longmans: London.

Hartley, G. S. & R. T. Brunskill (1958). Reflection of water drops from surfaces. In: *Surface Phenomena in Chemistry & Biology*, Danielli, J. F. *et al.*, pp. 214–23. Pergamon Press: London.

Hartley, G. S. & T. F. West (1969). *Chemicals for Pest Control*, 316 pp. Pergamon Press: London.

Hassan, E. (1977). *Major Insect and Mite Pests of Australian Crops*, 238 pp. Ento Press: Queensland.

Headley, J. C. (1972). Economics of agricultural pest control. *Ann. Rev. Entomol.* **17**, 273–86.

Hensley (1980). *See* Int. Congress of Entomology (1980).

Hercules Powder Co. (1960). *Cotton Insect Pests*, 44 pp. Hercules Powder Co.: Wilmington, Delaware.

Hill, A. F. (1952). *Economic Botany*, 560 pp. McGraw-Hill: New York.

Hill, D. S. (1974*a*). *Synoptic Catalogue of Insect and Mite Pests of Agricultural and Horticultural Crops*, 150 pp. Department of Zoology, Hong Kong University; Occasional Papers No. 1.

Hill, D. S. (1974*b*). Susceptibilities of carrot cultivars to carrot fly (*Psila rosae* (F.)). *Plant Pathology* **23**, 36–9.

Hill, D. S. & J. M. Waller (1982). *Pests and Diseases of Tropical Crops. Volume I: Principles and Methods of Control,* 175 pp. Longmans: London.

Hill, D. S. & J. M. Waller (1983). *Pests and Diseases of Tropical Crops. Volume II: Field Handbook*, Longmans: London.

Hinckley, A. D. (1963). *Trophic Records of some Insects, Mites, and Ticks in Fiji*, 116 pp. Dept. Agric., Fiji, Bull. No. 45.

Hinton, H. E. & A. S. Corbet (1955). *Common Insect Pests of Stored Food Products*, 3rd ed., 61 pp. Econ. Ser., no. 15, British Museum (NH): London.

HMSO (1964). *Review of the Persistent Organochlorine Pesticides.* HMSO: London.

HMSO (1981). *Approved Products for Farmers and Growers,* 331 pp. Agric. Chem. Approval Scheme.

Hocking, K. S. & C. Potter (1971). Problems on the use of insecticides in the tropics. *Proc. 6th Br. Insectic. Fungic. Conf.* **3**.

Hodgson, C. J. (1968). Soil application of systemic insecticides for control of woolly apple aphid (*Eriosoma lanigerum* (Hsm.)) in Rhodesia. *Bull. ent. Soc.* **58**, 73–82.

Hodgson, C. J. (1970). Pests of Citrus and their control. *PANS* **16**, 647–66.

Hodgson, C. J. & J. O. Whiteside (1969). Citrus in Rhodesia. Tech. Bull. no. 7: *Rhodesia Agric. J.*

Homeyer, B. (1970). Present state of soil insect pest control. *Pflanz.-Nachr. Bayer* **23**, 224–30.

Howe, R. W. (1957). A laboratory study of the Cigarette Beetle, *Lasioderma serricorne* (F.) (Col., Anobiidae) with a critical review of the literature on its biology. *Bull. ent. Res.* **48**, 9–56.

Howe, R. W. (1965). A summary of estimates of optimal and minimal conditions for population increase of some stored products insects. *J. Stored Prod. Res.* **1**, 177–84.

Huffaker, C. B. (ed.) (1971). *Biological Control*, 511 pp. Plenum: New York.

Huffaker, C. B., J. A. McMurtry & M. Van de Vrie (1969). The ecology of Tetranychid mites and their natural control. *Ann. Rev. Entomol.* **14**, 125–74.

Hughes, K. M. (1957). An annotated list and bibliography of insects reported to have virus diseases. *Hilgardia* **26**, 597–629.

Hussey, N. W., W. H. Read & J. J. Hesling (1969). *The Pests of Protected Cultivation*, 416 pp. Edward Arnold: London.

Imms, A. D. (1960). *A General Textbook of Entomology*, 9th ed., 886 pp. Methuen: London. Revised by O. W. Richards & R. G. Davies.

Imms, A. D. (1967). *Outlines of Entomology,* 5th ed., 224 pp. Methuen: London. Revised by O. W. Richards & R. G. Davies.

Ingram, W. R. (1965). An evaluation of several insecticides against berry borer and fruit fly in Uganda robusta coffee. *E. Afr. Agric. for. J.* **30**, 259–62.

Ingram, W. R. (1969). Observations on the pest status of bean flower thrips in Uganda. *E. Afr. Agric. for. J.* **34**, 482–4.

Ingram, W. R., J. R. Davies and J. N. McNutt (1970). *Agricultural Pest Handbook (Uganda)*, 50 pp. Govt. Printer: Entebbe.

Int. Congress of Entomology (XVI) (1980). *Abstracts*, 480 pp. Kyoto: Japan.

Int. Pest Control (1981*a*). Neem – pesticide potential. *Int. Pest Control,* Vol. 1981 (3), 68–70.

Int. Pest Control (1981*b*). *International Pesticide Directory*, suppl. to *Int. Pest Control*, Sept./Oct. 1981, 70 pp.

Ironside, D. A. (1973). *Insect Pests of Macadamia*, 11 pp. Advisory Lflt No. 1191, Div. Pl. Indust., Dept. Prim. Indust.: Queensland.

IRRI (1967). *The Major Insect Pests of the Rice Plant*, 729 pp. Johns Hopkins Press: Baltimore, Maryland.

Jacobson, M. (1965). *Insect Sex Attractants*, John Wiley: New York.

Jacobson, M. (1966). Chemical insect attractants and repellants. *Ann. Rev. Entomol.* **11**, 403–22.

Jacobson, M. & M. Beroza (1963). Chemical sex attractants. *Science* **140**, 1367–73.

Jacobson, M. & D. G. Crosby (1971). *Naturally Occurring Insecticides*, 585 pp. Dekker: New York.

Jameson, J. D. (1970). *Agriculture in Uganda*, 2nd ed., 395 pp. Oxford University Press: London.

Jeppson, L. R., H. H. Keifer & E. W. Baker (1975). *Mites Injurious to Economic Plants*, 614 pp. Univ. California Press: Berkeley.

Jepson, W. F. (1954). *A Critical Review of the World Literature on the Lepidopterous Stalk Borers of Tropical Graminaceous Crops*, 127 pp. CIE: London.

Jepson, W. F. (1956). The biology and control of the sugarcane chafer beetles in Tanganyika. *Bull. ent. Res.* **47**, 377–97.

Johnstone, D. R. (1970). High volume application of insecticide sprays in Cyprus Citrus. *PANS* **16**, 146–61.

Jones, F. G. W. & M. Jones (1974). *Pests of Field Crops*, 2nd Edition, 448 pp. Edward Arnold: London.

Jotwani, M. G. & W. R. Young (1971). Sorghum insect control – here's what's working in India. *World Farming*, 6–11.

Kennedy, J. S. (1966). Mechanisms of host plant selection. *Proc. Assoc. Appl. Biol.* 317–22.

Kennedy, J. S., M. F. Day & V. F. Eastop (1962). *A Conspectus of Aphids as Vectors of Plant Viruses*, 144 pp. CIE: London.

Kilgore, W. W. & R. L. Doutt (1967). *Pest Control – biological, physical, and selected chemical methods*, 477 pp. Academic Press: New York & London.

Kirkpatrick, T. W. (1966). *Insect Life in the Tropics*, 311 pp. Longmans: London.

Knipling, E. F. (1963). *Alternative Methods in Pest Control*. Symposium on New Developments and Problems in the Use of Pesticides, pp. 23–38. Fd. Nut. Bd., Nat. Acad. Sci.: Washington.

Knipling, E. F. & D. A. Spencer (1963). *Protection from Insect and Vertebrate Pests in Relation to Crop Production.* UN Conference on Application of Science and Technology for the Benefit of the Less Developed Areas, pp. 160–74. UN: Geneva.

Kranz, J., H. Schmütterer & W. Koch (1979). *Diseases, Pests, and Weeds in Tropical Crops*, 666 pp. John Wiley: Chichester.

Kring, J. B. (1972). Flight behaviour of aphids. *Ann. Rev. Entomol.* **17**, 461–92.

Krishna, K. & F. M. Weesner (1969–70). *Biology of Termites*: vol. 1, 598 pp. (1969); vol. 2, 643 pp. (1970). Academic Press: New York.

Kulman, H. M. (1971). Effects of insect defoliation on growth and mortality of trees. *Ann. Rev. Entomol.* **16**, 289–324.

La Brecque, G. C. & C. N. Smith (1968). *Principles of Insect Chemosterilisation*, 354 pp. Appleton-Century-Crofts: New York.

Laffoon, J. L. (1960). Common names of insects – approved by the Entomological Society of America. *Bull. Ent. Soc. Amer.* **6**, 175–211.

Lamb, K. P. (1974). *Economic Entomology in the Tropics*, 195 pp. Academic Press: London.

Lange, W. H., A. A. Grigarick, C. S. Davis, M. A. Miller, M. D. Miller, R. K. Washino, L. E. Rosenberg, R. L. Rudd & E. W. Jameson (1970). *Insects and Other Animal Pests of Rice*, Circ. 555, Calif. Agric. Expt. Sta. Ext. Serv., 32 pp.

Lavabre, E. M. (1961). *Protection des Cultures de Caféiers, Cacaoyers et autres Plantes Pérennes Tropicales*, 269 pp. Inst. Franc. Café Cacao: Paris.

Lavabre, E. M. (1966). *A Report on a Entomological Survey in Uganda*, 27 pp. Inst. Franc. Café Cacao: Paris.

Le Clerq, E. L., W. H. Leonard & A. G. Clark (1966). *Field Plot Technique*, 2nd ed., 373 pp. Burgess: Minneapolis.

Lee, K. E. & T. G. Wood (1971). *Termites and Soils*, 252 pp. Academic Press: London.

Leeuwangh, J. (1965). The biology of *Epigynopteryx stictigramma* (Hmps.) (Lepidoptera: Geometridae) a pest of coffee in Kenya. *J. ent. Soc. S. Afr.* **28**, 21–31.

Le Pelley, R. H. (1959). *Agricultural Insects of East Africa*, 307 pp. EA High Commission: Nairobi.

Le Pelley, R. H. (1968). *Pests of Coffee*, 590 pp. Longmans: London.

Lepesme, . . (1947). *Les insectes des palmiers*, Lechevalier: Paris.

Leston, D. (1970). Entomology of the cocoa farm. *Ann. Rev. Entomol.* **15**, 273–94.

Lever, R. J. A. W. (1969). *Pests of the Coconut Palm*, FAO Agric. Studies no. 77, 190 pp. FAO: Rome.

Lewis, T. (1973). *Thrips: their biology, ecology and economic importance*, 350 pp. Academic Press: London.

Li, Li-ying (1980). *See* Int. Congress of Entomology (1980).

Libby, J. L. (1968). *Insect Pests of Nigerian Crops*. Res. Bull. 269, 69 pp. University of Wisconsin: Madison, Wisconsin.

Lock, G. W. (1962). *Sisal*, 365 pp. Longmans: London. (2nd ed. 1969).

Long, W. H. & S. D. Hensley (1972). Insect pests of sugarcane. *Ann. Rev. Entomol.* **17**, 149–76.

Maas, W. (1971). *ULV Application and Formulation Techniques*, 164 pp. Philips-Duphar: Amsterdam.

McCallan, E. (1959). Some aspects of the geographical distribution of insect pests. *J. ent. Soc. S. Afr.* **22**, 3–12.

Mackerras, I. M. (ed.) (CSIRO) (1969). *The Insects of Australia*, 1029 pp. Melbourne University Press: Victoria.

McKinlay, K. S. & Q. A. Geering (1957). Studies of crop loss following insect attack on cotton in East Africa. I. Experiments in Uganda and Tanganyika. *Bull. ent. Res.* **48**, 833–49 (II. 851–66).

McKinley, D. J. (1971). An introduction to the use and preparation of artificial diets with special emphasis on diets for phytophagous Lepidoptera. *PANS* **17**, 421–4.

McNutt, D. N. (1963). The control of 'biting ants' *Macromischoides aculeatus* Mayr (Formicidae) on robusta coffee. *E. Afr. Agric. For. J.* **29**, 122–4.

McNutt, D. N. (1967). The White Coffee-borer (*Anthores leuconotus* Pasc.) (Col., Lamiidae): its identification, control and occurrence in Uganda. *E. Afr. Agric. for. J.* **32**, 469–73.

McNutt, D. N. (1976). *Insect Collecting in the Tropics*, 68 pp. COPR: London.

Madsen, H. F. (1971). Integrated control of the Codling Moth. *PANS* **17**, 417–20.

Madsen, H. F. & C. V. G. Morgan (1970). Pome fruit pests and their control. *Ann. Rev. Entomol.* **15**, 295–320.

Maeta & Kitamura (1980). *See* Int. Congress of Entomology (1980).

MAFF (1971). *Pesticides safety precautions scheme agreed between Government departments and industry*, 149 pp. MAFF: London.

MAFF (1981). *List of Approved Products and their uses for farmers and growers*, 331 pp. HMSO: London.

Maramorosch, K. & K. F. Harris (eds.) (1979). *Leafhopper Vectors and Plant Disease Agents*, 650 pp. Academic Press: New York.

Maramorosch, K. & K. F. Harris (eds.) (1981). *Plant Diseases and Vectors: Ecology and Epidemiology*, 360 pp. Academic Press: New York.

Marsh, R. W. (1969). Glossary of terms used in the application of crop protection measures. *Scient. Hort.* **21**, 147–55.

Martin, H. (1961). *Guide to the Chemicals Used in Crop Protection*, 4th ed., 387 pp. Canada Dept. of Agric.: London, Ont.

Martin, H. (1964). *The Scientific Principles of Crop Protection*, 384 pp. Edward Arnold: London.

Martin, H. (1969). *Insecticide and Fungicide Handbook for Crop Protection*, 3rd ed., 387 pp. Blackwell: Oxford.

Martin, H. (1970). *Pesticide Manual*, 2nd ed., 464 pp. Brit. Crop Prot. Council.

Martin, H. (1972*a*). *Pesticide Manual*, 3rd ed., 535 pp. Brit. Crop Prot. Council.

Martin, H. (1972*b*). *Insecticide and Fungicide Handbook for Crop Protection*, 4th ed., 415 pp. Blackwell: Oxford.

Martin, H. (1973). *The Scientific Principles of Crop Protection*, 6th ed. Edward Arnold: London.

Martin, H. & C. R. Worthing (1976). *Insecticide and Fungicide Handbook*, 5th ed. 427 pp. Blackwell: Oxford.

Massee, A. M. (1954). *The Pests of Fruit and Hops*, 3rd ed., 325 pp. Crosby Lockwood: London.

Materu, M. E. A. (1968). Biology and bionomics of *Acanthomia tomentosicollis* Stål. and *A. horrida* Germ. (Coreidae, Hemiptera) in Arusha area of Tanzania. Ph.D. thesis: University of E. Africa.

Matthews, G. A. (1977). C.d.a. – Controlled Droplet Application. *PANS* **23**, 387–94.

Matthews, G. A. (1979). *Pesticide Application Methods*, 334 pp. Longmans: London.

May, R. M. (ed.) (1976). *Theoretical Ecology – Principles and Applications*, 317 pp. Blackwell: Oxford.

Menzie, C. M. (1969). *Metabolism of Pesticides*. Bur. Sport, Fish, Wildlife; Sp. Sci. Rep. – Wildlife no. 127, 487 pp.

Menzie, C. M. (1972). Fate of pesticides in the environment. *Ann. Rev. Entomol.* **17**, 199–222.

Metcalf, C. L., W. P. Flint & R. L. Metcalf (1962). *Destructive and Useful Insects*, 1087 pp. McGraw-Hill: New York.

Metcalf, R. L. (ed.) (1957–68). *Advances in Pest Control Research*, vols. 1–8. Interscience: London & New York.

Metcalf, R. L. & W. H. Luckmann (eds.) (1975). *Introduction to Insect Pest Management*, 587 pp. John Wiley: New York.

Metcalf, J. R. (1972). An analysis of the population dynamics of the Jamaican sugarcane pest *Saccharosydne saccharivora* (Westw.) (Hom. Delphacidae). *Bull. ent. Res.* **62**, 73–85.

Miles, M. (1950). Studies of British Anthomyiid flies. I. Biology and habits of the Bean Seed Flies, *Chortophila cilicrura* (Rond.) and *C. tridactyla* (Rond.). *Bull. ent. Res.* **41**, 343–54.

Moran, V. C. (1968). Preliminary observations on the choice of host plants by adults of the citrus psylla, *Trioza erytreae* (Del Guercio) (Homoptera: Psyllidae). *J. ent. Soc. S. Afr.* **31**, 403–10.

Mound, L. A. (1965). An introduction to the Aleyrodidae of Western Africa (Homoptera). *Bull. Brit. Mus. Nat. Hist., Ent.* **17**, 113–60.

Mound, L. A. & S. H. Halsey (1977). *Whiteflies of the World: a systematic catalogue of the Aleyrodidae (Homoptera) with host plant and natural enemy data*, 336 pp. John Wiley: London.

Munro, J. W. (1966). *Pests of Stored Products*, 234 pp. Hutchinson: London.

Nat. Acad. Sci. US (1969). *Principles of Plant and Animal Pest Control*: Vol. 3, *Insect-pest management and control*, 508 pp.; Vol. 4, *Control of plant parasitic nematodes*, 172 pp. *Pub. Nat. Acad. Sci. US*, no. 1695 (Washington DC).

Nayar, K. K., T. N. Ananthakrishnan & B. V. David (1976). *General and Applied Entomology*, 589 pp. Tata McGraw-Hill: New Delhi.

Needham, J. G. (1959). *Culture Methods for Invertebrate Animals*, 509 pp. Dover: New York.

Nickel, J. L. (1979). *Annotated List of Insects and Mites Associated with Crops in Cambodia*, 75 pp. SEARCA: Philippines.

Nishida, T. & T. Torü (1970). *Handbook of field methods for research on rice stem borers and their natural enemies*, 132 pp. Blackwells: Oxford.

Nixon, G. E. J. (1951). *The Association of Ants with Aphids and Coccids*, 36 pp. CIE: London.

Noble-Nesbitt, J. (1970). Structural aspects of penetration through insects cuticles. *Pesticide Sci.* **1**, 204–8.

Nye, I. W. B. (1958). The external morphology of some of the dipterous larvae living in the Gramineae of Britain. *Trans. R. ent. Soc. London* **110**, 411–87.

Nye, I. W. B. (1960). The insect pests of graminaceous crops in East Africa. *Colonial Res. Studies no. 31.* HMSO: London.

O'Brien, R. D. (1967). *Insecticides, Action and Metabolism*, 332 pp. Academic Press: New York.

O'Brien, R. D. (ed.) (1970). *Biochemical Toxicology of Insecticides*, 218 pp. Academic Press: New York.

O'Connor, B. A. (1969). *Exotic Plant Pests and Diseases*, 424 pp. Noumea, New Caledonia: South Pacific Comm.

Odum, E. P. (1959). *Fundamentals of Ecology*, 2nd ed., 546 pp. Saunders: London.

Oei-Dharma, H. P. (1969). *Use of Pesticides and Control of Economic Pests and Diseases in Indonesia.* E. J. Brill: Leiden, Holland.

OILB (1970). Symposium OILB on borers of graminaceous plants. (Organisation Internationale de Lutte Biologique.) Published in *Entomophaga* **16** (1–2) (1971).

Oldfield, G. N. (1970). Mite transmission of plant viruses. *Ann. Rev. Entomol.* **15**, 343–80.

Oldroyd, H. (1958). *Collecting, Preserving and Studying Insects*, 327 pp. Hutchinson: London.

Oldroyd, H. (1968). *Elements of Entomology*, 312 pp. Weidenfeld & Nicolson: London.

Ordish, G. (1967). *Biological Methods in Crop Pest Control*, London. 242 pp.

Ordish, G. (1976). *The Constant Pest*, 240 pp. Peter Davies: London.

Ordish, G. *et al.* (1966). Current papers on integrated pest control. *PANS* A **12**, 35–72.

Ostmark, H. E. (1974). Economic insect pests of bananas. *Ann. Rev. Entomol.* **19**, 161–76.

Owen, D. F. (1966). *Animal Ecology in Tropical Africa*, 122pp. Oliver & Boyd: London.

Padwick, G. W. (1956). Losses caused by plant diseases in the colonies. *Phytopath. Papers*, no. 1, 60 pp. CMI Survey.

Painter, R. H. (1951). *Insect Resistance in Crop Plants*, MacMillan: New York.

Pan (1980). *See* Int. Congress of Entomology (1980).

Parkin, E. A. (1956). Stored Product Entomology (the assessment and reduction of losses caused by insects to stored foodstuffs). *Ann. Rev. Entomol.* **1**, 233–40.

Pathak, M. D. (1967). Significant developments in rice stem borer and leafhopper control. *PANS* **13**, 45–60.

Pathak, M. D. (1968). Ecology of common insect pests of rice. *Ann. Rev. Entomol.* **13**, 257–94.

Pathak (1980). *See* Int. Congress of Entomology (1980).

Pearson, E. O. (1958). *The Insect Pests of Cotton in Tropical Africa*, 355 pp. CIE: London.

Perring, F. H. & K. Mellanby (eds.) (1977). *Ecological Effects of Pesticides*, 193 pp. Linn. Soc. Symp. Series No. 5, Academic Press: London.

PESTDOC (1974). *Organism Thesaurus*, Vol. 1 *Animal Organisms*, 1317 pp. Ciba-Geigy: Basle, & Derwent Pub.: London.

Peterson, A. (1953). *Entomological Techniques*. Edwards Bros: Ann Arbor, Michigan.

Pfadt, R. E. (1962). *Fundamentals of Applied Entomology*, 668 pp. Macmillan: New York.

Pinhey, E. C. G. (1960). *Hawkmoths of Central and Southern Africa*, 139 pp. Longmans: London.

Pinhey, E. C. G. (1968). *Introduction to Insect Study in Africa*, 235 pp. Oxford University Press: London.

Poe, S. L. (1973). Tomato Pinworm, *Keiferia lycopersicella* (Walshingham) (Lepidoptera: Gelechiidae) in Florida. *Florida Dept. Agric., Ent. Circ. No. 131*, 2 pp.

Price Jones, D. & M. E. Solomon (1974). *Biology in Pest and Disease Control*, 398 pp. 13th Symp. Brit. Ecol. Soc., Blackwell: Oxford.

Proverbs, M. D. (1969). Induced sterilisation and control of insects. *Ann. Rev. Entomol.* **14**, 81–102.

Purseglove, J. W. (1968). *Tropical Crops. Dicotyledons*, vols. I & II, 719 pp. Longmans: London.

Purseglove, J. W. (1972). *Tropical Crops. Monocotyledons*, vol. I & II, 607 pp. Longmans: London.

Pury, J. M. S. de (1968). *Crop Pests of East Africa*, 227 pp. Oxford University Press: E. Africa.

Rabb, R. L. & F. E. Guthrie (1970). *Concepts of Pest Management*. Proceedings of a Conference held at N.C. State University at Raleigh, N.C., 25–27 March 1970, 242 pp. North Carolina State University: Raleigh, NC.

Rabb, R. L., F. A. Todd & H. C. Ellis (1976). *Tobacco Pest Management*, pp. 71–106. In: Apple & Smith (1976). *Integrated Pest Management*, Plenum Press: New York.

Rajamohan, N. (1976). Pest complex of sunflower – a bibliography. *PANS* **22**, 546–63.

Rao, B. S. (1965). *Pests of Hevea Plantations in Malaya*, 98 pp. Rubber Res. Inst.: Kuala Lumpur.

Rao, G. N. (1970). Tea pests in Southern India and their control. *PANS* **16**, 667–72.

Raw, F. (1959). Studies on the chemical control of cacao mirids *Distantiella theobroma* (Dist.) and *Sahlbergella singularis* Hagl. *Bull. ent. Res.* **50**, 13–23.

Raw, F. (1967). Some aspects of the wheat bulb fly problem. *Ann. appl. Biol.* **59**, 155–73.

Réal, P. (1959). Le cycle annuel de la cochenille *Dysmicoccus brevipes* Ckll., vectrice d'un 'wilt' de l'ananas en basse Cote d'Ivoire; son déterminisme. *Rev. Path. vég.* **38**, 3–111.

Reay, R. C. (1969). *Insects and Insecticides*, 152 pp. Oliver & Boyd: Edinburgh.

Reddy, D. B. (1968). *Plant Protection in India*, 454 pp. Allied Pub.: Bombay.

Richards, O. W. & R. G. Davies (1977). *Imm's General Textbook of Entomology,* 10th ed. Vol. 1. *Structure Physiology and Development*, 418 pp. Vol. 2. *Classification and Biology*, 1354 pp. Chapman & Hall: London.

Ripper, W. E. & L. George (1965). *Cotton Pests of Sudan*, 345 pp. Blackwell: Oxford.

Rose, D. J. W. (1972). Times and sizes of dispersal flights by *Cicadulina* species (Homoptera: Cicadellidae), vectors of Maize Streak Disease. *J. Anim. Ecol.* **41**, 495–506.

Rose, D. J. W. & C. J. Hodgson (1965). *Systates exaptus* Mshl. (Col., Curculionidae) and related species as soil pests of maize in Rhodesia. *Bull. ent. Res.* **56**, 303–18.

Rose, G. (1963). *Crop Protection*, 2nd ed., 490 pp. Leonard Hill: London.

Russell, G. E. (1978). *Plant Breeding for Pest and Disease Resistance*, 485 pp. Butterworths: London.

Sankaran, T. (1970). The oil palm bagworms of Sabah and the possibilities of their biological control. *PANS* **16**, 43–55.

Schmütterer, H. (1969). *Pests of Crops in Northeast and Central Africa*, 296 pp. G. Fischer: Stuttgart.

Schneider, D. (1961). The olfactory sense of insects. *Drapco Report* **8**, 135–46.

Schoohoven, L. M. (1968). Chemosensory basis of host plant selection. *Ann. Rev. Entomol.* **13**, 115–36.

Sharples, A. (1936). *Diseases and Pests of the Rubber Tree*, 480 pp. Macmillan: London.

SCI (1969). The effect of rain on plants, pests and pesticides. *Chemistry and Industry*, 1495–504.

Scopes, N. & M. Ledieu (eds.) (1979). *Pest and Disease Control Handbook*, BCPC Pub.: London.

Shillito, J. F. (1960). A biography of the Diopsidae (Diptera–Acalyptratae). *J. Soc. Bibl. Nat. Hist.* **3**, 337–50.

Shillito, J. F. (1971). The genera of Diopsidae (Insecta: Diptera). *Zool. J. Linn. Soc.* **50**, 287–95.

Shorey, H. H. (1976). *Animal Communication by Pheromones*, Academic Press: New York.

Short, L. R. T. (1963). *Introduction to Applied Entomology*, 235 pp. Longmans: London.

Simmonds, N. W. (1970). *Bananas*, 2nd ed., 512 pp. Longmans: London.

Singh, J. P. (1970). *Elements of Vegetable Pests*, 275 pp. Vora & Co.: Bombay.

Singh, S. R., H. F. van Emden & T. A. Taylor (eds.) (1978). *Pests of Grain Legumes: ecology and control*, 454 pp. Academic Press: London.

Skaife, S. H. (1953). *African Insect Life*, 387 pp. Longmans: London.

Smartt, J. (1976). *Tropical Pulses*, 348 pp. Longmans: London.

Smit, B. (1964). *Insects in Southern Africa – How to Control Them*, 399 pp. Oxford University Press: S. Africa.

Smith, C. N. (1966). *Insect Colonization and Mass Production.* Academic Press: New York.

Smith, D. (1973). Insect pests of Avocados. *Qld Agric. J.*, **99**, 645–53.

Smith, K. M. (1951). *Agricultural Entomology*, 2nd ed., 289 pp. Cambridge University Press: London.

Smith, R. F. & K. S. Hagen (1959). Impact of commercial insecticide treatments. *Hilgardia* **29**, 131–54.

Smith, R. F. & T. E. Mittler (etc.) (eds.) (1957–82). *Annual Review of Entomology*, vols. 1–27. Annual Reviews Inc.: California.

Soehardjan (1980). *See* Int. Congress of Entomology (1980).

Southwood, T. R. E. (1977). *The relevance of population dynamic theory to pest status.* pp. 35–54. In: Cherrett & Sagar (1977). *Origins of Pest, Parasite, Disease and Weed Problems*, Blackwell: Oxford.

Southwood, T. R. E. (1968). *Insect Abundance, a symposium*, 160 pp. R. ent. Soc.: London.

Southwood, T. R. E. (1978). *Ecological Methods*, 2nd ed. 524 pp. Chapman & Hall: London.

Southwood, T. R. E. & H. N. Comins (1976). A synoptic population model. *J. Anim. Ecol.* **45**, 949–65.

Spencer, K. A. (1973). *Agromyzidae (Diptera) of Economic Importance*, 418 pp. W. Junk: The Hague. (Vol. 9 of Series Entomologica.)

Stapley, J. H. & F. C. H. Gayner (1969). *World Crop Protection.* Vol. I, *Pests and Diseases*, 270 pp.; Vol. II, *Pesticides*, 249 pp. (K. A. Hassell).

Stern, V. M., R. F. Smith, R. van den Bosch & K. S. Hagen (1959). The integrated control concept. *Hilgardia* **29**, 81–101.

Stern, V. M. & R. van den Bosch (1959). Field experiments on the effects of insecticides. *Hilgardia* **29**, 103–30.

Storey, H. H. (1932). The inheritance by an insect vector of the ability to transmit a plant virus. *Proc. Roy. Soc.* (B) **112**, 46–69.

Storey, H. H. (1961). Vector relationships of plant viruses. *E. Afr. Med. J.* **38**, 215–20.

Strickland, A. H. (1967). Some problems in the economic integration of crop loss control. *Proc. 4th Br. Ins. Fung. Conf.* **2**, 478–91.

Strickland, A. H. (1971). The actual status of crop loss assessment. *EPPO Bull.* **1**, 39–51.

Stride, G. O. (1968). On the biology and ecology of *Lygus vosseleri* (Heteroptera: Miridae) with special reference to its host plant relationships. *J. ent. Soc. S. Afr.* **31**, 17–59.

Stride, G. O. (1969). Investigations into the use of a trap crop to protect cotton from attack by *Lygus vosseleri* (Heteroptera: Miridae). *J. ent. Soc. S. Afr.* **32**, 469–77.

Stroyan, H. L. G. (1961). Identification of aphids living on Citrus. *FAO Pl. Prot. Bull.* **9**, 30 pp.

Swaine, G. (1959). A preliminary note on *Helopeltis* spp. damaging cashew in Tanganyika Territory. *Bull. ent. Res.* **50**, 171–81.

Swaine, G. (1961). *Plant Pests of Importance in Tanganyika.* *Min. Agric. Bull.* no. 13, 44 pp. Min. Agric.: Tanganyika.

Sweetman, H. L. (1958). *Principles of Biological Control*, 560 pp. W. C. Brown: Iowa.

Tahori, A. S. (1971). *Pesticide Terminal Residues*, 374 pp. Butterworth: London.

Talhouk, A. M. S. (1969). *Insects and Mites Injurious to Crops in Middle Eastern Countries*, Monographien zur angew. Entomologie, Nr. 21, 230 pp. Verlag Paul Pary: Hamburg & Berlin.

Tapley, R. G. (1960). The white coffee borer, *Anthores leuconotus* Pasc., and its control. *Bull. ent. Res.* **51**, 279–301.

Taylor, T. A. (1964). Studies on Nigerian yam beetles. II. Bionomics and control. *J. W. Afr. Sci. Assoc.* **9**, 13–31.

Thomas, R. T. S. (1962). Checklist of Pests on Some Crops in West Irian (New Guinea). *Bull. Econ. Aff., Agric. Series no. 1*, 126 pp. Dept. Econ. Affairs: Hollandia.

Thompson, W. T. (1969). *The Ornamental Pesticide Application Guide*, 471 pp. Thompson: Fresno, California.

Thompson. *A Catalogue of the Parasites and Predators on Insect Pests.* Sec. I, Parasite Host Catalogue – Parts 1–11. Sec. II, Host Parasite Catalogue – Parts 1–5. Sec. III, Predator Host Catalogue. Sec. IV. Host Predator Catalogue. CAB (CIBC): London.

Trought, T. E. T. (1965). *Farm Pests*, 32 pp. Blackwell: Oxford.

Tunstall, J. P., G. A. Matthews & D. J. McKinley (1971). Tropical Crop Problems. The Introduction of Cotton Insect Control. *Proc. 6th Ins. Fung. Conf.* **3**, 6 pp.

Turnipseed, S. G. & M. Kogan (1976). Soybean entomology. *Ann. Rev. Entomol.* **21**, 247–82.

Tuttle, D. M. & E. W. Baker (1968). *Spider Mites of South-Western United States and a revision of the family Tetranychidae.* University of Arizona Press.

Tzanakakis, M. E. (1959). An ecological study of the Indian Meal Moth *Plodia interpunctella* (Hübner) with emphasis on diapause. *Hilgardia* **29**, 205–46.

Urquhart, D. H. (1961). *Cocoa*, 2nd ed., 293 pp. Longmans: London.

USDA (1968). *Suggested Guide for the use of insecticides to control insects affecting crops, livestock, households,* stored products, forests and forest products, 273 pp. US Dept. of Agric.

USDA (1952). *The Yearbook of Agriculture (1952) Insects*, 780 pp. US Dept. of Agric.

Uvarov, B. (1966). *Grasshoppers and Locusts*, Cambridge University Press: London.

Vanderplank, F. L. (1958). Studies on the Coconut pest, *Pseudotherapterus wayi* Brown (Coreidae), in Zanzibar. *Bull. ent. Res.* **49**, 559–84.

Van Emden, H. F., *et al.* (1969). The ecology of *Myzus persicae. Ann. Rev. Entomol.* **14**, 197–270.

Van Emden, H. F. (ed.) (1972). *Aphid Technology*, 344 pp. Academic Press: London.

Van Emden, H. F. (1974). *Pest Control and its Ecology*, 60 pp. Inst. Biol. Studies in Biology, No. 50 Edward Arnold: London.

Viggiani, G. (1965). La *Contarinia sorghicola* Coq. (Diptera, Cecidomyiidae) ed i suoi parassiti in Italia. *Boll. Lab. Ent. Agr. Portici* **23**, 1–36.

Wattanapongsiri, A. (1966). A revision of the genera *Rhyncophorus* and *Dynamis. Dept. Agric. Sci. Bull.* **1**, 1–328.

Walthers, H. J. (1969). Beetle transmission of plant viruses. *Adv. Virus Res.* **15**, 339–63.

Watson, M. A. & R. T. Plumb (1972). Transmission of plant-pathogenic viruses by aphids. *Ann. Rev. Entomol.* **17**, 425–52.

Watson, T. F., L. Moore & G. W. Ware (1975). *Practical Insect Pest Management*, 196 pp. W. H. Freeman & Co: San Francisco.

Webb, G. C. (1961). *Keys to the Genera of the African Termites.* Ibadan University Press: Ibadan.

Weber, N. A. (1972). Gardening ants: the Attines. *Mem. Amer. Phil. Soc.*, **92**, 1–146.

Weems, H. V. (1973). Citrus Whitefly, *Dialeurodes citri* (Ashmead) (Homoptera: Aleyrodidae). *Florida Dept. Agric., Ent. Circ. No. 128*, 2 pp.

Wheatley, G. A. (1971). Pest control in vegetables: some further limitations in insecticides for Cabbage Root Fly and Carrot Fly control. *Proc. 6th Br. Ins. Fung. Conf.* **2**, 386–95.

Wheatley, G. A. & T. H. Coaker (1969). Pest control objectives in relation to changing practices in agricultural crop production. *Tech. Econ. Crop Prot. Pest Control*, Mon. **36**, 42–55.

Wheatley, P. E. (1963). Laboratory studies of insecticides against the coffee leaf-miner *Leucoptera meyricki* Ghesq. (Lepidoptera, Lyonetiidae). *Bull. ent. Res.* **54**, 167–74.

Wheatley, P. E. (1964a). The Giant Looper. *Kenya Coffee*, 1–5.

Wheatley, P. E. (1964b). Field studies of insecticides against the coffee leaf-miner *Leucoptera meyricki* Ghesq. (Lepidoptera, Lyonetiidae). *Bull. ent. Res.* **55**, 193–203.

Wheatley, P. E. & T. J. Crowe (1967). *Pest Handbook. (The recognition and control of the more important pests of agriculture in Kenya.).* 33 pp. Govt. Printer: Nairobi.

Whellan, J. A. (1964). Some Rhodesian pests of entomological interest. *Span* 7, 3.

White, R. E. (1964). Injurious beetles of the genus *Diabrotica* (Coleoptera: Chrysomelidae). *Florida Dept. Agric., Ent. Circ. No. 27*, 2 pp.

Williams, D. J. (1969). The family-group of the scale insects (Hemiptera: Coccoidea). *Bull. Br. Mus. Nat. Hist., Ent.* **32**, 315–41.

Williams, D. J. (1970). The mealybugs (Homoptera, Coccidea, Pseudococcidae) of sugarcane, rice and sorghum. *Bull. ent. Res.* **60**, 109–88.

Williams, D. J. (1971). Synoptic discussion of *Lepidosaphes* Shimer and its allies with a key to genera (Homoptera, Coccoidea, Diaspididae). *Bull. ent. Res.* **61**, 7–11.

Williams, G. C. (1964). The life history of the Indian Meal Moth, *Plodia interpunctella* Hbn. (Lepidoptera: Phycitidae) in a warehouse in Britain and on different foods. *Ann. appl. Biol.* **53**, 459–75.

Williams, J. R. (1970). Studies on the biology, ecology, and economic importance of the sugarcane scale insect, *Aulacaspis tegalensis* (Zhnt.) (Diaspididae), in Mauritius. *Bull. ent. Res.* **60**, 61–95.

Williams, J. R., J. R. Metcalfe, R. W. Mungomery & R. Mathes (eds.) (1969). *Pests of Sugar Cane*, 568 pp. Elsevier: London & New York.

Wilson, A. (1969). *Further Review of Certain Persistent Organochlorine Pesticides in Great Britain*, 148 pp. HMSO: London.

Wilson, F. (1960). *A review of the Biological Control of Insects and Weeds in Australia and Australian New Guinea.*

Tech. Comm. no. 1, 102 pp. CIBC: Ottawa.

Wilson, F. (1971). *Biotic Agents of Pest Control as an Important Natural Resource.* Gooding Memorial Lecture, 12 pp. Cent. Assoc. Bee-keepers: London.

Wilson, J. W. (1931). The two-spotted mite (*Tetranychus telarius* L.) on *Asparagus plumosus. University Florida Agric. Expt. Sta. Bull. no. 234*, 20 pp.

Winteringham, F. P. W. (1969). Mechanisms of selective insecticidal action. *Ann. Rev. Entomol.* **14**, 409–41.

Wood, B. J. (1968). *Pests of Oil Palms in Malaysia and their Control*, 204 pp. Inc. Soc. of Planters: Kuala Lumpur.

Wood, B. J. (1971). The importance of ecological studies to pest control in Malaysian plantations. *PANS* **17**, 411–16.

Wood, D. L., R. M. Silverstein & M. Nakajima (1970). *Control of Insect Behaviour by Natural Products*, 346 pp. Academic Press: New York.

Wood, R. S. K. (ed.) (1971). Altering the resistance of plants to pests and diseases. *PANS* **17**, 240–57.

Worthing, C. R. (1979). *The Pesticide Manual: a world compendium* 6th ed. 655 pp. BCPC Pub.: London.

Wright, R. H. (1970). Some alternatives to insecticides. *Pesticide Sci.* **1**, 24–7.

Wyniger, R. (1962, 1968). *Pests of Crops in Warm Climates and their Control.* Supplement, *Control Measures*, 2nd ed., 555 & 162 pp. Basel, Switzerland.

Yamamoto, I. (1970). Mode of action of pyrethroids, nicotinoids and rotenoids. *Ann. Rev. Entomol.* **15**, 257–72.

Yasumatsu, K. & T. Toru (1968). Impact of parasites, predators, and diseases on rice pests. *Ann. Rev. Entomol.* **13**, 295–324.

Yathom, S. (1967). Control of the sorghum shoot fly in Israel. *Int. Pest Control* 9, 4.

Yunus, A. & A. Balasubramaniam (1975). *Major Crop Pests in Peninsular Malaysia*, 182 pp. Min. Agric. & Rur. Dev.: Malaysia.

Zimmerman, E. C. (1968). The *Cosmopolites* banana weevils (Coleoptera: Curculionidae: Rhynchophorinae). *Pacific Insects* **10**, 295–9.

Zur Strassen, R. (1960). Catalogue of South African Thysanoptera. *J. ent. Soc. S. Afr.* **23**, 330.

# Appendices

---

## A    List of pesticides cited and some of their trade names

---

Approved names are on the left; trade names follow in italic.
For further names refer to the Tropical Pest Management
(formerly PANS) Pesticide Index, 1981 edition, pp. 1–61.

### Chlorinated hydrocarbons

1. Aldrin    *Aldrex, Aldrite, Drinox, Toxadrin* etc.
2. BHC γ, (HCH)    *Lindane*
3. Chlorobenzilate    *Akar, Folbex*
4. DDT    many trade names
5. Dicofol    *Kelthane, Acarin, Mitigan*
6. Dieldrin    *Dieldrex, Alvit, Dilstan, Endosil* etc.
7. Endosulfan    *Thiodan, Cyclodan, Thifor, Thionex*
8. Endrin    *Endrex, Hexadrin, Mendrin*
9. Heptachlor    *Drinox, Heptamul, Velsicol*
10. Mirex    *Dechlorane*
11. Tetradifon    *Tedion, Duphar*
12. Tetrasul    *Animert V-101*
13. TDE    *DDD, Rhothane*
14. Amitraz    *Mitac, Taktic, Triatox, Baam, Azaform*

### Substituted phenols

15. Binapacryl    *Morocide, Acaricide, Ambox, Dapacryl*
16. DNOC    *Sinox, Dinitrol, Detal, Cresofin*
17. Pentachlorophenol    *Dowicide, Santophen, Pentacide*

### Organophosphorus compounds

18. Azinphos-methyl    *Gusathion, Benthion, Carfene*
19. Azinphos-methyl with demeton-S-methyl sulphone    *Gusathion MS*
20. Bromophos    *Brofene, Bromovur, Nexion, Pluridox*
21. Bromopropylate    *Acarol, Neoron*
22. Carbophenothion    *Trithion, Garrathion, Dagadip*
23. Chlorfenvinphos    *Birlane, Sapecron, Supona*
24. Chlorpyrifos    *Dursban, Lorsban*
25. Demephion    *Cymetox, Pyracide*
26. Demeton    *Systox, Systemox, Solvirex*
27. Demeton-S-methyl    *Metasystox 55, Demetox*
28. Diazinon    *Basudin, Diazitol, DBD, Neocide*
29. Dichlorvos    *Vapona, Nogos, Dedevap, Nuvan, Oko, Mafu*
30. Dimefox    *Terra-sytam, Hanane*
31. Dimethoate    *Rogor, Roxion, Dantox, Cygon*
32. Disulfoton    *Disyston, Murvin 50, Parsolin, Solvirex*
33. Ethion    *Embathion, Hylemox, Nialate, Rhodocide*
34. Ethoate-methyl    *Fitios*
35. Fenitrothion    *Folithion, Accothion, Sumithion*
36. Fenthion    *Baytex, Lebaycid, Queleton, Tiguvon*
37. Fonofos    *Dyfonate*
38. Formothion    *Anthio, Aflix*
39. Heptenophos    *Hostaquick, Ragadan*
40. Iodofenphos    *Nuvanol, Alfacron, Elocril*
41. Malathion    *Malathion, Malastan, Malathexo*
42. Mecarbam    *Murfotox, Pestan, Afos*
43. Menazon    *Saphizon, Saphicol, Sayfos, Aphex*
44. Methidathion    *Supracide, Ultracide*
45. Mevinphos    *Phosdrin, Menite, Phosfene*
46. Monocrotophos    *Nuvacron, Azodrin, Monocron*
47. Naled    *Dibrom, Ortho, Bromex*
48. Omethoate    *Folimat*
49. Oxydemeton-methyl    *Metasystox-R*
50. Oxydisulfoton    *Disyston-S*
51. Parathion    *Folidol, Bladan, Fosfex, Fosferno, Thiophos*
52. Parathion-methyl    *Dalf, Metacide, Folidol-M, Nitrox-80*

53. Phenisobromolate   *Acarol, Neoron*
54. Phenthoate   *Elsan, Cidial, Papthion, Tanone*
55. Phorate   *Thimet, Rampart, Granutox, Timet*
56. Phosalone   *Zolone, Embacide, Rubitox*
57. Phosmet   *Imidan, Appa, Prolate, Germisan*
58. Phosphamidon   *Dimecron, Dicron, Famfos*
59. Phoxim   *Baythion, Valexon, Volaton*
60. Pirimiphos-ethyl   *Primicide, Fernex, Primotec*
61. Pirimiphos-methyl   *Actellic, Actellifog, Blex*
62. Profenofos   *Curacron*
63. Prothoate   *Fac, Fostin, Oleofac, Telefos*
64. Quinomethionate   *Morestan, Erade, Forstan*
65. Schradan   *Sytam, Pestox 3*
66. TEPP   *Nifos T, Vapotone, Bladen, Fosvex, Tetron*
67. Terbufos   *Counter*
68. Tetrachlorvinphos   *Gardona, Ostabil, Rabon, Ravap*
69. Thiometon   *Ekatin, Intrathion*
70. Thionazin   *Nemafos, Zinophos, Nemasol*
71. Thioquinox   *Eradex, Eraditon*
72. Triazophos   *Hostathion*
73. Trichloronate   *Agritox, Agrisil, Phytosol, Fenophosphon*
74. Trichlorphon   *Dipterex, Anthon, Chlorofos, Tugon*
75. Tricyclohexyltin hydroxide   *Plictran*
76. Vamidothion   *Kilval, Vamidoate, Vation*

### Carbamates

77. Aldicarb   *Temik*
78. Bufencarb   *Bux*
79. Carbaryl   *Sevin, Carbaryl 85, Murvin, Septon*
80. Carbofuran   *Furadan, Curaterr, Yaltox*
81. Methiocarb   *Mesurol, Draza, Baysol*
82. Methomyl   *Lannate, Halvard, Nudrin*
83. Oxamyl   *Vydate, Tumex*
84. Pirimicarb   *Pirimor, Aphox, Fernos*
85. Promecarb   *Carbamult, Minacide*
86. Propoxur   *Baygon, Blattanex, Unden, Suncide*

### Miscellaneous compounds

87. Aluminium phosphide   *Phostoxin*
88. Copper acetoarsenite   *Paris Green*
89. Ethylene dibromide   *Bromofume, Dowfume-W* etc.
90. Lead arsenate   *Gypsine, Soprabel*

91. Mercurous chloride   *Calomel, Cyclostan*
92. Methyl bromide   *Bromogas, Embafume* etc.
93. Sulphur (Lime-sulphur)   *Cosan, Hexasul, Thiovit* etc.

### Natural organic compounds

94. Bioallethrin   *D-Trans, Esbiol*
94a. Cypermethrin   *Cymbush, Ripcord*
95. Nicotine   *Nicofume*
95a. Permethrin   *Ambush, Kalfil, Talcord*
96. Pyrethrins
97. Resmethrin   *Chryson, For-Syn, Synthrin*
98. Rotenone   *Derris, Cube, Rotacide*

### Organic oils

99. Petroleum oils (white oils)   *Volck*
100. Tar oils

### Biological compounds

101. *Bacillus thuringiensis*   *BTB-183, Thuricide, Biotrol*
102. *Heliothis* NPV   *Elcar*

### Insect growth regulators

103. Diflubenzuron   *Dimilin*
104. Methoprene   *Altosid, Manta, Kabat*

**Acaricide** Material toxic to mites (Acarina).

**Activator** Chemical added to a pesticide to increase its toxicity.

**Active ingredient (a.i.)** Toxic component of a formulated pesticide.

**Adherence** The ability of a material to stick to a particular surface.

**Adhesive (= Sticker)** Material added to increase pesticide retention; different commercial preparations of methyl cellulose are available for this purpose.

**Adjuvant** A spray additive to improve either physical or chemical properties (*see also* Supplement, Sticker, Adhesive, Spreader, Wetter and Emulsifier).

**Aedeagus** The male intromittent organ, or penis.

**Aerosol** A dispersion of spray droplets of diameter $0.1-5.0 \mu m$; usually dispersed from a canister.

**Aestivation** Dormancy during a hot or dry season.

**Agamic** Parthenogenetic reproduction; without mating.

**Agitator** A mechanical device in the spray tank to ensure uniform distribution of toxicant and to prevent sedimentation.

**Agroecology** The study of ecology in relation to agricultural systems.

**Allochthonous** Not aboriginal; exotic; introduced; acquired from elsewhere (opp. autochthonous).

**Allopatric** Having separate and mutually exclusive areas of geographical distribution (opp. sympatric).

**Anemophilous** Plants which are pollinated by the wind.

**Anionic surfactant** Salt of an organic acid, the structure of which determines its surface activity.

**Antibiosis** The resistance of a plant to insect attack by having, for example, a thick cuticle, hairy leaves, toxic sap, etc.

**Anti-feedant** A chemical possessing the property of inhibiting the feeding of certain insect pests.

**Anti-frothing agent** Material added to prevent frothing of the liquid in the spray tank.

**Approved product** Proprietary brand of pesticide officially approved by the Ministry of Agriculture, Fisheries and Food, UK.

**Arista** A large bristle, located on the dorsal edge of the apical antennal segment in the Diptera.

**Asymptote** The point in the growth of a population at which numerical stability is reached.

**Atomiser** Device for breaking up a liquid stream into very fine droplets by a stream of air.

**Atrophied** Reduced in size; rudimentary; vestigial.

**Attractant** Material with an odour that attracts certain insects; lure. Several proprietory lures are manufactured.

**Autecology** The study of a single species.

**Autochthonous** Aboriginal; native; indigenous; formed where found (opp. allochthonous).

**Autocide** The control of a pest by the sterile-male technique.

**Bait** Foodstuff used for attracting pests; usually mixed with a poison to form a poison bait.

**Band application** Treatment of a band of soil in row-crops, usually covering plant rows, with either sprays or granules.

**Biocide** A general poison or toxicant.

**Boom** (spray) Horizontal (or vertical) light frame carrying several spray nozzles.

**Brachypterous** Having short wings that do not cover the abdomen.

**Breaking** The separation of the phases from emulsion.

**Budworm** Common name in the USA for various tortricid larvae.

**Calling** A virgin female moth releasing sex pheromones to attract males for the purpose of mating.

**Carrier** Material serving as diluent and vehicle for the active ingredients; usually in dusts.

**Cationic surfactant** Material in which surface activity is determined by the basic part of a compound.

**Caterpillar** Eruciform larva; larva of a moth, butterfly, or sawfly.

**Chaetotaxy** The arrangement and nomenclature of the bristles on the insect exoskeleton, both adults and larvae.

**Chemosterilant** Chemical used to render an insect sterile without killing it.

**Chrysalis** The pupa of a butterfly.

**Climatograph** A polygonal diagram resulting from plotting temperature means against relative humidity.

**Clone** A group of identical individuals propagated vegetatively from a single plant.

**Coarctate pupa** A pupa enclosed inside a hardened shell formed by the previous larval skin.

**Cocoon** A silken case inside which a pupa is formed.

**Colloidal formulation** Solution in which the particle size is less than 6 $\mu$m in diameter, and the particles stay indefinitely dispersed.

**Commensalism** Two organisms living together and sharing food, both species usually benefiting from the association; a type of symbiosis.

**Community** The collection of different species and types of plants and animals, in their respective niches, within the habitat. 1. Closed: the habitat is completely colonized by plants; no areas of bare soil; strong competition for space. 2. Open: habitat is not completely colonized by plants; bare areas of soil; competition for space is thus reduced.

**Compatibility** The ability to mix different pesticides without physical or chemical interactions which would lead to reduction in biological efficiency or increase in phytotoxicity.

**Compressed** Flattened from side to side.

**Concentrated solution (c.s.)** Commercial pesticide preparation before dilution for use.

**Concentrate spraying** Direct application of the pesticide concentrate without dilution.

**Concentration** Proportion of active ingredient in a pesticide preparation, before or after dilution.

**Contact poison** Material killing pests by contact action, presumably by absorption through the cuticle.

**Control (noun)** Untreated subjects used for comparison with those given a particular crop protection treatment.

**Control (verb)** To reduce damage or pest density to a level below the economic threshold. 1. Legislative: the use of legislation to control the importation and to prevent any spread of a pest within a country. 2. Physical: the use of mechanical (hand picking, etc.) and physical methods (heat, cold, radiation, etc.) of controlling pests. 3. Cultural: regular farm operations designed to destroy pests. 4. Chemical: the use of chemical pesticides as smokes, gas, dusts, and sprays to poison pests. 5. Biological: the use of natural predators, parasites and disease organisms to reduce pest populations. 6. Integrated: the very carefully reasoned use of several different methods of pest control in conjunction with each other to control pests with a minimum disturbance to the natural situation.

**Cosmopolitan** A species occurring very widely throughout the major regions of the world.

**Costa** A longitudinal wing vein, usually forming the anterior margin (leading edge).

**Cover** Proportion of the surface area of the target plant on which the pesticide has been deposited.

**Crawlers** The active first instar of a scale insect.

**Cremaster** A hooked, or spine-like process at the posterior end of the pupa, often used for attachment (Lepidoptera).

**Crepuscular** Animals that are active in the twilight, pre-dawn and at dusk in the evenings.

**Crochets** Hooked spines at the tips of the prolegs of lepidopterous larvae.

**Crop hygiene** (= Phytosanitation) The removal and destruction of heavily infested or diseased plants from a crop so that they do not form sources of reinfestation.

**Cutworm** Larva of certain Noctuidae that lives in the soil, emerging at night to eat foliage and stems; serious pests of many crops as seedlings, and root crops.

**Deflocculating agent** Material added to a spray suspension to delay sedimentation.

**Defoliant** Spray which induces premature leaf-fall.

**Deposit (spray)** Amount and pattern of spray or dust deposited per unit area of plant surface.

**Deposit (dried)** Amount and pattern of active ingredient deposited per unit area of plant surface.

**Deposition velocity** Velocity at which the spray impinges on the target.

**Depressed** Flattened dorso-ventrally.

**Desiccant** Chemical which kills vegetation by inducing excessiv water loss.

**Diluent** Component of spray or dust that reduces the concentration of the active ingredient, and may aid in mechanical application but does not directly affect toxicity.

**Disinfect** 1. To free from infection by destruction of the pest or pathogen established in or on plants or plant parts. 2. To kill or inactivate pests or pathogens present upon the surface of plants or plant parts, or in the immediate vicinity (e.g. in soil).

**Dispersal** Movement of individuals out of a population (emigration) or into a population (immigration).

**Diurnal** Active during the daytime.

**Dormant** Alive but not growing; buds with an unbroken cover of scales; quiescent; inactive; a resting stage.

**Dose; dosage** Quantity of pesticide applied per individual, or per unit area, or per unit volume, or per unit weight.

**Drift** Spray or dust carried by natural air currents beyond the target area.

**Drop spectrum** Distribution, by number or volume of drops, of spray into different droplet sizes.

**Duster** Equipment for applying pesticide dusts to a crop.


**Ecdysis** The moulting (shedding of the skin) of larval arthropods from one stage of development to another — the final moult leading to the formation of the puparium or chrysalis.

**Ecesis** (= Oikesis) The establishment of an organism in a new habitat; accidental dispersal and establishment in a new area.

**Ecoclimate** Climate within the plant (crop) community.

**Ecology** The study of all the living organisms in an area and their physical environment.

**Economic damage** The injury done to a crop which will justify the cost of artifical control measures.

**Economic-injury level** The lowest population density that will cause economic damage.

**Economic pest** A pest causing a crop loss of about 5–10%, according to definition.

**Economic threshold** The pest population level at which control measures should be started to prevent the pest population from reaching the economic-injury level.

**Ecosystem** The interacting system of the living organisms in an area and their physical environment.

**Efficiency of a pest control measure** The more or less fixed reduction of a pest population regardless of the number of pests involved.

**Effectiveness of a pest control measure** This is shown by the number of pests remaining after control treatment.

**Elateriform larva** A larva resembling a wireworm with a slender body, heavily sclerotinized, with short thoracic legs and only a few body bristles.

**Elytron** The thickened forewing of the Coleoptera.

**Emergence** 1. The adult insect leaving the last nymphal skin, or pupal case. 2. Germination of a seed and the appearance of the shoot.

**Emigration** The movement of individuals out of a population.

**Emulsifiable concentrate (e.c.)** Liquid formulation that when added to water will spontaneously disperse as fine droplets to form an emulsion (= Miscible oil).

**Emulsifier** Spray additive which permits formation of a stable suspension of oil droplets in aqueous solution, or of aqueous solution in oil.

**Emulsion** A stable dispersion of oil droplets in aqueous solution, or vice cersa.

**Emulsion, Invert** Suspension of aqueous solution in oil.

**Encapsulation** Or microencapsulation: the encapsulation of a pesticide in a non-volatile envelope of gelatin, usually of minute size, for delayed release.

**Entomophagous** An animal (or plant) which feeds upon insects.

**Erinium** A growth of hairs in dense patches on plant leaves resulting from the attack of certain Eriophyidae (Acarina).

**Eruciform larva** Caterpillar; a larva with a cylindrical body, well-developed head, and with both thoracic legs and some abdominal prolegs.

**Exarate pupa** A pupa in which the appendages are free and not glued to the insect body.

**Exuvium** The cast skin of arthropods after moulting.


**Fecundity** Capacity to produce offspring (reproduce); power of a species to multiply rapidly.

**Filler** Inert component of pesticide dust or granule formulation.

**Flowability** Property of flowing possessed by dusts, colloids, liquids, and some pastes.

**Fluorescent tracer** Fluorescent material added to a spray to aid the assessment of spray deposits on plants.

**Formulation** 1. Statement of nature and amount of all constituents of a pesticide concentrate. 2. Method of preparation of a pesticide concentrate.

**Fossorial** Modified for digging; in the habit of digging or burrowing.

**Frass** Wood fragments made by a wood-boring insect, usually mixed with the faeces.

**Fumigant** Pesticide exhibiting toxicity in the vapour phase.

**Furrow application** Placement of pesticides with seed in the furrow at the time of sowing.

**Gall** An abnormal growth of plant tissues, caused by the stimulus of an animal or another plant.

**Generation** The period from any given stage in the life cycle (usually adult) to the same stage in the offspring.

**Granule** Coarse particle of inert material (pumice, Fuller's earth, rick husks) impregnated or mixed with a pesticide. Used mainly for soil application, but sometimes for foliar application (pumice formulation).

**Granule applicator** Machine designed to apply measured quantities of granules.

**Grease band** Adhesive material (e.g. resin in castor oil, or 'Sticktite') applied as a band around a tree to trap or repel ascending wingless female moths, or ants.

**Grub (White)** A scarabaeiform larva; thick-bodied, with a well-developed head and thoracic legs, without abdominal prolegs, usually sluggish in behaviour; general term for larvae of Coleoptera.

**Habitat** The place where plants and animals live, usually with a distinctive boundary (e.g. field, pond, sand-dune, rocky crevice).

**Hemelytron** The partly thickened forewing of Heteroptera.

**Hemimetabolous** Insects having a simple metamorphosis, like that in the Orthoptera, Heteroptera, and Homoptera.

**Herbivorous** Feeding on plants (phytophagous).

**Hibernation** Dormancy during the winter, or cold season.

**Hollow-cone** Spray jet with a core of air breaking to give drops in an annular pattern.

**Holometabolous** Insects having a complete metamorphosis, as in the Diptera, Hymenoptera, Coleoptera, Lepidoptera.

**Honey-dew** Liquid with high sugar content discharged from the anus of some Homoptera.

**Hornworm** A caterpillar with a dorsal spine or horn on the last abdominal segment − larvae of Sphingidae.

**Host** The organism in or on which a parasite lives; and the plant on which an insect feeds.

**Humectant** Material added to a spray to delay evaporation of the water carrier.

**Hyaline** Transparent.

**Hypermetamorphosis** A type of complete metamorphosis in which the different larval instars represent two or more different types of larvae.

**Hyperparasite** A parasite whose host is another parasite.

**Hythergraph** A polygonal diagram resulting from plotting temperature means against rainfall.

**Imago** The adult, or reproductive stage of an insect.

**Immigration** The movement of individuals into a population.

**Immune** Exempt from infection.

**Incompatible** Not compatible; incapable of forming a stable mixture with another chemical.

**Indicator** Marker.

**Inert** A material having no biological action.

**Infect** To enter and establish a pathogenic relationship with a plant (host); to enter and persist in a carrier.

**Infest** To occupy and cause injury to either a plant, soil or stored products.

**Injector** A device for ejecting a pesticide below the soil surface, or into the transport system of a tree.

**Insecticide** A toxin effective against insects.

**Instar** The form of an insect between successive moults; the first instar being the stage between hatching and the first moult.

**Intercropping** The growing of two crops simultaneously in the same field.

**Jet** Liquid emitted from a nozzle orifice (in USA = nozzle).

**Key pest** An important major pest species in the complex of pests attacking a crop, with a dominating effect on control practices.

**Lacquer** Pesticide incorporated into a lacquer or varnish to achieve slow release over a lengthy period of time.

**Larva** The immature stages of an insect, between the egg and p having a complete metamorphosis; the six-legged first insta the Acarina.

**Larvicide** Toxicant (poison) effective against insect larvae.

**$LC_{50}$** Lethal concentration of toxicant required to kill 50% of large group of individuals of one species.

**$LD_{50}$** Lethal dose of toxicant required to kill 50% of a large group of individuals of one species.

**Leaf area index (LAI)** Ratio of leaf surface to soil surface area, relation to utilization of solar energy for photosynthesis.

**Leaf miner** An insect which lives in and feeds upon the cells between the upper and lower epidermis of a leaf; these being larvae of Agromyzidae (Diptera), Lyonetiidae and Gracillariidae (Lepidoptera), Hispidae (Coleoptera), etc.

**Life table** The separation of a pest population into its different age components (e.g. eggs, larvae, pupae, adults).

**Looper** A caterpillar of the family Geometridae, with only one pair of abdominal prolegs (in addition to the terminal claspers), and which moves by looping its body.

**Macropterous** Large, or long-winged.

**Maggot** A vermiform larva, legless, without a distinct head capsule (Diptera).

**Miscible liquid (m.l.)** A formulation in which the technical product is dissolved in an organic solvent which is then, on dilution, dissolved in the water carrier.

**Mist blower** Sprayer producing a fine air-carried spray.

**Miticide** Preferably called Acaricide.

**Molluscicide** Toxicant effective against slugs and snails.

**Monoculture** The extensive cultivation of a single species of plant.

**Monophagous** An insect restricted to a single host plant species.

**Mortality** Population decrease factor; death rate.

**Mutualism** The symbiotic relationship between two organisms in which both parties derive benefit.

**Natality** Population 'increase' factor; birth rate.

**Necrosis** Death of part of a plant.

**Nematicide** Toxicant effective against nematodes (= eelworms).

**Nocturnal** Active at night.

**Non-ionic surfactant** A surfactant that does not ionize in solution and is therefore compatible with both anionic and cationic surfactants.

**Nozzle** 1. Air blast: nozzle using high velocity air to break up the spray liquid supplied at low pressure. 2. Anvil: nozzle in which the spray liquid jet strikes a smooth, solid surface at a high angle of incidence. 3. Cone (or swirl): nozzle in which the liquid emerges from the orifice with tangential velocity imparted by passage through one or more tangential or helical channels in the swirl chamber. 4. Hollow cone: nozzle in which spray jet has a core of air breaking to give drops in an annular pattern. 5. Fan nozzle: the aperture is an elongate horizontal slit, producing a fan-shaped spray pattern. 6. Deflector: nozzle in which a fan-shaped sheet of spray is formed by directing the liquid over a sharply inwardly curving surface.

**Nymph** The immature stage of an insect that does not have a distinct pupal stage; also the immature stages of Acarina that have eight legs.

**Obtect pupa** A pupa in which the appendages are more or less glued to the body surface (Lepidoptera).

**Oligophagous** (= Stenophagous) An animal feeding upon only a few, closely related, host plants; or it may be an animal parasite.

**Onisciform larva** A flattened platyform larvae, like a woodlouse in appearance.

**Orifice (nozzle) velocity** Velocity at which the spray leaves the nozzle orifice.

**Ovicide** Toxicant effective against insect or mite eggs.

**Oviparous** Reproduction by laying eggs.

**Paint gun** Type of small, hand-carried, air-blast machine.

**Pantropical** A species occurring widely throughout the tropical and subtropical parts of the world.

**Parasite** An organism living in intimate association with a living organism (plant or animal) from which it derives material essential for its existence while conferring no benefit in return.

**Parasitoid** An organism alternately parasitic and free-living; most parasitic Hymenoptera and Diptera fall into this category as usually only the larvae are parasitic.

**Parthenogenesis** Reproduction without fertilization; usually through eggs but sometimes through viviparity.

**Parts per million (ppm)** Proportion of toxicant present in relation to that of plant material on which it has been deposited. Usually applied in connection with the edible portion of a crop and its suitability for consumption.

**Pellet** Seed coated with inert material, often incorporating pesticides, to ensure uniform size and shape for precision drilling.

**Penetrant** Oil added to a spray to enable it to penetrate the waxy insect cuticle more effectively.

**Persistence** The term applied to chemicals that remain active for a long period of time after application.

**Pest** An animal or plant causing damage to man's crops, animals or possessions.

**Pest density** The population level at which a pest species causes economic damage.

**Pest management** The careful manipulation of a pest situation, after extensive consideration of all aspects of the life system as well as ecological and economic factors.

**Pest spectrum** The complete range of pests attacking a particular crop.

**Pesticide** A chemical which by virtue of its toxicity (poisonous properties) is used to kill pest organisms. A term of wide application which includes all the more specific applications – insecticide, acaricide, bactericide, fungicide, herbicide, molluscicide, nematicide, rodenticide, etc.

**Pheromone** (= Ectohormone). A substance secreted by an insect to the exterior causing a specific reaction in the receiving insects.

**Phytophagous** Herbivorous; plant eating.

**Phytosanitation** Measures requiring the removal or destruction of infected or infested plant material likely to form a source of reinfection or reinfestation. (See Crop Hygiene.) International phytosanitation refers to inspection of plants and seeds to prevent noxious pests and diseases from being brought into a country; often involves plant quarantine.

**Phytotoxic** A chemical liable to damage or kill plants (especially the higher plants), or plant parts.

**Planidium larva** A type of first instar larva in certain Diptera and Hymenoptera which undergoes hypermetamorphosis.

**Poison bait** An attractant foodstuff for insects, molluscs, or rodents, mixed with an appropriate toxicant.

**Polyphagous** An animal feeding upon a range of hosts.

**Pre-access interval** The interval of time between the last application of pesticide to an area and safe access to the area for domestic livestock, and man.

**Predisposition** Making a plant more susceptible to a pest or disease, usually as a result of genetic, cultural or environmental defects.

**Preference** The factor by which certain plants are more or less attractive to insects by virtue of their texture, colour, aroma or taste.

**Pre-harvest interval** The interval of time between the last application of pesticide and the safe harvesting of edible crops for immediate consumption.

**Pre-oviposition period** The period of time between the emergence of an adult female insect and the start of its egg laying.

**Pre-pupa** A quiescent stage between the larval period and the pupa; found in some Diptera and Thysanoptera.

**Preventative** A measure applied in anticipation of pest attack.

**Proleg** A fleshy abdominal leg found in caterpillars (Lepidoptera, sawflies), bearing a characteristic arrangement of crochets.

**Proprietory name** Distinguishing name given by the manufacturer to a particular formulated product.

**Protective clothing** Clothing to protect the spray operator from the toxic effects of crop protection chemicals. This may include rubber gloves, boots, apron, respirator, face mask, etc.

**Protonymph** The second instar of mites.

**Pterostigma** A thickened opaque or dark spot along the costal margin of the wing, near the tip (e.g. Odonata, Hymenoptera).

**Pupa** The stage between larva and adult in insects with complete metamorphosis; a non-feeding and usually inactive stage.

**Puparium** The case formed by the hardened last larval skin in which the pupa of Diptera is formed.

**Quarantine** All operations associated with the prevention of importation of unwanted organisms into a territory, or their exportation from it.

**Recruitment** The addition of new individuals to a population, usually by either birth or immigration.

**Redistribution** Movement of pesticide subsequent to the initial application to other parts of the plant, usually by rain.

**Repellant** A chemical which has the property of inducing avoidance by a particular pest.

**Residue** Amount of pesticide remaining in or on plant tissues (or in soil) after a given time, especially at harvest time.

**Resistance** The natural or induced capacity to avoid or repel attack by pests (or parasites). Also the ability to withstand the toxic effects of a pesticide or a group of pesticides, often by metabolic detoxification.

**Rodenticide** A toxicant effective against rodents.

**Roguing** The removal of unhealthy or unwanted plants from a crop.

**Rostrum** The beak or proboscis of Hemiptera.

**Run-off** The process of spray shedding from a plant surface during and immediately after application, when droplets coalesce to form a continuous film and surplus liquid drops from the surface.

**Scarabaeiform larva** A grub-like larva, with a thickened cylindrical body, well-developed head and thoracic legs, without abdominal prolegs, and sluggish in behaviour.

**Scavenger** An animal that feeds on dead plants and animals, on decaying matter, or on animal faeces.

**Secondary pest** Species whose numbers are usually controlled by biotic and abiotic factors which sometimes break down, allowing the pest to increase in numbers.

**Seed dressing** A coating (either dry or wet) of protectant pesticide applied to seeds before planting. Dry seed dressings are often physically stuck to the testa of the seed by a sticker such as methyl cellulose.

**Semi-looper** Caterpillar from the subfamily Plusiinae (Noctuidae) with two or three pairs of prolegs, which locomotes in a somewhat looping manner.

**Siphunculi** The paired protruding organs near the terminal end of the abdomen of Aphidoidea, also called cornicles, through which a waxy secretion is extruded.

**Slurry** Paste-like liquid used as a seed coating.

**Smoke** Aerial dispersal of minute solid particles of pesticides through the use of combustible mixtures.

**Soil sterilant** Toxicant added to, or injected into, soil for the purpose of killing pests and pathogens.

**Solid cone** Jet with air-core reduced to give a cone of spray droplets.

**Solvent** Carrier solution in which the pesticide (technical product) is dissolved to form the concentrate.

**Spray** 1. Air-carried: spray propelled to target in a stream of air. 2. Coarse: dispersion of droplets of mass median diameter over $200\,\mu$m. 3. Concentrate: undiluted commercial pesticide preparation. 4. Fine: dispersion of droplets of mass median diameter from $50-150\,\mu$m. 5. Floor: spray applied to the litter on the ground surface. 6. High-volume: over $1200$ l/ha on bushes and trees; over $700$ l/ha on ground crops (or over $400$ l/ha according to definition). 7. Low-volume: spray of $250-600$ l/ha on bushes and trees; $50-250$ l/ha on ground crops (or $5-400$ l/ha). 8. Median-volume: $600-1200$ l/ha on bushes and trees; $250-700$ l/ha on ground crops. 9. Mist: dispersion of droplets of $50-100\,\mu$m in diameter. 10. Ultra-low-volume: less than $50$ l/ha on ground crops; less than $250$ l/ha on trees and bushes (or less than $5$ l/ha according to definition).

**Spray angle** Angle between the sides of a jet leaving the orifice.

**Sprayer** Apparatus for applying pesticide sprays; not to be confused with 'Spray operator'.

**Spray operator** Person operating a sprayer, and applying a spray.

**Spread** Uniformity and completeness with which a spray deposit covers a continuous surface, such as a leaf or a seed.

**Spreader** Material added to a spray to lower the surface tension and to improve spread over a given area (= wetter).

**Spur** An articulated spine, often on a leg segment, usually the tibia. A serrulate tibial spur is characteristic of the Delphacidae (Homoptera).

**Stability** The ability of a pesticide formulation to resist chemical degradation over a period of time.

**Sticker** A material of high viscosity used to stick powdered seed dressings on to seeds; two commonly used stickers are paraffin and methyl cellulose. A solution of methyl cellulose can be added to a spray to increase retention on plant foliage.

**Stomach poison** A toxicant (poison) which operates by absorption through the intestine after having been injested by the insect, usually on plant material.

**Supplement (spray)** (= Adjuvant).

**Surfactant** (= Spreader; wetter).

**Susceptible** Capable of being easily infested or infected; not resistant.

**Swath** Width of target area sprayed at one pass.

**Symbiosis** The general term for two organisms that live together in a partnership, sometimes beneficial; includes commensalism, inquilinism, mutualism, and parasitism.

**Sympatric** Having the same, or overlapping, areas of geographical distribution.

**Synecology** The study of a particular community.

**Synergism** Increased pesticidal activity of a mixture of pesticides above that of the sum of the values of the individual components.

**Systematics** The classification of animal and plant species into their higher taxa; sometimes regarded as synonymous with taxonomy.

**Systemic** A pesticide absorbed through the plant surfaces (usually roots) and translocated through the plant vascular system.

**Taint** Unwanted flavour in fresh or processed food from a pesticide used on the growing crop.

**Target surface** The surface intended to receive a spray or dust application.

**Taxonomy** The laws of classification as applied to natural history; identification of plant and animal species.

**Technical product** The usual form in which a pesticide is prepared and handled prior to formulation; usually at a high level of purity (95–98%) but not completely pure.

**Tegmen** The thickened and leathery forewing in the Orthoptera and Dictyoptera.

**Tenacity** The property of a pesticide deposit or residue to resist removal by weathering.

**Tenacity index** Ratio of the quantity of residue per unit area at the end of a given period of weathering to that present at the beginning.

**Tolerance** Ability to endure infestation (or infection) by a particular pest (or pathogen) without showing severe symptoms of distress.

**Tolerance, permitted** Maximum amount of toxicant allowed in foodstuffs for human consumption.

**Toxicant** Poison, or chemical exhibiting toxicity.

**Toxicity** Ability to poison, or to interfere adversely with vital processes of the organism by physico-chemical means.

**Tracer** Additive to facilitate location of a deposit, by radio-active or fluorescent means.

**Translaminar** A pesticide which passes through from one surface of a leaf to the other (from lamina to lamina) through the leaf tissue.

**Translocation** The uptake of a pesticide into part of a plant body and its subsequent dispersal to other parts of the plant body.

**Trap crop** Crop of plants (sometimes wild plants) grown especially to attract insect pests, and when infested either sprayed or collected and destroyed. Trap plants usually grown between the rows of the crop plants or else peripherally.

**Trapping-out** The removal of individuals from a pest population, in significantly large numbers, by means of trapping (often using u.v. light traps).

**Triungulin larva** The active first instar larva of Meloidae (Coleoptera), and Strepsiptera.

**Vector** Organisms able to transmit viruses or other pathogens either directly or indirectly. Direct virus vectors include insects, mites and nematodes.

**Vermiform larva** A legless (apodous), headless (acephalic), worm-like larva typical of some Diptera.

**Vestigial** Poorly developed; degenerate; non-functional.

**Viviparous** Giving birth to living young (Aphidoidea).

**Volunteer** Crop plant growing accidentally from shed seed; not deliberately cultivated.

**Wireworm (Elateriform larva)** The larva of Elateridae (Coleoptera); long; slender, well-sclerotized, thoracic legs but no prolegs, and few setae.

This glossary was originally based in part upon one produced for the Horticultural Education Association, UK, by Mr R. W. Marsh, OBE, late of Long Ashton Research Station, Bristol, whose assistance is gratefully acknowledged.

### Killing

For specimens to be preserved dry: potassium cyanide, ethyl acetate, chloroform, ether, petrol, etc. Small insects are quickly killed by tobacco smoke. For preservation in fluids: the preservative or hot water.

### Labelling

Specimens without adequate labels are of little value. Data should include place, date and name of collector, clearly stated — e.g. 'Kabanyolo, Kampala, UGANDA. 12 Jan. 1972; J. Smith'. In remote districts state latitude and longitude rather than the name of a village that probably does not appear on any map. Notes such as 'boring sugarcanes', 'eating cacao leaves', 'at light', should be added when relevant. When it is desirable to add more on a label, make it quite clear to which specimens the descriptions refer.

Unnecessary words such as 'Insects found in . . .' should be avoided. Labels for tubes should be placed *inside* them, written in Indian ink or soft pencil on paper of good quality.

### Dried specimens

Most insects (e.g. grasshoppers, bugs, beetles, butterflies and moths, bees and wasps, flies) are usually pinned and dried. Only stainless steel insect pins, if obtainable, should be used in the humid tropics. If pinning is inconvenient, specimens can be preserved dry in pill-boxes, envelopes, etc., with packing of soft tissue paper, etc. to prevent movement. (Cotton-wool should not touch specimens since claws become entangled in it. Cellulose wool, however, is suitable for packing small Diptera, Hymenoptera, etc.)

After pinning, or packing, specimens should be protected from mould with liberal amounts of naphthalene (e.g. moth balls).

### Preservation in liquids

Alcohol 70% to 80%. Substitutes: methylated spirit (slightly diluted), strong rum, gin, Waragi or whisky. Bottles should not be more than half filled with specimens, to avoid excessive dilution of the preservative. Avoid formalin if possible. 80% alcohol to which 5% glycerine has been added is as good as anything. Soft-bodied insects (caterpillars, maggots, aphids, termites, etc.) should be preserved in liquids. Insects too small for pinning are often most conveniently preserved thus. (Almost all insects, except butterflies and moths, preserve well in liquids, though pinning is the usual practice with most of them.)

### Packing

Pins must be pushed deep into the cork of boxes, otherwise many of them come adrift. Large insects must be prevented from rotating on the pins by extra pins touching each side. Naphthalene should be firmly anchored with pins, or melted to adhere to the box. Boxes of pinned insects should be surrounded on all sides by at least 3 cm of thin wood shavings ('excelsior'), packed in a strong box sufficiently firmly to prevent all movement and to absorb shock. If all these precautions are adopted, delicate specimens survive mailing — otherwise considerable damage is almost inevitable.

For specimens in fluids, the only precautions necessary are to push corks firmly into bottles (sealing corks with paraffin is desirable, but not essential) and to pack securely put tissue paper in tubes of very delicate specimens. Cigarette and biscuit tins, wrapped in corrugated cardboard, make good containers for mailing bottles. Do not send corked tubes of alcohol by air.

### Living specimens

Although it is usually preferable either to dry specimens or preserve them in liquids for shipment, it is sometimes necessary to send living material. Live insects should be sent in suitable strong cardboard, wooden or tin boxes. When tins are used, the lid should be perforated with numerous small holes to prevent undue condensation of moisture. In general, larvae should be supplied with an adequate supply of food material, but if their food is of a succulent nature, it will probably decompose and it may be better to pack them loosely in damp moss. Many adult insects survive several days without food, but a little tissue paper or muslin, or in a dry climate some damp moss should be placed in the container with them.

For further information on collecting and preserving insects, with special reference to the tropics, the small book by McNutt is recommended (McNutt, D. N. (1976). *Insect Collecting in the Tropics*, 68 pp.; copies available from COPR, London, free to workers in Commonwealth countries, otherwise £1.30).

### Identification of insect pests

If the insects cannot be identified by the collector then in most countries the Ministry of Agriculture (or Department of Agriculture) will have a central laboratory with a named collection of local insect pests. Thus the newly collected pests can be identified with reasonable certainty by comparison with the named specimens in the regional collection.

As a last resort it may be necessary to send specimens to the taxonomic experts either at the Commonwealth Institute of Entomology, London, or the Smithsonian Institution, Washington DC. The appropriate addresses are as follows:

The Director,
Commonwealth Institute of Entomology,
c/o British Museum (Natural History),
Cromwell Road,
London, SW7 5BD,
England.

The Director,
Insect Identification Research Branch,
US Department of Agriculture,
c/o US National Museum,
Smithsonian Institution,
Washington DC (20560),
USA.

Parcels sent to either of these institutions should be sent by 'Air Parcel Post', and *not* by 'Air Freight' since this service involves heavy charges for customs clearance and delivery from the airport, and there is a considerable delay before delivery.

# D  Standard abbreviations

## Units and general abbreviations

| | |
|---|---|
| a.i. | active ingredient |
| °C | degrees Celsius |
| cm | centimetre |
| cv. | cultivar |
| e.c. | emulsifiable concentrate |
| g | gram |
| h | hour |
| ha | hectare |
| h.v. | high volume |
| i.r. | infra-red |
| kg | kilogram |
| km | kilometre |
| £ | pound sterling |
| l | litre |
| $LC_{50}$ | median lethal concentration |
| $LD_{50}$ | median lethal dose |
| l.v. | low volume |
| m | metre |
| mg | milligram |
| min | minute |
| ml | millilitre |
| mm | millimetre |
| pH | hydrogen ion concentration |
| post-em | post-emergence |
| ppm | parts per million |
| pre-em | pre-emergence |
| RH | relative humidity |
| s | second |
| sp. | species |
| spp. | species (plural) |
| ssp. | subspecies |
| sspp. | subspecies (plural) |
| $ | dollar |
| u.l.v. | ultra-low volume |
| u.v. | ultra-violet |
| var. | variety |
| vol. | volume |
| w.p. | wettable powder |
| w/w | weight for weight |

## Miscellaneous abbreviations

| | |
|---|---|
| BPH | Brown Planthopper of rice |
| BC | Biological control |
| BSI | British Standards Institute |
| Cda | Controlled droplet application |
| GV | Granulosis virus |
| IPM | Integrated pest management |
| EAG | Electroantennagram |
| HMSO | Her Majesty's Stationery Office, UK |
| OC | Organochlorine compounds |
| OP | Organophosphorous compounds |
| PHV | Polyhedrosis virus |
| PM | Pest management |
| SIRM | Sterile insect release method |

## Organizations

| | |
|---|---|
| ADAS | Agricultural Development and Advisory Service (formerly NAAS), MAFF, UK |
| ARC | Agricultural Research Council, UK |
| AVRS | Asian Vegetable Research Station, Taiwan |
| BM(NH) | British Museum (Natural History), London, UK |
| CAB | Commonwealth Agricultural Bureaux, Slough, UK |
| CIAT | Centre for International Tropical Agriculture, Cali, Colombia |
| CIBC | Commonwealth Institute of Biological Control (headquarters), West Indies |
| CIE | Commonwealth Institute of Entomology, London, UK |
| CIP | International Potato Centre, Lima, Peru |
| CIH | Commonwealth Institute of Helminthology, St Albans, UK |
| CMI | Commonwealth Mycological Institute, Kew, UK |
| COPR | Centre for Overseas Pest Research, London, UK |
| CSIRO | Commonwealth Scientific and Industrial Research Organization, Canberra, Australia |

| | |
|---|---|
| EAAFRO | East African Agricultural and Forestry Research Organization, Nairobi, Kenya |
| EPA | Environmental Protection Agency, Washington, USA |
| EPPO | European Plant Protection Organization, Paris, France |
| FAO | Food and Agricultural Organization of the United Nations, Rome, Italy |
| GCRI | Glasshouse Crops Research Institute, UK |
| IAC | International Agricultural Centre, Wageningen, Netherlands |
| ICRISAT | International Crops Research Institute for the Semi-arid Tropics, Hyderabad, India |
| IITA | International Institute of Tropical Agriculture, Ibadan, Nigeria |
| IRRI | International Rice Research Institute, Manila, Philippines |
| ICIPE | International Centre for Insect Physiology and Ecology, Nairobi, Kenya |
| MAFF | Ministry of Agriculture, Fisheries and Food, UK |
| MARDI | Malaysian Agricultural Research and Development Institute, Selangor, Malaysia |
| NAAS | National Agricultural Advisory Service (now ADAS), UK |
| NAPPO | North American Plant Protection Organisation, USA |
| NVRS | National Vegetable Research Station, Wellesbourne, UK |
| ODM | Ministry of Overseas Development, UK |
| PBI | Plant Breeding Institute, Cambridge, UK |
| PESTDOC | Derwent Pooled Pesticidal Literature Documentation |
| TPRI | Tropical Products Research Institute, London, UK |
| US AID | United States Aid for International Development, USA |
| USDA | United States Department of Agriculture, USA |
| WHO | World Health Organization, Geneva, Switzerland |
| WICSCBS | West Indies Central Sugarcane Breeding Station, Barbados, West Indies |
| WRO | Weed Research Organization, Oxford, UK |

# Index

*Major page references are in bold*

Autocide 52, 707
Avocado 507, 514
Avoidance 41, 45, 47
*Axiagastus campbelli* 548
*Ayyardia chaetophora* 556, 583
Azinphos-methyl 124, 152, 705
Azinphos-methyl + demeton-*S*-methyl sulphone 125, 152, 705
*Azochis gripusalis* 570

*Bacillus*
   *popiliae* 89
   *thuringiensis* 89, 150, 155, 706
Bacteria 89
*Bagrada*
   *cruciferarum* **261**, 524
   *hilaris* **261**, 524, 556, 605
   spp. 578, 634
Bagworms 66, 72, 284, 587, 593, 601
Baits 106, 707
*Balanogastris kolae* 588
*Balclutha viridis* 640
*Baldulus maidis* 598
*Balinus c-album* 642
*Baliothrips biformis* **271**, 638
Bamboo 507, 516
Bamboo Aphid 209, 516
Bamboo Borer 516
Bamboo Bug 255, 516
Bamboo Carpenter Bee 516
Bamboo Hispids 516
Bamboo Locust 516
Bamboo Longhorn 451, 516, 659
Bamboo Planthopper 516
Bamboo Star Scale 516

Bamboo Weevil 476, 516
*Bambusa*
   spp. 516
   *vulgaris* 516
Banana Aphid 203, 517, 571, 603, 666
Banana Fruit Caterpillar 518, 645, 668, 671
Banana Fruit Fly 518
Banana Fruit-scarring Beetle 459, 517, 518
Banana Lace Bug 259, 517, 666, 676
Banana Rust Thrips 518
Banana Scab Moth 326, 517, 599, 603
Banana Skippers 65, 517, 609
Banana Stem Borer 485, 517
Banana Stem Weevil 72, 485, 517
Banana Thrips 275, 517
Banana Weevil 64, 75, 91, 96, 475, 517, 603
Bananas 64, 507, 517
Band application 707
*Baris* spp. 627
Bark Beetles 418, 495
Barley (*see* Wheat)
Barley Fly 401, 599, 681
*Bathycoelia thalassina* 545
*Batocera*
   *horsfieldi* 678
   *rubus* **448**, 527, 602
   *rufomaculata* **450**, 602, 606, 645, 679
   spp. 74, 569, 584, 587, 606
*Batrachedra*
   *amydraula* 564
   *arenosella* 548
Bean
   Bovanist 583
   Field 519

French 519
Garbanzo (*see* Chickpea)
Hyacinth 583
Indian 583
Runner 519
Soya 654
Bean Aphid, Black 519
Bean Bruchid 454, 519, 688
Bean Fly 66, 78, 394, 519, 583, 620, 655
Bean Flower Thrips 278, 519, 577
Bean Pod Fly 521, 625, 646
Bean Seed Fly 402, 519, 598, 614, 619, 671, 682, 686
Beans 64, 507, 519
*Beauvaria* sp. 470
*Bedellia* spp. 663
Bees 16, 100
   Bumble 16
   Honey 16
   Leaf-cutter 16
   Mason 16
Beet Armyworm 376, 586, 614
Beetles 415–96
*Belippa laleana* 537
*Bemisia*
   *inconspicua* 525
   spp. 532
   *tabaci* **198**, 528, 532, 536, 555, 559, 567, 583, 611, 619, 646, 647, 662, 670, 673
Beniseed (*see* Sesame)
Ber (*see* Jujube)
Ber Fruit Fly 585
Ber Mealybug 585
Ber Scale 585
Ber Weevil 585
Betel Palm 507, 522
Betel-pepper 507, 523
Betelvine Bug 523

Betelvine Scale 523
*Betula* spp. 512
BHC, gamma (HCH) 119, 152, 705
*Bibio marci* 682
Binapacryl 123, 152, 705
Bioallethrin 147, 155, 706
Biocide 707
Biological control 36, 50, 79–98
Biological pesticides 149, 155, 706
*Biston suppressaria* 669
Biting Ant 408
*Bixadus sierricola* 553
Black Bean Aphid 519, 654
Black Borers 72, 416, 432, 588
Black Citrus Aphid 205, 588, 591, 594, 595, 665, 668
Black Cutworm 357, 568, 579, 612, 614, 615, 641, 655, 670, 674
Black Line Scale 239, 601, 609
Black Maize Beetle 78, 419, 598
Black Paddy Bug 267
Black Scale 230, 569, 572, 575, 613, 648, 665
Black Tea Thrips 274, 618, 645, 666, 667
Black Twig Borer 496, 573, 645
Black Wheat Beetle 419, 681
*Blatella germanica* 689
*Blatta orientalis* 689
*Blissus*
   *gibbus* 641
   *leucopterus* 599, 652, 682
Blister Beetles 442, 625, 654, 663

724

727

Lime-sulphur 146, 155, 706
*Limothrips cerealium* 683
*Liothrips oleae* 613
*Lipaphis erysimi* 524, 614, 680
*Liriomyza brassicae* 525
*Lissorhoptrus
    oryzophilus* **482,** 640
    spp. 482
Litchi 507, 591
*Litchi chinensis* 591
Litchi Gall Mite 592
Litchi Lantern Bug 190
Litchi Leaf Roller 592
Litchi Stink Bug 268, 591, 593, 637
*Lobesia botrana* 572
*Locris* spp. 181
*Locusta migratoria* sspp.
        **163,** 516, 547, 575, 578, 640, 652, 662, 685
Locust
    Australian Plague 685
    Bombay 166, 597, 604, 640, 652, 685
    Desert 167
    Mediterranean 685
    Migratory 163, 575, 578, 652, 667
    Red 164
    Spur-throated 556
Locusts 68, 163, 164, 166, 167, 640
*Lonchaea
    aristella* 570
    spp. 533
Longan 507, 593
Longhorn Beetles 72, 74, 417, 444, 569, 570, 587, 588, 600, 602, 606, 617, 618, 636, 657, 678, 684

*Longitarsus
    belagaumensis* 647
    *nigripennis* 624
Long-tailed Mealybug 600, 658
*Lophobaris piperis* 624
Loopers 711
*Lophodes
    miserana* 515
    *sinistraria* 596
Lophopidae 179, 194
*Lophosternus hugelii* 512
Loquat 507, 594
*Lotongus calathus* 548, 609
Low-volume mist blowers 113
Low-volume spraying 103, 110
Lucerne Crown Borer 579
Lycaenidae 282, 335
Lychee (*see* Litchi)
*Lycopersicum esculentum* 673
Lygaeidae 243, 249
*Lygaeus
    hospes* 587
    *pandurus* 629
*Lygocoris pabulinus* 510, 520, 622, 630, 633, 686
*Lygus
    oblineatus* 556, 622
    *rugulipennis* 634, 686
    spp. 520
    *viridanus* 511
*Lymantria
    mathura* 592
    *obfuscator* 513
Lymexylidae 74, 416, 434
Lyonetiidae 65


Macadamia 507, 595
Macadamia Cup Moth 595

Macadamia Felted Coccid 595
Macadamia Flower
    Caterpillar 595
Macadamia Kernel Grub 596
Macadamia Lace Bug 595
Macadamia Leaf Miner 595
Macadamia Nut Borer 592, 595
*Macadamia terhifolia* 595
Macadamia Tufted
    Caterpillar 596
Macadamia Twig Girdler 595
*Macalla moncusalis* 531
Mace 607
*Maconellicoccus hirsutus*
        **216,** 566, 572, 606, 643
*Macrocentrus homonae* 309
*Macromischoides aculeatus*
        **408,** 552
*Macropsis trimaculata* 630
Macropterous 711
*Macrosiphum
    avenae* 681
    *euphorbiae* 567, 590, 618, 633
    *fragariae* 681
    spp. 520, 575
*Macrotermes
    bellicosus* 173, 543, 544, 548
    spp. **172,** 173, 516, 644, 657, 685
Maggots 711
*Mahasena corbetti* **286,** 549, 608
Maize 64, 507, 597
Maize Aphid 204
Maize Beetle, Black 78
Maize Leafhopper 182, 578, 597, 598, 657
Maize Stalk Borer 360, 597, 651

Maize Tassel Beetle 68, 461, 598
Maize Webworm 324, 597, 605, 641, 652, 658, 682
Maize Weevil 70, 492, 598, 688
*Malacosoma indica* 509, 511, 513, 678
Malathion 131, 153, 705
*Maliarpha separatella* **323,** 639
*Mallodon downesi* 546
*Mallothrips indicus* 642
*Mamestra brassicae* 525
*Manduca
    quinquemaculata* 671, 674
    *sexta* 670, 675
*Mangifera indica* 600
Mango 507, 600
Mango Bark Borer 602
Mango Bud Mite 602
Mango Fruit Fly 389, 622
Mango Giant Mealybug 570, 580, 585, 601, 616, 621
Mango Hoppers 600, 648
Mango Leaf Weevil 602
Mango Red Spider Mite 602
Mango Scale 241
Mango Seed Weevil 493, 600
Mango Shoot Borer 592, 601
Mango Shoot Psyllid 600
Mango Soft Scale 601, 607
Mango Stone Weevil 493
Mango Twig Borers 600
Mango Webworm 601
Mango Weevil 70, 493, 602
Mango White Scale 601
*Manihot esculenta* 532
Manila Hemp 507, 603
Manioc (*see* Cassava)
*Marasmia trapezalis* **324,**

740